MW00581439

WATER DISTRIBUTION SYSTEM OPERATION AND MAINTENANCE

Fifth Edition

A Field Study Training Program

prepared by

Office of Water Programs
College of Engineering and Computer Science
California State University, Sacramento

in cooperation with the

National Environmental Training Association
(Now National Environmental, Safety & Health Training Association (NESHTA))

★★

Kenneth D. Kerri, Project Director

★★

for the

California Department of Health Services
Sanitary Engineering Branch
Standard Agreement #80-64652

and

U.S. Environmental Protection Agency
Office of Drinking Water
Grant No. T-901361-01-0

2005

NOTICE

This manual is revised and updated before each printing based on comments from persons using this manual.

FIRST EDITION—WATER SUPPLY SYSTEM OPERATION

First Printing, 1983	7,000

SECOND EDITION—WATER DISTRIBUTION SYSTEM O & M

First Printing, 1987	7,000
Second Printing, 1989	9,000
Third Printing, 1991	10,000

THIRD EDITION—WATER DISTRIBUTION SYSTEM O & M

First Printing, 1994	10,000
Second Printing, 1996	15,000
Third Printing, 1999	3,000

FOURTH EDITION—WATER DISTRIBUTION SYSTEM O & M

First Printing, 2000	15,000
Second Printing, 2002	25,000

FIFTH EDITION—WATER DISTRIBUTION SYSTEM O & M

First Printing, 2005	25,000
Second Printing, 2009	18,000

In recognition of the need to preserve natural resources, this manual is printed using recycled paper. The text paper is composed of 40% post-consumer waste and the cover is composed of 10% post-consumer waste. The Office of Water Programs will strive to increase its commitment to sustainable printing practices.

Copyright © 2005 by
California State University, Sacramento Foundation

ISBN 1-59371-020-8

OPERATOR TRAINING MATERIALS

OPERATOR TRAINING MANUALS AND VIDEOS IN THIS SERIES are available from the Office of Water Programs, California State University, Sacramento, 6000 J Street, Sacramento, CA 95819-6025, phone: (916) 278-6142, e-mail: wateroffice@csus.edu, FAX: (916) 278-5959, or website: www.owp.csus.edu.

1. *WATER DISTRIBUTION SYSTEM OPERATION AND MAINTENANCE,*

2. *WATER TREATMENT PLANT OPERATION,* 2 Volumes,

3. *SMALL WATER SYSTEM OPERATION AND MAINTENANCE,**

4. *UTILITY MANAGEMENT,*

5. *MANAGE FOR SUCCESS,*

6. *OPERATION OF WASTEWATER TREATMENT PLANTS,* 2 Volumes,

7. *SMALL WASTEWATER SYSTEM OPERATION AND MAINTENANCE,* 2 Volumes,

8. *OPERATION AND MAINTENANCE OF WASTEWATER COLLECTION SYSTEMS,* 2 Volumes,

9. *COLLECTION SYSTEMS: METHODS FOR EVALUATING AND IMPROVING PERFORMANCE,*

10. *ADVANCED WASTE TREATMENT,*

11. *INDUSTRIAL WASTE TREATMENT,* 2 Volumes,

12. *TREATMENT OF METAL WASTESTREAMS,* and

13. *PRETREATMENT FACILITY INSPECTION.*

* Other training materials and training aids developed by the Office of Water Programs to assist operators in improving small water system operation and maintenance and overall performance of their systems include the *SMALL WATER SYSTEMS VIDEO INFORMATION SERIES.* This series of ten 15- to 59-minute videos was prepared for the operators, managers, owners, and elected officials of very small water systems. The videos provide information on the responsibilities of operators and managers. They also demonstrate the procedures to safely and effectively operate and maintain surface water treatment systems, groundwater treatment systems, and distribution and storage systems. Other topics covered include monitoring, managerial, financial, and emergency response procedures for small systems. These videos are used with a *LEARNING BOOKLET* that provides additional essential information and references. The videos complement and reinforce the information presented in *SMALL WATER SYSTEM OPERATION AND MAINTENANCE.*

The Office of Water Programs at California State University, Sacramento, has been designated by the U.S. Environmental Protection Agency as a *SMALL PUBLIC WATER SYSTEMS TECHNOLOGY ASSISTANCE CENTER.* This recognition will provide funding for the development of training videos for the operators and managers of small public water systems. Additional training materials will be produced to assist the operators and managers of small systems.

PREFACE TO THE FIRST EDITION

The purposes of this water supply system field study training program are to:

1. Develop new qualified water supply system operators,

2. Expand the abilities of existing operators, permitting better service to both their employers and the public, and

3. Prepare operators for civil service and *CERTIFICATION EXAMINATIONS*.[1]

To provide you with the knowledge and skills needed to operate and maintain water supply systems as efficiently and effectively as possible, experienced water supply system operators prepared the material in each chapter of this manual.

Water supply systems vary from city to city and from region to region. The material contained in this program is presented to provide you with an understanding of the basic operation and maintenance aspects of your system and with information to help you analyze and solve operation and maintenance problems. This information will help you operate and maintain your system in a safe and efficient manner.

Water supply operation and maintenance is a rapidly advancing field. To keep pace with scientific and technological advances, the material in this manual must be periodically revised and updated. *THIS MEANS THAT YOU, THE OPERATOR, MUST RECOGNIZE THE NEED TO BE AWARE OF NEW ADVANCES AND THE NEED FOR CONTINUOUS TRAINING BEYOND THIS PROGRAM.*

BASIC TRAINING • ADVANCED TRAINING • CERTIFICATION EXAMINATION

The Project Director is indebted to the many operators and other persons who contributed to this manual. Every effort was made to acknowledge material from the many excellent references in the water supply field. Reviewers Leonard Ainsworth, Jack Rossum, and Joe Monscvitz deserve special recognition for their extremely thorough review and helpful suggestions. John Trax, Chet Pauls, and Ken Hay, Office of Drinking Water, U.S. Environmental Protection Agency, and John Gaston, Bill MacPherson, Bert Ellsworth, Clarence Young, Ted Bakker, and Beverlie Vandre, Sanitary Engineering Branch, California Department of Health Services, all performed outstanding jobs as resource persons, consultants and advisers. Larry Hannah served as Educational Consultant. Illustrations were drawn by Martin Garrity. Charlene Arora helped type the field test and final manuscript for printing. Special thanks are well deserved by the Program Administrator, Gay Kornweibel, who typed, administered the field test, managed the office, administered the budget, and did everything else that had to be done to complete this project successfully.

KENNETH D. KERRI
PROJECT DIRECTOR

1983

---

[1] *Certification Examination.   An examination administered by a state agency that water distribution system operators take to indicate a level of professional competence. In the United States, certification of water distribution system operators is mandatory.*

PREFACE TO THE SECOND EDITION

WATER DISTRIBUTION SYSTEM OPERATION AND MAINTENANCE is the title of this portion of the Second Edition. When the First Edition, **WATER SUPPLY SYSTEM OPERATION**, was developed, the objective was to produce one comprehensive training manual for water supply system operators. The resulting manual covered all aspects of water supply systems from the source of the water supply to the consumer's tap.

Operators using the First Edition, **WATER SUPPLY SYSTEM OPERATION**, expressed a valid concern that too much material was covered in one training manual. In response to this concern the manual was split approximately in half to produce the Second Edition as two manuals:

- **WATER DISTRIBUTION SYSTEM OPERATION AND MAINTENANCE**, and

- **SMALL WATER SYSTEM OPERATION AND MAINTENANCE**.

WATER DISTRIBUTION SYSTEM OPERATION AND MAINTENANCE describes the responsibilities of being an operator, storage facilities, distribution systems, water quality considerations, distribution system operation and maintenance, disinfection and safety. **SMALL WATER SYSTEM OPERATION AND MAINTENANCE** contains information on the responsibilities of being an operator, water sources, wells, small water treatment plants, disinfection, safety, lab procedures and a new chapter on setting water rates. The chapters on disinfection and safety are in both manuals because of their extreme importance. Both of these operator training manuals are equally important if you are responsible for a water supply system. If your agency is large enough to employ two or more crews of operators, the crews may specialize in a particular aspect of operation and maintenance. For example, one crew may deal primarily with wells and disinfection, while the other crew is assigned the responsibility of the distribution system. The splitting of the original manual, **WATER SUPPLY SYSTEM OPERATION**, into two manuals will allow operators to concentrate their studies on the subject areas related directly to their jobs. At the same time, it is hoped that operators will realize the importance of understanding all aspects of water supply systems and attempt to complete both of the manuals.

KENNETH D. KERRI
PROJECT DIRECTOR

1987

OBJECTIVES OF THIS MANUAL

Proper installation, inspection, operation, maintenance, repair, and management of water distribution systems have a significant impact on the operation and maintenance costs and effectiveness of the systems. The objective of this manual is to provide water distribution system operators with the knowledge and skills required to operate and maintain water systems effectively, thus eliminating or reducing the following problems:

1. Health hazards created by the delivery of unsafe water to the consumer's tap;

2. System failures that result from the lack of proper installation, inspection, preventive maintenance, surveillance, and repair programs designed to protect the public's investment in these facilities;

3. Tastes and odors caused by water distribution system problems;

4. Corrosion damages to pipes, equipment, tanks, and structures in the water distribution system;

5. Complaints from the public or local officials due to the unreliability or failure of the water distribution system to perform as designed; and

6. Fire damage caused by insufficient water and/or inadequate pressures at a time of need.

SCOPE OF THIS MANUAL

Operators with the responsibility for water storage and distribution systems will find this manual very helpful. This manual contains information on:

1. What water distribution system operators do,

2. Procedures for operating and maintaining clear wells and storage tanks,

3. Characteristics of distribution system facilities,

4. How to operate and maintain distribution systems,

5. How to maintain water quality in distribution systems,

6. Disinfection of new and repaired facilities as well as water delivered to consumers,

7. Techniques for recognizing hazards and developing safe procedures and safety programs, and

8. How to manage a water distribution system utility.

Material in this manual furnishes you with information concerning situations encountered by most water distribution system operators in most areas. These materials provide you with an understanding of the basic operational and maintenance concepts for water distribution systems and with an ability to analyze and solve problems when they occur. Operation and maintenance programs for water distribution systems will vary with the age of the system, the extent and effectiveness of previous programs, and local conditions. You will have to adapt the information and procedures in this manual to your particular situation.

Technology is advancing very rapidly in the field of operation and maintenance of water distribution systems. To keep pace with scientific advances, the material in this program must be periodically revised and updated. This means that you, the water distribution system operator, must be aware of new advances and recognize the need for continuous personal training reaching beyond this program. *TRAINING OPPORTUNITIES EXIST IN YOUR DAILY WORK EXPERIENCE, FROM YOUR ASSOCIATES, AND FROM ATTENDING MEETINGS, WORKSHOPS, CONFERENCES, AND CLASSES.*

USES OF THIS MANUAL

This manual was developed to serve the needs of operators in several different situations. The format used was developed to serve as a home-study or self-paced instruction course for operators in remote areas or persons unable to attend formal classes either due to shift work, personal reasons, or the unavailability of suitable classes. This home-study training program uses the concepts of self-paced instruction where you are your own instructor and work at your own speed. In order to certify that a person has successfully completed this program, objective tests and special answer sheets for each chapter are provided when a person enrolls in this course.

Also, this manual can serve effectively as a textbook in the classroom. Many colleges and universities have used this manual as a text in formal classes (often taught by operators). In areas where colleges are not available or are unable to offer classes in the operation of water distribution systems, operators and utility agencies can join together to offer their own courses using this manual.

Cities or utility agencies can use this manual in several types of on-the-job training programs. In one type of program, a manual is purchased for each operator. A senior operator or a group of operators are designated as instructors. These operators help answer questions when the persons in the training program have questions or need assistance. The instructors grade the objective tests, record scores, and notify California State University, Sacramento, of the scores when a person successfully completes this program. This approach eliminates any waiting while papers are being graded and returned by CSUS.

This manual was prepared to help operators operate and maintain their water distribution systems. Please feel free to use the manual in the manner which best fits your training needs and the needs of your operators. We will be happy to work with you to assist you in developing your training program. Please feel free to contact:

Project Director
Office of Water Programs
California State University, Sacramento
6000 J Street
Sacramento, California 95819-6025

Phone: (916) 278-6142
FAX: (916) 278-5959

ENROLLMENT FOR CREDIT AND CERTIFICATE

Students wishing to earn credits and a certificate for completing this course may enroll by contacting the Office of Water Programs, California State University, Sacramento, 6000 J Street, Sacramento, CA 95819-6025, (916) 278-6142. If you have already enrolled, the enrollment packet you were sent contains detailed instructions for completing and returning the objective tests. Please read these important instructions carefully before marking your answer sheets.

Following successful completion of each volume in this program, a Certificate of Completion will be sent to you. If you wish, the Certificate can be sent to your supervisor, the mayor of your town, or any other official you think appropriate. Some operators have been presented their Certificate at a City Council meeting, got their picture in the newspaper, and received a pay raise.

INSTRUCTIONS TO PARTICIPANTS IN HOME-STUDY COURSE

Procedures for reading the chapters and answering the questions are contained in this section.

To progress steadily through this program, you should establish a regular study schedule. For example, many operators in the past have set aside two hours during two evenings a week for study.

The study material is contained in eight chapters. Some chapters are longer and more difficult than others. For this reason, many of the chapters are divided into two or more lessons. The time required to complete a lesson will depend on your background and experience. Some people might require an hour to complete a lesson and some might require three hours; but that is perfectly all right. *THE IMPORTANT THING IS THAT YOU UNDERSTAND THE MATERIAL IN THE LESSON!*

Each lesson is arranged for you to read a short section, write the answers to the questions at the end of the section, check your answers against suggested answers; and then *YOU* decide if you understand the material sufficiently to continue or whether you should read the section again. You will find that this procedure is slower than reading a typical textbook, but you will remember much more when you have finished the lesson.

Some discussion and review questions are provided following each lesson in some of the chapters. These questions review the important points you have covered in the lesson. Write the answers to these questions in your notebook.

After you have completed the last chapter, you will find a final examination. This exam is provided for you to review how well you remember the material. You may wish to review the entire manual before you take the final exam. Some of the questions are essay-type questions, which are used by some states for higher-level certification examinations. After you have completed the final examination, grade your own paper and determine the areas in which you might need additional review before your next certification or civil service examination.

You are your own teacher in this program. You could merely look up the suggested answers at the end of the chapters or final exam or copy them from someone else, but you would not understand the material. Consequently, you would not be able to apply the material to the operation of your facilities nor recall it during an examination for certification or a civil service position.

YOU WILL GET OUT OF THIS PROGRAM WHAT YOU PUT INTO IT

SUMMARY OF PROCEDURE

OPERATOR (YOU)

1. Read what you are expected to learn in each chapter (the chapter objectives).

2. Read sections in the lesson.

3. Write your answers to questions at the end of each section in your notebook. You should write the answers to the questions just as you would if these were questions on a test.

4. Check your answers with the suggested answers.

5. Decide whether to reread the section or to continue with the next section.

6. Write your answers to the discussion and review questions at the end of each lesson in your notebook.

ORDER OF WORKING LESSONS

To complete this program you will have to work all of the lessons. You may proceed in numerical order, or you may wish to work some lessons sooner.

SAFETY IS A VERY IMPORTANT TOPIC. Everyone working in a water distribution system must always be safety conscious. Operators daily encounter situations and equipment that can cause a serious disabling injury or illness if the operator is not aware of the potential danger and does not exercise adequate precautions. For these reasons you may decide to work on Chapter 7, "Safety," early in your studies. In each chapter, *SAFE PROCEDURES ARE ALWAYS STRESSED.*

TECHNICAL CONSULTANTS

John Brady	Jim Sequeira
Gerald Davidson	R. Rhodes Trussell
Larry Hannah	Mike Young

NATIONAL ENVIRONMENTAL TRAINING ASSOCIATION REVIEWERS

George Kinias, Project Coordinator

E.E. "Skeet" Arasmith	Andrew Holtan	Rich Metcalf
Terry Engelhardt	Deborah Horton	William Redman
Dempsey Hall	Kirk Laflin	Kenneth Walimaa
Jerry Higgins		Anthony Zigment

PROJECT REVIEWERS

Leonard Ainsworth	Jerry Hayes	Joe Monscvitz	Gerald Samuel
Ted Bakker	Ed Henley	Angela Moore	Carl Schwing
Jo Boyd	Charles Jeffs	Harold Mowry	David Sorenson
Dean Chausee	Chet Latif	Theron Palmer	Russell Sutphen
Walter Cockrell	Frank Lewis	Eugene Parham	Robert Wentzel
Fred Fahlen	Perry Libby	Catherine Perman	James Wright
David Fitch	D. Mackay	David Rexing	Mike Yee
Richard Haberman	William Maguire	Jack Rossum	Clarence Young
Lee Harry	Nancy McTigue	William Ruff	

WATER DISTRIBUTION SYSTEM OPERATION AND MAINTENANCE
COURSE OUTLINE

Other similar operator training programs that may be of interest to you are our courses and training manuals on the operation and maintenance of small water systems, water treatment plant operation (two volumes), and utility management.

SMALL WATER SYSTEM OPERATION AND MAINTENANCE

(Wells, Small Treatment Plants, and Rates)

COURSE OUTLINE

WATER TREATMENT PLANT OPERATION, VOLUME I
COURSE OUTLINE

WATER TREATMENT PLANT OPERATION, VOLUME II
COURSE OUTLINE

UTILITY MANAGEMENT
COURSE OUTLINE

MANAGE FOR SUCCESS
COURSE OUTLINE

CHAPTER 1

THE WATER DISTRIBUTION SYSTEM OPERATOR

by

Ken Kerri

TABLE OF CONTENTS

Chapter 1. THE WATER DISTRIBUTION SYSTEM OPERATOR

OBJECTIVES

Chapter 1.   THE WATER DISTRIBUTION SYSTEM OPERATOR

At the beginning of each chapter in this manual you will find a list of OBJECTIVES. The purpose of this list is to stress those topics in the chapter that are most important. Contained in the list will be items you need to know and skills you must develop to operate, maintain, repair, and manage a water distribution system as efficiently and as safely as possible.

Following completion of Chapter 1, you should be able to:

1. Explain the type of work done by water distribution system operators,

2. Describe where to look for jobs in this profession, and

3. Find sources of further information on how to do the jobs performed by water distribution system operators.

CHAPTER 1.   THE WATER DISTRIBUTION SYSTEM OPERATOR

Chapter 1 is prepared especially for new operators or people interested in becoming water distribution system operators. If you are an experienced water distribution system operator, you may find some new viewpoints in this chapter.

In these training manuals the water supply system has been divided into two parts: (1) the water system, and (2) the water distribution system. The water system begins with the source of water such as groundwater or surface water. Wells, pumps, and treatment facilities are part of the water system which delivers water to the distribution system. The water distribution system consists of tanks and pipes which store and deliver treated water to consumers.

1.0   NEED FOR WATER DISTRIBUTION SYSTEM OPERATORS

People need safe and pleasant water to drink. Many sources of water are not directly suitable for drinking purposes. Engineers have designed water supply facilities to collect, store, transport, treat, and distribute water to people and industries. Once these facilities are built, the operators make the facilities do their intended job. Water distribution system operators operate, maintain, repair, and manage these facilities. Operators have the responsibility of ensuring that safe and pleasant drinking water is delivered to everyone's tap. Another responsibility is to be sure that adequate amounts of water and pressure are available during times of emergency, such as a fire. Cities and towns need qualified, capable, and dedicated operators to do these jobs.

The need for *RESPONSIBLE* water distribution system operators cannot be overstressed. You, as a water distribution system operator, have the responsibility for the health and well-being of the community you serve. Yes, you are responsible for the drinking water of your community and anytime you fail to do your job, you could be responsible for an outbreak of a waterborne disease, which could even result in death. As an operator, you do not want the knowledge that you were negligent in your duty and, as a result, were responsible for the death of a fellow human being.

QUESTIONS

Below are some questions for you to answer. You should have a notebook in which you can write the answers to the questions. By writing down the answers to the questions, you are helping yourself learn and retain the information. After you have answered all the questions, compare your answers with those given in the Suggested Answers section on page 10. Reread any sections you do not understand and then proceed to the next section. You are your own teacher in this training program, and YOU should decide when you understand the material and are ready to continue with new material.

1.0A   What do water distribution system operators do with water distribution facilities?

1.0B   What is the responsibility of water distribution system operators?

1.1   WHAT IS A WATER SUPPLY SYSTEM?

1.10   Sources of Water

A water supply system delivers water from its source to the ultimate consumer—homes, businesses, or industrial water users (Figure 1.1). The source of a water supply may be either groundwater or surface waters. Surface waters may be natural lakes, streams, rivers, or reservoirs behind dams. Groundwater is a water source below the ground surface.

Water is usually transported from its source through a transmission system consisting of either open channels (canals) or large-diameter pipes (called raw water conduits) to the water treatment plant. After treatment the water is usually pumped into transmission lines that are connected to a distribution grid system. From the grid system the water is delivered to individual service lines which serve private homes, businesses, factories, and industries. Groundwaters may not have transmission lines, but are often disinfected with chlorine and pumped directly to service storage and distribution systems.

1.11   Storage Facilities

Storage facilities may consist of large reservoirs behind dams (impoundments) or service storage reservoirs located at water treatment plants and/or at various places in distribution systems. Storage facilities for treated water at water treatment plants are called clear wells. Operational service storage tanks in distribution systems may be pressure tanks, elevated tanks, ground-level tanks or reservoirs, or underground facilities.

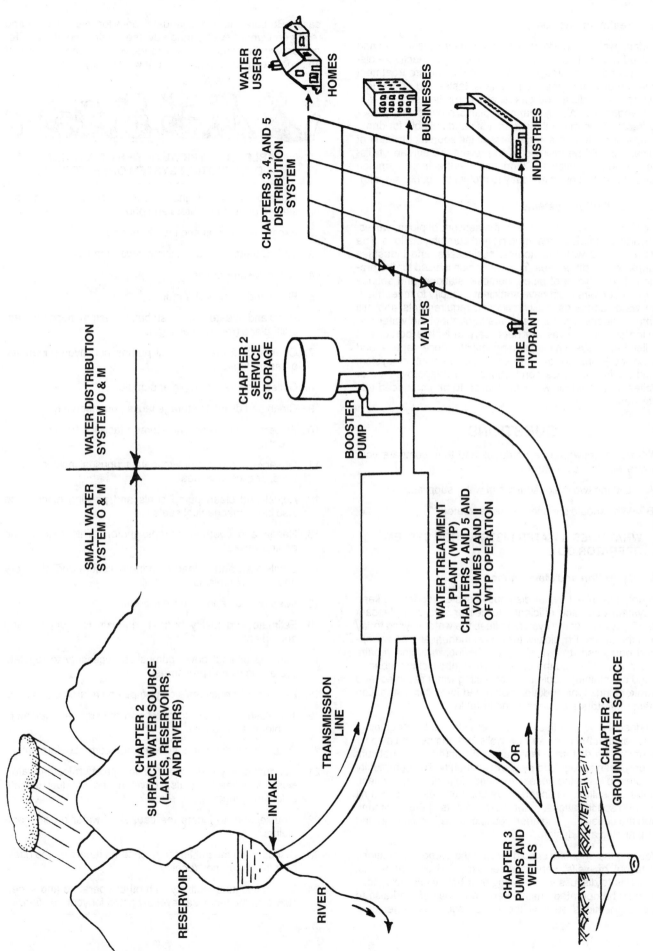

Fig. 1.1 Typical water supply system

1.12 Treatment Facilities

Surface waters require treatment to remove suspended and dissolved materials and also to remove, kill, or inactivate disease-causing organisms. Groundwater may require treatment for the removal of excessive hardness, taste- and odor-causing substances, dissolved gases, and impurities such as iron and manganese. Water treatment plants provide the necessary treatment to make the waters safe and suitable for drinking purposes. All waters, regardless of the source or treatment received, should be disinfected to prevent the spread of disease-causing organisms. Chlorination is the most common means of disinfection used today to protect the public's health.

1.13 Distribution Systems

A distribution system consists of a network of pipes, valves, fire hydrants, service lines, meters, and pumping stations. The system delivers water to homes, businesses, and industries for drinking and other uses. This water also is used for fire protection. The network of pipes, pumping stations, and service storage reservoirs must have sufficient capacity to meet maximum water demands plus firefighting requirements and still maintain adequate water pressures throughout the water distribution system. Valves are necessary to isolate portions of the distribution system for cleaning, maintenance, repairs, and making additions to the system. The distribution system should be free of cross connections with unapproved water supplies that could allow contamination to be introduced into the system.

QUESTIONS

Write your answers in a notebook and then compare your answers with those on page 10.

1.1A List the two major sources of water supplies.

1.1B Why should surface waters be treated?

1.2 WHAT DOES A WATER DISTRIBUTION SYSTEM OPERATOR DO?

1.20 Operation and Maintenance

Simply described, water distribution system operators keep the system operating efficiently to deliver safe and pleasant water. They inspect the system to keep the water flowing today and in the future. Physically, they have manual and power-operated equipment to help them do the job, including repairs and minor additions to the system. Other jobs include lubricating and maintaining equipment, collecting water samples, and recordkeeping. Typical duties performed by water distribution system operators are summarized in Table 1.1.

To describe your duties in detail, let's start at the beginning. Let us say that the need for new wells and pumps, and also for new or improved storage tanks and a distribution pipe network has long been recognized by your community. The community has voted to issue the necessary bonds to finance the project, and the consulting engineers have been requested to submit plans and specifications. In the best interests of the community and the consulting engineer, you should be in on the ground floor planning and design.

You and the engineer should discuss the proposed locations and layout of wells, service reservoirs, and pipe networks. Your job is to discuss with the engineer how these new facilities could be operated, maintained, and repaired, and also to make suggestions on how these jobs could be done more

easily. Be sure there is adequate room for maintenance and repair equipment even during adverse weather conditions. Together with the consulting engineers, you can be a member of an expert team able to advise your water utility.

TABLE 1.1 TYPICAL DUTIES OF A WATER DISTRIBUTION SYSTEM OPERATOR

1. Place barricades, signs, and traffic cones around work sites to protect operators and public.

2. Excavate trenches and install shoring.

3. Lay, connect, test, and disinfect water mains.

4. Tap into water mains.

5. Flush and clean water mains.

6. Read and update water distribution system maps and "as-built" plans (record drawings).

7. Operate and maintain well pumps and hydropneumatic pressure tanks.

8. Collect and transport water samples.

9. Clean and disinfect storage tanks and reservoirs.

10. Protect water mains and storage facilities from corrosion effects.

11. Observe pump motors to detect unusual noises, vibrations, or excessive heat.

12. Adjust and clean pump seals and packing glands and also clean mechanical seals.

13. Repair and overhaul pumps, motors, chlorinators, and control valves.

14. Safely load and unload chlorine cylinders and other dry and liquid chemicals.

15. Keep records and prepare reports.

16. Estimate and justify budget requests for supplies and equipment.

17. Start up or shut down pumps as necessary to regulate system flows and pressures.

18. Perform efficiency tests on pumps and related equipment.

19. Troubleshoot minor electrical and mechanical equipment problems and correct.

20. Detect hazardous atmospheres and correct before entry.

21. Conduct safety inspections, follow safety rules for waterworks facilities, and also develop and conduct tailgate safety meetings.

22. Troubleshoot to locate the causes of water quality complaints.

23. Discuss with the public their concerns regarding the quality of the water they receive.

24. Communicate effectively with other operators and supervisors on the technical level expected for your position.

Ultimately you want to operate your pumps and reservoirs in such a fashion that your consumers will always have sufficient, safe, and pleasant water at adequate pressures.

1.21 Supervision and Administration

In addition to operation and maintenance duties for your water distribution system, you may also be responsible for supervision and personnel. Chief operators or supervisors frequently have the responsibility of training new operators and should encourage all operators to strive for higher levels of certification.

As a water distribution system administrator, you may be in charge of recordkeeping. In this case, you will be responsible for operating and maintaining the system as efficiently as possible, keeping in mind that the primary objective is to deliver safe and pleasant drinking water to your consumers. Without adequate, reliable records of important phases of operation and maintenance, the effectiveness of your operation has not been documented (recorded). Also, accurate records are required by regulatory agencies in accordance with the Drinking Water Regulations of the Safe Drinking Water Act.

You may also be the budget administrator. Here you will be in the best position to give advice on budget requirements, management problems, and future planning. You should be aware of the necessity for additional expenditures, including funds for new facilities and enlargement of existing facilities, equipment replacement, and additional employees. You should recognize these and define them clearly for the proper officials. Early planning and budgeting of this type will contribute greatly to the continued smooth operation of your facility.

1.22 Public Relations

As an operator you are in the field of public relations and must be able to explain the purpose and operation of your facilities to civic organizations, school classes, representatives of news media, and even to city council members or directors of your district. Lots of people want to know about the large elevated storage tank or those "funny things" sticking out the top of an underground storage reservoir. One of the best results from a well-guided tour is gaining support from your city council and the public to obtain the funds necessary to run a good operation.

The overall appearance of your pump stations and elevated tanks indicates to the public the type of operation you maintain. If the facilities are dirty and rundown, in need of painting, and overgrown with weeds, you will be unable to convince the public that you are doing a good job. *YOUR RECORDS SHOWING THAT YOU ARE DELIVERING SAFE DRINKING WATER TO YOUR CONSUMERS WILL MEAN NOTHING TO VISITORS AND NEIGHBORS OF YOUR FACILITIES UNLESS YOUR FACILITIES APPEAR CLEAN AND WELL MAINTAINED.*

Another aspect of your job may be handling complaints. When someone contacts you complaining that their drinking water looks muddy, tastes bad, or smells bad, you may have a serious problem. Whenever someone complains, record all of the necessary information (name, date, location, and phone number) and have the complaint thoroughly investigated. Be sure to notify the person who complained of the results of your investigation and what corrective action was or will be taken.

1.23 Safety

Safety is a very important operator responsibility. Unfortunately, too many water distribution system operators take safety for granted. *YOU* have the responsibility to be sure that your facilities are a safe place to work and that everyone follows safe procedures. Following safe procedures is the most important responsibility of the operator with regard to safety. Everyone must follow safe procedures and understand why safe procedures must be followed at all times. All operators must be aware of safety hazards in and around water distribution facilities. Work in traffic and excavating for the installation or repair of pipes can be extremely hazardous. Explosive conditions can develop when painting the inside of an elevated tank if adequate ventilation is not provided. Most accidents result from carelessness or negligence. You should plan or be a part of an active safety program. Also, you may have the responsibility of training new operators and safe procedures must be stressed.

Clearly, today's water distribution system operator must be capable of doing many jobs—*AND DOING THEM ALL SAFELY.*

Write your answers in a notebook and then compare your answers with those on page 10.

1.2A   Why should water distribution system operators discuss proposed facilities with engineers?

1.2B   Why are adequate and reliable records very important?

1.2C   To whom might you have to explain the purpose and operation of your facilities?

1.2D   Why is the appearance of pump stations and elevated tanks and the grounds around them important?

1.2E   Why is safety important?

1.3 JOB OPPORTUNITIES

1.30 Staffing Needs

The water distribution system field is changing rapidly. New facilities and systems are being constructed, and old facilities and systems are being modified and enlarged to meet the water distribution demands of our growing population and industries. Towns, municipalities, special districts, and industries all employ water distribution system operators. Operators, maintenance personnel, foremen, managers, instrumentation experts, and laboratory technicians are sorely needed now and will be in the future. The Bureau of Labor Statistics (BLS)[1] predicts faster than average job growth in the water industry through the year 2006.

1.31 Who Hires Water Distribution System Operators?

Operators' paychecks usually come from a city, water agency or district, or a private utility company. The operator also may be employed by one of the many large industries that operate their own water distribution system. As an operator, you are always responsible to your employer for operating and maintaining an economical and efficient water distribution system. An even greater obligation rests with the operator because of the great number of people who drink the water from the distribution system. In the final analysis, the operator is really working for the people who depend on the operator to provide them with safe and pleasant drinking water.

1.32 Where Do Water Distribution System Operators Work?

Jobs are available for water distribution system operators wherever people live and need someone to deliver water to their homes, offices, or industrial processes. The different types and locations of water distribution systems offer a wide range of working conditions. From the mountains to the sea, wherever people gather together into communities, water distribution systems will be found. From a distribution system foreman or a computer control center operator at a complex municipal storage and distribution system to a one-person manager of a small town water distribution system, you can select your own special place in water distribution system operation.

1.33 What Pay Can a Water Distribution System Operator Expect?

In dollars? Prestige? Job satisfaction? Community service? In opportunities for advancement? By whatever scale you use, returns are mainly what you make them. If you choose a large municipality, the pay is good and advancement prospects are tops. Choose a small town and the pay may not be as good, but job satisfaction, freedom from time-clock hours, community service, and prestige may well add up to a more desirable outstanding personal achievement. If you have the ability and take advantage of the opportunities, you can make this field your career and advance to an enviable position. Many of

these positions are or will be represented by an employee organization that will try to obtain higher pay and other benefits for you. Total reward depends on you and how *YOU APPLY YOURSELF.*

1.34 What Does It Take To Be a Water Distribution System Operator?

DESIRE. First you must make the serious decision to enter this fine profession. You can do this with a high school or a college education. While some jobs will always exist for manual labor, the real and expanding need is for *QUALIFIED OPERATORS.* You must be willing to study and take an active role in upgrading your capabilities. New techniques, advanced equipment, and increasing use of complex instrumentation and computers require a new breed of water distribution system operator; one who is willing to learn today, and gain tomorrow, for surely your water distribution system will move toward newer and more effective operation and maintenance procedures. Indeed, the truly service-minded operator assists in adding to and improving the performance of the water distribution system on a continuing basis.

You can be a water distribution system operator tomorrow by beginning your learning today; or you can be a better operator, ready for advancement, by accelerating your learning today.

This training course, then, is your start toward a better tomorrow, both for you and for the public who will receive better water from your efforts.

QUESTIONS

Write your answers in a notebook and then compare your answers with those on page 10.

1.3A   Who hires water distribution system operators?

1.3B   What does it take to be a good water distribution system operator?

1.4 PREPARING YOURSELF FOR THE FUTURE

1.40 Your Qualifications

What do you know about your job or the job you'd like to obtain? Perhaps a little, and perhaps a lot. You must evaluate the knowledge, skills, and experience you already have and what you will need to achieve future jobs and advancement.

The knowledge, skills, abilities, and judgment required for your job depend to a large degree on the size and type of water distribution system where you work. You may work on a large, complex system serving several hundred thousand persons and employing a hundred or more operators. In this case, you are probably a specialist in one or more phases of the distribution system or storage reservoirs (such as being responsible for the valves).

On the other hand, you may operate and maintain a small water distribution system serving only a thousand people or even fewer. You may be the only operator for the entire system or, at best, have only one or two helpers. If this is the case, you must be a "jack-of-all-trades" because of the diversity of your tasks.

---

[1] *For additional information about future job opportunities, refer to the Bureau of Labor Statistics' website at www.bls.gov.*

1.41 Your Personal Training Program

Beginning on this page you are starting a training course that has been carefully prepared to help you improve your knowledge and skills to operate, maintain, and manage water distribution systems.

You will be able to proceed at your own pace; you will have the opportunity to learn a little or a lot about each topic. This training manual has been prepared this way to meet the various needs of water distribution system operators, depending on the size and type of system for which you are responsible. To study for certification and civil service exams, you may have to cover most of the material in this manual. You will never know everything about water distribution systems and the equipment and procedures available for operation and maintenance. However, you will be able to answer some very important questions about how, why, and when certain things happen in water distribution systems. You can also learn how to manage your water distribution system to provide a reliable supply of safe and pleasant drinking water to your customers while minimizing costs in the long run.

This training course is not the only one available to help you improve your abilities. Some state water utility associations, operator associations, vocational schools, community colleges, and universities offer training courses on both a short- and long-term basis. Many state, local, and private agencies have conducted training programs and informative seminars. Most state health departments can be very helpful in providing training programs or directing you to good programs.

Some libraries can provide you with useful journals and books on water distribution systems. Listed below are several very good references in the field of water distribution systems. Prices listed were those available when this manual was published; they will probably increase in the future.

1. *MANUAL OF INSTRUCTION FOR WATER TREATMENT PLANT OPERATORS (NEW YORK MANUAL)*. Obtain from Health Education Services, Inc., PO Box 7126, Albany, NY 12224. Price, $36.00, includes cost of shipping and handling.

2. *MANUAL OF WATER UTILITIES OPERATIONS (TEXAS MANUAL)*. Obtain from Texas Water Utilities Association, 1106 Clayton Lane, Suite 101 East, Austin, TX 78723-1093. Price to members, $22.85; nonmembers, $34.85; price includes cost of shipping and handling.

3. *WATER DISTRIBUTION OPERATOR TRAINING HANDBOOK*. Obtain from American Water Works Association (AWWA), Bookstore, 6666 West Quincy Avenue, Denver,

CO 80235. Order No. 20428. ISBN 1-58321-014-8. Price to members, $52.00; nonmembers, $76.00; price includes cost of shipping and handling.

4. *WATER SOURCES*. Obtain from American Water Works Association (AWWA), Bookstore, 6666 West Quincy Avenue, Denver, CO 80235. Order No. 1955. ISBN 1-58321-229-9. Price to members, $77.00; nonmembers, $113.00; price includes cost of shipping and handling.

5. *WATER TRANSMISSION AND DISTRIBUTION*. Obtain from American Water Works Association (AWWA), Bookstore, 6666 West Quincy Avenue, Denver, CO 80235. Order No. 1957. ISBN 1-58321-231-0. Price to members, $87.00; nonmembers, $128.00; price includes cost of shipping and handling.

6. *OPERATOR CERTIFICATION STUDY GUIDE: A GUIDE TO PREPARING FOR WATER TREATMENT AND DISTRIBUTION OPERATOR CERTIFICATION EXAMS*. Obtain from American Water Works Association (AWWA), Bookstore, 6666 West Quincy Avenue, Denver, CO 80235. Order No. 20517. ISBN 1-58321-287-6. Price to members, $52.00; nonmembers, $76.00; price includes cost of shipping and handling.

Throughout this manual we will be recommending American Water Works Association (AWWA) publications. Members of AWWA can buy some publications at reduced prices. You can join AWWA by writing to the headquarters office in Denver or by contacting a member of AWWA. Headquarters can help you contact your own state or regional AWWA Section. This professional organization can offer you many helpful training opportunities and educational materials when you join and actively participate with your associates in the field. We encourage you to visit AWWA's website at www.awwa.org.

1.42 Certification

Certification examinations are usually administered by state regulatory agencies or professional associations. Operators take these exams in order to obtain certificates which indicate a level of professional competence. You should continually strive to achieve higher levels of certification. Successful completion of this operator training program will help you achieve your certification goals.

1.5 ACKNOWLEDGMENTS

Many of the topics and ideas discussed in this chapter were based on similar work written by Larry Trumbull and Walt Driggs.

SUGGESTED ANSWERS

Chapter 1. THE WATER DISTRIBUTION SYSTEM OPERATOR

You are not expected to have the exact answers suggested for questions requiring written answers, but you should have the correct idea. The numbering of the questions refers to the section in the chapter where you can find the information to answer the questions. Answers to questions numbered 1.0A and 1.0B can be found in Section 1.0, "Need for Water Distribution System Operators."

Answers to questions on page 4.

1.0A   Water distribution system operators make sure water facilities do their intended job. Operators operate, maintain, repair, and manage these facilities.

1.0B   Water distribution system operators have the responsibility of ensuring that safe and pleasant drinking water is delivered to everyone's tap. Also adequate amounts of water and pressure must be available during times of emergency, such as a fire.

Answers to questions on page 6.

1.1A   The two major sources of water supplies are (1) groundwater, and (2) surface water.

1.1B   Surface water is usually treated to remove suspended and dissolved materials and to remove, kill, or inactivate disease-causing organisms.

Answers to questions on page 7.

1.2A   Water distribution system operators should discuss proposed facilities with engineers to find out how the engineer intends for these facilities to be operated, maintained, and repaired, and also to make suggestions on how these jobs could be done more easily. Be sure there is adequate room for maintenance and repair equipment even during adverse weather conditions.

1.2B   Adequate and reliable records are very important to document the effectiveness of your operation.

1.2C   Water distribution system operators must be able to explain the purpose and operation of their facilities to civic organizations, school classes, representatives of news media, and even to city council members or directors of their district.

1.2D   The appearance of pump stations and elevated tanks and the grounds around them indicates to the public the type of operation you maintain.

1.2E   Safety is a very important operator responsibility. Most accidents result from carelessness or negligence. Safe procedures must be stressed at all times.

Answers to questions on page 8.

1.3A   Water distribution system operators may be hired by cities, water agencies or districts, private utility companies, or industries.

1.3B   *DESIRE.* If you want to be a qualified water distribution system operator, you can do it.

CHAPTER 2

STORAGE FACILITIES

by

Dan Saenz

TABLE OF CONTENTS
Chapter 2.   STORAGE FACILITIES

OBJECTIVES

Chapter 2.   STORAGE FACILITIES

Following completion of Chapter 2, you should be able to:

1. Identify various types of storage facilities,

2. Determine suitable locations for facilities,

3. Inspect storage facilities,

4. Take a storage facility out of service and put it back on line,

5. Safely operate and maintain a storage facility,

6. Select protective coatings for a storage facility,

7. Apply interior and exterior protective coatings to a storage facility,

8. Collect samples from a storage facility,

9. Troubleshoot storage facility problems,

10. Protect a storage facility from corrosion,

11. Disinfect a storage facility, and

12. Maintain records for a storage facility.

PROJECT PRONUNCIATION KEY

by Warren L. Prentice

The Project Pronunciation Key is designed to aid you in the pronunciation of new words. While this key is based primarily on familiar sounds, it does not attempt to follow any particular pronunciation guide. This key is designed solely to aid operators in this program.

You may find it helpful to refer to other available sources for pronunciation help. Each current standard dictionary contains a guide to its own pronunciation key. Each key will be different from each other and from this key. Examples of the difference between the key used in this program and the *WEBSTER'S NEW WORLD COLLEGE DICTIONARY*[1] "Key" are shown below.

In using this key, you should accent (say louder) the syllable that appears in capital letters. The following chart is presented to give examples of how to pronounce words using the Project Key.

WORD	SYLLABLE				
	1st	2nd	3rd	4th	5th
acid	AS	id			
coliform	COAL	i	form		
biological	BUY	o	LODGE	ik	cull

The first word, *ACID*, has its first syllable accented. The second word, *COLIFORM*, has its first syllable accented. The third word, *BIOLOGICAL*, has its first and third syllables accented.

We hope you will find the key useful in unlocking the pronunciation of any new word.

Term	Project Key	Webster Key
acid	AS-id	aś id
coliform	COAL-i-form	kō′ lə fôrm
biological	BUY-o-LODGE-ik-cull	bī ə läj′ i kəl

---

[1] *The WEBSTER'S NEW WORLD COLLEGE DICTIONARY, Fourth Edition, 1999, was chosen rather than an unabridged dictionary because of its availability to the operator. Other editions may be slightly different.*

WORDS

Chapter 2.   STORAGE FACILITIES

AIR GAP

An open vertical drop, or vertical empty space, that separates a drinking (potable) water supply to be protected from another water system in a water treatment plant or other location. This open gap prevents the contamination of drinking water by backsi-phonage or backflow because there is no way raw water or any other water can reach the drinking water supply.

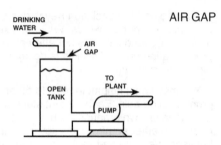

ALTITUDE VALVE   ALTITUDE VALVE

A valve that automatically shuts off the flow into an elevated tank when the water level in the tank reaches a predetermined level. The valve automatically opens when the pressure in the distribution system drops below the pressure in the tank.

ANODE (an-O-d)   ANODE

The positive pole or electrode of an electrolytic system, such as a battery. The anode attracts negatively charged particles or ions (anions).

AVAILABLE CHLORINE   AVAILABLE CHLORINE

A measure of the amount of chlorine available in chlorinated lime, hypochlorite compounds, and other materials that are used as a source of chlorine when compared with that of elemental (liquid or gaseous) chlorine.

AVERAGE DEMAND   AVERAGE DEMAND

The total demand for water during a period of time divided by the number of days in that time period. This is also called the average daily demand.

BACKFLOW   BACKFLOW

A reverse flow condition, created by a difference in water pressures, which causes water to flow back into the distribution pipes of a potable water supply from any source or sources other than an intended source. Also see BACKSIPHONAGE.

BACKSIPHONAGE   BACKSIPHONAGE

A form of backflow caused by a negative or below atmospheric pressure within a water system. Also see BACKFLOW.

BREAKPOINT CHLORINATION   BREAKPOINT CHLORINATION

Addition of chlorine to water until the chlorine demand has been satisfied. At this point, further additions of chlorine will result in a free chlorine residual that is directly proportional to the amount of chlorine added beyond the breakpoint.

CATHODE (KA-thow-d)   CATHODE

The negative pole or electrode of an electrolytic cell or system. The cathode attracts positively charged particles or ions (cations).

CATHODIC (ca-THOD-ick) PROTECTION   CATHODIC PROTECTION

An electrical system for prevention of rust, corrosion, and pitting of metal surfaces which are in contact with water or soil. A low-volt-age current is made to flow through a liquid (water) or a soil in contact with the metal in such a manner that the external electromo-tive force renders the metal structure cathodic. This concentrates corrosion on auxiliary anodic parts which are deliberately allowed to corrode instead of letting the structure corrode.

CENTRIFUGAL (sen-TRIF-uh-gull) PUMP   CENTRIFUGAL PUMP

A pump consisting of an impeller fixed on a rotating shaft that is enclosed in a casing, and having an inlet and discharge connection. As the rotating impeller whirls the liquid around, centrifugal force builds up enough pressure to force the water through the discharge outlet.

COLIFORM (COAL-i-form)   COLIFORM

A group of bacteria found in the intestines of warm-blooded animals (including humans) and also in plants, soil, air and water. Fecal coliforms are a specific class of bacteria which only inhabit the intestines of warm-blooded animals. The presence of coliform bacteria is an indication that the water is polluted and may contain pathogenic (disease-causing) organisms.

CROSS CONNECTION   CROSS CONNECTION

A connection between a drinking (potable) water system and an unapproved water supply. For example, if you have a pump moving nonpotable water and hook into the drinking water system to supply water for the pump seal, a cross connection or mixing between the two water systems can occur. This mixing may lead to contamination of the drinking water.

ELECTROMOTIVE FORCE (E.M.F.)   ELECTROMOTIVE FORCE (E.M.F.)

The electrical pressure available to cause a flow of current (amperage) when an electric circuit is closed. Also called VOLTAGE.

FLOAT ON SYSTEM   FLOAT ON SYSTEM

A method of operating a water storage facility. Daily flow into the facility is approximately equal to the average daily demand for water. When consumer demands for water are low, the storage facility will be filling. During periods of high demand, the facility will be emptying.

FOOT VALVE   FOOT VALVE

A special type of check valve located at the bottom end of the suction pipe on a pump. This valve opens when the pump operates to allow water to enter the suction pipe but closes when the pump shuts off to prevent water from flowing out of the suction pipe.

ION   ION

An electrically charged atom, radical (such as SO_4^{2-}), or molecule formed by the loss or gain of one or more electrons.

NAMEPLATE   NAMEPLATE

A durable metal plate found on equipment which lists critical operating conditions for the equipment.

PCBs   PCBs

Polychlorinated **B**iphenyls. A class of organic compounds that cause adverse health effects in domestic water supplies.

PEAK DEMAND   PEAK DEMAND

The maximum momentary load placed on a water treatment plant, pumping station or distribution system. This demand is usually the maximum average load in one hour or less, but may be specified as the instantaneous load or the load during some other short time period.

POSITIVE BACTERIOLOGICAL SAMPLE   POSITIVE BACTERIOLOGICAL SAMPLE

A water sample in which gas is produced by coliform organisms during incubation in the multiple tube fermentation test. See Chapter 11, Laboratory Procedures, "Coliform Bacteria," in *WATER TREATMENT PLANT OPERATION*, Volume I, for details.

PRIME   PRIME

The action of filling a pump casing with water to remove the air. Most pumps must be primed before start-up or they will not pump any water.

RESIDUAL CHLORINE   RESIDUAL CHLORINE

The concentration of chlorine present in water after the chlorine demand has been satisfied. The concentration is expressed in terms of the total chlorine residual, which includes both the free and combined or chemically bound chlorine residuals.

SET POINT   SET POINT

The position at which the control or controller is set. This is the same as the desired value of the process variable. For example, a thermostat is set to maintain a desired temperature.

STALE WATER   STALE WATER

Water which has not flowed recently and may have picked up tastes and odors from distribution lines or storage facilities.

TURBIDITY UNITS (TU)   TURBIDITY UNITS (TU)

Turbidity units are a measure of the cloudiness of water. If measured by a nephelometric (deflected light) instrumental procedure, turbidity units are expressed in nephelometric turbidity units (NTU) or simply TU. Those turbidity units obtained by visual methods are expressed in Jackson Turbidity Units (JTU) which are a measure of the cloudiness of water; they are used to indicate the clarity of water. There is no real connection between NTUs and JTUs. The Jackson turbidimeter is a visual method and the nephelometer is an instrumental method based on deflected light.

VOLTAGE

VOLTAGE

The electrical pressure available to cause a flow of current (amperage) when an electric circuit is closed. Also called ELECTROMOTIVE FORCE (E.M.F.).

WATER HAMMER

WATER HAMMER

The sound like someone hammering on a pipe that occurs when a valve is opened or closed very rapidly. When a valve position is changed quickly, the water pressure in a pipe will increase and decrease back and forth very quickly. This rise and fall in pressures can cause serious damage to the system.

CHAPTER 2.   STORAGE FACILITIES

2.0   PURPOSE OF STORAGE FACILITIES[2]

The main purpose of a water storage facility is to provide a sufficient amount of water to average or equalize the daily demands on the water supply system. The storage facility should be able to provide water for *AVERAGE*[3] and *PEAK DEMANDS*.[4] Also, the storage facility helps maintain adequate pressures throughout the entire system.

Other purposes of water storage include meeting the needs for fire protection, industrial requirements, and reserve storage. During a fire or other type of emergency, sufficient storage should be available to meet fire demands, as well as other demands, and also maintain system pressures. In some areas the water supply system may serve some type of industry. Storage requirements will depend on the type of industry and the flow and pressure demands of the industrial activities of each industrial facility served by the water supply system. Reserve storage requirements depend on standby requirements and alternate sources of water supply. Reserve requirements may be specified by fire insurance regulations. Reserve storage capacity may be provided to meet future growth and development demands of the area being served.

Reservoirs are storage facilities and may be of several different types. We often think of a reservoir as an open body of water contained by an earth-fill dam or a concrete dam. This chapter, however, will discuss various types of steel and concrete tanks (Figures 2.1 and 2.2) which are basically covered distribution system reservoirs for the storage of treated water and are commonly used in most water systems, especially small water systems.

2,000,000-Gallon Standpipe—Brown Deer, Wis.

1,700,000-Gallon Reservoir—Pleasant Hill, Calif.

500,000-Gallon Pedestal Spheroid—Warwick, R.I.

2,000,000-Gallon Turbospherical—Denton, Texas

Fig. 2.1   *Photos of typical storage facilities*
(Permission of Pittsburgh-Des Moines Corporation)

---

[2] *For additional information on storage facilities, see Chapter 3, "Reservoir Management and Intake Structures," in WATER TREATMENT PLANT OPERATION, Volume I, in this series of manuals.*

[3] *Average Demand.   The total demand for water during a period of time divided by the number of days in that time period. This is also called the average daily demand.*

[4] *Peak Demand.   The maximum momentary load placed on a water treatment plant, pumping station or distribution system. This demand is usually the maximum average load in one hour or less, but may be specified as the instantaneous load or the load during some other short time period.*

The requirements for a specific storage facility will depend upon a system's individual needs. To select a suitable type of storage facility, the answers to the following questions must be known:

1. What is the maximum-day use?

2. What is the maximum-hourly use?

3. What type of pressure will the facility be required to provide and maintain throughout the system?

4. What size will be necessary to fulfill the requirements for emergencies such as fire flow?

Water storage facilities are used to store water from wells or water treatment facilities at times when demands for water are low and to distribute the water during periods of high demand. Water storage facilities are found at one or more locations in areas closest to the ultimate users, where higher pressures are needed, and where land is available. The benefits of distribution system storage are:

1. Demands on the source of water, the pumping facilities, and the transmission and distribution mains are more nearly equalized and also the capacities of the tanks and other treatment facilities in the system need not be so large.

2. System flows and pressures are improved and stabilized, thus providing better service to the customers in the area.

3. Reserve water supplies are provided in the distribution system for emergencies such as firefighting and power outages.

QUESTIONS

Write your answers in a notebook and then compare your answers with those on page 45.

2.0A   What are the purposes of water storage facilities?

2.0B   List the benefits of distribution system storage.

2.1   TYPES OF STORAGE FACILITIES

The following types of storage facilities are common to a water system: clear wells (Figure 2.3), elevated tanks (Figure 2.4), standpipes (Figure 2.5), ground-level storage tanks (Figure 2.6), hydropneumatic or pressure tanks (Figure 2.7), and surge tanks (Figure 2.8). Let's study the definition of each tank and the differences among the tanks.

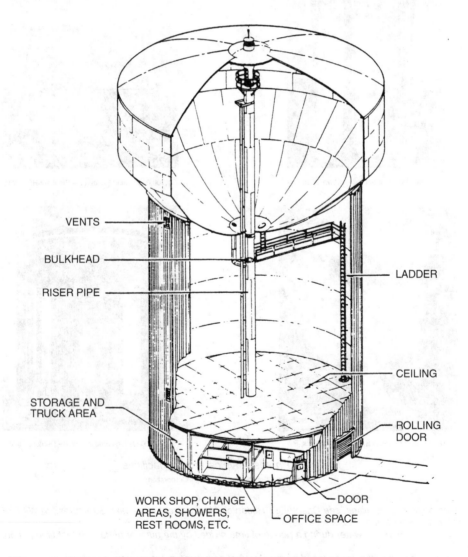

VENTS

BULKHEAD

RISER PIPE

LADDER

CEILING

STORAGE AND TRUCK AREA

ROLLING DOOR

WORK SHOP, CHANGE AREAS, SHOWERS, REST ROOMS, ETC.

DOOR

OFFICE SPACE

Fig. 2.2   Elevated tank (cross-sectional view)

(Permission of Pittsburgh-Des Moines Corporation)

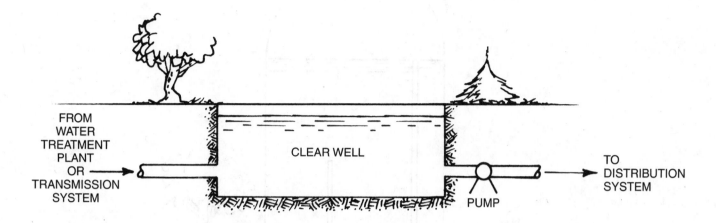

FROM WATER TREATMENT PLANT OR TRANSMISSION SYSTEM

CLEAR WELL

TO DISTRIBUTION SYSTEM

PUMP

Fig. 2.3   Clear well

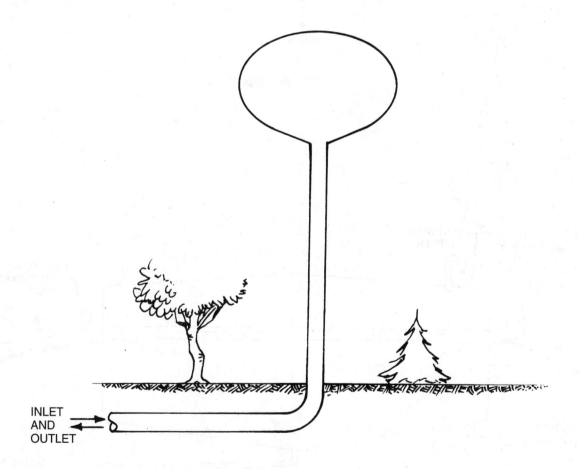

INLET AND OUTLET

Fig. 2.4   Elevated storage tank

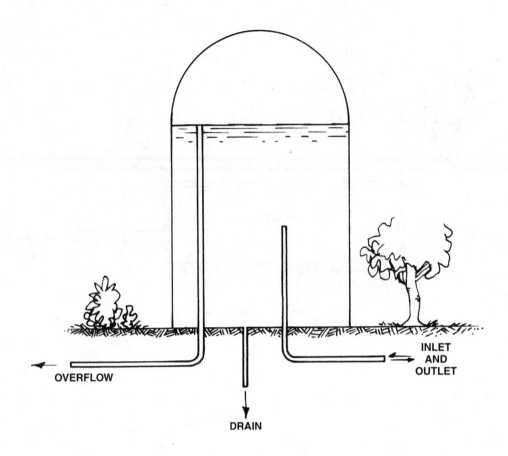

Fig. 2.5 Standpipe

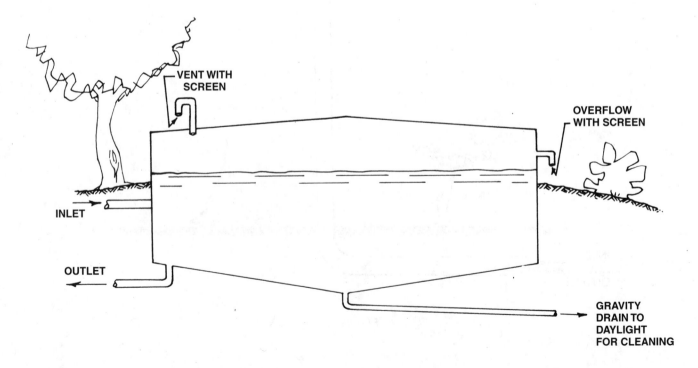

Fig. 2.6 Ground-level service storage reservoir

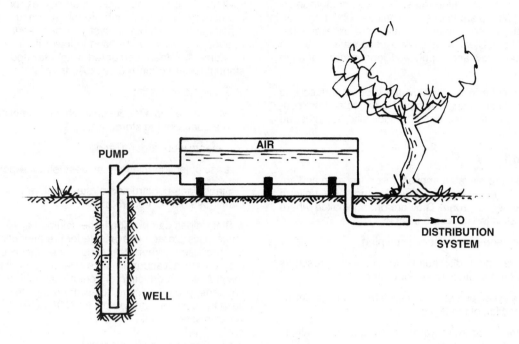

Fig. 2.7   Hydropneumatic or pressure tank

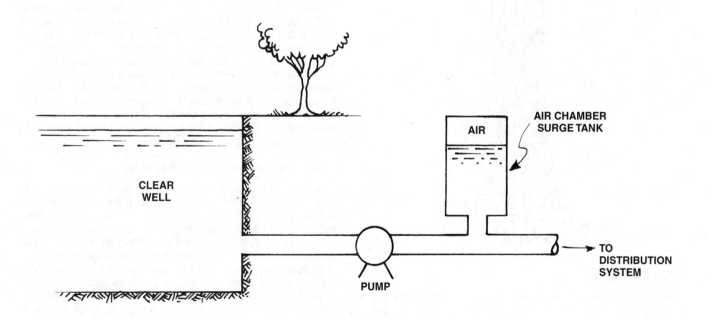

Fig. 2.8   Surge tank

2.10   Clear Wells

Clear wells are used for the storage of filtered water from a treatment plant. These storage tanks are of sufficient capacity to allow fairly constant filtration rates. This type of storage facility will allow a filter plant to operate continuously at an average flow rate and place neither fluctuating nor excessive demands on the filters. Clear wells store excess water when the demand for water is low and supply additional water when the demand is high.

Clear wells may be located below the ground surface and are often made of concrete. They must be protected from contamination and unauthorized entry. Periodically they must be inspected for leaks.

2.11   Elevated Tanks

Elevated tanks are elevated above the service zone and are used primarily to maintain an adequate and fairly uniform pressure to that service zone. They are often installed where the land is flat. Elevated tanks are used to:

1. Eliminate the need for continuous pumping,

2. Minimize variations in distribution system water pressures due to short-term shutdowns of power or pumps,

3. Equalize the water pressure in the distribution system by the proper location of the tanks,

4. Provide a small amount of water for storage (especially to meet demands such as fires), and

5. Reduce auxiliary power requirements.

One limitation of elevated tanks is that the pressure in the distribution system may vary with the water level in the tank.

2.12   Standpipes

Standpipes are tanks that stand on the ground and have a height greater than their diameter. For example, a tank that has the following dimensions is a standpipe—a 10-foot (3-m) diameter and a height of 20 feet (6 m). The bottom of a standpipe is on the ground, whereas the bottom of an elevated tank is above the ground.

Standpipes may be constructed of steel or concrete. They are usually located on high ground at or near a well field or at a point in the distribution system where equalizing storage is needed. Standpipes are used to lower fire insurance costs of consumers. When compared with elevated storage tanks, standpipes are preferred because they are:

1. Easier to maintain,

2. More accessible for observation and sampling to determine the quality of the stored water,

3. Safer to work around, and

4. Less objectionable from an aesthetic viewpoint.

Standpipes constructed of steel require more maintenance than those built of concrete.

Standpipes can provide large volumes of water at low pressures. This storage is available for fire protection if fire pumpers are used. When operating under these conditions, you may create a vacuum on the distribution system. This may result in the introduction of contamination to the mains. If this happens, you must issue boil-water orders to your consumers, flush the system, and sample for COLIFORMS[5] to be sure the water is safe to drink.

2.13   Ground-Level Reservoirs

Most ground-level reservoirs are constructed of concrete and are either circular or rectangular in shape. They may be buried in the ground or located on the ground surface. All reservoirs should be covered to reduce the possibility of contamination. Some concrete reservoirs are built with parks, parking lots, or tennis courts on top of them. These reservoirs are located above the service area to maintain the required pressures.

All storage facilities should be located above drainage areas and locations subject to flooding. Care must be taken to ensure that runoff water and debris cannot enter the reservoir and contaminate the water. Overflow and air vents should be screened so that birds, rodents, snakes, and debris cannot enter the reservoir. Vents must be adequate and never be blocked so air can flow freely without any obstruction. Properly functioning vents are essential to prevent a pressure developing in the reservoir when it fills or a vacuum when it empties or is drained. Reservoirs must be fenced to prevent access by vandals or other unauthorized persons.

2.14   Hydropneumatic or Pressure Tanks

A hydropneumatic tank contains a system in which a water pump is controlled by the air pressure in a tank partially filled with water. This type of facility is usually found in smaller water systems. Hydropneumatic tanks are used with a well or a booster pump in water supply systems that do not have a storage reservoir. The tank is used to maintain water pressures in the system and to control the operation of the pump. CAUTION: Hydropneumatic tanks are pressure vessels. Because of the high pressure in the tank, care must be used when operating and maintaining these facilities, especially the pressure relief valves.

---

[5] Coliform (COAL-i-form).   A group of bacteria found in the intestines of warm-blooded animals (including humans) and also in plants, soil, air and water. Fecal coliforms are a specific class of bacteria which only inhabit the intestines of warm-blooded animals. The presence of coliform bacteria is an indication that the water is polluted and may contain pathogenic (disease-causing) organisms.

Hydropneumatic tanks must contain the proper amount of air to be effective. The recommended water to air ratio is approximately two-thirds water to one-third air. The lack of sufficient air in a hydropneumatic tank will cause a rapid start-stop cycling of the pump, which may result in equipment damage. With insufficient air in a hydropneumatic tank, the pump will "short cycle," which means the pump will start and stop very quickly due to the rapid rise and fall of the pressure in the tank. In extreme cases, the pump may never reach full pumping speed, which can cause damage to the pump and the control system. In larger systems, WATER HAMMER[6] can be produced from the rapidly repeated start and stop cycles. An automatically operated air pump should be available to maintain the proper amount of air. The tank should have a sight glass so the water-air ratio in the tank can be observed.

A limitation of hydropneumatic or pressure tanks is that they do not provide much storage to meet demands during power outages. Also a very limited amount of time is available to do maintenance or repair work on wells, pumps, or water treatment equipment. If a pressure filter must be washed using water from a pressure tank, the amount of water remaining in storage and pressure in the system may become fairly low by the time the backwash job is completed. For additional information on hydropneumatic and pressure tanks, see Chapter 3, "Wells," Section 3.29, "Hydropneumatic Pressure Tank Systems," and Section 3.15, "Determination of Working Pressure," in SMALL WATER SYSTEM OPERATION AND MAINTENANCE.

2.15 Surge Tanks

Surge tanks are not necessarily storage facilities, but are used mainly to control water hammer or to regulate the flow of water. Their primary purpose is to protect pipelines against destruction due to excess surge pressure (water hammer). Surge tanks are filled with air and allow the pressure surge to flow into the air space instead of stressing the pipeline. Surge tanks do not absorb energy. Their function is simply to absorb the sudden surge (pressure of water) in the air space thus eliminating possible breaks in the distribution system pipelines. An insufficient amount of air in a surge tank will defeat its intended purpose and will allow surge problems in the distribution system.

QUESTIONS

Write your answers in a notebook and then compare your answers with those on page 45.

2.1A List the six common types of tanks found in water supply systems.

2.1B What is a clear well and its purpose?

2.1C What is the main purpose of an elevated storage tank?

2.2 SELECTION AND LOCATION OF STORAGE FACILITIES

The selection of the type of storage facility or tank depends on the system's individual needs and the type of terrain (land surface) where it is to be installed. Normally a clear well is located immediately downstream from a water treatment plant.

The capacity of the clear well should be sufficient to allow a water filtration plant to operate at a constant rate during periods of peak demands.

An elevated tank is used primarily to supply adequate pressure to a service area. A pump is used to lift water to the elevated tank which provides the head (ft) or pressure (psi) to the service zone. When the water level falls in the tank, a low level or pressure switch will activate an electric circuit which will start a pump. This pump will begin the fill cycle before the tank runs completely dry.

A standpipe may be a ground-level storage facility located at the highest point in a system or it may provide ground storage at a booster station.

A hydropneumatic tank is commonly used with small well systems. The well provides a source of water which is pumped to the tank. Air in the tank helps maintain pressure in the distribution system.

A surge tank is not a storage facility in the truest sense. Surge tanks are necessary to control the problem of surges and water hammer. The surge tank should always be located as close as possible to the activity which creates the water hammer. See the Appendix at the end of this chapter for a discussion on "Surge In Pipelines Carrying Liquids."

QUESTIONS

Write your answers in a notebook and then compare your answers with those on page 45.

2.2A The selection of the type of storage facility or tank depends on what two factors?

2.2B Surge tanks are commonly used for what purpose?

2.3 OPERATION[7]

2.30 Storage Tanks

Procedures for operating storage tanks will vary depending on the design of the facilities and the demands for water. Typical procedures could include (1) filling all storage tanks during periods of off-peak electrical demands, and/or (2) maintaining specified minimum pressures at certain critical points in the distribution system.

Normally, storage tanks are installed to supply water during periods of high water demand. Therefore, during periods of low water demand, the excess pumping capacity is used to fill the tanks before the next period of high water demand occurs. If off-peak electrical energy is available at reduced rates, the pumps can be operated during this period to fill the storage tanks.

All storage tanks should be operated according to the design engineer's and manufacturer's instructions. We must remember that in the design stages of any water system, there usually is a valid reason for the type of storage tank that was selected. Small changes in the distribution system, such as pipeline extensions or the addition of a few new services, usually will not change the requirements of the storage facility. Major system changes, such as larger size main lines or the

---

[6] Water Hammer. The sound like someone hammering on a pipe that occurs when a valve is opened or closed very rapidly. When a valve position is changed quickly, the water pressure in a pipe will increase and decrease back and forth very quickly. This rise and fall in pressures can cause serious damage to the system.

[7] For additional information on operation, see Chapter 5, "Distribution System Operation and Maintenance."

installation of multiple services, can change storage requirements and may require more tank storage.

Frequently the water supplier tries to maintain specified minimum target water pressures throughout the distribution system. Under these circumstances, the pump or pumps are operated whenever needed to maintain the desired pressures. Sometimes there may be several pumps available and they may be variable-speed pumps. The operator maintains pressures throughout the system by regulating the number of pumps operating and/or controlling the pump speed. Variable speed pumps can be very costly to operate.

Many water supply systems are instrumented and automated. Instruments are used to measure water pressures automatically throughout the distribution system and to measure water levels in service storage tanks. Whenever water pressures or water levels drop below minimum target levels, pumps will automatically be started. These pumps will stay on until pressures or water levels reach maximum levels and then the pumps will stop. In some systems there may be more than one minimum level. For example, pump number one may come on at one pressure or water level and pump number two may start at a lower level.

Operators of automated systems must inspect the measuring instruments (pressure gages and water levels) for proper measurement and must also be sure the pumps start and stop at the proper levels. (Sections 2.50, "Booster Pumps," and 2.52, "Gages," discuss the maintenance and troubleshooting of pumps and gages.)

A modification of normal operating procedures that can reduce costs is the "time-of-day" pumping concept. With this method, pumps are used only during the off-peak hours of electrical usage—namely during the hours between 10 p.m. and 6 a.m. Many electrical utilities offer a special lower rate for usage of electricity during off-peak hours.

Abnormal operating conditions include (1) excessive water demands, such as fire demands, (2) broken or out-of-service pumps, mains, or tanks, and (3) *STALE WATER*[8] in tanks creating tastes and odors in systems before design demand flows develop. Emergency plans must be developed in cooperation with the fire department that specify how operators should respond when a fire occurs. A serious fire will require that all pumps operate at full capacity. When all pumps are operating at full capacity, there is a significant danger of *BACKSIPHONAGE*[9] into the water supply. Whenever large

pumps are put on line or taken off line, care must be exercised so that surges (water hammer) do not burst pipes due to excessive pressure or collapse pipes due to negative pressures caused by pressure surge waves. These pumping conditions can also create electrical power surges which may cause pump motors to trip out (stop working) and produce even more serious problems.

Failures of pumps, mains, and tanks create serious problems. Emergency response plans must be developed *IN ANTICIPATION* of these events. Such planning could reveal the need for backup or alternative systems. Many water utility agencies have standby pumps and generators for use during emergencies. Know where you can obtain additional pumps, pipes, fittings, and other necessary supplies and equipment during emergencies. The purchase and storage of essential items *IN ADVANCE* may be appropriate. Temporary (and then permanent) repairs should be made as quickly as possible to avoid overloading the remaining facilities and to prevent additional breakdowns.

An emergency public relations program may be helpful to encourage consumers to reduce water usage during periods when facilities are overloaded. If the public understands the nature and seriousness of a crisis, they will usually cooperate.

New facilities are sometimes constructed and placed on line before a significant demand for water has developed. Under these circumstances, large quantities of water may remain in storage facilities for excessive time periods allowing tastes and odors to develop in the water. To avoid this problem, periodically allow a storage tank to empty and then refill. The tank could be emptied into the system and used by consumers. If this is not possible, the tank could be drained and the water donated for some use beneficial to the community. As a last resort, the water may simply have to be wasted.

QUESTIONS

Write your answers in a notebook and then compare your answers with those on page 45.

2.3A   Tank storage capacity can become inadequate due to what two factors?

2.3B   How are instruments used to operate water supply systems?

2.3C   What is the "time-of-day" pumping concept?

2.3D   List three abnormal conditions encountered by operators of water storage facilities.

---

[8] *Stale Water.   Water which has not flowed recently and may have picked up tastes and odors from distribution lines or storage facilities.*
[9] *Backsiphonage.   A form of backflow caused by a negative or below atmospheric pressure within a water system.*

2.31 Storage Levels

The main function of a distribution storage facility is to take care of daily demands and especially peak demands. Operators must be concerned with the amount of water in the storage facility at particular times of the day. Water levels drop during peak demands and rise during low demands. Most distribution systems establish a pattern which the operator should study in order to better anticipate system demands. Figure 2.9 shows a typical water usage flow curve over a 24-hour period. Ideally, extra water is supplied from storage during the hours that consumption is above average and water is delivered to fill the storage facility during the hours consumption is below average flows. The pattern for any particular system not only varies during the day as shown in Figure 2.9, but varies during different days of the week, especially on weekends and holidays. Demands for water also change during different times of the year because of varying weather conditions. Knowing these patterns, the operator can better anticipate and be ready for expected high-demand periods. The operator should know how high the water level should be each morning so that the system's demands will be met during the rest of the day.

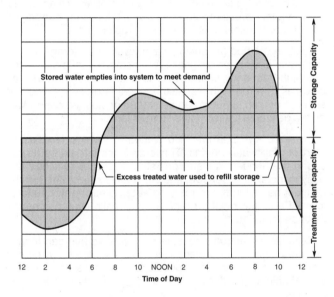

Fig. 2.9 Typical daily use (summer) flow curve
(Reprinted from *OPFLOW*, July 1978, by permission.
Copyright 1978, the American Water Works Association)

The volume of water in a reservoir can be readily determined if you know the water level. Water level indicators are, therefore, essential to successful reservoir operation. Devices range from a simple float connected to an indicator that goes up and down on a staff gage and which can be directly read (Figure 2.10), to telemetry equipment that transmits information concerning water levels to any distance and does it continuously or on demand. Recorders might also be used. Staff gages normally have water level readings in feet (or centimeters) and the level noted can be readily converted to the volume of water stored in the reservoir (see Example 1). Each

water level should be read at approximately the same time each day and the readings recorded. Checks should be made at other times of the day to determine whether any unusual demand conditions have occurred. If any significant increase in demand over that anticipated is noted, the operator should initiate an earlier increase of flow to the reservoir. Automatic water level regulation can be achieved using *ALTITUDE-CONTROL VALVES*.[10] These valves are designed to (1) prevent overflows from the storage tank or reservoir, or to (2) maintain a constant water level as long as water pressure in the distribution system is adequate. The simplest form of level control is a pressure-activated switch which turns a pump on and off. This switch is often sufficient for small water systems. The use of advance warning alarms to signal when water levels are too low or too high is recommended.

A number of routine checks need to be made at reservoirs depending on the type of equipment available. If advance warning alarms are used, they should be tested to ensure they will work when needed. In many reservoirs the water level is controlled automatically. Provisions should be made to allow the operator to test the system to be sure it is working properly. Time clocks are often used to operate valves or pumps that control water coming into the reservoir. They should occasionally be checked and reset if necessary. If they malfunction, the water level could vary too widely. Problems can also be caused by sticking or non-seating valves. The operator must be certain that all altitude and overflow valves in use are in good working condition.

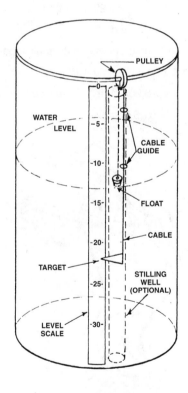

Fig. 2.10 Measuring water level in a tank

---

[10] *Altitude Valve. A valve that automatically shuts off the flow into an elevated tank when the water level in the tank reaches a predetermined level. The valve automatically opens when the pressure in the distribution system drops below the pressure in the tank.*

FORMULAS

To determine the volume of a storage facility, we need to know the dimensions of the facility. If the tank is rectangular, we need to know the length, width, and depth of water.

Volume, cu ft = (Length, ft)(Width, ft)(Depth, ft)

If the tank is circular, we need to know the diameter and depth of water.

Volume, cu ft = (0.785)(Diameter, ft)^2(Depth, ft)

To convert a volume from cubic feet to gallons, we multiply the volume in cubic feet by 7.48 gallons per cubic foot.

Volume, gal = (Volume, cu ft)(7.48 gal/cu ft)

EXAMPLE 1

Estimate the volume of water (cubic feet and gallons) in the tank shown in Figure 2.10. The tank is 20 feet in diameter and 35 feet deep. The target indicates that the depth of water is 22 feet and the water level indicates that the water is 13 feet from the top of the tank.

Known	Unknown
Diameter, ft = 20 ft	1. Volume, cubic feet
Depth, ft = 22 ft	2. Volume, gallons

1. Calculate the volume of water in the tank in cubic feet.

   Volume, cu ft = (0.785)(Diameter, ft)^2(Depth, ft)

   = (0.785)(20 ft)^2(22 ft)

   = 6,908 cu ft

2. Calculate the volume of water in the tank in gallons.

   Volume, gal = (Volume, cu ft)(7.48 gal/cu ft)

   = (6,908 cu ft)(7.48 gal/cu ft)

   = 51,672 gallons

Many operators prepare tables or charts that contain the volume in a tank for each foot in depth. These tables and charts could be prepared for the tank in Example 1 by reworking the problem for depths other than 22 feet. With these tables and charts, the operator records the depth in the tank, refers to the table or chart, and obtains the volume of water in the tank.

QUESTIONS

Write your answers in a notebook and then compare your answers with those on page 45.

2.3E   What is the main function of a distribution storage reservoir?

2.3F   Why should operators know how high the level in a storage tank should be each morning?

2.3G   How are water levels controlled in storage reservoirs?

2.32   Storage Level Controls

Several different types of control systems are used to monitor water levels in storage facilities and also to start and stop pumps. One of the simplest types of control systems uses electrodes mounted at various levels in the storage facility. These electrodes sense the rise and fall of the water level and start and stop the pumps as necessary.

Two problems may be encountered by operators of this type of control system. When the electrodes need replacement or repair, they can be difficult to reach. This problem can be solved by placing the electrodes at a location where access is easy. Another problem with probes is that they may corrode or become contaminated, thus requiring replacement.

Another method of sensing water levels in a storage facility is the use of ultrasonic signals. A transmitter sends a continuous high-frequency sound wave to a receiver. When the receiver is covered with water, the signal is broken. Once the signal is broken, a pump can be started or stopped depending on the desired situation in the storage tank.

Pressure switches are also placed in storage facilities to start or stop water pumps. These switches can be set to respond to changes in water pressure. Pressure switches are easy to install, inexpensive, and very easy to calibrate. Switches must be calibrated to respond at the proper water surface level. If a switch must be replaced, the calibration procedure must be repeated.

Solid-state electronic sensors are available that can measure the actual water surface level and start or stop a pump. Calibration of solid-state electronic sensors consists of dialing the set point to the desired start or stop water surface level for a pump.

Differential-pressure altitude valves are also used to regulate water surface levels in a storage facility. These valves shut off flow into a storage facility when it becomes full and reopen when the tank level drops to a predetermined level or the pressure in the distribution system drops to a predetermined pressure.

Pump stations are not always located in the same place as the storage facility. Under these conditions a signal must be transmitted from the storage facility to the pump or pumps. Also some type of indicator showing the level of water in the storage facility is desirable at the pump station. The transmission cable may be buried underground or an overhead wire may be used to transmit signals from a tank to a pump. Underground cable should be of the armored variety to protect it from rodents.

Radio frequencies are also used to transmit signals from sensors to pumps. Problems may develop if lightning or excessive voltage interferes or disrupts the signals. These improper signals may cause a pump to start or stop at the wrong time. Lightning protectors can be installed to prevent the disruption of signals.

Usually when a pump receives a signal to start, a light will come on in the control panel indicating that the pump is running. However, even when the pump running light is on, the pump may not be pumping water. To ensure that the pump is actually pumping water, a positive flow report-back signal should be installed on the control panel. This report-back signal may be a flowmeter, an ultrasonic flow switch, a pressure switch, or a check-valve with an arm-actuated mercury switch.

Most water utilities need only a small amount of instrumentation and controls. The best approach is to purchase high-quality instruments that will require little maintenance and operate for many years. If at all possible, the simple maintenance should be performed by the utility's operators.[11] *REMEMBER*, though, that only *TRAINED* and *QUALIFIED* operators should be permitted to repair electrical equipment. See Chapter 7, Section 7.37, "Working Around Electrical Units," for additional safety precautions.

QUESTIONS

Write your answers in a notebook and then compare your answers with those on page 45.

2.3H How can the depth of water in a storage facility be measured?

2.3I What problems are created by the use of electrodes to measure water depths?

2.3J What problem could occur when the pump running light is on?

2.33 Pumps

CENTRIFUGAL PUMPS[12] (Figure 2.11) cannot operate unless the impeller is submerged in water; *NEVER ATTEMPT TO START A CENTRIFUGAL PUMP UNTIL YOU KNOW THAT THE PUMP IS PROPERLY PRIMED*[13] (Figure 2.12). There are several ways to prime a pump so the impeller is submerged in water. A special primer pump may be used to pump water into the pump casing and submerge the impeller. A priming water tank or an auxiliary water supply may be used to prime the pump by adding water to the pump casing and bleeding off the air in the casing. Air may be removed from the pump casing by the use of an electric or hand-operated vacuum pump which causes water to flow into the suction pipe and pump casing. *FOOT VALVES*[14] (Figure 2.12) may be installed on the inlet of the pump suction pipe to keep the pump and suction line full of water. However, if the foot valve develops a leak, the pump will lose its prime. If the pump impeller is located below the surface of the water being pumped (such as a booster pump), the pump will not need to be primed.

When starting an electric motor-driven centrifugal pump, the following procedure is recommended:

1. Check the lubrication,

2. Prime the pump and make sure that the pump and suction piping are free of air,

3. Reduce electric current and water surge on start-up by starting the pump with the discharge valve closed or throttled,

4. Open the discharge valve slowly as soon as the pump is running,

5. Inspect the packing glands to see that water seals are functioning properly, and

6. Measure the running amperage and investigate any abnormal demands or changes from the usual readings.

Continuously operating water pumps should be inspected on a regular basis. Inspection should include observing and recording pump suction and discharge pressures, output flow, and electric current demands. Also check for excessive or abnormal noises, vibrations, heat, and odors. If the packing glands are leaking excessively, they should be tightened so that there is only a small amount of leakage. This small amount of leakage helps cool the pump shaft and reduce premature packing wear and scored shaft sleeves. Do not tighten too much and cause an increase in heating and/or damage to the pump. If a centrifugal pump does not operate as expected, refer to Section 2.50, "Booster Pumps," Table 2.2, "Centrifugal Pump Troubleshooting Chart," page 43.

QUESTIONS

Write your answers in a notebook and then compare your answers with those on page 45.

2.3K When is a pump primed?

2.3L How can a pump be primed?

2.3M How can the electric current and water surge be reduced when starting a centrifugal pump?

---

[11] *For additional information on instrumentation and controls, see Chapter 19, "Instrumentation," in WATER TREATMENT PLANT OPERATION, Volume II, in this series of manuals.*

[12] *Centrifugal (sen-TRIF-uh-gull) Pump.   A pump consisting of an impeller fixed on a rotating shaft that is enclosed in a casing, and having an inlet and discharge connection. As the rotating impeller whirls the liquid around, centrifugal force builds up enough pressure to force the water through the discharge outlet.*

[13] *Prime.   The action of filling a pump casing with water to remove the air. Most pumps must be primed before start-up or they will not pump any water.*

[14] *Foot Valve.   A special type of check valve located at the bottom end of the suction pipe on a pump. This valve opens when the pump operates to allow water to enter the suction pipe but closes when the pump shuts off to prevent water from flowing out of the suction pipe.*

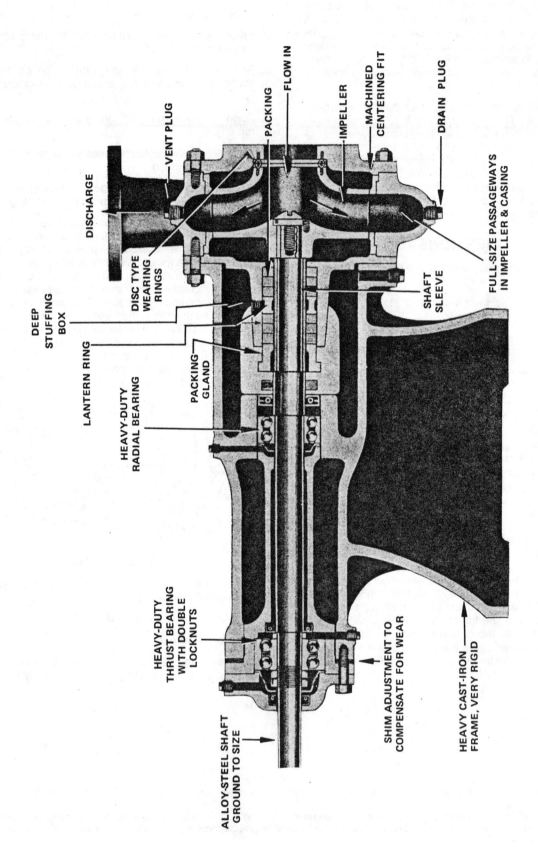

Fig. 2.11   Source water centrifugal pump

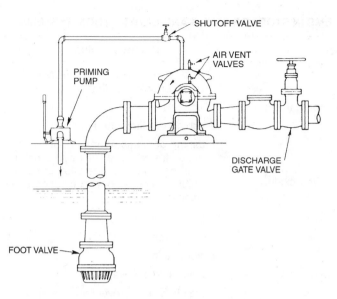

Fig. 2.12 Pump priming installation
(Reprinted from *WATER DISTRIBUTION OPERATOR TRAINING HANDBOOK*, by permission. Copyright 1976, the American Water Works Association)

2.34 Sampling

All storage facilities should be regularly sampled to determine the quality of water that enters and leaves the facility. Sampling data and visual observations can help you establish a routine for the periodic cleaning of the tank. Indicators that may help you decide when the tank will need cleaning are turbidities over an accepted standard (such as 1.0 *TURBIDITY UNITS*[15]), excessive color, tastes and odors, and *POSITIVE BACTERIOLOGICAL SAMPLES*[16] (which indicate the presence of bacterial contamination). Samples should be collected at the inlet and outlet to the storage facility as well as at various depths and cross sections to ensure measurement of water quality throughout the facility.

2.35 Troubleshooting

Water quality problems may be of the microbiological type, which could be caused by loss of chlorine residual, growth of bacteria, and direct entry of birds, rodents, snakes, and debris. Chemical water quality problems may be caused by leaching of chemicals from tank linings or coatings. Leaching of chemicals could cause taste and odor problems, and the quantity of disinfection by-products in the treated water could increase during storage. Common causes of physical water quality problems include settling and collection of sediment, rust, and chemical precipitates. Also dust, dirt, birds, snakes, and other animals could enter the storage facility.

Water quality in a storage facility could degrade due to excessive water age caused by low demands for water and short-circuiting within the distribution storage reservoir. Other causes of water quality degradation in the storage reservoir include poor design, inadequate maintenance, and improperly

applied and/or cured coatings and linings. Food-grade oil used in pumps can float to the water surface of a storage tank, then sink to the bottom and serve as food for a thriving microorganism population. This could require tank cleaning every two years.

If water quality problems develop, you must determine where the problem is coming from and what action will be taken to solve the problem. Table 2.1 is a summary of water quality problems that may develop in storage tanks and distribution systems. The table lists the problems, identifies possible causes, and proposes potential solutions. You will learn more about how to solve water quality problems as you continue through this manual. For additional information, see Chapter 4, "Water Quality Considerations in Distribution Systems," Section 4.52, "Quality Degradation in Storage Facilities."

In some areas where seasonal demands become very low, the chlorine residual may become depleted in the storage tank due to excessive detention time. If this is a problem, auxiliary chlorination stations may be needed in the distribution system or at secondary or booster pump stations to maintain desired chlorine residuals.

In addition to troubleshooting water quality problems, you should regularly inspect the storage facility. Be sure the gate to the facility is locked to keep out vandals and unauthorized persons. The access opening to the facility must be locked and provide a seal to keep out rain and debris. All vent, overflow, and drain screens must be in good condition. Carefully inspect the tank cover for any defects. Look at the water surface for debris and dead birds and rodents.

QUESTIONS

Write your answers in a notebook and then compare your answers with those on pages 45 and 46.

2.3N Which water quality indicators may help you decide when a storage tank needs cleaning?

2.3O List three possible causes of tastes and odors in a drinking water supply.

2.3P List three possible causes of positive coliform test results in a water storage tank.

---

[15] *Turbidity Units (TU). Turbidity units are a measure of the cloudiness of water. If measured by a nephelometric (deflected light) instrumental procedure, turbidity units are expressed in nephelometric turbidity units (NTU) or simply TU. Those turbidity units obtained by visual methods are expressed in Jackson Turbidity Units (JTU) which are a measure of the cloudiness of water; they are used to indicate the clarity of water. There is no real connection between NTUs and JTUs. The Jackson turbidimeter is a visual method and the nephelometer is an instrumental method based on deflected light.*

[16] *Positive Bacteriological Sample. A water sample in which gas is produced by coliform organisms during incubation in the multiple tube fermentation test. See Chapter 11, Laboratory Procedures, "Coliform Bacteria," in WATER TREATMENT PLANT OPERATION, Volume I, for details.*

TABLE 2.1   TROUBLESHOOTING WATER QUALITY PROBLEMS IN STORAGE TANKS AND DISTRIBUTION SYSTEMS

Problem	Possible Cause	Potential Solution
1. Tastes and Odors	High chlorine residual	Use *BREAKPOINT CHLORINATION*[a] or lower chlorine dosage
	Biological (algal) growth or microorganisms	Chlorinate
	Dead end in main or tank	Flushing or eliminate dead end
2. Turbidity	Silt or clay in suspension	Flushing of mains or proper operation of water treatment plant (proper coagulant, dosage, and operation of coagulation, flocculation, and filtration processes)
	Calcium carbonate	
	Aluminum hydrate, precipitated iron oxide	
	Microscopic organisms	
	Floc carryover	
3. Color	Decay of vegetable matter	Chlorination
	Microscopic organisms	Chlorination
4. Positive Coliform Results	Contaminated distribution system	Locate and remove source
	CROSS CONNECTION[b]	Install backflow prevention or *AIR GAP*[c] devices, flush, and temporarily increase chlorine dosage
	Negative pressure in main	Repair main, increase chlorine feed rate, flush system, and *SAMPLE*. Maintain a positive pressure in main (at least 5 psi, 0.35 kg/sq cm or 34.5 kPa)
	No or improper disinfection of new or repaired wells, reservoirs, or mains	Use proper disinfection procedures

[a] Breakpoint Chlorination.   Addition of chlorine to water until the chlorine demand has been satisfied. At this point, further additions of chlorine will result in a free chlorine residual that is directly proportional to the amount of chlorine added beyond the breakpoint.

[b] Cross Connection.   A connection between a drinking (potable) water system and an unapproved water supply. For example, if you have a pump moving nonpotable water and hook into the drinking water system to supply water for the pump seal, a cross connection or mixing between the two water systems can occur. This mixing may lead to contamination of the drinking water.

[c] Air Gap.   An open vertical drop, or vertical empty space, that separates a drinking (potable) water supply to be protected from another water system in a water treatment plant or other location. This open gap prevents the contamination of drinking water by backsiphonage or backflow because there is no way raw water or any other water can reach the drinking water supply.

2.4   MAINTENANCE[17]

2.40   Preventive and Corrective Maintenance

Maintenance is the necessary key in the efficient operation of any water system. Webster says that "maintenance is the upkeep of property or equipment." There are three types of maintenance—preventive, predictive, and corrective. Preventive maintenance is something that is done before some type of deterioration takes place. Painting puts a protective covering on the tank thus protecting it from deterioration (rust formation). Predictive maintenance attempts to predict when a failure might occur so corrective action can be performed before failure.

Corrective maintenance or repair is maintenance that is necessary when a problem already exists. An example is the replacing of deteriorated overflow and vent screens on a tank that have failed and do not serve the purpose of keeping birds or rodents out of the tank.

Problems will not disappear—they only get worse. For that reason alone, maintenance is important.

2.41   Painting

Several factors must be considered before painting the outside of a storage tank. Many water utility agencies try to paint the outside of their steel tanks once every five years. The time a coating lasts depends on: (1) *PROPER SURFACE PREPARATION*, (2) a good, durable paint, (3) good workmanship, (4) adequate drying and aging, and (5) proper maintenance (through periodic inspection and spot, partial, or complete removal of old paint and repainting as necessary).

A tank's interior coating will generally protect the interior for three to five years, depending on local conditions. Routine inspection is the best way to determine when a tank requires maintenance. New tanks or newly painted tanks should be inspected after one year of use. Otherwise, a tank should be drained, cleaned, and inspected once a year, depending on local conditions. Three types of inspections can be made:

1. A visual inspection, which is made from the roof hatch with the water level lowered to about half-full or less;

2. A detailed inspection, accomplished by draining the tank, washing it, and then inspecting the interior coating; and

---

[17] For additional information on maintenance, especially equipment maintenance, see Chapter 18, "Maintenance," in WATER TREATMENT PLANT OPERATION, Volume II, in this series of manuals.

3. A detailed inspection using divers and video cameras and then cleaning the tank with a device similar to a vacuum cleaner.

A visual inspection may be made in the odd numbered years and a detailed cleaning and inspection may be made in the even numbered years. For example, first anniversary (after painting)—visual inspection, and second anniversary—complete cleaning and inspection.

The best time of year to take a storage facility out of service for complete cleaning and inspection is during the period of lowest water consumption. For most water utilities in warmer climates, this will be during the period from October to April. However, this time period will vary depending on the climate and seasonal water demands. In very cold climates storage facilities are not taken out of service and painted in the winter.

When a tank is to be taken out of service, coordination and planning are necessary so that the distribution system will be able to meet the water demands of its customers. Here are some steps that may be followed for the draining of a storage facility for complete cleaning and inspection:

1. Make sure provisions are made to supply adequate water to the distribution system during shutdown. Be sure that customers who are most affected by the loss of available water from the tank in question will not experience shortages of water or inadequate pressures;

2. Secure (block) inlet line so that no water may enter while the tank is out of service for maintenance;

3. Draw the tank down until there is about 1 foot (0.3 m) of water left covering the bottom of the tank;

4. Secure (block) the discharge line so that no water will be used from the tank while it is being cleaned;

5. Collect a sample of the water remaining in the tank. Also collect a sample of silt/mud from the tank bottom for biological analysis. Analyze the mud/silt for snails and worms. If snails and/or worms are present, try to determine the source and eliminate it;

6. Drain and dispose of the remaining water and silt/mud;

7. Wash the interior tank walls with a water hose and brushes;

8. Inspect interior coating. Look for flaking, peeling, and rust.

The tank's interior inspection should be done by you or by a representative of a professional tank cleaning and painting service company. If you plan to hire someone to clean and recoat your tank, the following suggestions are recommended:

1. Verify the potential contractor's credentials, experience, license, and liability insurance,

2. Ask for references, and

3. Contact the references for their recommendations.

If substantial flaking, peeling, or rust are noted during inspection, the tank's interior should be prepared for repainting.

Preparing a tank's interior for repainting is a very important step if the new coat is to be of any value. Sandblasting is recommended. A brush-off blast (SSPC-SP 7/NACE No. 4)[18] may be used when minor deterioration has occurred. Where extensive deterioration is present, the near-white blast (SSPC-SP 10/NACE No. 2)[18] is recommended. Clean up the sand and wash down the interior of the tank to remove any particles of sand. The tank is now ready for painting. If the paint removed is a lead-base paint, special precautions must be taken during the sandblasting procedures and also for the collection and removal of sand and paint. Contact the appropriate regulatory agency in your area for details regarding the required procedures.

There are many interior paint and coating systems that can be used in water storage tanks. They are divided into two basic categories: (1) long-life coatings, and (2) short-life coatings. A short-life coating is any paint or material that will protect the interior surface of a tank up to approximately six years. A long-life coating is a coating or material that can protect the interior surface for up to ten years and possibly longer. Long-life coatings will cost from 30 to 50 percent more than a short-life coating. Since a long-life coating may last up to three times longer than a short-life coating, a long-life coating can reduce your maintenance cost over a period of 10 to 20 years.

SAFETY NOTICE

Before anyone ever enters a tank for any reason, these safety procedures must be followed:

1. First test the atmosphere in the tank for oxygen (should be 19.5 to 23.5 percent); then test for combustible and toxic gases. Contact your local safety equipment supplier for the proper types of atmospheric testing devices. These devices should have alarms that are activated whenever an unsafe atmosphere is encountered.

2. Provide adequate ventilation, especially when painting. A self-contained breathing apparatus may be necessary.

3. All persons entering a tank must wear a safety harness.

4. One person trained in tank rescue procedures, safety, and first aid must remain at the tank entrance observing the actions of all people in the tank. An additional person must be readily available to help the person at the tank entrance with any rescue operation.

Additional confined space safety procedures may also be required. Check with your local regulatory agency and see Chapter 7, "Safety," Section 7.63, "Confined Spaces," for information about permits and a description of some other confined space hazards and procedures.

---

[18] SSPC: The Society for Protective Coatings, 40 24th Street, 6th Floor, Pittsburgh, PA 15222-4656.

Care must be used in the selection of the tank's interior coating. Coatings fall into the category of *INDIRECT ADDITIVES* to the potable water supply and must be of an appropriate quality for this use. Accordingly, they must be nontoxic, must not release organic or inorganic compounds to the surrounding water in toxicologically (poisonous) significant amounts, and must not impart objectionable tastes or odors to the water.

Many states require that paints and coatings used in potable water systems be approved by the state. The National Sanitation Foundation publishes two standards to provide guidance to states regarding the acceptability of paints and coatings for use in potable water service (NSF Standard 60, Drinking Water Treatment Chemicals—Health Effects; NSF Standard 61, Drinking Water System Components—Health Effects; available from National Sanitation Foundation International, PO Box 130140, 789 N. Dixboro Road, Ann Arbor, MI 48113-0140, phone: (800) NSF-MARK ((800) 673-6275) or (734) 769-8010, e-mail: info@nsf.org, or website: www.nsf.org. Additionally, paints and coatings approved for use by the Food and Drug Administration (FDA) for continuous contact with aqueous (watery) foods as defined in the Code of Federal Regulations (21 CFR) are generally acceptable for use in potable water systems.

The paint or coating should also meet AWWA Specifications.[19] Products that contain lead or *PCBs*[20] are not recommended for potable water use. Coatings using trichloroethylene and tetrachloroethylene should also be avoided.

The best method of applying a coat of paint is by spraying. The cross-spraying method should be used. Spray an area to be painted with horizontal strokes. After the first area has been painted (a 6-foot (2-m) wide section) with one coat, move to one side of the first area. Spray another 6-foot (2-m) wide section. Spray half of the original section and three feet (one meter) of a new, unpainted area. Continue this procedure all the way around the tank. Repeat this procedure using vertical strokes and again overlapping half a section every time. This method not only provides an adequate film thickness, but also reduces the chances of holidays (unpainted spots). Cracks can develop in the coating if the coating is too thick. Coating failures usually appear on a welded seam. Coating repairs can be performed by divers under water fairly economically.

Drying time is of prime importance, especially if the paint job is going to be effective. *ALWAYS* assume that a tank's interior does not have sufficient ventilation. A forced-draft ventilation procedure should be used to dry (cure) the coating on a tank's interior. This ventilation will help speed up the drying time. Thorough drying is critical if the paint is to adhere (stick) properly to the tank's interior walls. If drying time is insufficient, problems with tastes and odors or the leaching of potentially toxic materials may occur. To be on the safe side, you may wish to double the drying (curing) time recommended by the paint manufacturer.

QUESTIONS

Write your answers in a notebook and then compare your answers with those on page 46.

2.4A   What is preventive maintenance?

2.4B   What is corrective maintenance?

2.4C   The time interval for painting the outside of steel tanks depends on what factors?

2.4D   During what time of the year should a water storage facility in a warm climate be taken out of service for complete cleaning and inspection?

2.4E   How is the interior surface of a tank prepared for repainting?

2.42   Corrosion Control[21]

Corrosion may occur either on the inside or outside of the metal surface of a storage facility. Many factors influence the rate of corrosion. The warmer the water, the faster the chemical corrosion reactions. If a water is corrosive, high water velocities will cause rapid pipe deterioration but with little metal pickup. When water velocities are low, there is a longer contact time between the water and the pipe and there could be a higher metal pickup (consumer complaints of red or dirty water). When two different metals come in contact, the difference in electrical potential causes a flow of current (electrons) which results in corrosion.

Substances in the water also influence the rate of corrosion. Dissolved oxygen in the water can increase corrosion rates. Carbon dioxide lowers the pH of a water which makes it more corrosive. Dissolved minerals (salts) in a water increase the rate of corrosion. Sulfate-reducing bacteria can cause increased corrosion.

When waters from different sources and of considerably different chemical content are mixed, the formation of a scale or coating may result. However, when a hard water is replaced with a soft water, corrosion will often result because the scale (protective coating) is dissolved by the soft water.

Painting either the outside or inside of a tank is a form of corrosion control. Usually when one thinks of corrosion control, interior corrosion or metal corrosion comes to mind. A coat of paint is the least expensive type of corrosion control; it is the most important part of any corrosion-control program.

---

[19] *D102-03 AWWA STANDARD FOR COATING STEEL WATER-STORAGE TANKS. See Section 2.9, "Additional Reading," for ordering information.*

[20] *PCBs or **P**oly**c**hlorinated **B**iphenyls.   A class of organic compounds that cause adverse health effects in domestic water supplies.*

[21] *For additional information on the causes of corrosion and methods of corrosion control, see Chapter 8, "Corrosion Control," in WATER TREATMENT PLANT OPERATION, Volume I, in this series of manuals.*

Methods of corrosion control include:

1. Metallic coatings such as zinc or aluminum to protect tank metals such as steel or aluminum (outside coating);

2. Nonmetallic coatings to protect tank metals. Appropriate coatings include coal tar enamels (bituminous), asphaltics, cement mortar, epoxy resins, vinyl resins and paints, coal tar-epoxy enamels, inorganic zinc, silicate paints, and organic zinc metals (be sure that any coating you use has been approved for use in potable water);

3. Chemicals could be added during the treatment of water which would deposit a protective coating or film on the tank's metal (calcium hydroxide (lime), sodium carbonate (soda ash), zinc orthophosphate, and silicate compounds, for example); and

4. Electrical control (*CATHODIC PROTECTION*[22]).

Rarely is one of these methods used by itself as a means of controlling corrosion. The addition of chemicals to water in order to deposit a protective coating is an excellent method of controlling corrosion. Any substance which promotes film formation on the surface to be protected when added to water is known as a corrosion inhibitor. Lime, soda ash, and caustic soda (sodium hydroxide) are chemicals that are used to help control corrosion. When added to water these chemicals will increase the pH. At a certain pH, depending on the temperature, total dissolved solids, calcium concentration, and alkalinity level of the water, the calcium carbonate equilibrium will be reached. For any pH above the equilibrium level, a thick coating of calcium carbonate will form on surfaces exposed to the water. This coating will protect the surfaces from corrosion. As long as the pH remains above the equilibrium level, the coating will not dissolve and will provide protection against corrosion.

Chemical inhibitors reduce the rate of corrosion by preventing dissolved oxygen from reaching the cathodic areas, thus inhibiting the rate-controlling cathodic reactions. Chemical inhibitors are applied to water entering a tank or pipeline by means of a continuously operated pump. The chemicals added to the water deposit a chemical film on the insides of the pipe or tank, thus protecting the interior. Sodium hexametaphosphate (polyphosphate) compounds are commonly used because they combine with (tie up) a variety of *IONS*[23] including iron, manganese, and calcium, thus preventing their build-

up as rust or scale. Polyphosphate is available either in a dry or liquid form. The liquid form is becoming more widely used because less equipment is needed for its application than for the dry form.

While polyphosphate compounds are commonly used, special pumps and safety equipment must be used for safe and proper application. Phosphate is usually fed at a rate of not more than 5 mg/*L*. This chemical is costly and monitoring the feed rate and resulting concentration in the potable water is essential. Chemical control methods should be used as a supplement to the proper choice of metals and metallic and nonmetallic coatings.

Cathodic protection (Figure 2.13) must not be regarded as a substitute for the use of a proper interior coating or the application of inhibitors, but should be used if necessary to supplement the other methods. Cathodic protection uses an electrical system for the prevention of corrosion and pitting of steel and iron surfaces in contact with water. A low-voltage current is made to flow through a liquid or soil in contact with the metal in such a manner that the external *ELECTROMOTIVE FORCE*[24] renders the metal structure *CATHODIC*[25] and transfers the corrosion to auxiliary sacrificial *ANODIC*[26] parts which are deliberately allowed to corrode instead of letting the water storage facility corrode. The key to successful use of cathodic protection is the replacement of the sacrificial anodes (usually made of magnesium or zinc) when they have been corroded away. These anodes (sacrificial) should be inspected (weighed to determine loss in weight due to corrosion) at least yearly and replaced when necessary. Do not energize cathodic protection on a re-coated water storage tank until after the first year's inspection following re-coating.

Cathodic protection systems not only prevent corrosion, but can reduce time and money spent cleaning and painting. Also the time required for a storage facility to be out of service for maintenance can be reduced. Cathodic protection of a water storage facility is effective only for those areas that are covered by water. The need for a high-quality coating is not eliminated by the installation of a cathodic protection system.

QUESTIONS

Write your answers in a notebook and then compare your answers with those on page 46.

2.4F   List five methods of corrosion control.

2.4G   What is a corrosion inhibitor?

2.4H   What is the key to a successful cathodic protection program?

2.43 Disinfection[27]

Disinfection is the inactivation or destruction of disease-producing organisms. New storage facilities and facilities that

---

[22] *Cathodic (ca-THOD-ick) Protection.   An electrical system for prevention of rust, corrosion, and pitting of metal surfaces which are in contact with water or soil. A low-voltage current is made to flow through a liquid (water) or a soil in contact with the metal in such a manner that the external electromotive force renders the metal structure cathodic. This concentrates corrosion on auxiliary anodic parts which are deliberately allowed to corrode instead of letting the structure corrode.*

[23] *Ion.   An electrically charged atom, radical (such as SO_4^{2-}), or molecule formed by the loss or gain of one or more electrons.*

[24] *Electromotive Force (E.M.F.).   The electrical pressure available to cause a flow of current (amperage) when an electric circuit is closed. Also called voltage.*

[25] *Cathode (KA-thow-d).   The negative pole or electrode of an electrolytic cell or system. The cathode attracts positively charged particles or ions (cations).*

[26] *Anode (an-O-d).   The positive pole or electrode of an electrolytic system, such as a battery. The anode attracts negatively charged particles or ions (anions).*

[27] *For additional information on disinfection, see Chapter 6, "Disinfection."*

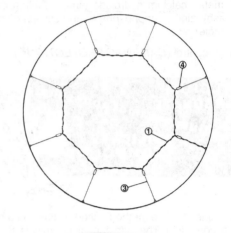

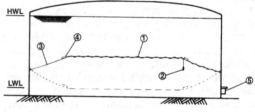

① PLATINIZED NIOBIUM ANODE
   TO PROTECT BOWL AREA

② SEGMENTED HIGH-SILICON
   CAST-IRON ANODES TO
   PROTECT WET RISER

③ COPPER-COPPER SULFATE
   REFERENCE ELECTRODE

④ STEEL EYE RING ANODE
   SUPPORT

⑤ STYROFOAM FLOAT

⑥ ANODE FEED WIRE

⑦ D. C. WIRING ENTERS
   TANK HERE

⑧ CONDUIT CONTAINING
   D. C. WIRING

⑨ T.A.S.C. AUTOMATIC
   CATHODIC PROTECTION
   POWER UNIT

Elevated Water
Storage Tanks

① PLATINUM NIOBIUM ANODE WIRE

② COPPER-COPPER SULFATE
   REFERENCE ELECTRODE

③ STRANDED PLASTIC ROPE

④ FLOTATION BLOCK

⑤ T.A.S.C. AUTOMATIC CATHODIC
   PROTECTION POWER UNIT

Ground-Level Water
Storage Tanks

PERMANODE ANODE SYSTEM

The only proven permanent anode system for water tanks subject
to winter icing conditions. Ten or more years of continuous cathodic
protection can be expected instead of the 9-10 months previously
provided by aluminum anode systems. This provides even greater
economics to water tank owners in Northern climates.

Fig. 2.13   Cathodic protection systems
(Courtesy of Harco Corporation, Medina, Ohio)

have been repaired, cleaned, or had cathodic protection installed must be disinfected. Always disinfect water storage facilities whenever there has been any opportunity for contamination.

Liquid chlorine (or gas), calcium hypochlorite, and sodium hypochlorite are commonly used as disinfectants. There are several methods used to disinfect a water storage facility. *NOTICE:* Before anyone enters a water storage facility for any reason, follow the safety procedures listed under the *SAFETY NOTICE* on page 33.

WARNING

BEFORE attempting to spray the interior of a tank, be sure there is adequate ventilation. If adequate ventilation is not available, the person doing the spraying must use a self-contained breathing apparatus.

ALWAYS wear protective clothing when spraying a tank interior. Protective clothing should consist of a rubberized suit (rain gear—rubber pants and coat), rubber boots, gloves, hat, and a face shield.

Whenever anyone enters a tank, they must wear a safety harness for rescue purposes. Someone must be at the tank entrance to observe the person in the tank and also to hold the safety rope attached to the harness being worn by the person in the tank. A method of communication using tugs on the safety rope must be established *BEFORE* anyone enters a tank. For example, one tug may mean help and two tugs means everything is OK.

In addition to the backup person at the tank entrance, a third person must be readily available to assist the person at the tank entrance.

NO ONE should enter a tank for any reason or under any circumstances without two people standing by for rescue purposes. Too often someone has entered a tank to rescue a friend and died also.

Spray or brush the interior of the storage facility with 200 mg/L (200 ppm) chlorine solution. The chlorine solution can be mixed in a crock or small tank. The application can be accomplished by using garden hoses with nozzles and pressure can be provided by a gasoline-driven pump. (Be sure to station the gasoline engine outside the tank to prevent the buildup of dangerous exhaust vapors.) A 200-mg/L chlorine con-

centration can be uncomfortable to work with inside of a storage facility. *ALWAYS PROVIDE ADEQUATE VENTILATION WHENEVER ANYONE IS INSIDE A TANK FOR ANY REASON.* The amount and type of ventilation depends on the situation.

To prepare a 200-mg/L solution using sodium hypochlorite, let's use a 100-gallon (380-L) crock of water and assume that the sodium hypochlorite has five percent *AVAILABLE CHLORINE.*[28] Sodium hypochlorite usually is available in the range from 5 to 15 percent available chlorine. The sodium hypochlorite must be thoroughly mixed with water to produce the chlorine solution.

FORMULAS

To determine the amount of sodium hypochlorite needed in pounds, you may use the following formula:

$$\text{Hypochlorite, lbs} = \frac{\left(\begin{array}{c}\text{Volume of Water,}\\ \text{Million Gallons}\end{array}\right)\left(\begin{array}{c}\text{Chlorine Conc,}\\ \text{mg/L}\end{array}\right)(8.34 \text{ lbs/gal})100\%}{\text{Available Chlorine, \%}}$$

Also, you should know that

100 gal of water = 0.0001 MG of water, and

1 gal of water = 8.34 lbs.

These are constants you should remember.

EXAMPLE 2

Prepare a 200-mg/L solution of chlorine using a 100-gallon crock (0.0001 million gallons) and a sodium hypochlorite solution containing five percent available chlorine. How many gallons of hypochlorite should be added to the 100 gallons of water in the crock?

Known	Unknown
Chlorine Conc, mg/L = 200 mg/L	Hypochlorite, gallons
Vol of Water, MG = 0.0001 MG	
Available Cl, % = 5%	

1. Calculate the pounds of hypochlorite needed to produce a chlorine concentration of 200 mg/L.

$$\text{Hypochlorite, lbs} = \frac{\left(\begin{array}{c}\text{Volume of Water,}\\ \text{Million Gallons}\end{array}\right)\left(\begin{array}{c}\text{Chlorine Conc,}\\ \text{mg/L}\end{array}\right)(8.34 \text{ lbs/gal})100\%}{\text{Available Chlorine, \%}}$$

$$= \frac{(0.0001 \text{ MG})(200 \text{ mg/L})(8.34 \text{ lbs/gal})(100\%)}{5\%}$$

$$= 3.34 \text{ lbs}$$

2. Determine the gallons of 5 percent sodium hypochlorite to be added to the 100 gallons of water.

$$\text{Hypochlorite, gallons} = \frac{\text{Hypochlorite, lbs}}{8.34 \text{ lbs/gal}}$$

$$= \frac{3.34 \text{ lbs}}{8.34 \text{ lbs/gal}}$$

$$= 0.40 \text{ gallon}$$

After the tank has been sprayed with the hypochlorite solution, allow it to stand unused for at least 30 minutes before filling. Fill the tank with distribution system water that has been

---

[28] *Available Chlorine.* *A measure of the amount of chlorine available in chlorinated lime, hypochlorite compounds, and other materials that are used as a source of chlorine when compared with that of elemental (liquid or gaseous) chlorine.*

treated with chlorine to provide a chlorine residual of 3 mg/L. Let the water in the tank stand for 3 to 6 hours. Take a bacterial sample of tank water in a sterile container and test for coliform bacteria (see Chapter 11, "Laboratory Procedures," "Coliform Bacteria," in *WATER TREATMENT PLANT OPERATION*, Volume I). Be sure to add enough sodium thiosulfate to the sampling bottle before it is sterilized to neutralize all chlorine in the water. After a sample is collected, shake the bottle and test to be sure there is no chlorine residual. After bacteriological tests prove negative (no coliforms), steps can be taken to put the tank back into service.

Another method of disinfecting a water storage tank is to fill the tank with water with a high enough chlorine concentration to produce a chlorine residual of 3 mg/L. This method of disinfection is used after the tank has been cleaned and the cathodic protection device has been serviced as necessary. The tank is inspected to be sure that all equipment and tools have been removed. Sufficient chlorine is then applied to the water entering the tank to end up with 3.0 mg/L of chlorine residual. The water in the tank is sampled for bacteriological tests 24 hours after the tank has been filled. After bacteriological tests prove negative (no coliforms), the tank is put back in service.

EXAMPLE 3

How many pounds of chlorine gas are needed to disinfect a one million-gallon water storage tank if the desired chlorine dose is 3.0 mg/L? Assume that there is no chlorine demand in the water used to fill the tank. Therefore, we will end up with a 3.0 mg/L chlorine residual.

Known	Unknown
Chlorine Conc, mg/L = 3.0 mg/L	Chlorine Gas, lbs
Vol of Water, MG = 1.0 MG	

Calculate the pounds of chlorine gas needed to produce a 3 mg/L chlorine residual in a one million-gallon storage tank.

$$\text{Chlorine Gas, lbs} = \left(\begin{matrix}\text{Vol of Water,}\\\text{Mil Gallons}\end{matrix}\right)\left(\begin{matrix}\text{Chlorine Conc,}\\\text{mg/L}\end{matrix}\right)(8.34 \text{ lbs/gal})$$

$$= (1.0 \text{ MG})(3.0 \text{ mg/L})(8.34 \text{ lbs/gal})$$

$$= 25.0 \text{ lbs Chlorine Gas}$$

A chlorinator can be installed to add chlorine to the water entering the storage tank or a separate chlorine solution feed line can be used to convey the chlorine solution to the one million-gallon storage tank. Weigh the 150-pound (68-kg) chlorine cylinder at the start. After the chlorine cylinder weight has dropped by 25 pounds (11.5 kg), you will know you have added enough chlorine. Be sure to mix the chlorine solution with the water filling the tank so the chlorine solution will be evenly mixed throughout the tank.

Another acceptable method of disinfecting a water storage tank is to fill the tank with potable water that has been treated to provide a chlorine residual of at least 10 mg/L after a six-hour contact period. This six-hour period is sufficient when chlorine or sodium hypochlorite has been applied to the water entering the tank at a uniform rate by the use of a portable chlorinator or a chemical feed pump (hypochlorinator). When sodium hypochlorite (a liquid) or calcium hypochlorite (a tablet) has been used either by pouring the liquid into the storage facility or by use of tablets that are dissolved by the water flowing into the tank, the water should have a 10-mg/L chlorine residual after a 24-hour contact period.

The last acceptable method described here requires that enough chlorine be added to the water to produce 50 mg/L of available chlorine when the storage tank is approximately five percent full. This solution of chlorine and water must be held in the tank for at least six hours. After six hours the storage tank is filled to the overflow level with potable water and held at this level for at least 24 hours.

After a water storage tank has been disinfected, the remaining mixture of chlorine and water must be properly disposed of. Any water with a chlorine residual of 2 mg/L or more either must be diluted with additional water or the chlorine must be properly neutralized before being discharged. Chlorine neutralization chemicals include sulfur dioxide (SO_2), sodium bisulfite ($NaHSO_3$), sodium sulfite (Na_2SO_3), or sodium thiosulfate ($Na_2S_2O_3 \cdot 5\ H_2O$). The amount of neutralizing chemical will depend on the chlorine residual and the type of neutralizing chemical. By mixing various amounts of a neutralizing chemical with the water in the tank, you can determine the amount of chemical that will just neutralize all of the chlorine residual.

A highly chlorinated water *MAY* be discharged to a sanitary sewer provided *PERMISSION IS OBTAINED* from the local agency that operates the wastewater collection system and treatment facilities. If the chlorinated water is discharged slowly and there is no opportunity for *BACKFLOW*[29] or *BACKSIPHONAGE*[30] between the wastewater collection system and the water storage tank, this method *MAY* be approved.

Chlorinated waters should not be discharged to any surface waters (storm drains, rivers, lakes) without obtaining approval of the appropriate water pollution control agencies.

NOTES:

1. When disinfecting a water storage facility for the first time (a newly purchased or constructed tank), the chlorine residual should be 50 mg/L (instead of 3 mg/L when placing a tank back in service) for 24 hours.

---

[29] *Backflow. A reverse flow condition, created by a difference in water pressures, which causes water to flow back into the distribution pipes or storage tank of a potable water supply from any source or sources other than the intended source. Also see BACKSIPHONAGE.*

[30] *Backsiphonage.   A form of backflow caused by a negative or below atmospheric pressure within a water system. Also see BACKFLOW.*

2. Small water storage tanks are commonly disinfected with a chlorine dosage of 50 mg/L. A chlorine residual of at least 10 mg/L must remain at the end of the 24-hour disinfection period. Large clear wells are usually disinfected with chlorine residuals of 3 mg/L. AWWA's standards for disinfection of tanks (see Section 2.9, AWWA Standard C652-02) list three disinfection methods. The important point is, *IF YOU RECEIVE NEGATIVE COLIFORM TEST RESULTS (NO COLIFORMS PRESENT), YOU HAVE ADEQUATELY DISINFECTED THE STORAGE FACILITY.*

3. Whenever you collect a sample for a bacteriological test (coliforms), be sure to use a sterile plastic or glass bottle that contains sufficient sodium thiosulfate to neutralize all of the chlorine residual.[31] The amount of sodium thiosulfate depends on whether the sample contains a chlorine residual of 3 mg/L or 50 mg/L. Add sufficient sodium thiosulfate to the sample bottle, sterilize the bottle, and then collect the sample for the coliform test. After the sample has been collected, shake the sample and measure the chlorine residual. If there is a chlorine residual, carefully add more sterilized sodium thiosulfate, shake sample, measure the chlorine residual, and continue repeating the procedures until there is no chlorine residual. A good practice is to discuss the project you are working on with your lab and have the lab sterilize sufficient sodium thiosulfate in the sample bottle when the bottle is sterilized.

Chlorine gas is used with large-capacity storage tanks because of the relatively low price of chlorine. Small storage tanks are treated with sodium hypochlorite or calcium hypochlorite. The guideline for use of chlorine gas versus sodium hypochlorite is: for any tank over 100,000 gallons (380 cu m) capacity, use chlorine gas; for any tank below 100,000 gallons (380 cu m) capacity, sodium hypochlorite may be easier to use.

After water in a storage tank has negative bacteriological results (no coliforms), it may be returned to service. To return a storage tank to service:

1. Empty the tank of chlorinated water. Discharge to a storm drainage system may be acceptable if you have a small volume of water with a high chlorine residual and there is sufficient dilution water in the drainage system. If you have a large volume of water with a low chlorine residual, gradually release this water into the distribution system, being careful not to produce too high a chlorine residual at a consumer's tap. Other alternatives include allowing the tank to sit while chlorine dissipates or adding sodium bisulfite to reduce the residual.

2. Open outlet valve from tank to the distribution system.

3. Make sure inlet valve to tank is open so that tank will *FLOAT ON SYSTEM*[32] (draw down and fill as needed).

4. If appropriate, remove interim method of providing water to distribution system (no longer needed because tank is back in service).

QUESTIONS

Write your answers in a notebook and then compare your answers with those on page 46.

2.4I List the three common types of disinfectants used to disinfect water storage facilities.

2.4J What are two methods of disinfecting a water storage tank?

2.4K What is the guideline for determining whether chlorine gas or sodium hypochlorite is used to disinfect a water storage tank?

2.44 Concrete Storage Facilities

Maintenance of concrete water storage facilities is very similar to steel tanks. Vents and overflows must be properly screened to keep out birds, rodents, and insects. After the first year in service and every other year thereafter, the facility should be drained, cleaned, and inspected. Special attention must be given to cracks which may allow either leakage or the inflow of contaminated water. Any material used for painting, coating, or repairs must be approved for use in potable water systems. New facilities and old facilities after inspection and repair must be properly disinfected. *ALWAYS FOLLOW SAFE PROCEDURES BEFORE ENTERING AND WHENEVER YOU MUST ENTER A CONCRETE STORAGE FACILITY (CONFINED SPACE).*

Whenever you inspect a concrete storage facility, be sure the facility is secure from vandals and unauthorized persons. Inspect the access to the facility and cover to be sure contaminated water cannot get inside. Look at the surface of the water to be sure there is no floating debris or dead birds or rodents.

WARNING: Never attempt to empty any underground tank when the water table is high. If the water table is high, an empty tank could float to the water surface like a cork. When this happens the tank could crack and inlet and outlet pipes could be damaged.

2.45 Inspection and Cleaning of Tanks

Commercial divers are being used by many water agencies to maintain the interior of water storage tanks. Divers are hired to inspect, clean, and do repairs under water, so the water storage tanks don't have to be drained (and consequently taken out of service) while the work is being done.

Commercial divers can videotape their inspection while in a water tank so operators will be able to see any problems discovered by divers during an underwater inspection. Other services provided by divers include underwater still photography, corrosion severity estimates, and corrosion and pit depths. Divers also clean the insides of tanks without having to drain a tank. Operators need to maintain a higher chlorine residual in the tank during the inspection and cleaning process to protect against bacterial contamination. Divers and equipment need to be disinfected with a solution containing a minimum of 200 mg/L chlorine before entering a tank.

---

[31] A 120-mL sample bottle requires 0.1 mL of ten percent sodium thiosulfate ($Na_2S_2O_3 \cdot 5 H_2O$) to neutralize about a 15 mg/L chlorine residual.

[32] *Float On System.* A method of operating a water storage facility. Daily flow into the facility is approximately equal to the average daily demand for water. When consumer demands for water are low, the storage facility will be filling. During periods of high demand, the facility will be emptying.

When hiring a commercial diver, use only qualified divers who have completed an Association of Commercial Diving Educators (ACDE) approved commercial dive training program. Do not use SCUBA-equipped or sport-trained divers in your potable water tank. Commercial divers are outfitted with a "dry" suit and surface-supplied diving equipment (an external air source that provides the lifeline of both air and information to and from the surface). The diver's "umbilical cord" includes the capability for voice communication between operators and the diver. Live video images can also be transmitted by this line, so that operators above water can see everything the diver sees. The American Water Works Association's (AWWA) current diving standard for potable water diving specifies surface-supplied diving equipment as well as the use of dry diving suits.

Safety must always be the ultimate concern. Commercial diving crews must be trained to assess onsite conditions and be equipped with all necessary equipment to perform routine jobs safely. A major concern is ladder safety and personal fall protection. Elevated tanks require special equipment and techniques to complete an inspection and cleaning job without incident.

Hiring a commercial diving contractor can save your water utility a lot of time, water, and money. The process can be simple, but requires some planning to be sure safety issues are resolved beforehand. If a tank has water in it, and there is safe access for a diver, cost-saving inspection, cleaning, and preventive maintenance can be performed by a commercial dive team.

2.46   Grounds

Storage facility grounds should be landscaped and kept clean. Well-kept landscaping is not only aesthetically desirable, but gives everyone the impression of a well-run operation.

2.47   Frozen Distribution Reservoirs
by Dick Krueger

When subzero temperature conditions exist for several days, ice formation may occur in both underground and elevated distribution storage reservoirs. In underground reservoirs ice formation is usually limited to surface ice on top of the water. In elevated tanks, icing may be more severe and thick accumulations on sidewalls have been observed. Damage to walls and structures may result from these accumulations of ice or falling ice inside or outside of the tank. Interior ladders, for example, may become distorted and either partially or totally tear away from the mounting brackets. As water freezes, expansion pressure may separate steel plates or panels and the tank diameter may be altered resulting in interior coating failure.

Ice formations can be minimized by continuously fluctuating reservoir water levels. The pumping and flow rates into and out of the reservoir should be adjusted to allow continuous water circulation and to prevent ice from becoming attached to walls and columns, thus being suspended above the surface level of the water. The use of automatic controls may have to be temporarily suspended, particularly during low-consumption hours when water levels remain constant for extended periods. Storage tanks can also be equipped with a small compressor for tank bubbler operation in order to circulate water in the tank and prevent the surface from freezing over.

In most cases the inflow of new water pumped into a reservoir coupled with the usual drawdown results in sufficient circulation to keep ice formation to a minimum. In elevated tanks, the water level should be varied by 50 percent every 24 hours. Water may have to be drained from a tank using a hydrant if pumping schedules, based on reduced winter water demand, do not allow continuous circulation in the tank. Freeze-up in elevated reservoirs may also be reduced by the installation of a frost jacket consisting of three inches (8 cm) of insulation covered with a metal banded jacket. In underground reservoirs, 25 percent fluctuations in water levels every 24 hours are sufficient to minimize icing. In arctic and subarctic areas, circulating water systems that continuously pump and reheat the water in the system are required to prevent severe icing. Water level indicators such as floats or electric sensors may be damaged when icing occurs. Float assemblies are highly susceptible to ice damage, including snapped steel float cables, crushed interior metal floats, and destruction of float guide pulleys or guide mounts. In reservoirs subject to recurrent ice formations, pressure gage devices are recommended as an alternative to float assemblies. Level measuring devices as well as level transmitters should be equipped with heaters to prevent damage or malfunction.

Air or overflow vents may also become obstructed by accumulations of ice and snow. The vent screens should be checked routinely in order to prevent blockage and ensure a free air circulation. In elevated reservoirs, the altitude valve that controls inflows must also be kept free of ice and snow.

Ice buildup on the outside of an elevated tank reservoir can erode paint surfaces, increase the load on structural supports, and create a falling ice hazard. Outer surface ice, which forms as a result of overflows or tower leakage, has actually caused some elevated tanks to collapse. In the Fall, the tower structure should be carefully inspected for structural defects and intentional overflows should be avoided when ice formation is possible.

If an excessive ice buildup inside an elevated distribution reservoir has occurred or the tank malfunctions due to icing, a number of steps can be taken to restore operation. The affected tank should be "valved off" and the distribution system pressurized using alternate means such as other storage tanks. Pressure relief valves should be attached at strategic locations if high pressure has built up. The pressure valves can be attached to fire hydrants and excess water can then be wasted to a storm sewer. All water pumps should be operated manually. The ice can be most effectively thawed using a steam generator equipped with 65 to 100 yards (60 to 90 meters) of hose. The hose can be inserted into one of the access hatches in the tank to thaw the ice. Usually, one day of steam thawing is adequate to restore water circulation in the tank. Any leaks or damage to the tank should be repaired before placing the reservoir back into service.

QUESTIONS

Write your answers in a notebook and then compare your answers with those on page 46.

2.4L   Does the maintenance of concrete water storage facilities differ much from steel tanks?

2.4M   What type of water level indicators could be damaged due to icing in a storage tank?

2.4N   What problems can be caused by ice buildup on the outside of an elevated tank?

2.48 Inspections

Inspections are a very important element of a preventive maintenance program. Inspections are conducted to determine the structural conditions of the storage reservoir, to identify any sanitary defects in the storage system, to evaluate the need for cleaning the storage facility, to determine maintenance needs and the effectiveness of the maintenance program, and also to identify any existing or potential water quality problems.

Inspections may be routine, periodic, or comprehensive. Routine inspections are part of the normal, daily routine and include a check of security items. Periodic inspections are more detailed than routine inspections and may require climbing the facility and looking inside. During these sanitary inspections, check the security at the site, site drainage, penetrations into the system from such sources as vents, hatches, taps, cathodic protection systems, and overflows. Also look for evidence of entrance into the storage tank by birds, animals, and snakes. Measure the depth of any sediment inside the tank. Inspect structural and coating systems to determine the condition of the exterior coating, concrete foundations and visible footings, structural components such as stiffeners and wind rods, ladders, vents, safety devices, interior coating or liner, cathodic protection system, overflow pipe, weir boxes, and bug screens. Wet inspections are conducted by divers or remotely operated equipment.

Routine inspections should be conducted daily or weekly, periodic inspections should be performed monthly or quarterly. Comprehensive inspections should be scheduled every three to five years maximum and possibly more often as indicated by water quality needs.

2.49 Emergency Maintenance

Examples of emergency maintenance include penetration of the storage facility due to vandalism, localized corrosion, or splits due to extensive metal loss. Stress failure in a plate or weld or structural collapse due to vacuum pressure will require an emergency response. Emergencies also are caused by high turbidity and/or bacteria from excessive sediment, animal contamination due to screen failure, or human contamination. Other emergencies include major rips or tears in floating covers, separation of the reservoir from inlet or outlet piping, and catastrophic failure of a concrete or steel reservoir caused by undermining of the foundation. Many of these emergency conditions are the direct result of having neglected to perform routine maintenance and inspections of water storage tanks.

Natural disasters such as earthquakes, floods, fires, and tornadoes also may create conditions requiring emergency maintenance of water storage facilities.

2.410 Cleaning Storage Tanks

Storage tanks may be cleaned by either out-of-service cleaning methods or in-service cleaning methods. Out-of-service cleaning consists of draining, washing, and disinfecting the tank before returning it to service. In-service cleaning uses divers or remotely operated equipment. This equipment is similar to a vacuum cleaner and it removes soft material from the bottom of the tank.

Frequency of cleaning storage tanks depends on sediment buildup, development of biofilms, and results from water quality monitoring. Covered facilities are cleaned every three to five years, but possibly more often based on the results of water quality monitoring and inspections. Uncovered storage facilities should be cleaned annually or possibly twice a year, if

needed. Uncovered facilities should be covered or replaced by covered facilities as soon as possible.

QUESTIONS

Write your answers in a notebook and then compare your answers with those on page 46.

2.4O Why are inspections of water storage facilities conducted?

2.4P What may be the cause of conditions that require emergency maintenance of a water storage facility?

2.4Q What are the two types of methods used to clean storage tanks?

2.5 RELATED EQUIPMENT

2.50 Booster Pumps (Figure 2.14)

A booster pump is used to increase the pressure in the mains on the discharge side of the main pumps or *TO SUPPLY WATER TO AN ELEVATED STORAGE TANK.* Booster pumps are also used to supply water to a service area at a higher elevation. Most booster pumps are connected with a solenoid switch to the altitude valve on the tank. When water falls below a predetermined amount, the solenoid is actuated and the booster pump begins to pump water to the tank. The operation and maintenance of the booster pump is critical to the effective operation of the elevated tank and the system it serves.

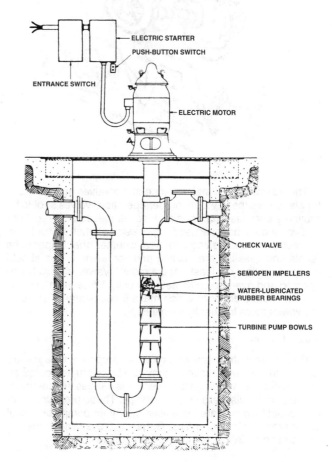

Fig. 2.14 Booster pump

(Reprinted from *WATER DISTRIBUTION OPERATOR TRAINING HANDBOOK,* by permission. Copyright 1976, the American Water Works Association)

Booster pumps are centrifugal pumps and care should be used in their operation. You need only be concerned with two critical points: installation and alignment. If the initial installation and the alignment are properly done, the pump should give few problems and require little maintenance other than routine checking of the packing and lubrication of bearings. Table 2.2 outlines procedures for troubleshooting centrifugal pumps.

2.51 Regulator Stations

A regulator station's main function is to reduce pressure and to maintain an even, acceptable pressure downstream from a high-pressure system. Regulator stations are frequently found in a water system using pumps or on a gravity system where the reservoir may have extreme pressure head (high above the service area of use). Regulator stations consist of a series of valves that reduce high pressures to lower, more acceptable working pressure. Water pressure over 100 psi (690 kPa or 7.0 kg/sq cm) should be regulated to a safe, usable pressure of 60 to 90 psi (414 to 621 kPa or 4.2 to 6.3 kg/sq cm). Where the excessive water pressure is not regulated, a pressure-reducing valve will be required on each consumer's service so that a safe, usable pressure can be obtained.

The operation and maintenance of the valves in the regulator station are the key to its proper operation. Valves should be routinely exercised (operated). Also, all valves in a regulator station should be exercised (operated) at their lowest and highest pressure settings to see whether the diaphragms (seals and gaskets) are holding and not leaking. This should be done routinely at least once a month. Valves should be inspected and overhauled at least annually. Water high in calcium may cause deposits which make it necessary to exercise the valves more frequently than this.

2.52 Gages

Gages are instruments that measure pressure. If gages are used in the correct locations, they can be useful to you as an indicator of your water system pressures. Gages are commonly used on the discharge side of pumps. These pressure readings could help indicate the efficiency of your pumps or loss of water pressure in your mains. Gages also are used in the operation and maintenance of regulator stations.

QUESTIONS

Write your answers in a notebook and then compare your answers with those on page 46.

2.5A   What is the purpose of a booster pump?

2.5B   What two points are critical to the successful operation of a booster pump?

2.5C   Regulator stations are used for what purpose?

2.5D   Where are gages commonly located?

2.6   RECORDS

Records provide necessary data for the effective operation of a water system. Records of equipment should include the *NAMEPLATE*[33] data, date of installation, start-up data, failures, and any type of service performed on the equipment. Results of tank inspections, condition of tank, maintenance performed, and all repairs or corrective action must be recorded. A list available to all involved operators of particular problems and the means used to resolve each problem is an essential part of any recordkeeping system. Performance data must be kept on your corrosion-control program. Also, the engineering plans and specifications of each type of tank structure in your system are part of your permanent records.

2.7   SAFETY

SAFETY IS EVERYONE'S RESPONSIBILITY FROM THE HEAD OF THE WATER UTILITY AGENCY TO EVERY WATER SUPPLY SYSTEM OPERATOR. However, you are the most important person when the subject of safety comes up. You have a dual responsibility. You are required to use safety equipment provided by your agency and you should be watchful for unsafe conditions and report them to your supervisor.

In working around water storage facilities, the following safety hazards should be of personal concern to you:

1. Make sure the area around the storage facility is clean and clear of debris.

2. Use care when climbing storage facility ladders or stairways. When using ladders, use a ladder safety belt. Remember: only one person at a time on a ladder. When a ladder is not in use, have the ladder guard in place so that unauthorized persons will not be able to use it. Be very careful using safety climbing devices that are equipped with a cable because cables can fail due to vibrations caused by winds.

   If a stairway is used to climb up to the storage facility, safety treads should be placed on stairs and a handrail should be used.

3. When entering a storage facility for inspection, test the atmosphere and make sure you have sufficient ventilation. Monitor the facility before entering to verify that you are not entering an oxygen-deficient atmosphere and that there are no toxic or explosive gases present (especially when applying coatings).

   Make sure adequate ventilation is available during inspection or maintenance in the facility. Use forced-draft ventilation where necessary for safety of personnel.

---

[33] *Nameplate.   A durable metal plate found on equipment which lists critical operating conditions for the equipment.*

TABLE 2.2   CENTRIFUGAL PUMP TROUBLESHOOTING CHART[a]

Problem	Possible Causes
Pump does not deliver water	1-4, 6, 11, 14, 16, 17, 22, 23, 48
Insufficient capacity delivered	2-11,14, 17, 20, 22, 23, 29-31
Insufficient pressure developed	5, 14, 16, 17, 20, 22, 29-31
Pump loses prime after starting	2, 3, 5-8, 11-13
Pumps require too much power	15-20, 23, 24, 26, 27, 29, 33, 34, 37
Stuffing box leaks excessively	1, 24, 26, 32-36, 38-40
Packing has short life	12, 13, 24, 26, 28, 32-40
Pump vibrates or is noisy	2-4, 9-11, 21, 23-28, 30, 35, 36, 41-47
Bearings have short life	24, 26-28, 35, 36, 41-47
Pump overheats and seizes	1, 4, 21, 22, 24, 27, 28, 35, 36, 41

Key to Possible Causes of Operational
Difficulties with Centrifugal Pumps

1. Pump not primed.

2. Pump or suction pipe not completely filled.

3. Suction lift too high.

4. Insufficient margin between suction pressure and vapor pressure.

5. Excessive amount of air or gas in liquid.

6. Air pocket in suction line.

7. Air leaks into suction line.

8. Air leaks into pump through stuffing boxes.

9. Foot valve too small.

10. Foot valve partially clogged.

11. Inlet of suction pipe insufficiently submerged.

12. Water-seal pipe plugged.

13. Seal cage improperly located in stuffing box, preventing sealing fluid from entering space to form seal.

14. Speed too low.

15. Speed too high.

16. Wrong direction of rotation.

17. Total head of system higher than design head of pump.

18. Total head of system lower than pump design head.

19. Specific gravity of liquid different from design.

20. Viscosity of liquid different from specifications.

21. Operation at very low capacity.

22. Parallel operation of pumps unsuitable for such operation.

23. Foreign matter in impeller.

24. Misalignment.

25. Foundations not rigid.

26. Shaft bent.

27. Rotating part rubbing on stationary part.

28. Bearings worn.

29. Wearing rings worn.

30. Impeller damaged.

31. Casing gasket defective, permitting internal leakage.

32. Shaft or shaft sleeves worn or scored at packing.

33. Packing improperly installed.

34. Incorrect type of packing for operating conditions.

35. Shaft running off center because of worn bearings or misalignment.

36. Rotor out of balance, resulting in vibration.

37. Gland too tight, resulting in no flow of liquid to lubricate packing.

38. Failure to provide cooling liquid to water-cooled stuffing boxes.

39. Excessive clearance at bottom of stuffing shaft and casing, causing packing to be forced into pump interior.

40. Dirt or grit in sealing liquid, leading to scoring of shaft or shaft sleeve.

41. Excessive thrust caused by mechanical failure inside pump or by failure of hydraulic balancing device, if any.

42. Excessive grease or oil in antifriction bearing housing or lack of cooling, causing excessive bearing temperature.

43. Lack of lubrication.

44. Improper installation of antifriction bearings (damage during assembly, incorrect assembly of stacked bearings, use of unmatched bearings as a pair).

45. Dirt getting into bearings.

46. Rusting of bearings due to water in housing.

47. Excessive cooling of water-cooled bearing, resulting in condensation in bearing housing.

48. Malfunction of foot valve.

---

[a]   *WATER DISTRIBUTION OPERATOR TRAINING HANDBOOK.* Obtain from American Water Works Association (AWWA), Bookstore, 6666 West Quincy Avenue, Denver, CO 80235. Order No. 20428. ISBN 1-58321-014-8. Price to members, $52.00; nonmembers, $76.00; price includes cost of shipping and handling.

Be sure to wear a safety harness. One person must be at the entrance watching you at all times. A third must be readily available to help with rescue operations. For additional information on how to safely enter a storage tank, see Section 7.63, "Confined Spaces."

Safety should be made part of your routine operation. Safety is a way of thinking that can help you operate and maintain your facilities more effectively.

2.8 ARITHMETIC ASSIGNMENT

A good way to learn how to solve arithmetic problems is to work on them a little bit at a time. In this operator training manual we are going to make a short arithmetic assignment at the end of every chapter. If you will work these assignments at the end of every chapter, you can easily learn how to solve waterworks arithmetic problems.

Turn to the Appendix "How to Solve Water Distribution System Arithmetic Problems," at the back of this manual and read the following sections:

1. *OBJECTIVES*,

2. A.0, *HOW TO STUDY THIS APPENDIX*,

3. A.1, *BASIC ARITHMETIC*,

4. A.2, *AREAS*,

5. A.3, *VOLUMES*,

6. A.10, *BASIC CONVERSION FACTORS*,

7. A.11, *BASIC FORMULAS*, and

8. A.12, *HOW TO USE THE BASIC FORMULAS*.

Solve all of the problems in Sections A.10, Addition; A.11, Subtraction; A.12, Multiplication; A.13, Division; A.14, Rules for Solving Equations; A.15, Actual Problems; A.2, *AREAS* (A.20, A.21, A.22, A.23, A.24, A.25, and A.26) and A.3, *VOLUMES* (A.30, A.31, A.32, A.33, and A.34) on an electronic pocket calculator. You should be able to get the same answers.

2.9 ADDITIONAL READING

1. *NEW YORK MANUAL*, Chapter 17,* "Protection of Treated Water."

2. *TEXAS MANUAL*, Chapter 16,* "Storage of Potable Water."

3. *WATER DISTRIBUTION OPERATOR TRAINING HANDBOOK*. Obtain from American Water Works Association (AWWA), Bookstore, 6666 West Quincy Avenue, Denver, CO 80235. Order No. 20428. ISBN 1-58321-014-8. Price to members, $52.00; nonmembers, $76.00; price includes cost of shipping and handling.

4. *WATER TRANSMISSION AND DISTRIBUTION*. Obtain from American Water Works Association (AWWA), Bookstore, 6666 West Quincy Avenue, Denver, CO 80235. Order No. 1957. ISBN 1-58321-231-0. Price to members, $87.00; nonmembers, $128.00; price includes cost of shipping and handling.

5. AWWA Standards on Storage. Obtain from American Water Works Association (AWWA), Bookstore, 6666 West Quincy Avenue, Denver, CO 80235.

   a. D102-03. *COATING STEEL WATER-STORAGE TANKS*, Order No. 44102. Price to members, $42.00; nonmembers, $61.00; price includes cost of shipping and handling.

   b. C652-02. *DISINFECTION OF WATER-STORAGE FACILITIES*, Order No. 43652. Price to members, $42.00; nonmembers, $61.00; price includes cost of shipping and handling.

* Depends on edition.

QUESTIONS

Write your answers in a notebook and then compare your answers with those on page 46.

2.6A   What information about equipment should be included in your records?

2.7A   What safety precautions should be exercised when working around a water storage facility?

Please answer the discussion and review questions next.

DISCUSSION AND REVIEW QUESTIONS

Chapter 2.   STORAGE FACILITIES

Write the answers to these questions in your notebook. The purpose of these questions is to indicate to you how well you understand the material in the chapter.

1. Why must care be exercised whenever large pumps are put on line or taken off line?

2. How can the development of tastes and odors be prevented in storage tanks in a new development with little demand for water?

3. What factors should be checked during the inspection of a continuously operating centrifugal pump?

4. How would you determine if a water storage tank needed cleaning?

5. What is the difference between preventive and corrective maintenance?

6. During inspection of a tank, how would you determine if the interior needed repainting?

7. How would you select a paint for the interior of a water storage tank?

8. How would you disinfect a water storage tank after it was taken out of service for inspection?

SUGGESTED ANSWERS

Chapter 2. STORAGE FACILITIES

Answers to questions on page 20.

2.0A The main purpose of a water storage facility is to provide a sufficient amount of water to average or equalize the demand on the water supply system. Also the storage facility is expected to help maintain adequate pressures throughout the entire system.

2.0B The benefits of distribution system storage include:

1. Demands on the source of water, the pumping facilities, and the transmission and distribution mains are more nearly equalized and also the capacities of the tanks and other treatment facilities need not be so great.
2. System flows and pressures are improved and stabilized.
3. Reserve water supplies are provided in the distribution system for emergencies such as firefighting and power outages.

Answers to questions on page 25.

2.1A The six common types of tanks found in water supply systems are: (1) clear wells, (2) elevated tanks, (3) standpipes, (4) ground-level storage tanks, (5) hydropneumatic tanks, and (6) surge tanks.

2.1B A clear well is used to store filtered water from a treatment plant. The purpose of a clear well is to allow fairly constant water filtration rates. Clear wells store excess water when the demand for water is low and supply additional water when the demand is high.

2.1C The main purpose of an elevated storage tank is to maintain an adequate and fairly uniform pressure to its service zone.

Answers to questions on page 25.

2.2A The selection of the type of storage facility or tank depends on the system's individual needs and the type of terrain where it is to be installed.

2.2B Surge tanks are commonly used to control the problem of surges and water hammer.

Answers to questions on page 26.

2.3A Major system changes, such as bigger size main lines or the installation of multiple services, can change storage requirements and make existing storage capacity inadequate.

2.3B Instruments operate water supply systems by measuring water supply system pressures and water levels in storage tanks. Pumps are automatically started and stopped by controls on the basis of measurements by instruments.

2.3C The "time-of-day" pumping concept suggests that pumps be operated during periods of off-peak electrical demands—namely, the hours between 10 p.m. and 6 a.m.

2.3D Three abnormal conditions encountered by operators of water storage facilities include: (1) excessive water demands, such as fire demands, (2) broken or out-of-service pumps, mains, or tanks, and (3) stale water creating tastes and odors.

Answers to questions on page 28.

2.3E The main function of a distribution storage reservoir is to take care of daily demands and especially peak demands.

2.3F Operators should know how high the level in a storage tank should be each morning so that the system's demands will be met during the rest of the day.

2.3G Water levels may be controlled in storage reservoirs either manually (by the operator) or by automatic control.

Answers to questions on page 29.

2.3H The depth of water in a storage facility can be measured by the use of floats, electrodes, ultrasonic signals, pressure switches, solid-state electronic sensors, and differential-pressure altitude valves.

2.3I Two problems may be encountered when using electrodes to measure water depths. The electrodes may corrode or become difficult to reach for replacement and repair.

2.3J Even when the pump running light is on, the pump may not be pumping water.

Answers to questions on page 29.

2.3K A pump is primed when the impeller is submerged in water.

2.3L A pump can be primed by use of (1) a special primer pump, (2) a priming tank or auxiliary water supply, and (3) removal of air by use of a vacuum pump.

2.3M The electric current and water surge can be reduced when starting a centrifugal pump by starting the pump with the discharge valve closed or throttled.

Answers to questions on page 31.

2.3N Water quality indicators that may help you decide when a water storage tank needs cleaning include: (1) turbidity, (2) color, (3) tastes and odors, and (4) positive bacteriological test results.

2.3O Four possible causes of tastes and odors in a water supply include: (1) high chlorine residual, (2) biological (algal) growths, (3) microorganisms, and (4) dead ends in mains or tanks.

2.3P Four possible causes of positive coliform test results in a water storage tank include: (1) contaminated distribution system, (2) cross connection, (3) negative pressure in a water main, and (4) no or improper disinfection of new or repaired wells, reservoirs, or mains.

Answers to questions on page 34.

2.4A Preventive maintenance is the repair or adjustment of equipment and facilities which is done before some type of deterioration takes place.

2.4B Corrective maintenance or repair is maintenance that is necessary when a problem already exists.

2.4C The time interval for painting the outside of steel tanks depends on (1) proper surface preparation, (2) a good, durable paint, (3) good workmanship, (4) adequate drying and aging, and (5) proper maintenance through periodic inspection and spot, partial, or complete removal of old paint and repainting as necessary.

2.4D The ideal time of year for taking a water storage facility out of service for complete cleaning and inspection is when the water consumption is lowest. For water utilities in warm climates, this time period is from October to April.

2.4E The interior surface of a tank is prepared for repainting by sandblasting.

Answers to questions on page 35.

2.4F Five methods of corrosion control include: (1) a coat of paint, (2) metallic coatings, (3) nonmetallic coatings, (4) chemicals for the treatment of water, and (5) electrical control (cathodic protection).

2.4G A corrosion inhibitor is any substance that promotes film formation when added to water.

2.4H The key to a successful cathodic protection program is the replacement of the sacrificial anodes when they have been corroded away. These anodes (sacrificial) should be inspected at least yearly and replaced when necessary.

Answers to questions on page 39.

2.4I The three common types of disinfectants used to disinfect water storage facilities are liquid chlorine (or gas), calcium hypochlorite, and sodium hypochlorite.

2.4J A water storage tank may be disinfected by:

1. Spraying the interior with a 200-mg/L chlorine solution, filling the tank with water with 3 mg/L chlorine residual, and allowing the water to remain in the tank for 3 to 6 hours;
2. Applying chlorine to the water entering the tank to produce a chlorine residual of 3.0 mg/L and holding the water in the tank for 24 hours;
3. Filling the tank with water that has been treated to provide a chlorine residual of at least 10 mg/L after a 6-hour contact period (24 hours for certain disinfectants); or
4. Filling the tank with water having 50 mg/L available chlorine when the tank is five percent full, holding for 6 hours, filling the tank, and then holding for 24 hours.

2.4K Chlorine gas is used to disinfect water storage tanks larger than 100,000 gallons (378,500 liters or 380 cu m) while sodium hypochlorite is used to disinfect smaller water storage tanks.

Answers to questions on page 40.

2.4L No, the maintenance of concrete storage facilities is very similar to steel tanks. Operators must inspect concrete tanks for cracks where water could leak out or contaminated water could flow into the storage facility.

2.4M Water level indicators such as floats or electric sensors may be damaged when icing occurs.

2.4N Problems that can be caused by ice buildup on the outside of an elevated reservoir include eroded paint surfaces, increased load on structural supports, and a falling ice hazard.

Answers to questions on page 41.

2.4O Inspections of storage facilities are conducted to determine the structural condition of the facility, to identify any sanitary defects in the storage system, to evaluate the need for cleaning the storage facility, to determine maintenance needs and the effectiveness of the maintenance program, and also to identify any existing or potential water quality problems.

2.4P Conditions requiring emergency maintenance may be caused by neglect of the storage facility or by natural disasters.

2.4Q Storage tanks may be cleaned by either out-of-service cleaning methods or in-service cleaning methods.

Answers to questions on page 42.

2.5A A booster pump is used to increase the pressure in the mains on the discharge side of the main pumps or to supply water to an elevated storage tank.

2.5B Two points that are critical to the successful operation of a booster pump are the initial installation and the alignment.

2.5C Regulator stations are used to reduce pressure and to maintain an even (low), acceptable pressure downstream from a high-pressure system.

2.5D Gages are commonly located on the discharge side of a pump. These pressure readings are used to indicate the efficiency of a pump and the loss of water pressure in mains. Gages are also used in the operation and maintenance of regulator stations.

Answers to questions on page 44.

2.6A Records regarding equipment should include the nameplate data, date of installation, start-up data, failures, and any type of service performed on the equipment.

2.7A Safety precautions that should be exercised when working around a water storage facility include:

1. Make sure area around the storage facility is clean and clear of debris,
2. Use care when climbing storage facility ladders and stairways, and
3. When entering a storage facility for inspection, test the atmosphere and make sure you have adequate ventilation.

APPENDIX

A. Surge in Pipelines Carrying Liquids

(Reproduced with the permission of Greer Hydraulics, Inc.,
6500 East Slauson Avenue, Commerce, CA 90040)

INTRODUCTION

Inertia is the tendency of matter to remain in its existing state of motion unless acted upon by outside forces. More energy is required to change steady state motion than to maintain it. Thus more energy is expended to increase or decrease flowing velocity in a pipeline than is necessary to maintain the liquid's steady state motion. One measure of the energy used to change liquid motion is pressure. If velocity is changed quickly by the application of much energy in a short period of time, the pressure change will be more significant than when the same amount of energy is expended over a longer time period. Thus, the energy applied to change pressure by 100 psi in one second is the same amount of energy as that applied for 10 seconds to change pressure by 10 psi.

BASIC CONCEPT

Any change from steady state conditions creates a temporary variation in pressure or flow called a hydraulic transient. Hydraulic transients are commonly called surge, shock, or water hammer. They are usually caused by opening, closing, or regulating valves, or by pumps starting or stopping. Their magnitude is a function of the: (1) change in flowing velocity, (2) liquid density, and (3) sound speed in the liquid and piping system.

These hydraulic transients may range in importance from a slight pressure and/or velocity change, to sufficiently high vacuum or pressure to collapse or burst pipes and fittings or damage pumps.

Surge arrestors are designed to control rapid velocity changes which may cause potentially dangerous pressure excursions.

This is accomplished by using a vessel charged with air or gas and connected to the pipeline carrying the liquid. This vessel has the capability to convert the kinetic energy of the moving liquid into stored potential energy when a liquid overpressure occurs. When a pump stops, the vessel gas expands and "pumps" needed fluid into the line to prevent the formation of vacuum and column separation.

It is this *hydropneumatic* feature that gives the surge arrestor its ability to control the *rate* of energy change.

An analogy can be drawn to develop an understanding of hydraulic transients and how the surge arrestor controls energy changes.

Consider that a railroad locomotive pushing railcars up a grade is similar to a pump pushing a column of liquid through a pipeline to an elevated reservoir.

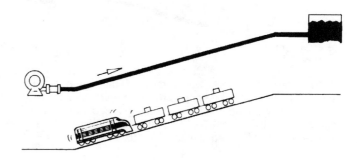

When the locomotive stops, the cars coast until stopped by friction and gravity.

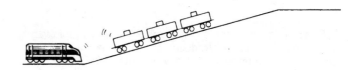

Gravity then causes the cars to roll back down the hill until they crash into the locomotive.

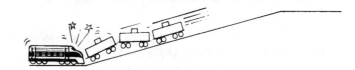

One can easily visualize the results. No crash would result if the railroad had a switch engine to move other cars from a sidetrack to the mainline behind the coasting cars to ease the reverse motion of the railcars back to the stopped locomotive.

And so it is with the pump-liquid pipeline system. When the pump stops, the liquid flows away from the pump, the liquid column separates and forms a vacuum downstream from the check valve until, under the influence of gravity and friction, it stops moving.

Then gravity causes it to reverse direction until it crashes into the closed check valve. The resulting shock may cause a simple annoyance or a catastrophic failure.

And again no shock would result if the pump-pipeline system is fitted with a hydropneumatic surge arrestor. It would force liquid into the pipeline replacing the pump flow for enough time to prevent liquid column separation and cushions reverse flow until equilibrium is established.

Shock in liquid systems may also be caused by rapid valve movement. This can be reduced by slower regulation of the valve action. Emergency valve requirements as in fire, deluge, and fuel transfer installations may not permit this type of regulation.

Liquid stopped by a quick closing valve would be like running the train head on into a large solid wall.

Bringing the train to a gradual halt by braking would parallel controlled (slower) valve closure.

A sidetrack up a hill would safely slow the train to a gradual stop without damage if the obstruction cannot be removed.

A fast acting valve could not cause damaging shock if a hydropneumatic surge arrestor is mounted upstream to "sidetrack," slow, and safely stop liquid movement.

Another place where energy is wasted and has a potential for damage is on start-up of pumps. This energy is easy to recognize in the railroad analogy. The locomotive must race its engine and spin its wheels to overcome inertia to get a long train of railcars moving.

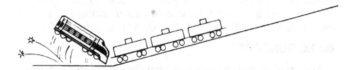

If the energy could be averaged by means of a large spring between the locomotive and the train, a smooth start would result.

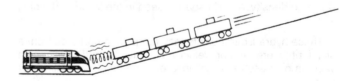

The high burst of energy required to overcome the inertia of liquid in a pipeline when a pump starts may similarly be averaged by the use of a hydropneumatic surge arrestor on the pump discharge.

HOW TO CONTROL SURGE?

Because surge is a complex problem, many solutions have been attempted. The following list compares some of these solutions.

Method	Advantages	Disadvantages	Application
None (ignore the problem)	No direct cost to plant; no fit-up problem.	Severe water hammer. Requires that pipe, supports, fittings be extra strong and designed for shock. Costly, high-pressure design required.	High-pressure steel lines.
Surge control valve	Easy to fit into plant.	Valve must be exercised regularly to ensure operation. Partially drains line. Reverses flow. Requires service and periodic maintenance. Needle valves and close internal tolerances may scale or plug up. Medium cost.	Transmission lines with smooth gradients and no service turnouts.
Back-spin pump with slow-closing discharge valve	Easy to fit into plant.	Pump and motor must be specially designed. Reverse flow may reach high velocity and cause damage. Not applicable for booster applications. Medium cost.	Transmission lines with smooth gradients and no service turnouts.
Flywheels	Good for large, short lines.	Requires reduced voltage starters. Keyways, couplings, bearings hard to maintain. Special pump and motor shaft requirements. High cost.	Transmission lines with smooth gradients and no service turnouts.
One-way surge tanks or standpipes	Open tank not a pressure vessel. Controlled upsurge and downsurge. Predictable performance.	Requires drawdown to tank level. Requires valve service. May be difficult to fit into plant. Medium cost.	Transmission lines with smooth gradients and no service turnouts.
Combination hydropneumatic tank with vacuum valve	May be buried. Reduces volume requirements for low-lift applications. Good for force main applications. Predictable performance.	Requires drawdown to tank elevation. May be difficult to fit into plant. Valve must be serviced. Medium cost.	Transmission lines with smooth gradients and no service turnouts.
Hydropneumatic surge arrestor	Minimum service. Controlled upsurge and downsurge. Predictable performance. Fail safe. Good for booster applications. Single pump check is only valve required. Permits use of thin-wall piping.	May be difficult to fit into plant. Becomes large for low-lift applications. Medium cost.	All including irregular profile and service turnouts.

CHAPTER 3

DISTRIBUTION SYSTEM FACILITIES

by

Sam Kalichman

and

Nick Nichols

TABLE OF CONTENTS

Chapter 3.   DISTRIBUTION SYSTEM FACILITIES

OBJECTIVES

Chapter 3. DISTRIBUTION SYSTEM FACILITIES

Following completion of Chapter 3, you should be able to:

1. Explain the purpose of a water distribution system,

2. Describe the importance of hydraulics in the performance of a distribution system,

3. Explain the purpose of distribution system storage and pumping facilities,

4. Identify various types of pipes and joints,

5. Properly and safely install pipe,

6. Identify and test various types of meters,

7. Identify the various types of backflow prevention devices, and

8. Determine the need for and install backflow prevention devices.

WORDS

Chapter 3. DISTRIBUTION SYSTEM FACILITIES

APPURTENANCE (uh-PURR-ten-nans)   APPURTENANCE

Machinery, appliances, structures and other parts of the main structure necessary to allow it to operate as intended, but not considered part of the main structure.

BACK PRESSURE   BACK PRESSURE

A pressure that can cause water to backflow into the water supply when a user's water system is at a higher pressure than the public water system.

BACKFLOW   BACKFLOW

A reverse flow condition, created by a difference in water pressures, which causes water to flow back into the distribution pipes of a potable water supply from any source or sources other than an intended source. Also see BACKSIPHONAGE.

BACKSIPHONAGE   BACKSIPHONAGE

A form of backflow caused by a negative or below atmospheric pressure within a water system. Also see BACKFLOW.

C FACTOR   C FACTOR

A value or factor used to indicate the smoothness of the interior of a pipe. The higher the C Factor, the smoother the pipe, the greater the carrying capacity, and the smaller the friction or energy losses from water flowing in the pipe. To calculate the C Factor, measure the flow, pipe diameter, distance between two pressure gages, and the friction or energy loss of the water between the gages.

$$\text{C Factor} = \frac{\text{Flow, GPM}}{193.75(\text{Diameter, ft})^{2.63}(\text{Slope})^{0.54}}$$

CATHODIC (ca-THOD-ick) PROTECTION   CATHODIC PROTECTION

An electrical system for prevention of rust, corrosion, and pitting of metal surfaces which are in contact with water or soil. A low-voltage current is made to flow through a liquid (water) or a soil in contact with the metal in such a manner that the external electromotive force renders the metal structure cathodic. This concentrates corrosion on auxiliary anodic parts which are deliberately allowed to corrode instead of letting the structure corrode.

CORPORATION STOP   CORPORATION STOP

A water service shutoff valve located at a street water main. This valve cannot be operated from the ground surface because it is buried and there is no valve box. Also called a corporation cock.

CORROSION   CORROSION

The gradual decomposition or destruction of a material by chemical action, often due to an electrochemical reaction. Corrosion may be caused by (1) stray current electrolysis, (2) galvanic corrosion caused by dissimilar metals, or (3) differential-concentration cells. Corrosion starts at the surface of a material and moves inward.

CROSS CONNECTION   CROSS CONNECTION

A connection between a drinking (potable) water system and an unapproved water supply. For example, if you have a pump moving nonpotable water and hook into the drinking water system to supply water for the pump seal, a cross connection or mixing between the two water systems can occur. This mixing may lead to contamination of the drinking water.

CURB STOP   CURB STOP

A water service shutoff valve located in a water service pipe near the curb and between the water main and the building. This valve is usually operated by a wrench or valve key and is used to start or stop flows in the water service line to a building. Also called a curb cock.

DYNAMIC PRESSURE DYNAMIC PRESSURE

When a pump is operating, the vertical distance (in feet) from a reference point (such as a pump centerline) to the hydraulic grade line is the dynamic head. Also see ENERGY GRADE LINE, STATIC HEAD, STATIC PRESSURE, and TOTAL DYNAMIC HEAD.

Dynamic Pressure, psi = (Dynamic Head, ft)(0.433 psi/ft)

ELECTROLYSIS (ee-leck-TRAWL-uh-sis) ELECTROLYSIS

The decomposition of material by an outside electric current.

ENERGY GRADE LINE (EGL) ENERGY GRADE LINE (EGL)

A line that represents the elevation of energy head (in feet) of water flowing in a pipe, conduit or channel. The line is drawn above the hydraulic grade line (gradient) a distance equal to the velocity head ($V^2/2g$) of the water flowing at each section or point along the pipe or channel. Also see HYDRAULIC GRADE LINE.

[SEE DRAWING ON PAGE 58]

ENTRAIN ENTRAIN

To trap bubbles in water either mechanically through turbulence or chemically through a reaction.

FRICTION LOSSES FRICTION LOSSES

The head, pressure or energy (they are the same) lost by water flowing in a pipe or channel as a result of turbulence caused by the velocity of the flowing water and the roughness of the pipe, channel walls, or restrictions caused by fittings. Water flowing in a pipe loses head, pressure or energy as a result of friction losses. Also see HEAD LOSS.

GALVANIZE GALVANIZE

To coat a metal (especially iron or steel) with zinc. Galvanization is the process of coating a metal with zinc.

GRADE GRADE

(1) The elevation of the invert (or bottom) of a pipeline, canal, culvert, or similar conduit.

(2) The inclination or slope of a pipeline, conduit, stream channel, or natural ground surface; usually expressed in terms of the ratio or percentage of number of units of vertical rise or fall per unit of horizontal distance. A 0.5 percent grade would be a drop of one-half foot per hundred feet of pipe.

HEAD LOSS HEAD LOSS

The head, pressure or energy (they are the same) lost by water flowing in a pipe or channel as a result of turbulence caused by the velocity of the flowing water and the roughness of the pipe, channel walls, or restrictions caused by fittings. Water flowing in a pipe loses head, pressure or energy as a result of friction losses. The head loss through a filter is due to friction losses caused by material building up on the surface or in the top part of a filter. Also see FRICTION LOSSES.

HIGH-LINE JUMPERS HIGH-LINE JUMPERS

Pipes or hoses connected to fire hydrants and laid on top of the ground to provide emergency water service for an isolated portion of a distribution system.

HYDRAULIC GRADE LINE (HGL) HYDRAULIC GRADE LINE (HGL)

The surface or profile of water flowing in an open channel or a pipe flowing partially full. If a pipe is under pressure, the hydraulic grade line is at the level water would rise to in a small vertical tube connected to the pipe. Also see ENERGY GRADE LINE.

[SEE DRAWING ON PAGE 58]

INVERT (IN-vert) INVERT

The lowest point of the channel inside a pipe, conduit, or canal.

MIL MIL

A unit of length equal to 0.001 of an inch. The diameter of wires and tubing is measured in mils, as is the thickness of plastic sheeting.

POTABLE (POE-tuh-bull) WATER POTABLE WATER

Water that does not contain objectionable pollution, contamination, minerals, or infective agents and is considered satisfactory for drinking.

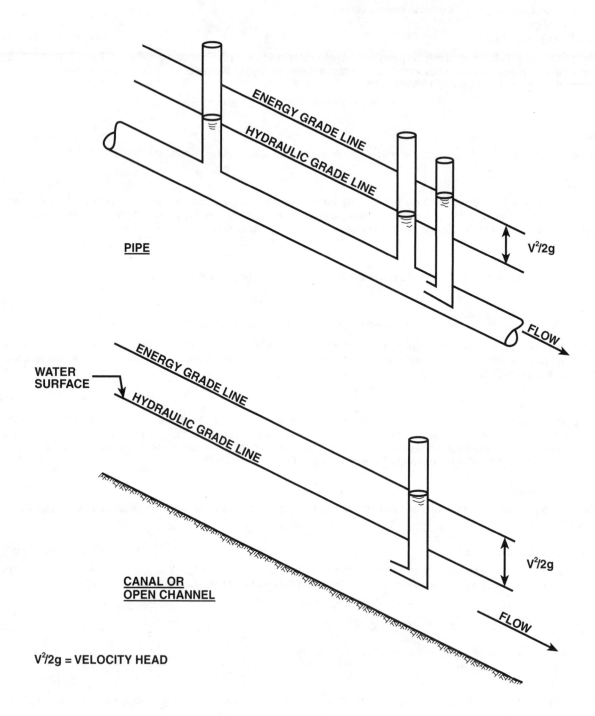

PIPE

ENERGY GRADE LINE

HYDRAULIC GRADE LINE

$V^2/2g$

FLOW

WATER
SURFACE

ENERGY GRADE LINE

HYDRAULIC GRADE LINE

$V^2/2g$

CANAL OR
OPEN CHANNEL

FLOW

$V^2/2g$ = VELOCITY HEAD

ENERGY GRADE LINE and HYDRAULIC GRADE LINE

PRESSURE HEAD   PRESSURE HEAD

The vertical distance (in feet) equal to the pressure (in psi) at a specific point. The pressure head is equal to the pressure in psi times 2.31 ft/psi.

PRESTRESSED   PRESTRESSED

A prestressed pipe has been reinforced with wire strands (which are under tension) to give the pipe an active resistance to loads or pressures on it.

SERVICE PIPE   SERVICE PIPE

The pipeline extending from the water main to the building served or to the consumer's system.

SLOPE   SLOPE

The slope or inclination of a trench bottom or a trench side wall is the ratio of the vertical distance to the horizontal distance or "rise over run." Also see GRADE (2).

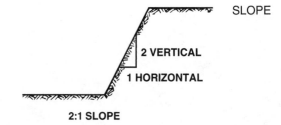

2:1 SLOPE

SLURRY (SLUR-e)   SLURRY

A watery mixture or suspension of insoluble (not dissolved) matter; a thin, watery mud or any substance resembling it (such as a grit slurry or a lime slurry).

SPOIL   SPOIL

Excavated material such as soil from the trench of a water main.

SPRING LINE   SPRING LINE

Theoretical center of a pipeline. Also, the guideline for laying a course of bricks.

STATIC HEAD   STATIC HEAD

When water is not moving, the vertical distance (in feet) from a specific point to the water surface is the static head. (The static pressure in psi is the static head in feet times 0.433 psi/ft.) Also see DYNAMIC PRESSURE and STATIC PRESSURE.

STATIC PRESSURE   STATIC PRESSURE

When water is not moving, the vertical distance (in feet) from a specific point to the water surface is the static head. The static pressure in psi is the static head in feet times 0.433 psi/ft. Also see DYNAMIC PRESSURE and STATIC HEAD.

SURGE CHAMBER   SURGE CHAMBER

A chamber or tank connected to a pipe and located at or near a valve that may quickly open or close or a pump that may suddenly start or stop. When the flow of water in a pipe starts or stops quickly, the surge chamber allows water to flow into or out of the pipe and minimize any sudden positive or negative pressure waves or surges in the pipe.

TAILGATE SAFETY MEETING   TAILGATE SAFETY MEETING

Brief (10 to 20 minutes) safety meetings held every 7 to 10 working days. The term *TAILGATE* comes from the safety meetings regularly held by the construction industry around the tailgate of a truck.

THRUST BLOCK   THRUST BLOCK

A mass of concrete or similar material appropriately placed around a pipe to prevent movement when the pipe is carrying water. Usually placed at bends and valve structures.

TOTAL DYNAMIC HEAD (TDH)   TOTAL DYNAMIC HEAD (TDH)

When a pump is lifting or pumping water, the vertical distance (in feet) from the elevation of the energy grade line on the suction side of the pump to the elevation of the energy grade line on the discharge side of the pump.

TUBERCULATION (too-BURR-cue-LAY-shun)   TUBERCULATION

The development or formation of small mounds of corrosion products (rust) on the inside of iron pipe. These mounds (tubercules) increase the roughness of the inside of the pipe thus increasing resistance to water flow (decreases the C Factor).

VELOCITY HEAD VELOCITY HEAD

The energy in flowing water as determined by a vertical height (in feet or meters) equal to the square of the velocity of flowing water divided by twice the acceleration due to gravity ($V^2/2g$).

WATER HAMMER WATER HAMMER

The sound like someone hammering on a pipe that occurs when a valve is opened or closed very rapidly. When a valve position is changed quickly, the water pressure in a pipe will increase and decrease back and forth very quickly. This rise and fall in pressures can cause serious damage to the system.

CHAPTER 3. DISTRIBUTION SYSTEM FACILITIES

(Lesson 1 of 5 Lessons)

3.0 PURPOSE OF DISTRIBUTION SYSTEMS

Water distribution typically consists of pipes, storage facilities, pumping stations, valves, fire hydrants, meters, and other *APPURTENANCES*[1] (Figure 3.1). Together they make up what can be considered a single operating unit working to achieve its intended purpose. That purpose is to deliver adequate quantities of water at sufficient pressures at all times under continually changing conditions while at the same time protecting water quality. Two major elements combine to accomplish this purpose. The first deals with the physical features of the system and relates to its ability to withstand the stresses imposed upon it, to protect the quality of the water it carries, and to deliver the supply in sufficient quantity through adequately sized facilities. The second element is the operation and maintenance of the system in a manner that preserves its integrity and the quality of the water it delivers. Of special importance is the maintenance of a *CONTINUOUS POSITIVE WATER PRESSURE* in the system under all conditions so as to protect the distribution system from the entrance of toxic and other undesirable substances. To properly understand and operate a distribution system, the operator must be thoroughly knowledgeable about both its physical and hydraulic characteristics. This chapter deals with the physical and hydraulic characteristics of a distribution system. The next chapter, Chapter 4, discusses the water quality aspects and the following chapter outlines the procedures to operate and maintain a distribution system.

QUESTIONS

Write your answers in a notebook and then compare your answers with those on page 153.

3.0A   List the major parts of a water distribution system.

3.0B   Why must a continuous positive water pressure be maintained in the distribution system at all times?

3.1 DISTRIBUTION SYSTEM HYDRAULICS

The major hydraulic concerns of a water distribution system are (1) will water flow, and (2) in what direction? Also, available pressures are of prime concern. Different agencies may use different values for target or desired minimum pressures. Generally, water should be delivered with a working minimum pressure of 35 pounds per square inch (psi) (241 kiloPascals or 2.5 kilograms per square centimeter) to all points in a distribution system. Twenty psi (138 kPa or 1.4 kg/sq cm) at a delivery point would be the absolute minimum. Excessive pressures (greater than 100 psi, 690 kPa or 7 kg/sq cm) will damage the customers' facilities and plumbing fixtures.

Accomplishing these hydraulic goals sometimes becomes very complex when a distribution system serves consumers living on high hills or in deep valleys. Another factor that must be considered is the fact that water flowing in a pipe or conduit will lose pressure due to pipe *FRICTION LOSS*.[2] Water pressure problems created by hills and valleys are a lesser concern than friction losses because pressure problems due to elevation can be anticipated and corrected when the distribution system is designed. However, the loss of pressure due to friction in a pipeline is not constant, being determined by the roughness of the interior of the pipe creating turbulence proportional to the velocity of water flowing in the pipe. This velocity is constantly changing with demand while pipe roughness can gradually increase over a period of time.

All pipes have a roughness that resists water flow and causes a drop in pressure under dynamic (flowing) conditions. The roughness is indicated by a factor called a *C FACTOR*.[3] The higher the C Factor, the smoother the inside of the pipe. Some pipe roughness is natural (due to the material used in the pipe construction); it will increase with deposits along the pipeline or pipe deterioration (corrosion).

To understand the hydraulics of a water supply system, there are several other technical terms you will need to master. Refer to the drawings on the next few pages if you have trouble picturing these concepts.

To begin with, imagine you are looking at one of the distribution pipes in a water supply system. If you were to drill a small hole in the top of the pipe and attach a vertical pipe to the hole, the water in the pipe would rise in the tube a distance equal to

---

[1] Appurtenance (uh-PURR-ten-nans).   Machinery, appliances, structures and other parts of the main structure necessary to allow it to operate as intended, but not considered part of the main structure.

[2] Friction Losses.   The head, pressure or energy (they are the same) lost by water flowing in a pipe or channel as a result of turbulence caused by the velocity of the flowing water and the roughness of the pipe, channel walls, or restrictions caused by fittings. Water flowing in a pipe loses head, pressure or energy as a result of friction losses.

[3] C Factor.   A value or factor used to indicate the smoothness of the interior of a pipe. The higher the C Factor, the smoother the pipe, the greater the carrying capacity, and the smaller the friction or energy losses from water flowing in the pipe. To calculate the C Factor, measure the flow, pipe diameter, distance between two pressure gages, and the friction or energy loss of water between the gages.

$$Slope = (Energy\ Loss)/Distance.$$

$$C\ Factor = \frac{Flow,\ GPM}{193.75(Diameter,\ ft)^{2.63}(Slope)^{0.54}}$$

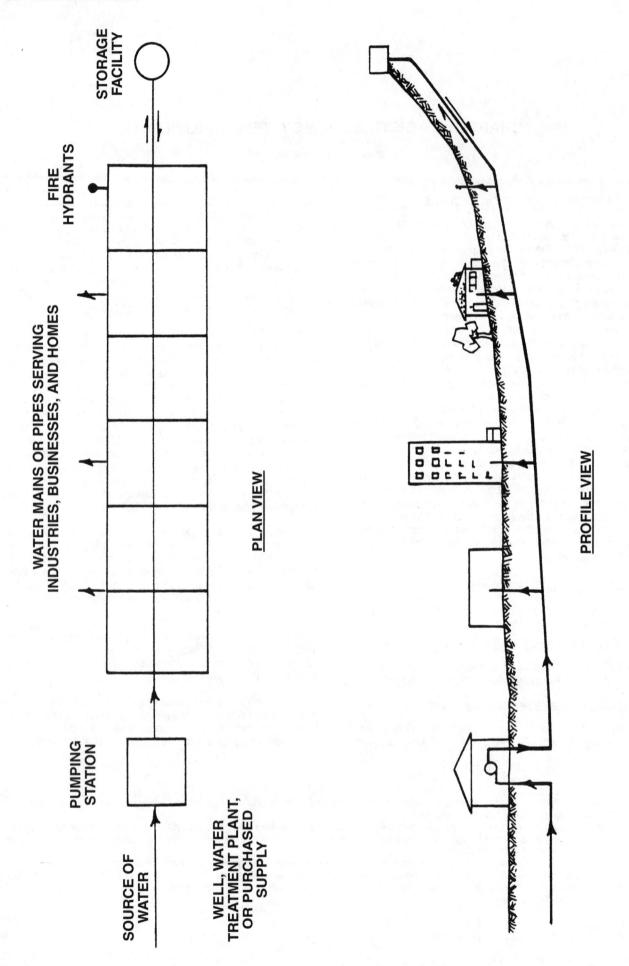

Fig. 3.1   Typical water distribution system facilities

the *PRESSURE HEAD*.[4] If you drilled several holes along the top of the pipe and inserted vertical tubes in each hole, the line connecting the top of the water in each tube would form the *HYDRAULIC GRADE LINE*[5] (also called the hydraulic gradient). The hydraulic grade line always slopes downward in the direction of flow in a pipe, regardless of the actual *SLOPE*[6] of the pipe (Figure 3.2).

When water changes its direction of flow in a pipe, the direction of the slope of the hydraulic grade line will also change. For example, if water is flowing into an elevated water storage reservoir, the hydraulic grade line will slope downward *TOWARD* the reservoir. When water is flowing out of a reservoir during periods of high demand, the hydraulic grade line will slope downward *AWAY* from the reservoir (Figure 3.3).

A major fire, leak, or other unanticipated high demand for water might draw the hydraulic grade line down to a point that a negative pressure (vacuum) is created at some service connections (Figure 3.4). This negative pressure could cause *BACKFLOW*[7] from a nonpotable source of water and contaminate the distribution system.

The hydraulic grade line is also affected by obstructions in the pipeline such as throttled valves (Figure 3.5). If pipe diameters are too small and/or flows are too great, the hydraulic grade line can drop very quickly and result in pressures that are too low. Remember, the pressures in a pipe are represented by the distance or size of the pressure head (the distance from the pressure gage to the hydraulic grade line). One way to evaluate your distribution system is to take pressure readings at various locations (such as fire hydrants) along a water main. By plotting the elevations of the hydraulic grade line on a profile drawing of the water main, you can locate obstructions (such as throttled or stuck valves) and other flow-limiting factors.

Other factors of hydraulic concern are thrust and *WATER HAMMER*.[8] Water flowing in a pipe exerts thrust at any point where the layout of the system either changes the direction of the flow, increases the velocity, or decreases or stops the flow. At these points, the pipes and fittings must be anchored and kept from moving or pulling apart by the use of *THRUST BLOCKS*[9] (Figure 3.6). Thrust blocks (or "kickers") distribute the thrust on the soil in such a manner that the thrust does not exceed the carrying capacity of the soil. Improperly installed thrust blocks can cause considerable damage to the pipe and fitting. Thrust blocks are not required on pipe where the joints are fixed (due to a mechanical anchorage such as welding) and cannot be pulled apart.

Water hammer is caused by surges in the pipeline resulting from the rapid increase or decrease in water flow. This usually results from turning pumps on or off or operating valves too rapidly. Water hammer exerts tremendous forces on the system and can be highly destructive. *SURGE CHAMBERS*[10] (Figure 3.7) are sometimes used to control water hammer. For a detailed description of surges and methods of controlling them, see Chapter 2, "Storage Facilities," Appendix, "Surge in Pipelines Carrying Liquids." Operators, firefighters, and anyone else authorized to open and close valves or start and stop pumps must be trained to slowly open and close valves. Large pumps should not be placed on line and started or stopped quickly.

FORMULAS

Frequently we determine the pressure in a main or at a fire hydrant by reading a pressure gage in psi (pounds per square inch). To convert a pressure reading in psi to a pressure head in feet, use the conversion factor 2.31 ft per psi. This means that a column of water 2.31 feet high would exert a pressure of one pound over one square inch.

Pressure Head, ft = (Pressure, psi)(2.31 ft/psi)

If we were given the pressure head in feet from a pressure gage, we could rearrange the above formula to determine the pressure in psi.

$$\text{Pressure, psi} = \frac{\text{Pressure Head, ft}}{2.31 \text{ ft/psi}}$$

or

Pressure, psi = (Pressure Head, ft)(0.433 psi/ft)

We obtained 0.433 psi/ft by solving for 1/2.31 ft/psi = 0.43 psi/ft. This term says that a column of water one foot high will exert a pressure of 0.433 pound over one square inch.

---

[4] *Pressure Head. The vertical distance (in feet) equal to the pressure (in psi) at a specific point. The pressure head is equal to the pressure in psi times 2.31 ft/psi.*

[5] *Hydraulic Grade Line (HGL). The surface or profile of water flowing in an open channel or a pipe flowing partially full. If a pipe is under pressure, the hydraulic grade line is at the level water would rise to in a small vertical tube connected to the pipe.*

[6] *Slope. The slope or inclination of a trench bottom or a trench side wall is the ratio of the vertical distance to the horizontal distance or "rise over run."*

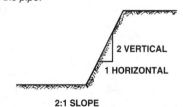

2 VERTICAL

1 HORIZONTAL

2:1 SLOPE

[7] *Backflow. A reverse flow condition, created by a difference in water pressures, which causes water to flow back into the distribution pipes of a potable water supply from any source or sources other than an intended source.*

[8] *Water Hammer. The sound like someone hammering on a pipe that occurs when a valve is opened or closed very rapidly. When a valve position is changed quickly, the water pressure in a pipe will increase and decrease back and forth very quickly. This rise and fall in pressures can cause serious damage to the system.*

[9] *Thrust Block. A mass of concrete or similar material appropriately placed around a pipe to prevent movement when the pipe is carrying water. Usually placed at bends and valve structures.*

[10] *Surge Chamber. A chamber or tank connected to a pipe and located at or near a valve that may quickly open or close or a pump that may suddenly start or stop. When the flow of water in a pipe starts or stops quickly, the surge chamber allows water to flow into or out of the pipe and minimize any sudden positive or negative pressure waves or surges in the pipe.*

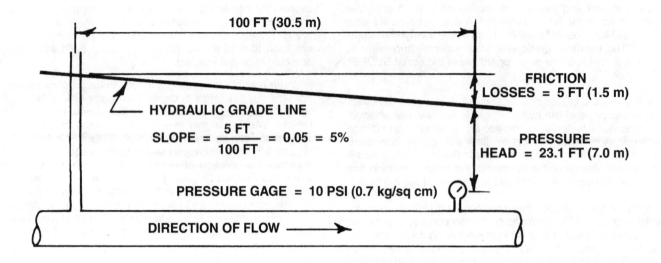

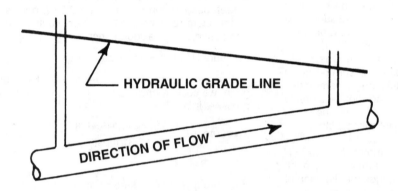

NOTE: The hydraulic grade line *ALWAYS* slopes downward
in the direction of flow.

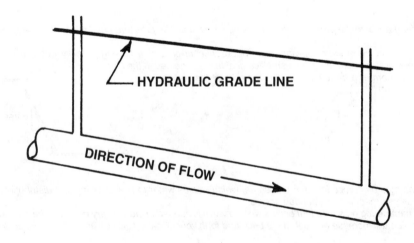

Fig. 3.2   Hydraulic grade lines

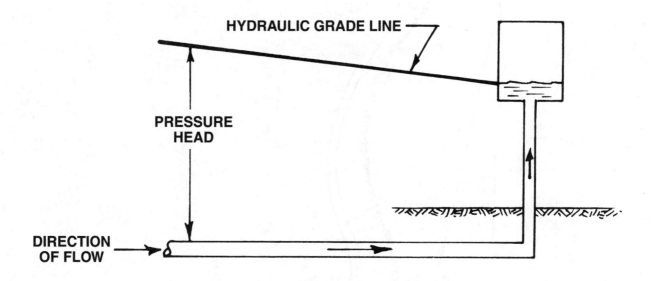

STORAGE RESERVOIR FILLING

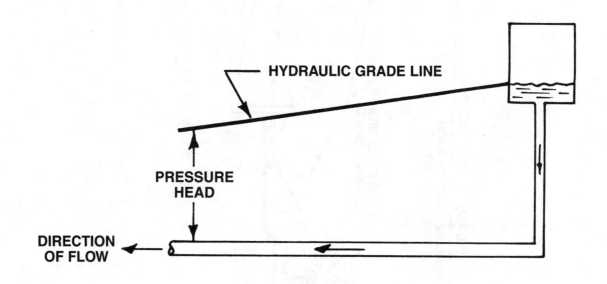

STORAGE RESERVOIR EMPTYING

Fig. 3.3 Storage reservoir hydraulic grade lines

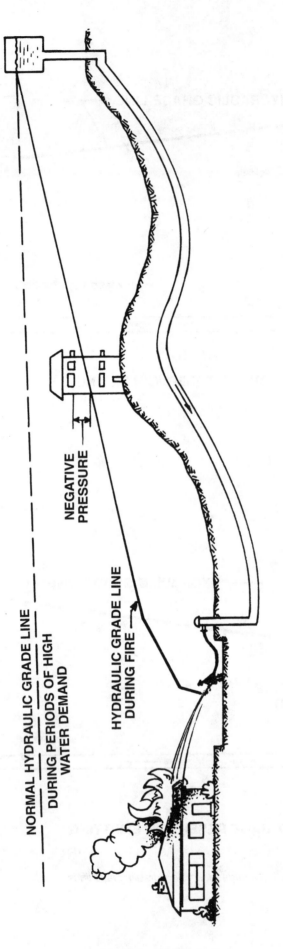

NORMAL HYDRAULIC GRADE LINE

DURING PERIODS OF HIGH WATER DEMAND

HYDRAULIC GRADE LINE DURING FIRE

NEGATIVE PRESSURE

NOTE: Any fixture or faucet above the hydraulic grade line will be exposed to a negative pressure.

Fig. 3.4   Negative pressure

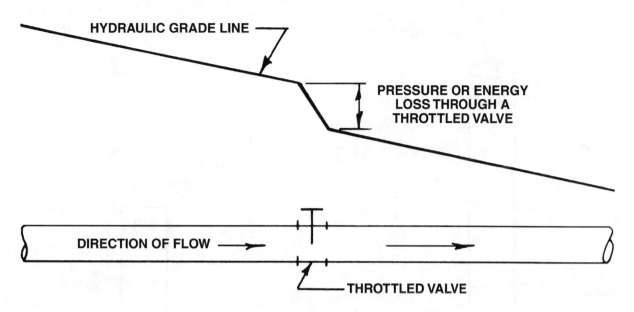

Fig. 3.5   *Throttled valve lowering hydraulic grade line*

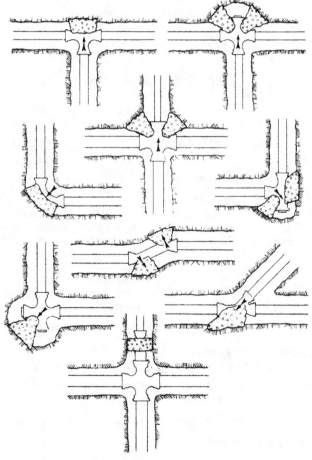

NOTE:
   1. The location of thrust blocks depends on the direction of
      thrust and type of fitting as shown above.
   2. Thrust blocks are not required if the ends of pipe are held
      tightly by mechancial anchorage such as a weld.

Fig. 3.6   *Location of thrust blocks*

(Permission of Johns-Manville Corporation,
a Subsidiary of Manville Corporation)

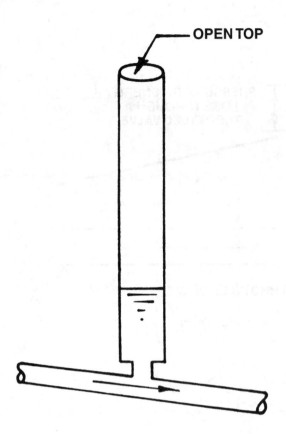

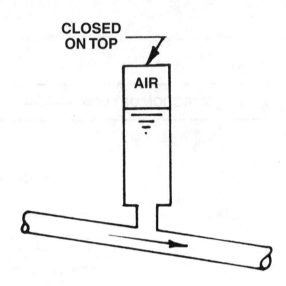

Fig. 3.7   *Types of surge chambers*

EXAMPLE 1

A pressure gage on a fire hydrant reads 40 psi. What is the pressure head in feet at the fire hydrant?

Known	**Unknown**
Pressure, psi = 40 psi	Pressure Head, ft

Calculate the pressure head in feet.

$$\text{Pressure Head, ft} = (\text{Pressure, psi})(2.31 \text{ ft/psi})$$
$$= (40 \text{ psi})(2.31 \text{ ft/psi})$$
$$= 92.4 \text{ ft}$$

EXAMPLE 2

A pressure gage on a fire hydrant reads 80 feet of pressure head. What is the pressure in psi? *NOTE: Pressure gages may read in psi and/or feet of head.*

Known	**Unknown**
Pressure Head, ft = 80 ft	Pressure, psi

Calculate the pressure in pounds per square inch (psi).

$$\text{Pressure, psi} = \frac{\text{Pressure Head, ft}}{2.31 \text{ ft/psi}}$$
$$= \frac{80 \text{ ft}}{2.31 \text{ ft/psi}}$$
$$= 34.6 \text{ psi}$$

or

$$\text{Pressure, psi} = (\text{Pressure Head, ft})(0.433 \text{ psi/ft})$$
$$= (80 \text{ ft})(0.433 \text{ psi/ft})$$
$$= 34.6 \text{ psi}$$

EXAMPLE 3 (ADVANCED PROBLEM)

Footnote 3 on page 69 shows the formula used to calculate the C Factor. If your calculator can calculate exponents, you can work this problem. A 24-inch diameter water main is carrying a flow of 3,000 GPM. Pressure gages installed 1,000 feet apart on the main indicate that the elevation of the pressure head at the upstream pressure gage is 101 feet and 100 feet at the downstream pressure gage. Calculate the C Factor for this pipe.

Known		**Unknown**
Flow, GPM	= 3,000 GPM	C Factor
Diameter, in	= 24 in	
Diameter, ft	= 2 ft	
Distance, ft	= 1,000 ft	
Energy Loss, ft	= 101 ft – 100 ft	
	= 1 ft	

1. Calculate the slope.

$$\text{Slope} = \frac{\text{Energy Loss, ft}}{\text{Distance, ft}}$$
$$= \frac{1 \text{ ft}}{1,000 \text{ ft}}$$
$$= 0.001 \text{ ft/ft}$$

2. Calculate the C Factor.

$$C\ Factor = \frac{Flow,\ GPM}{193.75(Diameter,\ ft)^{2.63}(Slope)^{0.54}}$$

$$= \frac{3,000\ GPM}{193.75(2.0\ ft)^{2.63}(0.001)^{0.54}}$$

$$= \frac{3,000}{(193.75)(6.19)(0.02399)}$$

$$= 104$$

QUESTIONS

Write your answers in a notebook and then compare your answers with those on page 153.

3.1A What problems may be created by excessive pressures in a distribution system?

3.1B What causes friction losses in pipes?

3.1C What conditions might cause thrust in a pipe?

3.1D Water hammer can be caused by what actions?

3.1E During a distribution system pressure test, a pressure gage attached to a faucet in a home read 30 psi. What is the pressure head in feet?

3.2 PERFORMANCE CONSIDERATIONS

Distribution systems and their components should be capable of carrying the required flow at the desired pressures to prevent the introduction of foreign substances and to minimize reactions between the water and parts of the system. Sufficient water must be available from the water sources and distribution reservoirs to supply adequately, dependably, and safely the total requirements of all users under maximum demand conditions.

Distribution system performance considerations include:

1. Demands for water throughout the entire system;

2. Firefighting water demands;

3. Elevation differences throughout the service area and the STATIC HEAD[11] conditions;

4. The most appropriate pipe for the system conditions;

5. Types of soils and depth to rock, groundwater, and frost levels;

6. Quality of the water supply which may affect the type of materials used;

7. Water volumes and pressures needed to minimize the effects of fires, earthquakes, floods, and sabotage;

8. Protection against unauthorized entry and vandalism;

9. Protection against the adverse effects of freezing weather; and

10. Potential service interruptions resulting from power supply, equipment, or structural failures.

The hydraulic adequacy of a distribution system is determined by the pressures that exist at various points in the system under the conditions of operation. While pressures must be high enough to serve the consumers and fire demand, excessive pressures will increase pump energy costs and may have adverse effects on some of the consumers' water-using devices.

Water should be delivered at a minimum desirable pressure of 35 psi (241 kPa or 2.5 kg/sq cm). An absolute minimum pressure is 20 psi (138 kPa or 1.4 kg/sq cm) to all points of the distribution system, and a maximum pressure is 100 psi (690 kPa or 7.0 kg/sq cm).[12] This recommended pressure range ensures sufficient pressure for normal use without risk of damage by excessive pressures to water heaters and other plumbing fixtures. In commercial districts, pressures of 75 psi (517 kPa or 5.25 kg/sq cm) and higher are desirable. If excessively high pressures (greater than 100 psi, 690 kPa or 7.0 kg/sq cm) cannot be avoided, pressure-reducing valves may be used (Figure 3.8).

The ideal system would rely completely on gravity to bring water from its source to the consumer at the desired pressures. In most systems, however, some pumping is required. Usually topography (hills and valleys) is such that water must be lifted to distribution reservoirs and elevated tanks. Also, booster pumps are often installed directly on mains to increase the available system pressures.

Pipes should be sufficiently large to carry the maximum quantity of water required at acceptable velocities. Too great a velocity can cause pressure drops and water hammer.

---

[11] Static Head. When water is not moving, the vertical distance (in feet) from a specific point to the water surface is the static head. (The static pressure in psi is the static head in feet times 0.433 psi/ft.)

[12] Maximum and minimum pressures vary from place to place and agency to agency. For example, some agencies have a minimum of 40 psi (2.8 kg/sq cm) and a maximum of 80 psi (5.6 kg/sq cm).

DIAPHRAGM
Separated upper chamber operating pressure from low chamber line pressure. Buna-N diaphragm standard; Viton available if required; all Nylon reinforced for high strength and long life.

BONNET
Four tapped ports for pilot piping. Center port for valve position indicator or valve actuated switches. Primed and painted like body.

VALVE SPRING
Stainless steel spring aids in closing the valve.

O-RING
Creates a static seal. No packing glands required, therefore breakaway friction is eliminated and valve will operate even at extremely low pressures.

DIAPHRAGM ASSEMBLY
The only moving part of the Model 65 valve. Ductile iron spool, seat retainer, diaphragm plate. Guided top and bottom by bronze or Teflon bushings.

VALVE SEAT
Buna-N or Viton compensates for wear on seating surface and maintains a drip-tight seal over extended service life.

BODY
Globe pattern 1-1/4 – 12": 250 lb. iron, 150 & 300 lb. steel, 150 lb. aluminum. Screwed ends 1-1/4 – 3" globe & angle. Iron & steel bodies epoxy primed inside and out with baked enamel exterior. Four tapped ports for pilot piping.

NOTE:
Basic valve can be used as a pressure relief, altitude control, or pressure-reducing valve depending on type of "brains" piped to valve.

SEAT RING
Bronze or stainless steel ring is replaceable and provides a lower guide for the stainless steel valve stem.

BASIC VALVE

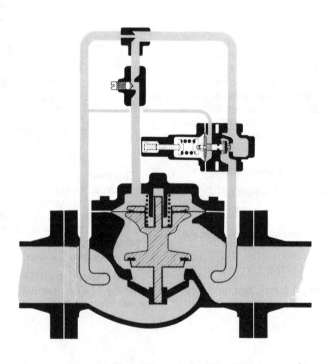

Fig. 3.8   Pressure-reducing valve
(Permission of OCV Control Valves)

Velocities of 2 to 4 feet per second (0.6 to 1.2 m/sec) at maximum flows are common and establish the size of water mains to be used. The minimum diameter of pipe commonly installed today is 6 inches (150 mm). Four-inch (100-mm) pipe should be the absolute minimum size used. Any pipe with a hydrant on the line must be at least 6 inches (150 mm) in diameter.

Avoid dead-end water mains (Figure 3.9) to prevent the development of tastes and odors. Any dead-end mains exceeding 1,000 feet (300 m) in length should be constructed of pipe at least 6 inches (150 mm) in diameter. For a dead end 2,000 feet (600 m) in length, the pipes should exceed 8 inches (200 mm). A hydrant, flushing valve, or blowoff is needed at the end of each dead-end water main for periodic flushing of the line.

The valve layout in a distribution system is very important. In general, valves on water mains of 12 inches (300 mm) in diameter and smaller should be located so that water main lengths of not more than 1,000 feet (300 m) can be isolated by valve closures. Fire hydrants should not be more than 500 feet (150 m) apart to avoid excessive *HEAD LOSS*[13] in fire hoses when fighting a fire.

Distribution reservoir capacity should be sufficient to meet the equalizing or operating storage demands, fire reserve, and emergency reserve. In pumped water supply systems, the cost of storage is balanced against cost of pumping. In all water supply systems, the cost of storage must be balanced against the cost of supply lines, increased fire protection, and more uniform pressures in the distribution system. Storage provides savings in energy consumption because tanks can be filled during periods of low water and energy demand, and the stored water used to meet peak water demands. Storage capacity of this type will eliminate frequent on-and-off pumping periods. Storage capacity is also an important factor when fire insurance rates are established.

Fire demand has a significant impact on the size of a system. Most of the components in a distribution system may need to be oversized to meet fire demands. While the total amount of water used for firefighting or prevention is small, a fire can put very high demands on a system for short periods of time. As fire demands are usually more than twice the normal user demands, facilities are often built to over twice the capacity needed for normal use. The proportional relationship between fire demand and daily demand will vary greatly due to the characteristics of the community. For instance, a community with heavy agricultural use or other use that can be stopped during an emergency might require little or no additional capacity for fire protection. On the other hand, a low-water-use industrial area that requires continuous service might require many times the daily water use for standby capacity for fire protection.

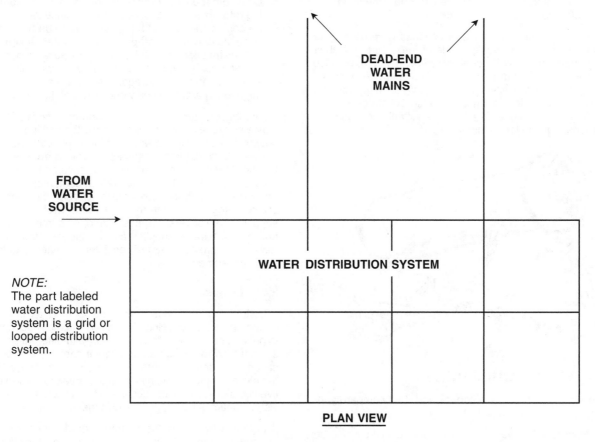

Fig. 3.9   Dead-end water mains (at top)

_____

[13] *Head Loss.   The head, pressure or energy (they are the same) lost by water flowing in a pipe or channel as a result of turbulence caused by the velocity of the flowing water and the roughness of the pipe, channel walls, or restrictions caused by fittings. Water flowing in a pipe loses head, pressure or energy as a result of friction losses. The head loss through a filter is due to friction losses caused by material building up on the surface or in the top part of a filter.*

QUESTIONS

Write your answers in a notebook and then compare your answers with those on page 153.

3.2A   What determines the hydraulic adequacy for a distribution system?

3.2B   Why are booster pumps sometimes installed directly on mains?

3.2C   Why should construction of dead-end water mains be avoided?

3.3   WATER TRANSMISSION SYSTEMS

Water is delivered from a system's source of supply through transmission lines to the distribution system. The source of supply might be a well, a treatment plant clear well, or a purchased water connection. Surface water supplies, for example streams or impounding reservoirs, and well supplies not meeting drinking water standards would normally be conveyed through a treatment plant before delivery to the distribution system. The source may or may not be under the jurisdiction of another agency. The location of the source determines whether the transmission mains are short or long, and whether the water is transported by gravity or by pumping. Wells provide system flexibility since they can be scattered within a distribution system. This can eliminate the need for long transmission mains and may also reduce the requirements for operational storage.

Normally, transmission mains or conduits have no service connections. They are usually made of large pipe with their size depending not only on peak daily flow demand, but also on available operational storage facilities and supplemental supplies such as wells within the system. While many different materials are used for transmission main construction, concrete with a steel cylinder is especially suitable because it is less expensive in the larger sizes.

QUESTIONS

Write your answers in a notebook and then compare your answers with those on page 153.

3.3A   How can the requirements for operational storage be reduced?

3.3B   What type of material is commonly used for large water transmission mains?

3.4   DISTRIBUTION STORAGE[14]

3.40   Purpose

A distribution reservoir is a storage facility connected to and serving the distribution system. This reservoir is used primarily to balance out fluctuations in demand that occur over short periods (several hours to several days), to provide local storage in case of an emergency such as a break in a main supply line or failure of a pumping plant, and to control the pressures in the service area.

The principal advantages of distribution storage are:

1. Water can be distributed and stored in advance of a need for water at one or more locations in a service area close to the ultimate user,

2. The demands on the sources of supply, production and treatment works, and transmission and distribution mains are more nearly equalized, and their sizes or capacities need not be so great,

3. System flows and pressures are improved and stabilized better to serve the customers throughout the service area,

4. Reserve supplies are provided in the distribution system for emergencies such as firefighting and power outages, and

5. Costs are reduced. Pumps can be used at a uniform pumping rate which results in cost savings since fewer or smaller pumps are needed. Other power costs can be reduced by pumping only during off-peak hours. Water can be pumped into distribution reservoirs during low-usage periods and discharged during periods of high water use. Also, with adequate operational storage, a small water treatment plant might be operated on only one or two shifts per day.

Distribution storage can provide other benefits. Reservoirs may also serve as sand traps to allow the settling of sand and suspended solids in the water. Chlorinated water may be held in covered distribution reservoirs to provide an extended contact time for effective disinfection. Uncovered distribution reservoirs are undesirable due to surface contamination. Water in uncovered distribution reservoirs should be rechlorinated before being returned to the distribution system. If the water supply is from several sources, with one source being of lower quality, the various supplies can be blended together in the reservoir to provide water of consistent and acceptable quality to the consumers.

3.41   Location

Distribution reservoirs should be located as close to the centers of use as possible. Normally, it is more advantageous to provide several smaller storage units in different parts of the system than to provide an equivalent capacity at a central location. The selection of the site or sites usually depends on the elevations available in or near the areas to be served, relative costs of security and access to the sites available, and the visual impact of a storage facility in these locations.

Figure 3.10 shows the hydraulic grade lines for a poorly located storage tank and a properly located tank. When the tank is located away from the area of use, the hydraulic grade line can drop too low and there will be times when adequate pressures can't be maintained in the distribution system.

---

[14] Also see Chapter 2, "Storage Facilities."

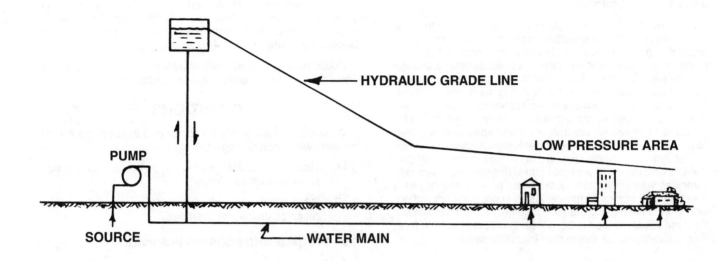

POORLY LOCATED STORAGE TANK

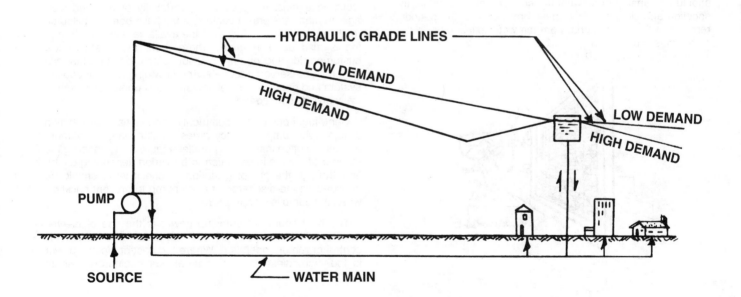

PROPERLY LOCATED STORAGE TANK

Fig. 3.10   Elevated storage tank locations

However, when the storage tank is located in the center of use, adequate pressures can be maintained even during periods of high demand.

3.42   Types of Storage

Distribution storage may consist of elevated structures or ground storage at high elevations or a combination of both. In areas that have hills sufficiently higher in elevation than most of the service area and in close-in locations, ground storage may be used. Where natural elevation is inadequate, elevated storage using standpipes and elevated tanks is more efficient. Ground storage is normally more economical than elevated storage and much larger capacities with fewer units can be provided. The useful capacity of standpipes and elevated tanks is limited to the volume of water stored above the elevation at which adequate pressure is created in the connected distribution system. In elevated tanks, this elevation generally coincides with the bottom of the water tank. In standpipes it may be much higher. Large variations in pressure are not desirable in distribution systems. Therefore, water level fluctuations are usually limited to 30 feet (9 m) or less in storage facilities controlling pressures in the distribution system.

3.43   Health and Water Quality Considerations

Open distribution reservoirs are not encouraged and are not approved in many states. Provide an adequate cover for all treated water reservoirs and try to locate and construct all reservoir openings in a manner that protects the water from contamination. Keep manholes and other means of entry securely locked. Cover all vents, drains, and overflow outlets with a fine mesh screen to prevent entry of birds, snakes, insects, and small animals. Take particular care to protect drains and overflows from *CROSS CONNECTIONS*[15] with any storm drain, sewer line, or other source of potential contamination. Drains should be capable of dewatering the reservoir to allow for inspection and cleaning. Also make provisions for the periodic removal of floating material from the water surface while the reservoir is in operation.

Care must be taken with underground facilities because of infiltration of groundwater or other contaminants. Ideally, the intake and outlet should be located on opposite sides to provide for proper circulation of water within the reservoir. However, a common pipe is often used for both the water inlet and outlet. Whenever a single reservoir is taken out of service, do not let the system water pressure within the service area drop below 20 psi (138 kPa or 1.4 kg/sq cm).

Public access to reservoirs should be restricted for safety and security reasons and to reduce vandalism.

QUESTIONS

Write your answers in a notebook and then compare your answers with those on page 153.

3.4A   How can the storage of water in covered distribution reservoirs aid disinfection?

3.4B   What precautions should be taken with distribution reservoir drains and overflows?

3.5   STORAGE AND BOOSTER PUMPING

Some water systems can rely completely on gravity to bring water from their source to the customer. Most communities, however, must do some pumping to get water from the source to the distribution system. Generally speaking, the topography of the distribution system is such that water must be lifted to the distribution reservoirs. Ground storage pumping to the system is used in some areas where topography does not permit the economical location of storage at the desirable hydraulic elevation of the system. The economy and desirability of pumped storage as compared to elevated storage must be determined for each individual system. Even if pumped storage is found to be more economical overall, its use should be weighed against the proven dependability of elevated storage. In many systems, booster pumps have been installed directly on the mains to increase the available pressure. Pumping is also used in areas where there are relatively few customers located on very high or remote points. They can generally be served adequately by providing a small booster system and pneumatic steel pressure tank operating automatically on pressure control.

Centrifugal pumps are commonly used in water distribution systems. As pump efficiency varies with the load, two or more pumps are often installed in parallel (Figure 3.11) enabling the number of pumps in operation to be varied depending on the flow through the pumping station. A check valve should be provided on the discharge of each pump to prevent backflow when the pump is stopped.

The best source of pumping power is the use of electric motors, which are compact and well adapted to automatic control or remote operation. Where the cost of electric power is too high, diesel motors, natural gas engines, or steam

---

[15] *Cross Connection.   A connection between a drinking (potable) water system and an unapproved water supply. For example, if you have a pump moving nonpotable water and hook into the drinking water system to supply water for the pump seal, a cross connection or mixing between the two water systems can occur. This mixing may lead to contamination of the drinking water.*

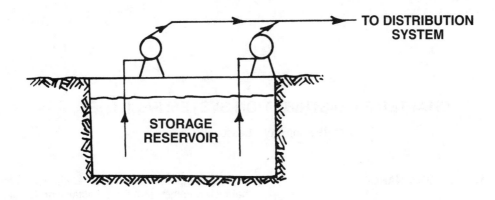

Fig. 3.11   Pumps installed in parallel

turbines are used. Standby power should always be provided to protect against possible failure of the main power source. Gasoline engines or motor generators are often used as standby power because of their low initial cost.

QUESTIONS

Write your answers in a notebook and then compare your answers with those on page 153.

3.5A   What type of pump is used for water distribution systems?

3.5B   What is the best source of power for distribution system pumps?

End of Lesson
1 of 5 on
DISTRIBUTION
SYSTEM
FACILITIES

Please answer the discussion and review questions next.

DISCUSSION AND REVIEW QUESTIONS

Chapter 3.   DISTRIBUTION SYSTEM FACILITIES

(Lesson 1 of 5 Lessons)

At the end of each lesson in this chapter you will find some discussion and review questions. The purpose of these questions is to indicate to you how well you understand the material in the lesson. Write the answers to these questions in your notebook before continuing.

1. What is the purpose of a water distribution system?

2. Why must a continuous positive water pressure be maintained in the distribution system at all times?

3. How are negative pressures created in a distribution system during a fire?

4. What problems may be encountered with distribution system underground storage reservoirs?

CHAPTER 3.   DISTRIBUTION SYSTEM FACILITIES

(Lesson 2 of 5 Lessons)

3.6   WATER MAINS AND APPURTENANCES

3.60   Layout

Water distribution system mains are laid out in grid, loop, or tree (branching) systems as shown in Figure 3.12. Most systems contain features of each type of layout, although the grid system is probably most common, particularly in large cities. Tree or branching systems produce dead-end lines which can cause taste and odor water quality problems. These lines must be frequently flushed and should be avoided wherever possible. Grid or loop systems permit greater water flow to an area when there is a fire or other source of high demand because water is being carried to any particular location from more than one direction. Unfortunately, some grid systems may include dead ends. A tree system consists of a single large main that decreases in size as it extends into the distribution area. Branches take off from the main at right angles with sub-branches from each branch. Branching patterns are more evident on the outskirts of a community, while grid patterns are found within the built-up portions of a community.

QUESTIONS

Write your answers in a notebook and then compare your answers with those on page 153.

3.60A   Dead-end water mains can cause what type of problem?

3.60B   List two advantages of grid or loop water distribution systems.

3.61   Pipe Features

The basic requirements of pipes for water distribution systems are adequate strength, durability, maximum corrosion resistance, and no adverse effect on water quality. Pipe strength must be sufficient to resist a variety of forces. Exterior forces such as backfill cover over the pipe, the weight of passing traffic, and other similar forces must be resisted. Pipes are also subjected to internal water pressures such as delivery pressure, surges, and water hammer. Pipes must also be durable and exist for a long time without deterioration. Good, durable pipe will retain a smooth inside surface with satisfactory flow characteristics throughout its service life. The inner surface of the pipe should not react with water and should resist corrosion. The exterior surface of the pipe also must be corrosion resistant. If the pipe is electrically conductive, then *CATHODIC PROTECTION*[16] may be needed. Tastes, odors, chemicals, or any other undesirable characteristics must not be imparted to the water by the pipe.

Also of importance is the ease with which the pipe can be handled and installed. Some installation features that are significant are weight, jointing characteristics, available sizes, and ease of tapping the line. The lighter the pipe, the easier it is to handle. Jointing is important not only for ease of installation but also with regard to any possible flexibility or deflection. Pipe should be available in all the commonly used sizes. If other lines (such as service lines) are to be attached to the pipe, the ease with which the pipe can be tapped is an important consideration.

The pressure rating of the pipe must be adequate to handle the pressures it will encounter in the system. Pressure ratings refer to the working pressure to be expected within a pipe and may take into account the internal pressure and a safety factor. Four classes of working pressure are commonly encountered—100, 150, 200, and 250 psi (690, 1,034, 1,380, and 1,724 kPa or 7, 10.5, 14, and 17.5 kg/sq cm).

A pipe may rupture or be crushed when subjected to external or internal pressures that exceed its ratings. Other common types of breakage are shear (cutting) and beam (bending) breakage. A shear break can occur when the earth moves, such as during an earthquake. Beam breakage may result when a pipe is unevenly supported along its length.

The size of a pipe must be sufficient to carry anticipated flows. The desirable minimum pipe diameter is 6 inches (150 mm) and pipes should certainly not be less than 4 inches (100 mm) in diameter.

QUESTIONS

Write your answers in a notebook and then compare your answers with those on page 153.

3.61A   List the forces water distribution system pipes must be capable of resisting.

3.61B   List the important installation features of water distribution pipes.

---

[16] *Cathodic (ca-THOD-ick) Protection.   An electrical system for prevention of rust, corrosion, and pitting of metal surfaces which are in contact with water or soil. A low-voltage current is made to flow through a liquid (water) or a soil in contact with the metal in such a manner that the external electromotive force renders the metal structure cathodic. This concentrates corrosion on auxiliary anodic parts which are deliberately allowed to corrode instead of letting the structure corrode.*

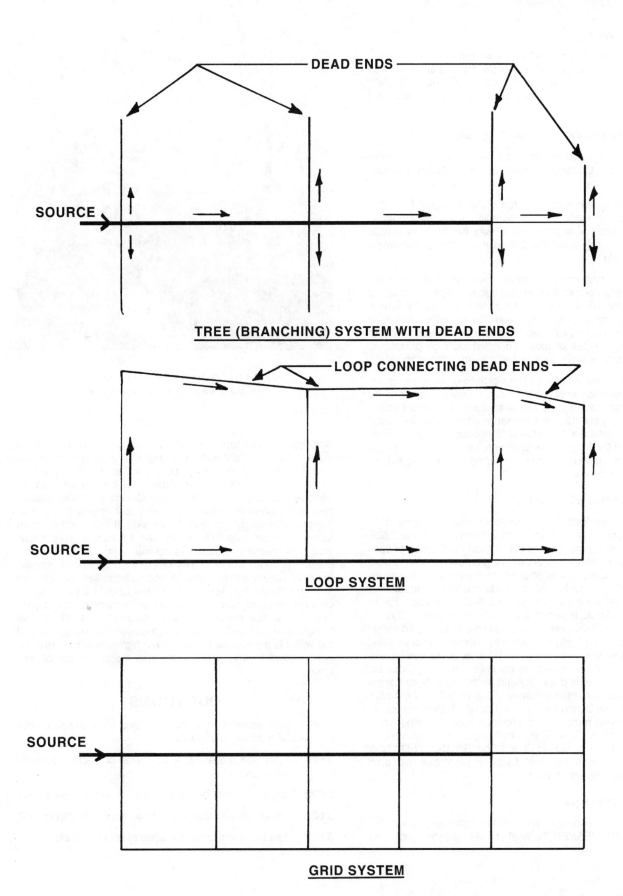

DEAD ENDS

SOURCE

TREE (BRANCHING) SYSTEM WITH DEAD ENDS

LOOP CONNECTING DEAD ENDS

SOURCE

LOOP SYSTEM

SOURCE

GRID SYSTEM

Fig. 3.12   Typical layouts of water distribution systems

3.62   Pipe Types

A wide variety of pipe materials is available for carrying water under pressure including ductile iron, steel, reinforced concrete, asbestos-cement, and plastic. *SERVICE PIPES*[17] are made of copper, plastic, iron, steel, asbestos-cement, and brass.

3.620   Ductile-Iron Pipe *(Figure 3.13, page 79)*

Ductile-iron pipe was introduced in 1955 as an improvement to cast-iron pipe. Currently, all iron pipe installed today is ductile-iron pipe.

Ductile-iron pipe is manufactured in sizes ranging from 2 to 54 inches (50 to 1,350 mm) in diameter. Transmission lines, distribution lines, or in-plant piping may be made of ductile iron. When properly coated and lined, it provides an extremely long service life.

Two types of cast-iron pipes may be encountered, gray cast-iron and ductile-iron pipe. Gray cast iron is tough and is easily tapped. Ductile iron is similar, but is stronger, less rigid, lighter, and offers corrosion resistance equal to or better than gray cast iron. Ductile iron will resist bending and twisting without breakage. Both types of pipe can withstand high pressures. Ductile-iron pipe is the only type being manufactured today.

Ductile-iron pipe is lined on the inside to prevent corrosion, including tuberculation (too-BURR-cue-LAY-shun) which is the pitting or growth of clumps of material that increase friction, thus reducing flows. Cement mortar or other available linings are used to produce a smoother pipe interior. A lined pipe may have one-half the friction loss of an unlined pipe. To reduce external pipe corrosion, bituminous coatings or a polyethylene wrap are commonly used.

3.621   Steel Pipe *(Figure 3.14)*

Steel pipe has long been popular in water systems. This pipe is available in sizes from $1/2$ to 144 inches (12.5 to 3,500 mm) in diameter. Steel pipe is lighter than ductile iron or concrete, strong, withstands high operating pressures, and can be designed in thickness and strength to meet the most severe conditions. This pipe has high tensile (pulling or stretching) strength, some flexibility, is easily installed and jointed, low in cost, readily welded together, and easily assembled, handled, and transported. However, steel pipe is more subject to corrosion and requires a lining or coating to maintain its service life. This pipe has low resistance to external pressures in the larger sizes. Though protective coatings and linings can be applied to increase its durability and service life, these linings are sometimes damaged because of the flexibility of the pipe as it bends under trench loads. Coatings and linings are usually of a bituminous material or a cement mortar. An epoxy lining is also sometimes used. The exterior of a steel pipe can be coated with epoxy or mastic, it could be *GALVANIZED*[18] or a protective wrap might be used. Cathodic protection may also be provided to prevent corrosion.

3.622   Concrete Pipe

Three types of concrete pipe can be obtained—steel cylinder (not *PRESTRESSED*[19]), steel cylinder (prestressed), and

Fig. 3.14   Coal tar enamel lined steel pipe
(Permission of Steel Plate Fabricators Association, Inc.)

non-cylinder (not prestressed). Concrete pipe is available in sizes 12 inches (300 mm) and larger. This pipe can actually be made in any size and has been produced in diameters up to 240 inches (6,000 mm). In large sizes it is usually less expensive than any other kind of pipe. Concrete pipe is durable, has good corrosion resistance and low maintenance features, and is easily installed. The inside of the pipe has been found to remain smooth for more than a hundred years of service. Prestressed concrete is strong, with normal working pressures up to 250 psi (1,724 kPa or 17.5 kg/sq cm). However, reinforced concrete that is not prestressed is not very strong and can only take up to 50 psi (345 kPa or 3.5 kg/sq cm) working pressure. The disadvantages of concrete pipe are that the pipe is heavy, may be hard to tap, needs special fittings, and may deteriorate in aggressive (corrosive) soils. If the water is aggressive (corrosive), a higher water pH may be encountered for one or two years after the installation of concrete pipe.

QUESTIONS

Write your answers in a notebook and then compare your answers with those on page 153.

3.62A   How can the external corrosion of ductile-iron pipe be reduced?

3.62B   Steel pipe may be lined with what types of materials?

3.62C   What types of coatings may be applied to steel pipe?

3.62D   List the major disadvantages of concrete pipe.

---

[17] *Service Pipe.   The pipeline extending from the water main to the building served or to the consumer's system.*
[18] *Galvanize.   To coat a metal (especially iron or steel) with zinc. Galvanization is the process of coating a metal with zinc.*
[19] *Prestressed.   A prestressed pipe has been reinforced with wire strands (which are under tension) to give the pipe an active resistance to loads or pressures on it.*

Fig. 3.13 Installation of ductile-iron pipe
(Courtesy of the Ductile Iron Pipe Association)

3.623   Asbestos-Cement Pipe

Asbestos-cement (AC) pipe was a relatively popular pipe material until people became concerned with the health hazard from breathing asbestos fibers. This pipe is made of asbestos fiber, silica sand, and cement. The asbestos fibers provide much of the pipe's strength. Asbestos is a mineral that is almost indestructible. This pipe material will not burn, deteriorate, or corrode, and has a high tensile strength. Available pipe sizes range from 4 to 42 inches (100 to 1,050 mm) in diameter. AC pipe is light and weighs approximately half as much as ductile-iron pipe of equal size and class. This pipe is easy to handle and join. Also, it costs less to transport because of its light weight. Since AC pipe is nonmetallic, it is completely immune to ELECTROLYSIS[20] and cannot tuberculate as do metal pipes. The pipe has a smooth, uniform bore with a high carrying capacity that holds up for a long time. The material may corrode if aggressive water conditions exist. AC pipe is easily tapped, cut, and machined in the field. One operator with lightweight cutting and machining tools can cut the pipe right on the job site.

The disadvantages of AC pipe are its low flexural strength (breaks when it is bent) in small sizes and its vulnerability to impact damage. AC pipe is difficult to locate when buried and cannot be thawed out using electrical methods because the material is nonconductive. Asbestos fibers have been known to leach out of the pipe in soft waters. People who work with AC pipe may be exposed to a health hazard if asbestos fibers are inhaled when the pipe is cut or machined. Respirators must be worn whenever there is a possibility that asbestos particles can become airborne when working with asbestos-cement pipe.

3.624   Plastic Pipe (Figure 3.15)

Plastic pipe is commonly used for distribution and service lines in water systems (Figure 3.15). Plastic pipe is used for water distribution mains in many parts of the country. Three types of plastic material are in general use: PVC (polyvinyl chloride), PE (polyethylene), and PB (polybutylene). The pipe is available in a wide range of sizes; however, PVC is the only one that goes up to 16 inches (400 mm) in diameter.

All plastic pipe has an exceptionally smooth interior surface and, therefore, low friction head losses. The pipe materials are chemically inert which means they are completely corrosion-free. PVC piping systems average 30 percent less in cost than steel pipe because of savings in installation and long-range corrosion-free maintenance.

PVC pipe's light weight makes movement by one person an easy task.

Fig. 3.15   Plastic pipe
(Permission of Uni-Bell Plastic Pipe Association)

---

[20] Electrolysis (ee-leck-TRAWL-uh-sis).   The decomposition of material by an outside electric current.

3.625 Types of Service Pipe

The service pipe is the pipeline extending from the water main to the building served or the customer's system. Normally, 3/4 to 1 1/2 inch (19 to 38 mm) pipe is used. The ability of the pipe material used to resist internal and external corrosion is the most important factor in the useful life of service lines. Most service lines use copper, PVC, PE, PB, or, in special cases, reinforced plastic pipe. Other materials are also used, such as ductile iron, black and galvanized wrought iron, steel, asbestos-cement, and brass. In some older systems, lead service lines still exist. These lines are no longer acceptable. Existing lead service lines should be replaced because of the potential for lead (which is a toxic material) to enter the water.

3.626 Summary

There are no rules that will tell you which type of pipe to select for a specific installation. Important considerations include cost of pipe and installation and also the corrosive conditions of the soil and water. Other important factors are the water pressures and external pressures the pipe must withstand.

Plastic pipe is very popular in small installations. Ductile-iron pipe is used in steep hills where pressure differences can be great. For pipes 24 inches (600 mm) and larger, reinforced concrete pipe or cement lined or coated steel pipe is often used.

QUESTIONS

Write your answers in a notebook and then compare your answers with those on page 154.

3.62E What are the disadvantages of asbestos-cement pipe?

3.62F List the three types of plastic material used for plastic pipe.

3.62G What are the most common materials used for service lines?

3.62H Why should existing lead service lines be replaced?

3.63 Joints

Joints connect two lengths of pipe. Different types of joints are used depending on the pipe materials, conditions of installation, and degree of joint deflection (change in direction or alignment) needed. Joints must provide a watertight seal. They should be easily installed, durable, and corrosion resistant.

3.630 Ductile-Iron Pipe

Push-on and mechanical joints are most often used for ductile-iron pipe. Flanged, restrained, flexible, and sleeve joints are also used.

Push-on joints are the most frequently used type of joints. They are used with bell and spigot pipe, which has an enlarged diameter or bell at one end and a spigot at the other end that fits into and is laid in the bell. Inside the bell there is a groove in which a heavy rubber gasket rests.

When the joint is made in the field, the gasket is inserted into the bell, the spigot end is coated with a lubricant film, and the spigot is pushed "home." In the larger pipe sizes, a jack is often used to push or pull the spigot into place. The assembly of a push-on joint is shown in Figure 3.16, page 82.

Mechanical joints (Figure 3.17) are easy to install and are flexible. The standard mechanical joint consists of (1) a specially shaped hub, (2) a ring or gasket shaped to fit the pipe's bell, and (3) an iron follower ring or gland which forces the gasket into and against the hub by means of bolts passing through lugs cast on the bell. The gaskets are generally plain rubber. Joint deflection of a few degrees is possible. Assembling a mechanical joint is very simple and requires only a ratchet wrench.

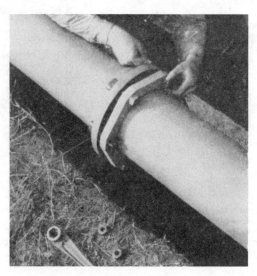

Slide gland into position, insert bolts, and tighten nuts.

Fig. 3.17 Mechanical joint
(Source: HANDBOOK OF CAST IRON PIPE)

The Dresser (Figures 3.18 and 3.19) and Victaulic couplings (Figure 3.20) are special types of mechanical joint couplings. The compression (Dresser) coupling, which joins plain-end ductile-iron pipe, provides a sleeve-type joint. A ductile-iron middle ring is used and a leakproof joint is obtained at each end of the middle ring by compressing two resilient gaskets with bolted follower rings. The gaskets absorb vibration and pipe movements. These joints allow for moderate deflection and are used to span short gaps.

Victaulic couplings are generally used on smaller pipe. In this type of coupling, each end of the pipe has a groove or shoulder that receives the sides of a trough-shaped metal housing in which there is a similarly shaped rubber gasket. A clamp fits over the grooves and makes the connection. The drawback of this coupling is that the groove weakens the pipe wall.

A flanged joint is one that is made by simply bolting the flanges of two pipes together with a resilient gasket between the flanges. Flanged joints are not flexible and are usually not installed when a pipe is buried.

Restrained joints are used where there is a lack of space to lock a joint in place to prevent movement, or where there is a possibility the soil behind a fitting will be disturbed. They are usually installed at the critical fitting and for a certain number of pipe lengths on each side of the fitting.

A flexible joint is any joint between two pipes that permits one of them to be moved without disturbing the other pipe. They are used at river crossings and other underwater installations and are similar to a ball and socket joint.

Assembly tool that is available from U.S. Pipe for assembling Tyton Joint fittings. The assembly tool consists of a bell yoke with chain attached, hinged pipe clamp, and a pair of handles.

1. Attach the pipe clamp to the pipe about twelve inches from the spigot end making sure that the swing bolt is on the top of the pipe. Tighten the swing bolt.

2. Place the yoke on the fitting behind the bell.

3. Place the two handles on the pivot pins of the pipe clamp and couple the chain to the handles.

4. Assemble the joint by pushing on the handles which are shown in position for the start of the assembly. The two handles operate independently helping to maintain the fitting in straight alignment throughout the assembly.

5. The completed assembly. Note that the second stripe is at the face of the bell.

Fig. 3.16   Assembly of a push-on type joint
(Permission of U.S. Pipe)

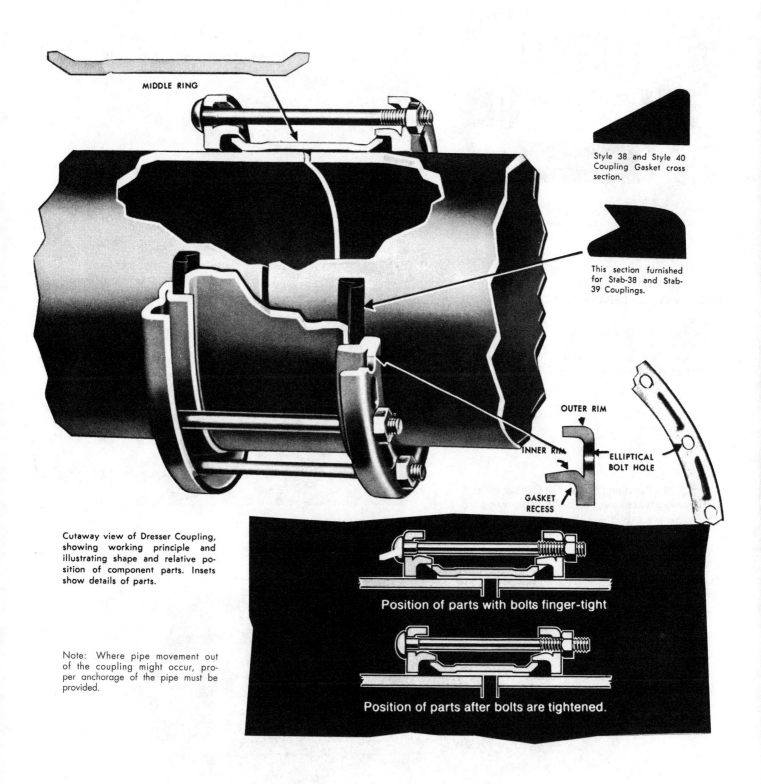

MIDDLE RING

Style 38 and Style 40 Coupling Gasket cross section.

This section furnished for Stab-38 and Stab-39 Couplings.

OUTER RIM

INNER RIM

ELLIPTICAL BOLT HOLE

GASKET RECESS

Cutaway view of Dresser Coupling, showing working principle and illustrating shape and relative position of component parts. Insets show details of parts.

Note: Where pipe movement out of the coupling might occur, proper anchorage of the pipe must be provided.

Position of parts with bolts finger-tight

Position of parts after bolts are tightened.

Fig. 3.18   Cutaway view of Dresser coupling
(Permission of Dresser Manufacturing Division, Dresser Industries, Inc.)

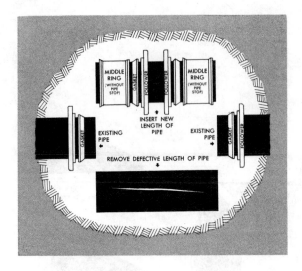

EASY WAY TO REPAIR A SECTION OF PIPE . . .

Sleeve-in a new section of pipe with Dresser Style 38 Couplings.
Diagram at left shows how Style 38 Couplings may be used to
make "cut-ins" into piping for repair. Valves or fittings can be inserted in
the same way. **With Stab-38" Couplings the whole coupling slips
on the pipe ends without disassembly.**

Fig. 3.19 Repairing pipe with a Dresser coupling
(Permission of Dresser Manufacturing Division, Dresser Industries, Inc.)

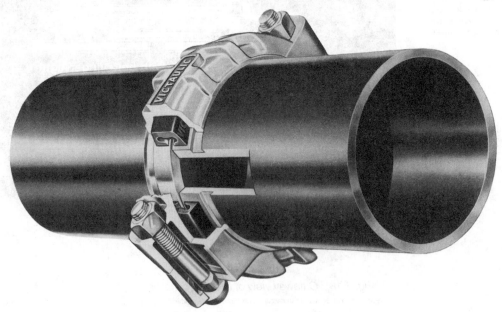

Fig. 3.20 Victaulic coupling used for ductile, grooved, iron pipe
(Permission of Victaulic Company of America)

3.631 Steel Pipe [21]

Steel pipe lengths can be joined together in the field by many different methods to produce rigid or flexible connections, as follows:

1. Bell and Spigot Lap Welded Joint,

2. Bell and Spigot Rubber Gasket Joint,

3. Harness Joint—Bell and Spigot,

4. Carnegie Shape Rubber Gasket Joint,

5. Butt-Welded Joint,

6. Butt Strap Joint for Welding,

7. Mechanically Coupled Joint, and

8. Flanged Joint for Bolting.

1. BELL AND SPIGOT LAP WELDED JOINT

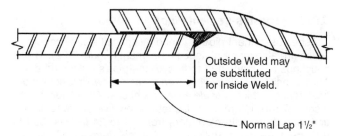

Outside Weld may be substituted for Inside Weld.

Normal Lap 1½"

The bell and spigot lap welded joint is widely used because of its flexibility, ease in forming and joining, watertightness, and simplicity. Small angle changes can be made in this joint. The joint may be welded on either the inside or outside with a small fillet weld.

2. BELL AND SPIGOT RUBBER GASKET JOINT

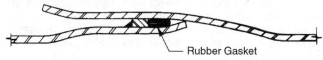

Rubber Gasket

a. Formed rubber gasket joint, usually applied to large-diameter water pipe.

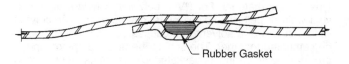

Rubber Gasket

b. Rolled-groove rubber gasket joint, usually applied to small-diameter water pipe.

Bell and spigot rubber gasket joints simplify laying the pipe and require no field welding. They permit flexibility, watertightness, lower installation costs, and elimination of bellholes.

3. HARNESS JOINT—Bell and Spigot Rubber Gasket

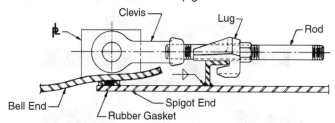

Clevis — Lug — Rod

Bell End — Spigot End — Rubber Gasket

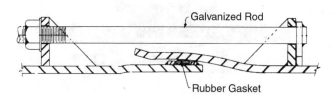

Galvanized Rod

Rubber Gasket

4. CARNEGIE SHAPE RUBBER GASKET JOINT

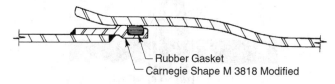

Rubber Gasket
Carnegie Shape M 3818 Modified

5. BUTT-WELDED JOINT

a. Single-V Butt-Welded

b. Double-V Butt-Welded

Butt-welded joints will develop full strength, but will require more care in cutting and fitting in the field if changes in alignment or profile occur frequently. This joint is not commonly used.

6. BUTT STRAP JOINT FOR WELDING

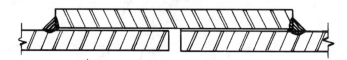

The butt strap is a closure joint used for joining ends of pipe when adjustments are required in the field.

---

[21] *This section reproduced from WELDED STEEL PIPE by permission of Steel Plate Fabricators Association, Inc.*

7. *MECHANICAL COUPLINGS*

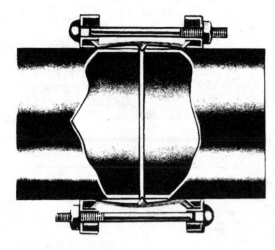

a. Sleeve type

b. Grooved and shouldered type

Mechanical couplings provide ease of installation and flexibility and are represented by the sleeve and clamp type of coupling.

8. *FLANGED JOINT*

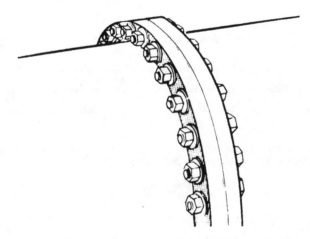

Flanged joints are not generally used for field joints on large-diameter steel pipe because of their high cost and lack of flexibility. They are advantageous, however, for special conditions, such as connections to flanged gate valves, bridge crossings, meters, and for field connections by unskilled labor.

QUESTIONS

Write your answers in a notebook and then compare your answers with those on page 154.

3.63A   List the basic requirements for a good pipe joint.

3.63B   What types of joints are most commonly used for ductile-iron pipe?

3.63C   What types of joints are used to join steel pipe?

3.632   *Steel Cylinder Concrete Pipe*

Steel cylinder concrete pipe is normally joined using a modified bell and spigot (Figure 3.21). The sealing element consists of a rubber gasket; steel rings in the bell and spigot are usually provided. The outside joint space is normally completely filled with cement mortar to protect the joint rings against external corrosion. Both the inner and outer space is filled with mortar. Concrete pressure pipe may be installed with or without a steel cylinder depending on the conditions.

3.633   *Asbestos-Cement Pipe*

Asbestos-cement pipe is joined by special sleeve couplings which are also made of asbestos-cement (Figure 3.22). These couplings are machined to fit over the pipe's machined ends. They consist of a sleeve grooved to receive two interior rubber gasket rings which provide a watertight seal. Couplings 3 to 12 inches (75 to 300 mm) in diameter allow five degrees deflection, and those with diameters from 14 to 24 inches (350 to 600 mm) permit a deflection of three degrees. Couplings should be able to take the same pressure as the pipe with which they are used and should be marked accordingly. The joint is normally jacked or pulled onto the pipe. In some pipe, the coupling comes already installed on one end; consequently, a push-on or slip-type joint can be used.

Cast-iron (ductile-iron) fittings including elbows, tees, crosses, reducers, valves, and hydrants are available with bells for connections to asbestos-cement pipe. Special adaptors are available to connect asbestos-cement pipe to ductile-iron pipe at either mechanical joint or poured bell fittings. Asbestos-cement pipe also can be connected to steel or plastic pipe by the use of adaptors.

3.634   *Plastic Pipe*

The jointing methods for plastic pipe vary depending on the type of pipe material. For PVC (polyvinyl chloride) pipe, solvent weld or bell and spigot joints may be used. The solvent weld consists of a solvent that causes both pieces of pipe to be cemented together. The seal in the bell and spigot pipe is provided by an O-ring. A gasketed joint is the usual type used for PVC pipe. The pipe may be an integral (attached) bell design or a separate sleeve-type coupling may be used. Schedule 80 and heavier wall PVC pipe may be threaded. PE (polyethylene) and PB (polybutylene) pipe are joined using a heat-fusion method. Threading of these materials is not recommended.

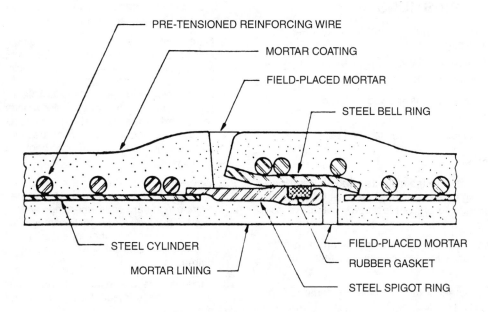

PRE-TENSIONED REINFORCING WIRE

MORTAR COATING

FIELD-PLACED MORTAR

STEEL BELL RING

STEEL CYLINDER

MORTAR LINING

FIELD-PLACED MORTAR

RUBBER GASKET

STEEL SPIGOT RING

Fig. 3.21   Steel cylinder concrete pipe modified bell and spigot coupling

(Reprinted from *WATER DISTRIBUTION OPERATOR TRAINING HANDBOOK*,
by permission. Copyright 1976, the American Water Works Association)

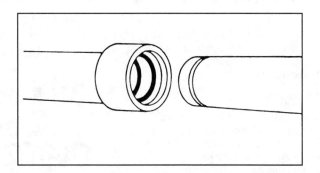

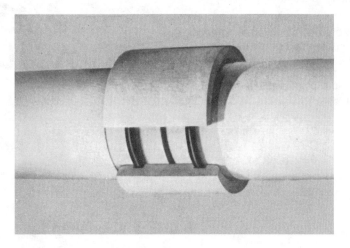

Fig. 3.22   Asbestos-cement pipe coupling

(Permission of Johns-Manville Corporation, a Subsidiary of Manville Corporation)

QUESTIONS

Write your answers in a notebook and then compare your answers with those on page 154.

3.63D   How are the joint rings for concrete pressure pipe protected against external corrosion?

3.63E   How much deflection is allowed in the couplings of asbestos-cement pipe?

3.63F   What types of joints are used for PVC pipe?

3.64   Pipe Protection

Pipe must be protected to maintain its strength, pipe performance and carrying capacity. Protection must be provided to control corrosion and to prevent leakage, rupture, or collapse.

A number of corrosion-control methods[22] are available. Some of the commonly used techniques and materials are listed below:

1. Nonmetallic pipe materials such as asbestos-cement, reinforced concrete, and plastic.

2. Nonmetallic coatings to protect metals such as coal tar enamels (bituminous), coal tar-epoxy enamels, asphaltics, cement mortar, epoxy resins, vinyl resins and paints, inorganic zinc silicate paints, and organic zinc paints. Pipe exteriors may also be protected with a loose polyethylene wrap or sleeve. Any protective coating must be continuous, without holes or breaks, or the corrosive attack will be concentrated on the exposed spots.

3. Metallic coatings to protect metals such as zinc (galvanizing) or aluminum.

4. Chemical treatment to deposit a protective coating or film on the pipe using calcium carbonate, sodium hexametaphosphates and silicates, pH adjustment, removal of dissolved oxygen, and removal of free carbon dioxide.

5. Electrical control using cathodic protection.

Often, a combination of the above methods must be used. The first choice is the proper use of noncorrosive metals and/or mechanical coatings. Next, chemical control methods may be tried as a supplementary measure. Finally, cathodic protection is tried if circumstances indicate that this method is appropriate.

Steel pipe is especially susceptible to corrosion and needs a great deal of protection. The inside and outside of the pipe are either protected with cement mortar, cold-applied mastic, or hot coal tar enamel. Small pipe up to four inches (100 mm) in diameter is galvanized. If steel pipe is exposed above ground, the outside is usually protected with paint.

In addition to corrosion control, other protection measures are needed to prevent pipe damage.

1. Care must be exercised when loading, unloading, and installing pipe. Pipe should not be dropped or allowed to strike other pipe. The bedding and backfilling material must be carefully chosen and placed in the trench. Asbestos-cement pipe is especially susceptible to puncture by sharp rocks or damage by heavy rocks which may be located directly next to the pipe. This is especially likely to occur if heavy loads or trucks drive over the pipe. Improper, nonuniform bedding will provide only poor support and weaken the pipe.

2. Where increased strength is needed, concrete encasement of a steel cylinder slipped over the pipe is often used. This is done in stream beds where heavy floods may be encountered. This type of protection may also be provided to a water or sewer line (or both) if they must be laid too close to one another. Special sewer pipe material may also be required.

3. Vacuum, pressure-relief, and pressure-reducing valves will prevent pipe damage from negative or excessively high pressures.

4. Thrust blocks are used to anchor pipes and fittings, and to keep them from moving or being pulled apart when there is a change in direction of the pipe, a sudden increase in velocity, or a sudden stoppage of flow.

QUESTIONS

Write your answers in a notebook and then compare your answers with those on page 154.

3.64A   List the most commonly used corrosion-control methods.

3.64B   How can steel pipe be protected from corrosion?

End of Lesson 2 of 5 on DISTRIBUTION SYSTEM FACILITIES

Please answer the discussion and review questions next.

---

[22] Also see Chapter 5, Section 5.75, "Corrosion Control," in this volume, and Chapter 8, "Corrosion Control," in WATER TREATMENT PLANT OPERATION, Volume I

DISCUSSION AND REVIEW QUESTIONS

Chapter 3.   DISTRIBUTION SYSTEM FACILITIES

(Lesson 2 of 5 Lessons)

Write the answers to these questions in your notebook before continuing. The question numbering continues from Lesson 1.

5. What are the basic requirements for water distribution system pipes?

6. What are the advantages of using ductile-iron pipe?

7. Discuss the advantages of asbestos-cement pipe.

8. Selection of a pipe joint is based on what factors?

9. Why must pipes be protected?

CHAPTER 3.   DISTRIBUTION SYSTEM FACILITIES

(Lesson 3 of 5 Lessons)

3.65   Pipe Installation

3.650   Locating Nearby Utility Lines

Prior to excavating, determine the location of all buried water, sewer, gas, power, telephone and cable TV lines, and storm drains in the work area. Many phone books list an "Underground Service Alert" or "One-Call" phone number which should be called *BEFORE* digging or drilling underground. If these underground lines are damaged during the construction or repairs, it could present serious problems to the community and even result in hazards to the workers. Detailed construction plans showing alignment, grade, and depth specifications for the nearby lines and structures are usually available, and you can ask the utility companies to physically locate their lines. To encourage cooperation, give copies of your preliminary alignment plans for water lines to the other utilities. Ideally, policies and procedures should be established among all public and private agencies involved in installing underground utilities in public right-of-ways (R/W) to coordinate use of the available space. Damage prevention programs benefit everyone. Agreement on standardized locations for lines will avoid conflicts between utilities and minimize service interruption for utility customers. A specific zone should be assigned for each utility's facilities within the street right-of-way. Standardized zones used by the City of Dallas, Texas, are shown in Figure 3.23 and Figure 3.24 shows another typical layout. After construction, as-built plans (record plans) must be prepared and maintained and made available to the other utilities as needed.

3.651   Water and Sewer Line Separation

An area of special concern is the separation between the water line being installed and existing nearby sewer lines. Water and sewer lines located close to each other present a hazard if leakage from the sewer line saturates the soil in the vicinity of the water line. If the water main suddenly becomes depressurized and there is no pressure or a negative pressure, contaminants in the saturated soil could be drawn into the water lines. Adequate separation between the two lines, therefore, is needed. Water mains parallel to sanitary sewers should be installed at least 10 feet (3 m) horizontally away from the sewers and at least one foot (0.3 m) higher than any nearby sanitary sewer line. They should never be laid in the same trench with sewer lines. Other wastewater facilities are also of concern. The water main should not be closer than 10 feet (3 m), and preferably 25 feet (7.5 m) horizontally from wastewater leach fields, cesspools, seepage pits, and septic tanks. If local conditions prevent the specific separation, special construction is required. Your local and state health departments may have specific guidelines.

3.652   Storage and Handling

Pipe must be handled very carefully to prevent damage when it is loaded, transported, and unloaded (Figure 3.25).

Pipe and other materials received should always be checked for damage upon arrival at the job site and again prior to installation. The best way to check metal pipe is by gently tapping each length of pipe with a hammer. If the pipe does not ring or hum clearly, the pipe is damaged and should be rejected.

Lined and coated pipe must be handled carefully to avoid damaging the protective material. Small-diameter pipe should be unloaded by using skids and ropes. Large-diameter pipe is unloaded by mechanical equipment (forklifts, cherry pickers, or front-end loaders with forks).

To prevent excessive handling of the pipe, it is desirable to unload pipe at the job site and place it as near as possible to where the pipe is to be used. Normally the pipe is placed along the trench on the side opposite the excavated material.

Be careful to protect pipe that is being stored. Stack it according to the manufacturer's directions and keep it off the ground so that it cannot be contaminated by flooding. Belled pipe should always be stacked with the bells overhanging so that the pipe lies flat. You can protect the ends of the pipes with paper covers. Do not remove the paper covers until the pipe is ready for installation. Plastic pipe stored outside longer than 30 days needs to be protected from the sun by covering it with canvas or other opaque material. All pipe fittings and gasket material should be properly stored so that they will be protected from contamination and kept as clean as possible.

QUESTIONS

Write your answers in a notebook and then compare your answers with those on page 154.

3.65A   What kinds of underground lines should be located prior to excavation?

3.65B   When water and sewer lines are not properly separated, what problems could develop?

3.65C   How would you check metal pipe for damage upon arrival at the job site and again prior to installation?

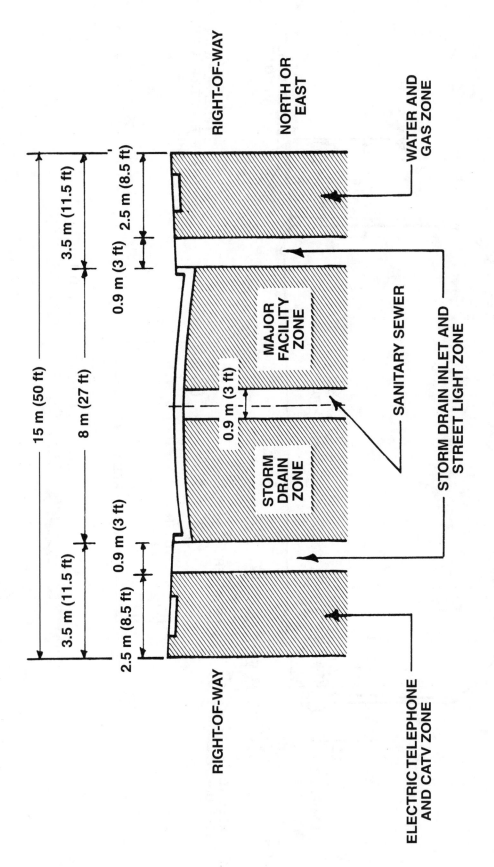

RIGHT-OF-WAY

NORTH OR EAST

WATER AND GAS ZONE

MAJOR FACILITY ZONE

SANITARY SEWER

STORM DRAIN INLET AND STREET LIGHT ZONE

STORM DRAIN ZONE

RIGHT-OF-WAY

ELECTRIC TELEPHONE AND CATV ZONE

15 m (50 ft)

3.5 m (11.5 ft)

2.5 m (8.5 ft)

0.9 m (3 ft)

8 m (27 ft)

0.9 m (3 ft)

3.5 m (11.5 ft)

0.9 m (3 ft)

2.5 m (8.5 ft)

Fig. 3.23   Typical utility locations in a minor street
(Adapted from Cowgill, J. E., "Coordinating Utility Location: A Defense Against Distribution System Damage," JAWWA, Vol. 73, No. 2, p. 87, February, 1981)

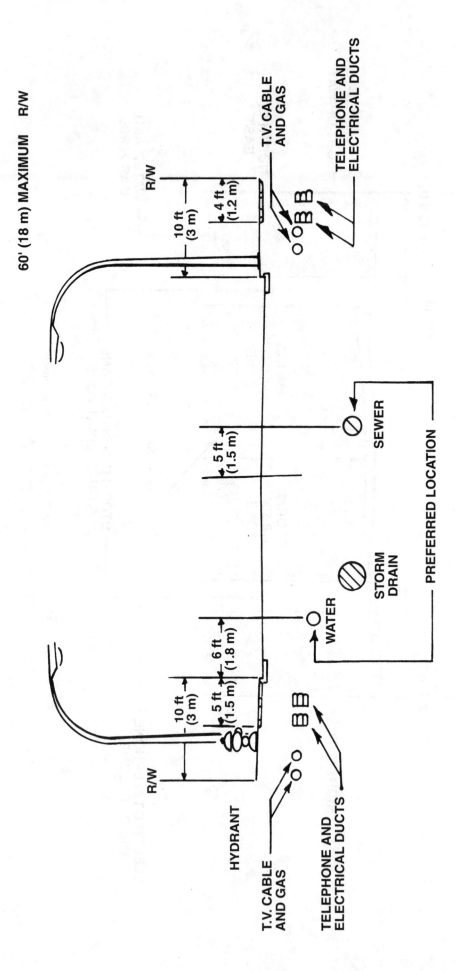

Fig. 3.24   *Typical utility locations*

(Adapted from *MODEL STANDARD LOCATIONS FOR UTILITIES IN PUBLIC RIGHTS OF WAY*, prepared by Uniform Ordinances and Practices Committee of the Southern California Chapter, American Public Works Association, 1975, Second Edition)

Forklift unloading (preferred)

Fig. 3.25   Unloading pipe
(Permission of Johns-Manville Corporation, a Subsidiary of Manville Corporation)

3.653 Excavation and Shoring

3.6530 EXCAVATION

When excavating a trench for water lines, you may first have to remove some type of surface paving material. Asphalt paving is cut with a wheel type of cutter, a clay spade, or a stomper. Concrete can be cut with power saws. The edges of all cuts must be smooth to reduce hazards to workers. All concrete and asphalt debris from the cut paving must be removed before the excavation starts because the debris cannot be used to refill the trench.

The actual excavation may be done by hand or machine. Backhoes, trenching machines, and power shovels are the equipment frequently used for excavation. The backhoe is most often used because it is usually readily available in any needed size, is easily operated, and can also be used to lay the pipe in the trench.

Beware of old, hazardous waste dump sites during trench excavation. If the soil looks different (color, composition, compaction) and/or smells different you may have encountered a former hazardous waste site. Try to find out what type of industries are located nearby or have used the site. If you think the soils contain hazardous materials or wastes, request a qualified soils lab to collect and analyze soil samples. If the soils contain a hazardous material, contact your local safety regulatory agency to determine what precautions must be taken to protect persons working in the trench.

The depth of the trench will depend on the desired grade line and local conditions. Trenches must be deep enough so that the pipe is protected from freezing, surface traffic, stream or river erosion, and possible future changes in the ground surface elevation. Rock must be excavated to a level six inches (150 mm) below the grade line of the pipe bottom. Water mains need at least 30 inches (750 mm) of cover over the top of the pipe. If water mains are not installed below the frost line, they must be protected to prevent freezing.

Generally, the trench should be no more than one to two feet (0.3 to 0.6 m) wider than the outside diameter of the pipe. This permits proper installation of the pipe and gives the workers enough room to do the work. The trench should be dug only as far ahead of pipe laying as necessary for a smooth operation.

Excavated soil is piled on the side of the trench between the trench and traffic and far enough away (2 to 5 feet (0.6 to 1.5 m)) so the crew can walk between the trench and the excavated material. The pipe is then laid out on the opposite side of the trench. Excavated rock and unstable material must be hauled away and not used for backfill.

3.6531 NEED FOR SHORING

Whenever excavation is necessary for repairs or new construction, shoring (Figures 3.26 and 3.27) may be necessary to protect operators from cave-ins. Shoring is a complete framework of wood and/or metal that is designed to support

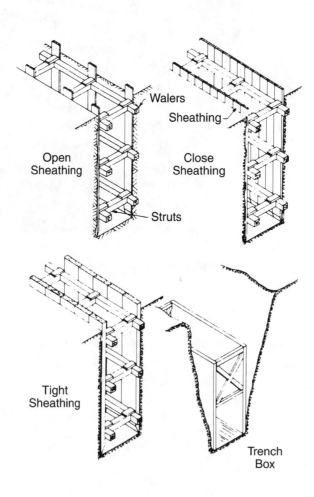

Fig. 3.26 Types of shoring

(Reprinted from *CONCRETE PIPE HANDBOOK*, by permission.
Copyright 1981, the American Concrete Pipe Association)

Hydraulic aluminum shoring devices

Trench boxes

Fig. 3.27   Shoring equipment
(Permission of Trench Shoring Company)

the walls of a trench. Sheeting is the solid material placed directly against the side of the trench. Either wooden sheets or metal plates might be used. Uprights are used to support the sheeting. They are usually placed vertically along the face of the trench wall. Spacing between the uprights varies depending on the stability of the soil. Stringers (or wales) are placed horizontally along the uprights. Trench braces are attached to the stringers and run across the excavation. The trench braces must be adequate to support the weight of the wall to prevent a cave-in. Different types of shores are described later in Section 3.6536. The need for shoring and shoring requirements depend on many factors, such as depth of trench, width of trench, type of soil (clay, loam, sand), soil conditions (compaction, moisture) and nearby activities that could cause vibrations.

Another consideration when determining shoring requirements is the length of time you expect the excavation to be open. If an excavation is going to be open for an extended period, you would want heavier bracing to provide additional protection against vibrations from traffic and rains that could cause flooding.

Shoring requirements are dictated by laws and codes and are strictly enforced by the regulatory agencies. Lack of shoring or shoring failure are the major causes of underground construction deaths. For example, seven people died and there were 67 lost-time injuries related to construction shoring problems in California during one year. How can we prevent these accidents? First, learn the shoring laws and codes, types of shoring, proper use, and then *APPLY THIS KNOWLEDGE*. Also, develop a safety program and organize an approved OSHA (Occupational Safety and Health Act) shoring class for operators and workers. The Williams-Steiger Occupational Safety and Health Act (OSHA) of 1970 contains shoring regulations, and they are strictly enforced. Penalties for violating the regulations include warnings, fines, and even prison sentences for flagrant or repeated violations. Contact your local safety office for all regulations pertaining to any excavations done by you or your crews.

3.6532   TYPICAL SHORING REQUIREMENTS

Typical shoring requirements include:

1. Shoring systems in trenches shall consist of uprights held rigidly opposite each other against the trench walls by jacks or horizontal cross members (cross braces) and, if required, longitudinal members (stringers) as specified below.

2. Uprights shall not exceed 15 degrees from the vertical. Uprights in trenches over 10 feet (3 m) deep shall be not less than 3-inch by 8-inch (7.5-cm by 20 cm) material, and shall be at least 2-inch by 8-inch (5-cm by 20-cm) material in trenches less than 10 feet (3 m) deep.

Uprights shall extend from above the top of the trench to as near the bottom as permitted by the material being installed, but not more than 2 feet (0.6 m) from the bottom.

3. Cross braces shall consist of steel screw-type trench jacks with a foot or base plate on each end of pipe or timbers placed horizontally and bearing firmly against uprights or stringers.

4. The minimum number of horizontal cross braces, either screw jacks or timbers, required for each pair of uprights shall be determined by the number of 4-foot (1.2-m) zones or segments into which the depth of trench may be divided. One horizontal cross brace shall be required for each of these zones, but in no case shall there be less than two cross braces or jacks. Trenches, the depths of which cannot be divided equally into these standard 4-foot (1.2-m) zones or segments, shall have an extra horizontal cross brace supplied for the short remaining zone, if such zone is greater than 2 feet (0.6 m). In no case, however, shall the vertical spacing of horizontal cross braces be spaced greater than 5 feet (1.5 m) center to center. Minor temporary shifting of horizontal cross bracing will be permitted, when necessary, for the lowering of materials into place. Be sure no one is allowed in the trench while the cross braces are temporarily shifted or moved. If the diameter of a pipe is greater than 4 feet (1.2 m), the lower cross braces require alternate considerations.

5. The dimensions and spacing of the elements of the shoring system shall be governed by the depth of the trench, type of soil encountered, and other special conditions of the site.

QUESTIONS

Write your answers in a notebook and then compare your answers with those on page 154.

3.65D   How is the depth of a trench determined?

3.65E   Shoring requirements depend on what factors?

3.65F   What is the number-one cause of cave-in deaths?

3.65G   What are the components of a typical trench shoring system?

3.65H   Uprights must extend how close to the bottom of a trench?

3.6533   RESPONSIBILITY

A "competent person" must be on the job site and in charge at all times. OSHA defines a "competent person" as a person capable of identifying existing predictable hazards in the surroundings, or working conditions which are unsanitary, hazardous or dangerous to employees, and who has authorization to take prompt corrective measures to eliminate the hazards. This person not only accepts responsibility, but is authorized to change any predetermined shoring system when soil conditions change. Material handling, traffic pattern changes, public relations, repair work excavation, backfill, and compaction are directed by the person in charge. Some of these jobs will have to be delegated, of course, or they won't be done effectively.

Excavations and the adjacent areas must be inspected on a daily basis by a "competent person" for evidence of potential cave-ins, protective system failures, hazardous atmospheres, or other hazardous conditions. The inspections are only required if an operator exposure is anticipated.

TAILGATE SAFETY MEETINGS[23] and preconstruction meetings are very important. Everyone on the job site must know exactly what to do and what the crew is expected to accomplish.

3.6534 PLACING OF SPOIL

While digging trenches less than 5 feet (1.5 m) in depth, the spoil (excavated material) should be placed 2 feet (0.6 m) or more back from the edge of the trench and on only one side of the trench. For trenches deeper than 5 feet (1.5 m), place the spoil 4 feet (1.2 m) or more back from the edge of the trench and on only one side of the trench. (*NOTE:* OSHA requires a minimum spoil setback of 2 feet (0.6 meter) from the edge of a trench. Some states require greater spoil setbacks depending on local conditions.) Try to keep the area between the trench and the spoil area clear of loose material. If not kept clear, rocks, tools, and pipe could be knocked into the excavation, or someone could trip on the equipment and fall into the trench. Keeping the area clear is not only good for the safety of operators, but reduces the chances of damage to new pipe or other conduits and utilities (which could cause thousands of dollars worth of damage). Also, keeping spoil away from the edge of the trench tends to reduce the load on the side walls, thus reducing the loads on the sheeting, uprights, stringers, and cross bracing.

3.6535 TYPICAL SHORING REGULATIONS

1. Operator Safety

All trenches 5 feet (1.5 m) in depth (4 ft (1.2 m) in some locations) and deeper must be effectively shored to protect operators against the hazard of moving ground. Trenches less than 4 feet (1.2 m) in depth also must be shored when examination indicates hazardous ground movement may be expected.

Special provisions must be made to prevent injury to operators while they install shoring. The use of special devices such as long-handled jacks will allow upper cross braces to be placed from the ground surface before operators enter the trench. At those points in deep trenches requiring additional braces, operators should then progress downward, protected by cross braces that have already been set firmly in place. Reverse this procedure when you remove the shoring.

2. Soil Conditions and Trench Location

The type of shoring you'll need depends on the soil conditions and the location of the trench. You cannot shore an excavation alongside a railroad track the same way as one in an open area because of vibrations created by passing trains. When the decision is made on the type of shoring to use, remember that soil conditions change and an alternate shoring system should be readily available so that little work time will be lost if a change must be made.

3. Sloping Excavation Walls

Instead of installing a shoring system, the sides or walls of an excavation may be sloped if a competent person determines that this method is more economical and the job can be done safely. If you decide to slope the walls of an excavation that would otherwise need shoring, excavate the side walls on a slope of $3/4$ foot horizontal to 1 foot vertical except where the instability of material requires a slope of greater than $3/4$:1 ($3/4$ vertical to 1 horizontal).

4. Access to Construction Work Area

Convenient and safe means must be provided for operators to enter and leave the excavated area. In trenches more than four feet deep, use a standard stairway, ladder, or ramp securely fastened in place at suitably guarded or protected locations where people are working. Ladders must be located within 25 feet (7.6 m) of any person working in the excavation. Ladders also must extend 3 feet (91 cm) above the surface of the excavation.

5. Pedestrian Safety

When working near curbs and sidewalks, keep a pedestrian way open on the sidewalks. If this area becomes covered with material and equipment, the pedestrians will be in the work area or walking across someone's yard. Neither one of these situations is acceptable. If necessary, erect barriers to keep pedestrians and children in safe, permissible areas. Be especially careful if children are playing or watching near the job site.

QUESTIONS

Write your answers in a notebook and then compare your answers with those on page 154.

3.65I What authority does the person supervising excavation work have?

3.65J Where should spoil or excavated material be placed in relation to the trench?

3.65K All trenches deeper than _____ feet must be shored.

3.65L How can operators be protected if shoring is not practical?

3.65M Why must a walkway be provided for pedestrians?

3.6536 TYPES OF SHORES

There are several types of shores that have been tested and approved by OSHA. Each type of shoring has its good and bad points. Selection of an appropriate type of shoring depends largely on soil conditions in your area.

1. Hydraulic Shores (Figure 3.28)

Hydraulic shores are used frequently due to their ease of installation and removal. They are usually not used on jobs

---

[23] *Tailgate Safety Meeting.   Brief (10 to 20 minutes) safety meetings held every 7 to 10 working days. The term TAILGATE comes from safety meetings regularly held by the construction industry around the tailgate of a truck.*

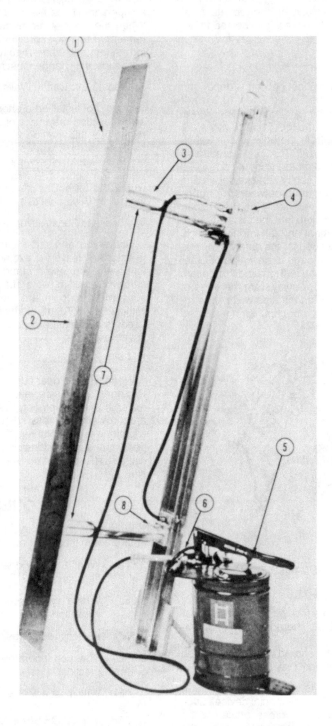

1. Aluminum alloy material
2. Flanged siderail
3. Hydraulic cylinder
4. Offset pins connecting siderails and hydraulic cylinders
5. Pump
6. Pressure gage
7. 4-foot (1.2-m) cylinder spacing
8. Cylinder pad

Fig. 3.28　Aluminum hydraulic cylinder
(Courtesy of SPEED-SHORE)

when they will have to be left installed in one segment of trench for longer than five days. This is due to the possibility of the hydraulic pressure bleeding off during a longer period of installation.

2. Screw Jacks (Figure 3.29)

Screw jacks are time consuming to install and usually are not used on large production jobs. They are inexpensive, however, and for minor repairs in smaller excavations, they are more practical than other types of shores. Screw jacks must be installed from the top of the trench down, and those who have used them know this is a slow process. Screw jacks are made of iron and come in two pieces. They are available with a 1 1/2- or 2-inch (38- to 50-mm) threaded shaft and wing nut that is used for adjustment to fit trench width. One end of the shaft is plain and the other has a ball-socket for a shoe that pushes against the shoring. The screw length on the 1 1/2-inch (38-mm) shaft varies from 10 to 18 inches (250 to 450 mm). The 2-inch (50-mm) shaft has a screw length of 18 inches (450 mm) only.

3. Air Shores (Figure 3.30)

Air shores are pneumatic shoring cross braces. They are not equipped with rails like hydraulic shores. They function like screw jacks and are used only as the cross brace which can be rapidly set to hold shoring members in place.

4. Solid Sheeting (Figure 3.31)

Occasionally it is necessary to excavate in materials that will not stand in a vertical position and where it is not practical to slope the trench. Under these circumstances, you must use a solid-sheeting method. Fully dimensional rough cut timbers are used. Minimum sizes are 2- x 8-inch (50- x 200-mm) rough timbers for trenches up to 8 feet (2.4 m) deep and 3- x 8-inch (75- x 200-mm) rough timber for trenches over 8 feet (2.4 m) deep. Metal plates can be used as solid sheeting if they are properly engineered.

The horizontal shoring members are generally referred to as "stringers" or "walers." When purchasing shoring material, buy the square 6" x 6" (15 cm x 15 cm) rough timber, or the 8" x 8" (20 cm x 20 cm) rough timber because finished dimensional timber is smaller and not as strong. Hydraulic shores, air shores, or screw jacks can be used as cross braces.

5. Cylinder Shoring (Figure 3.32)

Cylinder shoring can be made from 1/4-inch or 5/16-inch (6-mm or 8-mm) thick steel plates that are 4 feet (1.2 m) wide. These sheets are rolled into a cylinder with a 4-foot (1.2-m) inside diameter and are 4 feet (1.2 m) in height. The top and bottom of each cylinder are reinforced with a welded angle iron rolled to fit the cylinders. Each cylinder section weighs either 550 (1/4-inch (6-mm) plate) or 675 (5/16-inch (8-mm) plate) pounds (250 or 305 kg). This weight usually limits the use of cylinder shoring to street locations or areas where cranes or hydro lifts can be used to position the cylinders. Cylinders 4 feet (1.2 m) high and 4 feet (1.2 m) in diameter are easy to transport and handle, but are confining on large jobs.

6. Shield Shoring (Figure 3.33)

Shield shoring is used during excavation and installation of various types of water mains in areas and soil conditions that present a combination of problems. Shield shoring consists of a rectangular box that must be open on the top and bottom and usually the ends are open too. Size of shield depends on the size of water line being installed and the space needed to

install the pipe. Shields can vary in width from 30 inches (0.75 m) to 10 feet (3 m) and in length from 20 to 30 feet (6 to 9 m).

3.6537 SHORING SUMMARY

This section has identified the common types of shores and their uses. Selection of shores depends on several factors and the job you are trying to accomplish.

PROTECT YOUR OPERATORS AT ALL COST. Attend all trench safety lectures and seminars. Contact your own safety council and see what information is available and what is required. Be sure that everyone complies with all regulations. ALL TRENCHES AND EXCAVATIONS ARE UNSAFE UNTIL PROPERLY SHORED, USED, AND BACKFILLED.

QUESTIONS

Write your answers in a notebook and then compare your answers with those on page 154.

3.65N   List the different types of common shores.

3.65O   What should be your major concern during a shoring operation?

3.6538 BEDDING

Bedding is the material placed in the bottom of the trench to support the pipe. The trench bottom is often brought back up to grade with sand or soil to provide a suitable bedding. However, with the proper soil, pipe could be laid directly on the trench bottom. Special care must be taken to ensure that couplings will not rest on or settle down to the original trench bottom and also that the pipe is provided with even bedding for its entire length (Figure 3.34).

3.654 Pipe Laying and Jointing

Laying the pipe out for installation is called stringing. Small diameter pipe is laid by hand and should be strung along the trench. When belled pipe is used, point the bells in the direction that the work is proceeding except when the pipe is being laid downhill; then the bell should face uphill.

Never allow pipes to drop into the trench; carefully lower them into position. Pipes up to 12 inches (300 mm) in diameter can be lowered into the trench manually with two ropes, one tied to each end of the pipe (Figure 3.35). Larger pipe must be handled with power equipment. Cable slings may be used to lower pipe into trenches (Figure 3.36). Always move

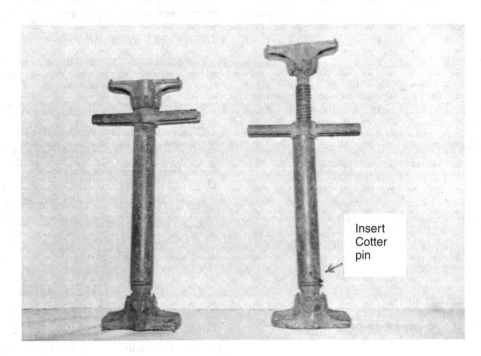

Screw jacks

Screw jacks with timber uprights

WARNING—Never use jacks for access ladders. Always use a ladder.

Fig. 3.29   Screw jacks

Fig. 3.30   Air shores

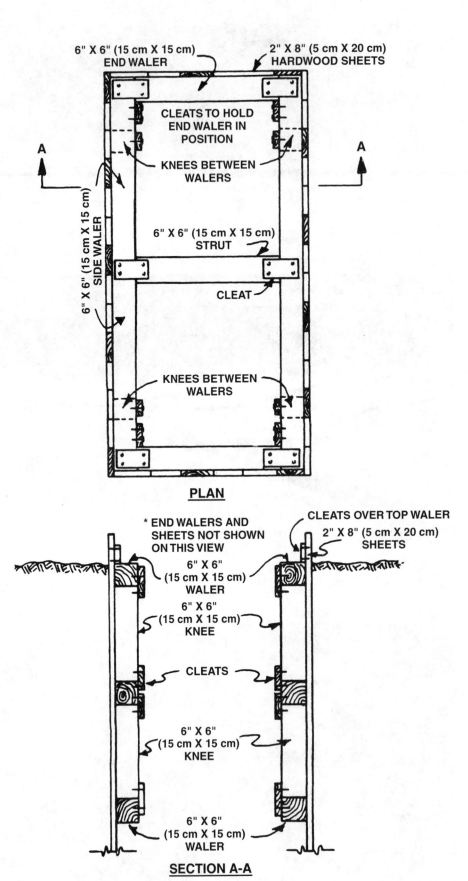

6" X 6" (15 cm X 15 cm)
END WALER

2" X 8" (5 cm X 20 cm)
HARDWOOD SHEETS

CLEATS TO HOLD
END WALER IN
POSITION

KNEES BETWEEN
WALERS

6" X 6" (15 cm X 15 cm)
SIDE WALER

6" X 6" (15 cm X 15 cm)
STRUT

CLEAT

KNEES BETWEEN
WALERS

PLAN

* END WALERS AND
SHEETS NOT SHOWN
ON THIS VIEW

CLEATS OVER TOP WALER

2" X 8" (5 cm X 20 cm)
SHEETS

6" X 6"
(15 cm X 15 cm)
WALER

6" X 6"
(15 cm X 15 cm)
KNEE

CLEATS

6" X 6"
(15 cm X 15 cm)
KNEE

6" X 6"
(15 cm X 15 cm)
WALER

SECTION A-A

Fig. 3.31   Solid sheeting

Cylinder shoring

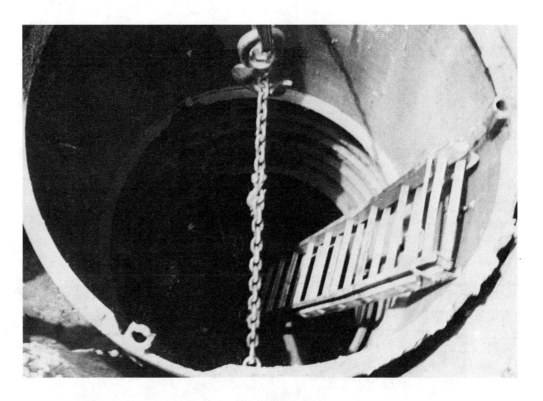

Ladder inside cylinder

Fig. 3.32   Cylinder shoring

Fig. 3.33   Shield shoring

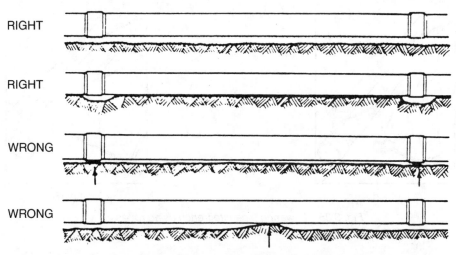

COUPLINGS—NEVER ALLOW COUPLINGS TO REST ON, OR TO SETTLE DOWN TO, ORIGINAL TRENCH BOTTOM.

PIPE—MAKE CERTAIN THAT PIPE BARREL IS GIVEN AN EVEN BEARING FOR ITS FULL LENGTH.

Fig. 3.34   Pipe bedding
(Reprinted from *WATER DISTRIBUTION OPERATOR TRAINING HANDBOOK*,
by permission. Copyright 1976, the American Water Works Association)

pipe very carefully so it will not be damaged. Special precautions must be taken to protect any interior and exterior pipe coatings.

Before placing pipe in the trench, inspect the inside and outside of each pipe. Remove all debris found inside the pipe. If there is any question about the cleanliness of the pipe, brush the inside to remove any loose dirt and then swab it with a strong hypochlorite solution. Keep all gaskets clean and dry. When work is not in progress and at night, the open ends of the installed pipe should be plugged tightly.

Before you begin jointing, carefully clean the ends of the pipes and the joint materials and smooth out any rough edges. Each type of pipe material requires special jointing procedures and all joints should be installed carefully in accordance with these procedures. One feature is common to most jointing methods: a distinctive sound is heard when the pipe is correctly pushed home. When rubber gaskets are used, check the completed joint with a feeler gage to make sure that there is proper seating. Only recommended lubricants are to be used because some lubricants could ruin gaskets or support bacterial growth.

The same methods used for jointing in straight pipe are also used for installing valves, fittings, and fire hydrants which are connected as the pipeline is laid.

3.655   Thrust Blocks (Figure 3.37)

Thrust forces in water mains are created where the pipeline changes direction or size, at dead ends, and when valves and hydrants are quickly closed. Thrust blocks are used at these points to prevent damage caused by internal pressures and to keep the pipeline intact. All tees, bends, caps, plugs, hydrants, or other fittings that change the direction of flow or stop the flow should be restrained or blocked to prevent leakage or separation of the joints. Thrust blocks made of concrete and steel reinforcement rods are often used. Wood or any other material that deteriorates is not acceptable. The concrete should be in contact only with the fitting, not with the pipe itself or the joint. Thrust blocks should rest against undisturbed solid soil for their required bearing area. The size of a thrust block depends on the water pressure, the size of the pipeline, the type of fitting in need of protection, and the type and character of soil that backs up the thrust block. Pipelines should not be pressure tested until the concrete in the thrust blocks has had time to set properly (at least five days).

3.656   Backfilling

After the pipe has been laid, jointed, and thrust blocks installed, the excavation is manually backfilled with select material. Compaction may be required depending on the trench conditions. Compaction around the pipe and above it, when required, is done gradually in stages. The first layer of backfill is compacted up to the horizontal centerline of the pipe. The backfill material must contain enough moisture for full compaction and consolidation under and on both sides of the pipe so that the pipe is supported without any hollow spaces in the soil. The purpose of tamping the soil around the pipe is to provide uniform support and bearing (Figure 3.38).

Frequently, the trench is backfilled and tamped except at the pipe joints which are left exposed until the pipe has been tested under water pressure. Following the water pressure test, the backfilling can be completed. Tamping of the backfill above the pipe can be done by hand or machine, or by applying water (jetting), depending on local conditions and convenience. If the trench is in a road, the completed backfill must meet the compaction requirements of the appropriate local agency.

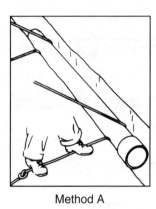 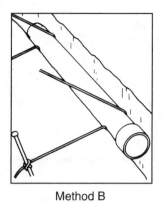

Method A   Method B

Fig. 3.35   Lowering pipe into trench using ropes
(Permission of the Johns-Manville Corporation, a subsidiary of the Manville Corporation)

Fig. 3.36   Use of cable sling to lower pipe into trench
(Permission of the Ductile Iron Pipe Research Association)

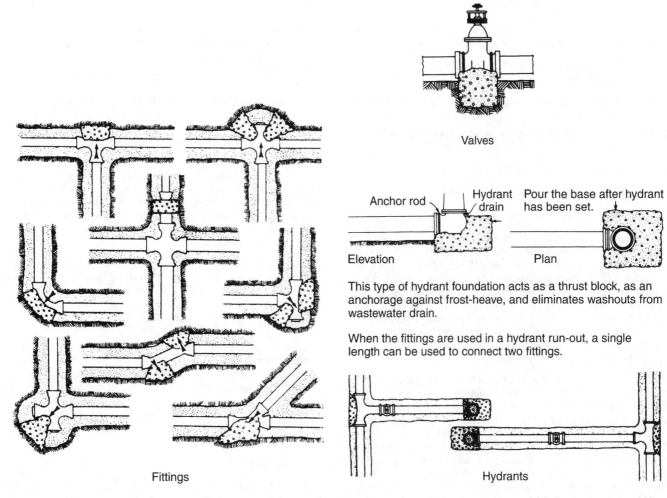

Valves

Anchor rod   Hydrant drain   Pour the base after hydrant has been set.

Elevation   Plan

This type of hydrant foundation acts as a thrust block, as an anchorage against frost-heave, and eliminates washouts from wastewater drain.

When the fittings are used in a hydrant run-out, a single length can be used to connect two fittings.

Fittings   Hydrants

Fig. 3.37   Thrust blocks
(Permission of the Johns-Manville Corporation, a subsidiary of the Manville Corporation)

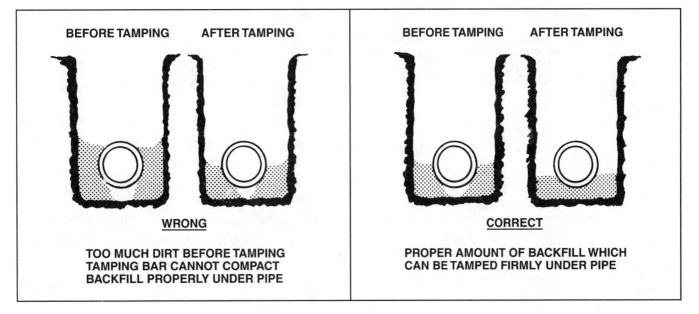

BEFORE TAMPING   AFTER TAMPING

WRONG

TOO MUCH DIRT BEFORE TAMPING
TAMPING BAR CANNOT COMPACT
BACKFILL PROPERLY UNDER PIPE

BEFORE TAMPING   AFTER TAMPING

CORRECT

PROPER AMOUNT OF BACKFILL WHICH
CAN BE TAMPED FIRMLY UNDER PIPE

Fig. 3.38   Tamping procedures
(Reprinted from *WATER DISTRIBUTION OPERATOR TRAINING HANDBOOK*,
by permission. Copyright 1976, the American Water Works Association)

QUESTIONS

Write your answers in a notebook and then compare your answers with those on pages 154 and 155.

3.65P   What precautions should be taken when laying pipe on bedding?

3.65Q   When laying pipe with a bell, the bell should point in what direction?

3.65R   How should pipes be prepared before they are joined together?

3.65S   Where should thrust blocks be installed?

3.65T   How deep is the first layer of backfill?

3.657   Pressure and Leakage Tests

Leakage tests are conducted after the trench has been partially backfilled. As mentioned, the joints are often left uncovered until after the test is completed. Each valved section of the main is filled slowly with water, with air in the pipe being released through CORPORATION STOPS[24] and hydrants. All air must be removed before the pressure testing starts, but do not fill the pipe too quickly. Excessive velocities could cause pipe movements when the water flows around fittings and thus cause leaks. Pipes should not be filled at a rate greater than one foot of pipe length per second.

After the pipe is full, it should sit idle for at least 24 hours before the test starts. Then, the pressure is brought up to a level at least 50 percent higher than the normal expected operating pressure, or 150 psi (1,034 kPa or 10.5 kg/sq cm), whichever is larger, and maintained for at least four hours. Any loss of pressure would indicate leakage. The amount of water needed to refill the pipe is measured. Leakage can also be determined by examination of the trench for visible leaks or by the use of a meter to detect any pipe flow. If any joints do show leakage, they are checked and adjusted or repaired. However, some leakage is to be expected, and the AWWA has specifications for allowable leakage for different types of pipe (Table 3.1). During and after the test the trench is also observed to determine if there is any pipe movement.

TABLE 3.1   ALLOWABLE LEAKAGE FROM PIPES [a]

Pipe Material	Pipe Length, ft [b]	Test Pressure, psi [c]	Allowable Leakage, GPD/mi-in [d,e]
Asbestos-Cement	13	150	30
Ductile Iron	18	150	23.3 [f]
Plastic	—	150	1.45 [g] or 1.88 [h]

[a] AWWA Standards for each type of pipe material.
[b] Multiply ft x 0.3 to obtain meters.
[c] Multiply psi x 0.07 to obtain kg/sg cm.
[d] Multiply GPD/mi-in x 0.926 to obtain LPD/km-cm.
[e] Allowable leakage in gallons per day per mile of pipe per inch of diameter.
[f] Allowable leakage for mechanical and push-on joints.
[g] Allowable leakage is 1.45 gallons (5.5 liters) per hour for each 100 joints for six-inch (150-mm) pipe.
[h] Allowable leakage is 1.88 gallons (7.1 liters) per hour for each 100 joints for eight-inch (200-mm) pipe.

FORMULAS

To determine the maximum flowmeter reading in gallons per minute, use the formulas shown below.

$$\text{Area, sq ft} = \frac{(0.785)(\text{Diameter, in})^2}{144 \text{ sq in/sq ft}}$$

$$\text{Flow, cu ft/sec} = (\text{Area, sq ft})(\text{Velocity, ft/sec})$$

and

$$\text{Flow, GPM} = (\text{Flow, cu ft/sec})(7.48 \text{ gal/cu ft})(60 \text{ sec/min})$$

The first formula calculates the area of a pipe in square feet when the diameter is given in inches. The second formula gives us the flow in cubic feet per second by multiplying the area in square feet times the velocity in feet per second. The last formula converts the flow from cubic feet per second to gallons per minute by converting the cubic feet to gallons and the seconds to minutes.

Leak tests on asbestos-cement and ductile-iron pipe report the results in terms of gallons of water leaked per day per miles of pipe length and inches of pipe diameter.

$$\text{Actual Leakage, GPD/mi-in} = \frac{\text{Leak Rate, GPD}}{(\text{Length, mi})(\text{Diameter, in})}$$

Leak tests on plastic pipe are reported in terms of gallons per hour for each 100 joints of pipe. Therefore, if the test section had 54 joints we would have 0.54 – 100 joints of pipe. This means we tested a little over half (0.54) of 100 joints.

$$\text{Actual Leakage, GPH/100 joints} = \frac{\text{Leak Rate, gal/hr}}{\text{Number of 100 joints}}$$

EXAMPLE 4

A 500-foot long section of a 12-inch diameter water main is being filled with water for a leakage test. If the pipe is filled at a flow rate of less than one foot of pipe length per second, what should be the maximum flowmeter reading in gallons per minute for filling the pipe?

Known		Unknown
Length, ft	= 500 ft	Flow, GPM
Diameter, in	= 12 in	
Fill Rate, ft/second	= 1 ft pipe length per second	

1. Calculate the maximum allowable flow in cubic feet per second.

$$\text{Flow, cu ft/sec} = (0.785)(\text{Diameter, ft})^2(\text{Fill Rate, ft/sec})$$
$$= (0.785)(1 \text{ ft})^2(1 \text{ ft/sec})$$
$$= 0.785 \text{ cu ft/sec}$$

2. Determine the maximum flowmeter reading in gallons per minute.

$$\text{Flow, GPM} = (\text{Flow, cu ft/sec})(7.48 \text{ gal/cu ft})(60 \text{ sec/min})$$
$$= (0.785 \text{ cu ft/sec})(7.48 \text{ gal/cu ft})(60 \text{ sec/min})$$
$$= 352 \text{ GPM}$$

Do not allow the flow through the meter to exceed 350 GPM.

---

[24] Corporation Stop.   A water service shutoff valve located at a street water main. This valve cannot be operated from the ground surface because it is buried and there is no valve box. Also called a corporation cock.

EXAMPLE 5

The normal expected operating pressure of a pipe is 120 psi. At what pressure should the pipe be tested if the test pressure is 50 percent higher than the normal expected operating pressure or 150 psi, whichever is larger?

Known	Unknown
Normal Pressure, psi = 120 psi	Test Pressure, psi
Test Pressure, psi = Normal + 50%	
or = 150 psi	

whichever is larger.

Calculate the test pressure.

$$\text{Test Pressure, psi} = \text{Normal Pressure, psi} + \frac{50\%(\text{Normal Pressure, psi})}{100\%}$$

$$= 120 \text{ psi} + \frac{(50\%)(120 \text{ psi})}{100\%}$$

$$= 120 \text{ psi} + 60 \text{ psi}$$

$$= 180 \text{ psi}$$

Use a test pressure of 180 psi because this is greater than 150 psi.

EXAMPLE 6

A 24-hour leak test is performed on 1,000 feet of 12-inch diameter asbestos-cement pipe with a test pressure of 150 psi. The pipes are 13 feet long. According to Table 3.1, the allowable leakage is 30 gallons per day per mile of pipe per inch of diameter. During the 24-hour test period, 20 gallons of water were added to maintain the 150 psi pressure. Did the pipe pass the leak test?

Known	Unknown
Test Duration, hr = 24 hr	1. Actual Leakage, GPD/mi-in
or = 1 day	2. Did pipe pass test?
Pipe Length, ft = 1,000 ft	
Pipe Diameter, in = 12 in	
Test Pressure, psi = 150 psi	
Pipe Section, ft = 13 ft	
Water Leaked, gal = 20 gal	
Allowable Leakage, GPD/mi-in = 30 GPD/mi-in	

1. Calculate the leakage rate in gallons per day.

$$\text{Leak Rate, GPD} = \frac{\text{Water Leaked, gal}}{\text{Test Duration, day}}$$

$$= \frac{20 \text{ gallons}}{1 \text{ day}}$$

$$= 20 \text{ GPD}$$

2. Convert pipe length from feet to miles.

$$\text{Pipe Length, mi} = \frac{\text{Pipe Length, ft}}{5,280 \text{ ft/mi}}$$

$$= \frac{1,000 \text{ ft}}{5,280 \text{ ft/mi}}$$

$$= 0.189 \text{ mi}$$

3. Determine the actual leakage during the test in gallons per day per mile of pipe per inch of diameter.

$$\text{Actual Leakage, GPD/mi-in} = \frac{\text{Leak Rate, GPD}}{(\text{Length, mi})(\text{Diameter, in})}$$

$$= \frac{20 \text{ GPD}}{(0.189 \text{ mi})(12 \text{ in})}$$

$$= 8.8 \text{ GPD/mi-in}$$

4. Did the pipe pass the leak test?

Yes. Since the actual leakage rate was 8.8 GPD/mi-in which is less than the allowable leakage rate of 30 GPD/mi-in, the pipe passed the leak test.

EXAMPLE 7

A 6-hour leak test is performed on 500 feet of 6-inch plastic pipe with a test pressure of 150 psi. The pipes are 10 feet long. According to Table 3.1, the allowable leakage is 1.45 gallons per hour for each 100 joints of 6-inch pipe. During the 6-hour test period, 12 gallons of water were added to maintain the 150 psi pressure. Did the pipe pass the leak test?

Known	Unknown
Test Duration, hr = 6 hr	1. Actual Leakage, gal per hr/100 joints
Pipe Length, ft = 500 ft	2. Did pipe pass test?
Pipe Diameter, in = 6 in	
Test Pressure, psi = 150 psi	
Pipe Section, ft = 10 ft	
Water Leaked, gal = 12 gal	
Allowable Leakage, gal per hr/100 joints = 1.45 gal per hr/100 joints	

1. Calculate the leakage rate in gallons per hour.

$$\text{Leak Rate, gal/hr} = \frac{\text{Water Leaked, gal}}{\text{Test Duration, hr}}$$

$$= \frac{12 \text{ gallons}}{6 \text{ hours}}$$

$$= 2 \text{ gallons per hour}$$

2. Determine the number of joints.

$$\text{Number of Joints} = \frac{\text{Pipe Length, ft}}{\text{Pipe Section, ft/joint}}$$

$$= \frac{500 \text{ ft}}{10 \text{ ft/joint}}$$

$$= 50 \text{ joints}$$

Since we have only 50 joints, and not 100 joints, we have 0.5(100 joints).

3. Determine the actual leakage during the test in gallons per hour per 100 joints.

$$\text{Actual Leakage, gal per hr/100 joints} = \frac{\text{Leak Rate, gal/hr}}{\text{Number of 100 joints}}$$

$$= \frac{2 \text{ gal/hr}}{0.5(100 \text{ joints})}$$

$$= 4 \text{ gal per hr/100 joints}$$

4. Did the pipe pass the leak test?

No. Since the actual leakage was 4 gal per hr/100 joints which is greater than the allowable leakage rate of 1.45 gal per hr/100 joints for 6-inch pipe, the pipe did not pass the test.

3.658   Main Disinfection

During the construction of a new water line, there is a lot of opportunity for contamination of the line even if special precautions have been taken. Therefore, effective disinfection is necessary before the line is placed into service. The effectiveness of disinfection depends largely on maintaining clean pipes and avoiding major contamination during construction or repairs. Pipe crews must be trained to be aware of this need.

Precautions must be taken to protect the interiors of pipes, fittings, and valves against contamination. When pipes are delivered to construction or repair sites they should be unloaded so as to minimize entrance of foreign material. Watertight plugs must be used to close all pipeline openings when pipe laying is stopped for any reason. Joints of all pipe in the trench must be completed before work is stopped. If any water gets into the trench, the plugs must remain in place until the trench is dry. Packing and sealing materials must not contain any contaminated material or any material capable of supporting growth of microorganisms and should be handled so as to avoid contamination. If there is dirt in the pipe from wet-trench construction, flooding by a storm, or an accident during construction which would probably not be removed by flushing, the pipes must be thoroughly cleaned. For more details on the disinfection of water mains, see Chapter 5, "Distribution System Operation and Maintenance," and Chapter 6, "Disinfection."

3.659   Safety

The safety of the public and the operators must be protected at all times during the installation of pipe. Barricades should be put up on the curb side of the trench to limit accidental access by pedestrians. Whenever possible, the excavated material from the trench should be piled on the street side of the main to form a barricade to keep traffic away from the trench. If this is not possible, barricades should be put up and moved along as the work progresses. Easily visible construction signs, guards, torches, and flaggers should be provided where needed for the protection of the public. In dense traffic, high signs are more visible. Children should be discouraged from playing near the job. At night, the excavated material, pipe, and fittings should be protected with flares and blinker lights.

Wherever the trenching destroys or interferes with the sidewalks used by pedestrians, wooden walkways at least 4 feet (1.2 m) wide should be provided. When open trenches interfere with traffic that must go through or driveways that must be used, steel plates over the trenches are required.

The need for shoring of the trenches will depend on the kind of soil, the depth of the trench, and the regulations of OSHA. Normally, unless the trench sides are sloped back at least 1: 2, a trench 5 feet (1.5 m) deep or more requires shoring.

All operators must wear gloves and hard hats. They should always stand clear of obvious safety hazard situations such as pipe being lifted by mechanical equipment and the swinging buckets of backhoes. Steel-toe shoes should be worn to protect your feet. During pipe grinding and polishing, safety goggles must be worn, and respirators are needed when cutting or machining asbestos-cement pipe.

QUESTIONS

Write your answers in a notebook and then compare your answers with those on page 155.

3.65U What water pressure is used for a leakage test?

3.65V During grinding and polishing work on pipe, what safety precautions must be taken?

End of Lesson 3 of 5 on DISTRIBUTION SYSTEM FACILITIES

Please answer the discussion and review questions next.

DISCUSSION AND REVIEW QUESTIONS

Chapter 3.   DISTRIBUTION SYSTEM FACILITIES

(Lesson 3 of 5 Lessons)

Write the answers to these questions in your notebook before continuing. The question numbering continues from Lesson 2.

10. Why should other utility companies be notified before any excavation work?

11. Why should water and sewer lines be adequately separated?

12. How should pipe be stored on the job site before installation?

13. Where should the excavated soil from a trench be piled?

14. How can operators in trenches be protected from cave-ins?

15. How can trenching accidents be prevented?

16. How can you determine the shoring requirements in your area?

17. How would you determine if there was leakage from a pipe during a leakage test?

CHAPTER 3.   DISTRIBUTION SYSTEM FACILITIES

(Lesson 4 of 5 Lessons)

3.66   Pipe Extensions

Water systems must grow with the community they serve. New pipes may extend directly out from the end of an existing main or stub out from the side of the main. To simplify the extension, a line valve may already have been installed. If so, existing services need not be interrupted and the new connections can be quickly and easily made. To add a new main or lateral into an existing main where no line valve was installed, cut a hole in the main using a tapping saddle or sleeve. In a wet tap, you will bolt a valve to the tapping saddle; the actual drilling of the pipe through the valve can be done with the pipe under pressure. Tapping a drained line is referred to as a dry tap. Wet taps are preferred because water service is not interrupted. Also, if the pipe is drained, there is a possibility of contamination because any time the pressure in a pipeline is removed, contamination may more easily enter that line.

The tapping sleeve used for a wet tap is a split sleeve as shown in Figure 3.39. One half of the sleeve has an opening with a flange or some other means by which a valve can be attached. Fasten the sleeve around the pipe and make a tight joint at each end. Fasten a valve of a size suitable for the proposed pipe extension to the sleeve opening. Bolt a tapping machine on the other side of this valve. The tapping machine (Figure 3.40) carries a cylindrical cutter (the same size as the proposed branch) which is enclosed in a watertight case. Open the valve, move the cutter through the valve to the pipe, and cut a hole through the pipe. Withdraw the cutter into its case and close the valve. Remove the tapping machine. The valve remains permanently in place and the pipeline extension can be connected to it.

QUESTIONS

Write your answers in a notebook and then compare your answers with those on page 155.

3.66A   How can a tap be installed in a water main under an operating water pressure?

3.66B   How can new service connections be installed in existing plastic mains?

3.67   Appurtenances

3.670   Valves

3.6700   TYPES OF VALVES

Valves are used for many purposes in a distribution system. They can shut off, turn on, and otherwise regulate the flow of water, reduce pressure, provide air and vacuum relief, blow off or drain water from parts of the system, and prevent backflow. Valves are one of the most important devices available in operating the water distribution system.

SHUTOFF AND FLOW REGULATION. Valves may be classified, according to the movement of the closure element, into two main categories: rotary and linear stroke valves. The rotary valves include plug, ball, and butterfly valves. The linear stroke valves include diaphragm, globe, gate, and check valves. These valves are shown in Figures 3.41 through 3.46.

In a rotary valve, the stem moves from the fully open to the fully closed position by rotating 90 degrees without a linear (up and down) motion of the closure element. Usually, a 90-degree rotation movement is used with a notch, arrow, or other indicator to show valve positions. In the ball valve (Figure 3.41), the movable part is a ball with a cylindrical hole bored through it. When the ball is in one position there is a straight passage through the valve, but when it is rotated 90 degrees the flow is blocked. Ball valves should be operated either fully open or fully closed. If you attempt to control flow with a ball valve, the seal around the ball and the valve body could be destroyed and render the valve useless.

The butterfly valve has a movable disc as large as the full diameter of the pipe. The disc rotates on a spindle or shaft in only one direction to either block or allow the flow of water. Figures 3.42 and 3.43 show butterfly valves.

Plug valves (Figure 3.41) may have a tapered or cylindrical plug with an opening through the side which can be turned to open, restrict, or close the flow. The smaller plug valves are used extensively as corporation stops on service lines.

Gate valves are the most common important type of linear stroke valves used. Their main application in a distribution system is to isolate sections of mains to permit emergency repairs without interruption of service to large numbers of customers. The stem in a linear stroke valve must turn around several times to move the closure element up or down from a fully

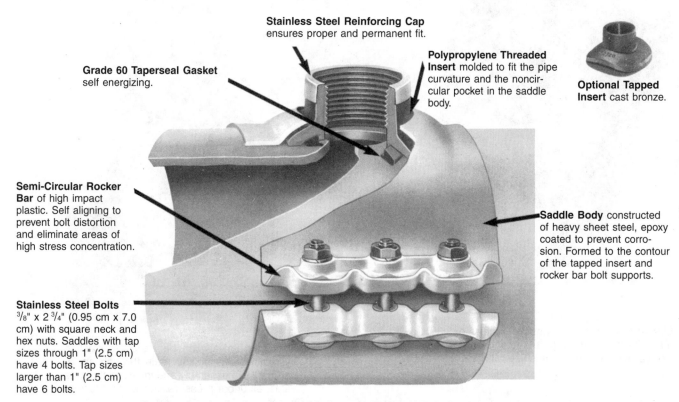

Stainless Steel Reinforcing Cap
ensures proper and permanent fit.

Grade 60 O-Ring maintains a positive hydraulic seal.

Heavy Duty Worm Drive Clamps of ALL STAINLESS STEEL construction for maximum corrosion resistance are geometrically positioned to provide uniform pressure balance on pipe and O-Ring seal.

Celcon M90 or Polypropylene Saddle Body is proportioned to fully encapsulate and protect pipe from undesirable stresses and deformation.

Saddles for plastic pipe sizes 1 inch through 4 inch (2.5 cm through 10 cm).

Stainless Steel Reinforcing Cap
ensures proper and permanent fit.

Grade 60 Taperseal Gasket
self energizing.

Polypropylene Threaded Insert molded to fit the pipe curvature and the noncircular pocket in the saddle body.

Optional Tapped Insert cast bronze.

Semi-Circular Rocker Bar of high impact plastic. Self aligning to prevent bolt distortion and eliminate areas of high stress concentration.

Saddle Body constructed of heavy sheet steel, epoxy coated to prevent corrosion. Formed to the contour of the tapped insert and rocker bar bolt supports.

Stainless Steel Bolts 3/8" x 2 3/4" (0.95 cm x 7.0 cm) with square neck and hex nuts. Saddles with tap sizes through 1" (2.5 cm) have 4 bolts. Tap sizes larger than 1" (2.5 cm) have 6 bolts.

Saddles for plastic pipe sizes 6 inch through 12 inch (15 cm through 30 cm).

Fig. 3.39   Tapping saddles for plastic pipes
(Permission of Rockwell International)

Drilling machine makes service connections to plastic mains by drilling the mains through a combined clamp and corporation stop.

Service connections can be made on any type of main under pressure by drilling through a corporation stop and service clamp.

Lateral branch main connections can be made by drilling through a tapping sleeve and valve with a drilling machine.

New gate valves can be added to existing systems by drilling the main through an inserting valve assembly and inserting the gate valve, all under pressure and without loss of water.

Fig. 3.40   Drilling and tapping machines
(Permission of Mueller Co. Copyright February 1971, Mueller Co.)

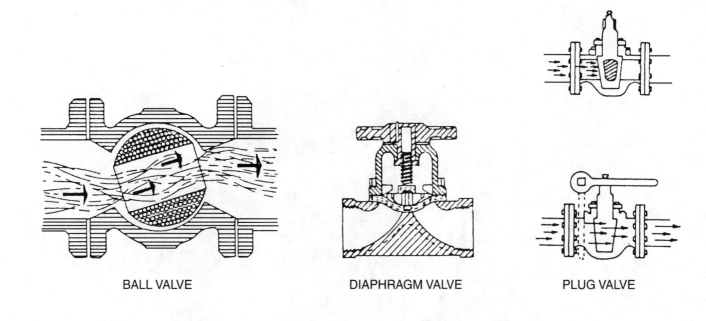

BALL VALVE   DIAPHRAGM VALVE   PLUG VALVE

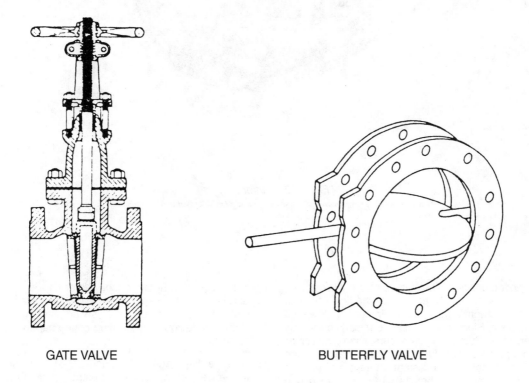

GATE VALVE   BUTTERFLY VALVE

Fig. 3.41   Valve types

(Reprinted from *WATER DISTRIBUTION OPERATOR TRAINING HANDBOOK*, by
permission. Copyright 1976, the American Water Works Association)

Fig. 3.42   Butterfly valve
(Permission of American Valve and Hydrant, A Division of American Cast Iron Pipe Company)

open to a fully closed position. The closure element is a sliding flat metal disc (or discs) slightly larger than the flow opening. The disc is moved at right angles to the direction of flow by a screw-operated stem. Since the disc can be raised completely out of the water, little reduction in flow occurs when gate valves are installed in a pipeline. The surface of a disc may be made of bronze or other corrosion-resistant materials. In some cases, the entire disc is made of bronze. Cutaway views of one gate valve model are shown in Figures 3.44 and 3.45.

Globe valves are very efficient in either flow or pressure regulation. They are used almost exclusively on pipes four inches (100 mm) in diameter or smaller, although they may be used for flow control in larger pipes. A section through a globe valve is shown in Figure 3.46. A tapered or flat disc made of rubber or leather is used. The disc is raised or lowered onto a seat as

in the common home faucet. Water flow is stopped when the disc contacts the seat. A needle design, which is used in some globe valves, provides precise adjustments of small flows. With this design, a protruding cone-shaped piece restricts or stops the flow.

In the diaphragm valve (Figure 3.41), a flexible piece inside the valve's body can be adjusted up or down using an attached stem. When the piece is depressed, the flow of water can be stopped or regulated. The piece is made of flexible material, such as rubber or leather, and is attached to the perimeter (outside edge) of the valve.

Summary information on the gate, butterfly, globe, ball, and plug valves is given in Table 3.2. The common size range, use, and main advantages and limitations are shown.

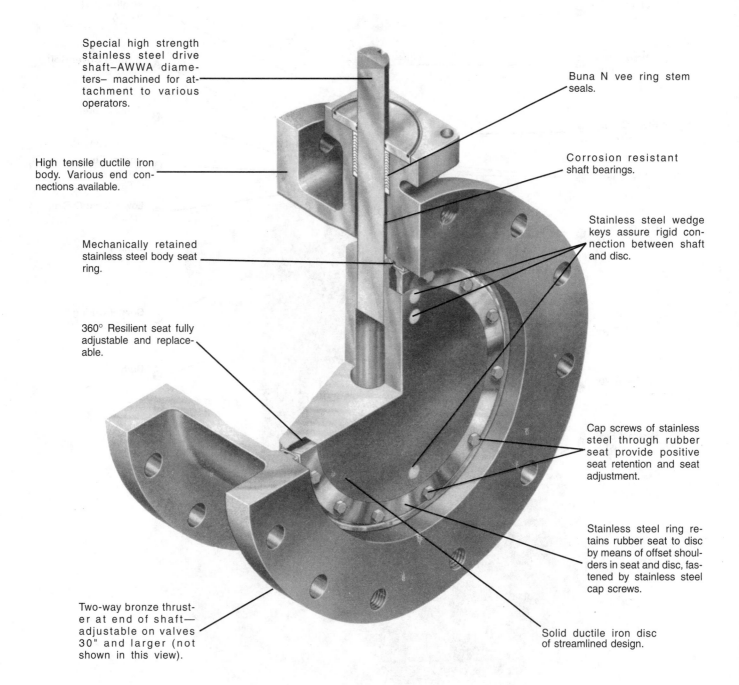

Special high strength stainless steel drive shaft—AWWA diameters— machined for attachment to various operators.

High tensile ductile iron body. Various end connections available.

Mechanically retained stainless steel body seat ring.

360° Resilient seat fully adjustable and replaceable.

Two-way bronze thruster at end of shaft— adjustable on valves 30" and larger (not shown in this view).

Buna N vee ring stem seals.

Corrosion resistant shaft bearings.

Stainless steel wedge keys assure rigid connection between shaft and disc.

Cap screws of stainless steel through rubber seat provide positive seat retention and seat adjustment.

Stainless steel ring retains rubber seat to disc by means of offset shoulders in seat and disc, fastened by stainless steel cap screws.

Solid ductile iron disc of streamlined design.

Fig. 3.43   Butterfly valve sectional view

(Permission of American Valve and Hydrant, A Division of American Cast Iron Pipe Company)

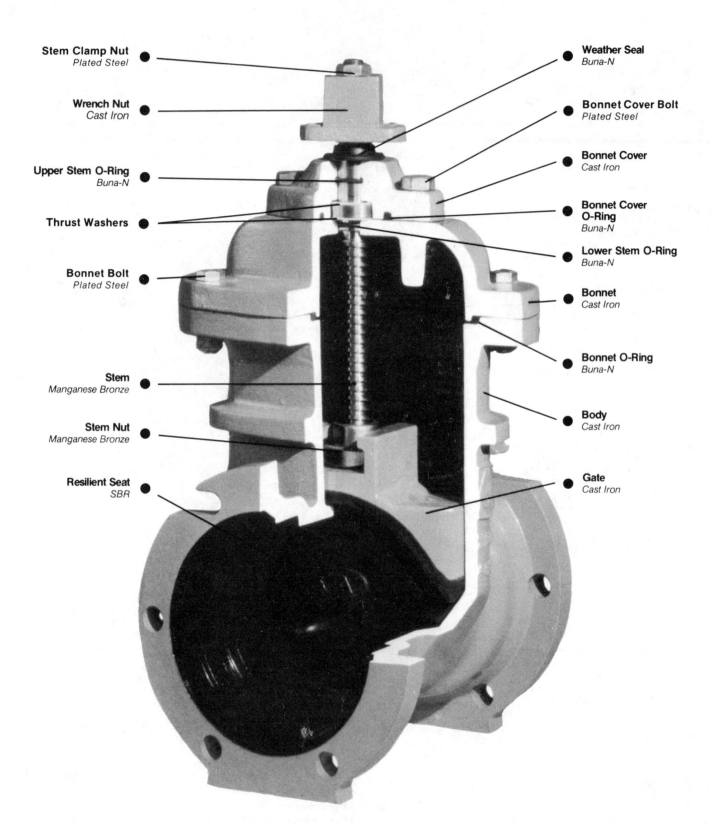

Stem Clamp Nut
Plated Steel

Wrench Nut
Cast Iron

Upper Stem O-Ring
Buna-N

Thrust Washers

Bonnet Bolt
Plated Steel

Stem
Manganese Bronze

Stem Nut
Manganese Bronze

Resilient Seat
SBR

Weather Seal
Buna-N

Bonnet Cover Bolt
Plated Steel

Bonnet Cover
Cast Iron

Bonnet Cover O-Ring
Buna-N

Lower Stem O-Ring
Buna-N

Bonnet
Cast Iron

Bonnet O-Ring
Buna-N

Body
Cast Iron

Gate
Cast Iron

Fig. 3.44   Gate valve parts
(Permission of American Valve and Hydrant, A Division of American Cast Iron Pipe Company)

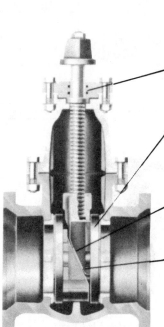

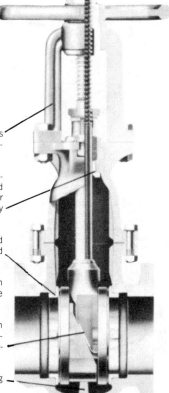

O Ring type stem seal with double O-Rings.

Upper wedge carries discs opposite their seats before wedges spread. Wedging pressure is released from backs of discs before they start to rise.

Radius faced upper wedge and transversely beveled faces on wedges provide horizontal and vertical equalization of wedging pressure on backs of discs.

No links or auxiliary means are necessary to hold parts in position.

One-piece yoke provides strength and rigidity, and assures alignment.

Bronze bonnet bushing provides seat for stem collar and permits repacking under pressure when valve is fully open.

Bronze seat rings screwed into body can be replaced while valve is in line.

Fully revolving discs seat in different positions each time valve is operated.

Nickel alloy insert faces in upper wedge prevent corrosion between wedging surfaces.

Bridge for positive wedging action upon contact.

NON-RISING STEM VALVE (NRS)

OUTSIDE SCREW AND YOKE VALVE (OSY)

Extra wide disc and seat ring faces provide large seating area.

Wedges and discs cannot be assembled incorrectly.

No links or auxiliary means are necessary to hold parts in position.

All working parts are perfectly plain with no pockets to collect sediment or prevent free and easy movement.

Discs are suspended by their center trunnions.

DISCS AND WEDGES—(Shown Assembled)

DISCS AND WEDGES—(Shown Separated)

SIMPLICITY AND RUGGEDNESS OF DESIGN ASSURES TIGHT CLOSING, LONG LIFE, AND EASY OPERATION

Fig. 3.45   Gate valve sectional views

(Permission of American Valve and Hydrant, A Division of American Cast Iron Pipe Company

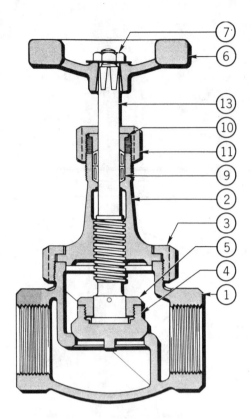

NO.	DESCRIPTION	MATERIAL	ASTM SPEC.
1	BODY	BRONZE	B-61
2	BONNET	BRONZE	B-61
3	BONNET RING	BRONZE	B-61
* 4	DISC	COPPER-NICKEL ALLOY (¾-3)	B-584 Alloy 976
5	DISC NUT	BRASS (¾-1)	B-16
		BRONZE (1¼-3)	B-62
6	HANDWHEEL	MALL. IRON	A-197
7	HANDWHEEL NUT	STEEL-ZINC PLATED	
9	PACKING	TEFLON® IMPREGNATED ASBESTOS	
10	PKG. GLAND	BRASS	
11	PKG. NUT	BRONZE	B-62
13	STEM	COPPER-SILICON ALLOY	B-371 Alloy 694

*Sizes 1/4-1/2 Integral with Stem

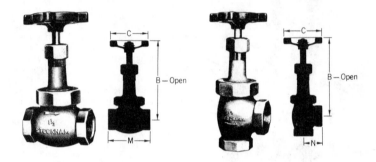

Globe Valve   Angle Valve

Fig. 3.46   Globe and angle valves
(Permission of Stockham Valves and Fittings)

TABLE 3.2   VALVE SUMMARY [a]

Valve Type	Common Size Range, in [b]	Particularly Adapted to	Main Advantages	Main Limitations
Gate valve	½-24	isolation service in pipe grids	low cost in small sizes; low friction loss; fairly good service life; ease of installation	high cost in large sizes; large valves are quite heavy; poor for throttling; should not be used where frequent operation is necessary
Butterfly valve	12-60	isolation and automatic control	low cost in larger sizes for normal service pressures; some types have very short lengths; ease of operation	higher friction loss than gate valve; must use care in selecting valve, particularly for line-break flow conditions; may cause problems when relining pipe; often leaks because of seat damage
Globe valve	½-24	isolation in small sizes, flow control in large sizes, pressure control	simple construction; dependable; can be used for throttling; good for pressure control; sediment or material unlikely to prevent complete closing	high friction loss; very heavy and expensive in large sizes
Ball and plug valves	¾-2	isolation	dependable; very low friction loss; slow shutoff characteristic minimizes closing surges; ease of operation	expensive; very heavy

[a] *WATER DISTRIBUTION OPERATOR TRAINING HANDBOOK.* Obtain from American Water Works Association (AWWA), Bookstore, 6666 West Quincy Avenue, Denver, CO 80235. Order No. 20428. ISBN 1-58321-014-8. Price to members, $52.00; nonmembers, $76.00; price includes cost of shipping and handling.

[b] Multiply inches x 2.5 to obtain centimeters.

QUESTIONS

Write your answers in a notebook and then compare your answers with those on page 155.

3.67A   List five types of valves.

3.67B   What is the main application of gate valves in a water distribution system?

3.67C   What are the two main uses of globe valves?

PRESSURE REGULATION. Valves that control water pressure operate by restricting flows (Figure 3.47). They are used to deliver water from a high-pressure system to a low-pressure system. The pressure downstream from the valve regulates the amount of flow. Usually, these valves are of the globe design and have a spring-loaded diaphragm that sets the size of the opening. An increase in the downstream pressure increases the pressure exerted against the diaphragm; as the spring is compressed it moves the valve element toward the seal, thereby decreasing the flow. When the downstream pressure decreases, the spring will open the valve element and permit more flow.

PRESSURE REDUCING AND PRESSURE SUSTAINING. For these functions, valves can be used separately or in combination. One valve can maintain a constant downstream pressure regardless of fluctuating demand while the other valve will hold the pressure at a predetermined minimum pressure. These valves are usually of a globe design controlled by a diaphragm with the diaphragm assembly being the only moving part. A resilient (material that springs back into shape) disc is held by a disc retainer and forms a tight seal with the valve seat when pressure is applied above the diaphragm flow. The open, closed, and in-between positions are controlled by small pilot valves and pressures sensed upstream and downstream of the valve. By sensing differential pressures across an orifice plate, the hydraulically operated disc globe valve can become an automatic flow-control valve.

DESCRIPTION

The Clayton Powertrol is a globe type valve, containing two operating chambers sealed from each other by a flexible, reinforced diaphragm. The lower chamber, or power chamber is also sealed from line pressure. Standard operation arrangement requires a four-way pilot control.

OPERATION

When operating pressure is applied above the diaphragm and removed from below, the valve closes tight. When these conditions are reversed, the valve opens wide. Rate of opening or closing is controlled by throttling the exhaust pressure. A locked position for any desired degree of opening can be accomplished.

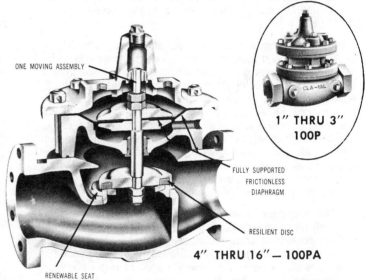

TIGHT CLOSING OPERATION

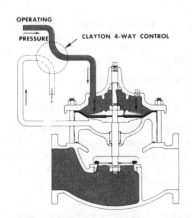

Valve closes drip-tight when operating pressure is applied above diaphragm and removed from below.

FULL OPEN OPERATION

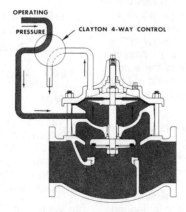

Valve opens wide when operating pressure is removed from chamber above diaphragm and applied to chamber below diaphragm. Flow in either direction is permitted through valve.

THROTTLING ACTION

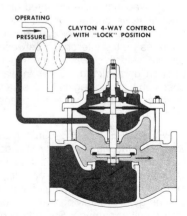

Valve holds any intermediate position when operating pressure is trapped above and below the diaphragm. Any controlling pilot valve can be used which provides four way action with a "lock" position in which no flow pilot occurs.

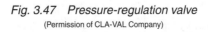

Fig. 3.47   Pressure-regulation valve
(Permission of CLA-VAL Company)

AIR AND VACUUM RELIEF. Both of these functions are usually combined in one valve unit. Air can cause serious pipeline problems. Air may get into the water through pumps, packing glands, leaky joints, or it may already be in the water as it enters the system. Once in the system air tends to collect at high points in the lines. These air pockets can increase the resistance to flow by 10 to 15 percent, and it is even possible for an air-lock condition to occur and stop the flow of water completely. Vacuum conditions sometimes develop when mains are being drained and can cause the collapse of pipelines, even of large steel pipes. Air relief valves are most effective when installed at high points in the line. A very simple type of air valve has no moving parts except a float, or the float may be connected to a hinged rubber stopper that seals the outlet (Figure 3.48). Normally, the outlet is sealed by either the float itself or the stopper. If air accumulates, the water level goes down and the float or stopper drops away from the outlet opening allowing the air to escape. After the air is released, the water level rises and the float or stopper reseals the outlet. These valves can combine three functions. They can allow large quantities of air to escape during the filling of a pipeline, permit air to enter a pipeline that is being drained, and allow *ENTRAINED*[25] air to escape while a line is operating under pressure.

Figure 3.48 shows various types of air valves and typical installations. Air and vacuum valves (Drawing 1) have a large orifice which permits great amounts of air to escape from a system while it is being filled. Air release valves (Drawing 2) have much smaller orifices than the air and vacuum valves. Air release valves release the small pockets of air which gather at the high points of a system after it is filled and under pressure. Combination air valves (Drawing 3) combine the features of both the air and vacuum valve and the air release valve. Drawing 4 shows air release valves installed on centrifugal pumps while Drawing 5 shows an air and vacuum valve on a vertical turbine pump.

Slow-closing air valves (Figure 3.49) are designed to eliminate critical shock conditions in those installations where operating conditions cause a regular air valve to slam shut. When pump discharge velocities exceed 10 feet per second (3 m/sec), a slow-closing air valve will prevent water hammer, muffle noise, and contain spillage. These valves consist of an air and vacuum valve mounted on top of a surge check unit. The air valve operates in the normal fashion allowing air to escape freely at any velocity (Figure A). The surge check unit operates on the interphase between the kinetic energy in the relative velocity flows of air and water so that air passes through unrestricted, but when water rushes into the surge check unit, a disc begins to close and reduces the rate of flow of water into the air valve by means of throttling holes in the disc (Figure B).

PRESSURE RELIEF. Surge pressures occur in systems whenever the water velocity is suddenly changed, either decreased or increased. Typical causes of sudden changes of velocity are quick opening or quick closing of a line valve or sudden starting or stopping of a pump. If this occurs, the built-up energy of the flowing water produces a high pressure that may result in damage. Some type of surge control is needed and although surge tanks are sometimes used, the simplest type of control is the use of pressure relief valves. These valves respond to pressure variations at their inlets. They open very rapidly at an increase in pressure above the set point of the control and discharge the air to the atmosphere.

They are most commonly installed on the side outlet of a tee at a specific point in a system.

VALVE PROTECTION. Air, vacuum, and pressure relief valves must be protected against possible contamination. Do not install air and vacuum relief valves in pits unless you are able to protect them against flooding. Screen air inlets and outlets to prevent insects and rodents from entering the valve. Pressure relief valves must never be directly connected to sewer or storm drains.

OTHER VALVE APPLICATIONS. Valves are also available for the uses described below.

Altitude valves are used mainly on supply lines to elevated tanks or standpipes. They close automatically when the tank is full and open when the pressure on the inlet side is less than that on the tank side of the valve. They control the high water level and prevent overflow.

Blowoff and drain valves are used primarily for blowing off accumulated sediment from low spots in the line and for dewatering lines or reservoirs for repairs or inspections.

Float-controlled valves control the level of water in storage tanks and may activate remote control equipment.

Backflow prevention valves will stop the flow of an unapproved and possibly contaminated supply into the domestic water system. These valves are discussed further in Section 3.8, "Customer Service."

Check valves are used in the discharge side of pumps to prevent backflow. For additional information on check valves, see Chapter 3, "Wells," Section 3.210, "Check Valves," in *SMALL WATER SYSTEM OPERATION AND MAINTENANCE*, and Chapter 18, "Maintenance," in *WATER TREATMENT PLANT OPERATION*, Volume II, in this series of manuals.

QUESTIONS

Write your answers in a notebook and then compare your answers with those on page 155.

3.67D   What type of valve design is used in pressure-regulating valves?

3.67E   What problems can be created when air gets into water?

3.67F   Air relief valves can serve what three functions?

3.67G   What is the purpose of blowoff valves?

3.6701 VALVE BOXES AND VAULTS (FIGURE 3.50)

Valve boxes provide access to buried valves. Their construction is similar to that of residential curb boxes. A valve key turns the valve stem from the top of the structure. Valve boxes are often made in two or more sections which telescope to give an adjustable length allowing the top of the box to be set at ground level.

Vaults are normally used for larger valves where access is needed for maintenance operations. Space is usually provided so that at least one person will be able to work on the valve or equipment. Valve vaults should be properly vented and drained. Inadequate draining may result in a flooded vault and a backsiphonage hazard. Special safety considerations must

---

[25] *Entrain.   To trap bubbles in water either mechanically through turbulence or chemically through a reaction.*

1. Air and Vacuum Valve

2. Air Release Valve

3. Combination Air Valves

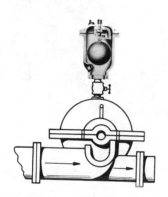

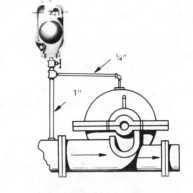

4. Air Release Valves for Centrifugal Pumps

5. Air Valve for Vertical Turbine Pumps

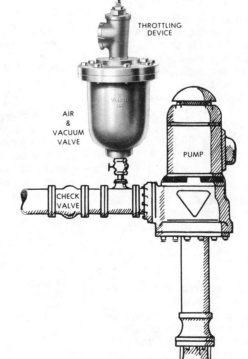

Fig. 3.48 Air valves

(Permission of APCO Valve and Primer Corporation)

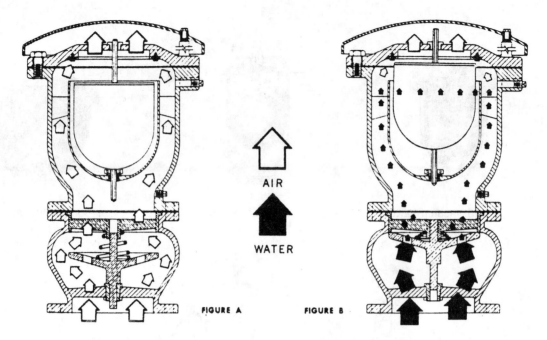

Fig. 3.49   Slow-closing air valve
(Permission of APCO Valve and Primer Corporation)

be taken into account with the deep vaults. Pump air into the vault continuously to provide adequate ventilation and use an atmospheric test meter to check for sufficient oxygen and harmful gases (combustible and toxic).

3.6702   DISTRIBUTION SYSTEM VALVE LAYOUT

Any water piping system should have valves in enough locations that relatively small parts of the system can be shut off for repairs or replacement without putting large areas out of service. Gate valves are normally used for this purpose. In general, provide enough valves so that repair work will not shut down a feeder main or a length of pipe greater than 1,000 feet (300 m). Some utilities use a maximum separation distance of 500 feet (150 m). A typical valving arrangement is shown in Figure 3.51. In general, valves should be located on all branches from feeder mains and between mains and fire hydrants. Three valves should be used at crosses and two at tees.

Valves should be located uniformly throughout the system in relation to street intersections (see Figure 3.51). Some utilities standardize the valve locations by putting them at the same corners of each intersection. This procedure makes it easier to find the valves in an emergency.

3.6703   VALVE OPERATION

Valves may be operated manually or by actuators. To operate a valve manually, use a wrench or turn a hand wheel on the valve. An actuator is a device designed to power-operate the closure element of a valve. Actuators can be assisted by water pressure or electrical power. In valves controlled by water pressures, the valve stem passes through a packing ring and is joined to a piston in a closed chamber. Water under pressure enters the chamber below the piston and any water in the chamber above the piston is permitted to escape. This lifts the piston and opens the valve. To close the valve, the pressure and release procedures are reversed. Solenoids and electric motors geared to the valve mechanism are also used to open and close valves.

QUESTIONS

Write your answers in a notebook and then compare your answers with those on page 155.

3.67H   Why are valve boxes installed?

3.67I   What problems could develop if a deep valve vault becomes flooded?

3.67J   How many valves should be located at pipe crosses?

3.67K   How are valves opened or closed?

3.671   Fittings

Pipe fittings are used for a number of different purposes in a water system. They connect pipes of the same size or different sizes, change the direction of flow, or stop the flow. Fittings can be obtained in most materials and sizes. Some can connect pipes made of dissimilar materials and others can connect pipes with different types of ends. Table 3.3 summarizes the various types of pipe fittings available.

TABLE 3.3   VARIOUS TYPES OF PIPE FITTINGS

CONNECTORS	CHANGE FLOW DIRECTION
Clamps	Elbows
Sleeves	Tees
Couplings	Bends
Adapters	
Expansion Joints	STOP FLOW
	Caps
	Plugs
REDUCERS	Blind Flanges
Reducing Flanges	
Tees	TAPPING
Crosses	Tapping Sleeves
Y-Laterals	Tapping Crosses
	Tapping Clamps

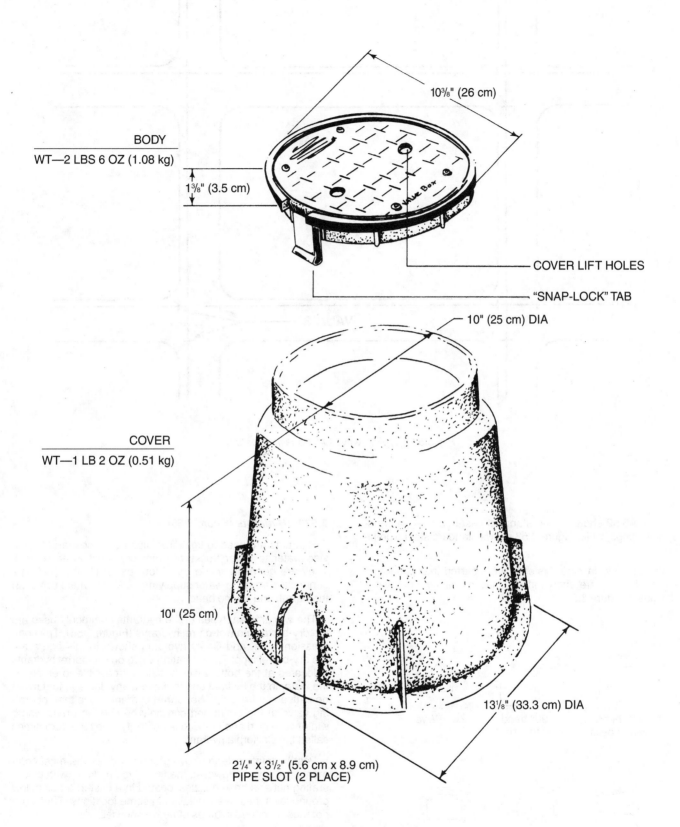

BODY
WT—2 LBS 6 OZ (1.08 kg)

10⅜" (26 cm)

1⅜" (3.5 cm)

COVER LIFT HOLES

"SNAP-LOCK" TAB

10" (25 cm) DIA

COVER
WT—1 LB 2 OZ (0.51 kg)

10" (25 cm)

13⅛" (33.3 cm) DIA

2¼" x 3½" (5.6 cm x 8.9 cm)
PIPE SLOT (2 PLACE)

Fig. 3.50   Valve box
(Permission of Hefner Plastics, Inc.)

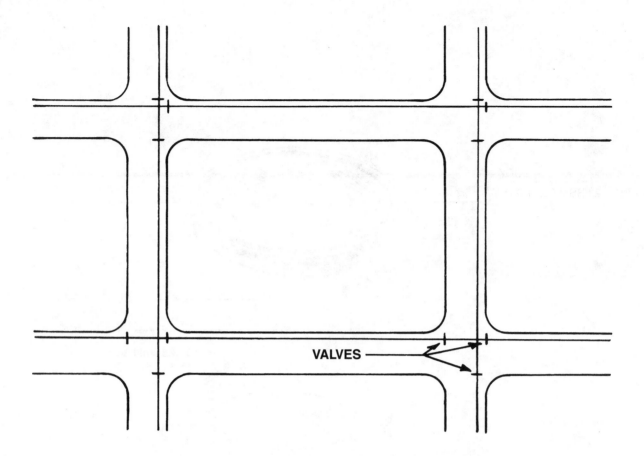

Fig. 3.51   *Typical valve locations in streets*

Figure 3.52 shows some standard cast-iron pipe bell-and-spigot fittings, while Figure 3.53 shows standard steel pipe fittings.

Valves used in pipe systems are inserted into couplings, some of which are shown in combination with various valve designs in Figure 3.54.

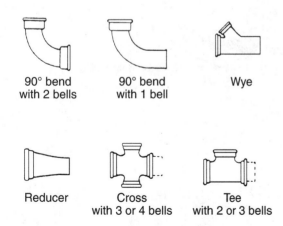

Fig. 3.52   *Standard bell-and-spigot fittings*

(Reprinted from *WATER-RESOURCES ENGINEERING*, Second Edition, by Linsley and Franzini, by permission. Copyright 1972, McGraw-Hill Book Company)

3.672 *Hydrants* (Figure 3.55)

Hydrants are used to fight fires, flush pipelines, and to provide water to water trucks or other construction equipment. They are generally made up of four parts: (1) the inlet pipe connection from the water supply main, (2) the main valve, (3) the barrel, and (4) the head.

The two different types of fire hydrants commonly used are the dry-barrel and wet-barrel hydrants (Figure 3.56). The compression, gate, and Corey hydrants shown in this figure are the dry-barrel type. The operating valve on dry-barrel hydrants is located at the bottom which makes it possible to empty all the water in the hydrant and to prevent any damage that might be caused by freezing. Wet-barrel hydrants have their operating valve at the outlet and can only be used in areas where winters are mild and freezing is not likely. They are sometimes called the California hydrant.

Post hydrants extend above ground and are the most commonly used type. However, the flush-type hydrant, with its operating nut and hose nozzles located in a cast-iron box below ground level, has been installed in some locations. These are not recommended in areas of heavy snowfall.

There are a number of other fire hydrant design features that are of interest. Two-, three-, and four-nozzle arrangements are used, with the two-nozzle, one-pumper connection arrangement being the most common. Nozzles are usually $2\frac{1}{2}$ inches (63 mm) in diameter with pumper connections having a diameter of $4\frac{1}{2}$ inches (11.3 mm). Protective caps over the

FITTINGS

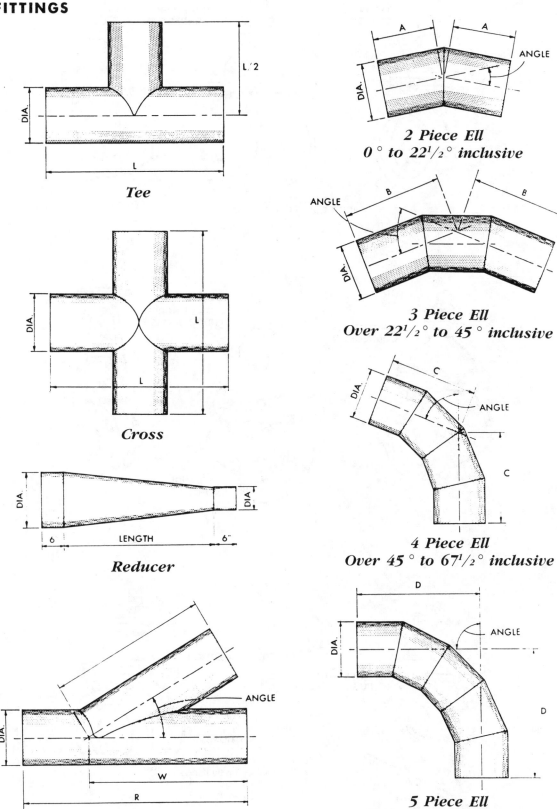

Tee

Cross

Reducer

Wye

2 Piece Ell
0 ° to 22¹/₂° inclusive

3 Piece Ell
Over 22¹/₂° to 45 ° inclusive

4 Piece Ell
Over 45 ° to 67¹/₂° inclusive

5 Piece Ell
Over 67¹/₂° to 90 ° inclusive

Fig. 3.53   Standard steel pipe fittings

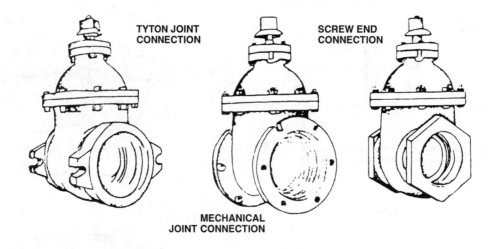

TYTON JOINT
CONNECTION

SCREW END
CONNECTION

MECHANICAL
JOINT CONNECTION

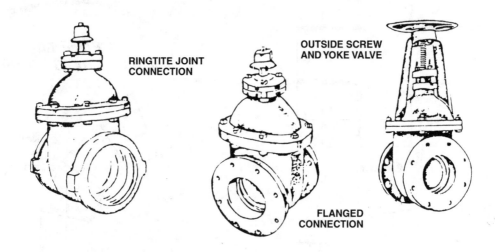

RINGTITE JOINT
CONNECTION

OUTSIDE SCREW
AND YOKE VALVE

FLANGED
CONNECTION

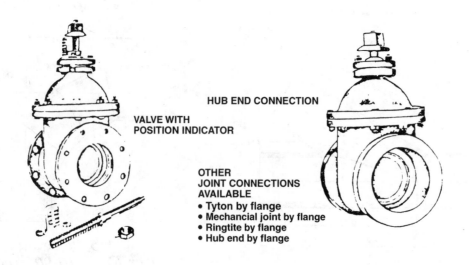

VALVE WITH
POSITION INDICATOR

HUB END CONNECTION

OTHER
JOINT CONNECTIONS
AVAILABLE
• Tyton by flange
• Mechancial joint by flange
• Ringtite by flange
• Hub end by flange

Fig. 3.54   Valve couplings

(Reprinted from *WATER DISTRIBUTION OPERATOR TRAINING HANDBOOK*, by
permission. Copyright 1976, the American Water Works Association)

① OPERATING NUT—cast one-piece bronze operating nut directly actuates hydrant rod. Extra-long, close tolerance threads assure long life and easy operation. Grease groove insures complete lubrication of threads and thrust bearing. Lead of the operating thread permits slow closing of the hydrant valve, eliminating water hammer.

② O-RINGS—seal lubrication chamber, eliminate stuffing box, assure dry top hydrant, reduce friction, prevent water from reaching the operating mechanism.

③ HYDRANT ROD—furnished in two sections of high tensile steel. Upper section has bronze sleeve where it passes through O-ring and has bronze stop nut to limit travel of rod. Lower section has a cadmium-plated spring with heavy bituminous coating to insure positive drain closure. Both sections are connected by cast iron coupling and pins. Rod directly operates hydrant valve and drains without need for auxiliary connections.

④ HYDRANT VALVE—consists of a hydrant valve top, hydrant valve bottom and special hydrant valve rubber of conical shape. Hydrant valve bottom is threaded and screwed to valve rod, protecting rod threads from corrosion. Thread locking compound prevents loosening in service. Hydrant valve seats against the bronze hydrant seat.

⑤ HYDRANT SEAT—made of special bronze, with accurately machined seat for hydrant valve and tapered drain ports.

⑥ HYDRANT SEAT DRAIN RING—securely held between barrel and base flanges, provides bronze-to-bronze threaded connection for hydrant seat. Serves as non-corrosive multiport drain channel.

⑦ DRAIN LEVER—rugged special bronze three-port drain performs dual function as carrier for drain lever washers and as wrench to remove working parts.

⑧ SOCKET HEAD PIPE PLUG—sealed grease chamber. No greasing required as long as rod is intact. Grease fitting can replace plug for greasing.

⑨ COVER—cast iron cover is quickly removed by unbolting two bolts. The word "open" and an arrow show direction to turn the operating nut.

⑩ HOUSING AND HOUSING COVER—retains operating nut and thrust bearing. Rugged construction supports pressure and operating forces.

⑪ THRUST BEARING—takes upward thrust when opening and closing hydrant valve and reduces operating torque. Bearing is lubricated through top of operating nut.

⑫ BARREL—large high-strength cast iron barrel is flanged at ground line with 8 bolts, permitting installation in eight different positions. Large barrel delivers water to nozzles with no perceptible pressure loss. A ductile iron lower barrel and flanges provide extra ruggedness.

⑬ NOZZLES—Bronze nozzles are streamlined and have micrometer checked threads.

⑭ BOLTS AND NUTS—all bolts and nuts are cadmium plated.

⑮ BASE—large, spherical-shaped base has no projections or cavities to obstruct flow or collect sediment.

All American B-62-B barrel castings are painted inside and outside with a special primer. Section below ground level is given two coats of special black asphaltum paint which resists corrosion from moisture and soil. Exterior surface above ground is painted with special primer and high-grade enamel to customer's color requirement.

American B-62-B hydrants can be furnished as a TRAFFIC type by the use of frangible bolts at the flange joint above the ground line. Should the hydrant be damaged by traffic, new bolts, rod coupling and gasket would be the only parts necessary to place the hydrant back into service.

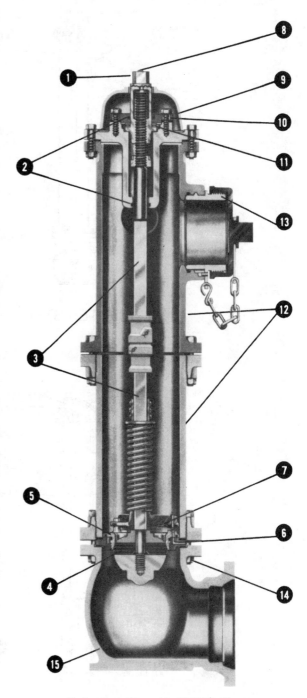

Fully complies with AWWA C-502.

Fig. 3.55   Typical fire hydrant

(Permission of American Valve and Hydrant, A Division of American Cast Iron Pipe Company)

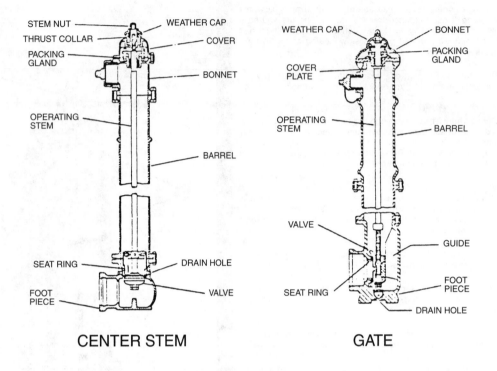

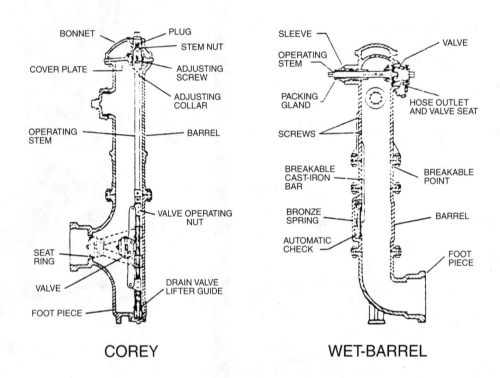

Fig. 3.56 Hydrant types

(Reprinted from *WATER DISTRIBUTION OPERATOR TRAINING HANDBOOK*, by permission. Copyright 1976, the American Water Works Association)

nozzle ends are needed to safeguard the nozzle threads. Hydrant designs generally permit lifting the valve and stem out of the barrel by removing the head. A shutoff valve should be provided on every hydrant lateral. Traffic safety is taken into account in the design of some barrels which have a breakaway point in case the hydrant is struck by a vehicle. If the barrel is broken, the valve stem is also designed to break.

Fire hydrants should not be more than 500 feet (150 m) apart to avoid excessive friction head loss in fire hoses. They are usually located much closer together in high value (business and commercial) districts. Preferably, they should be placed at intersections so they can be used in all directions from the corner.

QUESTIONS

Write your answers in a notebook and then compare your answers with those on page 155.

3.67L   List the uses of fittings.

3.67M   List the uses of hydrants.

3.67N   What are the four major parts of a hydrant?

End of Lesson 4 of 5 on DISTRIBUTION SYSTEM FACILITIES

Please answer the discussion and review questions next.

DISCUSSION AND REVIEW QUESTIONS

Chapter 3.   DISTRIBUTION SYSTEM FACILITIES

(Lesson 4 of 5 Lessons)

Write the answers to these questions in your notebook before continuing. The question numbering continues from Lesson 3.

18.   Why are wet taps preferred over dry taps on water distribution lines?

19.   List the purposes valves may be used for in a water distribution system.

20.   Surge pressures in water systems can be caused by what activities?

21.   How can air, vacuum, and pressure relief valves be protected against possible contamination?

22.   What safety precautions should be taken before entering a deep valve vault?

CHAPTER 3.   DISTRIBUTION SYSTEM FACILITIES

(Lesson 5 of 5 Lessons)

3.7   METERS

3.70   Purpose of Meters

The primary function of a meter is to measure and display the amount of water passing through it. A metered water system is one in which meters are used at all strategic points: on main supply lines, pumping stations, reservoir outlets, connections to other utility systems, and at each customer's service. Metering provides many benefits. The customer can be billed for the exact amount of water used, the amount of water produced can be determined, losses of water can be detected by comparing service meter and hydrant use meter readings with production meter readings, and capacities of pipelines can be determined. By determining costs, metering prevents waste through excessive use of water and thereby reduces water consumption. Meters are also used to provide accurate blending of waters of different quality so that the mix in the reservoirs is of the same constant quality. The information obtained from metering can show how efficiently the water utility is operating and can provide for system control.

3.71   Meter Selection

In selecting meters, look for the ability to measure and register your anticipated flow levels, ability to meet required capacity with minimum head loss, durability, ruggedness, precision of workmanship, ease of repair, availability of spare parts, freedom from irritating noise, a reasonable price, and a manufacturer with a good reputation. When selecting a brand of water meter for your utility, be sure to consider the availability of replacement meters and spare parts. Also try to determine the delivery time for the manufacturer to respond to an order for a needed part. Check with other utilities to verify that the manufacturer has the kind of warranty you want and will stand behind its products.

Sufficient accuracy is needed so that small meters do not register less than 98.5 percent nor more than 101.5 percent of the water passing through them. Large meters should accurately measure between 97 and 103 percent. Certain large meters are accurate within 1 percent. Generally, meters of 1 inch (25 mm) and smaller should have a head loss not exceeding 15 psi (103 kPa or 1.0 kg/sq cm), while the head loss of larger meters should not exceed 20 psi (138 kPa or 1.4 kg/sq cm).

QUESTIONS

Write your answers in a notebook and then compare your answers with those on page 155.

3.7A   At what strategic points should meters be used?

3.7B   How does metering reduce water consumption?

3.7C   List the important qualities you would look for when selecting a meter.

3.7D   Calculate the head loss through a meter in feet if the pressure loss is 15 psi (1 psi = 2.31 ft).

3.72   Meter Types

Basically, meters can be classified as small-flow meters, large-flow meters, and combination large/small-flow meters as shown below.

I. Small-flow meters (displacement type)

  A. Nutating-disc
  B. Piston

II. Large-flow meters (velocity type)

  A. Turbine
  B. Propeller
  C. Venturi
  D. Electronic
  E. Insertion

III. Combination meters (compound type)

3.720   Displacement Meters

Displacement-type meters are small-diameter meters (up to 2 inches (50 mm)) commonly used for customer services. For commercial services, however, they may be as large as 6 inches (15 mm) (Figure 3.57). This type of meter measures the flow by registering the number of times the meter chamber, whose volume is known, is filled and emptied. In this chamber, which is usually cylindrical in shape, a piston or disc goes through a certain cycle of motion that corresponds to a single filling and emptying of the chamber. This movement is transferred to a register.

The advantages of displacement meters are that they can measure wide variations of flow rate within their rated capacity and, in sizes up to 2 inches (50 mm), are accurate in registering low flows. The principal limitations are that in sizes greater than 2 inches (50 mm), the sensitivity at the lower rate of flow decreases so much that it is advisable to use another type of meter. Their capacity is limited and at high flows there is a high head loss. Also, large quantities of foreign matter or corrosion will either stop the meter entirely or cause its accuracy to decrease because of friction.

The different ways in which the disc or piston in the displacement meter moves have been used to categorize the various types of displacement meters. The two types normally encountered are the nutating-disc and the piston meters.

The nutating-disc meter is the most commonly used meter on small-diameter domestic services (Figures 3.57 and 3.58).

Fig. 3.57   Nutating-disc (nodding) water meters
(Permission of Neptune Water Meter Company)

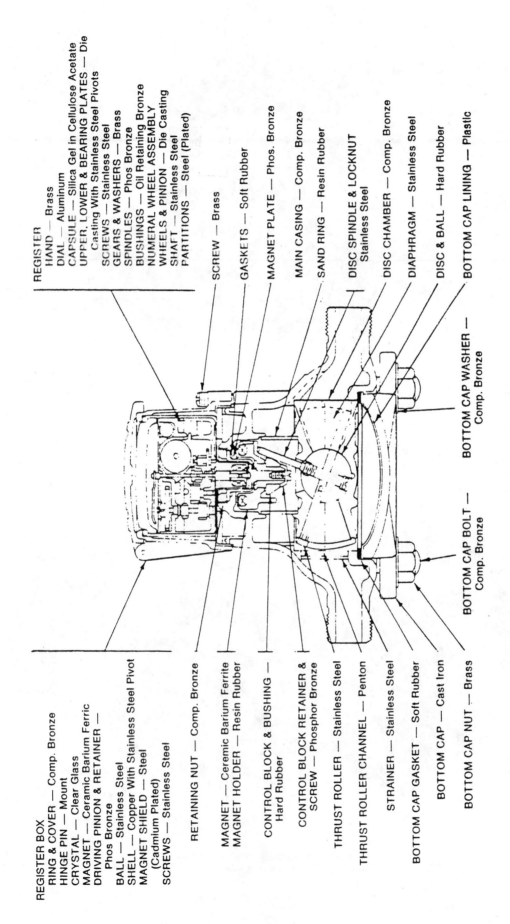

REGISTER
HAND — Brass
DIAL — Aluminum
CAPSULE — Silica Gel in Cellulose Acetate
UPPER, LOWER & BEARING PLATES — Die
Casting With Stainless Steel Pivots
SCREWS — Stainless Steel
GEARS & WASHERS — Brass
SPINDLES — Phos Bronze
BUSHINGS — Oil Retaining Bronze
NUMERAL WHEEL ASSEMBLY
WHEELS & PINION — Die Casting
SHAFT — Stainless Steel
PARTITIONS — Steel (Plated)

SCREW — Brass

GASKETS — Soft Rubber

MAGNET PLATE — Phos. Bronze

MAIN CASING — Comp. Bronze

SAND RING — Resin Rubber

DISC SPINDLE & LOCKNUT
Stainless Steel

DISC CHAMBER — Comp. Bronze

DIAPHRAGM — Stainless Steel

DISC & BALL — Hard Rubber

BOTTOM CAP LINING — Plastic

REGISTER BOX
RING & COVER — Comp. Bronze
HINGE PIN — Mount
CRYSTAL — Clear Glass
MAGNET — Ceramic Barium Ferric
DRIVING PINION & RETAINER —
Phos Bronze
BALL — Stainless Steel
SHELL — Copper With Stainless Steel Pivot
MAGNET SHIELD — Steel
(Cadmium Plated)
SCREWS — Stainless Steel

RETAINING NUT — Comp. Bronze

MAGNET — Ceramic Barium Ferrite
MAGNET HOLDER — Resin Rubber

CONTROL BLOCK & BUSHING —
Hard Rubber

CONTROL BLOCK RETAINER &
SCREW — Phosphor Bronze

THRUST ROLLER — Stainless Steel

THRUST ROLLER CHANNEL — Penton

STRAINER — Stainless Steel

BOTTOM CAP GASKET — Soft Rubber

BOTTOM CAP — Cast Iron

BOTTOM CAP NUT — Brass

BOTTOM CAP WASHER —
Comp. Bronze

BOTTOM CAP BOLT —
Comp. Bronze

Fig. 3.58   Section view of nutating-disc meter
(Reprinted from *WATER DISTRIBUTION OPERATOR TRAINING HANDBOOK*, by permission.
Copyright 1976, the American Water Works Association)

Nutating means nodding, which somewhat describes the action of the disc in the chamber of the meter. The disc may be flat or conical in shape and is made of hard rubber. When water flows, the tilted disc rotates. These meters are made with a bottom plate which breaks away automatically to prevent damage when the water freezes.

In the piston-type meter, water flows into a chamber and displaces a piston and the oscillatory circulating motion is transmitted to a register (Figure 3.59). Piston-type meters have a slightly greater head loss than nutating-disc meters.

3.721 Velocity-Type Meters

These meters are sometimes called current meters. They are found in sizes up to 36 inches (900 mm) and larger and actually measure the velocity of flow past a cross section of known area. Large quantities of water (within meter capacity) can be passed without any damage to their working parts and the meters are rugged and easy to maintain. Their rather low head loss is ideal for a meter that must pass high rates of flow; however, the meters become unreliable at low flow rates. Some low flows will pass through without being registered at all. This kind of meter is very satisfactory for main lines, pumps with continuous high flow rates and irrigation, golf course, industrial, and other high-flow uses.

Velocity meters include turbine (Figures 3.60 and 3.61) and propeller meters (Figure 3.62), as well as the Venturi, insertion type, and most electronic water meters.

In turbine or propeller meters, the rotors or propellers are turned by the flow of the water at a speed proportional to the velocity of water flow. This movement is then transmitted to a register. They are not designed for low flows or stop-and-go operation. Both types of meters are most useful in measuring continuously high flows, and both have low friction loss. The propeller meter may be installed within a section of pipe or it may be saddle mounted.

When main lines must be metered, it is desirable to have a meter that will not interfere with the flow of water if the meter fails. In case of fire, nothing must stop the effective flow of water. One principal type of meter that meets this requirement is the Venturi meter which is shown in Figure 3.63. A Venturi meter consists of an upstream reducer, a short throat piece, and a downstream expansion section which increases the diameter from the throat section to that of the downstream pipe. The amount of water passing through is metered by comparing the pressure at the throat and at a point upstream from the throat. Venturi meters are accurate over a large flow range and cause little friction loss.

An orifice plate can be used as an insertion-type meter. A thin plate with a circular hole in it is installed in the pipeline between a set of flanges (Figure 3.64). Flow is determined by comparing the upstream line pressure with the reduced pressure at the orifice restriction. Orifice plates are quite a bit less expensive than Venturi meters, occupy less space, but have more severe pressure losses and are somewhat less reliable.

QUESTIONS

Write your answers in a notebook and then compare your answers with those on pages 155 and 156.

3.7E   What is the major difference between small-flow meters and large-flow meters?

3.7F   List the two types of displacement meters.

3.7G   List the types of velocity meters.

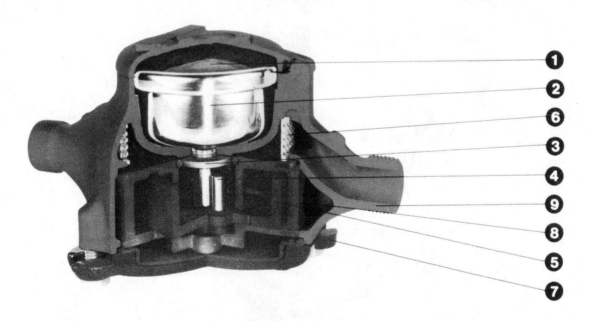

1. Heat-treated glass	4. Magnetic drive	7. Interchangeable bottom plate
2. Hermetically sealed register	5. Oscillating piston and piston roller	8. Measuring chamber
3. Register retainer	6. Cylindrical strainer	9. All cast bronze maincase

Fig. 3.59   Piston-type meter

(Courtesy of Rockwell International Corporation,
Municipal & Utility Division)

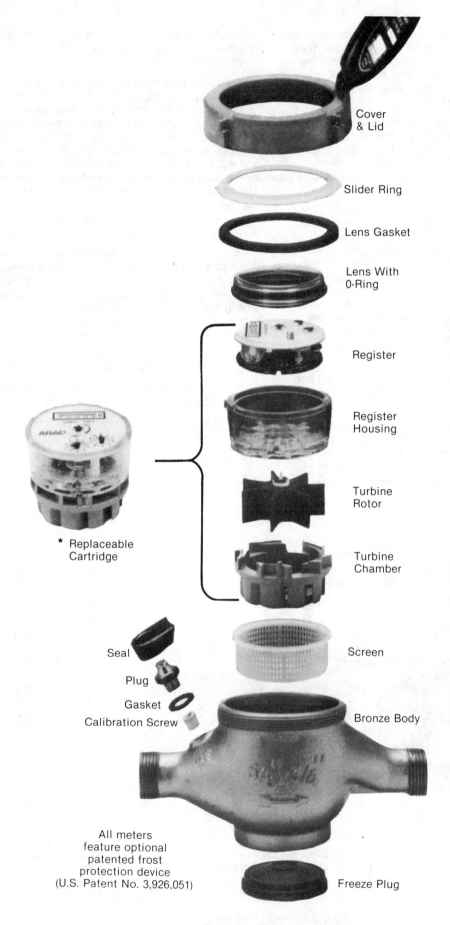

Cover
& Lid

Slider Ring

Lens Gasket

Lens With
0-Ring

Register

Register
Housing

Turbine
Rotor

Turbine
Chamber

Screen

Bronze Body

Freeze Plug

Seal

Plug

Gasket

Calibration Screw

* Replaceable
Cartridge

All meters
feature optional
patented frost
protection device
(U.S. Patent No. 3,926,051)

Fig. 3.60   Turbine water meter
(Permission of Neptune Water Meter Company)

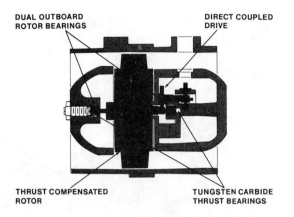

Fig. 3.61   Turbine water meter
(Permission of Western Water Meter)

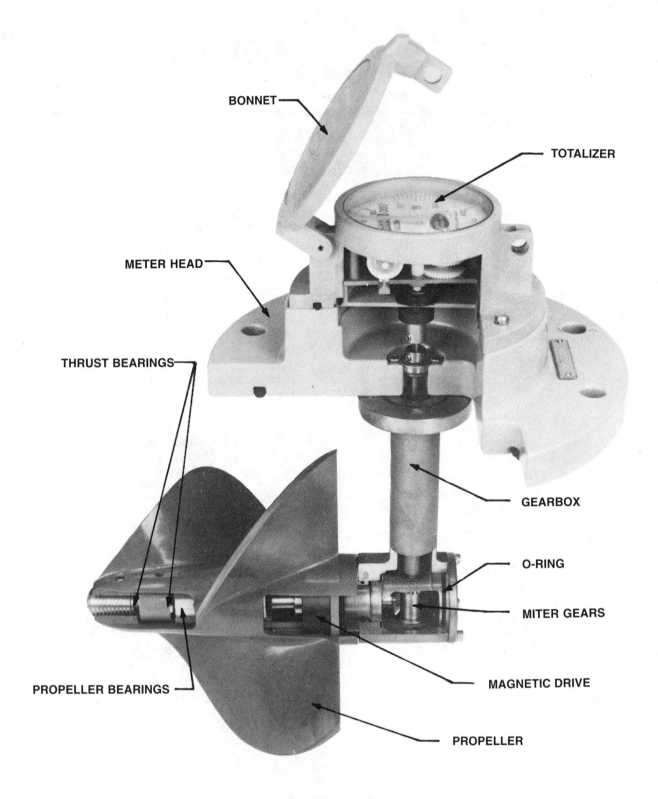

BONNET

TOTALIZER

METER HEAD

THRUST BEARINGS

GEARBOX

O-RING

MITER GEARS

MAGNETIC DRIVE

PROPELLER BEARINGS

PROPELLER

Fig. 3.62   Propeller meter
(Permission of Water Specialties Corporation)

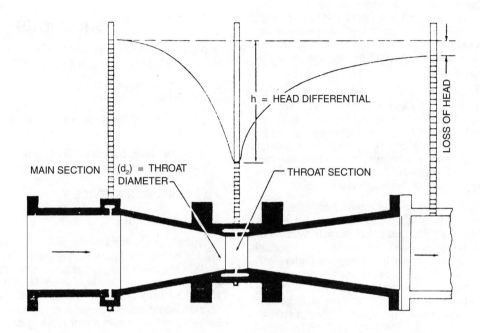

Fig. 3.63   Venturi meter

(Reprinted from *WATER DISTRIBUTION OPERATOR TRAINING HANDBOOK,* by permission.
Copyright 1976, the American Water Works Association)

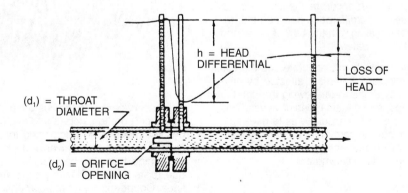

Fig. 3.64   Orifice plate meter

(Reprinted from *WATER DISTRIBUTION OPERATOR TRAINING HANDBOOK,* by permission.
Copyright 1976, the American Water Works Association)

3.722   Compound Meters

When water flow fluctuates widely, a compromise between the low- and high-flow meters is used. This compromise is a combination of the displacement and velocity-type meters and is called a compound meter. The displacement part of the meter records the low flows, and when the head loss through this part of the meter rises to a certain level, a compounding valve is actuated and permits the water to flow through the velocity portion of the meter. Compound meters are shown in Figures 3.65, 3.66, and 3.67.

Compound meters have special advantages in places where there are predominantly low or intermediate flows and only occasional high flows. This would include services to hotels, hospitals, factories, schools, apartment houses, commercial properties, and office buildings. They are best suited for locations with widely varying flows where accurate low-flow measurement is needed. However, where the majority of flows are moderate to high, a turbine meter is generally used.

The main advantage of compound meters, with the separate measuring chambers, is that they accurately measure flows from a fraction of a gallon up to the normal capacity of the pipeline. Even very small flows from leaks and individual faucets can be measured accurately. Compound meters can be subjected to high rates of flow for long time periods and are more rugged than displacement meters. Their limitations are that loss of head from friction is higher than in large-flow meters; during the changeover from low to high and high to low flows their accuracy drops; and they are large, cumbersome, and expensive.

3.723   Electronic Meters

Electronic meters include the magnetic and sonic types of meters. The magnetic meter is often called a "mag meter." Water flowing through a magnetic field induces a small electric current flow which is proportional to the water flow. The electric current produced is measured and mathematically changed to a measure of water flow.

Sonic meters contain sensors which are attached to the sides of a pipe. Sound pulses are sent alternately across the pipe in opposite diagonal directions. The frequency of the sound changes with the velocity of the water through which the sound waves must travel. An accurate measurement of water flow can be made using the difference between the frequency of the sound signal traveling with the flow of water and that traveling against the flow of water.

Electronic meters are generally highly accurate and there is no head loss. However, they are adversely affected by anything that distorts the velocity of the water flowing through the pipe such as elbows, pumps, and certain types of valves. This condition can make the meters so inaccurate as to be practically useless. The problem can be corrected, though, by leaving a distance of at least ten pipe diameters between an upstream obstruction or fitting and the flowmeter.

3.724   Proportional Meters

Another type of meter is the proportional meter in which a certain proportion of the total flow is diverted through a bypass meter and measured (Figure 3.68). The gears of the measuring bypass meter are adjusted to indicate on its register dial the total amount of water passing through the whole unit. The flows in the bypass line and the main pipe are proportional to the ratio of the areas of the bypass line and the main pipe. Therefore, a displacement or turbine meter which measures only the diverted flow can be calibrated to register the full amount of water in the main pipeline. The diversion of a portion of the flow through the bypass meter is accomplished by an orifice plate. This causes sufficient pressure differential to divert a portion of the water through the measuring meter. This type of meter is relatively accurate except for low flows. The meter is principally used to meter fire lines.

QUESTIONS

Write your answers in a notebook and then compare your answers with those on page 156.

3.7H   Under what type of flow conditions are compound meters installed?

3.7I   What are the two major types of electronic flowmeters?

3.7J   Fire lines may use what type of meter to measure flows?

3.8   CUSTOMER SERVICE

3.80   Installation

Typical service line installations are shown in Figure 3.69. In the top half of the figure (a) the installation includes a meter, while in the lower half (b) the meter is installed on the customer's premises or no meter is used. In cold climates the meter should be inside; however, the question of an inside versus an outside location of the meter is not as important as it once was because of the remote meter reading equipment that is available now.

The service line to the customer's system starts at the main with a corporation stop (shutoff valve) (see Figure 3.70). Various materials have been used for the service line including copper, plastic, and galvanized or lined steel, and iron pipe. Lead pipe has also been used, but is no longer acceptable due to its toxicity. Plastic pipe is frequently used for service lines. Copper pipe was popular until its price became too high. The service lines should be flexible. At its other end, the service line connects to another shutoff valve, the curb stop, which is attached to the meter. The meter is usually at the curb or property line. Finally, where a backflow contamination potential exists, backflow prevention devices are installed between the meter and the customer's line. Normally, the utility's responsibility for its service ends at the customer's side of the meter.

Service connections to the water main can be made at the time the main is installed or at any subsequent period. If a new

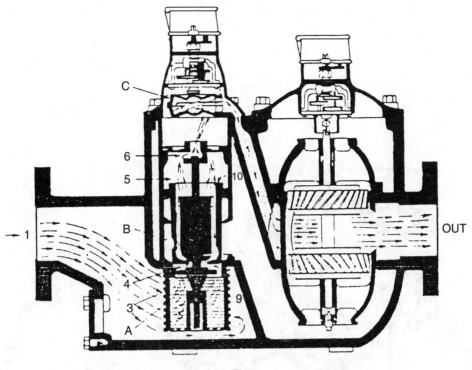

Low Flow

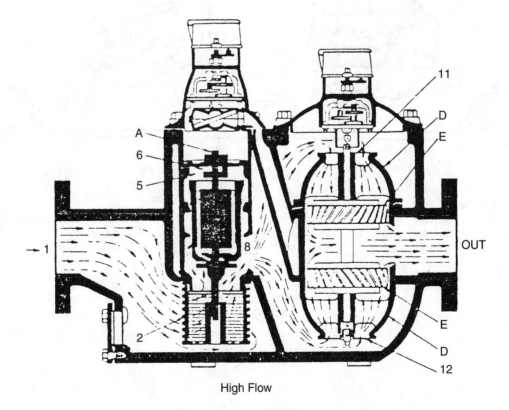

High Flow

Fig. 3.65   Flows through a compound meter

(Reprinted from *WATER DISTRIBUTION OPERATOR TRAINING HANDBOOK,* by permission.
Copyright 1976, the American Water Works Association)

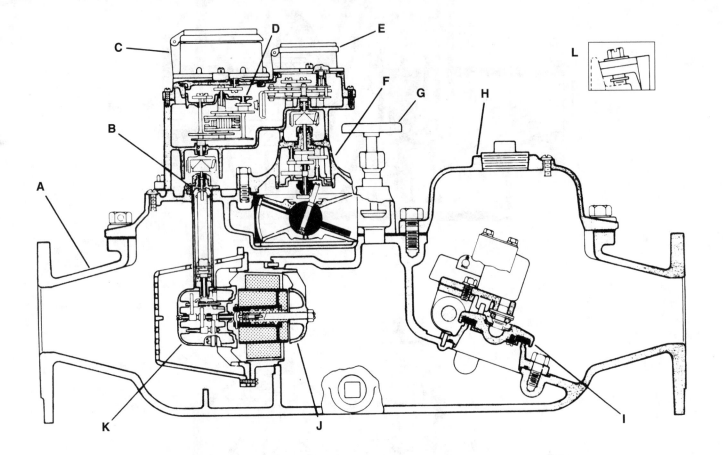

A. Main case
B. Stuffing box plate
C. Register and box
D. Combining drive

E. Test register and box
F. Disc section
G. Valve
H. Cover

I. Valve
J. Measure wheel and cage
K. Gear train
L. Clamps

Fig. 3.66   Sectional view of a compound meter
(Permission of Neptune Water Meter Company)

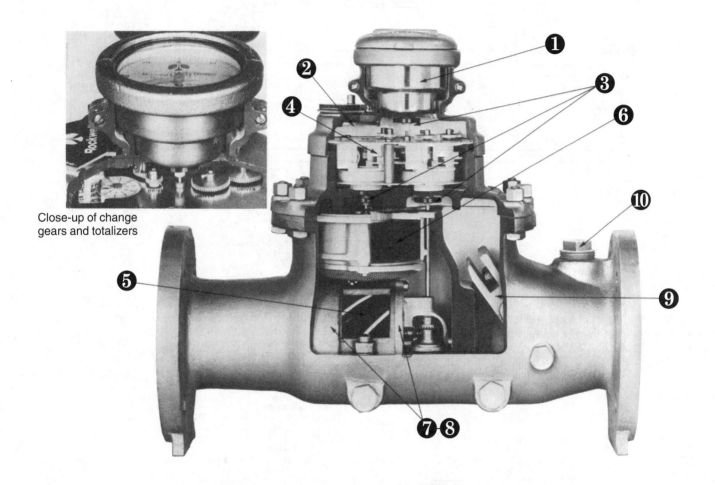

Close-up of change
gears and totalizers

1. Totalizing register
2. Bypass chamber register
3. Magnetic couplings
4. Sealed coordinator module
5. Bypass chamber
6. Turbine chamber
7. Magnetic suspension
8. Roller bearings
9. Automatic swing check valve
10. Test plug

Fig. 3.67   Cutaway views of a compound meter
(Permission of Rockwell International)

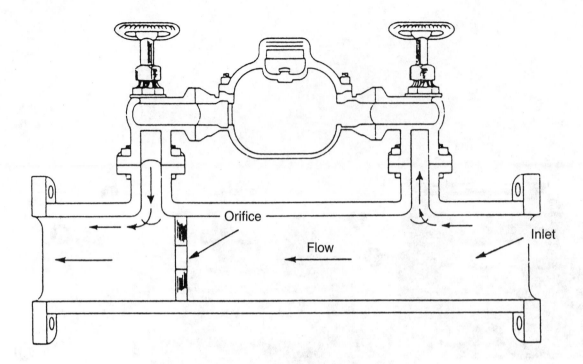

Fig. 3.68   Proportional meter

(Reprinted from *WATER DISTRIBUTION OPERATOR TRAINING HANDBOOK*, by permission.
Copyright 1976, the American Water Works Association)

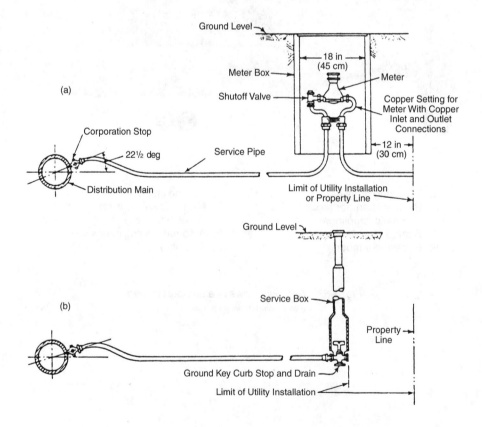

Fig. 3.69   Typical service line and meter box installation

(Reprinted from *A TRAINING COURSE IN WATER DISTRIBUTION, M8*, by permission.
Copyright 1976, the American Water Works Association)

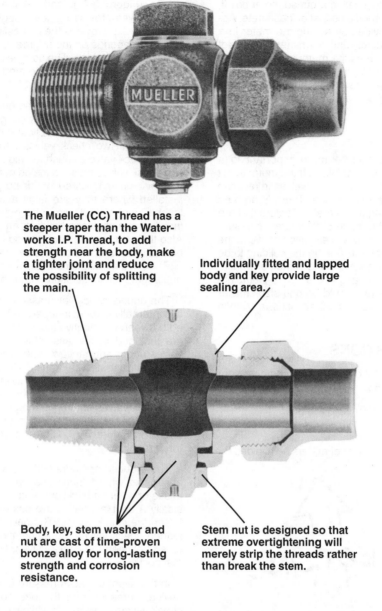

The Mueller (CC) Thread has a steeper taper than the Waterworks I.P. Thread, to add strength near the body, make a tighter joint and reduce the possibility of splitting the main.

Individually fitted and lapped body and key provide large sealing area.

Body, key, stem washer and nut are cast of time-proven bronze alloy for long-lasting strength and corrosion resistance.

Stem nut is designed so that extreme overtightening will merely strip the threads rather than break the stem.

Fig. 3.70   Corporation stop
(Permission of the Mueller Co. Mueller® Corporation Stops)

service line is to be installed after the street is paved over, the buried main must first be found. This information is usually in the utility's records and maps. If it is not available, electronic detection instruments are used. When the main is located, the pavement is cut and a narrow trench dug. Care should be taken in properly bedding and placing backfill material over the pipe to prevent any possible damage.

Another method for installing the service line is to use a mole for tunneling through the soil. This is a device shaped like an artillery shell that is activated by compressed air and tunnels through almost any kind of soil. The pipe to be installed is attached to the back of the mole and compressed air is fed through the pipe into the mole. This assembly is not easy to direct and steer. In some soils, a pipe pusher can be used to force pipe up to 4 inches (100 mm) in diameter for a distance of 50 to 150 feet (15 to 45 m). If an existing service line is being replaced, the new line is attached to the old one

and the whole assembly is pulled through with a backhoe or some other piece of motorized equipment.

3.81   Meters

The basic requirements of an acceptable meter installation, as recommended by the AWWA, are (1) that it be leaktight, (2) that it provide an upstream shutoff valve of high quality and with low pressure loss, (3) that it position the meter in a horizontal plane for optimum performance, (4) that it be reasonably accessible for service and inspection, (5) that it provide for easy reading either directly or with a remote-reading device, (6) that it be reasonably well protected against frost, mechanical damage, and tampering, and (7) that it not be an obstacle, or hazard, to customer or public safety. Mechanical damage can usually be avoided by placing the meter behind the curb and in a meter box. The meter box may be made of concrete, cast iron, or plastic. If it is plastic, metal particles are added to the plastic to facilitate locating a buried meter box if, by chance, maps and records should be lost or inaccurate. Additional protection can be provided by setting the meter in a copper yoke which connects rigidly and permanently the inlet and outlet pipes at the points where the meter is connected to them. This ensures proper spacing and alignment of the meter, cushions the meter against stresses and strains in the piping, and provides pipe support and electrical continuity if the meter is removed.

3.82 Meter Size

Meter selection and sizing can be more important to a system than all the meter repairs that follow. If the meter is not sized properly, it can be costly in repairs as well as in revenue lost for as long as that meter is in service. There is no hard and fast rule for selecting the right size meter. Factors that are taken into consideration are the range of flow rates; pressure at the service connection; differences in elevation between the street main and the highest water fixtures in the building to be served; and friction losses in the service line, meter, any backflow prevention devices installed, and in the house plumbing. A rule of thumb is that the meter should be one size smaller than the service line. In most residential services, a $^3/_4$-inch (18-mm) meter will be sufficient.

QUESTIONS

Write your answers in a notebook and then compare your answers with those on page 156.

3.8A   What is a corporation stop?

3.8B   What is a curb stop?

3.8C   How can mechanical damage of meters be avoided?

3.83 Backflow Prevention Devices

Backflow prevention devices, when used, are normally installed at the service connection directly downstream from the meter. Their purpose is, as indicated, to prevent backflow— the reverse flow of water of questionable quality from an unapproved water supply into a potable water system. Backflow may be caused by back pressure or backsiphonage, both of which result from a pressure differential. Back pressure results from a higher pressure in the unapproved water supply than in the domestic system, while backsiphonage takes place where there is a negative or reduced pressure in the domestic system.

Three basic backflow prevention devices are approved for use at service connections: (1) the double check valve assembly (Figure 3.71), (2) the reduced-pressure principle (RP) device (Figure 3.71), and (3) the air gap devices. Double check valves provide the least degree of protection, while an air gap device provides the highest degree. The type of device required depends on the degree of hazard and the probability of backflow occurring. The complexity of the piping on the premises and how difficult it is to modify the system are also important.

All devices should be located on the user's side of the service and as close to the user connection as is practical. The double check valve assembly and the reduced-pressure (RP) device have two check valves because the failure rate of a single check valve is much too high for reliable protection. The two check valves must be tested periodically and test cocks on the devices are required for this purpose. The devices must be installed where they are fairly accessible for testing and at least 12 inches (300 mm) above ground. If these devices are installed in a vault or a building, an adequate drain is required. Also they may not be enclosed so leaks and malfunctions can be easily observed. Double check valve and reduced-pressure (RP) device installations are shown in Figure 3.72 and an air gap separation installation is shown in Figure 3.73.

The double check valve assembly (Figure 3.74) consists of two internally loaded, independently acting check valves located between two tightly closing shutoff valves. Four test cocks are included in the assembly. The operation of the check valves under normal flow conditions and under backflow conditions is illustrated in Figure 3.75. The pressure of the incoming water keeps the check valves open, but when the on-premise pressure gets higher than the water system pressure, the check valves shut tight and prevent backflow. The relative locations of the check valves, gate valves, and test cocks are shown in Figure 3.76. Check valves should be tested every three months.

The reduced-pressure (RP) device (Figure 3.77) is installed at services where dangerous materials are present on the premises or where a backflow situation is likely. This device is basically a double check valve assembly with a relief valve between the check valves. The device consists of two internally loaded check valves, an automatic differential relief valve located between the check valves, two shutoff valves, and four test cocks.

In the event of back pressure (pressure downstream of the device increases and the direction of flow reverses), both check valves close and backflow is prevented. If there is leakage back into the zone between the two valves because the valves do not close tightly, the relief valve at the bottom opens and water is discharged to the atmosphere. If backsiphonage occurs with the supply pressure dropping to 2 psi (13.8 kPa or 0.14 kg/sq cm) or less, the relief valve will remain fully open. This will bring the pressure in the zone between the two check valves up to atmospheric pressure, thereby creating an air gap between the two check valves.

A properly designed air gap installation provides the most positive type of backflow prevention available. A vertical gap between the end of the pipe discharge and the top of the receiving tank is created, with the gap being twice the diameter of the discharge pipe, but no shorter than one inch (2.5 cm). With this system there is no possibility of backflow and correct installation can be observed immediately. An acceptable air gap separation installation is shown in Figure 3.73. All piping from the user connection to the receiving tank must be above grade and entirely visible unless the water supplier and health agency approve the burial of the line.

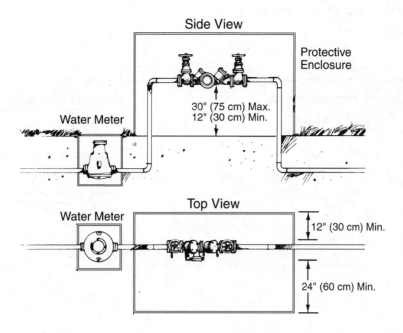

Reduced-pressure (RP) device

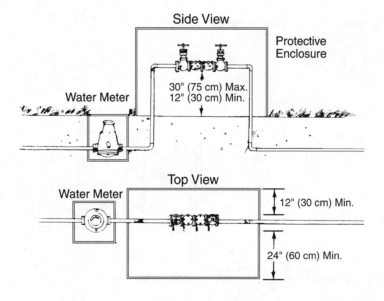

Double check valve

NOTE: Protective enclosures are not used by some agencies so any leaks or malfunctions can be easily observed.

Fig. 3.71   Backflow prevention devices
(Permission of Febco)

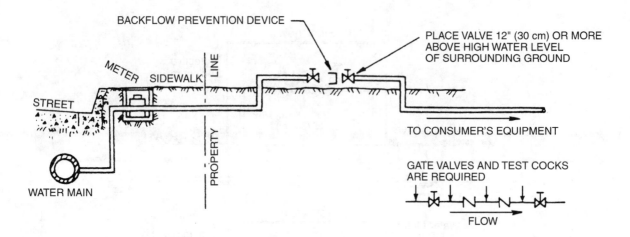

Fig. 3.72   Double check valve and RP device installation
(Permission of California Department of Health Services, Sanitary Engineering Branch)

TANK SHOULD BE OF SUBSTANTIAL CONSTRUCTION AND
OF A KIND AND SIZE TO SUIT CONSUMER'S NEEDS.
TANK MAY BE SITUATED AT GROUND LEVEL (WITH A
PUMP TO PROVIDE ADEQUATE PRESSURE HEAD) OR
BE ELEVATED ABOVE THE GROUND.

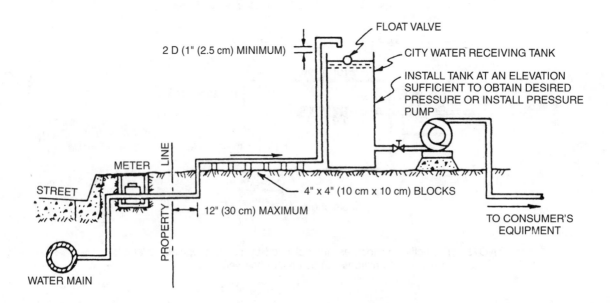

Fig. 3.73   Air gap separation installation
(Permission of California Department of Health Services, Sanitary Engineering)

¾" through 2" (1.9 cm through 5 cm)

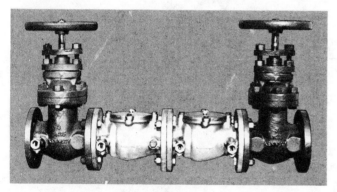

3" and 4" (7.5 cm and 10 cm)

6", 8", and 10" (15 cm, 20 cm, and 25 cm)

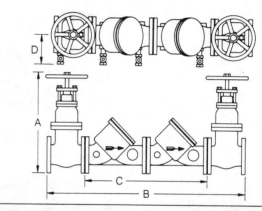

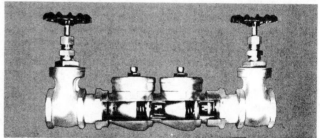

2" (5 cm) Cutaway

Fig. 3.74   Double check valve unit
(Permission of Febco)

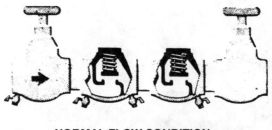

NORMAL FLOW CONDITION

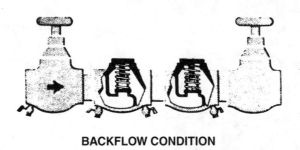

BACKFLOW CONDITION

Fig. 3.75   Normal and backflow conditions
(Permission of Febco)

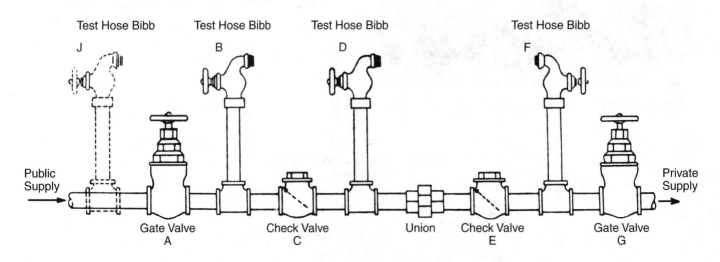

Fig. 3.76   Method of testing check valves

Dimensions and Weights

Size[a]	A	B	C	D	E	Wt/lbs[b]
1-1/2"	27-1/16"	7"	9"	5-3/4"	16-13/16"	90
2"	34-7/16"	12"	10"	6-1/4"	18-1/16"	138
2-1/2"	37-3/16"	12-1/2"	10-1/2"	7-1/2"	22-1/16"	280
3"	41-11/16"	14"	11-1/2"	8-1/16"	29-9/16"	285
4"	50-7/16"	17-3/8"	12-1/2"	11"	32-5/16"	460
6"	59-11/16"	21-1/4"	14"	14"	38-9/16"	775
8"	69-3/16"	26"	15"	18"	46-1/16"	1,270
10"	84-3/16"	30"	16"	22"	58-1/16"	1,720

[a] Multiply inches x 2.5 to obtain centimeters.
[b] Multiply pounds x 0.45 to obtain kilograms.

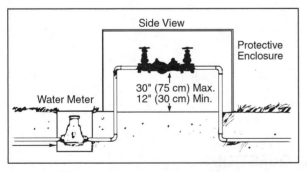

Approvals   Model 825 1-1/2" – 10" (3.8 cm – 25 cm)

ASSE	USC	IAMPO	SBCC	CSA	UL*

* 2-1/2" through 10" (6.4 cm through 25 cm)

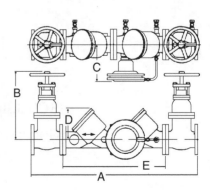

Fig. 3.77   *Reduced-pressure backflow preventer*
(Permission of Febco)

QUESTIONS

Write your answers in a notebook and then compare your answers with those on page 156.

3.8D   Where are backflow prevention devices normally installed when used?

3.8E   List the three basic backflow prevention devices that are approved for use at service connections.

3.8F   Why are two check valves required for the double check valve assembly and the reduced-pressure (RP) device?

3.8G   What is the distance of the vertical gap in an air gap device?

3.9   ARITHMETIC ASSIGNMENT

Turn to the Appendix "How to Solve Water Distribution System Arithmetic Problems," at the back of this manual and read the following sections:

1. A.4, *METRIC SYSTEM,*

2. A.5, *WEIGHT-VOLUME RELATIONS,* and

3. A.6, *FORCE, PRESSURE, AND HEAD.*

In Section A.13, "*TYPICAL WATER DISTRIBUTION SYSTEM PROBLEMS (ENGLISH SYSTEM),*" read and work the problems in Sections A.130, Flows, A.131, Chemical Doses, and A.132, Distribution System Facilities.

Check all of the arithmetic in these sections on an electronic calculator. You should be able to get the same answers.

3.10   ADDITIONAL READING

1. *NEW YORK MANUAL,* Chapter 3,* "Hydraulics and Electricity," and Chapter 17,* "Protection of Treated Water."

2. *TEXAS MANUAL,* Chapter 15,* "The Distribution System," and Chapter 17,* "Customer Meters."

3. *WATER DISTRIBUTION OPERATOR TRAINING HANDBOOK.* Obtain from American Water Works Association (AWWA), Bookstore, 6666 West Quincy Avenue, Denver, CO 80235. Order No. 20428. ISBN 1-58321-014-8. Price to members, $52.00; nonmembers, $76.00; price includes cost of shipping and handling.

4. *WATER TRANSMISSION AND DISTRIBUTION.* Obtain from American Water Works Association (AWWA), Bookstore, 6666 West Quincy Avenue, Denver, CO 80235. Order No. 1957. ISBN 1-58321-231-0. Price to members, $87.00; nonmembers, $128.00; price includes cost of shipping and handling.

* Depends on edition.

End of Lesson 5 of 5 on DISTRIBUTION SYSTEM FACILITIES

Please answer the discussion and review questions next.

DISCUSSION AND REVIEW QUESTIONS

Chapter 3.   DISTRIBUTION SYSTEM FACILITIES

(Lesson 5 of 5 Lessons)

Write the answers to these questions in your notebook. The question numbering continues from Lesson 4.

23. Distribution system metering can provide what benefits?

24. What are the advantages and limitations of displacement meters?

25. List the major differences between an orifice plate and a Venturi meter.

26. What are the advantages and limitations of electronic meters?

27. Why is the proper size of a meter important?

28. What causes backflow?

29. What factors are considered when selecting a backflow prevention device?

SUGGESTED ANSWERS

Chapter 3.   DISTRIBUTION SYSTEM FACILITIES

ANSWERS TO QUESTIONS IN LESSON 1

Answers to questions on page 61.

3.0A   Water distribution systems consist of pipes, storage facilities, pumping stations, valves, fire hydrants, meters, and other appurtenances.

3.0B   A continuous positive water pressure must be maintained in the distribution system at all times so as to protect the distribution system from the entrance of toxic and other undesirable substances.

Answers to questions on page 69.

3.1A   Excessive pressure may damage the customers' facilities and plumbing fixtures.

3.1B   Friction losses in pipes are caused by the roughness of the inside of the pipe creating turbulence proportional to the velocity of the flowing water.

3.1C   Thrust can be caused at any point where water changes direction, increases velocity, or the flow decreases or stops.

3.1D   Water hammer can be caused by the action of pumps suddenly starting or stopping or valves quickly opening or closing.

3.1E

Known	**Unknown**
Pressure, psi = 30 psi	Pressure Head, ft

Calculate the pressure head in feet.

Pressure Head, ft = (Pressure, psi)(2.31 ft/psi)

$$= (30 \text{ psi})(2.31 \text{ ft/psi})$$

$$= 69 \text{ ft}$$

Answers to questions on page 72.

3.2A   The hydraulic adequacy of a distribution system is determined by the pressures that exist at various points in the system under the conditions of operation.

3.2B   Booster pumps are installed directly on mains to increase the available system pressure.

3.2C   Dead-end water mains should be avoided to prevent the development of taste and odor problems.

Answers to questions on page 72.

3.3A   The requirements for operational storage can be reduced by the use of wells.

3.3B   Concrete with a steel cylinder is commonly used for large water transmission mains.

Answers to questions on page 74.

3.4A   Holding chlorinated water in covered distribution reservoirs provides an extended contact time for effective disinfection.

3.4B   Distribution reservoir drains and overflows should have no cross connections to any sewer line or other source of contamination and they should be screened.

Answers to questions on page 75.

3.5A   Centrifugal pumps are commonly used for water distribution systems.

3.5B   Electric motors are the best source of power for distribution system pumps because they are compact and well adapted to automatic control or remote operation.

ANSWERS TO QUESTIONS IN LESSON 2

Answers to questions on page 76.

3.60A   Dead-end water mains can cause taste and odor water quality problems.

3.60B   Two advantages of grid or loop water distribution systems include (1) fewer dead ends, and (2) a greater flow of water to an area when there is a fire or other source of high demand because water is being carried to any spot from more than one direction.

Answers to questions on page 76.

3.61A   Water distribution system pipes must be capable of resisting forces from (1) backfill, (2) weight of passing traffic, and (3) internal water pressures such as delivery pressure, surges, and water hammer.

3.61B   Important installation features of water distribution pipes include weight, jointing characteristics, available sizes, and ease of tapping the line.

Answers to questions on page 78.

3.62A   The external corrosion of ductile-iron pipe can be reduced by the application of bituminous coatings or a polyethylene wrap.

3.62B   Steel pipe may be lined with a bituminous material, cement mortar, or an epoxy.

3.62C   Steel pipes may be coated with epoxy or mastic, they can be galvanized, or a protective wrap could be used.

3.62D   The major disadvantages of concrete pipe are that the pipe is heavy, may be hard to tap, needs special fittings, and may deteriorate in aggressive (corrosive) soils.

Answers to questions on page 81.

3.62E The disadvantages of asbestos-cement pipe include its low flexural strength in small sizes and its vulnerability to impact damage. Because the material is nonconductive, it is also difficult to locate buried pipes or to thaw frozen pipes using electrical methods.

3.62F Plastic pipe may be made of PVC (polyvinyl chloride), PE (polyethylene), and PB (polybutylene).

3.62G The most common materials used for service lines are copper, PVC, PE, PB, or, in special cases, reinforced plastic pipe.

3.62H Existing lead service lines should be replaced because of the potential for toxic lead to enter the water.

Answers to questions on page 86.

3.63A Basic requirements for a good pipe joint include a watertight seal, easy installation, durability, and corrosion resistance.

3.63B Push-on and mechanical joints are most commonly used for ductile-iron pipe. Flanged, restrained, flexible, and sleeve joints are also used.

3.63C Steel pipe may be joined by bell and spigot lap welded joint, bell and spigot rubber gasket joint, harness joint, carnegie shape rubber gasket joint, butt-welded joint, butt strap joint for welding, mechanically coupled joint, and flanged joint for bolting.

Answers to questions on page 88.

3.63D The outside joint space is normally completely filled with cement mortar to protect the joint rings against external corrosion.

3.63E Couplings 3 to 12 inches (75 to 300 mm) in diameter allow five degrees deflection, and those with diameters from 14 to 24 inches (350 to 600 mm) permit a deflection of three degrees.

3.63F PVC pipe may be joined by solvent weld or bell and spigot joints.

Answers to questions on page 88.

3.64A The most commonly used corrosion-control methods include (1) use of nonmetallic pipe materials (asbestos-cement, reinforced concrete, plastic), (2) nonmetallic coatings, (3) metallic coatings, (4) chemical treatment, and (5) electrical control using cathodic protection.

3.64B Steel pipe can be protected from corrosion with cement mortar, cold-applied mastic, hot coal tar enamel or, in the case of piping up to four inches (100 mm) in diameter, it can be galvanized.

ANSWERS TO QUESTIONS IN LESSON 3

Answers to questions on page 90.

3.65A Prior to excavating, all buried water, sewer, gas, power, telephone and cable TV lines, and storm drains in the work area should be located.

3.65B Water and sewer lines should be separated to avoid possible contamination of water if leakage from sewer lines saturates the soil in the vicinity of the water line. Critical conditions develop when the water main is depressurized and there is no pressure or a negative pressure. Contaminants in the saturated soil could be drawn into the water lines.

3.65C Metal pipe can be checked for damage by gently tapping each length of pipe with a hammer. If the pipe does not ring or hum clearly, the pipe has been damaged.

Answers to questions on page 96.

3.65D Trenches must be deep enough so that the pipe is protected from freezing, surface traffic, stream or river erosion, and possible future changes in the ground surface elevation.

3.65E Shoring requirements depend on depth of trench, width of trench, type of soil, soil conditions, nearby activities, and regulatory laws and codes.

3.65F Lack of shoring or shoring failure is the number-one cause of cave-in deaths.

3.65G Shoring systems in trenches consist of some form of sheeting or liner, uprights held rigidly opposite each other against the trench walls by jacks or horizontal cross members (braces) and, if required, longitudinal members (stringers).

3.65H Uprights must extend as near to the bottom of the trench as permitted by the material being installed, but not more than 2 feet (0.6 m) from the bottom.

Answers to questions on page 97.

3.65I All excavation work shall at all times be under the immediate supervision of a competent person with authority to modify the shoring system or work methods as necessary to provide adequate safety.

3.65J Spoil or excavated material must be placed 2 feet (0.6 m) or more back from the edge of the trench and on only one side of the trench. If the trench is over 5 feet (1.5 m) deep, spoil must be placed more than 4 feet (1.2 m) back from the edge of the trench and on only one side of the trench.

3.65K All trenches deeper than 5 feet (1.5 m) must be shored. In some locations, all trenches deeper than 4 feet (1.2 m) must be shored.

3.65L If shoring is not practical, the sides or walls of an excavation may be sloped.

3.65M Walkways must be provided for pedestrians to keep them out of the way of workers and to protect them from injury.

Answers to questions on page 99.

3.65N The different types of common shores include (1) hydraulic, (2) screw jacks, (3) air, (4) solid sheeting, (5) cylinder, and (6) shield.

3.65O The major concern during a shoring operation is the protection of the operators. Assume all trenches and excavations are unsafe.

Answers to questions on page 108.

3.65P Precautions that should be taken when laying pipe on bedding include being sure that the couplings do not rest on or settle down to the original trench bottom and also that the pipe is provided with even bedding for its entire length.

3.65Q When laying pipe with a bell, the bell should point in the direction that the work is proceeding, except that when the pipe is being laid downhill, the bell should face uphill.

3.65R Before joining pipes together, brush the inside of the pipe to remove any loose dirt and swab with a strong hypochlorite solution if necessary. Carefully clean the ends of the pipes and the joint materials and smooth out the rough edges.

3.65S Thrust blocks should be installed to prevent leakage or separation of joints at all tees, bends, caps, plugs, hydrants, and other fittings that change the direction of flow or stop the flow.

3.65T The first layer of backfill should be compacted up to the horizontal center of the pipe.

Answers to questions on page 110.

3.65U The leakage test pressure is at least 50 percent higher than the normal expected operating pressure or 150 psi (1,034 kPa or 10.5 kg/sq cm), whichever is larger.

3.65V During grinding and polishing work on pipe, safety goggles must be worn.

ANSWERS TO QUESTIONS IN LESSON 4

Answers to questions on page 112.

3.66A In a wet tap, a valve is bolted to the tapping saddle and the actual drilling of the pipe through the valve is done with the pipe under pressure.

3.66B New service connections can be installed in existing plastic mains by using a drilling machine to drill into the main through a combined clamp and corporation stop.

Answers to questions on page 121.

3.67A Types of valves include plug, ball, and butterfly valves, and diaphragm, globe, gate, and check valves.

3.67B The main application of gate valves in water distribution systems is to isolate sections of mains to permit emergency repairs without interruption of service to a large number of customers.

3.67C The two main uses of globe valves are flow and pressure regulation. Globe valves are used to isolate sections of small-diameter pipe and for flow control in larger pipes.

Answers to questions on page 122.

3.67D Usually, pressure-regulating valves are of the globe design and have a spring-loaded diaphragm that adjusts the size of opening.

3.67E When air gets into water, the air tends to collect at high points in the lines. These air pockets can increase the resistance to flow by 10 to 15 percent, and it is even possible for an air-lock condition to occur and stop the flow of water completely.

3.67F Air relief valves (1) allow large quantities of air to escape during the filling of a pipeline, (2) permit air to enter a pipeline that is being drained, and (3) allow entrained air to escape while a line is operating under pressure.

3.67G Blowoff valves are used primarily for blowing off accumulated sediment from low spots in the line and for dewatering lines or reservoirs for repairs or inspections.

Answers to questions on page 124.

3.67H Valve boxes are installed to provide access to buried valves.

3.67I If a deep valve vault becomes flooded, a backsiphonage hazard could result.

3.67J Three valves should be installed at pipe crosses.

3.67K Valves may be operated manually, by water pressures (hydraulic or pneumatic), or electrically. Solenoids and electric motors geared to the valve mechanism are also used to open and close valves.

Answers to questions on page 131.

3.67L Fittings are used to connect pipes of the same size or different sizes, change the direction of flow, or stop the flow.

3.67M Hydrants are used for firefighting, flushing of pipelines, and providing water to water trucks or other construction equipment.

3.67N The four major parts of a hydrant are (1) the inlet pipe connection from the water supply main, (2) the main valve, (3) the barrel, and (4) the head.

ANSWERS TO QUESTIONS IN LESSON 5

Answers to questions on page 132.

3.7A Meters should be used on main supply lines, pumping stations, reservoir outlets, connections to other utility systems, and at each customer's service.

3.7B Metering reduces water consumption by determining costs and thus preventing waste through excessive use of water.

3.7C Important qualities to look for in a meter include ability to accurately measure and register your anticipated flow levels, ability to meet required capacity with minimum head loss, durability, ruggedness, precision of workmanship, ease of repair, availability of spare parts, freedom from irritating noise, a reasonable price, delivery time for spare parts, and a manufacturer with a good reputation.

3.7D

Known	Unknown
Head Loss, psi = 15 psi	Head Loss, ft

Convert the head loss from 15 psi to feet.

Head Loss, ft = (Head Loss, psi)(2.31 ft/psi)

$$= (15\ psi)(2.31\ ft/psi)$$

$$= 34.65\ ft$$

$$or = 35\ ft\ (approximately)$$

Answers to questions on page 135.

3.7E Small-flow meters are of the displacement type and large-flow meters are of the velocity type.

3.7F The two types of displacement meters are the piston and nutating-disc meters.

3.7G   The types of velocity meters include turbine, propeller, Venturi, insertion, and electronic.

Answers to questions on page 140.

3.7H   Compound meters are especially advantageous for places where there are predominantly low or intermediate flows and, only occasionally, high flows.

3.7I   The two major types of electronic flowmeters include the magnetic and sonic types of flowmeters.

3.7J   Proportional meters are used to measure flows in fire lines.

Answers to questions on page 146.

3.8A   A corporation stop is a water service shutoff valve located at a street water main.

3.8B   A curb stop is a water service shutoff valve located in a water service pipe near the curb and between the water main and the building.

3.8C   Mechanical damage of meters can be avoided by placing the meter behind the curb and in a meter box.

Answers to questions on page 152.

3.8D   Backflow prevention devices are normally installed at the service connection directly downstream from the meter.

3.8E   The three basic backflow prevention devices are (1) the double check valve assembly, (2) the reduced-pressure principle (RP) device, and (3) the air gap device.

3.8F   The double check valve assembly and the reduced-pressure (RP) device have two check valves because the failure rate of a single check valve is much too high for reliable protection.

3.8G   The distance of the vertical gap in an air gap device between the end of the pipe discharge and the top of the receiving tank should be twice the diameter of the discharge pipe, but no shorter than one inch (2.5 cm).

CHAPTER 4

WATER QUALITY CONSIDERATIONS IN DISTRIBUTION SYSTEMS

by

Sam Kalichman

TABLE OF CONTENTS

Chapter 4. WATER QUALITY CONSIDERATIONS IN DISTRIBUTION SYSTEMS

OBJECTIVES

Chapter 4.   WATER QUALITY CONSIDERATIONS IN DISTRIBUTION SYSTEMS

Following completion of Chapter 4, you should be able to:

1. Identify types of contaminants that could get into water distribution systems,

2. Identify and correct sources of contaminants in distribution systems, and

3. Identify and correct causes of water quality degradation in distribution systems.

WORDS

Chapter 4. WATER QUALITY CONSIDERATIONS IN DISTRIBUTION SYSTEMS

AEROBIC (AIR-O-bick)   AEROBIC

A condition in which atmospheric or dissolved molecular oxygen is present in the aquatic (water) environment.

AESTHETIC (es-THET-ick)   AESTHETIC

Attractive or appealing.

ANAEROBIC (AN-air-O-bick)   ANAEROBIC

A condition in which atmospheric or dissolved molecular oxygen is *NOT* present in the aquatic (water) environment.

BACK PRESSURE   BACK PRESSURE

A pressure that can cause water to backflow into the water supply when a user's water system is at a higher pressure than the public water system.

BACKFLOW   BACKFLOW

A reverse flow condition, created by a difference in water pressures, which causes water to flow back into the distribution pipes of a potable water supply from any source or sources other than an intended source. Also see BACKSIPHONAGE.

BACKSIPHONAGE   BACKSIPHONAGE

A form of backflow caused by a negative or below atmospheric pressure within a water system. Also see BACKFLOW.

CARCINOGEN (CAR-sin-o-JEN)   CARCINOGEN

Any substance which tends to produce cancer in an organism.

CHLORAMINES (KLOR-uh-means)   CHLORAMINES

Compounds formed by the reaction of hypochlorous acid (or aqueous chlorine) with ammonia.

COLIFORM (COAL-i-form)   COLIFORM

A group of bacteria found in the intestines of warm-blooded animals (including humans) and also in plants, soil, air and water. Fecal coliforms are a specific class of bacteria which only inhabit the intestines of warm-blooded animals. The presence of coliform bacteria is an indication that the water is polluted and may contain pathogenic (disease-causing) organisms.

CONTAMINATION   CONTAMINATION

The introduction into water of microorganisms, chemicals, toxic substances, wastes, or wastewater in a concentration that makes the water unfit for its next intended use.

CROSS CONNECTION   CROSS CONNECTION

A connection between a drinking (potable) water system and an unapproved water supply. For example, if you have a pump moving nonpotable water and hook into the drinking water system to supply water for the pump seal, a cross connection or mixing between the two water systems can occur. This mixing may lead to contamination of the drinking water.

ELECTROCHEMICAL REACTION   ELECTROCHEMICAL REACTION

Chemical changes produced by electricity (electrolysis) or the production of electricity by chemical changes (galvanic action). In corrosion, a chemical reaction is accompanied by the flow of electrons through a metallic path. The electron flow may come from an external source and cause the reaction, such as electrolysis caused by a D.C. (direct current) electric railway or the electron flow may be caused by a chemical reaction as in the galvanic action of a flashlight dry cell.

ENTRAIN   ENTRAIN

To trap bubbles in water either mechanically through turbulence or chemically through a reaction.

INTERFACE INTERFACE

The common boundary layer between two substances such as water and a solid (metal); or between two fluids such as water and a gas (air); or between a liquid (water) and another liquid (oil).

MACROSCOPIC (MACK-row-SKAWP-ick) ORGANISMS MACROSCOPIC ORGANISMS

Organisms big enough to be seen by the eye without the aid of a microscope.

NONPOTABLE (non-POE-tuh-bull) NONPOTABLE

Water that may contain objectionable pollution, contamination, minerals, or infective agents and is considered unsafe and/or unpalatable for drinking.

PALATABLE (PAL-uh-tuh-bull) PALATABLE

Water at a desirable temperature that is free from objectionable tastes, odors, colors, and turbidity. Pleasing to the senses.

POLLUTION POLLUTION

The impairment (reduction) of water quality by agricultural, domestic, or industrial wastes (including thermal and radioactive wastes) to a degree that has an adverse effect on any beneficial use of water.

POTABLE (POE-tuh-bull) WATER POTABLE WATER

Water that does not contain objectionable pollution, contamination, minerals, or infective agents and is considered satisfactory for drinking.

TRIHALOMETHANES (THMs) (tri-HAL-o-METH-hanes) TRIHALOMETHANES (THMs)

Derivatives of methane, CH_4, in which three halogen atoms (chlorine or bromine) are substituted for three of the hydrogen atoms. Often formed during chlorination by reactions with natural organic materials in the water. The resulting compounds (THMs) are suspected of causing cancer.

TUBERCLE (TOO-burr-cull) TUBERCLE

A protective crust of corrosion products (rust) which builds up over a pit caused by the loss of metal due to corrosion.

VORTEX VORTEX

A revolving mass of water which forms a whirlpool. This whirlpool is caused by water flowing out of a small opening in the bottom of a basin or reservoir. A funnel-shaped opening is created downward from the water surface.

CHAPTER 4.   WATER QUALITY CONSIDERATIONS IN DISTRIBUTION SYSTEMS

4.0   IMPORTANCE OF WATER QUALITY

"Water quality" is the general term used to describe the composite chemical, physical, and biological characteristics of a water supply. Whether the quality is high or low depends on the suitability of the water for a particular use. A water supply that is satisfactory in quality for one purpose might be entirely unsatisfactory for another. A domestic water supply is considered to be of good quality when it is free of disease-causing organisms and toxic chemicals, attractive in taste and appearance, of such a chemical composition that it may be distributed without undue corrosive or scale-forming effects on the water system, and will satisfy, to a maximum degree, the requirements of domestic and industrial consumers.

The quality of water received in the distribution system depends on the quality of the sources used and the type of treatment provided. The first order of quality control calls for prevention of contamination and pollution of the sources. If quality control at the source is not practiced, any breakdown in the treatment provided may result in delivery of questionable quality water (or even dangerous water) to the distribution system. The primary responsibility of the treatment plant is to produce a safe and *PALATABLE*[1] water; however, it should also condition the water to minimize its deterioration within the distribution system.

The importance of providing a water of acceptable quality is obvious. Service of a poor quality water could result in a range of consequences from the water not being acceptable to the consumer because of its appearance or taste, to illness or even death of some susceptible consumers. Each utility and, in turn, each operator working for that utility is responsible for the quality of the water served to its customers.

After a year-long study of water quality in distribution systems, Dr. John T. O'Connor of the University of Illinois reported:

"Perhaps the single greatest water quality problem which affects the daily lives of most American water consumers results from the deterioration of water quality within water distribution systems. In almost every circumstance, in the United States, the quality of the treated water put into the water distribution mains from the water purification plant is high. However, once the water has entered the distribution system, some fairly substantial changes can occur which the water purveyor (supplier) cannot readily control."

Water quality would not change in the distribution system if the system was inert to water, completely sealed off from any intrusion, and the water was completely stable and biologically sterile. Unfortunately, this situation does not exist, nor will it ever exist. A rather romanticized, but appropriate, statement was made at a British conference on problems associated with the distribution of drinking waters: "A distribution system is not just pipes and valves, but is like a living person—a dynamic, sensitive, individual that must be handled with great care."

QUESTIONS

Write your answers in a notebook and then compare your answers with those on page 176.

4.0A   Water quality is the general term used to describe which characteristics of a water?

4.0B   A domestic water supply is considered to be of good quality when it meets what guidelines or criteria?

4.0C   What are the consequences of delivering a poor quality water to the consumer?

4.1   WATER QUALITY STANDARDS

The waterworks industry is in the business of supplying one product—potable water. As in any industry, quality control of this product is necessary. To provide the needed quality control and ensure the acceptability of the product, water quality standards have been prepared and used. Some standards have been set by law and must be complied with, while others are merely guidelines. Water quality standards have changed considerably over the years. Increased knowledge of the substances in our water supplies and their effects has led to substantial revisions in both the types and the concentrations of substances allowable in water supplies.

---

[1] Palatable (PAL-uh-tuh-bull).   Water at a desirable temperature that is free from objectionable tastes, odors, colors, and turbidity. Pleasing to the senses.

Standards cannot possibly cover all potential chemicals. They can only include those constituents that are most likely to occur and that could be a significant problem. Existing water quality standards generally relate either to health or to aesthetics (general attractiveness or appeal). Mandatory limits have been set for substances that may result in a risk to the health of humans when present in concentrations above that limit and when continuously used for drinking or cooking purposes. Examples of contaminants in this category are arsenic, barium, chromium, lead, and mercury. Substances in the "aesthetics" category are those that may be objectionable to a large number of people, but are not generally hazardous to health. They include color, taste, odor, turbidity, iron, and manganese.

The United States National *PRIMARY*[2] Drinking Water Regulations limit the concentrations of substances that are harmful to health when consumed in drinking water. These limits are sometimes expressed as maximum contaminant levels (MCLs), but may also take the form of treatment techniques that a water treatment plant must use to ensure that the water is safe to drink. Primary standards have been set for many organic and inorganic chemicals, microorganisms, disinfection by-products, turbidity, and radionuclides.

National *SECONDARY*[2] Drinking Water Regulations, which are intended as guidelines for the states, have also been set for several organic and inorganic constituents, odor, color, pH, and total dissolved solids (TDS). These regulations are summarized on the poster inside the cover of this manual.

The Environmental Protection Agency is the federal agency responsible for regulating drinking water quality but states and local regulatory agencies also have a significant role in developing and implementing safe drinking water regulations. Each year EPA studies the effects of additional water constituents and frequently expands the list of regulated contaminants. Existing contaminant limits may also be modified if EPA's ongoing research reveals new information about the effects of a particular substance. The entire field of drinking water regulations is in a constant state of change. Operators are therefore urged to work closely with their state and local regulatory agencies to keep themselves informed of changes in drinking water requirements.

A more detailed discussion on water quality standards can be found in Chapter 2, "Water Sources and Treatment," in

SMALL WATER SYSTEM OPERATION AND MAINTENANCE. For additional information see *WATER TREATMENT PLANT OPERATION*, Volume II, Chapter 22, "Drinking Water Regulations."

QUESTIONS

Write your answers in a notebook and then compare your answers with those on page 176.

4.1A   Why have water quality standards been prepared and used by the waterworks industry?

4.1B   Drinking water quality standards can generally be categorized as being related either to _____ or to _____.

4.1C   MCL stands for what three words?

4.2 TYPES OF CONTAMINANTS

Contaminants found in water served from a distribution system may be chemical, physical, or biological in nature. They include, but are not restricted to, those substances already mentioned for which standards have been set.

Chemical contaminants are organic and inorganic substances which may be found as dissolved gases, in aqueous (liquid) solution, or in solid form in the water supply. These chemical substances range from simple dissolved gases to inorganic substances and to complex organic compounds. Inorganic substances that can be found in water include dissolved or suspended lead, mercury, fluoride, nitrate, and arsenic. Organics are chemical substances which are based on the element carbon. They may be natural substances of animal or vegetable origin such as the humic materials that originate from biological decomposition of plant life and animal remains. They can also be manmade synthetic compounds such as pesticides, herbicides, and *TRIHALOMETHANES.*[3]

The physical qualities of water are the ones that are recognized by the consumer and by which consumers judge the water being served to them. These include color, turbidity, taste, odor, and temperature. Other physical items of occasional concern include milky water due to air and *MACROSCOPIC ORGANISMS*[4] such as worms or bugs in the water.

Biological contaminants include the pathogenic (disease-causing) organisms such as bacteria, viruses, and intestinal parasites, and the non-pathogenic organisms such as iron bacteria, algal growths, slime growths, insects and insect larvae, nematodes (roundworms), mollusks such as snails and clams, crustacea such as water fleas and cyclops, and small animals and birds. Table 4.1 is a list of pathogenic organisms which can cause disease outbreaks when they contaminate water systems. Note that coliform bacteria are not listed in Table 4.1. Most coliform bacteria themselves are not pathogenic organisms but their presence in water indicates the potential presence of pathogens. However, *E. coli*, a type of coliform bacteria, has made people sick.

---

[2] *Primary standards deal with potentially harmful (toxic or disease) constituents; secondary standards deal with aesthetic constituents. Turbidity is an aesthetic concern; however, turbidity interferes with disinfection which is used to kill or inactivate disease-causing organisms.*

[3] *Trihalomethanes (THMs) (tri-HAL-o-METH-hanes).   Derivatives of methane, CH_4, in which three halogen atoms (chlorine or bromine) are substituted for three of the hydrogen atoms. Often formed during chlorination by reactions with natural organic materials in the water. The resulting compounds (THMs) are suspected of causing cancer.*

[4] *Macroscopic (MACK-row-SKAWP-ick) Organisms.   Organisms big enough to be seen by the eye without the aid of a microscope.*

TABLE 4.1 PATHOGENIC ORGANISMS (DISEASES) TRANSMITTED BY WATER

Bacteria
Salmonella (salmonellosis)
Shigella (bacillary dysentery)
Bacillus typhosus (typhoid fever)
Salmonella paratyphi (paratyphoid)
Vibrio cholerae (cholera)

Viruses
Enterovirus
Poliovirus
Coxsackie Virus
Echo Virus
Andenovirus
Reovirus
Infectious Hepatitis

Intestinal Parasites
Entamoeba histolytica (amoebic dysentery)
Giardia lamblia (giardiasis)
Ascaris lumbricoides (giant roundworm)
Cryptosporidium (cryptosporidiosis)

4.3 SOURCES OF CONTAMINANTS

Water served from the distribution system may have been of poor quality before it reached the system, or it may degrade in the distribution system itself. Operators must realize that the quality of the water coming out of the consumer's tap can also be affected by what happens to the water as it travels through the customer's own system.

Contamination of water can occur at the source of supply, at the treatment plant, in distribution system storage facilities, or even in water mains. Factors that may cause water quality degradation include cross connections, corrosion, biological growth and activity, high temperatures, unusual flows, time in system, dead ends, age of facilities, and operational procedures.

QUESTIONS

Write your answers in a notebook and then compare your answers with those on page 176.

4.2A What are trihalomethanes (THMs)?

4.2B What physical qualities of water are important to consumers?

4.3A Where might contamination of water occur?

4.3B How can water be degraded in the distribution system?

4.4 WATER SUPPLIED TO DISTRIBUTION SYSTEMS

4.40 Source of Supply

The quality of water entering the distribution system may have been adversely affected at the source or at the treatment plant. "Natural" source waters are never completely pure. During their precipitation (falling as rain or snow) and their passage over or through the ground, they can acquire a wide variety of dissolved or suspended impurities. Some of these impurities come from decaying vegetation or minerals in the soils. Other impurities result from the activities of people; waste discharges are of particular concern. Surface waters are exposed and especially susceptible to contamination. They can be degraded by domestic, industrial, and agricultural waste discharges and spills; humic materials such as decaying vegetation; refuse disposal; recreation; construction; animal activities; uncontrolled growths of algae and weeds; and radioactive fallout.

Groundwaters are less susceptible to contamination, but once they are contaminated, it normally takes a considerable time before they recover. This is due to the slow movement of the water in the ground. The groundwaters themselves may become contaminated or poorly constructed wells may contribute degraded water to the system. This degraded water may come from surface water moving above the groundwater table or from contaminants at or near the ground surface. Contamination may also result from leaching from nearby domestic, industrial, or agricultural wastewater operations or disposals; spills; humic material; flood waters; or passage through natural mineralized formations.

4.41 Treatment Plant

If a water treatment plant is not being operated effectively and reliably, the substances that are to be removed or altered can pass into the distribution system substantially unchanged. For example, a chlorinator breakdown could result in pathogenic organisms entering the distribution system. Chemicals used at water treatment facilities can also cause quality problems as when chlorination of water containing organic materials results in the formation of trihalomethanes (which are suspected *CARCINOGENS*[5]). Improper chlorination can cause objectionable tastes and odors; overdosing of alum during coagulation treatment can result in turbidity and deposits; and excessive use of manganese compounds for taste and odor control can result in a colored water.

QUESTIONS

Write your answers in a notebook and then compare your answers with those on page 176.

4.4A Why are "natural" source waters never completely pure?

4.4B How can surface waters become contaminated or degraded?

4.4C How can groundwaters become contaminated or degraded?

4.4D What problems can be created by improper chlorination in water treatment plants?

4.5 DISTRIBUTION SYSTEMS

Many conditions exist that could result in degradation of water quality in the distribution system. The system operator must be aware of them and take corrective or protective measures where possible. This section discusses the general

---

[5] *Carcinogen (CAR-sin-o-JEN). Any substance which tends to produce cancer in an organism.*

causes of quality degradation in distribution systems, particularly in water mains and storage facilities.

4.50   Causes of Quality Degradation in Distribution Systems

4.500   Cross Connections

A cross connection can be defined as an unprotected connection between any part of a water system used or intended to be used to supply water for drinking purposes and any source or system containing water or a substance that is not or cannot be approved as safe, wholesome, and potable for human consumption. Contamination resulting from *BACK-FLOW*[6] of unacceptable substances through cross connections to distribution systems has consistently caused more waterborne disease outbreaks in the United States than any other reported factor. Backflow can occur through a cross connection either by "backsiphonage" or "back pressure." Backsiphonage is a form of backflow caused by negative (below atmospheric) pressure in the water supply piping. Backsiphonage can develop from such causes as main breaks, inadequate source of supply capacity, undersized mains, unusual water demands, planned shutdowns for maintenance or repair, or the use of on-line booster pumps. Backflow caused by "back pressure" can occur when the user's water system is at a higher pressure than the public water system. Examples of sources of back pressure are services to premises where wastewater or toxic chemicals are handled under pressure, or where there are unapproved auxiliary supplies such as a private well.

4.501   Corrosion

Corrosion is the gradual deterioration or destruction of a substance or material by chemical action proceeding inward from the surface. Corrosion is frequently induced by *ELECTROCHEMICAL REACTIONS*.[7] Corrosion is of significant health and economic, as well as aesthetic, significance. Elevated levels of toxic or suspected toxic substances such as lead, cadmium, copper, zinc, asbestos, and certain organic compounds have been found in water being served from distribution systems as the result of corrosive action of water on the distribution system materials. Increased incidences of cardiovascular (heart) disease have been associated with consumption of a soft, corrosive water.

Perhaps the greatest water quality nuisance is the corrosion and precipitation of iron. "Red" water is the most common consumer complaint. The exact chemistry of how iron goes into solution and subsequently precipitates to yield rusty or "red" water is not fully understood. However, it appears that a number of different microorganisms are involved. Bacteria are rarely absent where iron is found in abundance in treated waters. The iron bacteria, *CRENOTHRIX*, are of special concern. These bacteria precipitate iron which forms the deposits in pipes, reduces carrying capacity, and produces color in water. The dead organisms also impart a disagreeable taste to the water. Improper potassium permanganate feed may also cause red-colored water.

Of the materials used for construction in water systems, metals are the most susceptible to corrosion; however, asbestos-cement pipe is also susceptible. Plastic (such as PVC or polyvinyl chloride) pipe is the least susceptible to corrosion. To prevent corrosion of cast-iron (ductile-iron) or steel pipes and to aid in the curing of cement-lined pipes, the inner side of such pipes may be lined with asphalt or coal tar materials. The types and durability of coatings for corrosion control are of special importance; however, some coatings can impart taste and odors to the water. Because of its importance and complexity, corrosion is discussed in detail in Chapter 8, "Corrosion Control," *WATER TREATMENT PLANT OPERATION*, Volume I.

4.502   Biological Growth and Activity (Biofilm)

Regardless of the efficiency of the treatment process, some microorganisms will enter the water and possibly interact with the distribution system. Geldreich and others (*OPFLOW*, Volume 23, No. 8, August 1997) with the EPA have reported that in waters with turbidity as low as 3.8 units, coliform bacteria have been observed to consistently survive 0.1 to 0.5 mg/L free chlorine residual after 30 minutes contact time. Some non-coliform organisms may be even more resistant. The survival and possible regrowth of these organisms may be affected by such factors as the amount of exposure to residual concentrations of chlorine, the amount of bacterial nutrients in the deposits, and water temperatures. The term "aftergrowth" has been used to describe the development of coliform organisms in distribution systems even though water delivered to the system meets bacterial standards and where other causes of contamination of the water appear unlikely.

Pockets of sediment deposited in the distribution system or pipe encrusted with chemical deposits (Figure 4.1) may form a protected habitat for these organisms. Such deposits often consist of silt, coagulants, precipitated chemicals, or products of corrosion. Some deposits have been found to contain bacterial populations of more than 100,000 organisms per gram.

Slimes are organic substances of a viscous nature formed from microbiological growth. Tastes and odors can be produced by slime growths of organisms that thrive on ammonia, iron, sulfide, and methane. The development of the iron bacteria *CRENOTHRIX* also results in slime growths.

Increasing levels of microbial activity can ultimately result in increased consumer complaints as well as accelerated corrosion and reduced flow from greater turbulence along the pipe

---

[6] *Backflow.   A reverse flow condition, created by a difference in water pressures, which causes water to flow back into the distribution pipes of a potable water supply from any source or sources other than an intended source.*

[7] *Electrochemical Reaction.   Chemical changes produced by electricity (electrolysis) or the production of electricity by chemical changes (galvanic action). In corrosion, a chemical reaction is accompanied by the flow of electrons through a metallic path. The electron flow may come from an external source and cause the reaction, such as electrolysis caused by a D.C. (direct current) electric railway or the electron flow may be caused by a chemical reaction as in the galvanic action of a flashlight dry cell.*

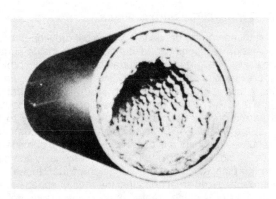

Fig. 4.1 Pipe encrusted with chemical deposits

walls. Taste and odor problems may result from the death and decay of organisms after heavy chlorination.

Biofilms are the result of a complex interaction among microorganisms. The organisms form microcolonies and secrete extracellular material that makes them highly resistant to biocides. A common example of a biofilm is the black staining around the bottom of a shower curtain. Biofilms may appear as a patchy mass within a pipe or as a uniform film inside a storage tank. The majority of microbial growth within a water distribution system occurs within the biofilms at the surface of the pipe and not in the water flowing in the pipe.

Biofilms will grow more easily on materials that supply nutrients, such as rubber gaskets or certain oils used in pumps. They need carbon, nitrogen, and phosphate in a ratio of 100:10:1. Iron is also an essential growth nutrient of limited supply in the water in a distribution system. Biofilm activity proceeds all year; however, the growth rate is faster in warmer water. TUBERCLE[8] crust contains high concentrations of nutrients for bacterial growth. Corrosion areas in iron pipe are often sites with the highest microbial activity.

Microbial control requires a good disinfectant residual. This can be accomplished with chlorine as either free chlorine, CHLORAMINES,[9] or chlorine dioxide. Chlorine is considered a more powerful disinfectant than chloramines. Chloramines are more persistent in a distribution system and will eventually penetrate farther into the system for a longer time. This persistence is why chloramines are more effective against troublesome biofilms. Free chlorine will attack biofilms and exhaust itself more quickly, whereas chloramines will continue to attack the biofilm and slowly break down the defenses of the biofilm. In chloraminated distribution systems, switching to free chlorine a few weeks before flushing each year can be an effective way to remove bacteria that become accustomed to chloramines.

The link between corrosion and biofilm control is a serious concern among distribution system operators. Iron corrosion control can be much more difficult to achieve than lead and copper corrosion control. When iron pipe corrodes, the accumulation of corrosion products on the pipe surface interferes with the ability of the disinfectant to penetrate the biofilm and inactivate the bacteria. In general, improved corrosion control

(pH, alkalinity, corrosion inhibitors) improves biofilm disinfection. There is concern that phosphate-based corrosion inhibitors will stimulate bacterial growth. Zinc orthophosphate has effectively controlled corrosion for compliance with the Lead and Copper Rule and also has been associated with improved microbial water quality. Because the corrosivity of water changes seasonally, some distribution system agencies are increasing the use of corrosion inhibitors during the warmer summer months when biofilm regrowth problems are most common.

QUESTIONS

Write your answers in a notebook and then compare your answers with those on page 176.

4.5A What is the most frequently reported cause of waterborne disease outbreaks in the United States?

4.5B What are the two causes of backflow through a cross connection?

4.5C How can corrosion cause health problems?

4.5D Slime growths can cause what problems in distribution systems?

4.5E Which form of chlorine is most effective against biofilms?

4.503 Temperature

Water temperature has three major impacts on system water quality: (1) higher temperatures tend to speed the rate of chemical reactions and increase biological growth rates; (2) biological decomposition may be intensified by summer temperatures; and (3) chlorine demand may be considerably greater so residuals won't carry as far in the distribution system in the summer.

4.504 Flow

Large variations in flow through a system may adversely affect water quality in three ways. First, changes in water velocity and flow reversals can result in sediments being stirred up and carried along until they reach the consumer. Second, low circulation and stagnant water can result in the growth of organisms, formation of sediments and corrosion products, depletion of oxygen, and increased tastes and odors. And third, turbulence can ENTRAIN[10] air into the supply causing "milky" water, which is objectionable to consumers.

4.505 Time in System

The "age" of water at a particular point in the distribution system can influence water quality. "Age" is defined as the total time from the entrance of the water into the system until it reaches a specific point of measurement. Water delivered to consumers close to the source might be only minutes old, while that received at remote parts of the system might have been in the system for several days or more depending on storage and demands. The longer the water is in the system, the more time is available for chemical and biological changes to take place.

---

[8] Tubercle (TOO-burr-cull). A protective crust of corrosion products (rust) which builds up over a pit caused by the loss of metal due to corrosion.

[9] Chloramines (KLOR-uh-means). Compounds formed by the reaction of hypochlorous acid (or aqueous chlorine) with ammonia.

[10] Entrain. To trap bubbles in water either mechanically through turbulence or chemically through a reaction.

4.506   Age of Facilities

As water mains and reservoirs become older, they require more and more maintenance. Gradual deterioration in protection against corrosion may lead to water quality problems. Ruptures and leaks in piping become more frequent. In older systems, it is not uncommon to find that the pipes and tanks were of poor construction when installed and perhaps had little or no protective coating even when new.

4.507   Operation

Careless or poor operating procedures can also result in water quality degradation. These include inadequate cross-connection control, poorly performed water quality monitoring and flushing programs, tolerance of low or negative pressures in the distribution system, insufficient surveillance to determine whether the protective features for storage facilities and mains are still adequate, inadequate disinfection after repair or installation of new facilities, and disregard for the hazards of installing facilities such as sewers in close proximity to water lines and reservoirs. The lack of trained, qualified, adequately paid operators contributes to the problem.

QUESTIONS

Write your answers in a notebook and then compare your answers with those on pages 176 and 177.

4.5F   High water temperatures can cause what problems to system water quality?

4.5G   What problems can be caused by low flows?

4.5H   How is water affected by the length of time it remains in a distribution system?

4.5I   What problems can develop as transmission and storage facilities grow older?

4.5J   How do water supply system operators contribute to the degradation of water quality in distribution systems?

4.51   Quality Degradation in Water Mains

In addition to the above considerations, quality deterioration can occur in mains because of pipe characteristics, material, construction, and location. The causes of such deterioration are discussed in the following paragraphs.

4.510   Large Interior Surface Areas

Substances can be deposited, biological growths attached, and corrosion reactions can take place on the inside surfaces of pipe walls. The large interior surface area available provides considerable space for these activities to take place. For example, 1 mile (1,600 km) of 6-inch (150-mm) diameter pipe has 8,300 sq ft (770 sq m) of inside pipe surface. Reduced flows permit greater chemical and biological activity to take place at the INTERFACE[11] where the pipe and water meet. This relatively quiet zone is capable of harboring considerable chemical or biological activity. Therefore, the quality conditions next to the pipe wall may be quite different from those in the main stream flow. In smaller mains, the ratio of surface area to volume of water is much greater, making the quality problems in these mains much more likely to be serious if chemical or biological activity occurs.

4.511   Dead Ends

The lack of circulation in a dead end creates nearly ideal conditions for degrading of water quality. The velocity is very low if not zero; the time of contact of water with the pipe and any deposits, encrustations, or slimes is long; there may be an accumulation of organic matter containing nutrients; organisms grow and use the available dissolved oxygen; and oxygen may be depleted, thus initiating ANAEROBIC[12] conditions which produce carbon dioxide, methane, and foul sulfide odors. Carbon dioxide may increase the corrosion potential. Chlorination treatment, if provided, may not even be effective because of increased demand by organics, biological forms, and corrosion products.

4.512   Pipe Material

Pipes generally are made of metal, asbestos-cement, or plastic. In metal and asbestos-cement pipe, corrosion can cause the release of substances from the piping materials into the water. With coal tar-lined or plastic pipe, however, the physical characteristics of the water are of minor importance; the concern is the possibility of leaching of material from the pipe or the lining by the water. To prevent corrosion, the insides of cement or metal pipes may be lined with asphalt or coal tar coatings. The potential for corrosion then relates to the integrity (durability) of these coatings.

Of special concern are systems that still have considerable unlined pipe in the ground and therefore are more susceptible to corrosion problems. When pipes of dissimilar metals are connected, there may be increased corrosion near the connection. Asbestos-cement pipe will deteriorate if aggressive water conditions exist, thus releasing increasing amounts of asbestos fibers. Corrosion has been discussed here to a limited degree and is discussed in more detail in Chapter 8, "Corrosion Control," in WATER TREATMENT PLANT OPERATION, Volume I, in this series of manuals.

4.513   Pipe Construction and Repair

There is ample opportunity for entry of contaminants during construction of new mains and repair of old mains. To ensure

---

[11] Interface.   The common boundary layer between two substances such as water and a solid (metal); or between two fluids such as water and a gas (air); or between a liquid (water) and another liquid (oil).

[12] Anaerobic (AN-air-O-bick).   A condition in which atmospheric or dissolved molecular oxygen is NOT present in the aquatic (water) environment.

the safety of the delivered water, proper protective, cleaning, and disinfection practices must be followed. When lines are being repaired, keep the hole dewatered to prevent possible contamination of the water line. After the line has been repaired, thoroughly flush the line downstream from the repair to remove any dirt and mud that may have entered the line. Flushing is important not only for health reasons, but to avoid consumer complaints. After flushing, the line should be disinfected before being returned to service. See Chapter 5, "Distribution System Operation and Maintenance," and Chapter 6, "Disinfection," for procedures on how to flush and disinfect water mains.

4.514 Proximity to Hazardous Facilities

When physical conditions or soil conditions prevent minimum separation distances from being met, water mains may be located adjacent to sewer lines, fuel lines, individual septic tanks and disposal systems, or tanks containing dangerous materials. If leakage from such hazardous facilities saturates the soil around the mains, the mains could become contaminated. When they are out of service for any reason and are not under pressure (or are under negative pressure) contamination may seep into the lines through cracked or inadequately sealed joints.

4.515 Appurtenances

Distribution system water quality can be adversely affected by improperly constructed or poorly located blowoffs or vacuum/air-relief valves. They may be located where they could become flooded and permit the entrance of contaminated water when they are open. A lack of blowoffs would make it difficult or impossible to correct a poor quality condition, or at least to give temporary relief. Air valves (or some form of manual air bleed valve) are important to relieve entrapped air and to help prevent milky water and surge problems when lines are being filled.

4.516 Operation

Hydrant testing and flushing programs can result in dirty water reaching consumers—and bring numerous complaints to the water utility. The dirty water results from high water velocities suspending material that has settled out and accumulated on the bottoms of water mains. For additional information on how to test hydrants, flush mains, and respond to dirty water complaints, see Chapter 5, "Distribution System Operation and Maintenance."

QUESTIONS

Write your answers in a notebook and then compare your answers with those on page 177.

4.5K What problems can develop at the interface where the distribution pipe and water meet?

4.5L Why may a chlorine residual not be effective at the dead end of a water main?

4.5M What is the major water quality concern regarding coal tar-lined pipe or plastic pipe?

4.5N How can water be protected from contaminants that enter the system during construction of new mains and repair of old mains?

4.5O What is the purpose of air valves (or some form of air-bleed valve)?

4.5P What problem might result from hydrant testing and flushing programs?

4.52 Quality Degradation in Storage Facilities

Tanks and reservoirs provide significant opportunity for pollution or contamination of the water they contain. Water in these facilities is not normally under pressure, the facility is never completely tight (except for pressure tanks), and there may be exposure to many sources of pollution. For a summary of problems, causes, and solutions, see Table 2.1 "Troubleshooting Water Quality Problems in Storage Tanks and Distribution Systems," page 32.

4.520 Uncovered Reservoirs

The use of distribution reservoirs that do not have covers is not encouraged and is often prohibited. Open reservoirs are the most vulnerable part of a water system (Figure 4.2). They are subject to contamination from vandalism; windblown and atmospheric contaminants; birds, animals, and rodents; and illegal bathing and fishing. Drownings have even occurred. Materials have been thrown into them either maliciously, carelessly, or for sabotage. An outbreak of salmonellosis in Alaska was attributed to sea gull contamination of a reservoir on a hill not far from where gulls had access to bay waters contaminated by raw domestic wastewater. Algae develop more readily in open reservoirs and may produce taste and odor problems. The larvae of the chironomid gnat are small worms and infestation of these worms in some distribution systems has led to severe consumer complaints. In most instances, the infestation started after deposition of gnat eggs in open reservoirs.

Note floating material having appearance of grease balls in lower pictures.

Fig. 4.2   Open reservoir

4.521   Inadequate Covers (Figures 4.3 and 4.4)

If covers are not properly maintained, animal, bird, and windblown contamination can enter through exposed openings in roofs or torn screening. A poorly constructed reservoir roof can often be worse than no roof at all in contributing contamination materials since bird droppings, dust, and other materials that have accumulated on a roof over a long seasonal dry period might be washed into the reservoir in a short period of time. If the vent areas on the roof are not protected, rain and windblown roof drainage debris may enter through these areas into the reservoir.

4.522   Below-Ground Construction (Figure 4.5)

Reservoirs that have part or all of their side walls below ground (cut and fill types) are subject to contamination from any materials leaching through the soils and getting into the reservoir vicinity if there is a failure (crack) in the side wall. Nearby septic systems or wastewater facilities are of special concern. Below-ground reservoirs are also more susceptible to flooding by surface storm waters and to windblown contamination. They are easier to vandalize since both the roof and the side wall vents are nearer to ground level and thus more accessible.

4.523   Other Sources

Improperly screened overflows may permit the entrance of small animals. Any cross connection between a drain or overflow and a sewer line is considered to be a serious hazard because it may result in a backflow of wastewater into the reservoir. Reservoirs that are not heavily used can be a source of taste, odor, and contamination problems.

4.524   Operation

Inadequate disinfection of reservoirs (or no disinfection at all) after construction or repairs may well result in degradation of water quality. If the water level in a reservoir is permitted to get too low, a *VORTEX*[13] will be formed in the water being discharged. Air may be entrained or materials that have settled in the bottom of the reservoir may be stirred up and get into the mains and the customers' service lines.

Reservoirs should be inspected at least twice a month. Remove any surface debris. In many cases the water quality of a storage facility is related to water clarity and surface debris.

4.53   Water Quality Monitoring

Water quality monitoring of distribution systems is important to identify when and where water quality changes occur in the system. Routine monitoring consists of collecting samples at remote locations in the system and testing for chlorine residual and coliforms. The minimum number of samples per month is based on the population served. When problems or complaints develop, a more detailed monitoring program is required to identify the cause and source of the problem. See Chapter 5, "Distribution System Operation and Maintenance," Section 5.2, "Water Quality Monitoring," for more details.

QUESTIONS

Write your answers in a notebook and then compare your answers with those on page 177.

4.5Q   How can open reservoirs become contaminated?

4.5R   Why can a poorly constructed reservoir roof be worse than no roof at all in terms of contamination?

4.5S   What are some of the problems associated with below-ground reservoirs?

4.5T   Why should the water level in storage facilities not be allowed to get too low?

4.6   TROUBLESHOOTING

This chapter has described the importance of water quality considerations in distribution systems. Types and sources of contaminants and causes of water quality degradation in water mains and storage facilities have been discussed. These water quality problems should not develop in properly designed, constructed, operated, and maintained water distribution systems.

Operators of distribution systems can maintain water quality in distribution systems by developing and implementing an effective distribution system operation and maintenance (O & M) program. Critical elements of an effective O & M program include:

- A comprehensive system surveillance and monitoring program;

- Developing and enforcing good main repair and replacement procedures, especially for disinfecting mains before placing them in service;

- Instituting a biofilm control program;

- Developing and implementing a unidirectional flushing program;

- Developing a regular schedule and program for inspecting and maintaining storage tanks and making sure there is adequate turnover of water in the tank (at least one-third of the water in the tank);

---

[13] *Vortex.   A revolving mass of water which forms a whirlpool. This whirlpool is caused by water flowing out of a small opening in the bottom of a basin or reservoir. A funnel-shaped opening is created downward from the water surface.*

Note openings in reservoir roof, lack of roofing paper, and general unsatisfactory conditions.

Fig. 4.3   Reservoir roof

Note rodent droppings next to pencil on concrete ledge inside reservoir.

Note deteriorated and rotting sills creating openings along sides into reservoir.

Fig. 4.4   Defective reservoir roof

Fig. 4.5   Reservoir with most of side wall below ground

- Developing and maintaining an active corrosion-control program;

- Instituting an annual cleaning and lining program;

- Developing a program to eliminate as many of the dead ends as possible in the distribution system;

- Developing a valve exercise program to ensure proper setting and operation of valves;

- Developing and implementing a backflow prevention program;

- Maintaining positive pressure on system at all times; and

- Thoroughly investigating and responding to all complaints.

Chapter 5, "Distribution System Operation and Maintenance," contains information on how to properly operate and maintain distribution systems. Also, procedures are outlined there on how to identify and locate the sources of problems causing water quality degradation and potential solutions to the problems.

4.7 ARITHMETIC ASSIGNMENT

Turn to the Appendix, "How to Solve Water Distribution System Arithmetic Problems," at the back of this manual and read all of Section A.7, *VELOCITY AND FLOW RATE*. Work all of the problems on your pocket calculator. You should be able to get the same answers.

4.8 ADDITIONAL READING

NEW YORK MANUAL, Chapter 17,* "Protection of Treated Water."

* Depends on edition.

Please answer the discussion and review questions next.

DISCUSSION AND REVIEW QUESTIONS

Chapter 4.   WATER QUALITY CONSIDERATIONS IN DISTRIBUTION SYSTEMS

Write the answers to these questions in your notebook. The purpose of these questions is to indicate to you how well you understand the material in the chapter.

1. What water quality criteria must a good quality domestic water supply meet?

2. What are the major sources of contaminants in a water distribution system?

3. How can water treatment plants contribute to water quality problems in a water distribution system?

4. What operational deficiencies can result in water quality degradation?

5. How may water quality be degraded in water mains?

6. How can water be degraded in dead ends?

7. What are some typical types of hazardous facilities that should not be located near water mains?

8. Under what conditions might water storage facilities contribute to water quality degradation?

SUGGESTED ANSWERS

Chapter 4.   WATER QUALITY CONSIDERATIONS IN DISTRIBUTION SYSTEMS

Answers to questions on page 163.

4.0A   Water quality is the general term used to describe the composite chemical, physical, and biological characteristics of a water.

4.0B   A domestic water supply is considered to be of good quality when it is free of disease-causing organisms and toxic chemicals, attractive in taste and appearance, of such a chemical composition that it may be distributed without undue corrosive or scale-forming effects on the water system, and will satisfy, to a maximum degree, the requirements of domestic and industrial consumers.

4.0C   The consequences of delivering a poor quality water to the consumer range from the water not being acceptable to the consumer because of its appearance or taste, to illness or even death of some susceptible consumers.

Answers to questions on page 164.

4.1A   Water quality standards have been prepared and used by the waterworks industry to provide the needed quality control and to ensure the acceptability of the product water.

4.1B   Drinking water quality standards can generally be categorized as being related either to *HEALTH* or *AESTHETICS*.

4.1C   MCL stands for **M**aximum **C**ontaminant **L**evel.

Answers to questions on page 165.

4.2A   Trihalomethanes (THMs) are derivatives of methane, CH_4, in which three halogen atoms (chlorine or bromine) are substituted for three of the hydrogen atoms. These may be formed during chlorination by reactions with natural organic materials in the water. The resulting compounds (THMs) are suspected of causing cancer.

4.2B   Physical qualities of water important to consumers include color, turbidity, taste, odor, and temperature. Other physical items of concern include milky water due to air and macroscopic organisms such as worms or bugs in the water.

4.3A   Contamination of water can occur at the source of supply, at the treatment plant, in distribution system storage facilities, or in water mains.

4.3B   Water can be degraded in the distribution system by cross connections, corrosion, biological growth and activity, high temperatures, unusual flows, time in system, dead ends, age of facilities, and operational procedures.

Answers to questions on page 165.

4.4A   "Natural" source waters are never completely pure because during their precipitation (falling as rain or snow) and their passage over or through the ground, they acquire a wide variety of dissolved or suspended impurities.

4.4B   Surface waters can become contaminated or degraded from domestic, industrial, or agricultural waste discharges and spills; humic materials such as decaying vegetation; refuse disposal; recreation; construction; animal activities; uncontrolled growths of algae and weeds; and radioactive fallout.

4.4C   Groundwaters may become contaminated or degraded by leaching from nearby domestic, industrial, or agricultural wastewater operations or disposals; spills; humic materials; flood waters; or passage through natural mineralized formations.

4.4D   Improper chlorination in water treatment plants can cause tastes and odors in waters delivered to consumers, and may produce trihalomethanes.

Answers to questions on page 167.

4.5A   Contamination through cross connections is the most frequently reported cause of waterborne disease outbreaks in the United States.

4.5B   Backflow through a cross connection can occur either by "backsiphonage" or "back pressure."

4.5C   Corrosion can cause health problems by increasing the levels of toxic or suspected toxic substances in water such as lead, cadmium, copper, zinc, asbestos, and certain organic compounds. Increased incidences of cardiovascular (heart) disease have been associated with consumption of soft, corrosive water.

4.5D   Slime growths can cause taste and odor problems in water distribution systems. Microbial activities can result in accelerated corrosion and reduced flow due to turbulence.

4.5E   The persistence of chloramines is why they are the most effective form of chlorine against biofilms.

Answers to questions on page 168.

4.5F   High water temperatures tend to speed the rate of chemical reactions and biological growth rates. Biological decomposition may also be intensified. Chlorine demand may be considerably greater so residuals may not carry as far into the distribution system.

4.5G   Low flows can contribute to the growth of organisms, formation of sediments and corrosion products, depletion of oxygen, and increased tastes and odors.

4.5H   The longer water is in the distribution system, the more time is available for chemical and biological changes to take place.

4.5I   As facilities grow older, protection against corrosion deteriorates and ruptures and leaks become more frequent.

4.5J   Careless or poor operating procedures by operators can result in water quality degradation. The lack of trained, qualified, adequately paid operators contributes to the problem.

Answers to questions on page 169.

4.5K   The interface where the pipe and water meet is a relatively quiet zone capable of harboring considerable chemical or biological activity. This activity could cause water quality degradation.

4.5L   A chlorine residual may not be effective at the dead end of a water main because of the increased chlorine demand by organics, biological forms, and corrosion products at that point.

4.5M   The major water quality concern in the case of coal tar-lined pipe or plastic pipe is the possible leaching of material from the pipe or the lining by the water.

4.5N   Contaminants that enter the system during construction of new mains and repair of old mains are removed by proper protective, cleaning, and disinfection practices.

4.5O   Air valves or some form of air-bleed valve are important to relieve entrapped air and to help prevent milky water and surge problems when lines are being filled.

4.5P   Hydrant testing and flushing programs can result in dirty water reaching consumers with resulting numerous complaints.

Answers to questions on page 171.

4.5Q   Open reservoirs are subject to contamination from vandalism; windblown and atmospheric contaminants; birds, animals, and rodents; and illegal bathing and fishing. Drowning has occurred; materials have been thrown into reservoirs; birds can carry salmonella bacteria into open reservoirs; and algae can cause taste and odor problems.

4.5R   A poorly constructed reservoir roof can be worse than no roof at all in contributing contamination materials since bird droppings, dust, and other materials that have accumulated on the roof over a long seasonal dry period might be washed into the reservoir in a short period.

4.5S   Below-ground reservoirs may be subject to contamination from materials leaching through soils and then through cracks or openings in the side wall. Below-ground reservoirs are also more susceptible to flooding by surface storm waters and to windblown contamination. They are easier to vandalize since both the roof and the side wall vents are nearer to ground level and thus more accessible.

4.5T   The water level in storage facilities should not be allowed to get too low because a vortex could be formed in the water being discharged. Air may be entrained or materials that have settled in the bottom of the reservoir may be stirred up and get into the mains and then into the customers' service lines.

CHAPTER 5

DISTRIBUTION SYSTEM OPERATION AND MAINTENANCE

by

Sam Kalichman

and

Nick Nichols

TABLE OF CONTENTS

Chapter 5.   DISTRIBUTION SYSTEM OPERATION AND MAINTENANCE

OBJECTIVES

Chapter 5.   DISTRIBUTION SYSTEM OPERATION AND MAINTENANCE

Following completion of Chapter 5, you should be able to:

1. Develop and conduct a water distribution system surveillance program,

2. Develop and conduct a water quality monitoring program for a water distribution system,

3. Develop and conduct a cross-connection control program,

4. Locate buried pipes and leaks,

5. Repair leaks,

6. Make pipe connections,

7. Flush pipes,

8. Clean pipes,

9. Contract out a pipe lining job,

10. Thaw frozen pipes and hydrants,

11. Test and read meters,

12. Disinfect mains and storage facilities,

13. Develop a recordkeeping system and keep accurate records,

14. Train operators to prepare for and respond to emergencies,

15. Effectively deal with the public,

16. Maintain the landscaped areas around distribution system facilities, and

17. Safely operate and maintain a water distribution system.

WORDS

Chapter 5.   DISTRIBUTION SYSTEM OPERATION AND MAINTENANCE

AUDIT, WATER   AUDIT, WATER

A thorough examination of the accuracy of water agency records or accounts (volumes of water) and system control equipment. Water managers can use audits to determine their water distribution system efficiency. The overall goal is to identify and verify water and revenue losses in a water system.

BACK PRESSURE   BACK PRESSURE

A pressure that can cause water to backflow into the water supply when a user's water system is at a higher pressure than the public water system.

BACKFLOW   BACKFLOW

A reverse flow condition, created by a difference in water pressures, which causes water to flow back into the distribution pipes of a potable water supply from any source or sources other than an intended source. Also see BACKSIPHONAGE.

BACKSIPHONAGE   BACKSIPHONAGE

A form of backflow caused by a negative or below atmospheric pressure within a water system. Also see BACKFLOW.

C FACTOR   C FACTOR

A factor or value used to indicate the smoothness of the interior of a pipe. The higher the C Factor, the smoother the pipe, the greater the carrying capacity, and the smaller the friction or energy losses from water flowing in the pipe. To calculate the C Factor, measure the flow, pipe diameter, distance between two pressure gages, and the friction or energy loss of the water between the gages.

$$\text{C Factor} = \frac{\text{Flow, GPM}}{193.75(\text{Diameter, ft})^{2.63}(\text{Slope})^{0.54}}$$

CAVITATION (CAV-uh-TAY-shun)   CAVITATION

The formation and collapse of a gas pocket or bubble on the blade of an impeller or the gate of a valve. The collapse of this gas pocket or bubble drives water into the impeller or gate with a terrific force that can cause pitting on the impeller or gate surface. Cavitation is accompanied by loud noises that sound like someone is pounding on the impeller or gate with a hammer.

COUPON   COUPON

A steel specimen inserted into water to measure the corrosiveness of water. The rate of corrosion is measured as the loss of weight of the coupon (in milligrams) per surface area (in square decimeters) exposed to the water per day. 10 decimeters = 1 meter = 100 centimeters.

HYDRAULIC GRADE LINE (HGL)   HYDRAULIC GRADE LINE (HGL)

The surface or profile of water flowing in an open channel or a pipe flowing partially full. If a pipe is under pressure, the hydraulic grade line is at the level water would rise to in a small vertical tube connected to the pipe.

HYDRAULIC GRADIENT   HYDRAULIC GRADIENT

The slope of the hydraulic grade line. This is the slope of the water surface in an open channel, the slope of the water surface of the groundwater table, or the slope of the water pressure for pipes under pressure.

PRIME   PRIME

The action of filling a pump casing with water to remove the air. Most pumps must be primed before start-up or they will not pump any water.

ROTAMETER (RODE-uh-ME-ter)   ROTAMETER

A device used to measure the flow rate of gases and liquids. The gas or liquid being measured flows vertically up a tapered, calibrated tube. Inside the tube is a small ball or bullet-shaped float (it may rotate) that rises or falls depending on the flow rate. The flow rate may be read on a scale behind or on the tube by looking at the middle of the ball or at the widest part or top of the float.

SCADA (ss-KAY-dah) SYSTEM   SCADA SYSTEM

Supervisory **C**ontrol **A**nd **D**ata **A**cquisition system. A computer-monitored alarm, response, control and data acquisition system used by drinking water facilities to monitor their operations.

SANITARY SURVEY   SANITARY SURVEY

A detailed evaluation and/or inspection of a source of water supply and all conveyances, storage, treatment and distribution facilities to ensure protection of the water supply from all pollution sources.

SUCTION LIFT   SUCTION LIFT

The *NEGATIVE* pressure [in feet (meters) of water or inches (centimeters) of mercury vacuum] on the suction side of the pump. The pressure can be measured from the centerline of the pump *DOWN TO* (lift) the elevation of the hydraulic grade line on the suction side of the pump.

TUBERCULATION (too-BURR-cue-LAY-shun)   TUBERCULATION

The development or formation of small mounds of corrosion products (rust) on the inside of iron pipe. These mounds (tubercles) increase the roughness of the inside of the pipe thus increasing resistance to water flow (decreases the C Factor).

VENTURI METER   VENTURI METER

A flow measuring device placed in a pipe. The device consists of a tube whose diameter gradually decreases to a throat and then gradually expands to the diameter of the pipe. The flow is determined on the basis of the difference in pressure (caused by different velocity heads) between the entrance and throat of the Venturi meter.

NOTE: Most Venturi meters have pressure sensing taps rather than a manometer to measure the pressure difference. The upstream tap is the high pressure tap or side of the manometer.

VENTURI METER

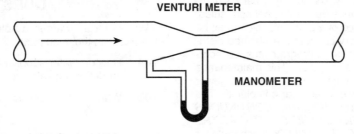

WATER AUDIT   WATER AUDIT

A thorough examination of the accuracy of water agency records or accounts (volumes of water) and system control equipment. Water managers can use audits to determine their water distribution system efficiency. The overall goal is to identify and verify water and revenue losses in a water system.

WATER HAMMER   WATER HAMMER

The sound like someone hammering on a pipe that occurs when a valve is opened or closed very rapidly. When a valve position is changed quickly, the water pressure in a pipe will increase and decrease back and forth very quickly. This rise and fall in pressures can cause serious damage to the system.

CHAPTER 5.   DISTRIBUTION SYSTEM OPERATION AND MAINTENANCE

(Lesson 1 of 4 Lessons)

5.0   NEED FOR SYSTEM OPERATION AND MAINTENANCE

Chapter 3 reviewed the design, construction, and installation of distribution system facilities, while Chapter 4 examined the water quality problems found in distribution systems and how they originate. In this chapter, consideration is given to operation and maintenance procedures needed to reliably ensure the delivery of sufficient water at adequate pressures and prevent or minimize water quality degradation. Operation and maintenance are ongoing daily concerns, in addition to emergency situations, as a never-ending job for water distribution system operators. Unless they are performed systematically, properly, and on a timely basis, problems in a system will get worse. The water supplier's management and operators have the legal responsibility for delivering a safe water to their customers and this is largely done through the conscientious operation and maintenance of the system facilities.

Both preventive and corrective maintenance are necessary. Preventive maintenance is that which is specifically scheduled; while corrective maintenance (repairs) is not scheduled, but is done after the appearance of a problem which must be corrected to continue satisfactory operation. A planned distribution system preventive maintenance program must be incorporated into every water utility's operation. This maintenance program ensures the safety of the water and results in fewer interruptions of service and fewer consumer complaints. The objective of maintenance is the *MAXIMUM CONTINUOUS SERVICE OF THE EQUIPMENT IN THE SYSTEM AT THE LOWEST POSSIBLE COST TO THE UTILITY.* A schedule for routine preventive maintenance should be prepared for *ALL* utility equipment. Operators should be required to follow the prepared schedule and document the work they do.

Also needed are proper and specific operational procedures for inspecting, monitoring, testing, repairing, and disinfecting a system, as well as for locating underground facilities and other activities that are done more or less on a routine basis. Weak links in the system must be detected and eliminated. System records must be sufficient to document the system facilities, their condition, the routine maintenance that should be done,

all maintenance that has been done, the problems that have been found, and corrective measures that were taken.

Routine and emergency operating procedures should be in writing and clear to all operators with the authority to act in emergencies (also see Section 5.12, "Emergency Planning"). Frequent training or refresher courses for all operators will help keep them familiar with the procedures and equipment for which they are responsible. Proper operation and maintenance will keep the system operating smoothly and will extend the useful life of the facilities.

Whenever it is necessary to work on equipment, be aware of the hazards associated with the job. *THINK SAFETY!* Lock out, block out, and tag any equipment that is being repaired or adjusted. Follow the safety guidelines presented in Section 7.37, "Working Around Electrical Units," to prevent accidental start-up of equipment or movement of parts that could seriously injure anyone working on the equipment.

QUESTIONS

Write your answers in a notebook and then compare your answers with those on page 287.

5.0A   How do water suppliers meet their legal responsibility for delivering safe water to customers?

5.0B   What is the objective of distribution system maintenance?

5.1   SYSTEM SURVEILLANCE

5.10   Purpose of System Surveillance

Surveillance of distribution systems is done for three reasons. First, to detect and correct any problems that are sanitary hazards; second, to detect and correct any significant deterioration of facilities or equipment used for the storage and transportation of the water supply; and third, to detect the encroachment of other utilities (sewer, power, gas, phone, cable TV). Some types of surveillance, such as checking for vandalism, are performed routinely, while others are done only under special circumstances, such as checking for damage after a storm. Generally, routine surveillance in a distribution system would only involve above-ground facilities such as reservoirs, pump stations, and valves. However, some less frequent surveillance can also be made of underground facilities as described later in this section.

Critical areas of the distribution system should be patrolled routinely so that the water utility will have an early warning of any adverse conditions that might appear. Any activity or situation that might endanger a water facility or the quality of the water must be investigated and reported without delay. Possible damage from floods, earthquakes, tornadoes, or fires needs to be looked into promptly.

Just as important as routine patrolling is having the field crews watch for actual or potential problems as they attend to their duties. The sooner corrective action is taken after a problem is found, the easier it usually is to make the correction. The longer water quality or facilities deteriorate, the more difficult the situation becomes.

5.11 Treated Water Storage Facilities

Storage facilities in the distribution system are the most obvious facilities needing surveillance. They are generally not under a positive pressure (except for pressure tanks) and are usually above or at ground level. They are, therefore, the most susceptible part of the distribution system to quality degradation from external sources. Open reservoirs are the most critical type of reservoir in this regard, with below-ground reservoirs next in line. Just about anything that one could think of has been found during inspections of reservoirs. For example, cans, bottles, papers, and even a pile of human fecal matter were found on a reservoir ledge inside a covered, 10 million gallon (38,000 cu m) reservoir. Finding small animals, dead or alive, in reservoirs is not uncommon.

Daily surveillance is recommended for trespassing, vandalism, dumping of trash, or swimming. Pay particular attention to fence and screen openings, roof damage, intact locks on reservoirs, and to manholes or doors. In one case, repeated high bacterial counts occurred and an inspection revealed that a roof manhole was open and several pigeons had entered and drowned in the water. If these daily inspections reveal any change in water color, odor, or turbidity or if any other evidence indicates the water may have been tampered with, take the facility out of service and contact the health department immediately to determine what further action should be taken. Some tests that can be quickly made are chlorine residual, conductance (for TDS), pH, and alkalinity. Bacteriological tests take longer to complete, but should also be made (for example, 24-hour coliform tests by the membrane filter method).

Inspect reservoir covers frequently to determine if they are watertight. If a roof develops leaks, any contaminants on the roof, such as bird droppings, will be washed into the reservoir. Even concrete roofs may develop cracks in time and permit leakage. Vent screens should be inspected to ensure they are in good condition. If the screens on the vents are torn or rust away, it is obvious that small animals can gain entry to the reservoir. Vent areas must be protected to prevent rain, wind-blown items (leaves, paper), roof drainage, debris and, to the extent possible, dust and dirt from entering the reservoir. Also check the screening on the vertical overflow pipes. In one case, the decomposed remains of two rodents were found in a tank which the rodents apparently had crawled into through an unscreened vertical overflow pipe some 20 ft (6 m) high.

Other reservoir inspections may be made at less frequent intervals. Weekly or more frequent inspections are advisable to note any algae, slime and/or worm growths, floating or settled materials, or deterioration (corrosion or dry rot) of roofing and wooden trusses.

After a heavy rain, inspect the reservoir for damage. If a below-ground reservoir is located where it could be affected by floods, an inspection should be made during such periods to determine whether or not storm waters are getting into the facility or erosion is undermining the structure.

At least once a year, the interior of each storage tank should be inspected to determine the condition of the interior coating and whether the tank needs to be washed out, completely cleaned, or recoated. Interior inspections may also reveal the presence of animals, birds, or debris in the tank. The condition of the vents can be seen quite well from the interior of the tank. Preferably every year, but certainly at least every other year, each reservoir should be drained and thoroughly inspected to determine if there is any significant leakage or corrosion and to observe the nature and amount of sediments on the floor or bottom. Cleaning of the reservoir usually takes place at this time. Divers are being used to inspect, clean, and repair tanks without draining them.

Routinely inspect any sewers located near below-ground reservoirs to determine if there is any noticeable leakage around them. The possibility exists that leaking wastewater could seep through (infiltrate) the ground and cracks in reservoir walls and reach the stored water. As part of the surveillance made after storms, be alert to possible overflow of wastewater from damaged or overloaded sewers to a reservoir area.

QUESTIONS

Write your answers in a notebook and then compare your answers with those on page 287.

5.1A   List the elements of surveillance that should be considered in a water distribution system surveillance program.

5.1B   Why should critical parts of a distribution system be patrolled on a routine basis?

5.1C   What items should be checked during a daily surveillance inspection of a reservoir?

5.1D   What items should be checked during weekly inspections of a reservoir?

5.12   Mains

Water utility operators should be on the lookout for signs of unauthorized construction activity or soil erosion on or near the utility's pipelines which may pose a physical threat to the line. All wastewater facilities must be kept a proper distance from the mains or meet special construction requirements as discussed in Chapter 3. An observer from the utility should be present during any digging or excavation (especially blasting) near the mains. Reasonable access to the main must always be maintained, and there should be no construction on the piping right-of-way unless it is authorized by the utility. During the routine work of the field crews, operators should be on the lookout for possible main leaks and report any wet spots, sunken areas, or other unusual conditions.

Even though water mains are located in the ground and out of sight, they still need to be monitored. Valuable information can be obtained by inspecting the condition of the inside of mains when repairs or additions are made to the system, or even by examining pipe cut-outs from main tapping operations. When *TUBERCULATION*[1] or other deposition is found, take samples to the laboratory for analysis or further study to help determine the cause of their formation and to develop remedial treatment measures. Another method for checking pipe conditions is to place test pipe specimens (*COUPONS*[2]) in the distribution system where they can be periodically removed, examined, weighed, photographed, and put back.

5.13   Valves and Blowoffs

Valve boxes and other valve appurtenances should be checked at least annually to determine if they are damaged, filled with earth, or covered over by pavement. Vacuum or air relief valves ("air-vacs") and blowoffs should be inspected after rains or floods to ensure that they are not submerged in drainage waters and that they work properly.

5.14   Customer Services

Meter readers should be alert for any problems noted during their rounds. A periodic check should be made of the condition of the meter or curb box. If a box has been displaced and parts of it project out of the ground, there is a danger of injury resulting from tripping. This has been one of the most frequent sources of lawsuits against utilities. Leaking services should also be noted and reported.

QUESTIONS

Write your answers in a notebook and then compare your answers with those on page 287.

5.1E   Why should field crews be on the lookout for any unauthorized construction activity or erosion near the utility's pipelines?

5.1F   How often should valves and valve boxes be inspected?

5.1G   When should air relief valves be inspected?

5.15   Vandalism

The normal security measures taken to prevent vandalism include locks on doors and gates, fences, lighting, posting signs, and patrolling. These measures will not stop vandalism, but should reduce it. These protective measures must not be ignored and allowed to deteriorate into uselessness (such as a fence with a hole in it).

If vandalism or illegal entry is found, make a thorough investigation to determine what damage might have been done and whether there is any possibility of a threat to the quality of the water supply. Check the appearance and odor of the water supply in the area. Record the condition of all locks, any questionable conditions in the areas where unauthorized entry apparently took place, the presence of hazardous material containers, missing items, and damaged equipment. Contact any neighbors in the area for additional information and request assistance to watch the facility. Then promptly report any damage or questionable conditions you have found to responsible supervisory personnel. If water quality may be affected, the health department should be called immediately. If the questionable part of the system can be isolated, this should be done. Notify customers and the health department if a decision is made to shut down part of the system.

As already noted, if there is any detectable change in color, odor, or turbidity of the water, or any other evidence that the water has been tampered with, the facility should be *IMMEDIATELY TAKEN OUT OF SERVICE*. Water quality tests that can be done promptly and would be indicative of suspected contamination should be promptly run. Tests that can be quickly made are turbidity, chlorine residual, conductance (for TDS), pH, and alkalinity. Bacteriological tests take longer to complete, but should also be made (for example, 24-hour coliform tests by the membrane filter method).

Fortunately, most acts of vandalism cause only superficial damage and do no serious harm. Water quality safeguards against minor vandalism are built into most systems and their operation. Existing chlorine residuals will assist in minimizing any threat from disease organisms. A large dilution factor is normally present which will also minimize any contamination threat. Routine sampling and continuous water quality monitoring assist in revealing any possible threat to water quality. For example, telemetering of chlorine residuals and turbidity can indicate water quality problems. Similarly, telemetering of pressures, flows, and reservoir levels can help detect operational problems early.

---

[1] Tuberculation.   The development or formation of small mounds of corrosion products (rust) on the inside of iron pipe. These mounds (tubercles) increase the roughness of the inside of the pipe thus increasing resistance to water flow (decreases the C Factor).

[2] Coupon.   A steel specimen inserted into water to measure the corrosiveness of water. The rate of corrosion is measured as the loss of weight of the coupon (in milligrams) per surface area (in square decimeters) exposed to the water per day. 10 decimeters = 1 meter = 100 centimeters.

Automatic controls will often compensate for or minimize certain acts of vandalism. The fact that most of the distribution system is underground and relatively inaccessible makes only a small portion of the system actually vulnerable. Reservoirs that may have been contaminated can usually be readily isolated as reserve supplies of water are often available from other sources. "Looped" distribution lines will allow the isolation of problem areas. Finally, the consumer can be counted upon to quickly report any perceived change in water appearance, taste, or odor.

If the vandal is still present and does not leave, notify the authorities (supervisor, police) immediately. *DO NOT ATTEMPT TO USE FORCE* on a vandal, even a child, because you or the vandal could be injured. After the episode is over, make a complete written report describing the vandalism (including photos) and its actual and potential effects on the physical system and water quality.

5.16   Telemetering

Some types of surveillance can also be accomplished by using remote indicating or recording systems. Remote control of a function usually involves telemetering. Quite simply, this involves the measuring, transmitting, and receiving of data at a distance by phone lines, radio, or microwave. Indicating or recording gages can be mounted at any convenient place, or a number of them can be put together on a central control panel and can show many different types of data, such as water elevations in reservoirs or tanks, rate of discharge of pumps, pressures in mains at any distant point, and rate of flow. Telemetered water quality data might also include residual chlorine, turbidity, and/or other water quality indicators. Strip charts or circular chart recorders may be used to record telemetered signals. Telemetering also can be used to start and stop pumps and open and close valves.

5.17   Other Surveillance

Water quality monitoring and cross-connection control, which are discussed in the following sections of this chapter, can also be an important part of a utility's distribution system surveillance program.

QUESTIONS

Write your answers in a notebook and then compare your answers with those on page 287.

5.1H   What should be done if vandalism or illegal entry is discovered?

5.1I   What water quality tests could be conducted immediately after an act of vandalism or illegal entry is discovered?

5.1J   What should an operator do if a vandal is discovered in a water distribution system facility?

5.1K   Give examples of information that can be telemetered.

5.2   WATER QUALITY MONITORING

5.20   Purpose of Water Quality Monitoring

A properly designed and meaningful water quality monitoring program is needed for each water utility. Monitoring is not just done to comply with the requirements of regulatory agencies, but also to:

1. Establish the safety and potability of the supply by determining if it meets standards,

2. Provide a record of the quality of the water served,

3. Assist in determining the source of any contamination that might have reached the system,

4. Check the quality of the water to be served after installation, cleaning, or repair of mains and reservoirs,

5. Detect any changes in quality between the source and the consumer's system (to determine if there is deterioration of the water after it has gone through the distribution system),

6. Answer consumer questions,

7. Verify or refute complaints,

8. Solve special problems such as those resulting from corrosion or encrustation, and

9. Assist in tracing the flow of water through the distribution system.

5.21   Monitoring Program

Many water quality monitoring programs are based on the requirements of the United States National Primary Drinking Water Regulations. A summary of these regulations, as prepared by the United States Environmental Protection Agency (USEPA) is given in the poster inserted inside the cover of this manual. This poster shows the health effects of the contaminants to be tested, maximum contaminant levels (MCLs), and monitoring requirements. The monitoring requirements given are the *MINIMUM* that must be followed. Water suppliers should not hesitate to take more samples and should collect as many samples as necessary to clearly prove the quality of water being served throughout the system or to solve a quality problem. Requirements for total trihalomethanes apply to community water systems that chlorinate their water supplies and serve 10,000 or more people. Generally, four samples per water treatment plant must be collected every three months. An MCL of 0.10 mg/*L* must be met based on a running annual average of the quarterly samples (see Examples 1, 2, and 3 at the end of this section).

NOTE: The MCL for total trihalomethanes was lowered to 0.080 mg/*L*; the new limit takes effect in December 2001 for public water systems serving more than 10,000 people, and in December 2003 for systems serving fewer than 10,000 people.

National Secondary Drinking Water Regulations have been established for contaminants that may adversely affect the aesthetic quality of the drinking water and result in consumer rejection of the water provided. The secondary standards are not federally enforceable, but are intended as guidelines for the states. The states have been encouraged to implement these regulations so that the public will not seek to obtain drinking water from potentially lower quality, higher risk sources. The regulations recommend that these contaminants be monitored at intervals no less frequent than the monitoring performed for the inorganic chemical contaminants listed in the National Primary Drinking Water Regulations. For additional information about the Secondary Drinking Water Regulations, see WATER TREATMENT PLANT OPERATION, Volume II, Chapter 22, "Drinking Water Regulations," Section 22.3, "Secondary Drinking Water Standards."

Federal and state regulations require monitoring water within the distribution system under three specific rules.

1. Total Coliform Rule. This rule controls the microbial water quality aspects by testing for coliform bacteria and chlorine residuals.

2. Lead and Copper Rule. This rule deals with the corrosivity of water distributed to homes with lead and copper plumbing. Water is tested for lead and copper in the dead ends of water mains and from the drinking water taps of homes when residents first get up in the morning (test stagnated samples at sites with lead and copper plumbing). Other water quality measurements associated with the Lead and Copper Rule include pH, alkalinity, and the residual of any corrosion inhibitor applied to the water.

3. Trihalomethane Rule. This rule is intended to limit chlorinated disinfection by-products. Samples are collected and tested for trihalomethanes (THMs) and chlorine residuals.

Additional routine sampling is needed where sanitary hazards exist or whenever levels of contaminants approach or exceed maximum contaminant levels. Establishment of a minimum monitoring program and compliance with the quality standards are not in themselves sufficient evidence that protection of the water in the distribution system is adequate. The overall operation and maintenance procedures used by the utility and results of inspections of the physical facilities (called SANITARY SURVEYS[3]) must be considered along with the quality analyses to evaluate the adequacy of protection of the supply.

Sampling should be done in accordance with accepted collection and transportation procedures. In addition, it will be necessary to test enough samples to provide an optimum amount of information on the quality of the water being supplied. Details on collection and transportation of samples are given in Chapter 11, "Laboratory Procedures," in WATER TREATMENT PLANT OPERATION, Volume I.

FORMULAS

To calculate an average measurement for a time period or a group of measurements or observations, add up all of the measurements and divide by the number of measurements.

$$\text{Average} = \frac{\text{Sum of All Measurements}}{\text{Number of Measurements}}$$

The Safe Drinking Water Act requires that total trihalomethanes be calculated as quarterly averages as shown above. When the running annual average exceeds the MCL of 0.10 mg/L, the value should be reported to the state within 48 hours. The running annual average is calculated by using the quarterly average for the quarter being considered and the three quarters immediately before the one being considered.

$$\text{Annual Running Average} = \frac{\text{Sum of Measurements}}{\text{Number of Measurements}}$$

Running averages are used to "smooth out" data. In the case of THMs, if one quarter was high and exceeded the MCL, the other three quarters could pull the annual running average MCL down and keep the water utility in compliance.

EXAMPLE 1

Four samples of water from a water distribution system were collected and analyzed for total trihalomethanes (TTHMs). Results of the analyses were as follows: 0.07 mg/L 0.05 mg/L, 0.04 mg/L, and 0.08 mg/L. What was the quarterly TTHM average in mg/L?

Known	Unknown
Results of TTHM Analyses	Quarterly TTHM Average, mg/L

Sample No.	Conc, mg/L
1	0.07
2	0.05
3	0.04
4	0.08

Formula for calculating average values:

$$\text{Average} = \frac{\text{Sum of All Measurements}}{\text{Number of Measurements}}$$

Calculate the quarterly TTHM average for the months of January, February, and March.

$$\text{Average TTHM, mg/}L = \frac{\text{Sum of Measurements, mg/}L}{\text{Number of Measurements}}$$

$$= \frac{0.07 \text{ mg/}L + 0.05 \text{ mg/}L + 0.04 \text{ mg/}L + 0.08 \text{ mg/}L}{4}$$

$$= \frac{0.24 \text{ mg/}L}{4}$$

$$= 0.06 \text{ mg/}L$$

---

[3] Sanitary Survey.   A detailed evaluation and/or inspection of a source of water supply and all conveyances, storage, treatment and distribution facilities to ensure protection of the water supply from all pollution sources.

EXAMPLE 2

Quarterly TTHM average values for year one are given. Calculate the annual average. First quarter (January, February, and March), 0.06 mg/L; second quarter (April, May, and June), 0.08 mg/L; third quarter (July, August, and September), 0.11 mg/L; and fourth quarter (October, November, and December), 0.09 mg/L.

Known	**Unknown**
Results of Quarterly TTHM Analyses	Annual TTHM Average, mg/L

Quarter	Conc, mg/L
1	0.06
2	0.08
3	0.11
4	0.09

Calculate the annual TTHM average for the year.

$$\text{Average TTHM, mg/}L = \frac{\text{Sum of Measurements, mg/}L}{\text{Number of Measurements}}$$

$$= \frac{0.06 \text{ mg/}L + 0.08 \text{ mg/}L + 0.11 \text{ mg/}L + 0.09 \text{ mg/}L}{4}$$

$$= \frac{0.34 \text{ mg/}L}{4}$$

$$= 0.085 \text{ mg/}L$$

EXAMPLE 3

Determine the annual running TTHM average using the results in Example 2 if the results during the next year were as follows: first quarter, 0.08 mg/L; second quarter, 0.09 mg/L; third quarter, 0.14 mg/L; and fourth quarter, 0.08 mg/L.

Known	**Unknown**
Results of Quarterly TTHM Analyses	Annual Running TTHM Average, mg/L

Year 1

Quarter	Conc, mg/L
1	0.06
2	0.08
3	0.11
4	0.09

Year 2

1	0.08
2	0.09
3	0.14
4	0.08

Calculate the annual running TTHM average for each quarter during the Year 2. Use the quarterly TTHM average for the quarter being considered and the three quarters immediately before the one being considered.

Year 2, Quarter 1

$$\text{Annual Running TTHM Average, mg/}L = \frac{\text{Sum of Measurements, mg/}L}{\text{Number of Measurements}}$$

$$= \frac{0.08 \text{ mg/}L + 0.09 \text{ mg/}L + 0.11 \text{ mg/}L + 0.08 \text{ mg/}L}{4}$$

$$= \frac{0.36 \text{ mg/}L}{4}$$

$$= 0.09 \text{ mg/}L$$

Year 2, Quarter 2

$$\text{Annual Running TTHM Average, mg/}L = \frac{0.09 \text{ mg/}L + 0.08 \text{ mg/}L + 0.09 \text{ mg/}L + 0.11 \text{ mg/}L}{4}$$

$$= \frac{0.37 \text{ mg/}L}{4}$$

$$= 0.093 \text{ mg/}L$$

Year 2, Quarter 3

$$\text{Annual Running TTHM Average, mg/}L = \frac{0.14 \text{ mg/}L + 0.09 \text{ mg/}L + 0.08 \text{ mg/}L + 0.09 \text{ mg/}L}{4}$$

$$= \frac{0.40 \text{ mg/}L}{4}$$

$$= 0.10 \text{ mg/}L$$

Year 2, Quarter 4

$$\text{Annual Running TTHM Average, mg/}L = \frac{0.08 \text{ mg/}L + 0.14 \text{ mg/}L + 0.09 \text{ mg/}L + 0.08 \text{ mg/}L}{4}$$

$$= \frac{0.39 \text{ mg/}L}{4}$$

$$= 0.098 \text{ mg/}L$$

SUMMARY OF RESULTS FOR YEAR 2

Quarter	Quarterly TTHM Average, mg/L	Annual Running TTHM Average, mg/L
1	0.08	0.090
2	0.09	0.093
3	0.14	0.100
4	0.08	0.098

QUESTIONS

Write your answers in a notebook and then compare your answers with those on page 287.

5.2A   Requirements for total trihalomethanes apply to community water systems that chlorinate their water supplies and serve _____ or more people.

5.2B   National Secondary Drinking Water Regulations apply to what types of contaminants?

5.2C   Under what conditions should more samples be collected than the minimum required by the National Primary Drinking Water Regulations?

5.22   Sampling Location

Ordinarily routine samples to analyze for primary and secondary contaminants are drawn from system sources of supply unless it appears there could be some change in constituent levels in the distribution system. Routine non-source sampling performed by utilities includes the following:

1. Bacteriological samples (for coliforms) are taken at points that are representative of the conditions within the distribution system. Be sure the sample bottles contain sodium thiosulfate to neutralize the chlorine residual. Measure the chlorine residual when the sample is collected.

2. Turbidity samples are taken at representative source entry points to the water distribution system.

3. For total trihalomethane sampling, 25 percent of the samples are collected at extreme ends of the distribution system and 75 percent at locations representative of population distribution.

4. Chlorine residual samples are frequently collected at the most remote locations of the distribution system to ensure that a chlorine residual exists throughout the entire system. Samples are also collected and tested for chlorine residual throughout the system.

5. Where tastes, odors, and color may be a problem, samples are often taken at representative or critical points in the distribution system. Such sampling may continue until there is adequate demonstration no problem exists. Critical points would include dead ends and other areas where water movement is slow, and areas where corrosion problems would be expected.

6. The Lead and Copper Rule requires that samples be collected at the consumers' taps. The samples must be collected from locations identified as "high-risk," including: (1) homes with lead solder installed after 1982, (2) homes with lead pipes, and (3) homes with lead service lines.

When samples are collected from the distribution system, they normally are taken from representative points throughout the system. Local conditions at a specific sample tap and in the piping connection to the main may make the sample point unrepresentative of the quality of the water being furnished to the consumers. The best and truest evaluation of water quality can be obtained from samples drawn *DIRECTLY* from the main at specially designed sampling stations.

Sampling taps must be properly located and protected to ensure accurate samples of water delivered to homes. Establishment by the utility of a specially designed distribution system sampling station, such as that in Figure 5.1, is recommended. The design should include a housing enclosing a single gooseneck sampling tap, a direct connection to the main being sampled with the shortest pipe possible (use pipe made of corrosion-resistant material), and a drip valve to drain the sampling tap when it is shut down.

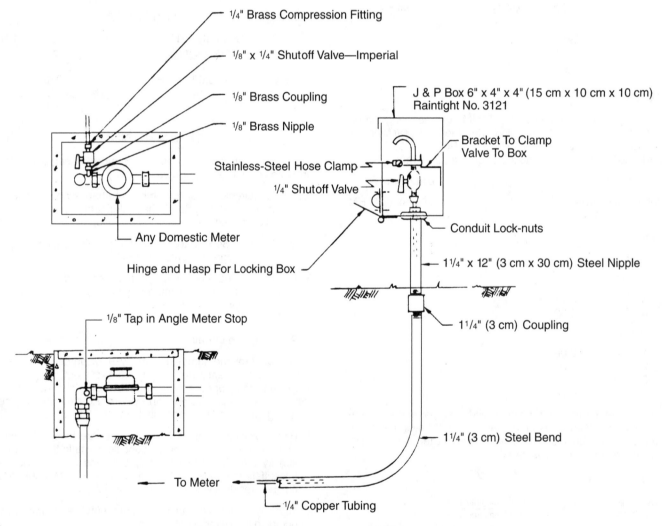

NOTE: Box should be located near a stationary object, such as power pole, for protection, or place sufficient concrete around riser below ground.

Fig. 5.1   Sampling station

(Source: *DISTRIBUTION SYSTEM BACTERIOLOGICAL SAMPLING AND CONTROL GUIDELINES,*
published by California-Nevada Section AWWA)

5.23 Bacteriological Sampling

Most of the effort in water quality monitoring programs is spent in collecting routine samples for bacteriological analysis. A detailed discussion on bacteriological sampling procedures is provided in Chapter 11, "Laboratory Procedures," *WATER TREATMENT PLANT OPERATION*, Volume I.

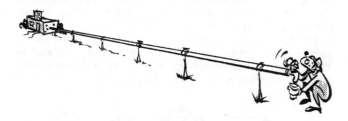

5.24 Chlorine Residual Testing

Many utilities attempt to maintain a chlorine residual throughout the distribution system. Chlorine is very effective in biological control and especially in elimination of coliform bacteria from the finished water. Adequate control of coliform "aftergrowth" is usually obtained only when chlorine residuals are carried to the farthest points of the distribution system. To ensure that this is taking place, make daily chlorine residual tests. A residual of about 0.2 mg/L measured at the extremities of the system is usually a good indication that a free chlorine residual is present in all other parts of the system. This small residual can destroy a small amount of contamination, so a lack of chlorine residual could indicate the potential presence of heavy contamination. If routine checks at a given point show measurable residuals, any sudden absence of a residual at that point should alert the water supplier to the possibility that a potential problem has arisen which needs prompt investigation. Immediate action that can be taken includes retesting for chlorine residual, then checking chlorination equipment, and finally searching for a source of contamination which could cause an increase in the chlorine demand.

5.25 Lead and Copper Monitoring and Sampling Procedures

The number of samples to be collected for lead and copper analysis is based on the size of the distribution system; the sampling frequency is every six months for the initial monitoring program. There are two monitoring periods each year, January to June, and July to December. If the system is in compliance, either as demonstrated by monitoring or after installation of corrosion control, a reduced monitoring frequency can be initiated. The number of sampling sites required based on system size for initial and reduced monitoring are listed in Table 5.1.

The samples are to be collected as "first draw" samples from the cold water tap in either the kitchen or bathroom, or from a tap routinely used for consumption of water if in a building other than a home. A first draw sample is defined as the first one liter of water collected from a tap which has not been used for at least six hours, but preferably unused no more than twelve hours. Faucet aerators should be removed prior to sample collection. The Lead and Copper Rule allows homeowners to collect samples for the utility as long as the proper sample collection instructions have been provided. The EPA specifically prohibits the utility from disputing the accuracy of a sample collected by a resident.

QUESTIONS

Write your answers in a notebook and then compare your answers with those on page 287.

5.2D   Samples are routinely collected from a water distribution system for what kinds of tests?

5.2E   Why do utilities attempt to maintain chlorine residuals throughout the distribution system?

5.2F   What should an operator do if a sudden absence of a chlorine residual is discovered in a water distribution system?

5.2G   When should samples for lead and copper be collected?

TABLE 5.1 SAMPLING SITES REQUIRED FOR LEAD AND COPPER ANALYSIS		
System Size (Population)	Sampling Sites Required (Base Monitoring)	Sampling Sites Required (Reduced Monitoring)
>100,000	100	50
10,001 - 100,000	60	30
3,301 - 10,000	40	20
501 - 3,300	20	10
101 - 500	10	5
≤100	5	5

Reduced Monitoring:

• All public water systems that meet the lead and copper action levels or maintain optimal corrosion-control treatment for two consecutive six-month monitoring periods may reduce the number of tap water sampling sites as shown in Table 5.1 and their collection frequency to once per year.

• All public water systems that meet the lead and copper action levels or maintain optimal corrosion-control treatment for three consecutive years may reduce the number of tap water sampling sites as shown in Table 5.1 and their collection frequency to once every three years.

5.3 CROSS-CONNECTION CONTROL

5.30 Importance of Cross-Connection Control

BACKFLOW[4] of contaminated water through cross connections into community water systems is not just a theoretical problem. As noted in Chapters 3 and 4, contamination through cross connections has consistently caused more waterborne disease outbreaks in the United States than any other reported factor. Inspections have often disclosed numerous unprotected cross connections between public water systems and other piped systems on consumers' premises which might contain wastewater; stormwater; processed waters (containing a wide variety of chemicals); and untreated supplies from private wells, streams, and ocean waters. Therefore, an effective cross-connection control program is essential.

Backflow results from either *BACK PRESSURE*[5] or *BACK-SIPHONAGE*[6] situations in the distribution system. Back pressure occurs when the user's water supply is at a higher pressure than the public water supply system (Figure 5.2). Typical locations where back pressure problems could develop include services to premises where wastewater or toxic chemicals are handled under pressure or where there are unapproved auxiliary water supplies such as a private well or the use of surface water or seawater for firefighting. Backsiphonage is caused by the development of negative or below atmospheric pressures in the water supply piping (Figure 5.3). This condition can occur when there are extremely high water demands (firefighting), water main breaks, or the use of on-line booster pumps.

The best way to prevent backflow is to permanently eliminate the hazard. Back pressure hazards can be eliminated by severing (eliminating) any direct connection at the pump causing the back pressure with the domestic water supply system. Another solution to the problem is to require an air gap separation device (Figure 5.6, page 200) where the water supply service line connects to the private system under pressure from the pump. To eliminate or minimize backsiphonage problems, proper enforcement of plumbing codes and improved water distribution and storage facilities will be helpful. As an additional safety factor for certain selected conditions, a double check valve (Figure 5.7, page 200) may be required at the meter.

5.31 Program Responsibilities

Responsibilities in the implementation of cross-connection control programs are shared by water suppliers, water users (businesses and industries), health agencies, and plumbing officials. The water supplier is responsible for preventing contamination of the public water system by backflow. This responsibility begins at the source, includes the entire distribution system, and ends at the user's connection. To meet this responsibility, the water supplier must issue (promulgate) and enforce needed laws, rules, regulations, and policies. Water service should not be provided to premises where the strong possibility of an unprotected cross connection exists. The essential elements of a water supplier cross-connection control program are discussed in the next section.

The water user is responsible for keeping contaminants out of the potable water system on the user's premises. When backflow prevention devices are required by the health agency or water supplier, the water user must pay for the installation, testing, and maintenance of the approved devices. The user is also responsible for preventing the creation of cross connections through modifications of the plumbing system on the premises. The health agency or water supplier may, when necessary, require a water user to designate a water supervisor or foreman to be responsible for the cross-connection control program within the water user's premises.

The local or state health agency is responsible for issuing and enforcing laws, rules, regulations, and policies needed to control cross connections. Also this agency must have a program that ensures maintenance of an adequate cross-connection control program. The protection of the system on the user's premises is provided where needed by the water utilities.

The plumbing agency (building inspectors) is responsible for the enforcement of building regulations relating to prevention of cross connections on the user's premises.

QUESTIONS

Write your answers in a notebook and then compare your answers with those on page 288.

5.3A What has caused more waterborne disease outbreaks in the United States than any other reported factor?

5.3B Who is usually responsible for the implementation of cross-connection control programs?

5.3C What is the water user's cross-connection control responsibility?

---

[4] *Backflow.   A reverse flow condition, created by a difference in water pressures, which causes water to flow back into the distribution pipes of a potable water supply from any source or sources other than an intended source. Also see BACKSIPHONAGE.*

[5] *Back Pressure.   A pressure that can cause water to backflow into the water supply when a user's water system is at a higher pressure than the public water system.*

[6] *Backsiphonage.   A form of backflow caused by a negative or below atmospheric pressure within a water system. Also see BACKFLOW.*

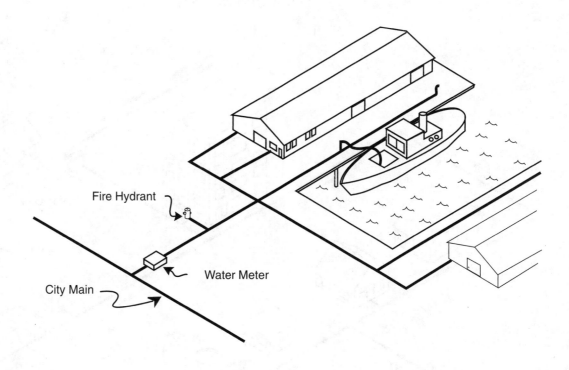

HYDRAULIC GRADIENT

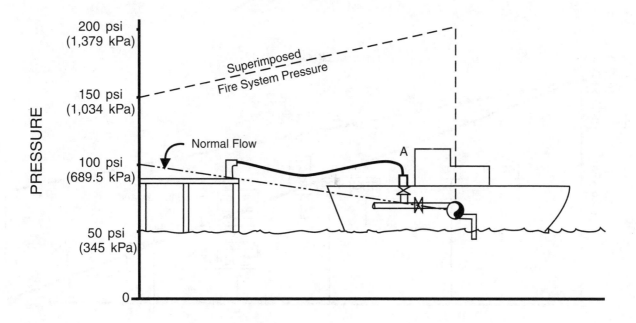

Fig. 5.2   Backflow due to back pressure

(Source: *MANUAL OF CROSS-CONNECTION CONTROL PROCEDURES AND PRACTICES*,
Sanitary Engineering Branch, California Department of Health Services, Berkeley, CA)

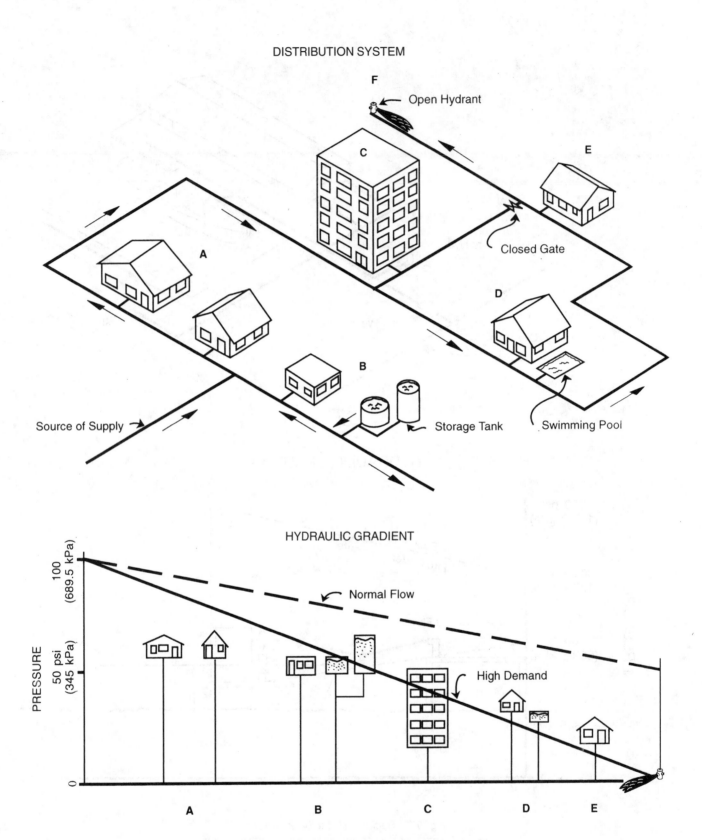

Fig. 5.3   *Backsiphonage due to extremely high water demand*

(Source: *MANUAL OF CROSS-CONNECTION CONTROL PROCEDURES AND PRACTICES,*
Sanitary Engineering Branch, California Department of Health Services, Berkeley, CA)

5.32   Water Supplier Program

The following elements should be included in each water supplier's cross-connection control program:

1. Enactment of an ordinance providing enforcement authority if the supplier is a governmental agency, or enactment of appropriate rules of service if the system is investor-owned.[7]

2. Training of personnel on the causes of and hazards from cross connections and procedures to follow for effective cross-connection control.

3. Listing and inspection or reinspection on a priority basis of all existing facilities where cross connections are of concern. A typical cross-connection survey form is shown in Figure 5.4.

4. Review and screening of all applications for new services or modification of existing services for cross-connection hazards to determine if backflow protection is needed.

5. Obtaining a list of approved backflow prevention devices and a list of certified testers, if available.

6. Acceptable installation of the proper type of device needed for the specific hazard on the premises.

7. Routine testing of installed backflow prevention devices as required by the health agency or the water supplier. Contact the health agency for approved procedures.

8. Maintenance of adequate records for each backflow prevention device installed, including records of inspection and testing. A typical form is shown in Figure 5.5.

9. Notification of each water user when a backflow prevention device has to be tested. This should be done after installation or repair of the device and at least once a year.

10. Maintenance of adequate pressures throughout the distribution system at all times to minimize the hazards from any undetected cross connections that may exist.

All field personnel should be constantly alert for situations where cross connections are likely to exist, whether protection has been installed or not. An example is a contractor using a fire hose from a hydrant to fill a tank truck for dust control or the jetting (for compaction) of pipe trenches. Operators should especially be on the lookout for illegal bypassing of installed backflow prevention devices.

5.33   Types of Backflow Prevention Devices

Different types of backflow prevention devices are available. The particular type of device most suitable for a given situation depends on the degree of health hazard, the probability of backflow occurring, the complexity of the piping on the premises, and the probability of the piping being modified. The higher the assessed risk due to these factors, the more reliable and positive the type of device needed. The types of devices normally approved are listed below according to the degree of assessed risk, with the type of device providing the greatest protection listed first. Only the first three devices are approved for use at service connections.

1. Air gap separation.

2. Reduced-pressure principle (RPP) device.

3. Double check valve.

4. Pressure vacuum breaker (only used for internal protection on the premises).

5. Atmospheric (non-pressure) vacuum breaker.

Figure 5.6 shows a typical air gap separation device and its recommended location. Figure 5.7 shows the installation of a typical double check valve backflow prevention device. These devices are normally installed on the water user's side of the connection to the utility's system and as close to the connection as practical. Figure 5.8 shows a typical installation of pressure vacuum breakers.

Only backflow prevention devices that have passed both laboratory and field evaluations by a recognized testing agency and that have been accepted by the health agency and the water supplier should be used.

5.34   Devices Required for Various Types of Situations

The state or local health agency should be contacted to determine the actual types of devices acceptable for various situations inside the consumer's premises. However, the types of devices generally acceptable for particular situations can be mentioned.

An air gap or a reduced-pressure principle (RPP) device is normally required at services to wastewater treatment plants, wastewater pumping stations, reclaimed water reuse areas, areas where toxic substances in toxic concentrations are handled under pressure, and premises having an auxiliary water supply that is or may be contaminated. The ultimate degree of protection is also needed in cases where fertilizer, herbicides, or pesticides are injected into a sprinkler system.

A double check valve device should be required when a moderate hazard exists on the premises or where an auxiliary supply exists, but adequate protection on the premises is provided.

Atmospheric and pressure vacuum breakers are usually required for irrigation systems; however, they are not adequate in situations where they may be subject to back pressure. If there is a possibility of back pressure, a reduced-pressure principle device is needed.

QUESTIONS

Write your answers in a notebook and then compare your answers with those on page 288.

5.3D   What items should be included on a cross-connection survey form?

5.3E   What factors should be considered when selecting a backflow prevention device?

5.3F   Air gap or reduced-pressure principle backflow prevention devices are installed under what conditions?

5.35   Typical Cross-Connection Hazards

5.350   Importance of Hazard Awareness

Water distribution system operators need to be aware of the types of hazardous chemicals that could enter the water distribution system as well as potential sources of these hazardous

---

[7] A typical ordinance is available in CROSS-CONNECTION CONTROL MANUAL, available from National Technical Information Service (NTIS), 5285 Port Royal Road, Springfield, VA 22161. Order No. PB91-145490. EPA No. 570-9-89-007. Price, $33.50, plus $5.00 shipping and handling per order.

CROSS-CONNECTION SURVEY FORM

Place:_____ Date:_____

Location:_____ Investigator(s)_____

Building Representative(s) and Title(s):

_____
_____

Water Source(s):_____
_____
_____

Piping System(s):_____
_____
_____

Points of Interconnection:_____
_____
_____
_____
_____

Special Equipment Supplied with Water and Source:

_____
_____
_____

Remarks or Recommendations:_____
_____
_____
_____
_____
_____

NOTE: Attach sketches of cross-connections found where necessary for
       clarity of description. Attach additional sheets for room-by-
       room survey under headings

       Room Number   Description of
                                        Cross-Connection(s)

Fig. 5.4 Typical cross-connection survey form

(From *CROSS-CONNECTION CONTROL MANUAL,* U.S. Environmental Protection Agency, Water Supply Division)

MAIL TO: Principal Sanitary Engineer
Department of Water and Power, City of Los Angeles
P.O. Box 111   Room A-18
Los Angeles, California 90051

CODE 03250
15M 7-77 P.O. 30582

RETURN NO LATER THAN:

MANUFACTURER	MODEL	SIZE	SERIAL NUMBER	SERVICE NUMBER

SERVICE ADDRESS:

LOCATION:

| 1 | 2 | 3 | 4 | 5 | 6 | 7 | 8 | 9 | 10 | 11 | 12 | 13 | 14 | 15 | 16 | 17 | 18 | 19 | 20 | 21 | 22 | 23 | 24 | 25 | 26 | 27 | 28 | 29 | 30 | 31 | 32 | 33 | 34 | 35 | 36 | 37 | 38 | 39 | 40 | 41 | 42 | 43 | 44 | 45 | 46 | 47 |

| 48 | 49 | 50 | 51 | 52 | 53 | 54 | 55 | 56 | 57 | 58 | 59 | 60 | 61 | 62 | 63 | 64 | 65 | 66 | 67 | 68 | 69 | 70 | 71 | 72 | 73 | 74 | 75 | 76 | 77 | 78 | 79 | 80 |

	CHECK VALVE #1		CHECK VALVE #2		CHECK VALVE #3		DIFFERENTIAL PRESSURE RELIEF VALVE	
INITIAL TEST	1. LEAKED*	☐	1. LEAKED	☐	1. LEAKED	☐	OPENED AT _____ LBS. REDUCED PRESSURE.	
	2. CLOSED TIGHT	☐	2. CLOSED TIGHT	☐	2. CLOSED TIGHT	☐	DID NOT OPEN	☐
REPAIRS	CLEANED	☐	CLEANED	☐	CLEANED	☐	CLEANED	☐
	REPLACED:		REPLACED		REPLACED		REPLACED:	
	DISC	☐	DISC	☐	DISC	☐	DISC:	
	SPRING	☐	SPRING	☐	SPRING	☐	UPPER	☐
	GUIDE.	☐	GUIDE	☐	GUIDE	☐	LOWER	☐
	PIN RETAINER	☐	PIN RETAINER	☐	HINGE PIN	☐	SPRING	☐
	HINGE PIN	☐	HINGE PIN	☐	SEAT	☐	DIAPHRAGM:	
	SEAT	☐	SEAT	☐	DIAPHRAGM	☐	LARGE:	
	DIAPHRAGM	☐	DIAPHRAGM	☐			UPPER	☐
							LOWER	☐
	OTHER, DESCRIBE	☐	OTHER, DESCRIBE	☐	OTHER, DESCRIBE	☐	SMALL	☐
							SEAT:	
							UPPER	☐
							LOWER	☐
							SPACER:	
							LOWER	☐
							OTHER DESCRIBE	☐
FINAL TEST	CLOSED TIGHT	☐	CLOSED TIGHT	☐	CLOSED TIGHT	☐	OPENED AT _____ LBS. REDUCED PRESSURE	

The above report is certified to be true.

INITIAL TEST BY _____   CERTIFIED TESTER NO. | | | | | | DATE | MO. | DAY | YR. |

REPAIRED BY _____   DATE _____

FINAL TEST BY _____   CERTIFIED TESTER NO. | | | | | | | MO. | DAY | YR. |

Fig. 5.5   Typical backflow prevention device maintenance record form

(Permission of Department of Water and Power, City of Los Angeles)

TANK SHOULD BE OF SUBSTANTIAL CONSTRUCTION AND
OF A KIND AND SIZE TO SUIT CONSUMER'S NEEDS.
TANK MAY BE SITUATED AT GROUND LEVEL (WITH A
PUMP TO PROVIDE ADEQUATE PRESSURE HEAD) OR
BE ELEVATED ABOVE THE GROUND.

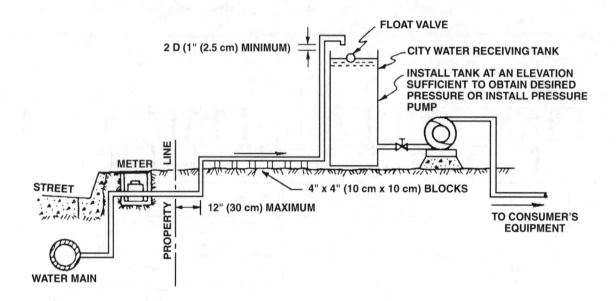

Fig. 5.6   Typical air gap separation

(From MANUAL OF CROSS-CONNECTION CONTROL PRACTICES AND PROCEDURES, Sanitary Engineering Branch, California Department of Health Services, Berkeley, CA)

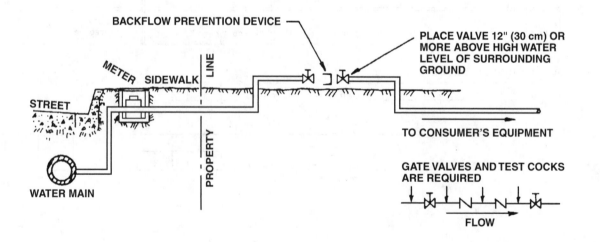

Fig. 5.7   Typical double check valve backflow prevention device

(From MANUAL OF CROSS-CONNECTION CONTROL PRACTICES AND PROCEDURES, Sanitary Engineering Branch, California Department of Health Services, Berkeley, CA)

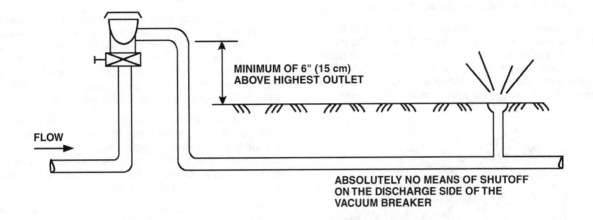

MINIMUM OF 6" (15 cm)
ABOVE HIGHEST OUTLET

FLOW

ABSOLUTELY NO MEANS OF SHUTOFF
ON THE DISCHARGE SIDE OF THE
VACUUM BREAKER

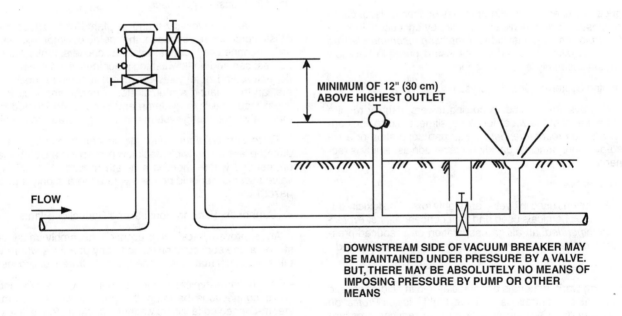

MINIMUM OF 12" (30 cm)
ABOVE HIGHEST OUTLET

FLOW

DOWNSTREAM SIDE OF VACUUM BREAKER MAY
BE MAINTAINED UNDER PRESSURE BY A VALVE.
BUT, THERE MAY BE ABSOLUTELY NO MEANS OF
IMPOSING PRESSURE BY PUMP OR OTHER
MEANS

Fig. 5.8 Typical installations of atmospheric (top) and pressure (bottom) vacuum breakers

(From *MANUAL OF CROSS-CONNECTION CONTROL PRACTICES AND PROCEDURES*, Sanitary Engineering Branch, California Department of Health Services, Berkeley, CA)

chemicals. Operators need to know the types of pollutants or contaminants that are used by consumers that could threaten the public health of consumers if they entered the distribution system. A knowledge of typical industries that have contaminated public water systems through cross connections will help operators protect the quality of the water delivered to consumers through the distribution system.

5.351 Hazardous Chemicals Used by Consumers

Chemicals used by industries may create a hazard to the public when the chemicals are used on site, enter the industry's water distribution system, and then may enter the public water distribution system through a cross connection. This section describes some typical industries and the protection recommended to protect public water distribution systems.

● Agriculture

Agriculture uses many different types of chemicals for different purposes. Toxic chemicals are used by agriculture in fertilizers, herbicides, and pesticides. *Protection recommended:* an air gap separation or a reduced-pressure principle backflow prevention assembly is recommended.

● Cooling Systems (Open or Closed)

Cooling systems, including cooling towers, usually require some treatment of the water for algae, slime, or corrosion control. *Protection recommended:* an air gap separation or a reduced-pressure principle backflow prevention assembly is recommended.

● Dye Plants

Most solutions used in dyeing are highly toxic. The toxicity depends on the chemicals used and their concentrations. *Protection recommended:* an air gap separation or a reduced-pressure principle backflow prevention assembly is recommended.

● Plating Plants

In plating work, materials are first cleaned in acid or caustic solutions at concentrations that are highly toxic. *Protection recommended:* an air gap separation or a reduced-pressure principle backflow prevention assembly is recommended.

● Steam Boiler Plants

Most boiler plants will use some form of boiler feedwater treatment. The chemicals typically used for this purpose include highly toxic compounds. *Protection recommended:* an air gap separation or a reduced-pressure principle backflow prevention assembly is recommended.

5.352 Industries With Cross-Connection Potential

A knowledge of industries that may have cross connections to a public water distribution system will help operators prevent the occurrence of cross connections. The following inspection guidelines list typical hazards and recommend cross-connection protection.

● Auxiliary Water Systems

An auxiliary water system is a water supply or source that is not under the control or the direct supervision of the water purveyor. An approved backflow prevention assembly must be installed at the service connection of the water purveyor to any premises where there is an auxiliary water supply or system, even though there is no connection between the auxiliary water supply and the public potable water system.

Typical auxiliary water systems include water used in industrialized water systems; water in reservoirs or tanks used for firefighting purposes; irrigation reservoirs; swimming pools; fish ponds; mirror pools; memorial and decorative fountains and cascades; cooling towers; and baptismal, quenching, washing, rinsing, and dipping tanks.

Protection recommended: an air gap separation or a reduced-pressure principle backflow prevention assembly is recommended where there is a health hazard. A double check valve assembly should be used where there is only a pollution hazard.

● Beverage Bottling and Breweries

An approved backflow prevention assembly must be installed on the service connection to any premises where a beverage bottling plant is operated or maintained and water is used for industrial purposes.

The hazards typically found in plants of this type include cross connections between the potable water system and steam-connected facilities; washers, cookers, tanks, lines, and flumes; can and bottle washing machines; lines in which caustic, acids, and detergents are used; reservoirs, cooling towers, and circulating systems; steam generating facilities and lines; industrial fluid systems and lines; water-cooled equipment; and firefighting systems, including storage reservoirs.

Protection recommended: an air gap separation or a reduced-pressure principle backflow prevention assembly is recommended where there is a health hazard. A double check valve assembly should be used where there is only a pollution hazard.

● Canneries, Packing Houses, and Reduction Plants

An approved backflow prevention assembly must be installed at the service connection to any premises where vegetable or animal matter is canned, concentrated, or processed.

The hazards typically found in plants of this type include cross connections between the potable water system and steam-connected facilities; washers, cookers, tanks, lines, and flumes; reservoirs, cooling towers, and circulating systems; steam generating facilities and lines; industrial fluid systems and lines; firefighting systems, including storage reservoirs; water-cooled equipment; and tanks, can and bottle washing machines, and lines.

Protection recommended: an air gap separation or reduced-pressure principle backflow prevention assembly is recommended.

- Chemical Plants—Manufacturing, Processing, or Treatment

An approved backflow prevention assembly shall be installed on the service connection to any premises where there is a facility requiring the use of water in the industrial process of manufacturing, storing, compounding, or processing chemicals. This will also include facilities where chemicals are used as additives to the water supply or in processing products. Cross connections may be numerous because of the intricate piping. The severity of these cross connections varies with the toxicity of the chemical used.

The hazards typically found in plants of this type include cross connections between the potable water system and formulating tanks and vats, decanter units, extractor/precipitators, and other processing units; reservoirs, cooling towers, and circulating systems; steam generating facilities and lines; firefighting systems, including storage reservoirs; water-cooled equipment; hydraulically operated equipment; equipment under hydraulic tests; pressure cookers, autoclaves, and retorts; and washers, cookers, tanks, flumes, and other equipment used for storing, washing, cleaning, blanching, cooking, flushing, or for the transmission of food, fertilizers, or wastes.

Protection recommended: an air gap separation or a reduced-pressure principle backflow prevention assembly is recommended where there is a health hazard.

- Dairies and Cold Storage Plants

An approved backflow prevention assembly shall be installed on the service connection to any premises on which a dairy, creamery, ice cream, cold storage, or ice manufacturing plant is operated or maintained, provided such a plant has on the premises an industrial fluid system, wastewater handling facilities, or other similar source of contamination that, if cross connected, would create a hazard to the public system.

The hazards typically found in these types of plants include cross connections between the potable water system and reservoirs, cooling towers, and circulation systems; steam generating facilities and lines; water-cooled equipment and tanks; can and bottle washing machines; and lines in which caustics, acids, detergents, and other compounds are circulated for cleaning, sterilizing, and flushing.

Protection recommended: an air gap separation or a reduced-pressure principle backflow prevention assembly is recommended where there is a health hazard. A double check valve assembly should be used where there is only a pollution hazard.

- Film Laboratories

An approved backflow prevention assembly must be installed on each service connection to any premises where a film laboratory, processing, or manufacturing plant is operated or maintained. This does not include darkroom facilities.

The hazards typically found in a plant of this type include cross connections between the potable water system and tanks, automatic film processing machines, and water-cooled equipment that may be connected to a sewer, such as compressors, heat exchangers, and air conditioning equipment.

Protection recommended: an air gap separation or a reduced-pressure principle backflow prevention assembly is recommended.

- Hospitals, Medical Buildings, Sanitariums, Morgues, Mortuaries, Autopsy Facilities, and Clinics

An approved backflow prevention assembly must be installed on the service connection to any hospital, medical buildings, and clinics. The hazards typically found in this type of facility include cross connections between the potable water system and contaminated or sewer-connected equipment; water-cooled equipment that may be sewer-connected; reservoirs, cooling towers, and circulating systems; and steam generating facilities and lines.

Protection recommended: an air gap separation or a reduced-pressure principle backflow assembly must be installed on the service connection to any hospital, mortuary, morgue, or autopsy facility, or to any multistoried medical building or clinic.

- Laundries and Dye Works (Commercial Laundries)

An approved backflow prevention assembly must be installed on each service connection to any premises where a laundry or dyeing plant is operated or maintained.

The hazards typically found in plants of this type include cross connections between the potable water system and laundry machines having under rim or bottom outlets; dye vats in which toxic chemicals and dyes are used; water storage tanks equipped with pumps and recirculating systems; shrinking, bluing, and dyeing machines with direct connections to circulating systems; retention and mixing tanks; wastewater pumps for priming, cleaning, flushing, or unclogging purposes; water-operated wastewater sump ejectors for operational purposes; sewer lines for the purpose of disposing of filter or softener backwash water from cooling systems; reservoirs, cooling towers, and circulating systems; and steam generating facilities and lines.

Protection recommended: an air gap separation or a reduced-pressure principle backflow prevention assembly is recommended.

- Marine Facilities and Dockside Watering Points

The actual or potential hazard to the utility's water system created by any marine facility or dockside watering point must be individually evaluated. The basic risk to a domestic water system is the possibility that contaminated water can be pumped into the domestic water system by the fire pumps or other pumps aboard a ship. The additional risk of dockside water facilities located on fresh water or diluted salt water is that if backflow occurs, it can be more easily digested because of the lack of salty taste.

Protection recommended: minimum system protection for marine installations may be accomplished in one of the following ways: (1) water connections directly to vessels for any purpose must have a reduced-pressure backflow prevention assembly installed at the pier hydrants; (2) where water is delivered to marine facilities for fire protection only, and no auxiliary water supply is present, all service connections should be protected by a reduced-pressure principle backflow prevention assembly; (3) water delivered to a marine repair facility should have a reduced-pressure principle backflow prevention assembly; (4) water delivered to small boat moorages that maintain hose bibbs on a dock or float should have a reduced-pressure principle backflow prevention assembly installed at the user connection and a hose connection vacuum breaker on all hose bibbs; and (5) water for fire protection aboard ship, connected to dockside fire hydrants, must not be taken aboard from fire hydrants unless the hydrants are on a fire system separated from the domestic system by an approved reduced-pressure principle backflow prevention assembly.

- Metal Manufacturing, Cleaning, Processing, and Fabricating Plants

An approved backflow prevention assembly must be installed on the service connection to any premises where metals are manufactured, cleaned, processed, or fabricated and the process involves used water and/or industrial fluids. This type of facility may be operated or maintained either as a separate function or other facility, such as an aircraft or automotive manufacturing plant.

The hazards typically found in a plant of this type include cross connections between the potable water and reservoirs, cooling towers, and circulating systems; steam generating facilities and lines; plating facilities involving the use of highly toxic chemicals; industrial fluid systems; tanks, vats, or other vessels; water-cooled equipment; tanks, can and bottle washing machines, and lines; hydraulically operated equipment; and equipment under hydraulic tests.

Protection recommended: an air gap separation or a reduced-pressure principle backflow prevention assembly is recommended where there is a health hazard. A double check valve assembly should be used where there is only a pollution hazard.

- Multistoried Buildings

Multistoried buildings may be broadly grouped into the following three categories in terms of their internal potable water systems:

1. Using only the service pressure to distribute the potable water throughout the structure, and with no internal potable water reservoir,

2. Using a booster pump to provide potable water directly to the upper floors, and

3. Using a booster pump to fill a covered roof reservoir from which there is a down-feed system for the upper floors.

Considerable care must be exercised to prevent the use of the suction-side line to these pumps from also being used as the takeoff for domestic, sanitary, laboratory, or industrial uses on the lower floors. Pollutants or contaminants from equipment supplied by takeoffs from the suction-side line may be easily pumped throughout the upper floors. In each of these systems it is probable that there is one or more takeoffs for industrial water within the building. Any loss of distribution main pressure will cause backflow from these buildings' systems unless approved backflow prevention assemblies are properly installed.

Protection recommended: an air gap separation or a reduced-pressure principle backflow prevention assembly where there is a health hazard; a reduced-pressure principle backflow prevention assembly when takeoffs for lower floor sanitary facilities are connected to the suction side of booster pump(s); and a double check valve assembly where there is a non-health hazard. The suction pressure on booster pumps should be limited to prevent drawing water from adjacent unprotected premises.

- Oil and Gas Production, Storage, or Transmission Properties

An approved backflow prevention assembly shall be installed at the service connection to any premises where animal, vegetable, or mineral oils and gases are produced, developed, processed, blended, stored, refined, or transmitted in a pipeline, or where oil or gas tanks are maintained. An approved backflow prevention assembly must be installed at the service connection where an oil well is being drilled, developed, operated, or maintained; or where an oxygen, acetylene, petroleum, or other manufactured gas production or bottling plant is operated or maintained.

The hazards typically found in plants of this type include cross connections between the potable water system and steam boiler lines; mud pumps and mud tanks; oil well casings; dehydration tanks and outlet lines from storage and dehydration tanks; oil and gas tanks; gas and oil lines; reservoirs, cooling towers, and circulating systems; steam generating facilities; industrial fluid systems; firefighting systems, including storage reservoirs; water-cooled equipment; hydraulically operated equipment and equipment under hydraulic tests.

Protection recommended: an air gap separation or a reduced-pressure principle backflow prevention assembly is recommended.

- Paper and Paper Product Plants

An approved backflow prevention assembly must be installed on the service connection to any premises where a paper or paper products plant (wet process) is operated or maintained.

The hazards typically found in a plant of this type include cross connections between the potable water system and pulp, bleaching, dyeing, and processing facilities that may be contaminated with toxic chemicals; reservoirs, cooling towers, and circulating systems; steam generating facilities and lines; industrial fluid systems and lines; water-cooled equipment; and firefighting systems, including storage reservoirs.

Protection recommended: an air gap separation or a reduced-pressure principle backflow prevention assembly is recommended.

- Plants or Facilities Handling Radioactive Material or Substances

An approved backflow prevention assembly must be installed at the service connection to any premises where radioactive materials or substances are processed in a laboratory or plant, or where they may be handled in such a manner as to create a potential hazard to the water system, or where there is a reactor plant.

Protection recommended: an air gap separation or a reduced-pressure principle backflow prevention assembly is recommended.

- Restricted, Classified, or Other Closed Facilities

An approved backflow prevention assembly must be installed on the service connection to any facility that is not readily accessible for inspection by the water purveyor because of military secrecy requirements or other prohibitions or restrictions.

Protection recommended: an air gap separation or a reduced-pressure principle backflow prevention assembly is recommended.

- Solar Domestic Hot Water Systems

An approved backflow prevention assembly must be installed on the service connection to any premises where there is a solar domestic hot water system.

The hazards typically found in a solar domestic system include cross connections between the potable water system and heat exchangers, tanks, and circulating pumps. Contamination can occur when the piping or tank walls of the heat exchanger between the potable hot water and the transfer medium begin to leak.

Protection recommended: the recommended protection depends on whether there is a possible health hazard or a nonhealth hazard. In either case a reduced-pressure principle backflow prevention assembly will always provide safe protection.

- Water Hauling Equipment

An approved backflow prevention assembly must be installed on any portable spraying or cleaning units that have the capability of connection to any potable water supply that does not contain a built-in approved air gap.

The hazards typically found in water hauling equipment include cross connections between the potable water system and tanks contaminated with toxic chemical compounds used in spraying fertilizers, herbicides, and pesticides; water hauling tanker trucks used in dust control; and other tanks on cleaning equipment.

Protection recommended: an air gap separation or a reduced-pressure principle backflow prevention assembly is recommended.

5.353   Maintenance and Testing Procedures

Backflow prevention programs must require the owners of buildings and facilities with backflow prevention devices to maintain the devices within their buildings and facilities in good working condition. Also, qualified persons must periodically test the backflow preventers for satisfactory performance.

5.354   Acknowledgment

Material in this section was obtained from *RECOMMENDED PRACTICE FOR BACKFLOW PREVENTION AND CROSS-CONNECTION CONTROL* (M14), available from American Water Works Association (AWWA), Bookstore, 6666 West Quincy Avenue, Denver, CO 80235. Order No. 30014. ISBN 1-58321-288-4. Price to members, $69.00; nonmembers, $99.00; price includes cost of shipping and handling.

QUESTIONS

Write your answers in a notebook and then compare your answers with those on page 288.

5.3G   List five typical industries that use hazardous chemicals.

5.3H   What type of cross-connection control device is recommended where there is a health hazard?

5.3I   Who should test backflow preventers for satisfactory performance?

5.4   SYSTEM PRESSURES

A main concern in operation of a water distribution system is the maintenance of a *CONTINUOUS POSITIVE PRESSURE* at all times to all customers. This is necessary not only to meet the flow needs of the customers, but to prevent contamination from backflow. The water supplier should try to maintain a minimum pressure of 35 psi (241 kPa or 2.45 kg/sq cm) at all points in the distribution system, with an absolute minimum of 20 psi (138 kPa or 1.4 kg/sq cm) even during fire flows. In commercial districts, pressures of 75 psi (517 kPa or 5.3 kg/sq cm) and higher are used, but the delivered pressure should

not be more than 100 psi (690 kPa or 7.0 kg/sq cm). Pressures in this range (35 to 100 psi, 241 to 690 kPa, or 2.4 to 7.0 kg/sq cm) are sufficient for normal use without risk of damage to water-using facilities in the distribution system or on the consumer's premises. Excessive pressures could cause such damage as well as unintentional higher water usage. Where high main pressures cannot be avoided (on transmission mains), pressure-reducing valves are normally used.

Promptly investigate low-pressure complaints from consumers and make corrections where necessary. To determine the extent of a reported problem, use pressure gages or portable mechanical or battery-powered pressure recorders which can record data for a 24-hour period or longer by connecting to the consumer's system.

Pressure problems have many possible causes. A faulty pressure regulator might have a control water strainer that needs cleaning, or the regulator itself may be in need of repair. Sometimes, purposefully or otherwise, a line valve is left closed or partly closed causing loss of pressure, especially during higher flows. Low pressures can be caused by system lines being too small which will cause high velocities. High velocities occur when there is a large use of water due to unusually high consumer demands, firefighting, or a major leak. While these problems are out of the operator's control, they can be eased by maintaining adequate reserves in distribution storage. Although about 35 percent or more of the storage capacity should normally be reserved for fire demand and emergency use, excessive peak hour or peak day demands can still seriously drain the available storage. Storage reservoirs should never be allowed to run too low. Activating a well or an interagency connection or increasing pumping to provide more water might help maintain adequate pressure. A potential source of an emergency water supply is older reservoirs that have been taken out of service when the physical growth of the service area has left them at too low an elevation to be useful; however, they have been kept available for emergency purposes. By slightly modifying the piping and using portable pumps, these older reservoirs can be used during emergencies.

Pressure problems can also be caused by pump failure. If automatic pump controls fail, the pumps must be operated with manual overrides during peak periods. If the pumps themselves fail or the power fails, the availability of self-powered pumps and generators is invaluable. Distribution system operators should have the basic knowledge to make prompt repairs on pumps that have failed and the necessary repair parts should be readily available. Operators should have access to emergency equipment and rental equipment with the authority to act.

QUESTIONS

Write your answers in a notebook and then compare your answers with those on page 288.

5.4A   Why should a distribution system be under a continuous positive pressure at all times?

5.4B   What problems may be caused by excessive distribution system pressures?

5.4C   What are the possible causes of low water distribution system pressures?

5.5   DISTRIBUTION STORAGE

Surveillance of distribution storage facilities was discussed at the beginning of this chapter. Deficiencies found during surveillance should be taken care of as soon as possible. Any condition that might affect the quality of the water in the reservoir must be corrected immediately. When leaving the reservoir site either after surveillance or completion of other work, be particularly careful to leave all locks in a locked position and make sure the reservoir is completely secured.

The main function of a distribution system storage reservoir is to take care of daily demands as well as peak demands. Operators must be concerned with the amount of water in the storage tanks at particular times of the day. Try to have the tank full or nearly full before the start of the peak demand period. When the peak demand period is over, the water level in the tank should not have dropped below a minimum target level. Fill the tank during the low-demand period. See Chapter 2, "Storage Facilities," Section 2.3, "Operation," for procedures on how to operate distribution storage facilities.

QUESTIONS

Write your answers in a notebook and then compare your answers with those on page 288.

5.5A   What is the main function of a distribution system storage reservoir?

5.5B   When should a distribution storage reservoir be full?

5.6   PUMPS

A comprehensive preventive maintenance program is necessary for all pump installations. Without such a program, the continued satisfactory operation of the pumps will be threatened and may result in unsatisfactory system performance. In many systems, a lengthy pump outage will severely strain the ability of the system to continue to serve sufficient water. Written detailed inspection, operation, and maintenance proce-

dures should be available at convenient locations for all operators responsible for these duties. The statement, "When all else fails, follow directions" was never more true than for pump operation. Manufacturers' directions should be closely followed.

Pumps must be firmly installed to prevent problems from operating noise levels and alignment during the life of the equipment. Alignment procedures for the pump and driver are critical. There should be no strain on the pump case from the suction and discharge piping connections.

The starting, stopping, and regulation of pumping equipment can be achieved manually, automatically, or semi-automatically. There has been an increasing use of sensing and automatic control devices. The variables sensed may be water elevation, water pressure, quantity of water pumped, or any combination of the three. Long distance (remote) sensing operations can be set up using leased telephone circuits, radio signals, or pressure communicated through the water in the pipelines to transmit control system signals. The entire pumping function in a complex distribution system can be operated and controlled from a single point or control station or from multiple points.

As pump operating conditions vary widely, it is impossible to provide a preventive maintenance schedule that would apply to all systems. Regular inspections must be planned and followed. Inspection schedules are often based on the number of expected pump operating hours since the last inspection. A record of the inspections and of the pump maintenance performed must be kept to ensure that the necessary scheduled procedures are being followed.

Inspection and preventive maintenance procedures will cover many types of operations. Some of the more important procedures are listed below.

1. Observe and record pump pressures and output (flow), and the pump's current (electricity) demands.

2. Regularly check for excessive or abnormal noise, vibration, heat, and odor.

3. Provide grease and oil lubrication in accordance with the manufacturer's instructions. Lubrication should never be overdone as the addition of too much grease will cause a bearing to overheat, and too much oil will result in foaming. Proper lubrication for pump bearings cannot be overemphasized. The conditions of operation will determine how often a bearing should be greased.

4. Check bearing temperatures once per month with a thermometer. If the bearings are found to be running too hot, it could be the result of too much lubricant. Every three months, check the lubricated bearings for "saponification." This is a soapy or foamy condition of the lubricant which usually results from water or other fluid infiltrating past the bearing shaft seals. If the grease appears as a white foamy

substance, flush the bearings with kerosene or solvent, clean thoroughly, and repack with the type of lubricant recommended by the manufacturer.

5. Listen for any bearing noise. Usually a ball bearing will give audible notice of impending failure. This early warning will give you time to plan a shutdown for maintenance.

6. Tighten the packing glands if they are leaking excessively. The tightening should be sufficient to permit only a small amount of leakage, but not result in an increase in packing-follower heating. Packing glands should never be tightened to the point where there is no leakage since this will cause undue packing wear and even a scored shaft or sleeve. Check the leakage rate daily. When the packing wears or is compressed so that the gland cannot be tightened farther, install a new set of rings. While mechanical seals require higher initial cost than packing, long-term costs may be lower.

7. Inspect the pump *PRIMING*[8] system. A priming system must be used to prevent the pump from running dry if the pump operates with a *SUCTION LIFT.*[9] To protect against loss of prime, check valves or foot valves on the intake or suction piping of a pump are a necessity. Use only clean water for the priming. To facilitate priming, ejectors or vacuum pumps are also available. As the pumps are also prone to air binding, they are usually provided with manual or automatic vent valves on top of the pump casing. One type of priming pump installation is shown in Figure 5.9.

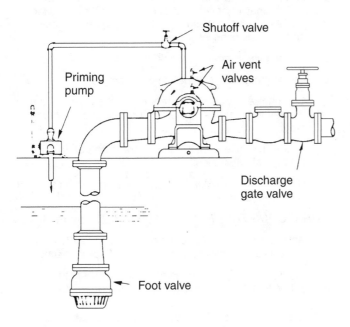

Fig. 5.9   Pump priming installation

(Reprinted from *WATER DISTRIBUTION OPERATOR TRAINING HANDBOOK*, by permission. Copyright 1976, the American Water Works Association)

8. Routinely operate internal combustion-driven pumps and generators on standby for 15 minutes once per week. Also check any automatic pump controls. If they fail during an emergency, the pumps must be operated with manual overrides.

9. Check pump alignment periodically to guard against premature bearing and coupling wear. Alignment should be checked when the pump is cold, and again when a unit has run long enough for it to reach the proper operating temperature. Alignment on new installations can change considerably in a short period of time. Daily checks should be made until a stable operation has been established.

Since pumps and controls are usually powered by electric energy, the operator should have some working knowledge of electric circuits and circuit testing instruments if effective maintenance is to be provided. Although the operator is usually not expected to be an expert in this field, an understanding of the basics of electricity is important if for no other reason than to learn how to avoid electric shock.

An electric motor failure can have serious consequences. The principal causes of such a failure are overheating caused by overloading, a locked rotor, rapid cycling, loss of cooling due to clogged or impeded ventilation, or low water level in a submersible pump casing. When overheating occurs, power is usually shut off by protective devices installed in motor circuits. Become familiar with these devices and learn where and how to reset them after finding the problem, if any, which caused the tripping of the circuit.

A common problem found in pump stations with a high rate of motor failure is voltage imbalance or unbalance. Unlike a single-phase condition, all three phases are present but the phase-to-phase voltage is not equal in each phase.

Voltage imbalance can occur in either the utility side or the pump station electrical system. For example, the utility company may have large single-phase loads (such as residential services) which reduce the voltage on a single phase. This same condition can occur in the pump station if a large number of 120/220 volt loads are present. Slight differences in voltage can cause disproportional current imbalance; this may be six to ten times as large as the voltage imbalance. For example, a two percent voltage imbalance can result in a 20 percent current imbalance. A 4.5 percent voltage imbalance will reduce the insulation life to 50 percent of the normal life. This is the reason a dependable voltage supply at the motor terminals is critical. Even relatively slight variations can greatly increase the motor operating temperatures and burn out the insulation.

It is common practice for electrical utility companies to furnish power to three-phase customers in open delta or wye configurations. An open delta or wye system is a two-transformer bank that is a suitable configuration where *LIGHTING LOADS ARE LARGE AND THREE-PHASE LOADS ARE LIGHT.* This is the exact opposite of the configuration needed by most pumping facilities where *THREE-PHASE LOADS ARE LARGE.* (Examples of three-transformer banks include

---

[8] *Prime.   The action of filling a pump casing with water to remove the air. Most pumps must be primed before start-up or they will not pump any water.*

[9] *Suction Lift.   The NEGATIVE pressure [in feet (meters) of water or inches (centimeters) of mercury vacuum] on the suction side of the pump. The pressure can be measured from the centerline of the pump DOWN TO (lift) the elevation of the hydraulic grade line on the suction side of the pump.*

Y-delta, delta-Y, and Y-Y.) In most cases, three-phase motors should be fed from three-transformer banks for proper balance. The capacity of a two-transformer bank is only 57 percent of the capacity of a three-transformer bank. The two-transformer configuration can cause one leg of the three-phase current to furnish higher amperage to one leg of the motor, which will greatly shorten its life.

Operators should acquaint themselves with the configuration of their electric power supply. When an open delta or wye configuration is used, operators should calculate the degree of current imbalance existing between legs of their polyphase motors. If you are unsure about how to determine the configuration of your system or how to calculate the percentage of current imbalance, *ALWAYS* consult a qualified electrician. *CURRENT IMBALANCE BETWEEN LEGS SHOULD NEVER EXCEED 5 PERCENT UNDER NORMAL OPERATING CONDITIONS* (NEMA Standards MGI-14.35).

Loose connections will also cause voltage imbalance as will high-resistance contacts, circuit breakers, or motor starters. Motor connections at the circuit box should be checked frequently (semiannually or annually) to ensure that the connections are tight and that vibrating wires have not rubbed through the insulation on the conductors. Measure the voltage at the motor terminals and calculate the percentage imbalance (if any) using the procedures below.

Another serious consideration for operators is voltage fluctuation caused by neighborhood demands. A pump motor in near perfect balance (for example, 3 percent unbalance) at 9:00 AM could be as much as 17 percent unbalanced by 4:00 PM on a hot day due to the use of air conditioners by customers on the same grid. Also, the hookup of a small market or a new home to the power grid can cause a significant change in the degree of current unbalance in other parts of the power grid. Because energy demands are constantly changing, water system operators should have a qualified electrician check the current balances between legs of their three-phase motors at least once a year.

Do not rely entirely on the power company to detect unbalanced current. Complaints of suspected power problems are frequently met with the explanation that all voltages are within the percentages allowed by law and no mention is made of the percentage of current unbalance which can be a major source of problems with three-phase motors. A little research of your own can pay large benefits. For example, a small water company in Central California configured with an Open Delta system (and running three-phase unbalances as high as 17 percent as a result) was routinely spending $14,000 a year for energy and burning out a 10-HP motor on the average of every 1.5 years (six 10-HP motors in 9 years). After consultation, the local power utility agreed to add a third transformer to

each power board to bring the system into better balance. Pump drop leads were then rotated, bringing overall current unbalances down to an average of 3 percent, heavy-duty three-phase capacitors were added to absorb the prevalent voltage surges in the area, and computerized controls were added to the pumps to shut them off when pumping volumes got too low. These modifications resulted in a saving in energy costs the first year alone of $5,500.00.

FORMULAS

Percentage of current unbalance can be calculated by using the following formulas and procedures:

$$\text{Average Current} = \frac{\text{Total of Current Value Measured on Each Leg}}{3}$$

$$\text{\% Current Unbalance} = \frac{\text{Greatest Amp Difference from the Average}}{\text{Average Current}} \times 100\,\%$$

PROCEDURES

A. Measure and record current readings in amps for each leg. (Hookup 1.) Disconnect power.

B. Shift or roll the motor leads from left to right so the drop cable lead that was on terminal 1 is now on 2, lead on 2 is now on 3, and lead on 3 is now on 1. (Hookup 2.) Rolling the motor leads in this manner will not reverse the motor rotation. Start the motor, measure and record current reading on each leg. Disconnect power.

C. Again shift drop cable leads from left to right so the lead on terminal 1 goes to 2, 2 goes to 3, and 3 to 1. (Hookup 3.) Start pump, measure and record current reading on each leg. Disconnect power.

D. Add the values for each hookup.

E. Divide the total by 3 to obtain the average.

F. Compare each single leg reading to the average current amount to obtain the greatest amp difference from the average.

G. Divide this difference by the average to obtain the percentage of unbalance.

H. Use the wiring hookup that provides the lowest percentage of unbalance.

CORRECTING THE THREE-PHASE POWER UNBALANCE

Example: Check for current unbalance for a 230-volt, 3-phase 60-Hz submersible pump motor, 18.6 full load amps.

Solution: Steps 1 to 3 measure and record amps on each motor drop lead for Hookups 1, 2, and 3 (Figure 5.10).

	Step 1 (Hookup 1)	Step 2 (Hookup 2)	Step 3 (Hookup 3)
(T_1)	DL_1 = 25.5 amps	DL_3 = 25 amps	DL_2 = 25.0 amps
(T_2)	DL_2 = 23.0 amps	DL_1 = 24 amps	DL_3 = 24.5 amps
(T_3)	DL_3 = 26.5 amps	DL_2 = 26 amps	DL_1 = 25.5 amps
Step 4	Total = 75 amps	Total = 75 amps	Total = 75 amps
Step 5	Average Current =	Total Current =	$\dfrac{75}{3}$ = 25 amps
		$\dfrac{\text{Total Current}}{\text{3 readings}}$	
Step 6	Greatest amp difference from the average:	(Hookup 1) = 25 − 23 = 2	
		(Hookup 2) = 26 − 25 = 1	
		(Hookup 3) = 25.5 − 25 = 0.5	
Step 7	% Unbalance	(Hookup 1) = 2/25 x 100 = 8	
		(Hookup 2) = 1/25 x 100 = 4	
		(Hookup 3) = 0.5/25 x 100 = 2	

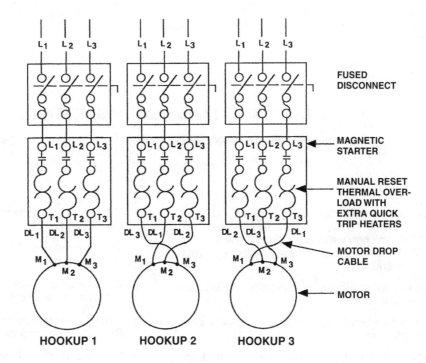

Fig. 5.10   Three hookups used to check for current unbalance

As can be seen, Hookup 3 should be used since it shows the least amount of current unbalance. Therefore, the motor will operate at maximum efficiency and reliability on Hookup 3.

By comparing the current values recorded on each leg, you will note the highest value was always on the same leg, L_3. This indicates the unbalance is in the power source. If the high current values were on a different leg each time the leads were changed, the unbalance would be caused by the motor or a poor connection.

If the current unbalance is greater than 5 percent, contact your power company for help.

ACKNOWLEDGMENT

Material on unbalanced current was provided by James W. Cannell, President, Canyon Meadows Mutual Water Company, Inc., Bodfish, CA. His contribution is greatly appreciated.

For an excellent summary of pump troubleshooting procedures, see Chapter 2, Table 2.2, "Centrifugal Pump Troubleshooting Chart," page 43. For additional information on pump maintenance, see Chapter 18, Section 18.2, "Mechanical Equipment," in *WATER TREATMENT PLANT OPERATION*, Volume II, in this series of manuals.

QUESTIONS

Write your answers in a notebook and then compare your answers with those on page 288.

5.6A   Why are pump alignment procedures for the pump and driver critical?

5.6B   List the important pump inspection and preventive maintenance procedures.

5.6C   What types of checks should be performed on bearings?

5.6D   When should new pump packing rings be installed?

End of Lesson 1 of 4 Lessons
on
DISTRIBUTION SYSTEM
OPERATION and MAINTENANCE

Please answer the discussion and review questions next.

DISCUSSION AND REVIEW QUESTIONS

Chapter 5.   DISTRIBUTION SYSTEM OPERATION AND MAINTENANCE

(Lesson 1 of 4 Lessons)

At the end of each lesson in this chapter you will find some discussion and review questions. The purpose of these questions is to indicate to you how well you understand the material in the lesson. Write the answers to these questions in your notebook before continuing.

1. What steps should a water distribution utility take in preparation for emergencies?

2. Why should water distribution systems have a system surveillance program?

3. What problems can an operator expect from open distribution storage reservoirs?

4. How frequently should a storage tank be cleaned?

5. How could you determine the condition (smoothness) of the inside of a pipe?

6. What normal security measures should be taken to prevent vandalism?

7. Under what conditions should a water supply facility be immediately taken out of service?

8. Why are the states encouraged to implement the National Secondary Drinking Water Regulations?

9. What items should be included in a water supplier's cross-connection control program?

10. How does a voltage imbalance affect electric motors?

CHAPTER 5.   DISTRIBUTION SYSTEM OPERATION AND MAINTENANCE

(Lesson 2 of 4 Lessons)

5.7   WATER MAINS AND APPURTENANCES

5.70   Pipe

5.700   Pipe Maintenance

Pipe maintenance is performed to prevent leakage, maintain or restore the pipe's carrying capacity, maintain proper water quality conditions in the pipe, and prolong the effective life of the pipe. A pipe's useful life can be greatly extended if it is properly maintained and rehabilitated. The type of maintenance carried out includes repairing leaks and breaks, flushing, cleaning, disinfecting, and relining.

Pipes deteriorate on the inside because of water corrosion and erosion and on the outside because of corrosion from aggressive soil moisture. Even under the best of conditions, pipe can be weakened and damaged with time. All types of metal, concrete, and asbestos-cement pipe are subject to some deterioration. This deterioration may be revealed as a loss of water carrying capacity, leaks, or degradation of water quality. Loss of water carrying capacity can result from corrosion, pitting, tuberculation, deposition of sediment, and slime growth.

5.701   Locating Leaks

Leak detection programs are an effective means for some water utilities to reduce operating and maintenance costs. If a leak detection crew can reduce the flow of leaks and produce cost savings greater than the cost of maintaining the field crew, then the leak detection program is economically justified. Leak detection programs can also be justified in terms of the early detection and repair of leaks while they are small, before serious failure occurs with resulting property damage, crew overtime, delays of other projects, and similar problems. Also a water shortage may require an effective leak detection program.

Leaks may originate from any weakened joint or fitting connection or from a damaged or corroded part of the pipe. Leaks are undesirable not only because they waste water, but because they can undermine pavements and other structures. Another undesirable effect of leaks is that the leak soaks the ground surrounding the pipe and, in the event that pressure is lost in the pipe, the water, combined now with dirt and other contaminants, may backflow into the pipe.

The total amount of leakage in a distribution system is affected by a number of factors. Improper pipe installation (bedding, backfill, misalignment) can result not only in weakened joint and fitting connections, but also in damage to the pipe itself and to any corrosion-protection measures that have been provided. The pipe's durability, strength, and corrosion resistance will vary with the type of pipe material used. Protection against corrosion is an important maintenance activity which will minimize leakage. The older the pipe, the more time there has been for corrosion to act and for the pipe system to

be weakened, especially at joints and fittings. The longer the pipe system, the more "opportunity" (joints) there is for leaks to occur. Aggressive water and soil will accelerate corrosion in susceptible pipe. Systems with higher pressures produce more leakage. The vibration caused by traffic loading may cause damage to buried pipes. Soil movement due to changes in moisture, frost heave, and earthquakes causes damage to pipes. In all of these cases, the total amount of leakage is also affected by the type of soil surrounding the leaking pipes. In coarse soils (sands) the leakage may continue for a long time without detection, whereas in finer soils (clays) the leaks show up sooner on the surface, which means they may be detected sooner.

The process of locating a leak is often not easy and sometimes becomes a troublesome and frustrating experience. Methods used to locate leaks include direct observation as well as use of sounding rods, listening devices, and data from a waste control study. The checklist in Figure 5.11 identifies a variety of steps that can be taken to determine whether leaks are occurring and to locate the sources of leaks.

The simplest method of leak detection is to search for and locate wet spots which might indicate the presence of a leak. Sometimes these are reported by the system's customers. However, even if a damp spot is found, it does not necessarily mean that the leak can be easily found. The leak may be located directly below the damp area or it may be 50 yards (46 m) or more away. Often the leak is someplace other than where it would be expected because water will follow the path of least resistance to the ground surface.

After the general location of the leak has been determined, a probe may be used to find it exactly. This probe is a sharp-pointed metal rod that is thrust into the ground and then pulled up for inspection. If the rod is moist or muddy, the line of the leak is being followed. *BE CAREFUL YOU DON'T PROBE INTO AN ELECTRICAL CABLE.*

Listening devices are made up of sound-intensifying equipment that is used in a systematic fashion to locate leaks. The simplest listening device is a steel bar held against the pipe or

Checklist for Leak Detection

1. Using sound-intensifying instruments, listen on fire hydrants, valves, meters, mains, and services. Do this periodically.

2. If leak sounds are heard, conduct a detailed investigation by listening on each meter in the area of the leak sound. Meters are convenient points for making contact with the underground piping system. Listening on the meter allows you to check the meter coupling and curb stop for leakage. Sounds heard at a meter may be a leak on the service or on the street main.

3. If meters are widely spaced, listen over the main at closely spaced intervals with sound-intensifying instruments to locate leaks in the main.

4. Have meter readers listen on services. Develop incentive programs to encourage meter readers to report leaks.

5. Inspect sewer manholes and catch basins for unusual amounts of clear water running in the sewer or coming through joints in the manhole.

6. Check all stream crossings for water bubbling up through the streambed or for the stream to be carrying a much larger volume of water than normal.

7. Check out sudden increases in metered consumption. This could indicate a service line leak.

8. Investigate complaints from customers who report hearing water running in their house piping. This may be caused by a service leak, by a leak in the neighbor's service, or by a leak in the main.

9. Investigate complaints of low pressure in the distribution system. This could indicate that a large leak has occurred. This condition may be reported by customers or by the fire department.

10. Check for commercial, industrial, and residential unmetered use.

11. Check for use from unmetered fire services and private yard hydrants.

12. Review policy on unmetered use: Is public use (for parks, street cleaning, and so forth) unmetered? Review policy on allowing contractors and others to fill tank trucks from hydrants without the water being metered.

13. Install meters in public buildings, churches, hospitals, schools, parks, municipal golf courses, pollution control plants, fire service, or anywhere unmetered water is used.

14. Meter all blowoffs of water from the distribution system.

15. Monitor the metered ratios. (Compare readings from the master meter at the plant, which measures all water entering the distribution system, to all metered readings from customers for the same period of time.)

Fig. 5.11   Checklist for leak detection

(Source: Pitotmeter Associates, Consulting Engineers. Reprinted from *PACIFIC MOUNTAIN NETWORK NEWS*)

valve. The device is moved in the direction of increasing sound until the leak is found. Patented leak detectors use audiophones to pick up the sound of escaping water. Some types of test rods can be driven into the ground and held against the pipe or held against above-ground hydrants and valves. Listening sticks are sometimes equipped with electronic controls giving the operator the ability to select different frequencies to reduce background noise. Different types of leak detection equipment are shown in Figures 5.12, 5.13, and 5.14.

Another method for locating leaks is the use of a leak noise correlator. This type of instrument locates leaks by noise intensity and the time it takes for the leak sound to travel to a pair of microphones placed on fittings (fire hydrants or stop valves) on each side of a suspected leak. Leak correlators are fairly accurate in locating a leak. However, they are of limited use in systems with reduced noise levels such as low-pressure systems and/or pipes made of materials that absorb sound (for example, concrete pipes). Also, leak noise correlators are of limited use in systems with relatively few fittings.

The amount of water lost from the distribution system through leakage is only one component of the system's total water losses. Losses due to illegal connections also occur, and meter malfunctions that produce incorrect readings could give the appearance that higher or lower quantities of water have been used. The total amount of water lost from a distribution system from all sources is often referred to as "unaccounted for water" or UFW. The UFW is the difference between the total amount of water produced and the total amount of water consumed. The amount of unaccounted for water lost by a distribution system is usually determined by conducting a water audit.

A water audit is a thorough examination of the accuracy of water agency records or accounts (volumes of water) and system control equipment. Water managers can use audits to determine their water distribution system efficiency. The overall goal is to identify and verify water and revenue losses in a water system. This allows the water utility to select and implement programs to reduce water and revenue losses. Such examinations must be performed annually to update the results of an earlier audit.

Benefits resulting from a water audit can be significant, including:

1. *Reduced Water Losses*—Conducting a leak detection project will identify and locate system leakage. Upon repair of the leaks, water savings will result. Savings are also realized in reduced power costs to deliver water and reduced chemicals to treat water.

2. *Financial Improvement*—A water audit and leak detection program can increase revenues from customers who have been undercharged, lower costs of wholesale supplies, and reduce treatment and pumping costs.

3. *Increased Knowledge of the Distribution System*—The added familiarity of the distribution system gained during a water audit and leak detection project helps a utility to respond more quickly to such emergencies as main breaks.

4. *More Efficient Use of Existing Supplies*—Reducing water losses will help stretch existing supplies to meet increased needs. This could help defer the construction of new water facilities, such as a new well, reservoir, or treatment plant.

5. *Reduced Property Damage*—Improved maintenance of a water distribution system can reduce the likelihood of property damage and better safeguard public health and safety.

6. *Improved Public Relations*—The public appreciates seeing that its water systems are being maintained. Field teams carrying out water audit and leak detection tasks and doing repair and maintenance work make a favorable impression.

7. *Reduced Legal Liability*—Conducting a water audit and leak detection project provides better information for protection against expensive lawsuits.

Waste control or water audit studies are usually conducted when no specific reason can be found for a significant water loss in the system. Routine comparisons of water production and use should be made to determine the amount of unaccounted for or "lost" water. The amount of unaccounted for

Fig. 5.12   Leak detection instrument

(Permission of Heath Consultants, Inc.)

Fig. 5.13   Leak detection
(Permission of Heath Consultants, Inc.)

water that is acceptable depends largely on the conditions in each system. In some systems there is concern when the loss exceeds 10 percent of the water produced while in other systems there is little concern until 20 percent of the water produced cannot be accounted for. The amount of unaccounted for water is affected by leaks, pressures, efficiency of meter maintenance, and the attention given to leakage reduction and unauthorized uses of water. Leakage has already been discussed. Higher pressures not only result in more leakage, but increase the underregistration of meters. Meters also tend to underregister as they get older and if they are not being properly maintained. Utilities that pay attention to finding and eliminating hidden losses of their supply will naturally have a more efficient and cost-effective operation.

Waste control studies basically involve flow measurements beginning at the source of supply and working out into the system. The system's source and master meters are checked out first. Then separate areas ("districts") are isolated by making appropriate valve manipulations. All the water to the isolated district is made to pass through a single pipe where the flow is measured with a meter. Usually, there is an average of about 18 miles (29 km) of mains per district, but this could vary from 4 to 46 miles (6 to 75 km), depending on the situation. If the flow through the pipe is found to be greater than normal, the isolated district is divided into smaller districts to narrow in on the source of the abnormal use, leaks, or wastage. Measurements of water consumption are made in the late evening hours because they are more likely to indicate leakage, waste, or abnor-

Fig. 5.14   Leak detector

(Permission of Metrotech)

QUESTIONS

Write your answers in a notebook and then compare your answers with those on page 288.

5.70A   What types of pipe maintenance are performed by distribution system operators?

5.70B   What causes pipe deterioration?

5.70C   Where can leaks originate?

5.70D   How are probes used to locate leaks?

5.70E   Why are waste control studies (water audits) conducted?

5.702   Locating Pipes

Ideally, all pipe in a water system would be in standardized, easy to find locations and maps would be available showing precisely where the installation was made. Unfortunately, this ideal situation rarely exists. Other measures must often be

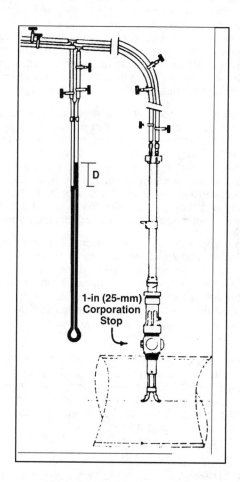

Fig. 5.15   Modified pitotmeter

(Reprinted from AWWA *WATER DISTRIBUTION TRAINING COURSE,* by permission. Copyright 1962, the American Water Works Association)

mal use of water without the complications presented by normal daytime use. Those areas where unexplained high consumption rates are found are further investigated in detail by visual and sounding methods. Surface areas can be observed for wet spots or depressions, and valves, hydrants, and services can be sounded for typical audible vibrations produced by water escaping from defects in the underground piping.

The most convenient flow measurement device used in studies of this type is the pitotmeter. This device is a reversible (water velocity) pitot (PEA-toe) tube that can be inserted into a main through a one-inch (25-mm) corporation cock (Figure 5.15). The differential pressure (D) caused by the velocity of the water flowing past the openings to the tubes is calibrated to the average velocity of the water flowing in the main. Recording devices can be used with it to record continuous flow rates over time periods of 12, 24, or 48 hours. This type of meter has a field accuracy of ± 2 percent and can measure flows over a wide range from a few thousand gallons per day to over 200 MGD (750 ML/day).[10]

---

[10] *750 megaliters per day or 750 million liters per day.*

used to locate the pipes while searching for a leak, when a new connection is to be made, or if an excavation must be made in the area to locate or install other nearby facilities. Even when good records are available, an accurate determination of the underground facilities should be made in the field. Numerous devices have been developed to expedite finding buried pipe. One of these devices is shown in Figure 5.16. Electronic pipe finders consist basically of a portable radio-direction-finder receiver. The transmitter induces an electromagnetic field into any buried metallic object within its range. As the receiver is carried across a pipe location, the induced electromagnetic field is detected and produces an audible tone in the earphones used and a deflection on a visual indicating instrument. Both the position and depth of the buried pipe can usually be determined. Another device used is a stainless-steel-tipped shaft, which is pushed into the ground to locate the pipe. For future location of nonmetallic pipe, a metallic tracer tape (wire), which is detectable by electronic finders, is put on top of the pipe before it is covered.

Locating buried nonmetallic pipes, such as plastic pipes, that do not have metallic tracer tape (wire) can be very challenging for operators. Some operators can locate buried pipe with water flowing in the pipe by using two wire coat hangers. They straighten out the two coat hangers and place a 90-degree bend in each hanger about 6 inches from the end. Holding the 6-inch pieces loosely in each hand and pointing the long pieces in front of their body, they walk toward the suspected location of the buried pipe. When they cross over the pipe, the coat hangers will rotate and align over the buried pipe.

Technology is developing sound-generating devices that could be effective in locating buried nonmetallic pipe. These devices are connected to the pipe, a water meter, a faucet, or a fire hydrant. They use a transmitter to generate a traceable sound that is created by tapping the pipe and a receiver to detect the sound as it moves through the pipe. Thus, the receiver is capable of identifying the location of the buried pipe. This technique may only be effective for about 100 feet because nonmetallic pipe is not a good conductor of sound.

Another approach is the attachment of a type of butterfly valve that is rapidly shut when water is flowing through the pipe. A "water hammer" is generated that moves back and forth down the pipe and can be detected by the receiver. This procedure can detect the water hammer for about 100 feet. A limitation of this procedure is that the water hammer could cause leaks or breaks in the pipe unless properly used by trained persons.

The cost of purchasing a transmitter and receiver may be prohibitive for many water systems. However, a contractor with this equipment may be available to locate the buried nonmetallic pipe. Be sure that the vendor or contractor guarantees the results and performance you desire, in writing, before agreeing to pay for equipment or services. Do not pay unless performance is in agreement with the contract and meets your expectations.

As a last resort for locating buried nonmetallic pipe, cut open the pipe at both known ends. Clean an electrical fish tape with a chlorine-soaked rag. Run the electrical fish tape down the line and trace the fish tape to locate the buried pipe. Another approach could be to insert a "pig" with an electronic transmitter into the line and trace the pig above ground. Remove the pig after the line has been traced.

5.703 Repairing Leaks and Pulled Services

After the leak has been located, route traffic around the work area and take any other necessary safety precautions *BEFORE* starting excavation. Also try to locate other underground utilities *BEFORE* starting any excavation. Many phone books list in the "White" pages an "Underground Service Alert" or "One-Call" phone number which should be called *BEFORE* digging or drilling underground.

A maintenance crew excavates and uncovers the leaky pipe. Sometimes a dewatering pump is also needed to drain the hole before it is possible to work around the pipe. If the main must be isolated before the repairs are made, notify all affected consumers in advance and give them an estimated length of time that the main will be out of service. Leaks at joints or splits or breaks may have to be repaired. Shoring may be required depending on depth of pipe and soil conditions. A good slogan to remember is, *"WHEN IN DOUBT, SHORE IT."*[11] The simplest repair is accomplished by using repair clamps (Figure 5.17). These devices are short, cylindrical pieces usually made of one piece of pipe or in two halves and bolted together or otherwise fastened around a pipe, covering a break, or making a joint between two pipes. For long splits or other defects, clamps up to 18 inches (450 mm) long with rubber or lead gaskets are available. Repair of cracks and breaks in steel pipe is often done by welding. Joint leaks in caulked bell and spigot joints can be stopped by recaulking, by clamps (Figure 5.17), or the entire joint may be removed and replaced. If the main was taken out of service and drained during the repair job, flush and disinfect the main and test a sample for coliforms before placing the main back in service.

Occasionally a contractor digging another utility trench will hit a service line and pull the corporation cock out of the main line. To repair a break of this type, follow the procedures in this section and also in Section 5.74, "Main Breaks." Be sure to notify all consumers who will be without water how long they can expect to be without water. Some agencies notify the consumers before any valves are closed to isolate the break while others close the valves first. If extensive damage is occurring (erosion, washouts, and flooding) as a result of the break, close the valves as soon as possible.

QUESTIONS

Write your answers in a notebook and then compare your answers with those on pages 288 and 289.

5.70F Why does the location of a buried pipe sometimes have to be determined?

5.70G How can buried nonmetallic pipes be located?

5.70H How can a leak in the wall of a pipe be repaired?

5.704 *Making Pipe and Service Connections*

Every water utility agency should have a policy regarding the installation of services and setting of meters to ensure the consumer of an effective installation and to avoid problems in the future. Perhaps the most satisfactory policy is for the agency to install all services and meters. If the workload is too great for agency crews, all contractors and plumbers should be notified of required procedures and fittings. Work done in new subdivisions is usually performed by contractors. All in-

---

[11] See Chapter 3, "Distribution System Facilities," Section 3.653, "Excavation and Shoring," for details.

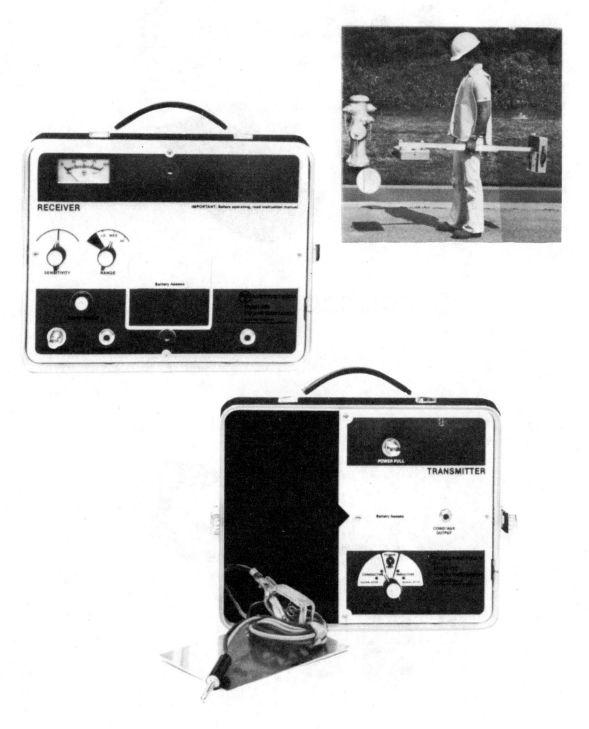

Fig. 5.16   Devices for locating buried pipe
(Permission of Metrotech)

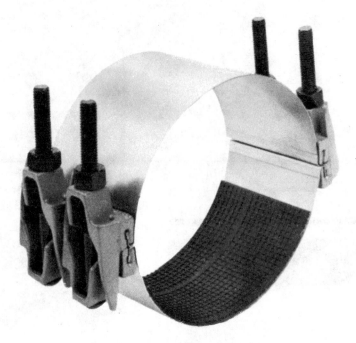

Pipe clamp for repairing pin holes, cracks, bruises, fractures, holes, and other damage in any type of pipe

Bell joint clamp for repairing or preventing leaks in cast-iron bell and spigot caulked or rubber ring joints

Fig. 5.17   Leak repair clamps

(Courtesy of Rockwell International Corporation, Municipal & Utility Division)

stallations should be inspected and tested for leaks by the agency *BEFORE* the installation is covered with backfill.

Connections from existing pipe are often made to a new main or a new service pipe. Making service connections, from a street main to a home, is one of the most frequently performed jobs in a water system. Ideally, when new mains are installed, the connections for the service pipe are made prior to pressurizing the main.

The most common method used in making a connection is "wet tapping" where the connection is made with the main under water pressure. A corporation stop is directly inserted into the main (ductile-iron and PVC thick-wall pipe) using a tapping machine. Although it is called a tapping machine, this device allows three operations to be performed on a main under pressure and should more properly be referred to as a drilling, tapping, and inserting machine.

To insert a tap into a main under pressure, the first step is to excavate down to and around the main. Install shoring if necessary. Clean the main. Install saddle and tapping equipment. Be sure the saddle and equipment are tight.

A combined drill and tap is used to first drill a hole into the pipe (Figure 5.18a). Then, a tap is inserted into the main to thread the hole (Figure 5.18b). Next, still using the tapping machine, the threaded inlet of a corporation stop is threaded into the hole (Figure 5.18c). Wet taps are frequently made with a clamp and corporation stop. Finally, the service line is connected to the corporation stop fitting to be activated when the corporation stop is turned on.

Large service connections are usually made using wet tapping with the water mains under pressure. This is much more convenient for both the water utility and the customer. No customer is out of water, and the job can be done at the convenience of the water utility. Wet tapping also eliminates complaints of dirty water which frequently result when shutting down a section of main. In addition, wet tapping avoids loss of considerable amounts of water.

If you do not have the capabilities for wet tapping, you will have to use a dry tapping machine after closing off the valves and emptying that part of the main that will be tapped. In this operation, attach a service clamp around the main and then thread the corporation stop into it (Figure 5.19).[12] With the corporation stop in an open position, attach the drilling device to the corporation stop's outlet threads and drill a hole or make a shell-cut through the wall of the main. After the main is cleaned, bolt a tapping sleeve to it. A tapping sleeve is a split sleeve or clamp in one-half of which is an opening with a flange or other means of attaching a valve. Figure 5.20 is a picture of one type of tapping sleeve used. Next, attach a tapping valve (a permanent valve installation) to the sleeve outlet (Figure 5.21a). Then attach the drilling machine and adapter to the valve outlet flange (Figure 5.21b). With the tapping valve open, advance the cutter and drill a hole into the pipe; retract the cutter and close the tapping valve (Figure 5.21c).

After removing the drilling machine, attach a new lateral to the valve outlet flange and activate it by opening the valve.

5.705 Special Investigations

Deficiencies in service can be recognized and corrected in advance and potential failures can be detected in advance by making pressure, flow, hydraulic grade line, and pipe *C FACTOR*[13] surveys. Pressure surveys can indicate the hydraulic efficiency of the system in meeting normal requirements. These surveys are performed by attaching a pressure gage to a hose nozzle or a hydrant or other available connection, opening the valve, and reading the pressure in the main (Figure 5.22). The best location to obtain pressure readings is the closest possible location to the main being tested. Pressure gages often receive rough handling and should be tested and recalibrated periodically to ensure accurate readings. If a pressure record covering a longer period of time is desired, a recording pressure gage is used to evaluate changing conditions. The gage is connected to some type of recorder such as a strip chart. Pressure readings are of special interest during periods of maximum demands.

Pressures may fall off in some parts of the system due to an increase in water use by consumers, leaks, obstructions, or diminished carrying capacity in the pipes. Pressure readings can be taken at various points in the system to follow out the changes in pressure and locate the problem areas. Results of pressure tests combined with rate-of-flow (fire) information are very useful in locating the source of a pressure problem.

Use flow tests to determine the efficiency and adequacy of a distribution system in transmitting water and to measure the amount of water available from hydrants for firefighting. This information is of particular importance during peak demand periods. Flow tests are usually made from hydrants using a pitot gage. Hydrant flow tests are made by measuring flow from one or more hydrants and at the same time noting the change in pressures at a nearby hydrant when going from no flow to full flow.

HYDRAULIC GRADIENT[14] tests are used to determine the ability of the distribution system to maintain adequate pressures throughout the system. Since the purpose of these tests is to find any weaknesses in the system, they are normally run during peak periods of delivery when such weaknesses are most evident. Plotting the pressure measurements at various points in the system will give you a visual picture of the losses of head between these points as the water is distributed through the system.

Pipe interior roughness coefficient (C Factor) tests will demonstrate whether or not friction losses in the pipe are increasing and whether the ability of the pipe to transmit water is being seriously hindered. Isolate the pipe under study as much as possible and close even the service connections if you can. Using two points a known distance apart, determine the hydraulic gradient and measure the flow (normal or induced) through the pipeline. Using these data, calculate the pipe roughness coefficient (C Factor). Once the coefficient is

---

[12] *Some agencies use this type of equipment for wet tapping.*

[13] *C Factor. A factor or value used to indicate the smoothness of the interior of a pipe. The higher the C Factor, the smoother the pipe, the greater the carrying capacity, and the smaller the friction or energy losses from water flowing in the pipe. To calculate the C Factor, measure the flow, pipe diameter, distance between two pressure gages, and the friction or energy loss of the water between the gages.*

$$C \text{ Factor} = \frac{Flow, \ GPM}{193.75(Diameter, \ ft)^{2.63}(Slope)^{0.54}}$$

[14] *Hydraulic Gradient. The slope of the hydraulic grade line. This is the slope of the water surface in an open channel, the slope of the water surface of the groundwater table, or the slope of the water pressure for pipes under pressure.*

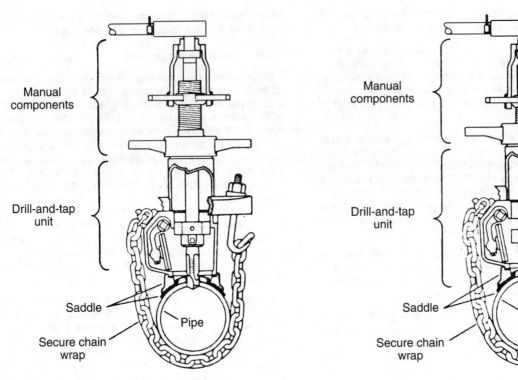

Manual components

Drill-and-tap unit

Saddle

Pipe

Secure chain wrap

a. With the combined drill-and-tap unit, first drill a hole into the main.

Manual components

Drill-and-tap unit

Saddle

Pipe

Secure chain wrap

b. After the hole is drilled, insert the tap.

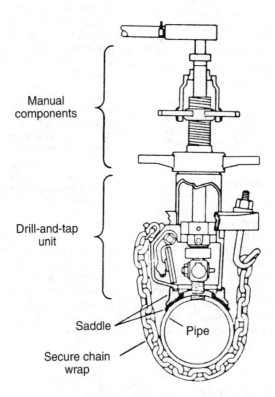

Manual components

Drill-and-tap unit

Saddle

Pipe

Secure chain wrap

c. Again, with the drill-and-tap unit, insert the threaded inlet of a corporation cock. The service can then be activated.

Fig. 5.18   Wet tapping

(Art reproduced by permission of Mueller Company)

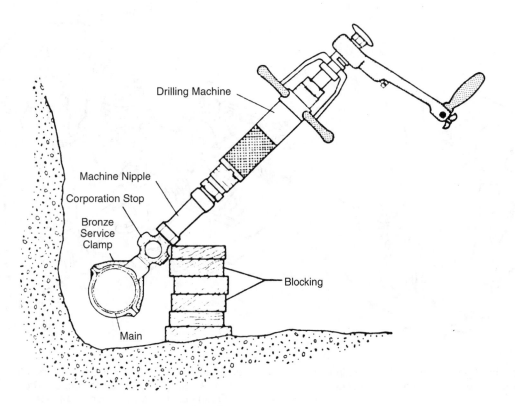

Fig. 5.19 Dry tapping

(Art reproduced by permission of Mueller Company)

Fig. 5.20 Tapping sleeves

(Courtesy of Rockwell International Corporation, Municipal & Utility Division)

a. Clean the main and then bolt the tapping sleeve to it. Attach tapping valve to sleeve outlet.

b. Attach the drilling machine and adapter to valve outlet flange. Position support blocks. Open tapping valve, advance cutter, and drill hole inside of pipe inside the sleeve.

c. Retract cutter and then close tapping valve. Remove drilling machine and attach new line or lateral. Open tapping valve, activating new line.

Fig. 5.21   Drilling machine

(Art reproduced by permission of Mueller Company)

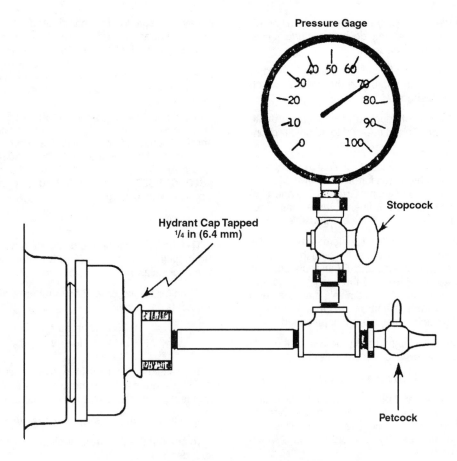

Pressure Gage

Stopcock

Hydrant Cap Tapped
¼ in (6.4 mm)

Petcock

Fig. 5.22   Gage assembly for measuring hydrant pressures
(Reprinted from *AWWA DISTRIBUTION MANUAL*, by permission. Copyright by the American Water Works Association)

known, head losses under other flows can be determined for the particular section of pipe. For procedures on how to calculate the C Factor, see Chapter 3, Example 3, page 68, or the Arithmetic Appendix at the end of this manual, Section A.132, "Distribution System Facilities," Example 11, page 561.

QUESTIONS

Write your answers in a notebook and then compare your answers with those on page 289.

5.70I   What is wet tapping?

5.70J   Why are pressure surveys of a water distribution system conducted?

5.70K   How can an operator obtain a pressure record covering a long period of time at one location?

5.706   Pipe Flushing

Flushing is done to clean out distribution pipelines by removing any impurities or sediment that may be present in the pipe. Routine flushing of dead-end lines is often necessary to avoid taste and odor complaints. Many operators use flushing as a short-term solution to distribution system problems. Flushing is commonly practiced after receiving water quality complaints (red water, sand and grit, tastes and odors, cloudy (air) water, and something swimming), when the water in the

system appears to have become contaminated, and to clean newly installed or repaired mains prior to disinfection.

Flushing may remove deposits, encrustations, sediments, and other materials. Deposits that have settled out and accumulated in pipelines may result in taste, odor, and turbidity problems. Encrustations may restrict the water flow. Sand, rust, and biological materials cause quality problems and are not uncommon in pipelines. The needed frequency of routine flushing can usually be determined by customer complaints and the types of material found during the flushing procedure. Flushing should not be considered the only solution to distribution system water quality problems. The water utility should always try to prevent water quality degradation through proper design, operation, and treatment.

Water mains should be flushed before consumers start complaining about poor water quality. Flushing should be conducted during periods of low water demand (spring or fall) when the weather is suitable. Prior planning and good communications will allow the flushing crew to conduct the flushing operation quickly and without confusion. Flushing crews consist of two operators. The following procedures are recommended for flushing operations.

1. Preplan an entire day's flushing using the available distribution system maps. Consider flushing at night between 9:00 PM and 5:00 AM to minimize any inconvenience to customers. Night operations encounter little traffic, but

traffic must be made aware of the operation if it will be affected by the flushing. Warning devices include lights, traffic cones, barricades, and flaggers.

2. Determine where sections of mains are to be flushed at one time, the valves to be used, and the order in which the pipelines will be flushed.

3. Start at or near a source of supply and work outward into the distribution system. This is referred to as unidirectional flushing.

4. Ensure that an adequate amount of flushing water is available at sufficiently high pressures. A *MINIMUM* flushing velocity of 2.5 ft/sec (5 ft/sec preferred) (0.75 and 1.50 m/sec) should be used. Do not flush a large main supplied by a single smaller one if a choice is possible.

5. Prior to flushing the mains, notify all customers who will be affected of the dates and times of the flushing through billing, newspapers, and local radio and TV announcements. Explain the intent and objective of the flushing program. Notify individuals who might be on dialysis machines and also hospitals, restaurants, laundromats, and others who might be affected while the mains are being flushed.

6. Isolate the section to be flushed from the rest of the system. Close the valves slowly to prevent water hammer.

7. Open the fire hydrant or blowoff valve slowly.

8. Direct flushing water away from traffic, pedestrians, and private lots (Figure 5.23). Avoid erosion damage to streets, lawns, and yards by the use of tarpaulins and lead-off discharge devices. Try to avoid flooding, which can cause traffic problems.

9. Open hydrant fully for a period long enough (5 to 10 minutes) to stir up the deposits inside the water main. Usually lines are flushed for at least 30 minutes.

10. Ensure that system pressures in nearby areas do not drop below 20 psi (138 kPa or 1.5 kg/sq cm).

11. Record all pertinent data regarding the flushing operation as well as a description of the appearance and odor of the water flushed. Figure 5.24 is a sample flushing log sheet that can also be used for a pipeline cleaning operation.

12. Collect two water samples from each flowing hydrant, one in the beginning (about 2 to 3 minutes after the hydrant was opened) and the second sample when the discolored water turns clear (just before closing the hydrant). These samples allow a check on the water quality for certain basic water quality indicators (iron, chlorine residual, turbidity) and the development of water quality trends for comparison purposes.

13. After the flushing water becomes clear, slowly close the hydrant or blowoff valves.

14. In areas where the water does not become completely clear, the operator should use judgment as to the relative color and turbidity and decide when to shut down. A water sample in a clear glass bottle will allow the operator to visually observe the color from time to time.

15. Mark closed valves on a map when they are closed and erase marks after the valves are reopened. Do this promptly and do not depend on memory.

16. After flushing one section of pipe, move on to the next section to be flushed and repeat the same procedures.

Contact your water pollution control agency to determine if flushed water needs to be dechlorinated. In many communities,

it may be illegal to allow chlorinated water to flow into a storm drain or surface waters. Devices are available that can be attached to a fire hydrant or a fire hose to deflect the flow or dissipate the energy in flushed water. Some of these devices have a space where tablets can be inserted that contain a chemical that neutralizes chlorine in the flushed water. Use National Sanitation Foundation-approved dechlorination chemicals (NSF Standard 60, Drinking Water Treatment Chemicals—Health Effects; NSF Standard 61, Drinking Water System Components—Health Effects; available from National Sanitation Foundation International, PO Box 130140, 789 N. Dixboro Road, Ann Arbor, MI 48113-0140, phone: (800) NSF-MARK ((800) 673-6275) or (734) 769-8010, e-mail: info@nsf.org, or website: www.nsf.org.

FORMULAS

The formulas needed to calculate a desired flowmeter reading in gallons per minute (GPM) to flush a water main are the same formulas we've used before.

To calculate the cross-sectional area of a pipe when the diameter is given in inches,

$$\text{Area, sq ft} = \frac{(0.785)(\text{Diameter, in})^2}{144 \text{ sq in/sq ft}}$$

To calculate the flow in a pipe in cubic feet per second (CFS), we need to know the area in square feet and the velocity in feet per second. This is the familiar formula, Q = AV.

Flow, CFS = (Area, sq ft)(Velocity, ft/sec)

To convert the flow from cubic feet per second (CFS) to gallons per minute, GPM,

Flow, GPM = (Flow, cu ft/sec)(7.48 gal/cu ft)(60 sec/min)

EXAMPLE 4

A 15-inch diameter water main is to be flushed at a velocity of 5 ft/sec. What should be the reading on the flowmeter in gallons per minute?

Known	**Unknown**
Diameter, in = 15 in	Flow, GPM
Velocity, ft/sec = 5 ft/sec	

FORMULA (Q = AV)

Flow, CFS = (Area, sq ft)(Velocity, ft/sec)

1. Calculate the cross-sectional area of the pipe in square feet.

$$\text{Area, sq ft} = (0.785)(\text{Diameter, in})^2$$
$$= \frac{(0.785)(15 \text{ in})^2}{144 \text{ sq in/sq ft}}$$
$$= 1.23 \text{ sq ft}$$

2. Determine the flow in the pipe in cubic feet per second (CFS).

Flow, CFS = (Area, sq ft)(Velocity, ft/sec)
$$= (1.23 \text{ sq ft})(5 \text{ ft/sec})$$
$$= 6.15 \text{ CFS}$$

3. Calculate the flowmeter reading in gallons per minute (GPM).

Flow, GPM = (Flow, cu ft/sec)(7.48 gal/cu ft)(60 sec/min)
$$= (6.15 \text{ cu ft/sec})(7.48 \text{ gal/cu ft})(60 \text{ sec/min})$$
$$= 2,760 \text{ GPM}$$

Direct flushing water away from traffic, pedestrians,
underground utility vaults, and private lots.

Deflection tubes keep water and swabs from going into traffic on a busy street.
Without the chain, vibration will break tube.

Fig. 5.23   Diverting flushing water

Date	Time	Location	Press. Zone	Size of Main		Swabs	P Pitot Press, psi	d Disch. Opening, in	Q Flush. Rate, GPM	V Flush. Velocity, FPS	Time Req'd. to Clear, min.	Flushed Water Description
				D Dia, in	L Length, ft	# of Runs # per Run						

Q = Flushing rate in GPM

d = Diameter of nozzle or opening in inches

P = Pitot gage pressure at nozzle or opening in psi

$Q = 26.8\, d^2 \sqrt{P}$

V = Flushing velocity in main in FPS

D = Diameter of main being flushed in inches

$V = \dfrac{0.409\, Q}{D^2}$

Fig. 5.24 Main flushing and swabbing log

QUESTIONS

Write your answers in a notebook and then compare your answers with those on page 289.

5.70L   How is the frequency of routine flushing determined?

5.70M   Why should valves be closed slowly during a flushing operation?

5.70N   List the information that should be recorded during a flushing operation.

5.707   Pipe Cleaning (Swabs and Pigs)

Mechanical cleaning devices are often used to clean pipes if flushing does not provide relief from water quality problems or from problems in maintaining the carrying capacity. Foam swabs, pigs (Figure 5.25), and pressurized air have been successfully used to remove loose sediments and soft scales from mains. Pigs have also been used to flush new mains prior to disinfection. Scrapers or brushes may have to be used in mains with hardened scales or extensive tuberculation, and they are usually used prior to relining. Of the available de-

vices, foam swabs and pigs are the easiest and most effective to use. Pipe cleaning projects should produce improved pipe carrying capacity and a reduction of power (and cost) to pump the water.

Swabs are made of polyurethane foam; both soft and hard grade forms are available. Some agencies purchase sheets of foam and cut out the desired size. They are efficient in removing loose sediments, soft scales, and slimes. Use different col-

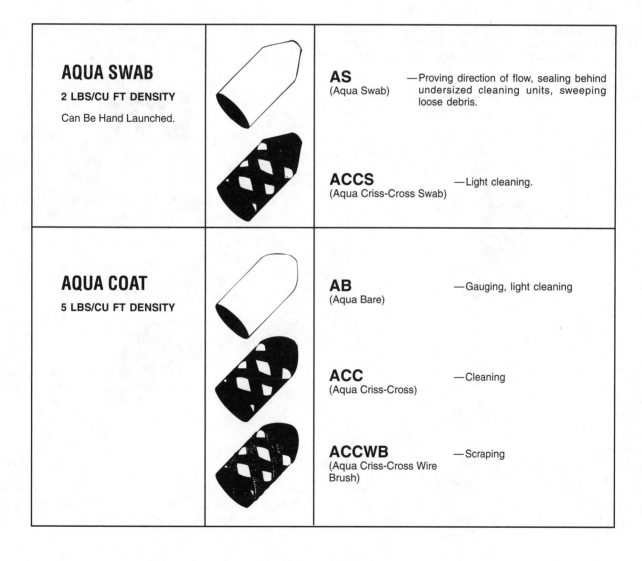

Fig. 5.25   Types of swabs and pigs

(Permission of Girard Industries)

ored swabs or some other means of identifying swabs. All swabs inserted in mains must be retrieved. Pigs are also made of polyurethane foam, but are much heavier in weight, harder, and less flexible than swabs. They are bullet-shaped and come in various grades of flexibility and roughness.

Generally, if loose sediments and soft scales in the pipe are to be removed without disturbing hardened encrustations, swabs are used. If you also want to improve the carrying capacity of the main, but not remove all the existing encrustations, then pigs should be used. However, the use of pigs is more likely to result in leaks at a later date.

A mixture of air and water can effectively clean small mains up to 4 inches (100 mm) in diameter. Air is introduced into the upstream end of the pipe from a compressor of the same type used for pneumatic tools. Spurts of water mixed with the air will remove all but the toughest scale.

The use of compressible foam swabs and pigs provides flexibility in their insertion and removal. The entry and exit points used for smaller size mains are fire hydrants, air valves, blowoffs, wyes, and tees (Figure 5.26). Figures 5.27 and 5.28 show the insertion of a pig and a swab into fire hydrants. In larger mains, a section of pipe may be removed and a wye inserted in its place at the entry and exit points to allow insertion, launching, and exiting of the swabs and pigs.

The routine procedures used for cleaning pipe are very similar to those used for flushing except that services to customers will have to be shut off during cleaning. Temporary servic-es may have to be installed to some customers if there is a critical need for a continuing water supply (such as a hospital). The water used may be that from the upstream main or may be from an external pressurized water source. A swab flushing plan is shown in Figure 5.29. A typical main cleaning operation is shown in Figure 5.30. Both flushing and swabbing operations usually start near the beginning of the system and move outward toward the ends of the system.

First insert the swab into the section of main to be cleaned (the section is already isolated). A pressurized water source such as a nearby fire hydrant can be used to launch the swab. Also a water truck with a pump can provide the pressurized water. Open the valve at the exit.

Pass the swab through the main at a speed of 2 to 4 ft/sec (0.6 to 1.2 m/sec). Using velocities in this range, up to 4,000 feet (1,200 m) of pipe can be effectively cleaned before the swab wears down to a size smaller than the main. The entire operation may thus require 10 to 20 swabs. Typically, 2 to 3 runs are made using 4 to 5 swabs in each run. Continue the cleaning until the water behind the swabs emerging at the exit clears up within one minute. All swabs inserted into and ejected from the main must be accounted for. Figure 5.31 shows swabs being flushed from an exit hydrant.

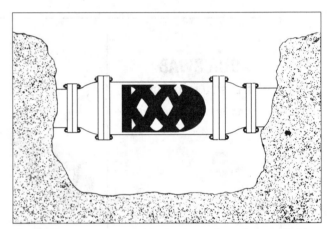

Figure 2 shows how an oversized spool can be coupled into the line for launching.

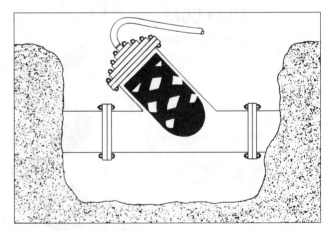

Figure 3 shows Aqua Pigs being introduced into the line through a standard "Y" section. Regular "T" sections can also be used.

Figure 1 shows how a fire hydrant can be used to launch the Aqua Pig, by removing the internal valve assembly (including the seat) and attaching a swage reducer.

Fig. 5.26   Methods of inserting and launching pigs

(Permission of Girard Industries)

Fire hydrant "pig launcher"

Insertion of pig through fire hydrant

Fig. 5.27   Insertion of pig

Insertion of foam swabs into a hydrant

If you have the screw-type hydrant in your system with the bleeders on the side that you cannot plug, you can make up a hydrant such as this with steel pipe and a reducer, moving it from job to job.

Fig. 5.28   Insertion of swab

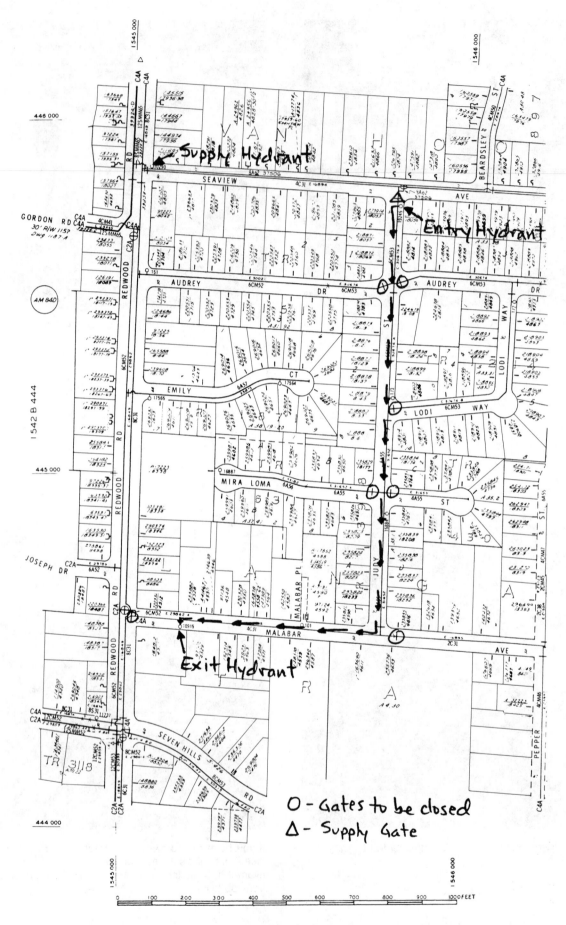

Fig. 5.29 Swab flushing plan

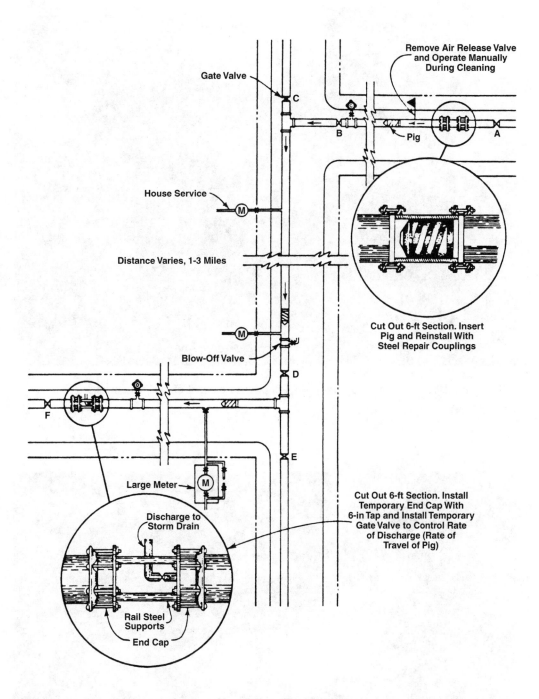

Valves A and B are closed. Cut pipe and insert pig. Open valve A and fire hydrant as needed to move pig; when pig approaches, close hydrant and valves C and D and service lines. Valve B is opened and flow at blow-off is regulated to move pig; when pig approaches blow-off, valves E and F and service lines are closed. Open valve D and regulate flow using valve on end cap. Close valve D, remove end cap and pig, flush thoroughly, and replace pipe section.

Fig. 5.30   Typical main cleaning operation

(Reprinted from *JOURNAL AMERICAN WATER WORKS ASSOCIATION*, Volume 60, Number 8 (August 1968), by permission. Copyright 1968, American Water Works Association)

Dirty water flushed from main just prior to the exit of the foam swabs

Foam swabs after being flushed from fire hydrant

Fig. 5.31   Foam swabs leaving system

Before starting any cleaning job, determine how to dispose of or remove the water and deposits discharged from the cleaned water main. If the water is discharged onto a street or the ground, be sure that the drainage is proper and adequate. When the water flows down a street, sandbags or a screened barrier may be used to catch the solids yet allow the water to flow down a storm drain.

The procedures to follow for cleaning a water main using pigs or swabs are as follows:

1. Isolate the line to be cleaned. Be sure that those customers requiring temporary services have enough water.

2. Be sure that all valves in the section to be cleaned are fully opened.

3. Turn on the water and verify the direction of flow.

4. Run a full-sized bare swab through the main to prove the direction of flow.

5. Run a bare squeegee unit through the main. Measure the diameter of the unit upon exiting and introduce a crisscross type unit into the main that will just fit the "true" opening. Run a full-sized bare swab behind the crisscross unit to ensure a tight seal. Continue this process until a unit is discharged from the main in reusable condition.

6. Increase the size of the crisscross pigs in one-inch (25-mm) increments until the units that measure the same as the pipe inside diameter are being used. For pipes with a buildup of hard scale, such as carbonates, crisscross wire pigs can be used on the final pass.

7. Run a full-sized bare swab to sweep out any loose debris.

To obtain the best possible cleaning results, be sure to:

1. Flush thoroughly after each pig run;

2. Avoid applying more than two wire-brush pigs on the final pass (this prevents overcleaning);

3. Launch the pigs from fire hydrants for mains of 8 inches (200 mm) or smaller, or from concentric reducers, pipe couplings, spools, eccentric reducers, in-line launchers, or by hand; and

4. Have an operator with experience in proper main cleaning procedures help you the first time you attempt to clean a main. This is good practice to avoid stuck, lost, or damaged pigs or swabs.

After the cleaning operation is completed, flush and disinfect (chlorinate) the main. When the main is reactivated, flush service lines and remove any temporary services.

Portions of the material in this section were reproduced from *DISTRIBUTION MAIN FLUSHING AND CLEANING*, published by the California-Nevada Section of the American Water Works Association.

QUESTIONS

Write your answers in a notebook and then compare your answers with those on page 289.

5.70O Foam swabs are efficient in removing what types of materials from water mains?

5.70P Where would you insert foam swabs and pigs into a small water main to be cleaned?

5.70Q What is the recommended speed range for a swab cleaning a water main?

5.70R What should be done, before use, after a water main is cleaned with a swab?

5.708 *Cement-Mortar Lining* (Figure 5.32)[15]

Nearly all new metal mains are protected by cement lining; however, this has not always been the case. Many operators still have to operate and maintain a considerable distance of unlined pipe. As time goes by, the condition of those water mains susceptible to corrosion or tuberculation becomes progressively worse. Cement-mortar lining of metal water mains in place has proven to be very valuable where:

1. The pipe carrying capacity has been seriously decreased because of tuberculation,

2. Corrosion products are released into the water in unacceptable amounts, and

3. Abnormal leakage through poor joints or holes in the pipe wall exist.

Water mains of all pipe diameters over four inches (100 mm) have been successfully relined, but greater cost effectiveness is to be expected in relining the larger pipes (18 inches (450 mm) or larger). Lining of pipe just after it has been cleaned is often recommended. Cleaning alone is frequently found to be only a temporary solution as tuberculations and corrosion tend to recur at a faster rate after cleaning.

If metal water mains are corroded, cement-mortar lining may be used to protect the pipe and improve the carrying capacity of the water main. Usually transmission mains larger than 18 inches (450 mm) are lined. The first step is to inspect the mains for external corrosion. If the mains are badly corroded on the outside, they will need to be replaced. Lining the inside will not solve the problem of preventing eventual collapse of the main.

If inspection indicates that external corrosion is not a problem, the next step is to inspect the inside of the main to determine if internal corrosion is a problem. Operators should personally inspect the inside of the main by the use of a TV camera, by crawling through the main, or by being pulled through on a sled. Look for any dips in the main such as where the main had to go under a sewer or other obstruction.

---

[15] *For additional information, see AWWA STANDARD FOR CEMENT-MORTAR LINING OF WATER PIPELINES IN PLACE—4 INCHES (100 mm) AND LARGER, C602-00. Obtain from American Water Works Association (AWWA), Bookstore, 6666 West Quincy Avenue, Denver, CO 80235. Order No. 43602. Price to members, $42.00; nonmembers, $61.00; price includes cost of shipping and handling. Other standards include C104/A21.4-03, CEMENT-MORTAR LINING FOR DUCTILE-IRON PIPE AND FITTINGS FOR WATER, price to members, $42.00; nonmembers, $61.00; price includes cost of shipping and handling, and C205-00, CEMENT-MORTAR PROTECTIVE LINING AND COATING FOR STEEL WATER PIPE—4 INCHES (100 mm) AND LARGER—SHOP APPLIED, price to members, $42.00; nonmembers, $61.00; price includes cost of shipping and handling.*

1 Bypass lines are installed to provide water service to residents and businesses during rehabilitation process.

2 Small excavations are sufficient for access to the pipeline. Spacing between openings varies according to job conditions.

3 Shoring of trench prior to work on pipeline.

4-4A Six-foot (1.8 m) steel nipple is removed by cutting torch to provide access to interior of pipeline. This section is cleaned and lined separately, later reinstalled.

5 Tuberculation and scale are removed from the pipe by steel scraper blades drawn through pipeline by winch before cement-mortar lining operations begin.

6 Emerging at end of run, squeegee assembly removes water and debris freed by steel scrapers.

Fig. 5.32   Photos showing the steps in the cement-mortar lining process

(Permission of Ameron)

7-7A Cement mortar is mixed and pumped through hose to Spunline equipment in trench.

9 Trowel assembly follows Spunline unit providing a smooth surface finish.

8 Attaching cement-mortar carrying hose to Spunline equipment. Spinning centrifugal head of unit in foreground sprays cement mortar on inside wall as unit travels through pipe.

10 Inspection of pipe after cement-mortar lining has been completed.

11 Typical pipe lining street scene. Minimum access requirements permit unhampered traffic flow.

Fig. 5.32   Photos showing the steps in the cement-mortar lining process (continued)

(Permission of Ameron)

Dips may cause problems for the contractor's cleaning and lining equipment. If the main is corroded, then the procedures outlined in the remainder of this section should be followed to cement line the main, test it, and return the main to service.

Plan to do the lining job during the time of the year when demands for water are low and the weather is suitable. Determine which sections of the main will be lined. The sections usually run from valve to valve and are around 500 to 600 feet (140 to 180 m) long. Locate the holes where the lining equipment will be inserted into and removed from the main on both sides of a valve.

Shut off each section of the main and determine if anyone is out of water. Make provisions to provide water to all consumers while the main is out of service during the lining project. New mains may have to be installed or special temporary, above-ground lines may have to be provided (Figure 5.32-1). While each section is isolated, look for any leaking valves. Sometimes old valve seats will be tuberculated and won't shut tight. These valves will have to be repaired or replaced before the lining job is started to prevent leakage from damaging the mortar before it is cured. Once everyone is ensured of an adequate supply of water, the actual lining of the main can get started. This work is usually performed by an outside contractor because lining requires special equipment and specially trained personnel.

A 5- or 6-foot (1.5- or 1.8-m) section of the top of the main is removed at each end of a section for the insertion and removal of the lining equipment (Figure 5.32-4). Pump or drain dry the section of the main to be lined. Inspect the main and look for any dips that must be drained. A scraper (Figure 5.32-5) is pulled through the main to remove the scale, tubercules, and any old tar or other lining. Do not attempt to run a scraper past a valve because the valve seats could be badly damaged. Next an oversized squeegee is pulled through the main to clean out all remaining debris and water (Figure 5.32-6). Inspect the main to be sure it is clean. Any "live tar" (tar firmly sticking to the main) may be allowed to remain and can be covered with the mortar lining.

Insert the cement-mortar lining equipment into the main. The mortar is pumped into a header (Figure 5.32-7, 8, and 9) which applies the mortar to the side of the pipe. The mortar is sprayed by the "centrifugal method," from a rapidly rotating "head," onto the wall of the pipe as the machine carrying the spray equipment travels through the pipe. The slower the equipment moves, the thicker the lining. A troweling cone follows the mortar applicator and trowels the mortar smooth. Figure 5.33 shows a cement-mortar lining process for both smaller and larger diameter pipe.

At specified intervals some contractors will skip (not line) a short section. If there is a problem with the mortar mix, the mortar could fall off the pipe and ruin all the new lining in the entire section. This is a difficult mess to clean up. By skipping short sections, the length of failure can be confined and thus reduce the size of the cleanup job and length of section which must be redone.

After a section has been lined, close the section tight and allow the mortar to cure. After 24 hours of curing, inspect the entire section. The skipped sections should be hand lined with mortar at this time. Also all valves, tees, reducers, and wyes should be hand lined. Replace the 5- or 6-foot (1.5- or 1.8-m) section which was cut out at each end of the section. A 6-inch (150-mm) hand hole is cut on the top of each of these short sections. Use this hole to hand line the section. Finally the 6-inch (150-mm) hand hole is lined, reinserted, and sealed in place.

Pressure test the section of the main that was lined to be sure there are no leaks. Disinfect the new section by following the procedures for disinfecting mains as outlined in Section 5.8, "Field Disinfection."

QUESTIONS

Write your answers in a notebook and then compare your answers with those on page 289.

5.70S   Why are water mains lined with cement?

5.70T   Why must metal water mains be inspected on the outside before considering cement-mortar lining?

5.70U   How are the insides of water mains inspected before a lining job?

5.709   *Thawing*
by Don Thomas

Frozen pipe can be a very serious problem in frigid climates. Difficulties are usually experienced with service lines rather than water mains because the service lines have less ground cover over them and are smaller. Also, since there is little or no flow through service lines at night, they are more likely to freeze unless water in the lines is kept moving and wasted. If a metallic pipe is frozen, electrical thawing is a quick and relatively inexpensive method to start the water flowing again. However, electric thawing can cause property damage to household electrical wiring and appliances even when done by trained and experienced operators. Operators must realize that electrical thawing can be dangerous not only to the property where the thawing equipment is located, but to other houses on the block and sometimes to houses farther away. Fires have been caused in some communities almost two blocks from the location of the equipment. For this reason many communities have codes that prohibit the use of electric thawing equipment. Due to this problem, hot water thawing is the preferred method, although it is considerably slower. These methods are explained in the following paragraphs.

A.   Electrical Thawing of Frozen Water Services

1. Obtain information from the utility office responsible for thawing.

    The operator will receive a priority list of lines to be thawed from the Thawing Supervisor. The operator will then check the thawing card records (service cards) as to service shutoff measurements and ground wire locations, if any. This information should be on or attached to existing service cards. Also, the operator should check to see if service is in a cathodically protected area; this would mean the service is isolated from the water main.

For Pipelines 4 Inches (100 mm) Through 36 Inches (914 mm) in Diameter

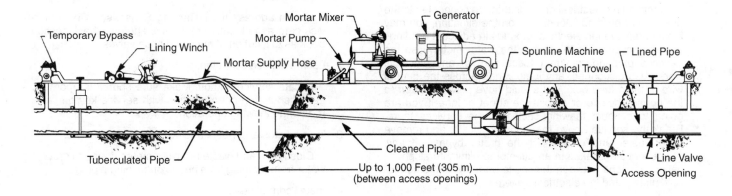

For Larger Pipelines to 264 Inches (6.7 m) in Diameter

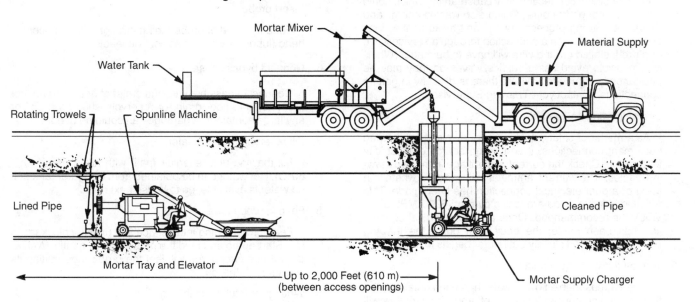

Fig. 5.33   Drawings showing cement-mortar lining process

(Permission of Ameron)

2. Locate section of frozen pipe.

Upon arriving at given address, determine whether the service is frozen on private property or on utility's side. After locating service valve, try to operate to a half on/half off position. Listen on service to determine if there is water flowing from main to service valve. If water is flowing, you should hear water flowing out of service valve drain hole. *NOTE:* Some service valves have no drain holes, but you still have to check. There is no way of determining whether there is a drain hole or not.

3. Remove electrical grounds and disconnect lines (Figure 5.34).

Remove any electrical grounds that are connected to the pipe in that building. Depending on the situation, you may have to remove the electrical grounds in adjacent buildings and may even have to remove the water meter. Use the services of a Journeyman Electrician for any electrical work. Also disconnect any jump wiring across the neutral wire at main fuse box. You should have power shut off to the houses beside or across the street or lane if you are going to connect thawing cables to the services there. Break the service above the stop and waste, and remove water meter and disconnect the meter bypass, if any. These steps are taken in an attempt to eliminate all physical connections between the line to be thawed, the household piping, and the electrical system.

4. Connect cable to line.

Clean the service pipe at the point where the thawing power is to be connected (use wire brush or emery cloth). A poor electrical connection may cause arcing which could cut holes through the pipe. Arcing also wastes energy and slows the thawing process. If you can't make connections between houses, then a connection through a service valve or a pre-installed ground wire will have to be made. (Wherever possible avoid using service valves as they may be permanently damaged.) Where there is difficulty in getting connections, contact your Thawing Supervisor.

5. Use of thawing equipment.

As a matter of safety, double check all electrical thawing equipment connections, then turn on source of electricity for thawing. Check the current and voltage in the two thawing cables. The current ammeter will indicate whether a good or a poor electrical connection has been made. This will also indicate if you are losing current. 600 AMPS at 12 VOLTS is recommended. Once a good electrical connection has been made, the equipment doing the thawing should be watched closely until water begins to flow.

6. Turn off thawing equipment.

When a reasonable flow of water has been obtained, the thawing equipment can be turned off and the water flow will finish thawing the line.

7. Reconnect ground wires and service lines.

Reconnect all disconnected wires and water lines and water meters. To prevent refreezing of the thawed line, have consumer let the water run until the danger of freezing has passed. If consumer is on meter, notify meter department of present meter reading so consumer is charged flat rate for duration of potential freeze up. The consumer is issued a directive as to when to discontinue running water.

8. Complete paperwork.

Complete frozen service sheet including the address and the name of the party to whom the thawing service will be charged (Figure 5.35). The consumer should be charged when service is frozen on private property. Figure 5.36 is a copy of a letter describing "Utility policy" to customers. The utility is responsible when service is frozen on utility property. Turn in paperwork to Thawing Supervisor as soon as possible.

B. Thawing Frozen Service Lines with a Hot Water Thawing Unit (Figure 5.37)

1. Obtain address.

Obtain address from Thawing Supervisor. Check service card as to location of service shutoff. Check section book as to location of main and main valves.

2. Locate service curb stop.

Locate service shutoff. Make sure shutoff is operable. Leave shutoff in open position with service key on valve rod during thawing operation.

3. Turn off stop and waste and connect to service line.

Close stop and waste (at water meter, if any) to prevent water from draining from house or building piping.

4. Install control head.

Install control head onto service line. Make a tight joint with a compression fitting. Some fitting variations may be required depending upon size and type of service line.

5. Insert probe.

Feed free end of probe tubing through gland on control head (lubricate gland to assist free feeding).

6. Connect bypass hose.

Connect bypass hose to side outlet of head and anchor firmly to sink, floor drain, or reservoir, such as a pail or any large container. A garbage can could be used.

7. Fill reservoir with hot water.

Fill thawing unit reservoir (pail) with hot water (140°F, 60°C). Hot water can be obtained from hot water tank or if no water is available, go to house next door.

8. Feed in probe tube.

Feed probe tubing into service line until ice blockage is felt. Start pump and feed the probe tube in gently. (*NOTE: do not use force.*) You will feel the probe tube cutting its way through the ice.

9. When probe tubing is through ice.

When probe tubing breaks through the ice, water will pass through the bypass hose rapidly, indicating that the line has been thawed. Tighten the gland nut to prevent excess leakage. Place footage marker clip on tubing at this point to establish location of problem area. Water pressure will force the tube back through the gland.

10. Pull back probe and close service valve.

Close service valve only after tubing has been pulled clear so the service valve will not cut off tubing.

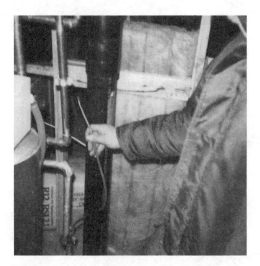

STEP #3—Ground wire removed from service line.

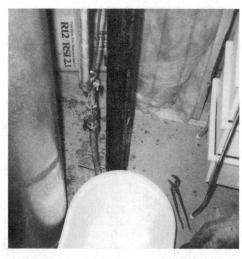

STEP #3—Service line disconnected above stop and waste.

STEP #4—Thawing cable clamped onto service key on service rod.

STEP #4—Thawing cable clamped onto service key on ground wire cable at main cock.

STEP #4—Thawing cables clamped onto service keys when two services grounded on side of isolating main cock. Variation.

STEP #5—Aircraft energizer used as power source. 600 AMP 12 volts recommended.

Fig. 5.34   Procedures to electrically thaw frozen water services

CITY OF CALGARY

WATERWORKS DIVISION

FROZEN SERVICE

(PLEASE PRINT)

NAME   Karen Brooks

ADDRESS   8344 Bowness Road NW

PHONE   286-8686   METER READING   00653

METER TAG NO.   1478910

IF CHARGE TO OTHER PARTY:

NAME   Ron Brooks

ADDRESS   4640 B - 83st   NW

PHONE   286-4783

COST OF WORK PERFORMED: $ 85.00

DATE: 19/01/05

Karen Brooks
(Signature)

COMMERCIAL _____   RESIDENTIAL ____✔____

PROPERTY ____✔____   CITY _____

OW Burton
(Operator)

Fig. 5.35   Frozen service sheet

OCCUPANT:

DUE TO THE DEPTH OF FROST IN THE SOIL THIS YEAR YOUR WATER SERVICE HAS FROZEN. IN ORDER TO PREVENT THIS FROM OCCURRING AGAIN, WE RECOMMEND THAT A TAP BE RUN STEADY, 24 HOURS A DAY (ABOUT THE SIZE OF A PENCIL ½" in DIAMETER).

BECAUSE OF THE INCREASED NUMBER OF FROZEN SERVICES AND RISING COSTS, THE CITY OF CALGARY WATERWORKS DIVISION WILL BE CHARGING A SERVICE FEE OF $85.00 PER HOUR (MINIMUM ONE (1) HOUR) FOR REPEATED CALLS TO THAW OUT WATER SERVICES. SERVICES FROZEN ON PRIVATE PROPERTY WILL BE CHARGED $85.00 PER HOUR FOR ALL CALLS.

NOTE: REGARDLESS OF TEMPORARY PERIODS OF WARM WEATHER, CONTINUE TO RUN THE WATER UNTIL YOU ARE NOTIFIED BY THE WATERWORKS DIVISION TO DISCONTINUE RUNNING WATER.

METERED CUSTOMERS' WATER BILLS WILL BE ADJUSTED ACCORDINGLY DURING THIS PERIOD.

SHOULD YOU REQUIRE FURTHER INFORMATION PLEASE CONTACT THE WATERWORKS AT 268-4904.

YOUR COOPERATION IN THIS REGARD WILL BE MOST APPRECIATED.

WATERWORKS DIVISION
THE CITY OF CALGARY

Fig. 5.36   City policy regarding frozen service

HOT WATER THAWING UNIT

STEP #3
   Disconnect service line below stop and waste.

STEP #4 and #5
   Install control head, feed in probe tube.

STEP #6
   Bypass return into reservoir.

STEP #7
   Hot water fed into pump.

STEP #9
   Thawing complete. Probe tube removed.

Fig. 5.37   *Procedures used to thaw lines with a hot water thawing unit*

11. Reconnect service line.

Remove probe tubing completely and reconnect service line as quickly as possible. Turn on service valve and stop and waste and let water run at a steady flow (size of a pencil) to prevent refreezing. If consumer is on a water meter, notify the meter department of present reading so consumer is charged flat rate for duration of potential freeze up. Consumer is issued a directive when to discontinue running water.

12. Complete required paperwork.

Complete Frozen Service Sheet (Figure 5.35) as to address and to whom the thawing service is to be charged. The consumer pays when the service is frozen on the property. The utility is responsible where the service is frozen on utility property. Turn in paperwork to Thawing Supervisor as soon as possible.

C. Steam Thawing Service Valves, Main Valves and Hydrants

1. Steam thawing equipment (Figure 5.38).

Portable steam boilers of about 8 to 10 horsepower (6 to 7.6 kW) are commonly used for thawing. These boilers may be mounted vertically on a truck and have a working pressure of 120 psi (827 kPa or 8.4 kg/sq cm). These units are completely self-contained and can be mounted on one-ton, dual-wheeled trucks.

2. Thawing valves.

Thawing of main and service valves is done with approximately 8-foot (2.5-m) long, ³/₈-inch (10-mm) diameter metal tube attached to the one-inch (25-mm) steam hose. The tube is fed down the valve casing to the valve. The time required to thaw a frozen valve will depend on size and depth of frost. Two hours is not an unreasonable time to thaw a 10- or 12-inch valve (250- or 300-mm valve). This method could also be used to blow debris, small rocks, and mud out of valve casings.

3. Thawing of fire hydrants.

Steam boilers are equipped with a 30-foot (10-m) long, 1-inch (25-mm) diameter steam hose. This hose is used for thawing and also for siphoning the water out of a hydrant body.

The hydrant is thawed through a hydrant outlet if it is a pumper hydrant. The pumper outlet could be used also.

a. Make sure hydrant is in OFF position before thawing process.

IMPORTANT NOTE: NEVER STAND IN FRONT OF ANY HYDRANT OUTLET

b. Insert steam hose into hydrant through hydrant outlet. Siphon water out first, then thaw to bottom of hydrant and make sure the hydrant is operable.

c. If hydrant is not draining, siphon out all the water to prevent re-freezing. Notify the Hydrant Repair Foreman if you have problems operating the hydrant or if the hydrant is not draining properly. A *"FIRE USE ONLY"* sign may have to be attached to one of the hydrant outlets until the hydrant is repaired.

4. Steam thawing frozen water mains.

a. The frozen section of the main to be thawed should be isolated (turned off at main valve).

b. Expose the main by excavating at 130-foot (40-m) intervals and cut out a section (approximately 3 feet (1 meter)). Push the steam hose up the main with the aid of a 1-inch (25-mm) steel snake.

c. After the thawing process is completed each way from the cut-out sections, turn on main valves very slightly to make sure water is flowing freely; turn off main valves and reconnect the cut-out section with approved couplings.

d. After the main is reconnected, make certain the main is put on the flushing program. This procedure is used to thaw large services.

D. Flushing the Water Mains to Prevent Freezing

A water main flushing program should be developed and carried out each winter to prevent freezing of water mains in problem areas where mains have frozen in the past. The need for flushing is based on whether mains have frozen in the past, the depth of frost penetration, the type of soil, the depth of the main, and the flow in the main. Another important factor is the temperature of the water in the main.

1. Spot check problem areas.

Spot check problem areas for evidence of ice or slush in water mains. This should be done when records indicate severe frost penetration.

2. Dispatch crews.

Once the need for the flushing program has occurred, dispatch flushing crews to assigned locations with properly marked maps or flushing diagrams. Be sure crews notify their flushing foremen and/or waterworks dispatcher of their flushing locations. The flushing foreman and/or dispatcher should be notified in case any complaints from consumers are received regarding loss of water pressure, no water, sanitary sewer backup into home (sewer being flushed into), or any other problem that could be caused by the flushing operation.

3. After crew arrives at flushing location.

a. Check all main valves, hydrant control valves, hydrants, and washouts to ensure all are in operating condition.

b. If any valves or hydrants are frozen and cannot be thawed with a tiger torch, notify the flushing foreman or contact steam boiler crew to thaw valves or hydrants.

c. Check catch basins and manholes that will be used to make sure that water will flow away freely. If any sewer installation is found frozen and cannot be used and there are no alternatives, notify flushing foreman to have Sewer Division thaw them out.

Truck-mounted steam boiler

Water tank, diesel fuel tank, and steam hose

Thawing main valve

Thawing hydrant and siphoning out water

Fig. 5.38   Steam thawing unit

4. Prepare for flushing (Figure 5.39).

   a. Operate main valves to positions indicated on flushing diagram.

   b. Attach gate valve in closed position to designated hydrant as per flushing diagram. Make sure all other outlets on hydrant are on tight. Open hydrant slowly to full open position, then open gate valve slowly to let out the air from hydrant.

      NOTE: Make sure hydrant is in full open position so water cannot escape through drain hole, thereby causing the drainage bed to freeze.

   c. Attach 2½-inch (63-mm) fire hose to gate valve and attach nozzle to other end of hose. Secure fire hose nozzle to catch basin or to manhole step so it will not whip when under pressure (Figure 5.40).

   d. Open gate valve slowly to full open position.

   e. Either visually or by feel with hand on hose, determine if ice is present in water main during flushing.

      NOTE: Ice may be present at start due to poor hydrant drainage or ice being formed in hydrant body.

      Allow 15 minutes of flow before test. When determining the presence of ice by feel, grip the hose firmly with bare hand. If ice is present you will feel it pass through. Also, by placing an ear to the hose you could hear a rattling sound as ice passes through, or by using a pail to catch the water flow, you can visually determine the presence of ice or slush. Record all results on reverse side of flushing diagram.

5. Operate main valves.

   a. Operate main valves as per flushing diagram. Operate valves slowly.

   b. Flush each phase for a minimum of 30 minutes or until no ice is evident. Record results of each phase.

   c. Continue flushing and recording results until each phase is complete.

   d. After all phases have been completed, return all valves to open position unless otherwise noted.

      NOTE: Do not leave open valves that are indicated to be closed for pressure separation.

   e. *IMPORTANT*

      Make sure all valves that should be open are open to maintain water circulation through water main and to avoid dead ends. Still water in dead ends could cause the water main to re-freeze quickly.

   f. Any valves that are frozen in a closed position must be referred to the steam boiler crew to be thawed out and opened as soon as possible.

   g. Ensure that plastic discs are reinserted under valve lid in top boxes to prevent snow, gravel, and other debris from filling up casing, thus making it difficult to get at the rod.

6. Shut off hydrant (labeled Step 6 on Figure 5.40).

   a. Turn off gate valve slowly to avoid creating water hammer.

   b. Disconnect and drain fire hose, remove nozzle, and roll up hose so no water remains in hose to eventually freeze up.

   c. Shut off hydrant slowly to again avoid water hammer. At this time open gate valve to allow air in so hydrant can drain. If hydrant is draining, there will be suction at gate valve when hand is placed over outlet.

   d. If hydrant is not draining, pump out by using the 12-volt pump with suction and discharge hoses attached. If any problem with pumping occurs, notify flushing foreman to obtain steam boiler to siphon water out.

   e. If a problem occurs with operation of hydrant, refer to hydrant repair foreman. Provide foreman with make and location of hydrant. Also request the repairs to be made as soon as possible because they will be required for flushing.

   f. Ensure all gate valves and fittings, hose couplings, and nozzles are free of ice and are not damaged. Any frozen or damaged fittings will hinder the next hookup, and also could cause leaks creating hazardous conditions.

   g. All gate valves and pumps should be kept in a warm place to prevent them from freezing and splitting.

   h. If any water has been spilled on sidewalks or roadways, flushing foreman should be notified so the location can be sanded or salted.

   i. Ensure all manhole lids or catch basin grates that were disturbed have been replaced and secured.

   j. Clean up flushing site. Pick up all traffic signs and warning devices that were used. Check to make sure all tools have been picked up, are in good working order, and are in their proper place or container.

7. Proceed to next flushing location.

   Proceed to next location. Notify flushing foreman and repeat procedure. At end of shift, notify flushing foreman if any minor problems have to be looked into.

QUESTIONS

Write your answers in a notebook and then compare your answers with those on page 289.

5.70V   Why is the freezing of service lines a greater problem than with mains?

5.70W   How can a frozen water main be thawed?

5.70X   What precaution should be exercised when thawing pipe using electrical thawing methods?

End of Lesson 2 of 4 Lessons on DISTRIBUTION SYSTEM OPERATION and MAINTENANCE

Please answer the discussion and review questions next.

STEP #4 - (a)
  Operate main valves as per flushing diagram.

STEP #4 - (c)
  Attach hose to gate valve.

STEP #4 - (b)
  Attach closed gate valve to hydrant, release air pressure.

STEP #4 - (b)
  Attached closed gate valve.

Fig. 5.39   Preparing to flush water mains to prevent freezing

STEP #4 - (c)
    Attach nozzle to hose.

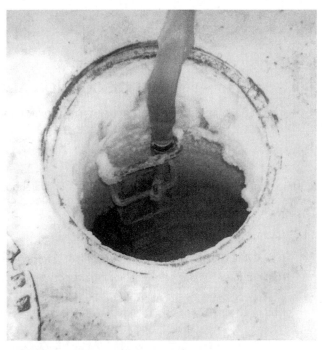

STEP #4 - (c)
    Secure nozzle to manhole or catch basin.

STEP #4 - (c)
    Secure nozzle to manhole or catch basin.

STEP #6 - (b)
    Shut off hydrant. Disconnect fire hose and drain.

Fig. 5.40   Flushing water mains to prevent freezing

DISCUSSION AND REVIEW QUESTIONS

Chapter 5.   DISTRIBUTION SYSTEM OPERATION AND MAINTENANCE

(Lesson 2 of 4 Lessons)

Write the answers to these questions in your notebook before continuing. The question numbering continues from Lesson 1.

11. Why must pipes be maintained?

12. Why should water utilities have a leak detection program?

13. What are the advantages of wet tapping over dry tapping?

14. How can potential distribution system failures be recognized in advance?

15. What information can be obtained from pipe roughness coefficient (C Factor) tests?

16. What problems may be created if distribution system lines are not regularly flushed?

17. If flushing fails to clean pipes, what alternative methods might be effective?

18. How is water provided to consumers while a water main is being lined?

CHAPTER 5.   DISTRIBUTION SYSTEM OPERATION AND MAINTENANCE

(Lesson 3 of 4 Lessons)

5.71   Valves

Distribution system shutoff valves are provided primarily to isolate small areas for emergency maintenance. Most of these valves, therefore, suffer from lack of operation rather than from wear. A comprehensive program of inspection, exercising, and maintenance of valves on a regular basis can help water utilities avoid potentially serious problems when the need to use a valve arises.

Operators should know *EXACTLY* where to go to shut off any valves at any time in case of a line break or other emergency. When breaks occur in water mains, crews often experience problems in *FINDING* valves whose locations are marked incorrectly on system maps. Other problems include valves that won't close or open after they are located. Time is often wasted looking for valves and after finding them getting them to work. The same devices used to locate mains are used to locate valves that may be lost or buried under earth or snow.

Routine valve inspections should be conducted and the following tasks performed:

1. Verify the accuracy of the location of the valve boxes on the system map (if incorrect, *CHANGE THE MAP*),

2. After removing the valve box cover, inspect the stem and nut for damage or obvious leakage,

3. Close the valve fully, if possible, and record the number of turns to the fully closed position,

4. Reopen the valve to reestablish system flows, and

5. Clean valve box cover seat. Sometimes covers on valve boxes will come off when traffic passes over them due to dirt in the seat.

Exercising (opening and closing) a valve should be done at the same time the valve inspection is made. Some manufacturers recommend that a valve stem never be left in a fully open position. They recommend that after fully opening a valve, back off the stem by one turn. Be careful closing valves because if some valves are closed too tight, damage to the valve or valve seat could result and cause the valve to leak.

Conditions of each system will determine how often the valves should be exercised, but in general it is recommended that all valves be exercised at least once a year. Planned exercising of valves verifies valve location, determines whether or not the valve works, and extends valve life by helping to clean encrustations from the valve seats and gates. Any valves that do not completely close or open should be replaced. Valves that leak around the stems should be repacked. To determine that a valve is closed, an aquaphone or other listening device can be used. Valves should be exercised in both directions (fully closed and fully opened) and the number of turns and direction of operation recorded. Valves operating in a direction opposite to that which is standard for the system need to be identified and this fact recorded. The condition of the valve packing, stem, stem nut, and gearing should be noted. A timely maintenance program should be initiated to correct any problems found during the inspection and exercising.

Two types of hydraulic problems can occur while operating a valve, cavitation (CAV-uh-TAY-shun) and water hammer. Cavitation results when a partial vacuum (voids) occurs on the downstream side of a valve and a small section of the pipe is filled with low-pressure vapor pockets. These pockets will collapse downstream (implode) and in doing so create a mechanical shock that causes small chips of metal to break away from the valve surfaces. A noisy or vibrating valve may be an indication that cavitation is occurring and the valve may eventually have to be replaced if cavitation is permitted to go on indefinitely. Water hammer is caused by closing a valve too quickly. The water flow is suddenly stopped, shock waves are generated, and the resulting large pressure increases throughout the system (even though very brief) can result in significant damage. Water hammer can be prevented by always closing the valves slowly, regardless of size or type.

Valves can be operated either manually or by a power actuator. Manual operation of large valves not only can be backbreaking labor, but is a slow process and, therefore, time consuming and costly. Power equipment (Figures 5.41 and 5.42) is available which will cut valve operating time considerably. Most types of power equipment are portable, fast, and efficient and can be powered by a portable air compressor, an electric generator, or a gas engine. A power valve operator can also be used to accurately count the number of turns to open or close a valve.

Two of the most important factors in maintaining distribution system valves are the availability of current and correct maps of the distribution system. A portion of a typical distribution

Fig. 5.41   How powered valve operators work

(Permission of E. H. Wachs Company)

system map for valves and hydrants is shown in Figure 5.43. Each utility should use this type of map, verify often that it is accurate, and keep the map up to date by immediately recording any changes such as replacements or additions. Some water departments equip their service trucks with "gate books" which carry all of the pertinent valve information including location, direction of turning to close, and number of turns required.

Maintaining current records is as important as maintaining current maps. A typical two-sided valve record form is shown in Figure 5.44. The location of a valve is obtained from a controlled survey bench mark or permanent reference point. The make of valve is important because different makes have different operating characteristics. The use of a simple valve numbering system keyed to up-to-date maps is recommended. This procedure has proven to be quite helpful in locating valves rapidly and in communicating with others about particular valves.

Road improvements require constant attention from water distribution system operators to ensure that valves are not lost. Valve boxes can be graded out or covered with pavement. The center lines of roads, curb lines, and right-of-way lines used as reference points for locating valves can be changed. Changed measurements must be noted on valve record forms.

Corrosion is a problem for valves in some areas and can cause failure of bonnet and packing gland bolts. This is apparent when stem leakage occurs or when a valve is closed and the bonnet separates from the body. Stainless-steel bolts can be used for replacement, and the valve should be encased in polyethylene wrap.

Valves left closed in error can cause severe problems in a distribution system. Construction and maintenance crews operate valves as they do their work. Contractors and plumbers sometimes operate valves without permission. Separate pressure zones in distribution systems may be established by closing valves, thus increasing the possibility of problems related to the incorrect use of valves. Unexplained problems with pressure and excessive operation of pumps in a given area have been traced to valves left closed or open in error. When crews change shifts during a project, valve closures and openings information must be exchanged. Crew chiefs must be sure all valves are restored to proper positions whenever anyone discovers a valve in the wrong position.

Repairing in-line (installed) gate valves is a difficult task. If repairs are needed, proper advance planning is important. The valves needing repairs must be located. The valves that will be used to isolate a damaged valve must be in good operating condition. The necessary repair parts must be obtained in advance. When ordering repair parts, be sure to include the size, make, direction of opening, year of manufacture, and other pertinent information in order to ensure that the proper repair parts will be received.

Until the valve is isolated and opened up, it is difficult to determine what part of the valve is damaged. Therefore, make sure that all replacement parts are available before isolating the necessary section of the water main, excavating the valve, and making the repairs.

Most valves are located along roadways, and operators who locate, exercise, or dig up valves for repair are exposed to traffic hazards. Routine work is usually done during daylight, but

Fig. 5.42   Portable and truck-mounted powered valve operators

(Permission of E. H. Wachs Company)

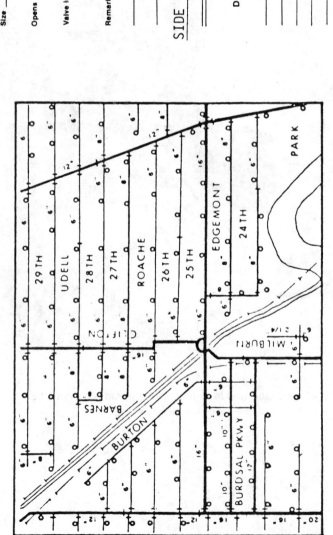

SIDE 1

Valve Record

Location _____   No. _____

Make _____

Type _____

Size _____   Operating nut _____

Opens _____   No. of turns _____

Valve in _____   Route No. _____

Remarks: _____

SIDE 2

Date	Opens	No. of turns	Stem'	Packing	Nut	Box or manhole	Valve is now	Check by	Remarks

Fig. 5.44   Typical (two-sided) valve record form
(Reprinted from OPFLOW, by permission. Copyright 1977, The American Water Works Association)

Fig. 5.43   Portion of a valve and hydrant map with street names and main sizes given
(Reprinted from OPFLOW, by permission. Copyright 1977, The American Water Works Association)

° Hydrant
+ Valve

traffic must be warned at all times. Motorists must be notified in advance of blocked lanes or work alongside of traveled lanes. This can be done by using high-level warning signs, barricades with lights for night work, traffic cones, warning flags, and flaggers. Repair crew vehicles with flashers can be positioned to alert traffic and to provide physical protection for the crew from oncoming traffic at the work site. Supervisors should hold a job-site meeting with operators to explain the task, the equipment to be used for the job, the hazards, the safety procedures to follow, and the safety equipment needed. See Chapter 7, Section 7.5, "Working in Streets," for more safety information.

For additional information on valve maintenance, see Chapter 18, "Maintenance," Section 18.26, "Valves," in *WATER TREATMENT PLANT OPERATION*, Volume II, in this series of manuals.

QUESTIONS

Write your answers in a notebook and then compare your answers with those on page 289.

5.71A How can water utilities avoid serious valve operating problems?

5.71B Why should operators be able to quickly find valves?

5.71C Why should valves be exercised regularly?

5.71D How can you tell if cavitation is occurring at a valve?

5.72 Fire Hydrants

The different types of hydrants available are described in Chapter 3, "Distribution System Facilities." Operators responsible for hydrant inspections should be thoroughly familiar with the various types of hydrants used in their system. If difficult questions arise about the fire hydrant or its operation, a good source of information is the hydrant supplier. Contact the supplier whenever necessary to obtain descriptive literature, operation and maintenance instructions, parts manuals, or assistance on particular problems.

In general, fire hydrants should be inspected and maintained twice a year. These operations are often done in the spring and the fall. However, each hydrant should also be inspected after each use. Inspect dry-barrel hydrants after use, especially during freezing weather, to ensure that the drain remains open when the hydrant is not in use.

A good source of information on fire hydrants is AWWA's Manual M17, *INSTALLATION, FIELD TESTING, AND MAINTENANCE OF FIRE HYDRANTS.*[16] The operator is referred to this manual for detailed procedures.

Some general inspection and maintenance procedures used for hydrants include:

1. Inspect for leakage and make corrections when necessary.

2. Open hydrant fully, checking for ease of operation.

3. Flush hydrant to waste (take care to direct flow).

4. Remove all nozzle caps and inspect for thread damage from impact or cross threading. Wire-brush the nozzle and cap threads. Clean and lubricate outlet nozzle threads, preferably with a dry graphite-base lubricant, and check for ease of operation. Be sure that the outlet nozzle cap gaskets are in good condition.

5. Replace caps, tighten with a spanner wrench, then back off on the threads slightly so that the caps will not be excessively tight but will leave sufficient frictional resistance to prevent removal by hand.

6. Check for any exterior obstruction that could interfere with hydrant operation during an emergency.

7. Check dry-barrel hydrants for proper drainage.

8. Clean exterior of hydrant and repaint if necessary.

9. Be sure that the auxiliary valve is in the fully opened position.

10. If a hydrant is inoperable, tag it with a clearly visible marking to prevent loss of time by firefighting crews if an emergency should arise before the hydrant is repaired. Immediately report the condition of this fire hydrant to your fire department.

11. Prepare a record of your inspection and maintenance operations and any repair work. A recommended Hydrant Maintenance Report and Master Record are shown in Figure 5.45.

Everyone working with hydrants should be aware that when operating a dry-barrel hydrant, it must be opened completely so that the drain will become *FULLY* closed. If this is not done, the drain will remain partially open and water seeping through it could saturate the drain field and result in hydrant damage from freezing.

Hydrants can be partially protected against freezing by covering them with a box which can be quickly removed when the hydrant must be used. To keep hydrants from freezing, (those that won't drain in the winter due to frozen conditions), insert in the hydrant propylene glycol or some other nontoxic substance that won't freeze or cause water quality problems. Frozen hydrants may be thawed by the same methods used for thawing pipe. Electric current or hot water thawing can be used. Live steam injected through a hose into the hydrant barrel is a relatively quick, inexpensive, and effective method of thawing.

Standardization of hydrants minimizes the requirement for stocking parts, simplifies repair procedures, and allows replacing only defective parts. Every water utility should keep a basic stock of repair parts on hand for immediate use. If spare parts are not readily available, *YOUR* community's fire protection system could be jeopardized.

---

[16] *Obtain from American Water Works Association (AWWA), Bookstore, 6666 West Quincy Avenue, Denver, CO 80235. Order No. 30017. ISBN 0-89867-460-3. Price to members, $57.00; nonmembers, $84.00; price includes cost of shipping and handling.*

HYDRANT MAINTENANCE REPORT

XYZ Water Utility _____ Hydrant No _____

Location _____

Caps   Missing _____ Replaced _____ Greased _____

Chains   Missing _____ Replaced _____ Freed _____

Paint   O.K. _____ Repainted _____

Oper. Nut   O.K. _____ Greased _____ Replaced _____

Nozzles   O.K. _____ Caulked _____ Replaced _____

Valve & Seat   O.K. _____ Replaced _____

Packing   O.K. _____ Tightened _____ Replaced _____

Drainage   O.K. _____ Corrected _____

Flushed _____ Minutes _____ Nozzle Open

Pressure   Static _____ Residual _____ Flow _____ gpm

Branch Valve   Condition _____

Any Other Defects _____

_____

_____

_____

_____

Inspected _____ By _____

Defects Corrected _____ By _____

FIRE HYDRANT MASTER RECORD
XYZ WATER UTILITY

Manufacturer _____ Date _____ Hydrant No _____

Type _____ MVO _____ Inlet _____

Bury _____ Hose Nozzle Size _____ Thread Type _____

Pumper Nozzle Size _____ Thread Type _____

Installed by _____ Date _____ W/O No. _____ Cost _____

Operating Nut _____ Turns to Open _____

Location _____ Line Static Pressure _____

Date	Inspected	Tested	Repaired	Painted	Opened by	Cost	Remarks

Avenue   Property Line

Right of Way

Water Main—Size/Type

N

In order to carry out a meaningful inspection and maintenance program, it is essential to record the location, make, type, size, and date of installation of each hydrant.

Fig. 5.45   Hydrant maintenance report and master record forms

(Reprinted from *OPFLOW*, by permission. Copyright
1981, The American Water Works Association)

The repair job most often performed on fire hydrants is replacing main valves. Therefore, try to keep an ample supply of main valves in your stockroom. Other important items to keep on hand are drain parts, seat rings, stem seals and packing, and "traffic-damage" repair kits. These items should be stocked for each of the various types and sizes of hydrants in your system. The number of parts to keep on hand depends on the past experience of your water utility.

Fire hydrant vandalism causes serious problems for water utilities. Illegally opened fire hydrants cause damage due to flooding and washouts. Consumers can suffer damages due to a lack of water and pressure. Water may not be available to fight fires when needed and the loss of water results in a loss of revenue for the water utility.

Fire hydrant caps or guards can be installed on the tops of fire hydrants to eliminate fire hydrant vandalism.

Fire hydrants are usually the only part of the distribution system regularly seen by the general public. Frequent painting of hydrants creates a favorable impression and is, therefore, an excellent public relations tool.

QUESTIONS

Write your answers in a notebook and then compare your answers with those on page 289.

5.72A   What information can be obtained from a fire hydrant supplier?

5.72B   When should fire hydrants be inspected?

5.72C   List some general fire hydrant inspection and maintenance procedures.

5.72D   What happens when a dry-barrel hydrant is operated with the valve only partially open?

5.73   Meters

5.730   Testing of Meters

Water meters can over- or underregister because of wear, deposits, or turbulence resulting from valves and fittings. Overregistration rarely occurs. Each utility should establish a schedule for periodic meter testing based on meter use, water quality, age of meter, cost of testing, and water revenue loss. The age of a meter reflects the degree of wear of the meter parts; wear also increases when the water is corrosive or abrasive. The potential revenue loss from inaccurate meters, which almost always underregister, must also be considered.

5.731 Test Procedures

Small meters should be tested once every five to ten years, and large ones every one to four years. New meters should also be tested before installation, although a survey of customer metering practices showed that 68 percent of the utilities surveyed did not test their meters before they were first put into use.

One type of meter testing installation used in a utility's shops is shown in Figure 5.46. Field testing is usually done only to meters larger than two inches (50 mm) in size. One method used in the field is to connect a calibrated test meter in series with the meter to be tested and then compare the readings of the two meters. Most of the larger meters have provisions for attaching a test meter. The smaller meters and some of the larger ones are generally tested by running a measured volume of water through the meter and then comparing the reading to the known volume used. Some larger meters have built-in calibration checks, such as a *VENTURI METER*[17] with a test head which simulates pressure differential from a Venturi tube.

Most modern water meters have sealed registers and easily changed measuring chambers. These meters are usually maintained and repaired by the utility. Older style meters with heads that must be worked on should be returned to the manufacturer for maintenance and repair if they are not being replaced.

FORMULAS

Domestic water service meters are called totalizing water service meters because they measure and record the total volume of water that passes through the meter to the consumer. Totalizing water service meters are tested at a *GIVEN RATE OF FLOW* as measured by a *ROTAMETER*[18] in gallons per minute. The rotameter measures the actual flow in gallons per minute (GPM) while the water service meter measures the *TOTAL VOLUME IN GALLONS*. Using the meter testing installation shown in Figure 5.46, we can determine the accuracy of a totalizing meter at both low and high flows as indicated by the rotameter.

To determine the actual volume of water that passes through a totalizer, calculate the volume of water that flows into a cylinder during the test time period.

$$\text{Volume, cu ft} = (\text{Area, sq ft})(\text{Height, ft})$$

and

$$\text{Volume, gal} = (\text{Volume, cu ft})(7.48 \text{ gal/cu ft})$$

To calculate the actual rate of flow through the meter, divide the actual volume in gallons by the time in minutes.

$$\text{Actual Flow, GPM} = \frac{\text{Actual Volume, gal}}{\text{Time, min}}$$

Accuracy of meters is determined by dividing the observed meter values (rotameter or totalizer) by actual values and multiplying by 100 percent.

$$\text{Meter Accuracy, \%} = \frac{(\text{Observed Value})(100\%)}{\text{Actual Value}}$$

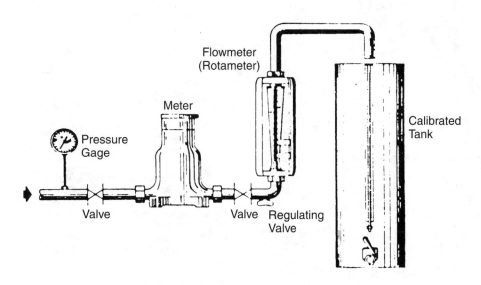

Fig. 5.46   Meter testing installation

(Reprinted from *WATER DISTRIBUTION OPERATOR TRAINING HANDBOOK*, by permission. Copyright 1976, the American Water Works Association)

---

[17] *Venturi Meter.   A flow measuring device placed in a pipe. The device consists of a tube whose diameter gradually decreases to a throat and then gradually expands to the diameter of the pipe. The flow is determined on the basis of the difference in pressure (caused by different velocity heads) between the entrance and throat of the Venturi meter.*

[18] *Rotameter (RODE-uh-ME-ter).   A device used to measure the flow rate of gases and liquids. The gas or liquid being measured flows vertically up a tapered, calibrated tube. Inside the tube is a small ball or bullet-shaped float (it may rotate) that rises or falls depending on the flow rate. The flow rate may be read on a scale behind or on the tube by looking at the middle of the ball or at the widest part or top of the float.*

EXAMPLE 5

A water service meter is tested in the installation shown in Figure 5.46. The calibrated tank is 2 feet in diameter. The flowmeter (rotameter) reads 1.5 GPM. The discharge valve on the tank is closed. A stopwatch is started when the water level in the tank goes past the 1.00-foot mark and is stopped when the water level goes past the 3.00-foot mark. The stopwatch reads 30 minutes and 55 seconds. What should the flowmeter read in GPM and what should be the total volume of water recorded during the test? The meter reading at the start of the test (stopwatch) was 112,474 gallons and was 112,520 gallons at the end of test.

Known	Unknown
Tank Diameter, ft = 2.0 ft	1. Actual Volume, gal
Flowmeter, GPM (Rotameter) = 1.5 GPM	2. Actual Flow, GPM
Water Level, ft = 1.0 ft (start)	
Water Level, ft = 3.0 ft (stop)	
Time, min, sec = 30 min, 55 sec	
Service Meter, gal = 112,474 gal (start)	
Service Meter, gal = 112,520 gal (stop)	

1. Calculate the volume of water measured by the cylinder in cubic feet.

$$\text{Volume, cu ft} = (0.785)(\text{Diameter, ft})^2(\text{Stop Level, ft} - \text{Start Level, ft})$$

$$= (0.785)(2 \text{ ft})^2(3.0 \text{ ft} - 1.0 \text{ ft})$$

$$= 6.28 \text{ cu ft}$$

2. Convert cylinder volume from cubic feet to gallons to find actual volume of water in gallons.

$$\text{Volume, gal} = (\text{Volume, cu ft})(7.48 \text{ gal/cu ft})$$

$$= (6.28 \text{ cu ft})(7.48 \text{ gal/cu ft})$$

$$= 47 \text{ gallons}$$

3. Determine volume of water measured by flowmeter in gallons.

Meter Stop Reading, gal =	112,520 gal
Meter Start Reading, gal =	112,474 gal
Recorded Flow, gal =	46 gal

NOTE: Since meter volume is slightly less than the volume measured in the cylinder, the meter underregistered.

4. Calculate the actual flow rate in gallons per minute.

$$\text{Actual Flow, GPM} = \frac{\text{Actual Volume, gal}}{\text{Time, min}}$$

$$= \frac{47 \text{ gallons}}{30 \text{ min} + \left(\dfrac{55 \text{ sec}}{60 \text{ sec/min}}\right)}$$

$$= \frac{47 \text{ gallons}}{30 \text{ min} + 0.92 \text{ min}}$$

$$= \frac{47 \text{ gallons}}{30.92 \text{ min}}$$

$$= 1.52 \text{ GPM}$$

NOTE: Since the actual flow is slightly different than the meter flow, the meter is not perfect.

5.732   *Accuracy Requirements*

Table 5.2 shows the accuracy requirements for new meters. In most instances, the meter underregisters slightly to favor the customer.

TABLE 5.2   ACCURACY REQUIREMENTS FOR NEW METERS [a,b]

AWWA Standard Designation and Type of Meter	At Normal Test Flow, percent	At Minimum Test Flow, percent
C700-Displacement Type	98.5 – 101.5	95
C701-Turbine Type	98.0 – 102.0	Not required
C702-Compound Type [c]	97.0 – 103.0	95
C703-Fire Service Type [c]	97.0 – 103.0	95
C704-Propeller Type	98.0 – 102.0	95

[a] AWWA Standards, *OPFLOW*, Vol. 3, No. 8, August 1977, p. 3.
[b] There is no AWWA standard for the accuracy of repaired meters, but the practice at many utilities is to require the same accuracy as for new meters at the normal test flow and at least 90 percent at the minimum test flow.
[c] The accuracy during the "change-over period," which is defined in the applicable standards, shall not be less than 85 percent.

EXAMPLE 6

If a totalizing water service meter reads 46 gallons and the actual flow calculated by the use of a calibrated tank was 47 gallons, did the meter meet the accuracy requirements for a displacement type of meter as listed for new meters in Table 5.2?

Known	Unknown
Meter Reading, gal = 46 gal	Does the meter meet accuracy requirements?
Actual Volume, gal = 47 gal	

Calculate the accuracy of the meter.

Meter Accuracy, % $= \dfrac{\text{(Meter Reading, gal)}(100\%)}{\text{Actual Volume, gal}}$

$= \dfrac{(46 \text{ gal})(100\%)}{47 \text{ gal}}$

$= 97.9\%$

NOTE: Since the meter accuracy (97.9%) is not within the accuracy required (98.5 to 101.5%), the meter does not meet the accuracy requirements.

EXAMPLE 7

If a rotameter reads 1.5 GPM and the actual flow calculated by the use of a calibrated tank was 1.52 GPM, what was the accuracy of the meter?

Known	**Unknown**
Meter Reading, GPM = 1.5 GPM	Meter Accuracy, %
Actual Flow, GPM = 1.52 GPM	

Calculate the accuracy of the meter.

Meter Accuracy, % $= \dfrac{\text{(Meter Reading, GPM)}(100\%)}{\text{Actual Flow, GPM}}$

$= \dfrac{(1.5 \text{ GPM})(100\%)}{1.52 \text{ GPM}}$

$= 98.7\%$

5.733   Meter Registers and Readouts

Meter registers may be mechanically or magnetically driven. In the older type of mechanically driven registers, a series of gears within the meter transmits the rotary motion of a disc or piston to the register. Magnetic drive units are now available, and older mechanical drive meters can be updated and converted to magnetic drive units. The newer meters use magnetic couplings to drive the registers which eliminates many mechanical parts. The mechanically driven register has a number of disadvantages in that it has more moving parts, fogs up when set in the ground, and corrodes. Magnetically driven registers are better protected, being normally hermetically sealed (airtight) and somewhat tamper-proof. The magnetic drive gets rid of the conventional stuffing box to greatly reduce the friction in the mechanism.

The total flow measured by meters may be read directly from the registers on the meters or may be read remotely. The flow rate can be read from gages, either indicating or recording, which can be mounted at any convenient location. Strip chart or circular chart recorders may be used to provide continuous permanent records. Metering operations can become quite complex. Customer meter readings can be converted into electrical data, transmitted by telemetry via telephone lines from the meter to a central utility office, and then translated by a data processor or recording device that compiles the meter-reading information. The compiled readings may then be transferred manually or automatically to the billing office for further manipulation.

Service meters may be installed and read either indoors or outdoors. Indoor locations are often used in cold weather areas; however, they can present a problem of access to the meter reader and can be an annoyance to the owner or resi-dent. This has led to the use of remote-reading registers which are installed on the same premises as the meter, but in a location where they can be easily observed. With this type of operation, the time required to read meters has been greatly reduced. Remote-reading meters are becoming more popular.

QUESTIONS

Write your answers in a notebook and then compare your answers with those on page 290.

5.73A   A meter testing program should be based on what facts?

5.73B   What type of water meter is usually maintained and repaired by the utility?

5.73C   How are meter registers driven?

5.734   Maintenance

Like any other mechanical device, meters need regular servicing to maintain their efficiency. Their many moving parts and bearing surfaces always eventually result in some degree of wear. Wear, corrosion, and deposition result in inaccurate registration. Overregistration rarely occurs. Instead, the usual result of meter deterioration is underregistration which is of serious concern because it results in a loss of revenue.

Some water should be sealed in a meter when a meter is taken out of service in the field. This can be done with slip-in plugs or protective caps, which will also protect the threads on the connections. Do not allow the inside of a meter to dry out or allow anything to get inside the meter because either situation can change the way a meter measures water flow. An electrical jumper should be installed around a meter whenever working on a meter to protect you from electric shock.

Meter maintenance and repair consists essentially of dismantling and thoroughly cleaning the meter and inspecting all of its parts. The step-by-step procedures used for a nutating disc meter are:

1. Remove meter from service and take to shop,

2. Test meter for accuracy,

3. Dismantle the meter,

4. Inspect the various parts for excessive wear and corrosion, pitting, or distortion,

5. Replace all defective and badly worn parts,

6. Thoroughly clean all parts to be used,

7. Ensure that the gear train (if any) runs freely,

8. Check the action of the disc in the chamber of the main casing before and after assembly,

9. Use a new meter as a comparison standard for tolerances and clearances, and

10. Retest the meter for accuracy.

The cost of meter maintenance, especially for large meters, can be quite small compared to the revenues that would be gained in the long term. Each utility should select the best method of meter maintenance for its own situation, determine the most economical test frequency, and stick to their planned program.

5.735   Meter Reading

Every operator who may have to read a meter should have a thorough understanding of how to do it correctly. The customer should be paying a fair bill for the amount of water being used. Also, the customer may be able to read the meter and the operator had better be correct if questioned. An alert meter reader can often spot underregistering of meters by a quick comparison with past readings. Meter stoppages should be noted immediately at the time of the meter reading and reported.

Figure 5.47 shows a water meter with a circular register face. As shown on the face, the meter records in cubic feet. In reading a meter of this type, start with the scale showing the highest number (10,000,000) and then read each scale around the register successively until the lower numbered scale (10) is read. Note that in some scales the hands turn in a clockwise direction, while in others they move counterclockwise. If a hand is between two numbers, the lower number is used. If the hand is close to a number, the hand on the next scale should be looked at to see if it is past the zero (0) point. If that hand is not on or past zero (0), the hand being read is still between the two numbers and the lower number on the dial should be read. If the adjacent (next lower) hand is on or past zero (0), the higher number is used.

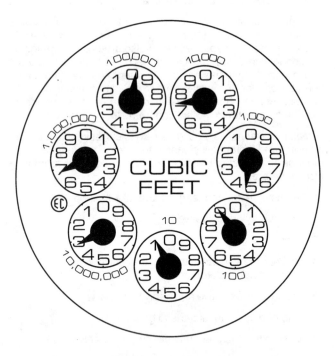

Fig. 5.47   Water meter

(Reprinted from *WATER DISTRIBUTION OPERATOR TRAINING HANDBOOK*, by permission. Copyright 1976, the American Water Works Association)

When a hand makes a complete revolution on its dial, the amount of water that has passed through the meter equals the number designated above that dial. One revolution on the 10 dial is 10 cubic feet and one revolution on the 10,000,000 dial is 10 million cubic feet. For the 10 dial, the numbers correspond to 1 through 9 cubic feet, and for the 10,000,000 dial, the numbers are 1 million through 9 million cubic feet. Other meters might read in gallons or in metric units depending on what is shown on the face. Some meters have multipliers shown on the meter or the dial face such as "10X" or "100X."

To determine the correct total, the reading must be multiplied by the multiplier.

The most important consideration in getting an accurate reading is that the proper number of digits be determined. As a check, the final meter reading should be between the numbers shown on the largest numbered scale read and the next largest. An error in the higher numbered scales will be much more significant than one in the lower numbered scales. Never make the assumption that the previous reading was correct.

Many meters have digital readouts. This means that you merely read and record the numbers shown by the meter. Figure 5.48 is a digital readout totalizing water meter.

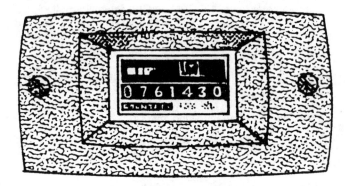

Totalizing receiver

Fig. 5.48   Digital readout totalizing water meter

QUESTIONS

Write your answers in a notebook and then compare your answers with those on page 290.

5.73D   What causes inaccurate meter registration?

5.73E   List the procedures used for meter maintenance and repair.

5.73F   What is the meter reading for the meter shown in Figure 5.47?

5.74   Main Breaks

Breaks in water mains can occur at any time and every agency must have an established, written response plan. A break may be obvious, such as water spouting from a main as a result of a traffic accident, an earthquake, or a washout. At other times, consumers may complain of a lack of pressure or no water at all and the underground break will have to be located.

Sometimes a main will fail under the middle of a paved street and water will be flowing out from under the outside edges of the pavement. When this happens, locate the section of broken main in the center of the pavement. Poke holes through the pavement over the main until the point where the greatest water discharge is located. Don't do any poking in areas where underground electrical or other types of cables may be buried.

After a break has been located, determine which valves will have to be closed to isolate the break. A good policy *BEFORE* shutting off any valves is to notify every consumer involved that they will be out of water for an estimated length of time. The purpose of this advance notification is to allow consumers to make any necessary preparations. For example, water-cooled refrigeration equipment must be shut off before the valves are closed. If extensive damage is being caused by the break (flooding and/or washouts), you will have to close the valves and isolate the section as soon as possible and maybe even before notifying all consumers.

After the valves are closed, the trash pump can be used to drain the hole. A backhoe or other equipment can be used to dig down to the break. Before anyone enters the hole, determine the necessary shoring needed. Use the appropriate shoring (see Section 3.653, "Excavation and Shoring," in Chapter 3). Remove the damaged section of pipe. Remove as much silt and debris as possible from the remaining sections of the main by flushing or other methods. Replace the damaged section of pipe and/or valves using clamps and other fittings. Flush the entire section that was isolated using hydrants or drains. Disinfect the system by following the procedures for disinfecting mains as outlined in Section 5.8, "Field Disinfection." Figure 5.49 (page 260) is a "Water Main Failure Report" used by some agencies to evaluate the cause of breaks and to justify preventive measures. This figure lists the common causes of water main breaks.

5.75 Corrosion Control

Whenever water mains are drained for any reason they should be inspected for signs of corrosion. Also pieces of pipe removed during tapping operations or sections of pipe removed for repair or replacement should be examined for indications of corrosion.

Some agencies insert *COUPONS*[19] (Figure 5.50) in water mains to serve as an indication of the corrosiveness of water and the rate of corrosion of the water mains. The rate of corrosion is measured by the loss of weight of the coupon between weighing time intervals. Be sure to scrape off all encrustations before weighing the coupons.

Another indication of corrosion problems is consumer complaints about red water. By marking on a map the location of consumer complaints, you can identify those water mains where the corrosion problem is the most serious.

There are numerous ways to control corrosion and frequently several methods may be used at the same time. Chemicals can be added to water at a treatment plant to help control the corrosion of water mains. Water mains can be lined with protective coatings to control corrosion (see Section 5.708, "Cement-Mortar Lining"). Cathodic protection is the use of an electrical system to prevent corrosion.

Water mains must be protected from both internal corrosion and external corrosion. For detailed procedures on how to control corrosion, see Section 3.64 "Pipe Protection," in this manual and Chapter 8, "Corrosion Control," in *WATER TREATMENT PLANT OPERATION*, Volume I, of this series of manuals.

5.76 Treatment for Control of Corrosion

A water's pH/alkalinity combination determines that water's tendency to cause pipe corrosion. Corrosion-control treatment often involves adjusting pH and alkalinity to make water less corrosive.

The inhibitors used to control corrosion act by forming a protective coating over the site of corrosion activity, thus "inhibiting" corrosion. The success of inhibitor addition depends on the ability of the inhibitor to provide a continuous coating throughout the distribution system.

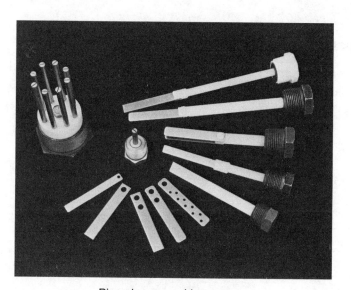

Pipe plug assembly coupons

(Permission of Metal Samples Co., Inc.)

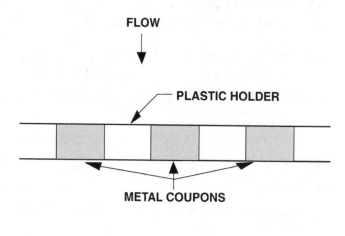

Fig. 5.50 Coupon

---

[19] *Coupon.   A steel specimen inserted into water to measure the corrosiveness of water. The rate of corrosion is measured as the loss of weight of the coupon (in milligrams) per surface area (in square decimeters) exposed to the water per day. 10 decimeters = 1 meter = 100 centimeters.*

WATER MAIN FAILURE REPORT
FIELD DATA FOR MAIN BREAK EVALUATION

DATE OF BREAK: _____   TIME: _____   A.M. _____   P.M.

TYPE OF MAIN: _____   SIZE _____   JOINT _____   COVER _____ FT. _____ IN.

THICKNESS AT POINT OF FAILURE _____ INCH.

NATURE OF BREAK: Circumferential ☐   Longitudinal ☐   Circumferential & Longitudinal ☐   Blowout ☐   Joint ☐

Split at Corporation ☐   Sleeve ☐   Miscellaneous _____ ☐
(describe)

APPARENT CAUSE OF BREAK: Water Hammer (surge) ☐   Defective Pipe ☐   Corrosion ☐   Deterioration ☐

Improper   Excessive   Differential   Temp
Bedding ☐   Operating Pressure ☐   Settlement ☐   Change ☐   Contractor ☐   Misc. _____ ☐
(describe)

STREET SURFACE: Paved ☐   Unpaved ☐   TRAFFIC: Heavy ☐   Medium ☐   Light ☐

TYPE OF STREET SURFACE _____   SIDE OF STREET: Sunny ☐   Shady ☐

TYPE OF SOIL _____   RESISTIVITY _____ ohm/cm

ELECTROLYSIS INDICATED: Yes ☐   No ☐   CORROSION: Outside ☐   Inside ☐

CONDITIONS FOUND: Rocks ☐   Voids ☐   PROXIMITY TO OTHER UTILITIES _____

DEPTH OF FROST _____ INCH   DEPTH OF SNOW _____ INCH

OFFICE DATA FOR MAIN EVALUATION

WEATHER CONDITIONS: PREVIOUS TWO WEEKS _____

SUDDEN CHANGE IN AIR TEMP?   Yes ☐   No ☐   TEMP. _____ °F.   RISE _____ °F.   FALL _____ °F.

WATER TEMP.: SUDDEN CHANGE: Yes ☐   No ☐   TEMP. _____ °F.   RISE _____ °F.   FALL _____ °F.

SPEC. OF MAIN _____   CLASS OR THICKNESS _____   LAYING LENGTH _____ FT.

PREVIOUS BREAK
DATE LAID _____   OPERATING PRESSURE _____ PSI.   REPORTED _____

INITIAL INSTALLATION DATA:

TRENCH PREPARATION: Native Material _____   ☐ Sand Bedding   ☐ Gravel Bedding ☐
(describe type)

BACKFILL: Native Material ☐ DESCRIBE _____   Bank Run Sand & Gravel ☐

Gravel ☐   Sand ☐   Crushed Rock ☐   OTHER _____

SETTLEMENT: Natural ☐   Water ☐   Compactors ☐   Vibrators ☐   OTHER _____
(describe)

ADDITIONAL DATA FOR LOCAL UTILITY USE

LOCATION OF BREAK _____   MAP NO. _____

REPORTED BY _____

DAMAGE TO PAVING AND/OR PRIVATE PROPERTY _____

_____

REPAIRS MADE (Materials, Labor, Equipment) _____

_____

REPAIR DIFFICULTIES (If Any) _____

_____

INSTALLING CONTRACTOR _____

Fig. 5.49   Water main failure report

Phosphates are by far the most common inhibitors used in water treatment for corrosion control. Silicates have a more limited application and may be most suitable for small systems with iron and manganese problems. The control of lead using a phosphate inhibitor occurs when a lead-phosphate compound is formed.

Orthophosphates are primarily used to control lead, not copper. Under optimal conditions, orthophosphate treatment is usually more effective in reducing lead than pH/alkalinity adjustment. The optimal pH for orthophosphate treatment is between 7.2 to 7.8. The orthophosphate residual must be maintained continuously throughout the distribution system. Systems that have raw water naturally in the optimal pH range from 7.2 to 7.8 can treat with orthophosphate alone, provided the water has sufficient alkalinity for a stable pH.

Optimal corrosion-control treatment minimizes lead and copper concentrations at users' taps while ensuring that the treatment does not cause the water system to violate any drinking water regulations. When applying corrosion-control treatment, a substantial amount of time may elapse between the time treatment changes are made and the detection of reduced lead and copper by the analysis of tap water samples. Be very cautious because by changing the chemistry of water, conditions may get worse before they get better.

When using corrosion inhibitors, distribution system operators must be aware of how the inhibitors affect many other aspects of water treatment and delivery and the potential effects of corrosion inhibitors on consumers. Some of the side effects of applying corrosion inhibitors include:

- Phosphate-based inhibitors may stimulate the growth of biofilms in the distribution system which may deplete disinfection residuals within the distribution system;

- Consumer complaints regarding red water, dirty water, color, and sediment may result from the action of the inhibitor on existing corrosion by-products within the distribution system;

- The use of zinc orthophosphate may present problems for wastewater facilities with zinc or phosphorus limits in their NPDES Permits;

- Users with specific water quality needs, such as health care facilities, should be advised of any treatment changes at the plant;

- The use of silicates may reduce the useful life of domestic hot water heaters due to "glassification" because silicates precipitate rapidly at higher water temperatures; and

- For systems using alum or other products containing aluminum, the aluminum will bind with the orthophosphate at a ratio of 1:4 and will interfere with efforts to maintain an effective inhibitor film. For example, if the target orthophosphate residual is 0.5 mg/L and the water has 0.1 mg/L of aluminum, only 0.1 mg/L of the orthophosphate will be available to coat the pipes while 0.4 mg/L will be bound to the aluminum. To solve this problem, aluminum measurements should be taken at the entry to the distribution system, multiplied by four, and then added to the desired operating orthophosphate residual to obtain an effective residual of orthophosphate needed for corrosion control in the distribution system.

QUESTIONS

Write your answers in a notebook and then compare your answers with those on page 290.

5.74A   List the possible causes of a water main break.

5.74B   How can the hole where a water main break is located be drained?

5.75A   Why are coupons installed in water mains?

5.76A   How do chemical inhibitors control corrosion in water distribution systems?

End of Lesson 3 of 4 Lessons on DISTRIBUTION SYSTEM OPERATION and MAINTENANCE

Please answer the discussion and review questions next.

DISCUSSION AND REVIEW QUESTIONS

Chapter 5.   DISTRIBUTION SYSTEM OPERATION AND MAINTENANCE

(Lesson 3 of 4 Lessons)

Write the answers to these questions in your notebook before continuing. The question numbering continues from Lesson 2.

19. Why are valves installed in distribution systems?

20. How can frozen fire hydrants be thawed?

21. How would you determine the frequency of a meter testing program?

22. What are the limitations of mechanically driven meters?

CHAPTER 5.   DISTRIBUTION SYSTEM OPERATION AND MAINTENANCE

(Lesson 4 of 4 Lessons)

5.8   FIELD DISINFECTION

5.80   Need for Disinfection

Field disinfection of distribution system mains and reservoirs is necessary when they are new, after repairs are made, or whenever there is any possibility of contamination. Under any of these conditions, inadequate disinfection could result in a waterborne-disease outbreak somewhere in the community with your utility agency being legally liable.

5.81   Disinfection of Mains[20]

During the construction of a new water line, or after an extensive repair (involving dewatering), there is a small but real opportunity for contamination of the line, even if special precautions, as described in Chapter 3, have been taken. Therefore, effective disinfection is necessary before the line is placed into service.

The disinfecting agent most often used is chlorine, which is available in three chemical forms:

1. Liquid chlorine contains 100 percent available chlorine and is packaged in 100-pound, 150-pound, or one-ton (45-, 68-, or 909-kg) steel cylinders. Special equipment and controls, trained operators, and close attention to safety practices are needed when chlorine gas is used.

2. Sodium hypochlorite is a liquid solution. This form of chlorine contains approximately 5 to 15 percent available chlorine and comes in one-quart to five-gallon (1.0- to 20-liter) containers. Precautions must be taken to prevent deterioration of the hypochlorite solution. Store hypochlorite in a cool, dark location and use as soon as possible.

3. Calcium hypochlorite is a dry material containing approximately 65 percent available chlorine; it comes in powder, granular, or tablet form. Calcium hypochlorite is relatively soluble in water and, therefore, adaptable to solution feeding. Storage conditions must be controlled to prevent deterioration or reaction with combustible chemicals or materials.

Disinfection is commonly accomplished using the tablet, continuous feed, or slug methods, depending on the type of chlorine used and specific job conditions.

Tablets are best suited for short sections (a few hundred feet) and small-diameter lines (24 inches (600 mm) or less). Since preliminary flushing cannot be performed, tablets cannot be used unless the main is initially kept clean and dry. As the pipe is laid in the trench, tablets are placed in each pipe section by securely attaching them with an approved adhesive

to the top of the pipe. Sufficient tablets are added to provide a dose of 25-50 mg/L chlorine (see Table 5.3). The line is filled slowly (less than one foot (0.3 m) of pipe per second) to prevent the tablets from being flushed away. The chlorinated water is allowed to remain in the line for at least 24 hours. Check the chlorine residual during the test to be sure the residual does not drop below 25 mg/L. Sometimes the tablets will be washed to the far end of the pipe when the pipe is being filled with water. For this reason, samples should be collected from the entire length of pipe and the chlorine residual measured to be sure the chlorine is uniformly distributed throughout the pipe.

TABLE 5.3   PIPE DISINFECTION USING TABLETS[a]

| Pipe Diameter, in[d] | Number of Five-gram Hypochlorite Tablets Required for a Dose of 25 mg/L[b] | | | | |
| | Length of Pipe Section, ft[c] | | | | |
	13 or less	18	20	30	40
4	1	1	1	1	1
6	1	1	1	2	2
8	1	2	2	3	4
10	2	3	3	4	5
12	3	4	4	6	7
16	4	6	7	10	13

[a]   AWWA Standard for Disinfecting Water Mains, ANSI/AWWA C651-92.
[b]   Based on 3.25 grams available chlorine per tablet, any portion of tablet required rounded to next higher number.
[c]   Multiply feet x 0.3 to obtain meters.
[d]   Multiply inches x 2.5 to obtain centimeters.

---

[20] *AWWA STANDARD FOR DISINFECTING WATER MAINS, C651-99. Obtain from American Water Works Association (AWWA), Bookstore, 6666 West Quincy Avenue, Denver, CO 80235. Order No. 43651. Price to members, $42.00; nonmembers, $61.00; price includes cost of shipping and handling.*

In the continuous-feed method, preliminary flushing at not less than 5 ft/sec (1.5 m/sec) is required. A chlorine solution containing not less than 25 mg/L free chlorine is injected into the pipe through a corporation cock or other fitting. The solution is injected as the line is being filled. Table 5.4 gives the amount of chlorine required for each 100 feet (30 m) of pipe of various diameters. At the end of a minimum 24-hour period, properly treated water will have a residual of not less than 10 mg/L free chlorine in all portions of the main.

TABLE 5.4   CHLORINE REQUIRED TO PRODUCE 25 mg/L CONCENTRATION IN 100 FEET (30 m) OF PIPE[a]

Pipe Diameter, in [b]	100 Percent Chlorine, lb [c]	1 Percent Chlorine Solution, gal [d]
4	0.013	0.16
6	0.030	0.36
8	0.054	0.65
10	0.085	1.02
12	0.120	1.44
16	0.217	2.60

[a] AWWA Standard for Disinfecting Water Mains, ANSI/AWWA C651-92.
[b] Multiply inches x 2.5 to obtain centimeters.
[c] Multiply pounds x 454 to obtain grams.
[d] Multiply gallons x 3.785 to obtain liters.

The slug method is especially advantageous for use with long, large-diameter mains as it reduces the volume of heavily chlorinated water to be flushed to waste and results in significant savings in chlorine costs. Place calcium hypochlorite (granules or tablets) in the main during construction in quantities shown in Table 5.5. This initial chlorine dose will meet the initial chlorine demand. Fill the main completely to remove all air pockets, then flush.

TABLE 5.5   PIPE DISINFECTION USING GRANULES[a]

Pipe Diameter, in [c]	Ounces of Calcium Hypochlorite Granules to be Placed at Beginning of Main and at Each 500-ft Interval [b] Calcium Hypochlorite Granules, oz [d]
4	0.5
6	1.0
8	2.0
12	4.8
16 and larger	8.0

[a] AWWA Standard for Disinfecting Water Mains, ANSI/AWWA C651-92.
[b] 500 feet is equal to 150 meters.
[c] Multiply inches x 2.5 to obtain centimeters.
[d] Multiply ounces x 28.4 to obtain grams.

Next, using the continuous-feed method, dose the water to produce and maintain a chlorine concentration of at least 100 mg/L free chlorine. Apply the chlorine continuously for a long enough time period to produce a slug (solid column of water) of highly chlorinated water that will slowly move through the main and expose all interior surfaces of the main to a chlorine concentration of 100 mg/L as it is being applied and in the main as the slug moves along.

If the free chlorine residual in the slug drops below 50 mg/L, stop the flow, add more chlorine to the slug to increase the residual to 100 mg/L and continue. Try to maintain a three-hour contact time as the entire slug moves through the main. When the chlorine slug flows past fittings and valves, operate related valves and hydrants in order to disinfect pipe branches and appurtenances.

After the slug has passed through the main, flush out the chlorinated water until the chlorine concentration in the water leaving the main is no higher than that commonly found in the system. When the disinfection procedures are completed, collect bacteriological samples. Do not put the main into service until the samples are found to be negative for coliform organisms. If positive coliform samples are found, reflush the main and resample. If samples are still positive, the main must be rechlorinated and resampled until satisfactory results are obtained. The 24-hour membrane filter test is commonly used to test for coliform bacteria because test results are available more quickly than with other test methods.

Repair of mains under pressure presents little danger of contamination and disinfection is not required. However, when mains are wholly or partially dewatered, they must be disinfected. In wet excavations, large quantities of hypochlorite are applied to open trench areas to lessen the danger of contamination. The interior of all pipes and fittings used in making the repair must be swabbed or sprayed with a one percent hypochlorite solution before they are installed.

The most practical way of removing contamination introduced during repairs is by thorough flushing. Where it can be done, a section of main, in which the break is located, should be isolated and all service connections shut off. Then the section should be flushed and chlorinated using the slug method as described for new main disinfection except that the dose may be increased to as much as 300 mg/L and the contact time reduced to as little as 15 minutes.

After the chlorination has been completed, flushing is resumed and continued until any discolored water is eliminated and the water is free of noticeable chlorine odor. The main may be returned to service prior to completion of bacteriological testing so that the time customers are out of water will be minimized. Samples should be taken on each side of the main break if the direction of flow in the main was not known at the time of the break. If positive samples are found, daily sampling must be continued until two consecutive samples are negative.

For more details on the disinfection of water mains, see Chapter 6, "Disinfection."

FORMULAS

To determine the flow in a pipe in gallons per minute, we usually have to calculate the flow in cubic feet per second. The flow in cubic feet per second (CFS) is determined by multiplying the area of the pipe in square feet times the velocity. We have referred to this formula as Q = AV.

$$\text{Area, sq ft} = \frac{(0.785)(\text{Diameter, in})^2}{144 \text{ sq in/sq ft}}$$

We divided by 144 sq in/sq ft to convert the calculated area from square inches to square feet.

Flow, CFS = (Area, sq ft)(Velocity, ft/sec)

or

Flow, GPM = (Flow, cu ft/sec)(7.48 gal/cu ft)(60 sec/min)

EXAMPLE 8

A 6-inch diameter water main is to be flushed at 4 ft/sec before disinfection. What should be the reading on the flow-meter in gallons per minute?

Known	Unknown
Diameter, in = 6 in	Flow, GPM
Velocity, ft/sec = 4 ft/sec	

1. Calculate the flow in cubic feet per second.

Flow, CFS = (Area, sq ft)(Velocity, ft/sec)

$$= \frac{(0.785)(6 \text{ in})^2(4 \text{ ft/sec})}{144 \text{ sq in/sq ft}}$$

$$= 0.785 \text{ cu ft/sec}$$

2. Convert flow in cubic feet per second to GPM.

Flow, GPM = (Flow, cu ft/sec)(7.48 gal/cu ft)(60 sec/min)

$$= (0.785 \text{ cu ft/sec})(7.48 \text{ gal/cu ft})(60 \text{ sec/min})$$

$$= 352 \text{ gal/min}$$

QUESTIONS

Write your answers in a notebook and then compare your answers with those on page 290.

5.8A   Why should distribution system mains and reservoirs be disinfected?

5.8B   List three forms of chlorine available for disinfecting a water main.

5.8C   When using tablets to disinfect a water main, how long should the chlorinated water remain in the line?

5.8D   What should be done when the disinfection procedures for a water main are completed?

5.82   Disinfection of Storage Facilities

A detailed description of procedures to disinfect storage facilities is given in AWWA C652-02, *AWWA STANDARD FOR DISINFECTION OF WATER-STORAGE FACILITIES*,[21] and in Chapter 6, "Disinfection." If the water agency contracts out this work, the contract should specify use of the appropriate AWWA standard.

Distribution system storage facilities must be disinfected when they are newly installed before being placed in or returned to service or when they are taken out of service (dewatered) for any reason and whenever the possibility exists that the water has become contaminated.

Before disinfection is started, cleaning of the facilities is necessary. All materials not belonging in the empty tank must first be removed. Then, all interior surfaces of the facility are cleaned thoroughly using a high-pressure water jet, scrubbing, sweeping, or some equivalent method. Water and dirt that have accumulated from the cleaning operation are discharged or otherwise removed before disinfection starts.

The same forms of chlorine used for main disinfection can be used for disinfecting storage facilities. These are liquid chlorine, sodium hypochlorite solution, calcium hypochlorite solution, and calcium hypochlorite granules or tablets. Table 5.6 shows the amounts of chemicals required to provide various initial chlorine concentrations in 100,000 gallons (378,500 L) of water.

TABLE 5.6   AMOUNTS OF CHEMICALS REQUIRED TO GIVE VARIOUS CHLORINE CONCENTRATIONS IN 100,000 GALLONS (378,500 L) OF WATER [a]

Desired[b] Chlorine Dose in Water, mg/L	Pounds of Liquid Chlorine Required[c]	Gallons of Sodium Hypochlorite Required[d]			Pounds of Calcium Hypochlorite Required[c]
		5% Available Chlorine	10% Available Chlorine	15% Available Chlorine	65% Available Chlorine
2	1.7	3.9	2.0	1.3	2.6
10	8.3	19.4	9.9	6.7	12.8
50	42.0	97.0	49.6	33.4	64.0

[a] From AWWA Standard C652-92, *DISINFECTION OF WATER-STORAGE FACILITIES*.
[b] Amounts of chemicals are for initial concentrations of available chlorine. Allowance may need to be made for chlorine depletion where low concentrations are held for extended time periods.
[c] Multiply pounds x 0.454 to obtain kilograms.
[d] Multiply gallons x 3.785 to obtain liters.

Three methods of chlorination are described in the AWWA standard referred to in Table 5.6. The operator should decide which of these methods is most suitable considering the availability of materials and equipment, training status of the operators who will do the disinfection, and safety considerations.

In Chlorination Method 1 of the AWWA Standard, the storage facility is filled to the overflow level with potable water and enough chlorine is added to provide a free chlorine residual in the full facility of not less than 10 mg/L at the end of the required retention period. When chlorine gas is used, the water entering the storage facility is chlorinated uniformly by a chlorinator. If sodium hypochlorite is used, it is poured into the storage facility as it begins filling, when the depth of the water is between 1 and 3 feet (0.3 and 1 m). If calcium hypochlorite is used, granules or tablets are poured into or placed in the storage facility before water is discharged into it. If chlorine gas is used, the required retention period is not less than six hours. When sodium hypochlorite or calcium hypochlorite is used, a minimum (chlorine contact) period of 24 hours is required. After the required retention period, the free chlorine residual in the water is reduced to not more than 2 mg/L to minimize possible complaints when the water is served. This can be done by completely draining and refilling the facility, or by additional holding time, and/or by blending with potable water having a low chlorine concentration.

In Chlorination Method 2, the surfaces of all parts of the storage facility, which would be in contact with the water, are thoroughly coated with a solution of 200 mg/L free available chlorine using brushes or spray equipment. Potable water is used to fill the reservoir after at least 30 minutes of contact

---

[21] *Obtain from Customer Services, AWWA, 6666 West Quincy Avenue, Denver, CO 80235. Order No. 43652; price to members, $42.00; nonmembers, $61.00; price includes cost of shipping and handling.*

time after application of the chlorine solution. Better disinfection can be achieved if the tank is then filled with distribution system water that has been treated with chlorine to provide a chlorine residual of 3 mg/L. Let the water in the tank stand for 3 to 6 hours.

In Chlorination Method 3, the storage facility is filled to approximately five percent of the total storage volume. Sufficient chlorine is then added to provide a 50 mg/L available chlorine solution which is held for a period of not less than six hours. The facility is then filled to the overflow level and the chlorinated water is held for at least 24 hours. At the end of the 24-hour holding period, there must be a minimum of 2 mg/L free chlorine residual.

In all cases, water is not delivered to the distribution system until bacteriological tests are negative for coliform organisms and the water is of acceptable aesthetic quality. The 24-hour membrane filter test for coliform is commonly used in this case because results are available sooner than with the multiple-tube fermentation (MPN) method. If coliform bacteria are found, repeat samples must be taken until two consecutive samples are negative. The water should also be tested to be certain there is no offensive odor or color.

If the chlorinated water is discharged to the environment and there is a possibility it may be toxic to fish and other aquatic life, the chlorine residual must be neutralized using a reducing agent. Also chlorinated water should not be discharged to a sewer if there is any possibility of a chlorine residual remaining in the wastewater when it reaches a treatment plant. Table 5.7 shows the amounts of reducing chemicals required to neutralize various residual chlorine concentrations in 100,000 gallons (378,500 L) of water. Disposal of highly chlorinated water may require a permit from a regulatory agency.

For additional detailed procedures on how to disinfect storage facilities, see Chapter 6, "Disinfection."

FORMULAS

When disinfecting either a water main or a storage facility, we need to know the volume to be disinfected in million gallons and desired chlorine dose in milligrams per liter.

$$\text{Chlorine Needed, lbs} = (\text{Volume, M Gal})(\text{Chlorine Dose, mg}/L)(8.34 \text{ lbs/gal})$$

If a hypochlorite solution is used to provide the chlorine needed, we need to know the percent of chlorine in the hypochlorite solution to calculate the gallons of hypochlorite required.

$$\text{Chlorine Needed, lbs} = \frac{(\text{Hypochlorite, gal})(8.34 \text{ lbs/gal})(\text{Hypochlorite, \%})}{100\%}$$

or

$$\text{Hypochlorite, gal} = \frac{(\text{Chlorine Needed, lbs})(100\%)}{(8.34 \text{ lbs/gal})(\text{Hypochlorite, \%})}$$

EXAMPLE 9

A 50,000-gallon storage tank is to be disinfected with a chlorine solution of 50 mg/L. How many pounds of chlorine (gas) will be needed?

Known		Unknown
Tank Volume, gal	= 50,000 gal	Chlorine Needed, lbs
Tank Volume, Million gal	= 0.050 M gal	
Chlorine Conc, mg/L	= 50 mg/L	

Formula for calculating pounds of chlorine needed.

$$\text{Chlorine, lbs} = (\text{Tank Volume, M Gal})(\text{Chlorine, mg}/L)(8.34 \text{ lbs/gal})$$

Calculate the pounds of chlorine needed.

$$\text{Chlorine, lbs} = (\text{Tank Volume, M Gal})(\text{Chlorine, mg}/L)(8.34 \text{ lbs/gal})$$

$$= (0.050 \text{ M Gal})(50 \text{ mg}/L)(8.34 \text{ lbs/gal})$$

$$= 21 \text{ lbs}$$

EXAMPLE 10

How many gallons of five percent hypochlorite solution will be needed to disinfect the storage tank in Example 9 (where 21 pounds of chlorine gas was needed)?

Known		Unknown
Chlorine Needed, lbs	= 21 lbs	Hypochlorite, gallons
Hypochlorite, %	= 5%	

Formula for calculating pounds of available chlorine.

$$\text{Chlorine, lbs} = (\text{Hypochlorite, gal})(8.34 \text{ lbs/gal})\left(\frac{\text{Hypochlorite, \%}}{100\%}\right)$$

Calculate gallons of hypochlorite needed.

$$\text{Hypochlorite, gal} = \frac{(\text{Chlorine, lbs})(100\%)}{(8.34 \text{ lbs/gal})(\text{Hypochlorite, \%})}$$

$$= \frac{(21 \text{ lbs})(100\%)}{(8.34 \text{ lbs/gal})(5\%)}$$

$$= 50 \text{ gallons}$$

TABLE 5.7   AMOUNTS OF CHEMICALS REQUIRED TO NEUTRALIZE VARIOUS RESIDUAL CHLORINE CONCENTRATIONS IN 100,000 GALLONS (378,500 L) OF WATER[a]

Residual Chlorine Concentration, mg/L	Pounds of Chemical Required			
	Sulfur Dioxide (SO$_2$)	Sodium Bisulfite (NaHSO$_3$)	Sodium Sulfite (Na$_2$SO$_3$)	Sodium Thiosulfate (Na$_2$S$_2$O$_3 \cdot$ 5 H$_2$O)
1	0.8	1.2	1.4	1.2
2	1.7	2.5	2.9	2.4
10	8.3	12.5	14.6	12.0
50	41.7	62.6	73.0	60.0

[a] From AWWA Standard C652-92, *DISINFECTION OF WATER-STORAGE FACILITIES.*

QUESTIONS

Write your answers in a notebook and then compare your answers with those on page 290.

5.8E   When should distribution system storage facilities be disinfected?

5.8F   How can the interior surfaces of a storage facility be cleaned before disinfection?

5.8G   After the required retention (chlorine contact) period for disinfecting a storage facility is over, how is the free chlorine residual in the water reduced to less than 2 mg/*L*?

5.8H   List the reducing agents that could be used to neutralize a free chlorine residual in water that has been used to disinfect a storage facility.

5.9 RECORDS

5.90 Importance of Records

In too many systems the importance and value of keeping good records are not recognized. Where this occurs, the operators may not know (or be able to find) the construction details of important facilities, where they are located, or what shape they are in. The need for a good record system regardless of the size or complexity of the water system cannot be overemphasized.

Records are needed for many reasons. In general, they promote the efficient operation of the water system. Records can remind the operator when routine operation or maintenance is necessary and help ensure that schedules will be maintained and no needed operation or maintenance will be overlooked or forgotten. Records are the key to an effective maintenance program. They are also needed for those regulatory agencies that require submission of periodic water quality and operational records. Records can be used to determine the financial health of the utility, provide the basic data on the system's property, and prepare monthly and annual reports. Another reason for keeping accurate and complete records of system operations is the legal liability of the utility. Such records are required as evidence of what actually occurred in the system. Good records can help if the utility is threatened with litigation. Records also assist in answering consumer questions or complaints. Finally, clear, concise records are required to effectively meet future operational needs, that is, for planning purposes.

Records should be tailored to meet the demands of the particular system and only records known to be useful should be kept. Operators should determine what type of information will be of value for their system and then prepare maps, forms, or other types of records on which the needed information can be easily recorded and clearly shown. Records should be prepared as if they will be kept indefinitely. In fact some will be kept for a long time, while other records will not. Records should be put into a filing system that can be easily used and understood by everyone concerned, readily accessible, and protected from damage in a safe environment. The nonpermanent records should be disposed of in accordance with a disposal schedule set up for the different types of records maintained. Good recordkeeping tips can be found throughout these manuals. For additional information on recordkeeping from the viewpoint of the distribution system manager, see Chapter 8, Section 8.12, "Recordkeeping."

5.91 Maps

Comprehensive maps and sectional plats are used by most utilities. The comprehensive map provides an overall view of the entire distribution system. Important structures are shown, including water intakes, treatment plants, wells, reservoirs, mains, hydrants, and valves. The preferred scale for comprehensive maps is 500 feet per inch (60 m/cm), while the maximum scale recommended is 1,000 feet per inch (120 m/cm). Sectional plats will show various portions of the comprehensive map in much more detail (Figure 5.51). The scale varies from 50 to 200 feet per inch (6 to 24 m per cm). Standard symbols are used to indicate different items on these and other utility maps. Some common symbols are shown in Figure 5.52.

Valve and hydrant maps as shown in Figure 5.53 may cover the same area as up to four sectional maps. They not only show valve and hydrant locations, but provide such information as the direction to open the valve, the number of turns to open, the model type, and the installation date. A valve intersection plat with the valves identified by letters is shown in Figure 5.54.

Other types of maps are also used. Leak survey maps are usually another modification of sectional or valve maps and, where regular leak survey work is conducted, these maps show the valves to be closed and the areas to be isolated. Leak frequency maps show the locations where leaks have been found and pinpoint problem areas. Customer complaints, when indicated on a map, are also very helpful in showing problem areas and the need for follow-up investigations. Locations of leaks and complaints may be marked on wall-mounted maps by pins with colored heads for ease in interpreting location, size, and type of problems.

All maps and drawings should show constructed facilities "as built" (or record drawings). If there was any change whatsoever from the construction plans, the maps and plans used by the utility should show this change. Whenever modifications are made, plans should be changed to show the details of the modification, the date of the modification, and who recorded the modification on the plans. The modified drawings are called "as-builts" or "record drawings."

Plan and profile maps are engineering drawings that show the depth of pipe, pipe location both vertically and horizontally, and the correct distance from a reference starting point. Operators occasionally need to use this type of map. Water gradient contour maps are prepared by taking pressure readings

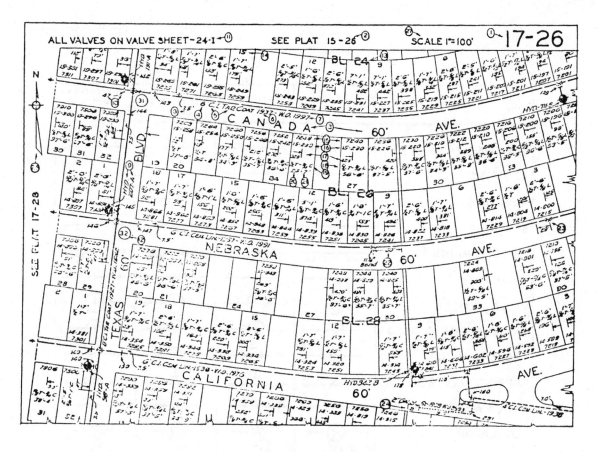

Fig. 5.51   Portion of sectional plat

(Reprinted from *WATER DISTRIBUTION TRAINING COURSE*, by permission.
Copyright, the American Water Works Association)

during peak use periods over the system, calculating hydraulic gradients, placing these gradients on the comprehensive map, and drawing in gradient contour lines. They can be used to indicate problem areas with insufficient or excessive pressures.

The geographic information system (GIS) is a computer program that combines mapping with detailed information about the physical structures within geographic areas. To create the database of information, "entities" within a mapped area, such as streets, manholes, line segments, and lift stations, are given "attributes." Attributes are simply the pieces of information about a particular feature or structure that are stored in a database. The attributes can be as basic as an address, manhole number, or line segment length, or they may be as specific as diameter, rim invert, and quadrant (coordinate) location. Attributes of a main line segment might include engineering information, maintenance information, and inspection information. Thus an inventory of entities and their properties is created. The system allows the operator to periodically update the map entities and their corresponding attributes.

The power of a GIS is that information can be retrieved geographically. An operator can choose an area to look at by pointing to a specific place on the map or outlining (windowing) an area of the map. The system will display the requested section on the screen and show the attributes of entities located on the map. A printed copy may also be requested. Figure 5.55 shows a GIS-generated map. This example shows data

from an inflow/infiltration analysis, including pipe attributes, hydraulic data, and selected engineering data. The example also shows a map of the system. In most cases computer-based maintenance management system (CMMS) software has the ability to communicate with geographic information systems so that attribute information from the collection system can be copied into the GIS.

A GIS can generate work orders in the form of a map with the work to be performed outlined on the map. This minimizes paper work and gives the work crew precise information about where the work is to be performed. Completion of the work is recorded in the GIS to keep the work history for the area and entity up to date. Reports and other inquiries can be requested as needed, for example, a listing of all line segments in a specific area could be generated for a report.

In many areas GISs are being developed on an area-wide basis with many agencies, utilities, counties, cities, and state agencies participating. Usually a county-wide base map is developed and then all participants provide attributes for their particular systems. For example, information on the sanitary sewer collection system might be one map layer, the second map layer might be the water distribution system, and the third layer might be the electric utility distribution system. In addition to sharing databases with CMMSs, GISs generally now also have the ability to operate smoothly with computer-aided design (CAD) systems.

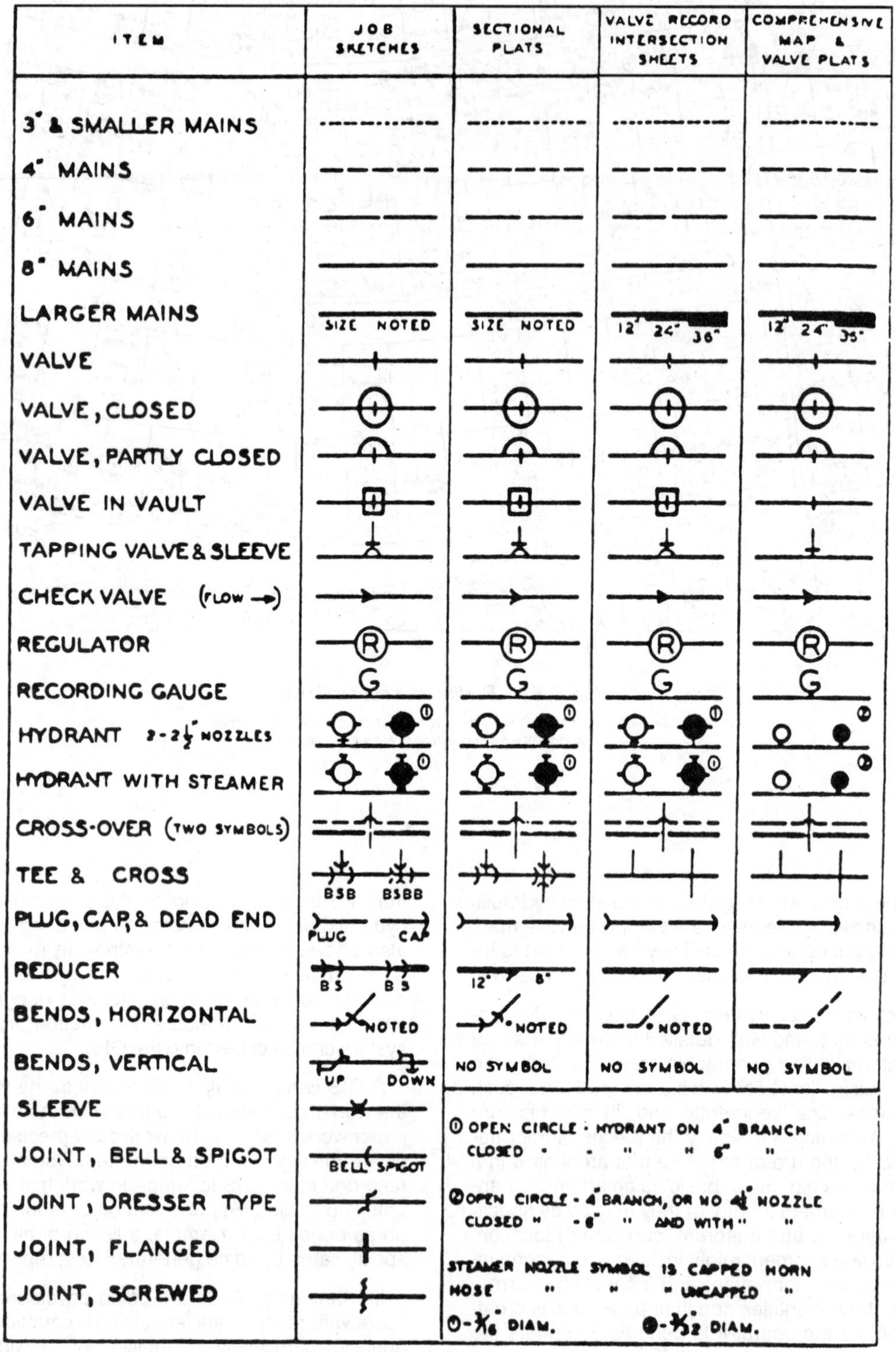

Fig. 5.52   Common map symbols

(Reprinted from WATER DISTRIBUTION OPERATOR TRAINING HANDBOOK,
by permission. Copyright 1976, the American Water Works Association)

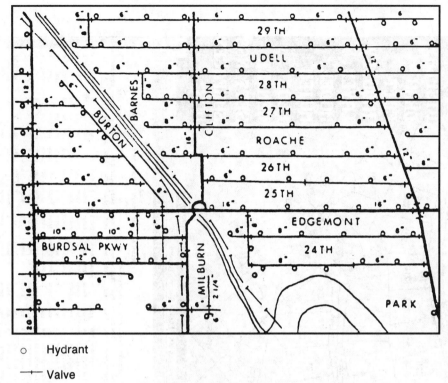

o   Hydrant

┼   Valve

Fig. 5.53   Valve and hydrant map

(Reprinted from *WATER DISTRIBUTION OPERATOR TRAINING HANDBOOK*,
by permission. Copyright 1976, the American Water Works Association)

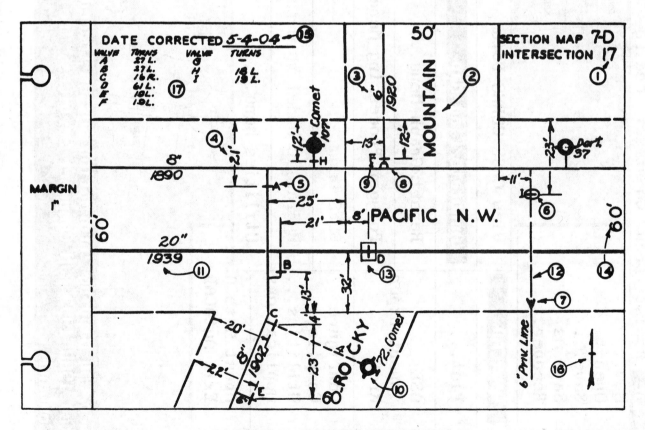

Fig. 5.54   Valve intersection plat

(Reprinted from *WATER DISTRIBUTION TRAINING COURSE*, by permission.
Copyright, the American Water Works Association)

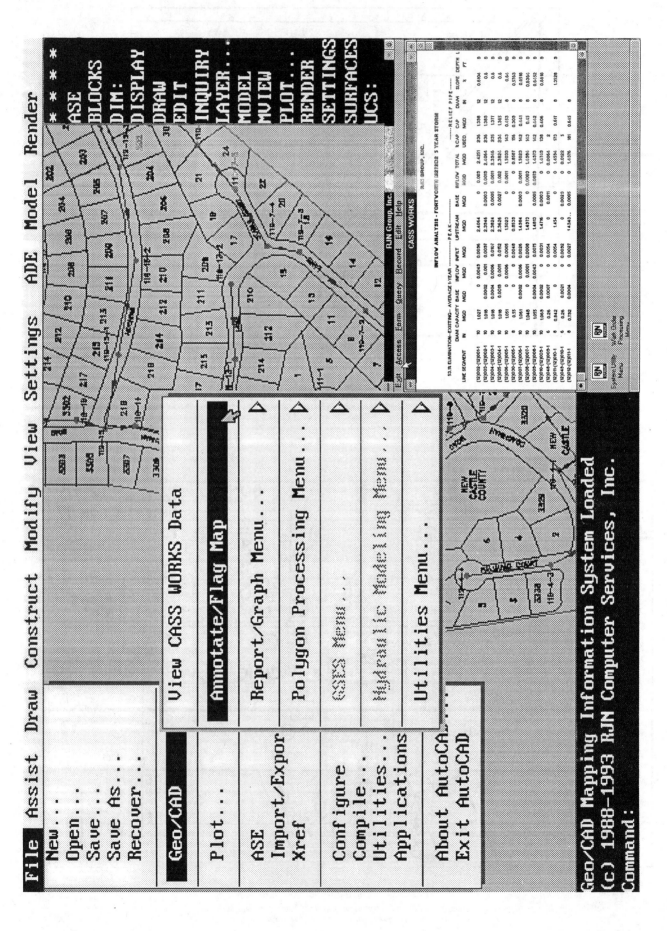

Fig. 5.55 Typical Geographic Information System (GIS) map

5.92 Types of Records

Written records should be kept on all facilities in the distribution system. These records should describe the facility, its construction, date installed, repair and maintenance work done, manufacturer, and condition during latest inspection. As new facilities are installed, they should be included in the existing record system. Records should also note those facilities retired from service. Other types of records to be kept include:

1. Results of water quality monitoring,
2. Cross-connection control,
3. Main flushing,
4. Main cleaning,
5. Consumer complaints,
6. Disinfection,
7. Pressure surveys,
8. Leaks,
9. Engineering reports,
10. Any operations done in compliance with the request of the health department, and
11. A daily log kept on any and all unusual events occurring, such as equipment malfunction, unusual weather conditions, and natural or manmade disasters.

Each public water system is required by the Primary Drinking Water Regulations to maintain records of water quality analyses, written reports, variances or exemptions, and actions taken to correct violations of the regulations. The lengths of time these records must be kept are summarized in Table 5.8.

TABLE 5.8 MINIMUM TIME PERIODS FOR RETAINING RECORDS

Record	Minimum Years of Retention
Bacteriological analyses	5
Chemical analyses	10
Lead and Copper Monitoring Rule	12
Written reports such as sanitary surveys, engineering reports	10[a]
Variances or exemptions	5[b]
Action taken to correct violation	3[c]

[a] Following completion of surveys and reports.
[b] Following expiration of variance or exemption.
[c] After last action with respect to violation.

Important distribution system operational records include technical and maintenance information on pumps, meters, valves, and other equipment. Information on design, capacities, and required maintenance should be kept with each major piece of equipment. Each piece of equipment should have its own file containing manufacturer's instructions, repairs made, and the schedule to be followed for routine maintenance. Each valve and hydrant should have a permanent file card such as that shown in Figure 5.56 and an inspection report form such as that shown in Figure 5.57.

Every utility should have a work order system which provides that all additions or changes to, or removal of, any of the system facilities be accomplished using approved work orders. Examples of some work order forms used are shown in Figures 5.58 and 5.59. The work order system should be devised to account properly for all of the physical plant facilities. The work order sketch (Figure 5.59), when completed, is used to revise the distribution maps to record additions, changes, or removals.

QUESTIONS

Write your answers in a notebook and then compare your answers with those on page 290.

5.9A What problems can develop if good records are not kept?

5.9B What is the purpose of a comprehensive map?

5.9C What information is shown on valve and hydrant maps?

5.9D What information should be recorded on the "as built" plans (record drawings) whenever a change or modification is made to the system?

5.10 SCADA SYSTEMS

SCADA (pronounced ss-KAY-dah) stands for **S**upervisory **C**ontrol **A**nd **D**ata **A**cquisition system. This is a computer-monitored alarm, response, control and data acquisition system used by operators to monitor and adjust their treatment processes and water distribution systems.

A SCADA system collects, stores, and analyzes information about all aspects of operation and maintenance, transmits alarm signals when necessary, and allows fingertip control of alarms, equipment, and processes. SCADA provides the information that operators need to solve minor problems before they become major incidents. As the nerve center of a water distribution system, the SCADA system allows operators to enhance the efficiency of their water distribution system by keeping the operators fully informed and fully in control.

In water applications, SCADA systems monitor levels, pressures, and flows and also operate pumps, valves, and alarms. They monitor temperatures, speeds, motor currents, pH, chlorine residuals, and other operating guidelines, and provide control as necessary. SCADA also logs event and analog signal trends and monitors equipment operating time for maintenance purposes.

Applications for SCADA systems include water distribution system control and monitoring, water treatment plant control monitoring, and other related applications. SCADA systems can vary from merely data collection and storage to total data

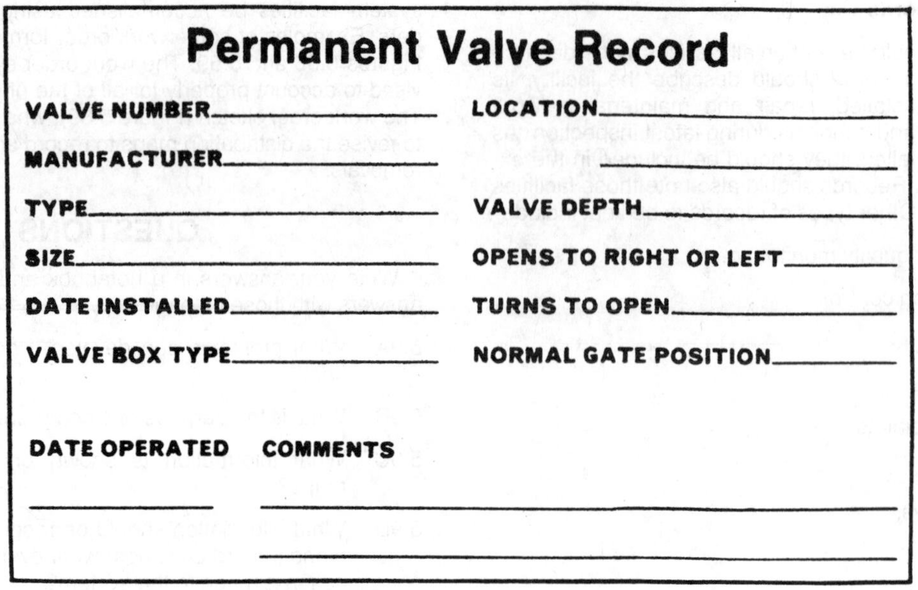

Permanent Valve Record

VALVE NUMBER ____________ LOCATION ____________

MANUFACTURER ____________

TYPE ____________ VALVE DEPTH ____________

SIZE ____________ OPENS TO RIGHT OR LEFT ____________

DATE INSTALLED ____________ TURNS TO OPEN ____________

VALVE BOX TYPE ____________ NORMAL GATE POSITION ____________

DATE OPERATED COMMENTS

Fig. 5.56 Typical valve record form

(Reprinted from *OPFLOW*, January 1979, by permission. Copyright 1979, the American Water Works Association)

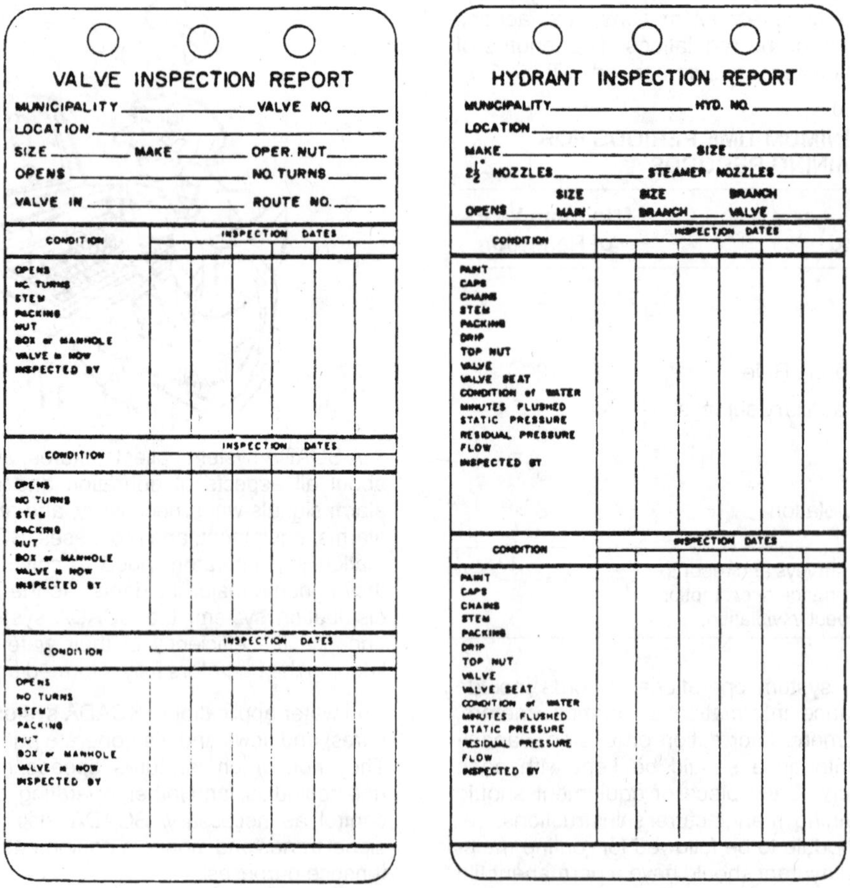

VALVE INSPECTION REPORT

MUNICIPALITY ____________ VALVE NO. ______
LOCATION ____________
SIZE ______ MAKE ______ OPER. NUT ______
OPENS ____________ NO. TURNS ______
VALVE IN ____________ ROUTE NO. ______

CONDITION	INSPECTION DATES					
OPENS NO. TURNS STEM PACKING NUT BOX or MANHOLE VALVE is NOW INSPECTED BY						

CONDITION	INSPECTION DATES					
OPENS NO TURNS STEM PACKING NUT BOX or MANHOLE VALVE is NOW INSPECTED BY						

CONDITION	INSPECTION DATES					
OPENS NO TURNS STEM PACKING NUT BOX or MANHOLE VALVE is NOW INSPECTED BY						

HYDRANT INSPECTION REPORT

MUNICIPALITY ____________ HYD. NO. ______
LOCATION ____________
MAKE ____________ SIZE ______
2½" NOZZLES ______ STEAMER NOZZLES ______
OPENS ______ SIZE MAIN ______ SIZE BRANCH ______ BRANCH VALVE ______

CONDITION	INSPECTION DATES					
PAINT CAPS CHAINS STEM PACKING DRIP TOP NUT VALVE VALVE SEAT CONDITION of WATER MINUTES FLUSHED STATIC PRESSURE RESIDUAL PRESSURE FLOW INSPECTED BY						

CONDITION	INSPECTION DATES					
PAINT CAPS CHAINS STEM PACKING DRIP TOP NUT VALVE VALVE SEAT CONDITION of WATER MINUTES FLUSHED STATIC PRESSURE RESIDUAL PRESSURE FLOW INSPECTED BY						

Fig. 5.57 Valve and hydrant inspection report forms

(Reprinted from *WATER DISTRIBUTION TRAINING COURSE*, by permission. Copyright, the American Water Works Association)

INVESTMENT WORK ORDER

FOR PROPERTY ADDITIONS
AND IMPROVEMENTS

Investment Work Order No.	A-80
Charge to Budget (Item C Below) $	1,652.65

The Metropolitan Water Company
Name of Company

Apply to Budget Item No.	1A-2

Metropolitan City
Location

File No.	-443

DESCRIPTIVE TITLE: Install 400 feet of 6"   pipe on
Main Street south from First Avenue.

PURPOSE: Give complete explanation here, not by letter. Attach extra sheet if necessary. Sketch must accompany changes in distribution system, buildings, pumping equipment and piping. If contribution or refundable deposit toward the cost is to be obtained, state amount thereof, from whom it will be received, and if an agreement is to be entered into submit the signed agreement. If State regulations, franchises, municipal contracts or company rules are involved, state extent.

This installation required to replace a 3/4" galvanized pipe
installed in 1926, which has become inadequate due to corrosion
and an increased number of customers.

Pressure on Proposed Mains _____60_____ psi   Traffic ___light___   Type of Paving ___none___

SUMMARY ESTIMATE OF COST		RETURN ON INVESTMENT	
A. Total estimated cost (per reverse) $ 1,652.65		1. Total investment cost (Item E) . . . . $ 1,625.65	
B. Less: Contributions (non-refundable)$		2. Increase in revenue:	
C. Net direct cost to Company $ 1,652.65		2a. _____ Consumers @ $ _____ . . $_____	
D. Service and meter installations $		2b. _____ Consumers @ $ _____ . . _____	
E. Total cost $ 1,652.65		2c. _____ Hydrants @ $ _____ . . _____	
		2d. _____ . . _____	
		3. Total increase in revenue . . . . . . $_____	
		4. Decrease in operating expenses . . . . _____	
		5. Total—Lines 3 and 4 . . . . _____	
If any preliminary investigation costs have been incurred		6. Increase in expenses:	
for this project, to what account were they charged?		6a. Operating, maintenance and general	
None		taxes, _____ % of Line 3 . . . . . $_____	
		6b. Depreciation _____ % of Line 1 . . . _____	
		7. Total increase in expenses . . . . . . _____	
Does the proposed work replace any Company owned		8. Income before income taxes—Line 5 minus 7 _____	
property? ___Yes___		9. Income taxes _____ % of Line 8 . . . _____	
If yes, what number is assigned to the related Retirement		10. Income available for return—Line 8 minus 9 $_____	
Order? R - 31		11. Rate of return—Line 10 ÷ Line 1 . . . _____ %	
In what political subdivision will new property be located? Metropolitan City			

PREPARATION AND APPROVAL

LOCAL AND DIVISION OFFICES	Date	GENERAL OFFICE	Date
Prepared By:			
Local Manager:			
Budget Approval:			
Division Comptroller:			
Division Manager:			

Fig. 5.58 *Typical investment work order form*

(Reprinted from *WATER DISTRIBUTION TRAINING COURSE*, by permission.
Copyright, the American Water Works Association)

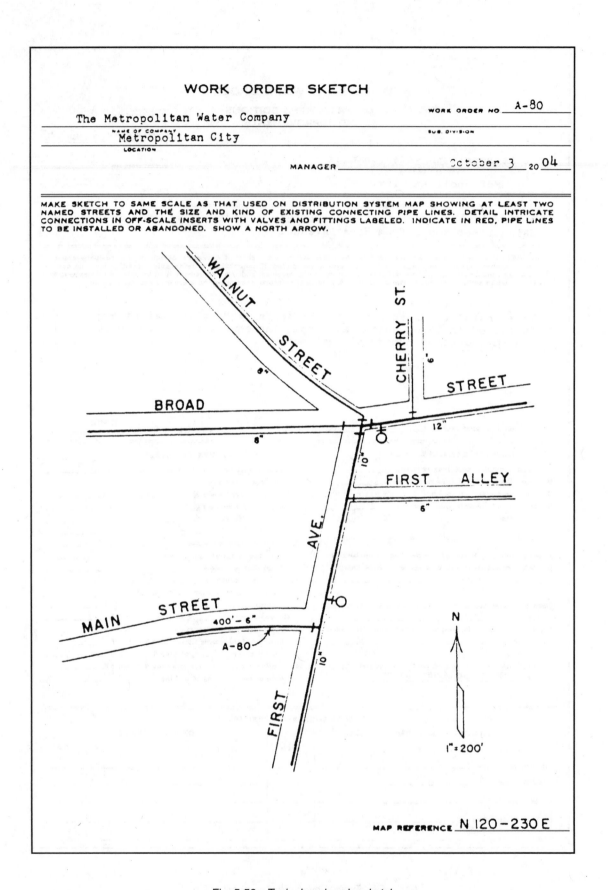

WORK ORDER SKETCH

The Metropolitan Water Company
NAME OF COMPANY
Metropolitan City
LOCATION

WORK ORDER NO. A-80

SUB. DIVISION

MANAGER _____ October 3 20 04

MAKE SKETCH TO SAME SCALE AS THAT USED ON DISTRIBUTION SYSTEM MAP SHOWING AT LEAST TWO NAMED STREETS AND THE SIZE AND KIND OF EXISTING CONNECTING PIPE LINES. DETAIL INTRICATE CONNECTIONS IN OFF-SCALE INSERTS WITH VALVES AND FITTINGS LABELED. INDICATE IN RED, PIPE LINES TO BE INSTALLED OR ABANDONED. SHOW A NORTH ARROW.

MAP REFERENCE N 120-230 E

Fig. 5.59 Typical work order sketch

(Reprinted from *WATER DISTRIBUTION TRAINING COURSE*, by permission.
Copyright, the American Water Works Association)

analysis, interpretation, and process control. Please refer to Chapter 8, Section 8.94, for a more detailed description of SCADA systems.

QUESTIONS

Write your answers in a notebook and then compare your answers with those on page 290.

5.10A   What does SCADA stand for?

5.10B   What does a SCADA system do?

5.11 EQUIPMENT AND STORES

Tools and equipment for distribution system operation and maintenance must be carefully selected according to the needs of the utility. The following points should be considered:

1. Cost and efficiency (including any resulting reduction in labor costs),

2. Shorter service interruptions,

3. Ability to make the job easier for the operators (a morale factor), and

4. Versatility of tools or equipment for use in different types of operation and maintenance tasks.

System requirements should be carefully analyzed to determine what is needed and how much should usually be kept on hand. A continuing stock inventory should be kept with the stock properly classified, described, and stored where it is easily located. Physical inventories (counts) should be periodically made, once or twice each year. An efficient record system is needed to provide for advance ordering of depleted items. The required items must be available on demand. Arrangements have to be made for proper handling and storing of supplies, equipment, parts, and other needed items.

The general types of equipment used include various hand tools such as picks, shovels, sledges, caulking tools, and valve and hydrant wrenches. Also used are trench pumps, lighting equipment for night work, rubber boots, rain gear, mobile air compressors, power valve operators, excavators, pipe and leak locators, electric generators, ladders, rope, jacks, tamping and valve inserting tools, pipe pushers (Figure 5.60), mechanical joints, couplings, safety and first-aid equipment, signs, barricades, warning signals, and traffic cones. An emergency truck containing all of this needed equipment, ready to be sent out at a moment's notice, should be available. This properly staffed emergency vehicle should be able to handle a variety of system repairs and operation.

These and other supplies, equipment, materials, and parts represent an appreciable part of the assets of a utility. Every water utility should also keep a basic stock of pipe and repair parts available for immediate use.

Motorized equipment is often categorized as (1) general purpose service trucks, (2) excavators (backhoes), (3) cranes of all types, (4) miscellaneous work equipment, and (5) specialized vehicles. Specialized vehicles include trucks used for meter maintenance, fire hydrant and valve maintenance, and emergency vehicles. Tractor backhoes have become the standard in most trenching operations. Pipe laying operations can proceed with the use of a motorized crane without interrupting excavation operations. Tractors equipped with front loader attachments can efficiently backfill excavations after pipe laying has been completed. Mechanical earth compactors or vibrating attachments for hydraulic tractors along with jetting will ensure a compact trench backfill.

QUESTIONS

Write your answers in a notebook and then compare your answers with those on page 291.

5.11A   What factors should be considered when selecting tools and equipment for distribution system operation and maintenance?

5.11B   How frequently should physical inventories be made?

5.12 EMERGENCY PLANNING

Every water utility should have an emergency operations plan relating to disasters such as earthquakes, floods, hurricanes, tornadoes, and even war. Normally, this type of plan does not include "everyday" routine disruptions of service. The utility should have a separate plan that outlines what must be done when significant "routine" problems such as the following occur:

1. A key water main breaks,

2. An extended electrical power interruption occurs,

3. A vitally needed well pump motor or bearing "burns out,"

4. A key chlorinator or treatment unit fails, and

5. Test results or direct observations indicate the water supply has become contaminated.

Operators must recognize that these types of problems can also result from a disaster so some of the discussions in this section will apply to such "routine" problems. During a disaster, all of the above problems and more could take place at once with the entire system affected.

An emergency operations plan has been defined simply as: "Under disaster conditions, *WHO* does *WHAT* and *WHEN* with the existing resources." The operations plan should be concise, giving guidance quickly at the time of need. If there is too much detail, the whole operation can get bogged down. The best plan is only a guide to action, but may be specific in some critical details. The plan must recognize that under emergency conditions, trained operators must react quickly and make on-the-spot decisions which must be appropriate to each level of responsibility. The emergency plan must be widely distributed, but also updated at least annually and when changes take place in staffing, equipment, and facilities. All water utility operators must be familiar with the written plan and its location. If the plan is to be workable, effective training of personnel on a periodic basis (once a year) is required.

Fig. 5.60   Pipe pusher

(Permission of Trojan Manufacturing Company, Inc.)

During a disaster, the water utility has the following prime responsibilities:

1. Provide water for firefighting, drinking, and sanitation,

2. Prevent unnecessary loss of stored water, and

3. Restore the integrity of the entire water system as soon as possible.

The first step in preparing for a disaster is to form a disaster organization. The disaster organization's staff and teams within the utility should be designated with alternates specified. An alerting list is needed with telephone numbers. The responsibilities of each individual with channels of command should be clearly defined. Related nearby utilities and local civil defense and military authorities should be contacted to establish liaison channels, to determine their own plans and what support is available from them, and to obtain possible help in the utility's planning. "Mutual aid agreements" should be made where appropriate providing for exchange or assignment of personnel, equipment, and materials. The plan should contain work and home telephone numbers for all personnel of the disaster and mutual aid organizations, civil defense, police and fire departments, and state and local health departments.

An Emergency Operations Plan is developed by the disaster organization by first assuming a certain type of emergency and the resulting effects. The type of natural disaster a water utility might encounter depends on where it is located. Utilities on the West Coast can anticipate earthquakes or floods, on the East Coast hurricanes are not uncommon, while the Central United States is threatened by tornadoes. An estimate is made of the facilities, personnel, and equipment capabilities remaining after the disaster, and the requirements of the community. The capabilities are then matched to the requirements. The next three steps become the operations plan. First, priorities by which the requirements can be met are specified. Next, the best ways of using the available resources are designated. Finally, specific tasks are determined and are assigned to the assumed surviving or available personnel.

A vulnerability assessment should be made of the entire system to determine where the weaknesses are if the disaster being considered took place and what kind of improvements are feasible to reduce the weaknesses found. To do this, the separate components of the system are identified and described, including sources, collection works, transmission system, treatment facilities, distribution system, personnel, power supply, materials and supplies, and communications. The existing emergency plans and mutual aid agreements are included in the assessment. Then, the effects of the assumed disaster on each component are estimated. Under the assumed conditions, water is allocated for firefighting, potable, sanitary, decontamination, industrial, and livestock uses. The capability of the system to meet these requirements is then estimated and critical, suspect components identified.

Using the above information, the actual plan of operation during the emergency can be developed. Priorities are specified and the best apparent way of using the available resources outlined. A determination is made on how to allocate water under the assumed conditions for the uses specified above. Guidelines are prepared for water allowances, priorities, rationing (if appropriate), and time-phasing of estimated water requirements. Next, procedures are established for emergency treatment, pumping, and distribution of water and for stations for service of emergency water. Restoring a system after a disaster may take weeks or longer and so it may be necessary to bring in hauled potable water and put it into intact facilities, or deliver it from tank trucks or other means directly to consumers. The public must be kept informed of the availability of potable water.

The vulnerable parts of the system, personnel, or equipment that were found from the vulnerability assessment can be reduced by a number of means. Facilities can be strengthened and, where appropriate, duplicated; alternative operation procedures can be made available; auxiliary booster pumps, power and disinfection facilities can be obtained; stockpiles of materials and supplies can be increased; additional repair equipment can be obtained; and emergency procedures, communications, and training given to disaster organization personnel can be improved.

Those operators who would like to have further information on this subject are referred to AWWA Manual M19, *EMERGENCY PLANNING FOR WATER UTILITIES*.[22] Also see Chapter 8, Section 8.10, "Emergency Response," in this training manual.

QUESTIONS

Write your answers in a notebook and then compare your answers with those on page 291.

5.12A Utilities should develop plans for responding to what types of routine disruptions which are normally not considered when developing disaster plans?

5.12B Define an Emergency Operations Plan.

5.12C List the three basic steps of an Emergency Operations Plan.

5.13 PUBLIC RELATIONS

Every employee of a utility is responsible for customer relations and for positive customer communication. Everything an employee does from meter reading, installing pipelines, repairing lines, and flushing hydrants to answering consumer complaints has an effect on the public's image of the utility. System operators are often the most visible of the system's employees and at such times they can be considered as being in the front line of public relations. Some operators come into contact with

[22] *Obtain from American Water Works Association (AWWA), Bookstore, 6666 West Quincy Avenue, Denver, CO 80235. Order No. 30019. ISBN 1-58321-135-7. Price to members, $69.00; nonmembers, $99.00; price includes cost of shipping and handling.*

the public incidentally, while others have direct intentional contact such as when dealing with consumer complaints. The system operator, therefore, must be aware of and take into account the responsibilities of public relations. Some water suppliers have recognized this fact and now call their meter readers water service representatives.

All employees communicate with the utility's customers in one way or another even if they do not realize it. If an employee operates an official vehicle in a reckless manner, appears to be idle while others are busy, does not know the answer to a question when asked, or is surly or uncooperative, the utility has communicated with its customers, but certainly not in the way it should have.

This added responsibility of representing the water supplier to the public means the operator must be careful to present a neat, efficient, knowledgeable, and cooperative image. Work habits should not encourage public criticism. Any encounter with the public, regulatory agencies, or municipal officials should be viewed as an opportunity to provide valid information and to correct misunderstandings. All operators working for a water agency should have a basic knowledge of the system facilities and the system's operation. The public normally expects operators to have this knowledge and an operator appears to be uninformed and disinterested if simple questions about the agency cannot be answered.

All complaints should be treated as legitimate and investigated as soon as possible. If the person with the complaint is loud and abusive, the operator should not retaliate in kind. Instead, the operator should show concern, listen carefully and calmly, offer to look into the complaint, and help correct any problems found. Questions should be asked as necessary to make certain the problem is understood. Every effort should be made to give the customer an immediate, clear, and accurate answer to the problem in nontechnical language. If the answer is not known, it should be referred to the proper person in the organization. Always inform the person with a problem what will be done to correct the situation or indicate when someone will call back with additional information. After the problem is corrected, a follow-up phone call or postcard to be sure everything is satisfactory or appropriate.

Sometimes it's only through complaints from its consumers that the water utility learns the service being given is not satisfactory. Complaint records should be reviewed periodically by responsible operators and they need to be especially looked at when numerous complaints start coming in. As previously mentioned in this chapter, if complaints are displayed on a map with the use of colored pins, a pattern might be detected which could indicate the area affected and where the investigation should center. By doing this routinely, the utility might be able to catch a problem at an early stage.

A customer's complaint may at times result from a problem that is originating on the customer's own premises. For example, a customer may complain about lack of water pressure such as often results from deficiencies in the customer's own plumbing. Pressure readings should be taken in the house, first without any flow so as to obtain a reading that should be close to the pressure in the system main. Then a pressure test is made with other taps in the house flowing to show the pressure drop was due to some obstruction or inadequate plumbing. Often, this may be due to the shutoff valve on the house water line being turned off or partially off. If this is not the cause of failure, the owner must usually make corrections to the plumbing. Replacing the service line from the meter to the house is often sufficient; however, occasionally repairs to the plumbing in the house may also be necessary.

Another problem that is not uncommon is red or dirty water which could also be due to deficiencies in the consumer's own lines. The cause of the problem can be demonstrated by taking samples from the customer's tap and comparing them with samples of water collected from the utility's service line just upstream of the consumer's system. Another indication of the source is that if no other home in the area has, or has recently had, this problem, it almost certainly is coming from the consumer's plumbing.

The operator may also receive complaints that some persons in a home are becoming ill, with this emphatically attributed to the water they are drinking. These complaints should be thoroughly investigated by the utility agency and also referred to the health department. Sometimes the consumer's doctor says that the water supply is at fault (or so the consumer claims). If this takes place, the doctor should be referred to the health officer. Often, a bacteriological sample is collected directly from the consumer's home. If water quality test results are negative and the customer's neighbors and the rest of the community have no problems, it is very doubtful that the utility's supply is the culprit.

To record all the necessary information, many agencies use a complaint form such as the one shown in Figure 5.61.

Some utilities go directly to their customers to find out how the customers feel about the system and its operation. Figure 5.62 was prepared as part of an overall plan of utility self-evaluation, consumer opinion, follow-up with consumer, and development of a capital construction improvement program. The questionnaire was designed to ask questions a layperson could understand and was successfully field tested in two water utilities. From the responses received, problems can be identified and priorities can be designated as to which problems can be solved soon and which must await solution at some later date.

For an excellent discussion on how to solve problems causing complaints, see Section 10.111, "Investigating Complaints," in *WATER TREATMENT PLANT OPERATION*, Volume I, of this series of manuals. The section describes the more common types of consumer complaints, investigation procedures, and the possible causes for complaints.

WESTERN MUNICIPAL WATER DISTRICT
WATER SYSTEM COMPLAINT

AREA: _____   DATE: _____

FROM: _____   TAKEN BY: _____

_____   ☐ WMWD OFFICE

_____   ☐ OPERATIONS CENTER

TELEPHONE _____   ☐ OTHER: _____

COMPLAINT: ☐ TURBIDITY   ☐ LOW PRESSURE

☐ TASTE   ☐ ORGANISMS   ☐ HIGH PRESSURE

☐ ODOR   ☐ NOISE   ☐ OTHER: _____

☐ COLOR   ☐ NO WATER

INVESTIGATION: (LIST ALL APPLICABLE ITEMS)

CAUSE: ☐ PUMP FAILURE   ☐ CUSTOMER LINE STOPPAGE

☐ EMERGENCY   ☐ DEAD-END MAIN   ☐ PIPE TOO SMALL

☐ WATER OUTAGE   ☐ WATER SURGES   ☐ OTHER: _____

☐ POWER OUTAGE   ☐ REGULATOR FAILED   _____

☐ POWER OUTAGE   ☐ REGULATOR FAILED

EXPLAIN: _____

_____

_____

_____

_____

CORRECTION: ☐ UNDER INVESTIGATION

☐ NO ACTION REQ'D.   ☐ PLANNED SHUTDOWN

☐ CONTROLS REPAIRED   ☐ CUSTOMER NOT NOTIFIED

☐ CUSTOMER TO CORRECT   ☐ OTHER: _____

☐ MAIN FLUSHED

EXPLAIN: _____

_____

_____

_____

ADDITIONAL INFORMATION: _____

_____

_____

_____

INVESTIGATORS: _____   DATE: _____

DISTRIBUTION: ORIGINAL TO OPERATIONS CENTER
            COPY TO CHIEF ENGINEER

Fig. 5.61   Sample water system complaint and investigation form

SAMPLE QUESTIONNAIRE

(A low-cost format for general use by systems of all sizes)

Please rate us. Where you rate us low we'll try to improve. Or explain why we don't think we can. Please return this rating form when you pay your bill.

1. Do you think you have good water service generally? (circle one)

    Always   Usually   Half the time   Occasionally   Never

2. Is your water pressure too high?

    Always   Usually   Half the time   Occasionally   Never

3. Is your water pressure too low?

    Always   Usually   Half the time   Occasionally   Never

4. Is your water dirty or off-color?

    Always   Usually   Half the time   Occasionally   Never

5. Do you avoid sprinkling in summer because water costs so much?

    Always   Usually   Half the time   Occasionally   Never

6. Have you had to complain about your service in the last two years?

    Yes   Can't recall   No

7. Do you know who to call if you have a complaint?   Yes   No

8. Are you treated courteously when you talk to water utility people?

    Always   Usually   Half the time   Occasionally   Never

9. Do you get prompt action when you ask them for something?

    Always   Usually   Half the time   Occasionally   Never

10. Is your water bill too high for the service you get?

    Always   Usually   Half the time   Occasionally   Never

11. Please tell us on the back of this sheet if you have a specific complaint or comment.

Thanks for your help. Do NOT sign your name unless you have a complaint and want someone to call you.

We'll let you know the outcome of this survey.

Fig. 5.62   Water utility evaluation form

(Source: *THE SAFE DRINKING WATER HANDBOOK FOR WATER SYSTEM OPERATORS*, prepared by the American Water Works Association)

For additional information about setting up and administering a public relations program for a water agency, see Chapter 8, Section 8.7, "Public Relations."

QUESTIONS

Write your answers in a notebook and then compare your answers with those on page 291.

5.13A   What kind of an image should an operator present to the public?

5.13B   How would you attempt to determine the cause of a customer's complaint regarding low water pressure?

5.14   LANDSCAPING

5.140   Purpose of Landscaping

Landscaping enhances the appearance of water supply and treatment facility sites. When people drive by well pumping sites, water storage facilities, and booster pumping stations, many may have preconceived thoughts. A well-manicured lawn and trimmed shrubs, along with a variety of ground cover, will provide a positive "first impression." Take pride in your facilities and it will be apparent to all who drive by or visit your facilities.

When facilities are designed and built, landscaping is a very important part of the construction. Usually landscaping is well done by professionals. Our responsibility is to maintain that investment and appearance.

The following paragraphs will shed light on how to best achieve adequate landscape appearance and yard maintenance.

5.141   Irrigating

If your facility is one of many that enjoy a beautiful green lawn, you know the importance of proper watering and water coverage. If your sprinkler system was installed by a competent private contractor, chances are it was engineered to provide adequate lawn coverage. A good indication that coverage is inadequate is a nonuniform appearance of the lawn. The sprinkler drops the greatest amount of water near the head and very little at its outer limits. The small amount of water doesn't penetrate very deep and the grass roots are very shallow. The weak grass allows for weeds, such as dandelion, sorrel, and plantain, that can take less water than typical lawn grasses. On the other hand, spots that are subject to too much water invite sedge, crabgrass, chickweed, and annual bluegrass. The best answer to unequal water distribution is to alternate the sprinklers and wetting pattern.

When and how often do you water? Grass shows its need for water first by loss of resilience. When you walk on grass needing water, there is no spring back. Next, the color chang-

es from a fresh green to a dull gray-green. The grass tops turn brown and die in the next stage. Ideally you should water before the no-spring back sign begins. After living with a lawn for a while you can sense that timing. Deep watering once a week is recommended. This causes plants to root deeper. As water is drawn from the soil by the grass roots, more air enters the ground and a better growth environment is created. Long slow watering will allow for good penetration. A three-hour application is quite common. Depending upon the type of soil you have, water penetration will be approximately one inch (25 mm) during the proper watering period.

Some of the reasons for not using light, frequent waterings are:

1. Greater water use,

2. Shallow rooting of grass,

3. Encouragement of shallow-rooted weeds,

4. Greater soil compaction, and

5. Rapid buildup of salinity due to lack of leaching.

5.142   Controlling Weeds

If we always grew only what we sowed originally in a lawn, the task of keeping a lawn would be easier. The trouble is that soils are generally full of dormant weed seeds waiting for the right conditions to germinate. Weed seed is also present in lawn seed mixtures to a small degree. Wind and birds help bring in new seeds. We will always have weeds. The question is how to get rid of them. There is a wide variety of chemical weed killers (herbicides) available. Most herbicides are selective and control a specific type of weed. They fall into two classes, broadleaf and grassy.

The term "broadleaf" is used to describe non-grassy weeds. A few of the familiar ones are: dandelion, curly dock, chickweed, bur clover, oxalis, knotweed, and English daisy. The hormone-type weed killers such as 2,4D and 2,4,5-TP are the best known of the herbicides. They kill broadleaf plants by speeding growth so the plants literally grow themselves to death. Depending on the species, weeds may require one to four treatments to obtain a good kill. This type of chemical is absorbed by the leaves and carried through the system to the roots. Both tops and roots are killed. Weeds treated with chemicals show curled and twisted stems in the first stage. Finally, the roots are expanded and ruptured and the entire plant dies. Select a weed killer that kills the entire plant. Burning the leaf may indicate a kill, but a vigorous root system could survive and cause regrowth.

The grassy weeds are also all too familiar: Bermuda grass, quack grass, nutgrass, velvet grass, and orchard grass. One thing to remember if you are applying a chemical to control grassy weeds: it can also kill the lawn grasses. Therefore, it is best to do controlled spot killing in the affected areas. Spot application of dalapon or amitrol will generally take care of local grasses. Be sure to follow label directions closely. If your lawn is overcome by grassy weeds, it may be necessary to use a soil fumigant. The fumigant chemically kills almost everything in the soil and leaves it sterile. Three commonly used fumigants are Vapam, Mylone, and calcium cyanamid. If applying the chemical yourself, follow directions closely.

If you are faced with a large weed-control problem, hire a professional. There are commercial companies that are knowledgeable and capable of handling your problem. When faced with weeds that grow along chain-link fences or soil around buildings, a light application of diesel fuel will do the trick. All you need is a handheld pressure sprayer to ensure adequate coverage. Do your spraying on calm days without wind. The objective is to kill only the selected weeds and grasses. When the wind kicks up, discontinue application until the wind dies down again. If using diesel fuel be sure to consider the potential fire hazard as well as any possible storm water runoff and long-term cumulative contamination.

5.143 Fertilizing

Grass has a continual need for nitrogen. Nature does a poor job of supplying that nutrient in quantities necessary to keep plants growing thick, green, and dense. If a lawn turns yellow or pale green or if grass becomes thin and weeds come in, it is probably an indication that nitrogen is needed. If you apply a fertilizer containing nitrogen and the lawn does not respond, then suspect other trouble such as disease, lack of air, or grubs and other larvae.

When you purchase a fertilizer, the three-number formula such as 6-4-2 indicates first the percentage of nitrogen, second the percentage of phosphorus, and third the percentage of potassium. Consider the phosphorus and potassium as a bonus to lawn feeding. The ever-important nitrogen comes in several forms. Some types are slow acting and may take a number of weeks or months for the nitrogen to be fully used by the lawn. Other forms are fast acting and immediate results can be seen in just a few days. When reading labels, remember not all organics are slow acting and not all inorganics are fast acting. The rate of availability grades down from immediate to gradual over a period of months.

Here are the meanings of the words on the fertilizer label that refer to nitrogen:

NITRATE. The form of nitrogen that is available to the plant as is, regardless of temperatures.

AMMONICAL OR AMMONIC. Available to plants when converted by bacteria to nitrate. Speed of conversion depends upon soil temperatures.

ORGANIC. Describes sludge, cottonseed meal, and any other materials that must first be broken down by bacteria.

UREA. A synthetic organic that water and the enzyme urease change immediately to inorganic ammonia. The ammonia is then converted to nitrate.

UREA FORM OF UREA FORMALDEHYDE. A nitrogen fertilizer that has been specially compounded for slow release.

How often do you feed a lawn? The color of the grass can show nitrogen need. An easy method used to determine need is the number of times per week or month that the lawn needs cutting. At the start, if cuttings were done every five to eight days and then extended to eight to fourteen days, more fertilizer is needed. Depending on the fertility of the lawn soil, grass will show a loss of color in four to eight weeks after fertilization. Thus, the feed-when-it-needs-it method could require fertilizer from three to twelve times a year. If fertilizing is a problem, then feed the lawn adequately in the spring and fall. This procedure may not be quite as effective as desirable, but the lawn should do quite well.

How much do I feed the lawn? A rule of thumb is one pound (0.5 kg) actual nitrogen for every 1,000 square feet (100 square meters) of lawn per month (this applies only to the growing months). To figure actual nitrogen needs, take the percentage stated on the label times the weight of the fertilizer. For example a 20-pound (10-kg) bag (it could be liquid) containing 15 percent nitrogen will yield approximately 3 pounds (1.5 kg) of nitrogen or enough to feed 3,000 square feet (300 sq m) for a month (20 pounds (10 kg) times 0.15 equals 3.0 pounds (1.5 kg) nitrogen).

Fertilizer should be distributed as evenly as possible. The most effective applicator is a hopper spreader. Four common mistakes when using the spreader are:

1. You cannot turn an open hopper inside a previous turn. The square corner will allow for excessive fertilizer application at that point.

2. You cannot stop and start with the hopper open. Walking at an unsteady pace will do the same thing.

3. Do not overlap with fertilizer.

4. Do not make 180-degree turns. Since one wheel is still driving the sifter, dispersal on the turn is very erratic.

After applying fertilizer, water the lawn heavily. Most fertilizers require a soaking to get the action started. This also prevents burning the grass blades.

5.144 Mowing and Pruning

Mowing and pruning are very important. Lawns must be mowed and edged at regular intervals to maintain their appearance. Trees and shrubs should be kept pruned so they won't interfere with the operation and maintenance of the plant, as well as to appear neat.

5.145 Surface Water Drainage

Depending on the location of your plant, drainage of storm water runoff and other surface water runoff can be very important. No one wants unwanted water running into buildings or flooding roadways and walkways. Provisions to handle this water should have been made when the plant was designed. Maintenance of these facilities requires keeping all curb drains and curb inlets free of leaves and other debris. Also, the storm drain lines must be clear and ready to handle any runoff.

Sump pumps in the drainage system should be included in a regular preventive maintenance program, depending on the type, size, and frequency of use of the pumps.

QUESTIONS

Write your answers in a notebook and then compare your answers with those on page 291.

5.14A How do weed seeds get in a lawn?

5.14B How could you tell if a lawn needs nitrogen?

5.15 ARITHMETIC ASSIGNMENT

There are four additional types of distribution system problems that have not been covered in the previous material in this manual. These problems are encountered by operators in the operation and maintenance of distribution systems and you should know how to work them.

FORMULAS

1. Trench Excavation

   When excavating a trench for a pipeline, you need to be able to calculate the volume of soil removed in cubic yards.

   Volume, cu ft = (Length, ft)(Width, ft)(Depth, ft)

   and

   $$\text{Volume, cu yd} = \frac{\text{Volume, cu ft}}{27 \text{ cu ft/cu yd}}$$

2. Volume of Flow

   If you are given a flowmeter reading in gallons per minute (GPM), you can calculate the total flow volume if you know the length of time.

   Flow Volume, gal = (Flow Rate, gal/min)(Time, min)

   If you are given the volume of a tank and the flow rate in gallons per minute (GPM), you can calculate the time required to fill a tank.

   $$\text{Time Required, min} = \frac{\text{Tank Volume, gal}}{\text{Flow Rate, gal/min}}$$

3. Reservoir Storage

   Earlier in this manual we calculated the volume of water in a tank or reservoir using the following formulas:

   a. Rectangular

      Volume, cu ft = (Length, ft)(Width, ft)(Depth, ft)

   b. Cylindrical

      Volume, cu ft = (0.785)(Diameter, ft)^2(Depth, ft)

   To convert the volume from cubic feet to gallons, we multiplied by 7.48 gal/cu ft.

   Volume, gal = (Volume, cu ft)(7.48 gal/cu ft)

   Many operators calculate the volume of water in gallons per each foot of water depth in their storage reservoirs. With this constant number, they measure the depth of water in feet and multiply this depth by the constant in gallons per foot to determine the volume in gallons.

For a cylindrical tank,

Volume, gal = (Area, sq ft)(Depth, ft)(7.48 gal/cu ft)

The area times 7.48 gal/cu ft will give us a constant and then we multiply this constant by the depth to obtain the volume in gallons. With this constant we can also develop an elevation (depth) versus storage curve so that for any depth of water we obtain the storage or volume in gallons from the curve.

4. Measurement of Flow from a Hydrant (Figure 5.63)

   When flushing water mains, the flow can be estimated by measuring the pipe diameter (D) or diameter (D) of the water jet flowing out of the hydrant, the distance or height (H) the water drops down to the ground and the distance (L) the water shoots out from the hydrant before hitting the ground. All three of the values must be measured in inches.

   $$\text{Flow, GPM} = \frac{(2.83)(\text{Diameter, in})^2(\text{Length, in})}{\sqrt{\text{Height, in}}}$$

EXAMPLE 11

A trench 3 feet wide, 1,400 feet long, and 8 feet deep is to be excavated for a water main. How many cubic feet and also cubic yards of soil have to be excavated for the water main?

Known	Unknown
Length, ft = 1,400 ft	1. Volume Excavated, cu ft
Width, ft = 3 ft	2. Volume Excavated, cu yd
Depth, ft = 8 ft	

1. Calculate the volume excavated in cubic feet.

   Volume, cu ft = (Length, ft)(Width, ft)(Depth, ft)

   = (1,400 ft)(3 ft)(8 ft)

   = 33,600 cu ft

2. Calculate the volume excavated in cubic yards.

   $$\text{Volume, cu yd} = \frac{\text{Volume, cu ft}}{27 \text{ cu ft/cu yd}}$$

   $$= \frac{33,600 \text{ cu ft}}{27 \text{ cu ft/cu yd}}$$

   = 1,244 cu yd

EXAMPLE 12

A flowmeter indicates a flow rate of two gallons per minute. How many gallons of water will flow through the meter in two hours?

Known	Unknown
Flow Rate, GPM = 2 GPM	Flow Volume, gal
Time, hr = 2 hr	

Calculate the total volume of water in gallons.

Flow Volume, gal = (Flow Rate, gal/min)(Time, hr)(60 min/hr)

= (2 gal/min)(2 hr)(60 min/hr)

= 240 gal

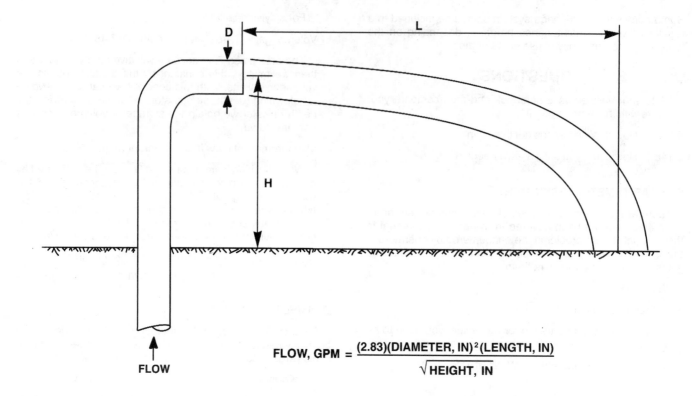

$$\text{FLOW, GPM} = \frac{(2.83)(\text{DIAMETER, IN})^2(\text{LENGTH, IN})}{\sqrt{\text{HEIGHT, IN}}}$$

Fig. 5.63   Measurements for estimating flow

EXAMPLE 13

A chemical solution tank is 4 feet in diameter and 5 feet deep. How long will it take to fill this tank when water flows in at a rate of 2 gallons per minute?

Known	**Unknown**
Diameter, ft = 4 ft	Time to Fill, min
Depth, ft = 5 ft	
Flow Rate, GPM = 2 GPM	

1. Calculate the tank volume in gallons.

$$\begin{aligned}\text{Tank Volume,} \atop \text{gal} &= (0.785)(\text{Diameter, ft})^2(\text{Depth, ft})(7.48 \text{ gal/cu ft}) \\ &= (0.785)(4 \text{ ft})^2(5 \text{ ft})(7.48 \text{ gal/cu ft}) \\ &= 470 \text{ gal}\end{aligned}$$

2. Calculate the time to fill the tank in minutes.

$$\begin{aligned}\text{Time to Fill, min} &= \frac{\text{Tank Volume, gal}}{\text{Flow Rate, gal/min}} \\ &= \frac{470 \text{ gal}}{2 \text{ gal/min}} \\ &= 235 \text{ min}\end{aligned}$$

3. Convert 235 minutes to hours and minutes.

$$\begin{aligned}\text{Time to Fill} &= \frac{235 \text{ min}}{60 \text{ min/hr}} \\ &= 3.917 \text{ hr} \\ &= 3 \text{ hr} + (0.917 \text{ hr})(60 \text{ min/hr}) \\ &= 3 \text{ hr} + 55 \text{ min}\end{aligned}$$

EXAMPLE 14

A cylindrical water storage tank is 10 feet in diameter and 20 feet high. Calculate the gallons of water per foot of depth and then the total gallons when the water is 5 feet deep.

Known	**Unknown**
Diameter, ft = 10 ft	1. Volume, gal/ft
Height, ft = 20 ft	2. Volume, gal
Depth, ft = 5 ft	(for Depth = 5 ft)

1. Calculate the volume of water in the tank per foot of depth.

$$\begin{aligned}\text{Volume, gal} &= (\text{Area, sq ft})(\text{Depth, ft})(7.48 \text{ gal/cu ft}) \\ \text{Volume, gal/ft} &= (0.785)(\text{Diameter, ft})^2(7.48 \text{ gal/cu ft}) \\ &= (0.785)(10 \text{ ft})^2(7.48 \text{ gal/cu ft}) \\ &= 587 \text{ gal/ft}\end{aligned}$$

2. Calculate the volume of water in the tank in gallons when the water is 5 feet deep in the tank.

Volume, gal = (Volume, gal/ft)(Depth, ft)

$$= (587 \text{ gal/ft})(5 \text{ ft})$$

$$= 2,935 \text{ gal}$$

EXAMPLE 15

Using the 10-foot diameter tank that is 20 feet high from the previous example, and the constant, 587 gal/ft, prepare an elevation (depth) versus storage curve for the tank.

Known		Unknown
Diameter, ft	= 10 ft	Depth versus storage
Height, ft	= 20 ft	
Constant, gal/ft	= 587 gal/ft	

To prepare the depth versus storage curve, determine the storage in gallons for three or more water depths. In this example we will use depths of 5 ft, 10 ft, 15 ft, and 20 ft.

Volume, gal = (Constant, gal/ft)(Depth, ft)

For 5 ft

Volume, gal = (587 gal/ft)(5 ft)

$$= 2,935 \text{ gal}$$

For 10 ft

Volume, gal = (587 gal/ft)(10 ft)

$$= 5,870 \text{ gal}$$

For 15 ft

Volume, gal = (587 gal/ft)(15 ft)

$$= 8,805 \text{ gal}$$

For 20 ft

Volume, gal = (587 gal/ft)(20 ft)

$$= 11,740 \text{ gal}$$

SUMMARY OF RESULTS

Depth, ft	Storage, gal
5	2,935
10	5,870
15	8,805
20	11,740

To prepare depth versus storage curve, plot the summary of results as shown in Figure 5.64. Draw a line of best fit (a straight line in this case) through the plotted points. If we measure a depth of water of 12 feet, we could go to the curve and obtain a storage volume of 7,050 gallons. The larger the curve, the more accurate the readings that can be obtained.

EXAMPLE 16

Estimate the flow from a hydrant if the water is flowing out a pipe 2 inches in diameter. The water drops 36 inches before hitting the ground a distance of 60 inches away from the hydrant (see Figure 5.63).

Known		Unknown
Diameter, in	= 2 in	Flow, GPM
Height, in	= 36 in	
Length, in	= 60 in	

Estimate the flow from the hydrant in gallons per minute.

$$\text{Flow, GPM} = \frac{(2.83)(\text{Diameter, in})^2(\text{Length, in})}{\sqrt{\text{Height, in}}}$$

$$= \frac{(2.83)(2 \text{ in})^2(60 \text{ in})}{\sqrt{36 \text{ in}}}$$

$$= \frac{(2.83)(4)(60)}{6}$$

$$= 113 \text{ GPM}$$

Turn to the Appendix, "How to Solve Water Distribution System Arithmetic Problems," at the back of this manual and read all of Section A.8, *PUMPS*.

In Section A.13, *TYPICAL WATER DISTRIBUTION SYSTEM PROBLEMS (ENGLISH SYSTEM)*, read and work the problems in Section A.133, Distribution System Operation and Maintenance.

Check all of the arithmetic in these two sections on an electronic pocket calculator. You should be able to get the same answers.

5.16   ADDITIONAL READING

1. *NEW YORK MANUAL*, Chapter 3,* "Hydraulics and Electricity," and Chapter 17,* "Protection of Treated Water."

2. *TEXAS MANUAL*, Chapter 15,* "The Distribution System," and Chapter 17,* "Customer Meters."

3. *WATER DISTRIBUTION OPERATOR TRAINING HANDBOOK*. Obtain from American Water Works Association (AWWA), Bookstore, 6666 West Quincy Avenue, Denver, CO 80235. Order No. 20428. ISBN 1-58321-014-8. Price to members, $52.00; nonmembers, $76.00; price includes cost of shipping and handling.

4. *WATER TRANSMISSION AND DISTRIBUTION*. Obtain from American Water Works Association (AWWA), Bookstore, 6666 West Quincy Avenue, Denver, CO 80235. Order No. 1957. ISBN 1-58321-231-0. Price to members, $87.00; nonmembers, $128.00; price includes cost of shipping and handling.

5. *CROSS-CONNECTION CONTROL MANUAL*. Obtain from National Technical Information Service (NTIS), 5285 Port Royal Road, Springfield, VA 22161. Stock No. PB91-145490. EPA No. 570-9-89-007. Price, $33.50, plus $5.00 shipping and handling per order.

6. *WATER AUDITS AND LEAK DETECTION* (M36). Obtain from American Water Works Association (AWWA), Bookstore, 6666 West Quincy Avenue, Denver, CO 80235. Order No. 30036. ISBN 1-58321-018-0. Price to members, $64.00; nonmembers, $94.00; price includes cost of shipping and handling.

7. *RECOMMENDED PRACTICE FOR BACKFLOW PREVENTION AND CROSS-CONNECTION CONTROL* (M14). Obtain from American Water Works Association (AWWA), Bookstore, 6666 West Quincy Avenue, Denver, CO 80235. Order No. 30014. ISBN 1-58321-288-4. Price to members, $69.00; nonmembers, $99.00; price includes cost of shipping and handling.

* Depends on edition.

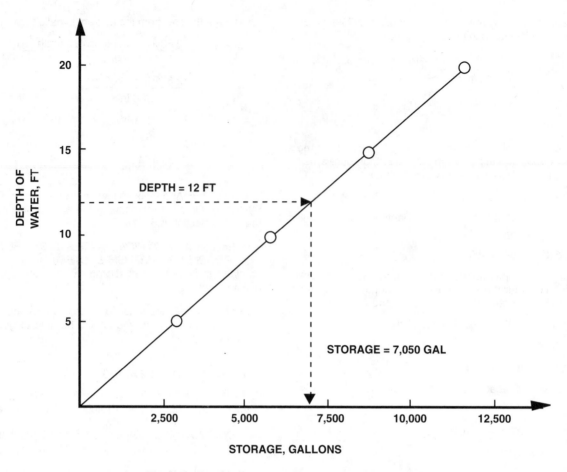

Fig. 5.64 Depth versus storage curve for Example 15

5.17 ACKNOWLEDGMENTS

The authors of these three chapters (3, 4, and 5) on distribution systems wish to thank the American Water Works Association for graciously granting permission to reproduce material from the many excellent AWWA publications, including the operator publication *OPFLOW*. Also thanks are given to the many manufacturers who granted us permission to use their material.

End of Lesson 4 of 4 Lessons on DISTRIBUTION SYSTEM OPERATION and MAINTENANCE

Please answer the discussion and review questions next.

DISCUSSION AND REVIEW QUESTIONS

Chapter 5.   DISTRIBUTION SYSTEM OPERATION AND MAINTENANCE

(Lesson 4 of 4 Lessons)

Write the answers to these questions in your notebook. The question numbering continues from Lesson 3.

23. When should distribution system mains and reservoirs be disinfected?

24. Why is the slug method of chlorine disinfection advantageous for use with long, large-diameter mains?

25. What factors would you consider when selecting a method for using chlorine to disinfect a storage facility?

26. Why should good records be prepared and kept?

27. Why should utilities have a work order system?

28. What are the prime responsibilities of a water utility during a disaster?

29. How should complaints be handled?

30. How would you respond to customers complaints that the water they are drinking is making them ill?

SUGGESTED ANSWERS

Chapter 5.   DISTRIBUTION SYSTEM OPERATION AND MAINTENANCE

ANSWERS TO QUESTIONS IN LESSON 1

Answers to questions on page 186.

5.0A   Water suppliers meet their *LEGAL* responsibility to deliver safe water to their customers through the conscientious operation and maintenance of the system facilities.

5.0B   The objective of distribution system maintenance is the maximum continuous service of the equipment in the system at the lowest possible cost to the utility.

Answers to questions on page 187.

5.1A   Water distribution system surveillance elements that should be considered include vandalism, storm damage, reservoirs, underground and above-ground storage tanks, pump stations, and valves.

5.1B   Critical parts of a distribution system should be patrolled on a routine basis to provide the utility with an early warning of any adverse conditions that might appear.

5.1C   Daily surveillance of a reservoir should include a check to see if there has been any trespassing, vandalism, dumping of trash, or swimming. Particular attention should be paid to fence and screen openings, roof damage, locks on reservoirs, and to manholes or doors. Reservoir roofs should be watertight and vent screens should cover openings.

5.1D   Weekly inspections of reservoirs should note any algae, slime and/or worm growths, floating or settled materials, or deterioration of roofing and wooden trusses.

Answers to questions on page 188.

5.1E   Field crews should be on the lookout for any unauthorized construction activity or soil erosion near the utility's pipelines because these activities could damage the water distribution system, especially buried pipes.

5.1F   Valves and valve boxes should be inspected annually.

5.1G   Air relief valves should be inspected after rains or floods to ensure that they are not submerged in drainage water and that they work properly.

Answers to questions on page 189.

5.1H   If vandalism or illegal entry is discovered, make a thorough investigation to determine what damage might have been done and whether there is any possibility of a threat to the quality of the water supply. Record the condition of all locks, any questionable conditions in the areas where unauthorized entry apparently took place, presence of hazardous material containers, missing items, and damaged equipment.

Report your findings to the appropriate authorities (supervisor and police).

5.1I   Water quality tests that can be quickly made are turbidity, chlorine residual, conductance (for TDS), pH, and alkalinity. Tests should also be done for coliforms.

5.1J   If an operator discovers a vandal in a water distribution system facility, the authorities (supervisor and police) should be immediately notified. The operator should not attempt to use force on a vandal because the operator could be injured.

5.1K   Information that can be telemetered includes water elevations in reservoirs or tanks, pump discharge rates, pressures in mains, and rate of flow. Water quality data can include residual chlorine and turbidity. Telemetering can also be used to start and stop pumps and to open and close valves.

Answers to questions on page 191.

5.2A   Requirements for total trihalomethanes apply to community water systems that chlorinate their water supplies and serve 10,000 or more people.

5.2B   National Secondary Drinking Water Regulations apply to contaminants that may adversely affect the aesthetic quality of the drinking water and result in consumer rejection of the water provided.

5.2C   More samples than the minimum required would be collected where sanitary hazards exist or when levels of contaminants approach or exceed maximum contaminant levels.

Answers to questions on page 193.

5.2D   Samples are routinely collected from water distribution systems and tested for (1) coliform bacteria, (2) turbidity, (3) total trihalomethanes, (4) chlorine residuals, (5) physical quality (tastes and odors), and (6) lead and copper.

5.2E   Utilities attempt to maintain chlorine residuals throughout the system to protect the public health, control coliform aftergrowths, and to serve as an indicator that the water is free from contamination.

5.2F   The sudden absence of a chlorine residual in a water distribution system should alert the operator to the possibility that a problem has arisen which needs prompt investigation. Possible action includes retesting for chlorine residual, checking chlorination equipment, and searching for a source of contamination.

5.2G   Lead and copper samples are to be collected as "first draw" samples from the cold water tap in either the kitchen or bathroom, or from a tap routinely used for consumption of water if in a building other than a home.

Answers to questions on page 194.

5.3A   Contamination through cross connections has consistently caused more waterborne disease outbreaks in the United States than any other reported factor.

5.3B   Responsibilities in the implementation of cross-connection control programs are shared by water suppliers, water users, health agencies, and plumbing officials.

5.3C   The water user's cross-connection control responsibility is to keep contaminants out of the potable water system on the user's premises. The user also has the responsibility to prevent the creation of cross connections through modifications of the plumbing system on the premises.

Answers to questions on page 197.

5.3D   Items that should be included on a cross-connection survey form include (1) location, (2) date, (3) names of investigator(s) and building representative(s), (4) water source(s), (5) piping system(s), (6) points of interconnection, (7) special equipment, and (8) remarks or recommendations.

5.3E   Factors that should be considered when selecting a backflow prevention device include the (1) degree of the health hazard, (2) probability of backflow occurring, and (3) complexity of the piping on the premises and the probability of its modification.

5.3F   Air gap or reduced-pressure principle backflow prevention devices are installed at wastewater treatment plants, wastewater pumping stations, reclaimed water reuse areas, areas where toxic substances in toxic concentrations are handled under pressure, premises having an auxiliary water supply that is or may be contaminated, and in locations where fertilizer, herbicides, or pesticides are injected into a sprinkler system.

Answers to questions on page 205.

5.3G   Typical industries that use hazardous chemicals include agriculture, cooling systems, dye plants, plating plants, and steam boiler plants.

5.3H   An air gap separation or a reduced-pressure principle backflow prevention assembly is recommended where there is a health hazard.

5.3I   Qualified persons should test backflow preventers for satisfactory performance.

Answers to questions on page 206.

5.4A   Distribution systems should always be under a continuous positive pressure at all times to meet the needs of the customers and to prevent contamination from backflow.

5.4B   Excessive distribution system pressures could cause damage to water-using facilities and unintentional higher water use.

5.4C   Possible causes of low water distribution system pressures include pipes being too small or velocities being too high as a result of excessive demands by consumers, fires, or leaks; improperly adjusted valves; and blocked regulator screens.

Answers to questions on page 206.

5.5A   The main function of a distribution storage reservoir is to meet daily water demands as well as peak demands.

5.5B   Distribution system reservoirs should be full at the start of the peak demand period.

Answers to questions on page 209.

5.6A   Pump alignment procedures are critical because there should be no strain on the pump case from the suction and discharge piping connections.

5.6B   Important pump inspection and preventive maintenance procedures include:

1. Observe and record pump pressures and output (flow), and current demands,
2. Check for excessive or abnormal noise, vibration, heat, and odor,
3. Provide necessary grease and oil lubrication,
4. Check bearing temperatures,
5. Listen for any bearing noise,
6. Tighten packing glands if they are leaking excessively,
7. Inspect the pump priming system,
8. Routinely operate internal combustion-driven pumps and generators on standby, and
9. Check pump alignment.

5.6C   Bearings should be checked for temperature, noise, and proper lubrication.

5.6D   New pump packing rings should be installed when the packing is worn or compressed so that the gland cannot be tightened farther.

ANSWERS TO QUESTIONS IN LESSON 2

Answers to questions on page 215.

5.70A   Types of pipe maintenance performed by distribution system operators include repairing, flushing, cleaning, disinfecting, and relining.

5.70B   Pipes deteriorate on the inside because of water corrosion and erosion and on the outside due to corrosion from aggressive soil moisture.

5.70C   Most leaks originate from a weakened joint or fitting connection or from a damaged or corroded part of a pipe.

5.70D   A probe is a sharp-pointed metal rod that is thrust into the ground and then pulled up for inspection. If the rod is moist or muddy, the line of the leak is being followed.

5.70E   Waste control studies are usually conducted when no specific reason can be found for a significant water loss in the system.

Answers to questions on page 216.

5.70F   Buried pipes may have to be located to find leaks, to make a new connection, or if an excavation must be made in the area to locate or install other facilities.

5.70G Buried nonmetallic pipes can be located by (1) driving stainless-steel-tipped shafts into the ground until they hit the pipe, and (2) placing a tracer tape which is detectable by electronic finders on top of the pipe before it is covered.

5.70H A leak in the wall of a pipe can be repaired by using a repair clamp. Repair clamps are short, cylindrical pieces usually made in two halves and bolted together or otherwise fastened around the leaking spot.

Answers to questions on page 223.

5.70I Wet tapping is a procedure for making a service connection while the water main is under pressure.

5.70J Pressure surveys are conducted in a water distribution system to indicate the hydraulic efficiency of the system in meeting normal requirements.

5.70K A pressure record covering a long period of time at one location can be obtained by using a recording pressure gage. The gage is connected to some type of recorder such as a strip chart.

Answers to questions on page 227.

5.70L The frequency of routine flushing can usually be determined by customer complaints and the types of material found during the flushing procedure.

5.70M Valves should be closed slowly during a flushing operation to prevent water hammer.

5.70N Information that should be recorded during a flushing operation includes date, time, location, pressure zone, size of main (diameter, length), flushing rate (GPM), flushing velocity (ft/sec), time required to clear, and description of the flushed water (appearance and odor).

Answers to questions on page 233.

5.70O Foam swabs are efficient in removing loose sediments, soft scales, and slimes from water mains.

5.70P The entry and exit points for foam swabs and pigs in a small water main are fire hydrants, air valves, blow-offs, wyes, and tees.

5.70Q The recommended speed for a swab cleaning a water main is 2 to 4 feet per second (0.6 to 1.2 m per sec).

5.70R After a water main is cleaned with a swab, the main should be flushed and disinfected (chlorinated). When the main is reactivated, service lines should be flushed and any temporary services removed.

Answers to questions on page 236.

5.70S Water mains are cement lined to prevent corrosion and tuberculation, to increase capacity, to improve water quality, and to reduce leakage through poor joints and holes.

5.70T Metal water mains must be inspected on the outside before considering cement-mortar lining to be sure outside corrosion is not so bad that the mains must be replaced.

5.70U The insides of water mains are inspected before a lining job by the use of a TV camera, by crawling through the main, or by being pulled through on a sled.

Answers to questions on page 245.

5.70V Freezing problems with service lines are usually greater than with mains because service lines have less ground cover over them and are smaller.

5.70W Frozen water mains can be thawed by the use of electrical thawing methods, hot water, or by blowing steam from a portable boiler into the frozen pipes.

5.70X When using electrical thawing methods, remove any electrical grounds that are connected to the pipe in that building.

ANSWERS TO QUESTIONS IN LESSON 3

Answers to questions on page 253.

5.71A Water utilities can avoid serious valve problems by a comprehensive program of valve inspection, exercising, and maintenance.

5.71B Operators must be able to find valves in order to shut off any valve at any time in case of a line break or other emergency.

5.71C Valves should be exercised to (1) determine whether or not the valve works, (2) extend valve life, and (3) help clean encrustations from the valve seats and gates.

5.71D A noisy or vibrating valve may be an indication that cavitation is occurring at a valve.

Answers to questions on page 254.

5.72A Hydrant suppliers can furnish operators with information on hydrant operation, maintenance, parts manuals, and solving particular problems.

5.72B Fire hydrants should be inspected twice a year and after each use.

5.72C Fire hydrant inspection and maintenance procedures include:

1. Inspect for leakage and make corrections when found,
2. Open hydrant fully, checking for ease of operation,
3. Flush hydrant,
4. Inspect, clean, and lubricate threads,
5. Replace caps, but not too tight,
6. Check for any exterior obstructions,
7. Check dry-barrel hydrants for drainage,
8. Clean exterior of hydrant and repaint if necessary
9. Be sure auxiliary valve is fully open,
10. Tag inoperable hydrants, and
11. Record inspection, maintenance, and repair work performed.

5.72D If a dry-barrel hydrant is operated when the valve is not fully opened, the drain will be partially open. Water will seep through the drain, saturate the drain field, and could result in hydrant damage due to freezing.

Answers to questions on page 257.

5.73A A meter testing program should be based on meter use, water quality, age of meter, and cost of testing. The potential revenue loss from inaccurate meters must also be considered.

5.73B Most modern water meters have sealed registers and easily changed measuring chambers. These meters are usually maintained and repaired by the utility.

5.73C Meter registers may be mechanically or magnetically driven.

Answers to questions on page 258.

5.73D Inaccurate meter registration results from wear, corrosion, and deposition.

5.73E Meter maintenance and repair consists essentially of dismantling and thoroughly cleaning the meter and inspecting all of its parts. The main procedures are:

1. Remove meter from service and take to shop,
2. Test meter for accuracy,
3. Dismantle the meter,
4. Inspect the various parts of the meter for excessive wear and corrosion, pitting or distortion,
5. Replace all defective and badly worn parts,
6. Thoroughly clean all parts to be used,
7. Ensure that the gear train (if any) runs freely,
8. Check the action of the disc in the chamber of the main casing before and after assembly,
9. Use a new meter as a comparison standard for tolerances and clearances, and
10. Retest the meter for accuracy.

5.73F The correct reading for the meter shown in Figure 5.47 is 3,697,491 cu ft.

Answers to questions on page 261.

5.74A Water main breaks can be caused by traffic accidents, earthquakes, washouts, water hammer, defective pipe, corrosion, and improper bedding.

5.74B After the valves are closed, the hole where the break is located can be drained by the use of a trash pump.

5.75A Coupons are steel specimens inserted in water mains to indicate the rate of corrosion of metal pipe. By weighing the specimens, the loss of weight over time is a measure of the corrosion rate.

5.76A The inhibitors used to control corrosion act by forming a protective coating over the site of corrosion activity, thus "inhibiting" corrosion.

ANSWERS TO QUESTIONS IN LESSON 4

Answers to questions on page 264.

5.8A Distribution system mains and reservoirs should be disinfected to prevent waterborne-disease outbreaks with the utility agency being legally liable.

5.8B The three forms of chlorine available for disinfecting a water main are (1) liquid chlorine, (2) sodium hypochlorite, and (3) calcium hypochlorite.

5.8C When using tablets to disinfect a water main, the chlorinated water should remain in the line for at least 24 hours.

5.8D When the disinfection procedures for a water main are completed, bacteriological samples are collected and tested for coliform bacteria.

Answers to questions on page 266.

5.8E Distribution system storage facilities should be disinfected before being placed in or returned to service, when they are newly installed, or when they are taken out of service for any reason and the circumstances are such that there could be contamination of water.

5.8F The interior surfaces of a storage facility are cleaned before disinfection by first removing all materials not belonging in the empty tank. Then all interior surfaces of the facility are cleaned thoroughly using a high-pressure water jet, scrubbing, sweeping, or some equivalent method.

5.8G The free chlorine residual in the water can be reduced to less than 2 mg/L by (1) completely draining and refilling the facility, (2) having additional holding time and blending with potable water having a low chlorine concentration, or (3) by the addition of reducing chemicals to neutralize the chlorine.

5.8H Reducing agents that could be used to neutralize a free chlorine residual include sulfur dioxide (SO_2), sodium bisulfite ($NaHSO_3$), sodium sulfite (Na_2SO_3), and sodium thiosulfate ($Na_2S_2O_3 \cdot 5\,H_2O$).

Answers to questions on page 271.

5.9A If good records are not kept, operators may not know the construction details of important facilities, where they are located, or what shape they are in. Also the utility may lose damage suits because of poor records.

5.9B The comprehensive map provides an overview of the entire distribution system.

5.9C Valve and hydrant maps show valve and hydrant locations, the direction to open the valve, the number of turns to open, the model type, and the installation date.

5.9D Whenever changes or modifications are made, the "as-built" plans (record drawings) should be changed to show the details of the modification, the date of the modification, and who recorded the modification on the plans.

Answers to questions on page 275.

5.10A SCADA stands for **S**upervisory **C**ontrol **A**nd **D**ata **A**cquisition system.

5.10B A SCADA system collects, stores, and analyzes information about all aspects of operation and maintenance, transmits alarm signals when necessary, and allows fingertip control of alarms, equipment, and processes.

Answers to questions on page 275.

5.11A Factors that should be considered when selecting tools and equipment for distribution system operation and maintenance include:

1. Cost and efficiency (including any reduction in labor costs),
2. Shorter service interruptions,
3. Ability to make the job easier for the operators (a morale factor), and
4. Versatility of tools or equipment for use in different types of operation and maintenance jobs.

5.11B Physical inventories should be made once or twice a year.

Answers to questions on page 277.

5.12A Utilities should develop plans for the following routine disruptions of service:

1. A key water main breaks,
2. An extended electrical power interruption occurs,
3. A vitally needed well pump motor or bearing "burns out,"
4. A key chlorinator or treatment unit fails, and
5. Test results or direct observations indicate the water supply has become contaminated.

5.12B An Emergency Operations Plan has been defined simply as: "Under disaster conditions, *WHO* does *WHAT* and *WHEN* with the existing resources."

5.12C The three basic steps of an Emergency Operations Plan are:

1. Priorities by which the requirements can be met are specified,
2. The best ways of using available resources are designated, and
3. Specific tasks are determined and are assigned to the assumed surviving or available personnel.

Answers to questions on page 281.

5.13A The operator should present to the public the image of a neat, efficient, knowledgeable, and cooperative individual.

5.13B To determine the cause of a customer's complaint regarding low pressures:

1. Take pressure readings in the customer's house without any flow (taps closed) so as to obtain the approximate system pressure, and
2. Turn on the taps in the house and take another pressure reading.

A drop in pressure readings could indicate plumbing problems in the house.

Answers to questions on page 283.

5.14A Weed seeds get into a lawn from the soil, lawn seeds, birds, and the wind.

5.14B A lawn needs nitrogen if it turns yellow or pale green or if the grass becomes thin and weeds start growing.

CHAPTER 6

DISINFECTION

by

Tom Ikesaki

NOTICE

Drinking water rules and regulations are continually changing. Two major new laws were signed by the president in December 1998: the Disinfectant/Disinfection By-Products (D/DBP) Rule and the Interim Enhanced Surface Water Treatment Rule (IESWTR). Several other drinking water laws are being developed and are expected to be signed into law over the next two or three years. The regulations described in this chapter are current in 2002 as this manual is being prepared for publication. Please see Section 6.01, "Safe Drinking Water Laws," for information about recently passed laws and anticipated future regulations.

Keep in contact with your state drinking water agency to obtain the rules and regulations that currently apply to your water distribution system agency. For additional information or answers to specific questions about the regulations, phone EPA's toll-free Safe Drinking Water Hotline at (800) 426-4791.

TABLE OF CONTENTS

Chapter 6.   DISINFECTION

OBJECTIVES

Chapter 6.   DISINFECTION

Following completion of Chapter 6, you should be able to:

1. Disinfect new and existing wells,

2. Disinfect pumps, mains, and storage facilities,

3. Calculate chlorine dosage,

4. Determine hypochlorinator and chlorinator settings,

5. Operate and maintain hypochlorinators,

6. Operate and maintain chlorinators,

7. Troubleshoot chlorination systems, and

8. Conduct a chlorine safety program.

WORDS

Chapter 6.   DISINFECTION

ACUTE HEALTH EFFECT   ACUTE HEALTH EFFECT

An adverse effect on a human or animal body, with symptoms developing rapidly.

AMPEROMETRIC (am-PURR-o-MET-rick)   AMPEROMETRIC

A method of measurement that records electric current flowing or generated, rather than recording voltage. Amperometric titration is a means of measuring concentrations of certain substances in water.

AMPEROMETRIC (am-PURR-o-MET-rick) TITRATION   AMPEROMETRIC TITRATION

A means of measuring concentrations of certain substances in water (such as strong oxidizers) based on the electric current that flows during a chemical reaction. Also see TITRATE.

BACTERIA (back-TEAR-e-ah)   BACTERIA

Bacteria are living organisms, microscopic in size, which usually consist of a single cell. Most bacteria use organic matter for their food and produce waste products as a result of their life processes.

BREAKPOINT CHLORINATION   BREAKPOINT CHLORINATION

Addition of chlorine to water until the chlorine demand has been satisfied. At this point, further additions of chlorine will result in a free chlorine residual that is directly proportional to the amount of chlorine added beyond the breakpoint.

BUFFER CAPACITY   BUFFER CAPACITY

A measure of the capacity of a solution or liquid to neutralize acids or bases. This is a measure of the capacity of water for offering a resistance to changes in pH.

CARCINOGEN (CAR-sin-o-JEN)   CARCINOGEN

Any substance which tends to produce cancer in an organism.

CHLORAMINES (KLOR-uh-means)   CHLORAMINES

Compounds formed by the reaction of hypochlorous acid (or aqueous chlorine) with ammonia.

CHLORINATION (KLOR-uh-NAY-shun)   CHLORINATION

The application of chlorine to water, generally for the purpose of disinfection, but frequently for accomplishing other biological or chemical results (aiding coagulation and controlling tastes and odors).

CHLORINE DEMAND   CHLORINE DEMAND

Chlorine demand is the difference between the amount of chlorine added to water and the amount of residual chlorine remaining after a given contact time. Chlorine demand may change with dosage, time, temperature, pH, and nature and amount of the impurities in the water.

Chlorine Demand, mg/*L* = Chlorine Applied, mg/*L* – Chlorine Residual, mg/*L*

CHLORINE REQUIREMENT   CHLORINE REQUIREMENT

The amount of chlorine which is needed for a particular purpose. Some reasons for adding chlorine are reducing the number of coliform bacteria (Most Probable Number), obtaining a particular chlorine residual, or oxidizing some substance in the water. In each case a definite dosage of chlorine will be necessary. This dosage is the chlorine requirement.

CHLOROPHENOLIC (klor-o-FEE-NO-lick)   CHLOROPHENOLIC

Chlorophenolic compounds are phenolic compounds (carbolic acid) combined with chlorine.

CHLORORGANIC (klor-or-GAN-ick) CHLORORGANIC

Organic compounds combined with chlorine. These compounds generally originate from, or are associated with, life processes such as those of algae in water.

CHRONIC HEALTH EFFECT CHRONIC HEALTH EFFECT

An adverse effect on a human or animal body with symptoms that develop slowly over a long period of time or that recur frequently.

COLIFORM (COAL-i-form) COLIFORM

A group of bacteria found in the intestines of warm-blooded animals (including humans) and also in plants, soil, air and water. Fecal coliforms are a specific class of bacteria which only inhabit the intestines of warm-blooded animals. The presence of coliform bacteria is an indication that the water is polluted and may contain pathogenic (disease-causing) organisms.

COLORIMETRIC MEASUREMENT COLORIMETRIC MEASUREMENT

A means of measuring unknown chemical concentrations in water by measuring a sample's color intensity. The specific color of the sample, developed by addition of chemical reagents, is measured with a photoelectric colorimeter or is compared with "color standards" using, or corresponding with, known concentrations of the chemical.

COMBINED AVAILABLE RESIDUAL CHLORINE COMBINED AVAILABLE RESIDUAL CHLORINE

The concentration of residual chlorine that is combined with ammonia, organic nitrogen, or both in water as a chloramine (or other chloro derivative) and yet is still available to oxidize organic matter and help kill bacteria.

COMBINED RESIDUAL CHLORINATION COMBINED RESIDUAL CHLORINATION

The application of chlorine to water to produce combined available chlorine residual. This residual can be made up of monochloramines, dichloramines, and nitrogen trichloride.

DPD (pronounce as separate letters) DPD

A method of measuring the chlorine residual in water. The residual may be determined by either titrating or comparing a developed color with color standards. DPD stands for N,N-diethyl-p-phenylene-diamine.

DISINFECTION (dis-in-FECT-shun) DISINFECTION

The process designed to kill or inactivate most microorganisms in water, including essentially all pathogenic (disease-causing) bacteria. There are several ways to disinfect, with chlorination being the most frequently used in water treatment. Compare with STERILIZATION.

EDUCTOR (e-DUCK-ter) EDUCTOR

A hydraulic device used to create a negative pressure (suction) by forcing a liquid through a restriction, such as a Venturi. An eductor or aspirator (the hydraulic device) may be used in the laboratory in place of a vacuum pump. As an injector, it is used to produce vacuum for chlorinators. Sometimes used instead of a suction pump.

EJECTOR EJECTOR

A device used to disperse a chemical solution into water being treated.

ENTERIC ENTERIC

Of intestinal origin, especially applied to wastes or bacteria.

ENZYMES (EN-zimes) ENZYMES

Organic substances (produced by living organisms) which cause or speed up chemical reactions. Organic catalysts and/or biochemical catalysts.

FREE AVAILABLE RESIDUAL CHLORINE FREE AVAILABLE RESIDUAL CHLORINE

That portion of the total available residual chlorine composed of dissolved chlorine gas (Cl_2), hypochlorous acid (HOCl), and/or hypochlorite ion (OCl^-) remaining in water after chlorination. This does not include chlorine that has combined with ammonia, nitrogen, or other compounds.

HTH (pronounce as separate letters) HTH

High **T**est **H**ypochlorite. Calcium hypochlorite or $Ca(OCl)_2$.

HEPATITIS (HEP-uh-TIE-tis) HEPATITIS

Hepatitis is an inflammation of the liver caused by an acute viral infection. Yellow jaundice is one symptom of hepatitis.

HYDROLYSIS (hi-DROLL-uh-sis)   HYDROLYSIS

(1)   A chemical reaction in which a compound is converted into another compound by taking up water.

(2)   Usually a chemical degradation of organic matter.

HYPOCHLORINATION (HI-poe-KLOR-uh-NAY-shun)   HYPOCHLORINATION

The application of hypochlorite compounds to water for the purpose of disinfection.

HYPOCHLORITE (HI-poe-KLOR-ite)   HYPOCHLORITE

Chemical compounds containing available chlorine; used for disinfection. They are available as liquids (bleach) or solids (powder, granules, and pellets) in barrels, drums, and cans. Salts of hypochlorous acid.

IDLH   IDLH

Immediately **D**angerous to **L**ife or **H**ealth. The atmospheric concentration of any toxic, corrosive, or asphyxiant substance that poses an immediate threat to life or would cause irreversible or delayed adverse health effects or would interfere with an individual's ability to escape from a dangerous atmosphere.

MPN (pronounce as separate letters)   MPN

MPN is the **M**ost **P**robable **N**umber of coliform-group organisms per unit volume of sample water. Expressed as a density or population of organisms per 100 mL of sample water.

NITROGENOUS (nye-TRAH-jen-us)   NITROGENOUS

A term used to describe chemical compounds (usually organic) containing nitrogen in combined forms. Proteins and nitrates are nitrogenous compounds.

ORTHOTOLIDINE (or-tho-TOL-uh-dine)   ORTHOTOLIDINE

Orthotolidine is a colorimetric indicator of chlorine residual. If chlorine is present, a yellow-colored compound is produced. This reagent is no longer approved for chemical analysis to determine chlorine residual.

OXIDATION (ox-uh-DAY-shun)   OXIDATION

Oxidation is the addition of oxygen, removal of hydrogen, or the removal of electrons from an element or compound. In the environment, organic matter is oxidized to more stable substances. The opposite of REDUCTION.

OXIDIZING AGENT   OXIDIZING AGENT

Any substance, such as oxygen (O_2) or chlorine (Cl_2), that will readily add (take on) electrons. The opposite is a REDUCING AGENT.

PALATABLE (PAL-uh-tuh-bull)   PALATABLE

Water at a desirable temperature that is free from objectionable tastes, odors, colors, and turbidity. Pleasing to the senses.

PATHOGENIC (PATH-o-JEN-ick) **ORGANISMS**   PATHOGENIC ORGANISMS

Organisms, including bacteria, viruses or cysts, capable of causing diseases (giardiasis, cryptosporidiosis, typhoid, cholera, dysentery) in a host (such as a person). There are many types of organisms which do *NOT* cause disease. These organisms are called non-pathogenic.

PHENOLIC (fee-NO-lick) **COMPOUNDS**   PHENOLIC COMPOUNDS

Organic compounds that are derivatives of benzene.

POSTCHLORINATION   POSTCHLORINATION

The addition of chlorine to the plant effluent, *FOLLOWING* plant treatment, for disinfection purposes.

POTABLE (POE-tuh-bull) **WATER**   POTABLE WATER

Water that does not contain objectionable pollution, contamination, minerals, or infective agents and is considered satisfactory for drinking.

PRECHLORINATION   PRECHLORINATION

The addition of chlorine at the headworks of the plant *PRIOR TO* other treatment processes mainly for disinfection and control of tastes, odors, and aquatic growths. Also applied to aid in coagulation and settling.

PRECURSOR, THM (pre-CURSE-or)

Natural organic compounds found in all surface and groundwaters. These compounds *MAY* react with halogens (such as chlorine) to form trihalomethanes (tri-HAL-o-METH-hanes) (THMs); they *MUST* be present in order for THMs to form.

REAGENT (re-A-gent)

A pure chemical substance that is used to make new products or is used in chemical tests to measure, detect, or examine other substances.

REDUCING AGENT

Any substance, such as base metal (iron) or the sulfide ion (S^{2-}), that will readily donate (give up) electrons. The opposite is an OXIDIZING AGENT.

REDUCTION (re-DUCK-shun)

Reduction is the addition of hydrogen, removal of oxygen, or the addition of electrons to an element or compound. Under anaerobic conditions (no dissolved oxygen present), sulfur compounds are reduced to odor-producing hydrogen sulfide (H_2S) and other compounds. The opposite of OXIDATION.

RELIQUEFACTION (re-LICK-we-FACK-shun)

The return of a gas to the liquid state; for example, a condensation of chlorine gas to return it to its liquid form by cooling.

RESIDUAL CHLORINE

The concentration of chlorine present in water after the chlorine demand has been satisfied. The concentration is expressed in terms of the total chlorine residual, which includes both the free and combined or chemically bound chlorine residuals.

ROTAMETER (RODE-uh-ME-ter)

A device used to measure the flow rate of gases and liquids. The gas or liquid being measured flows vertically up a tapered, calibrated tube. Inside the tube is a small ball or bullet-shaped float (it may rotate) that rises or falls depending on the flow rate. The flow rate may be read on a scale behind or on the tube by looking at the middle of the ball or at the widest part or top of the float.

SAPROPHYTES (SAP-row-FIGHTS)

Organisms living on dead or decaying organic matter. They help natural decomposition of organic matter in water.

STERILIZATION (STARE-uh-luh-ZAY-shun)

The removal or destruction of all microorganisms, including pathogenic and other bacteria, vegetative forms and spores. Compare with DISINFECTION.

TITRATE (TIE-trate)

To *TITRATE* a sample, a chemical solution of known strength is added drop by drop until a certain color change, precipitate, or pH change in the sample is observed (end point). Titration is the process of adding the chemical reagent in small increments (0.1 – 1.0 milliliter) until completion of the reaction, as signaled by the end point.

TOTAL CHLORINE RESIDUAL

The total amount of chlorine residual (value for residual chlorine, including both free chlorine and chemically bound chlorine) present in a water sample after a given contact time.

TRIHALOMETHANES (THMs) (tri-HAL-o-METH-hanes)

Derivatives of methane, CH_4, in which three halogen atoms (chlorine or bromine) are substituted for three of the hydrogen atoms. Often formed during chlorination by reactions with natural organic materials in the water. The resulting compounds (THMs) are suspected of causing cancer.

TURBIDITY (ter-BID-it-tee)

The cloudy appearance of water caused by the presence of suspended and colloidal matter. In the waterworks field, a turbidity measurement is used to indicate the clarity of water. Technically, turbidity is an optical property of the water based on the amount of light reflected by suspended particles. Turbidity cannot be directly equated to suspended solids because white particles reflect more light than dark-colored particles and many small particles will reflect more light than an equivalent large particle.

CHAPTER 6.   DISINFECTION

(Lesson 1 of 2 Lessons)

6.0   PURPOSE OF DISINFECTION

6.00   Making Water Safe for Consumption

Our single most important natural resource is water. Without water we could not exist. Unfortunately, safe water is becoming very difficult to find. In the past, safe water could be found in remote areas, but with population growth and related pollution of waters, there are very few natural waters left that are safe to drink without treatment of some kind.

Water is the universal solvent and therefore carries all types of dissolved materials. Water also carries biological life forms which can cause diseases. These waterborne *PATHOGENIC ORGANISMS*[1] are listed in Table 6.1 Most of these organisms and the diseases they transmit are no longer a problem in the United States due to proper water protection, treatment, and monitoring. However, many developing regions of the world still experience serious outbreaks of various waterborne diseases.

TABLE 6.1   PATHOGENIC ORGANISMS (DISEASES) TRANSMITTED BY WATER

Bacteria
Salmonella (salmonellosis)
Shigella (bacillary dysentery)
Bacillus typhosus (typhoid fever)
Salmonella paratyphi (paratyphoid)
Vibrio cholerae (cholera)

Viruses
Enterovirus
Poliovirus
Coxsackie Virus
Echo Virus
Andenovirus
Reovirus
Infectious Hepatitis

Intestinal Parasites
Entamoeba histolytica (amoebic dysentery)
Giardia lamblia (giardiasis)
Ascaris lumbricoides (giant roundworm)
Cryptosporidium (cryptosporidiosis)

One of the cleansing processes in the treatment of safe water is called disinfection. Disinfection is the selective destruction or inactivation of pathogenic organisms. Don't confuse disinfection with sterilization. Sterilization is the complete destruction of all organisms. Sterilization is not necessary in water treatment and is also quite expensive. (Also note that disinfection does not remove toxic chemicals which could make the water unsafe to drink.)

6.01   Safe Drinking Water Laws

In the United States, the U.S. Environmental Protection Agency is responsible for setting drinking water standards and for ensuring their enforcement. This agency sets federal regulations which all state and local agencies must enforce. The Safe Drinking Water Act (SDWA) and its amendments contain specific maximum allowable levels of substances known to be hazardous to human health. In addition to describing maximum contaminant levels (MCLs), these federal drinking water regulations also give detailed instructions on what to do when you exceed the maximum contaminant level for a particular substance.

In 1986, Congress amended the SDWA. The amendments set deadlines for the establishment of maximum contaminant levels (MCLs), placed greater emphasis on enforcement, authorized penalties for tampering with drinking water supplies, and mandated the complete elimination of lead from drinking water. In addition, the SDWA amendments placed considerable emphasis on the protection of underground drinking water sources.

The Surface Water Treatment Rule (SWTR) is an important set of drinking water regulations promulgated by the Environmental Protection Agency in 1989. The SWTR requires disinfection of all surface water supply systems as protection against exposure to viruses, bacteria, and *Giardia*. Table 6.2 shows an example of the regulations for *COLIFORM*[2] bacteria, which are supposed to be killed or inactivated by disinfection. Table 6.3 lists the number of coliform samples required based on the population served.

Drinking water regulations are constantly changing. The Interim Enhanced Surface Water Treatment Rule (IESWTR), the Long Term 1 Enhanced Surface Water Treatment Rule, and the Disinfection/Disinfection By-Products (D/DBP) Rule are now in effect. The goal of the IESWTR is to increase public protection from illness caused by the *Cryptosporidium* organism. The D/DBP Rule, which applies to water systems using a disinfectant during treatment, limits the amount of certain potentially harmful disinfection by-products that may remain in drinking water after treatment.

---

[1] Pathogenic (PATH-o-JEN-ick) Organisms.   Organisms, including bacteria, viruses or cysts, capable of causing diseases (giardiasis, cryptosporidiosis, typhoid, cholera, dysentery) in a host (such as a person). There are many types of organisms which do NOT cause disease. These organisms are called non-pathogenic.

[2] Coliform (COAL-i-form).   A group of bacteria found in the intestines of warm-blooded animals (including humans) and also in plants, soil, air and water. Fecal coliforms are a specific class of bacteria which only inhabit the intestines of warm-blooded animals. The presence of coliform bacteria is an indication that the water is polluted and may contain pathogenic (disease-causing) organisms.

TABLE 6.2   MICROBIOLOGICAL STANDARDS [a, b]

Maximum Contaminant Level Goal (MCLG):   zero

Maximum Contaminant Level (MCL):

1. Compliance based on presence/absence of total coliforms in sample, rather than an estimate of coliform density.

2. MCL for system analyzing at least 40 samples per month: no more than 5.0 percent of the monthly samples may be total coliform-positive.

3. MCL for systems analyzing fewer than 40 samples per month: no more than 1 sample per month may be total coliform-positive.

---

[a] See Chapter 22, "Drinking Water Regulations," *WATER TREATMENT PLANT OPERATION*, Volume II, and the poster provided with this manual for more details.
[b] See Chapter 11, "Laboratory Procedures," *WATER TREATMENT PLANT OPERATION*, Volume I, for details on how to do the coliform bacteria tests (membrane filter (MF), multiple tube fermentation (MPN), presence-absence, Colilert, and Colisure).

**MONITORING AND REPEAT SAMPLE FREQUENCY AFTER A
TOTAL COLIFORM-POSITIVE ROUTINE SAMPLE**

No. Routine Samples/Month	No. Repeat Samples [a]	No. Routine Samples Next Month [b]
1/mo or fewer	4	5/mo
2/mo	3	5/mo
3/mo	3	5/mo
4/mo	3	5/mo
5/mo or greater	3	Table 6.3

---

[a] Number of repeat samples in the same month for each total coliform-positive routine sample.
[b] Except where state has invalidated the original routine sample, or where state substitutes an on-site evaluation of the problem, or where the state waives the requirement case by case.

For more information about the requirements of the Safe Drinking Water Act (SDWA), refer to the poster that was included with this manual. Also see Chapter 22, "Drinking Water Regulations," in *WATER TREATMENT PLANT OPERATION*, Volume II, in this series of operator training manuals.

QUESTIONS

Write your answers in a notebook and then compare your answers with those on page 356.

6.0A   What are pathogenic organisms?

6.0B   What is disinfection?

6.0C   Drinking water standards are established by what agency of the United States government?

6.0D   MCL stands for what words?

6.1   FACTORS INFLUENCING DISINFECTION

6.10   pH

The pH of water being treated can alter the efficiency of disinfectants. Chlorine, for example, disinfects water much faster at a pH around 7.0 than at a pH over 8.0.

6.11   Temperature

Temperature conditions also influence the effectiveness of the disinfectant. The higher the temperature of the water, the more efficiently it can be treated. Water near 70 to 85°F (21 to 29°C) is easier to disinfect than water at 40 to 60°F (4 to 16°C). Longer contact times are required to disinfect water at lower temperatures. To speed up the process, operators often simply use larger amounts of chemicals. Where water is exposed to the atmosphere, the warmer the water temperature the greater the dissipation rate of chlorine into the atmosphere.

6.12   Turbidity

Under normal operating conditions, the turbidity level of water being treated is very low by the time the water reaches the disinfection process. Excessive turbidity will greatly reduce the efficiency of the disinfecting chemical or process. Studies in water treatment plants have shown that when water is filtered to a turbidity of one unit or less, most of the bacteria have been removed.

The suspended matter itself may also change the chemical nature of the water when the disinfectant is added. Some types of suspended solids can create a continuing demand for the chemical, thus changing the effective germicidal (germ killing) properties of the disinfectant.

**TABLE 6.3 TOTAL COLIFORM SAMPLING REQUIREMENTS
ACCORDING TO POPULATION SERVED**

Population Served		Minimum Number of Routine Samples Per Month[a]	Population Served		Minimum Number of Routine Samples Per Month[a]
25	to 1,000 [b]	1 [c]	59,001	to 70,000	70
1,001	to 2,500	2	70,001	to 83,000	80
2,501	to 3,300	3	83,001	to 96,000	90
3,301	to 4,100	4	96,001	to 130,000	100
4,101	to 4,900	5	130,001	to 220,000	120
4,901	to 5,800	6	220,001	to 320,000	150
5,801	to 6,700	7	320,001	to 450,000	180
6,701	to 7,600	8	450,001	to 600,000	210
7,601	to 8,500	9	600,001	to 780,000	240
8,501	to 12,900	10	780,001	to 970,000	270
12,901	to 17,200	15	970,001	to 1,230,000	300
17,201	to 21,500	20	1,230,001	to 1,520,000	330
21,501	to 25,000	25	1,520,001	to 1,850,000	360
25,001	to 33,000	30	1,850,001	to 2,270,000	390
33,001	to 41,000	40	2,270,001	to 3,020,000	420
41,001	to 50,000	50	3,020,001	to 3,960,000	450
50,001	to 59,000	60	3,960,001	or more	480

[a] A noncommunity water system using groundwater and serving 1,000 persons or fewer may monitor at a lesser frequency specified by the state until a sanitary survey is conducted and the state reviews the results. Thereafter, noncommunity water systems using groundwater and serving 1,000 persons or fewer must monitor in each calendar quarter during which the system provides water to the public, unless the state determines that some other frequency is more appropriate and notifies the system (in writing). In all cases, noncommunity water systems using groundwater and serving 1,000 persons or fewer must monitor at least once/year.

A noncommunity water system using surface water, or groundwater under the direct influence of surface water, regardless of the number of persons served, must monitor at the same frequency as a like-sized community public system. A noncommunity water system using groundwater and serving more than 1,000 persons during any month must monitor at the same frequency as a like-sized community water system, except that the state may reduce the monitoring frequency for any month the system serves 1,000 persons or fewer.

[b] Includes public water systems which have at least 15 service connections, but serve fewer than 25 persons.

[c] For a community water system serving 25 to 1,000 persons, the state may reduce this sampling frequency if a sanitary survey conducted in the last five years indicates that the water system is supplied solely by a protected groundwater source and is free of sanitary defects. However, in no case may the state reduce the sampling frequency to less than once/quarter.

6.120 Organic Matter

Organics found in the water can consume great amounts of disinfectants while forming unwanted compounds. *TRIHAL-OMETHANES*[3] are an example of undesirable compounds formed by reactions between chlorine and certain organics. Disinfecting chemicals often react with organics and *RE-DUCING AGENTS*[4] (Section 6.13). Then, if any of the chemical remains available after this initial reaction, it can act as an effective disinfectant. The reactions with organics and reducing agents, however, will have significantly reduced the amount of chemical available for this purpose.

6.121 Inorganic Matter

Inorganic compounds such as ammonia (NH_3) in the water being treated can create special problems. In the presence of ammonia, some oxidizing chemicals form side compounds causing a partial loss of disinfecting power. Silt can also create a chemical demand. It is clear, then, that the chemical properties of the water being treated can seriously interfere with the effectiveness of disinfecting chemicals.

6.13 Reducing Agents

Chlorine combines with a wide variety of materials, especially reducing agents. Most of the reactions are very rapid, while others are much slower. These side reactions complicate the use of chlorine for disinfection. The demand for chlorine by reducing agents must be satisfied before chlorine becomes available to accomplish disinfection. Examples of inorganic reducing agents present in water that will react with chlorine include hydrogen sulfide (H_2S), ferrous ion (Fe^{2+}), manganous ion (Mn^{2+}), and the nitrite ion (NO_2^-). Organic re-

---

[3] Trihalomethanes (THMs) (tri-HAL-o-METH-hanes). *Derivatives of methane, CH_4, in which three halogen atoms (chlorine or bromine) are substituted for three of the hydrogen atoms. Often formed during chlorination by reactions with natural organic materials in the water. The resulting compounds (THMs) are suspected of causing cancer.*

[4] Reducing Agent. *Any substance, such as base metal (iron) or the sulfide ion (S^{2-}), that will readily donate (give up) electrons. The opposite is an oxidizing agent.*

ducing agents in water also will react with chlorine and form chlorinated organic materials of potential health significance.

6.14 Microorganisms

6.140 Number and Types of Microorganisms

Microorganism concentration is important because the higher the number of microorganisms, the greater the demand for a disinfecting chemical. The resistance of microorganisms to specific disinfectants varies greatly. Non-spore-forming bacteria are generally less resistant than spore-forming bacteria. Cysts and viruses can be very resistant to certain types of disinfectants.

6.141 Removal Processes

Pathogenic organisms can be removed from water, killed, or inactivated by various physical and chemical water treatment processes. These processes are:

1. *COAGULATION.* Chemical coagulation followed by sedimentation and filtration will remove 90 to 95 percent of the pathogenic organisms, depending on which chemicals are used. Alum usage can increase virus removals up to 99 percent.

2. *SEDIMENTATION.* Properly designed sedimentation processes can effectively remove 20 to 70 percent of the pathogenic microorganisms. This removal is accomplished by allowing the pathogenic organisms (as well as nonpathogenic organisms) to settle out by gravity, assisted by chemical floc.

3. *FILTRATION.* Filtering water through granular filters is an effective means of removing pathogenic and other organisms from water. The removal rates vary from 20 to 99+ percent depending on the coarseness of the filter media and the type and effectiveness of pretreatment.

4. *DISINFECTION.* Disinfection chemicals such as chlorine are added to water to kill or inactivate pathogenic microorganisms.

The first three processes are covered in Chapter 4, "Small Water Treatment Plants," in the manual on *SMALL WATER SYSTEM OPERATION AND MAINTENANCE.* The fourth, disinfection, is the subject of this chapter.

NOTE: Drinking water regulations may require the calculation of log removals for inactivation of Giardia cysts, viruses, or particle counts. For two examples of how to make these calculations, see *WATER TREATMENT PLANT OPERATION*, Volume I, Arithmetic Appendix, Section A.17, "Calculation of Log Removals."

6.15 Disinfection Considerations

Several different types of disinfectants are used to disinfect water that is delivered to consumers through distribution systems. Free chlorine is the most commonly used distribution system disinfectant. Monochloramine is the second most common disinfectant and its use is becoming more frequent. Monochloramine residuals decrease more slowly in distribution systems than free chlorine. Chlorine dioxide is used in some systems, but its use is decreasing due to regulations. Mixed oxidants are produced on site by a new disinfection system that generates a combination of disinfectants—free chlorine, chlorine dioxide, and ozone. Ozone and ultraviolet (UV) systems are used to disinfect water at water treatment plants, but do not produce a disinfection residual in the water distributed by distribution systems.

Distribution system operators must maintain a sufficient and effective disinfectant residual in the distribution system to prevent contamination of the drinking water being distributed. Some possible sources of contamination include new main installations, cross connections, and main breaks. Microbial growths (biofilms) may develop in distribution systems as a result of contamination from any of these sources. Monochloramine is more effective than free chlorine in controlling biofilms. Also the absence of a chlorine residual is an indication that contamination could have entered the distribution system.

Monochloramine is a more stable disinfectant than free chlorine; therefore, it is easier to maintain a disinfectant residual in a distribution system using monochloramine. Also, lower total trihalomethane (TTHM) concentrations are produced when using monochloramine and chlorine dioxide than when using free chlorine. TTHMs will increase with contact time in the distribution system when using free chlorine, but TTHMs are stable (unaffected by contact time) when using monochloramine. One disadvantage of using monochloramine, however, is that monochloramine removal is required for persons undergoing dialysis treatment.

Disinfectants may be the cause of tastes and odors and other aesthetic considerations in distribution system drinking water. For example, free chlorine may produce a chlorinous taste and odor plus a by-product taste and odor. Dichloramine causes taste and odor problems and is a poor disinfectant. (The solution to dichloramine problems may be to increase the chlorine dose.) By-products from chlorine dioxide may produce undesirable odors. Free chlorine and chlorine dioxide may increase copper corrosion as compared to monochloramine. Monochloramine and chlorine dioxide have been implicated in the deterioration of some types of rubber products.

Ultimately the selection of a distribution system disinfectant is a complex process that is based on the water quality and the competing needs of microbial protection from a disinfectant, disinfectant by-product minimization, aesthetics, and economics. Every disinfectant has different advantages and limitations that need to be carefully monitored, controlled, and evaluated to produce the best possible water quality for consumers.

QUESTIONS

Write your answers in a notebook and then compare your answers with those on page 356.

6.1A How does pH influence the effectiveness of disinfection with chlorine?

6.1B How does temperature of the water influence disinfection?

6.1C What two factors influence the effectiveness of disinfection on microorganisms?

6.1D What are possible sources of drinking water contamination in distribution systems?

6.2 PROCESS OF DISINFECTION

6.20 Purpose of Process

The purpose of disinfection is to destroy harmful organisms. This can be accomplished either physically or chemically.

Physical methods may (1) physically remove the organisms from the water, or (2) introduce motion that will disrupt the cells' biological activity and kill or inactivate them.

Chemical methods alter the cell chemistry causing the microorganism to die. The most widely used disinfectant chemical is chlorine. Chlorine is easily obtained, relatively cheap, and most importantly, leaves a *RESIDUAL CHLORINE*[5] that can be measured. Other disinfectants are also used. There has been increased interest in disinfectants other than chlorine because of the *CARCINOGENIC*[6] compounds that chlorine may form (trihalomethanes or THMs).

This chapter will focus primarily on the use of chlorine as a disinfectant. However, let's take a brief look first at other disinfection methods and chemicals. Some of these are being more widely applied today because of the potential adverse side effects of chlorination.

6.21   Agents of Disinfection

6.210   *Physical Means of Disinfection*

A. *ULTRAVIOLET RAYS* can be used to destroy pathogenic microorganisms. To be effective, the rays must come in contact with each microorganism. The ultraviolet energy disrupts various organic components of the cell causing a biological change that is fatal to the microorganisms.

   This system has not had widespread acceptance because of the lack of measurable residual and the cost of operation. Currently, use of ultraviolet rays is limited to small or local systems and industrial applications. Ocean-going ships have used these systems for their water supply.

B. *HEAT* has been used for centuries to disinfect water. Boiling water for about 5 minutes will destroy essentially all microorganisms. This method is very energy intensive and thus very expensive. The only practical application is in the event of a disaster when individual local users are required to boil their water.

C. *ULTRASONIC WAVES* have been used to disinfect water on a very limited scale. Sonic waves destroy the microorganism by vibration. This procedure is not yet practical and is very expensive.

6.211   *Chemical Disinfectants (Other Than Chlorine)*

A. *IODINE* has been used as a disinfectant in water since 1920, but its use has been limited to emergency treatment of water supplies. Although it has long been recognized as a good disinfectant, iodine's high cost and potential physiological effects (pregnant women can suffer serious side effects) have prevented widespread acceptance. The recommended dosage is two drops of iodine (tincture of iodine which is 7 percent available iodine) in a liter of water.

B. *BROMINE* has been used only on a very limited scale for water treatment because of its handling difficulties. Bromine causes skin burns on contact. Because bromine is a very reactive chemical, residuals are hard to obtain. This also limits its use. Bromine can be purchased at swimming pool supply stores.

C. *BASES* such as sodium hydroxide and lime can be effective disinfectants but the high pH leaves a bitter taste in the finished water. Bases can also cause skin burns when left too long in contact with the skin. Bases effectively kill all microorganisms (they sterilize rather than just disinfect water). Although this method has not been used on a large scale, bases have been used to sterilize water pipes.

D. *OZONE* has been used in the water industry since the early 1900s, particularly in France. In the United States it has been used primarily for taste and odor control. The limited use in the United States has been due to its high costs, lack of residual, difficulty in storing, and maintenance requirements.

   Although ozone is effective in disinfecting water, its use is limited by its solubility. The temperature and pressure of water being treated regulate the amount of ozone that can be dissolved in the water. These factors tend to limit the disinfectant strength that can be made available to treat the water.

   Many scientists claim that ozone destroys all microorganisms. Unfortunately, significant residual ozone does not guarantee that a water is safe to drink. Organic solids may protect organisms from the disinfecting action of ozone and increase the amount of ozone needed for disinfection. In addition, ozone residuals cannot be maintained in metallic conduits for any period of time because of ozone's reactive nature. The inability of ozone to provide a residual in the distribution system is a major drawback to its use. However, recent information about the formation of trihalomethanes by chlorine compounds has resulted in renewed interest in ozone as an alternative means of disinfection.

---

[5] *Residual Chlorine.   The amount of free and/or available chlorine remaining after a given contact time under specified conditions.*
[6] *Carcinogen (CAR-sin-o-JEN).   Any substance which tends to produce cancer in an organism.*

QUESTIONS

Write your answers in a notebook and then compare your answers with those on page 356.

6.2A   List the physical agents that have been used for disinfection.

6.2B   List the chemical agents that have been used for disinfection, other than chlorine.

6.2C   What is a major limitation to the use of ozone?

6.22   Chlorine (Cl_2)

6.220   Properties of Chlorine

Chlorine is a greenish-yellow gas with a penetrating and distinctive odor. The gas is two-and-a-half times heavier than air. Chlorine has a very high coefficient of expansion. If there is a temperature increase of 50°F (28°C) (from 35°F to 85°F or 2°C to 30°C), the volume will increase from 84 to 89 percent. This expansion could easily rupture a cylinder or line full of liquid chlorine. For this reason no chlorine container should be filled to more than 85 percent of its volume. One liter of liquid chlorine can evaporate and produce 450 liters of chlorine gas.

Chlorine by itself is nonflammable and nonexplosive, but it will support combustion. When the temperature rises, so does the vapor pressure of chlorine. This means that when the temperature increases, the pressure of the chlorine gas inside a chlorine container will increase. This property of chlorine must be considered when:

1. Feeding chlorine gas from a container, and

2. Dealing with a leaking chlorine cylinder.

6.221   Chlorine Disinfection Action

The exact mechanism of chlorine disinfection action is not fully known. One theory holds that chlorine exerts a direct action against the bacterial cell, thus destroying it. Another theory is that the toxic character of chlorine inactivates the *ENZYMES*[7] which enable living microorganisms to use their food supply. As a result, the organisms die of starvation. From the point of view of water treatment, the exact mechanism of chlorine disinfection is less important than its demonstrated effects as a disinfectant.

When chlorine is added to water, several chemical reactions take place. Some involve the molecules of the water itself, and some involve organic and inorganic substances suspended in the water. We will discuss these chemical reactions in more detail in the next few sections of this chapter. First, however, there are some terms associated with chlorine disinfection that you should understand.

When chlorine is added to water containing organic and inorganic materials, it will combine with these materials and form chlorine compounds. If you continue to add chlorine, you will eventually reach a point where the reaction with organic and inorganic materials stops. At this point, you have satisfied what is known as the *"CHLORINE DEMAND."*

The chemical reactions between chlorine and these organic and inorganic substances produce chlorine compounds. Some of these compounds have disinfecting properties; others do not. In a similar fashion, chlorine reacts with the water itself and produces some substances with disinfecting properties. The total of all the compounds with disinfecting properties *PLUS* any remaining free (uncombined) chlorine is known as the *"CHLORINE RESIDUAL."* The presence of this measurable chlorine residual is what indicates to the operator that all possible chemical reactions have taken place and that there is still sufficient *"AVAILABLE RESIDUAL CHLORINE"* to kill or inactivate the microorganisms present in the water supply.

Now, if you add together the amount of chlorine needed to satisfy the chlorine demand and the amount of chlorine residual needed for disinfection, you will have the *"CHLORINE DOSE."* This is the amount of chlorine you will have to add to the water to disinfect it.

Chlorine Dose, mg/L = Chlorine Demand, mg/L + Chlorine Residual, mg/L

where

Chlorine Demand, mg/L = Chlorine Dose, mg/L – Chlorine Residual, mg/L

$$\text{Chlorine Residual, mg/}L = \text{Combined Chlorine Forms, mg/}L + \text{Free Chlorine, mg/}L$$

6.222   Reaction With Water

Free chlorine combines with water to form hypochlorous and hydrochloric acids:

Chlorine + Water $\leftrightarrows$ Hypochlorous Acid + Hydrochloric Acid

$$Cl_2 + H_2O \leftrightarrows HOCl + HCl$$

Depending on the pH, hypochlorous acid may be present in the water as the hydrogen ion and hypochlorite ion (Figure 6.1).

Hypochlorous Acid $\leftrightarrows$ Hydrogen Ion + Hypochlorite Ion

$$HOCl \leftrightarrows H^+ + OCl^-$$

In solutions that are dilute (low concentration of chlorine) and have a pH above 4, the formation of HOCl (hypochlorous acid) is most complete and leaves little free chlorine (Cl_2) existing. The hypochlorous acid is a weak acid and hence is poorly dissociated (broken up into ions) at pH levels below 6. Thus any free chlorine or hypochlorite (OCl^-) added to water will immediately form either HOCl or OCl^-; the species formed is thereby controlled by the pH value of the water. This is extremely important since HOCl and OCl^- differ in disinfection ability. HOCl has a much greater disinfection potential than OCl^-. Normally in water with a pH of 7.3, 50 percent of the chlorine present will be in the form of HOCl and 50 percent in the form of OCl^-. The higher the pH level, the greater the percent of OCl^-.

6.223   Reaction With Impurities in Water

Most waters that have been processed still contain some impurities. In this section we will discuss some of the more common impurities that react with chlorine and we will examine the effects of these reactions on the disinfection ability of chlorine.

---

[7] *Enzymes (EN-zimes). Organic substances (produced by living organisms) which cause or speed up chemical reactions. Organic catalysts and/or biochemical catalysts.*

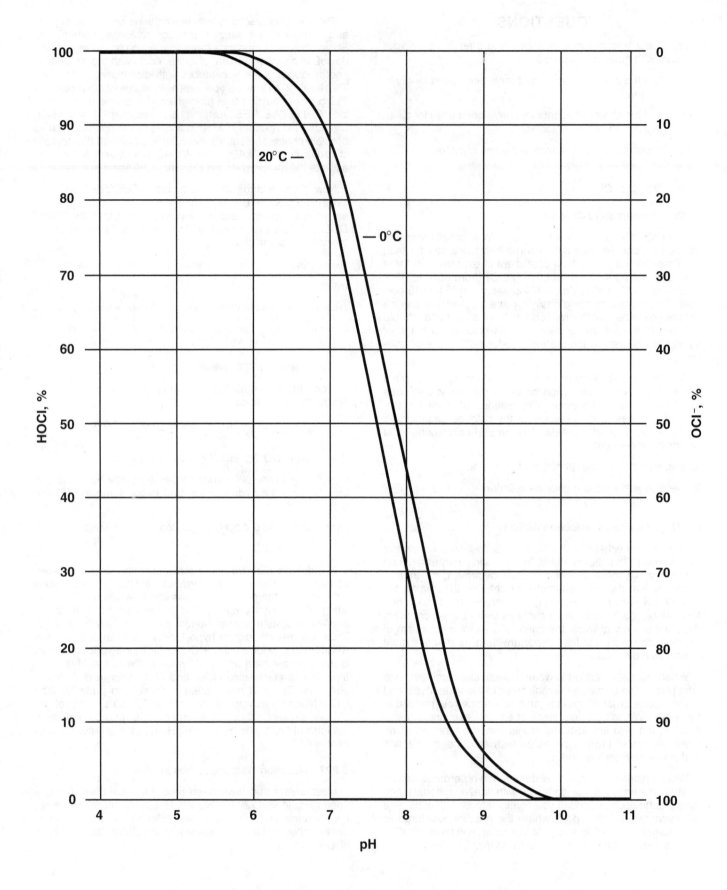

Fig. 6.1   Relationship between hypochlorous acid (HOCl),
hypochlorite ion (OCl⁻), and pH

A. Hydrogen sulfide (H_2S) and ammonia (NH_3) are two inorganic substances that may be found in water when it reaches the disinfection stage of treatment. Their presence can complicate the use of chlorine for disinfection purposes. This is because hydrogen sulfide and ammonia are what is known as *REDUCING AGENTS*. That is, they give up electrons easily. Chlorine reacts rapidly with these particular reducing agents producing some undesirable results.

Hydrogen sulfide produces an odor which smells like rotten eggs. It reacts with chlorine to form sulfuric acid and elemental sulfur (depending on temperature, pH, and hydrogen sulfide concentration). Elemental sulfur is objectionable because it can cause odor problems and will precipitate as finely divided white particles which are sometimes colloidal in nature.

The chemical reactions between hydrogen sulfide and chlorine are as follows:

$$\text{Hydrogen Sulfide} + \text{Chlorine} + \text{Oxygen Ion} \rightarrow \text{Elemental Sulfur} + \text{Water} + \text{Chloride Ions}$$

$$H_2S + Cl_2 + O^{2-} \rightarrow S\downarrow + H_2O + 2\ Cl^-$$

The chlorine required to oxidize hydrogen sulfide to sulfur and water is 2.08 mg/L chlorine to 1 mg/L hydrogen sulfide. The complete oxidation of hydrogen sulfide to the sulfate form is as follows:

$$\text{Hydrogen Sulfide} + \text{Chlorine} + \text{Water} \rightarrow \text{Sulfuric Acid} + \text{Hydrochloric Acid}$$

$$H_2S + 4\ Cl_2 + 4\ H_2O \rightarrow H_2SO_4 + 8\ HCl$$

Thus, 8.32 mg/L of chlorine are required to oxidize one mg/L of hydrogen sulfide to the sulfate form. Note that in both reactions the chlorine is converted to the chloride ion (Cl^- or HCl) which has no disinfecting power and produces no chlorine residual. In waterworks practice we always chlorinate to produce a chlorine residual; therefore, the second reaction (complete oxidation of hydrogen sulfide) occurs before any chlorine residual develops in the water being treated.

When chlorine is added to water containing ammonia (NH_3), it reacts rapidly with the ammonia and forms *CHLORAMINES*.[8] This means that less chlorine is available to act as a disinfectant. As the concentration of ammonia increases, the disinfectant power of the chlorine drops off at a rapid rate.

B. When organic materials are present in water being disinfected with chlorine, the chemical reactions that take place may produce suspected carcinogenic compounds (trihalomethanes). The formation of these compounds can be prevented by limiting the amount of prechlorination and by removing the organic materials prior to chlorination of the water.

QUESTIONS

Write your answers in a notebook and then compare your answers with those on page 356.

6.2D How is the chlorine dosage determined?

6.2E How is the chlorine demand determined?

6.2F List two inorganic reducing chemicals with which chlorine reacts rapidly.

6.23 Hypochlorite (OCl^-)

6.230 Reactions With Water

The use of hypochlorite to treat potable water achieves the same result as chlorine gas. Hypochlorite may be applied in the form of calcium hypochlorite ($Ca(OCl)_2$) or sodium hypochlorite ($NaOCl$). The form of calcium hypochlorite most frequently used to disinfect water is known as **H**igh **T**est **H**ypochlorite (**HTH**). The chemical reactions of hypochlorite in water are similar to those of chlorine gas.

CALCIUM HYPOCHLORITE

$$\text{Calcium Hypochlorite} + \text{Water} \rightarrow \text{Hypochlorous Acid} + \text{Calcium Hydroxide}$$

$$Ca(OCl)_2 + 2\ H_2O \rightarrow 2\ HOCl + Ca(OH)_2$$

SODIUM HYPOCHLORITE

$$\text{Sodium Hypochlorite} + \text{Water} \rightarrow \text{Hypochlorous Acid} + \text{Sodium Hydroxide}$$

$$NaOCl + H_2O \rightarrow HOCl + Na(OH)$$

Calcium hypochlorite (HTH) is used by a number of small water supply systems. A problem occurs in these systems when sodium fluoride is injected at the same point as the hypochlorite. A severe crust forms when the calcium and fluoride ions combine.

6.231 Difference Between Chlorine Gas and Hypochlorite Compound Reactions

The only difference between the reactions of the hypochlorite compounds and chlorine gas is the "side" reactions of the end products. The reaction of chlorine gas tends to lower the pH (increases the hydrogen ion (H^+) concentration) by the formation of hydrochloric acid which favors the formation of hypochlorous acid ($HOCl$). The hypochlorite tends to raise the pH with the formation of hydroxyl ions (OH^-) from the calcium or sodium hydroxide. At a high pH of around 8.5 or higher, the hypochlorous acid ($HOCl$) is almost completely dissociated to the ineffective hypochlorite ion (OCl^-) (Figure 6.1). This reaction also depends on the *BUFFER CAPACITY*[9] (amount of bicarbonate, HCO_3^-, present) of the water.

$$\text{Hypochlorous Acid} \leftrightarrows \text{Hydrogen Ion} + \text{Hypochlorite Ion}$$

$$HOCl \leftrightarrows H^+ + OCl^-$$

---

[8] *Chloramines (KLOR-uh-means).* Compounds formed by the reaction of hypochlorous acid (or aqueous chlorine) with ammonia.

[9] *Buffer Capacity.* A measure of the capacity of a solution or liquid to neutralize acids or bases. This is a measure of the capacity of water for offering a resistance to changes in pH.

6.24   Chlorine Dioxide (ClO₂)

6.240   Reaction in Water

Chlorine dioxide may be used as a disinfectant. Chlorine dioxide does not form carcinogenic compounds that may be formed by other chlorine compounds. Also it is not affected by ammonia, and is a very effective disinfectant at higher pH levels. In addition, chlorine dioxide reacts with sulfide compounds, thus helping to remove them and eliminate their characteristic odors. Phenolic tastes and odors can be controlled by using chlorine dioxide.

Chlorine dioxide reacts with water to form chlorate and chlorite ions in the following manner:

$$\text{Chlorine Dioxide} + \text{Water} \rightarrow \text{Chlorate Ion} + \text{Chorite Ion} + \text{Hydrogen Ions}$$

$$2\,ClO_2 + H_2O \rightarrow ClO_3^- + ClO_2^- + 2\,H^+$$

6.241   Reactions With Impurities in Water

A. INORGANIC COMPOUNDS

Chlorine dioxide is an effective *OXIDIZING AGENT*[10] with iron and manganese and does not leave objectionable tastes or odors in the finished water. Because of its oxidizing ability, chlorine dioxide usage must be monitored and the dosage will have to be increased when treating waters with iron and manganese.

B. ORGANIC COMPOUNDS

Chlorine dioxide does not react with organics in water. Therefore, there is little danger of the formation of potentially dangerous trihalomethanes.

QUESTIONS

Write your answers in a notebook and then compare your answers with those on page 356.

6.2G   How do chlorine gas and hypochlorite influence pH?

6.2H   How does pH influence the relationship between HOCl and OCl⁻?

6.25   Breakpoint Chlorination[11]

In determining how much chlorine you will need for disinfection, remember you will be attempting to produce a certain chlorine residual in the form of *FREE AVAILABLE RESIDUAL CHLORINE*.[12] Chlorine in this form has the highest disinfecting ability. *BREAKPOINT CHLORINATION* is the name of this process of adding chlorine to water until the chlorine demand has been satisfied. Further additions of chlorine will result in a chlorine residual that is directly proportional to the amount of chlorine added beyond the breakpoint. Public water supplies are normally chlorinated *PAST THE BREAKPOINT.*

Take a moment here to look at the breakpoint chlorination curve in Figure 6.2. Assume the water being chlorinated contains some manganese, iron, nitrite, organic matter, and ammonia. Now add a small amount of chlorine. The chlorine reacts with (oxidizes) the manganese, iron, and nitrite. That's all that happens—no disinfection and no chlorine residual (Figure 6.2, points 1 to 2). Add a little more chlorine, enough to react with the organics and ammonia; *CHLORORGANICS*[13] and *CHLORAMINES*[14] will form. The chloramines produce a combined chlorine residual—a chlorine residual combined with other substances so it has lost some of its disinfecting strength. Combined residuals have rather poor disinfecting power and may cause tastes and odors (points 1 to 3).

With just a little more chlorine the chloramines and some of the chlororganics are destroyed (points 3 to 4). Adding just one last amount of chlorine we get *FREE AVAILABLE RESIDUAL CHLORINE* (beyond point 4)—free in the sense that it has not reacted with anything and available in that it *CAN* and *WILL* react if need be. Free available residual chlorine is the best residual for disinfection. It disinfects faster and without the "swimming pool" odor of combined residual chlorine. Free available residual chlorine begins to form at the breakpoint; the process is called *BREAKPOINT CHLORINATION.* In water treatment plants today it is common practice to go "past the breakpoint." This means that the treated water will have a low chlorine residual, but the residual will be a very effective disinfectant because it is in the form of *FREE AVAILABLE RESIDUAL CHLORINE.*

CAUTION: Ammonia must be present to produce the breakpoint chlorination curve from the addition of chlorine. Sources of ammonia in raw water include fertilizer in agricultural runoff and discharges from wastewater treatment plants. High-quality raw water without any ammonia will not produce a breakpoint curve. Therefore, if there is no ammonia present in the water, and chlorinated water smells like chlorine, and a chlorine residual is present, DO NOT ADD MORE CHLORINE.

Let's look more closely at some of the chemical reactions that take place during chlorination. When chlorine is added to waters containing ammonia (NH_3), the ammonia reacts with hypochlorous acid (HOCl) to form monochloramine, dichloramine, and trichloramine. The formation of these chloramines depends on the pH of the solution and the initial chlorine-ammonia ratio.

---

[10] *Oxidizing Agent.*   Any substance, such as oxygen (O_2) or chlorine (Cl_2), that will readily add (take on) electrons. The opposite is a reducing agent.

[11] *Breakpoint Chlorination.*   Addition of chlorine to water until the chlorine demand has been satisfied. At this point, further additions of chlorine will result in a free chlorine residual that is directly proportional to the amount of chlorine added beyond the breakpoint.

[12] *Free Available Residual Chlorine.*   That portion of the total available residual chlorine composed of dissolved chlorine gas (Cl_2), hypochlorous acid (HOCl), and/or hypochlorite ion (OCl–) remaining in water after chlorination. This does not include chlorine that has combined with ammonia, nitrogen, or other compounds.

[13] *Chlororganic (klor-or-GAN-ick).*   Organic compounds combined with chlorine. These compounds generally originate from, or are associated with, life processes such as those of algae in water.

[14] *Chloramines (KLOR-uh-means).*   Compounds formed by the reaction of hypochlorous acid (or aqueous chlorine) with ammonia.

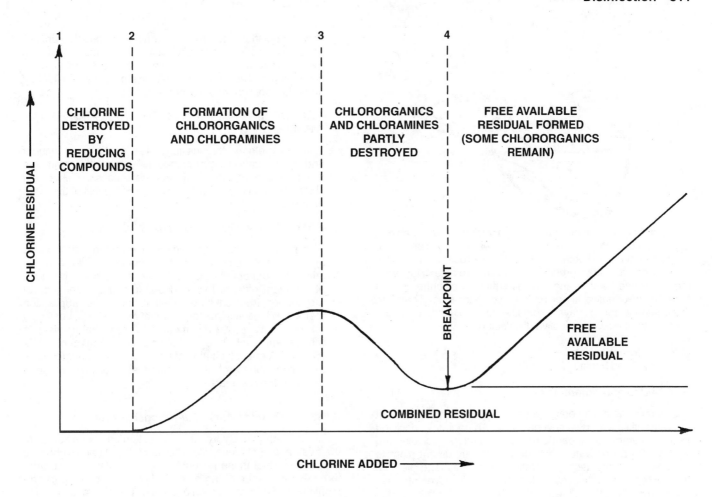

1	2		3	4	
CHLORINE DESTROYED BY REDUCING COMPOUNDS	FORMATION OF CHLORORGANICS AND CHLORAMINES		CHLORORGANICS AND CHLORAMINES PARTLY DESTROYED	FREE AVAILABLE RESIDUAL FORMED (SOME CHLORORGANICS REMAIN)	

Fig. 6.2   Breakpoint chlorination curve

Ammonia + Hypochlorous Acid → Chloramine + Water

$NH_3 + HOCl \rightarrow NH_2Cl + H_2O$ Monochloramine

$NH_2Cl + HOCl \rightarrow NHCl_2 + H_2O$ Dichloramine

$NHCl_2 + HOCl \rightarrow NCl_3 + H_2O$ Trichloramine[15]

At the pH levels usually found in water (pH 6.5 to 9.5), monochloramine and dichloramine exist together. At pH levels below 5.5, only dichloramine exists. Below pH 4.0, trichloramine is the only compound found. The mono- and dichloramine forms have definite disinfection powers and are of interest in the measurement of chlorine residuals. Dichloramine has a more effective disinfecting power than monochloramine. However, dichloramine is not recommended as a disinfectant because of taste and odor problems. Chlorine reacts with *PHENOLIC COMPOUNDS*[16] and salicylic acid (both are leached into water from leaves and blossoms) to form *CHLOROPHENOL*[17] which has an intense medicinal odor. This reaction goes much slower in the presence of monochloramine.

Historically some water treatment plants added ammonia (aqua ammonia) to the filter effluent when they chlorinated (postchlorination) to produce chloramines. The reason for this practice was that chloramines produced very long-lasting chlorine residuals. Today plants are considering adding ammonia because chloramines do not form trihalomethanes. A major limitation of using chloramine residuals is that chloramines are a less effective disinfectant than free chlorine residuals. Another potential problem is that chloramines may cause tastes and odors in the treated water. Studies have indicated a dosage of three parts chlorine to one part ammonia (3:1) will form monochloramines. Monochloramines form combined residual chlorine (rising part of breakpoint curve in Figure 6.2) and apparently do not produce tastes and odors. Trichloramines form during the oxidation of combined residual materials (Figure 6.2) and will cause tastes and odors. Although chloramines are nontoxic to healthy humans, they can have a debilitating (weakening) effect on individuals with renal disease who must undergo kidney dialysis.

---

[15] *More commonly called nitrogen trichloride.*

[16] *Phenolic (fee-NO-lick) Compounds.   Organic compounds that are derivatives of benzene.*

[17] *Chlorophenolic (klor-o-FEE-NO-lick).   Chlorophenolic compounds are phenolic compounds (carbolic acid) combined with chlorine.*

In plants where trihalomethanes (THMs) are not a problem, sufficient chlorine is added to the raw water (prechlorination) to go "past the breakpoint." The chlorine residual will aid coagulation, control algae problems in basins, reduce odor problems in treated water, and provide sufficient chlorine contact time for an effective kill of pathogenic organisms. Therefore the treated water will have a very low chlorine residual, but the residual will be a very effective disinfectant.

6.26   Chloramination
by David Foust

6.260   Use of Chloramines

Chloramines have been used as an alternative disinfectant by water utilities for over seventy years. An operator's decision to use chloramines depends on several factors, including the quality of the raw water, the ability of the treatment plant to meet various regulations, operational practices, and distribution system characteristics. Chloramines have proven effective in accomplishing the following objectives:

1. Reducing the formation of trihalomethanes (THMs) and other disinfection by-products (DBPs),

2. Maintaining a detectable residual throughout the distribution system,

3. Penetrating the biofilm (the layer of microorganisms on pipeline walls) and reducing the potential for coliform regrowth,

4. Killing or inactivating HETEROTROPHIC[18] plate count bacteria, and

5. Reducing taste and odor problems.

6.261   Methods for Producing Chloramines

There are three primary methods by which chloramines are produced: (1) preammoniation followed by later chlorination, (2) addition of chlorine and ammonia at the same time (concurrently), and (3) prechlorination/postammoniation.

1. PREAMMONIATION FOLLOWED BY LATER CHLORINATION

In this method, ammonia is applied at the rapid-mix unit process and chlorine is added downstream at the entrance to the flocculation basins. This approach usually produces lower THM levels than the postammoniation method.

Preammoniation to form chloramines (monochloramine) does not produce phenolic tastes and odors, but this method may not be as effective as postammoniation for controlling tastes and odors associated with DIATOMS[19] and anaerobic bacteria in source waters.

2. CONCURRENT ADDITION OF CHLORINE AND AMMONIA

In this method, chlorine is applied to the plant influent and, at the same time or immediately thereafter, ammonia is introduced at the rapid-mix unit process. Concurrent chloramination produces the lowest THM levels of the three methods.

3. PRECHLORINATION/POSTAMMONIATION

In prechlorination/postammoniation, chlorine is applied at the head of the plant and a free chlorine residual is maintained throughout the plant processes. Ammonia is added at the plant effluent to produce chloramines. Because of the longer free chlorine contact time, this application method will result in the formation of more THMs, but it may be necessary to use this method to meet the disinfection requirements of the Surface Water Treatment Rule (SWTR). A major limitation of using chloramine residuals is that chloramines are less effective as a disinfectant than free chlorine residuals.

6.262   Chlorine to Ammonia-Nitrogen Ratios

After a method of chloramine application has been selected, the best ratio of chlorine to ammonia-nitrogen (by weight) and the desired chloramine residual for each system must be determined. A dosage of three parts of chlorine to one part ammonia (3:1) will form monochloramines. This 3:1 ratio provides an excess of ammonia-nitrogen which will be available to react with any chlorine added in the distribution system to boost the chloramine residual.

Higher chlorine to ammonia-nitrogen weight ratios such as 4:1 and 5:1 also have been used successfully by many water agencies. However, the higher the chlorine to ammonia-nitrogen ratio, the less excess ammonia will be available for rechlorination. Some agencies have found it necessary to limit the amount of excess available ammonia to prevent incomplete NITRIFICATION.[20]

Monochloramines form combined residual chlorine (rising part of curve in Figure 6.3) as the chlorine dose is increased in the presence of ammonia. As the chlorine dose increases, the combined residual increases and excess ammonia decreases. The maximum chlorine to ammonia ratio that can be achieved is 5:1. At a chlorine dose above the 5:1 ratio, the combined residual actually decreases and the total ammonia-nitrogen also begins to decrease as it is oxidized by the additional chlorine. Dichloramines form during this oxidation and may cause tastes and odors. As the chlorine dose is further increased, breakpoint chlorination will eventually occur. Trichloramines are formed past the breakpoint and also may form tastes and odors. As with breakpoint chlorination, further additions of chlorine will result in a chlorine residual that is proportional to the amount of chlorine added beyond the breakpoint.

---

[18] Heterotrophic (HET-er-o-TROF-ick).   Describes organisms that use organic matter for energy and growth. Animals, fungi and most bacteria are heterotrophs.

[19] Diatoms (DYE-uh-toms).   Unicellular (single cell), microscopic algae with a rigid (box-like) internal structure consisting mainly of silica.

[20] Nitrification (NYE-truh-fuh-KAY-shun).   An aerobic process in which bacteria reduce the ammonia and organic nitrogen in water into nitrite and then nitrate.

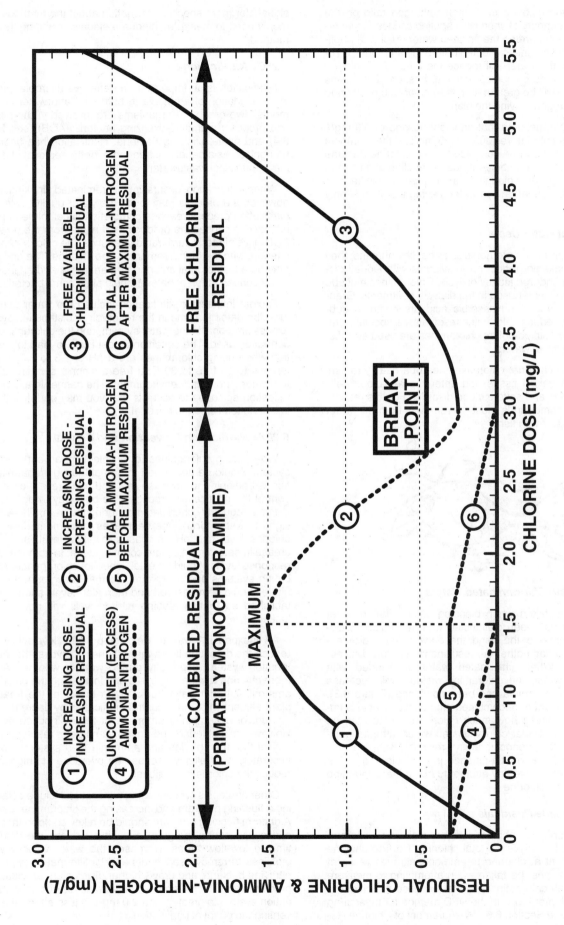

Fig. 6.3   Typical chloramination dose-residual curve

Calculating the chlorine to ammonia-nitrogen ratio on the basis of actual quantity of chemicals applied can lead to incorrect conclusions regarding the finished water quality. In applications in which chlorine is injected before the ammonia, chlorine demand in the water will reduce the amount of chlorine available to form the combined residual. In such applications the *applied* ratio will be greater than the *actual* ratio of chlorine to ammonia-nitrogen leaving the plant.

As an example, assume that an initial dosage of 5.0 mg/L results in a free chlorine residual of 3.5 mg/L at the ammonia application point; it can be concluded that a chlorine demand of 1.5 mg/L exists. If ammonia-nitrogen is applied at a dose of 1.0 mg/L, the applied chlorine to ammonia-nitrogen ratio is 5:1, whereas the actual ratio in water leaving the plant is only 3.5:1.

6.263   Special Water Users

Although chloramines are nontoxic to healthy humans, they can have a weakening effect on individuals with kidney disease who must undergo kidney dialysis. Chloramines must be removed from the water used in the dialysis treatments. Granular activated carbon and ascorbic acid are common substances used to reduce chloramine residuals. All special water users should be notified before chloramines are used as a disinfectant in municipal waters.

Also, like free chlorine, chloramines can be deadly to fish. They can damage gill tissue and enter the red blood cells causing a sudden and severe blood disorder. For this reason, all chloramine compounds must be removed from the water prior to any contact with fish.

6.264   Blending Chloraminated Waters

Care must be taken when blending chloraminated water with water that has been disinfected with free chlorine. Depending on the ratio of the blend, these two different disinfectants can cancel each other out resulting in very low disinfectant residuals. When chlorinated water is blended with chloraminated water, the chloramine residual will decrease after the excess ammonia has been combined (Figure 6.3). Knowing the amount of uncombined ammonia available is important in determining how much chlorinated water can be blended with a particular chloraminated water without significantly affecting the monochloramine residual. Knowing how much uncombined ammonia-nitrogen is available is also important before you make any attempt to boost the chloramine residual by adding chlorine.

6.265   Chloramine Residuals

When measuring combined chlorine residuals (chloramines) in the field, analyze for total chlorine. No free chlorine should be present at chlorine to ammonia-nitrogen ratios of 3:1 to 5:1. Care must be taken when attempting to measure free chlorine with chloraminated water because the chloramine residual will interfere with the DPD method for measuring free chlorine. (See Section 6.5, "Measurement of Chlorine Residual," for more specific information about the methods commonly used to measure chlorine residuals, including the DPD method.)

6.266   Nitrification

Nitrification is an important and effective microbial process in the oxidation of ammonia in both land and water environments. Two groups of organisms are involved in the nitrification process: ammonia-oxidizing bacteria (AOB) (see Figure 6.4) and nitrite-oxidizing bacteria. Nitrification has been well recognized as a beneficial treatment for the removal of ammonia in municipal wastewater.

When nitrification occurs in chloraminated drinking water, however, the process may lower the water quality unless the nitrification process reaches completion. Incomplete or partial nitrification causes the production of nitrite from the growth of AOB. This nitrite, in turn, rapidly reduces free chlorine and can interfere with the measurement of free chlorine. The end result may be a loss of total chlorine and ammonia and an increase in the concentration of heterotrophic plate count bacteria.

Factors influencing nitrification include the water temperature, the detention time in the reservoir or distribution system, excess ammonia in the water system, and the chloramine concentration used. The conditions most likely to lead to nitrification when using chloramines are a pH of 7.5 to 8.5, a water temperature of 25 to 30°C, a free ammonia concentration in the water, and a dark environment. The danger in allowing nitrification episodes to occur is that you may be left with very low or no total chlorine residual.

6.267   Nitrification Prevention and Control

When using chloramines for disinfection, an early warning system should be developed to detect the signs that nitrification is beginning to occur so that you can prevent or at least control the nitrification process. The best way to do this is to set up a regularly scheduled monitoring program. The warning signs to watch for include decreases in ammonia level, total chlorine level and pH, increases in nitrite level, and an increase in heterotrophic plate count bacteria. In addition, action response levels should be established for chloraminated distribution systems and reservoirs. Normal background levels of nitrite should be measured and then alert levels should be established so that increasing nitrite levels will not be overlooked.

An inexpensive way to help keep nitrite levels low is to reduce the detention times through the reservoirs and the distribution system, especially during warmer weather. Adding more chlorine to reservoir inlets and increasing the chlorine to ammonia-nitrogen ratio from 3:1 up to 5:1 at the treatment plant effluent will further control nitrification by decreasing the amount of uncombined ammonia in the distribution system. However, at a chlorine and ammonia-nitrogen ratio of 5:1, it is critical that the chlorine and ammonia feed systems operate accurately and reliably because an overdose of chlorine can reduce the chloramine residual.

Other strategies for controlling nitrification include establishing a flushing program and increasing the chloramine residual. A uniform flushing program should be a key component of any nitrification control program. Flushing reduces the detention time in low-flow areas, increases the water velocity within pipelines to remove sediments and biofilm that would harbor nitrifying bacteria, and draws higher disinfectant residuals into problem areas. Increasing the chloramine residual in the distribution system to greater than 2.0 mg/L is also effective in preventing the onset of nitrification.

Nitrification is a biological process caused by naturally occurring ammonia-oxidizing bacteria. These bacteria feed on free ammonia and convert it to nitrite and then nitrate. They thrive in covered reservoirs during warm summer months and are very resistant to chloramine disinfection. The by-products of their biological breakdown can support the growth of coliform bacteria.

THE NITRIFICATION PROCESS

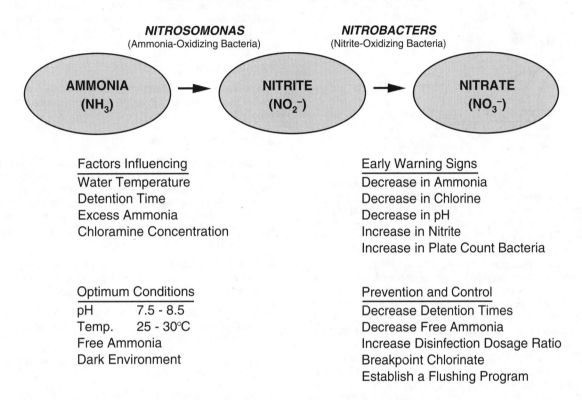

Fig. 6.4   The nitrification process

QUESTIONS

Write your answers in a notebook and then compare your answers with those on page 356.

6.2I   What is breakpoint chlorination?

6.2J   An operator's decision to use chloramines depends on what factors?

6.2K   What are the three primary methods by which chloramines are produced?

6.2L   Why is the *applied* chlorine to ammonia-nitrogen ratio usually greater than the *actual* chlorine to nitrogen ratio leaving the plant?

6.2M   Incomplete nitrification causes the production of nitrite which produces what problems in disinfection of water?

6.27   Chlorine Residual Testing

6.270   Importance

Many small system operators attempt to maintain a chlorine residual throughout the distribution system. Chlorine is very effective in biological control and especially in elimination of coliform bacteria that might reach water in the distribution

system through cross connections or leakage into the system. A chlorine residual also helps to control any microorganisms that could produce slimes, tastes, or odors in the water in the distribution system.

Adequate control of coliform "aftergrowth" is usually obtained only when chlorine residuals are carried to the farthest points of the distribution system. To ensure that this is taking place, make daily chlorine residual tests. A chlorine residual of about 0.2 mg/L measured at the extreme ends of the distribution system is usually a good indication that a free chlorine residual is present in all other parts of the system. This small residual can destroy a small amount of contamination, so a lack of chlorine residual could indicate the presence of heavy contamination. If routine checks at a given point show measurable residuals, any sudden absence of a residual at that point should alert the operator to the possibility that a potential problem has arisen which needs prompt investigation. Immediate action that can be taken includes retesting for chlorine residual, then checking chlorination equipment, and finally searching for a source of contamination which could cause an increase in the chlorine demand.

6.271   Chlorine Residual Curve

The chlorine residual curve procedure is a quick and easy way for an operator to estimate the proper chlorine dose, especially when surface water conditions are changing rapidly such as during a storm.

Fill a CLEAN five-gallon bucket from a sample tap located at least two 90-degree elbows (or where chlorine is completely mixed with the water in the pipe) AFTER the chlorine has been injected into the pipe. Immediately measure the chlorine residual and record this value on the "time zero" line of your record sheet (Figure 6.5). This is the initial chlorine residual. At 15-minute intervals, vigorously stir the bucket using an up and down motion. (A large plastic spoon works well for this purpose.) Collect a sample from one or two inches below the water surface and measure the chlorine residual. Record this chlorine residual value on the record sheet. For at least one hour, collect a sample every 15 minutes, measure the chlorine residual, and record the results to indicate the "chlorine demand" of the treated water. Plot these recorded values on a chart or graph paper as shown on Figure 6.5. Connect the plotted points to create a chlorine residual curve. If the chlorine residual after one hour is not correct (about 0.2 mg/L), increase or decrease the initial chlorine dose so the final chlorine residual will be approximately at the desired ultimate chlorine residual in the water distribution system. Repeat this procedure until the desired TARGET initial chlorine residual will achieve the desired chlorine residual throughout the distribution system.

Precautions that must be taken when performing this test include being sure the five-gallon plastic test bucket is clean and only used for this purpose. A new bucket does not need to be used for every test, but the bucket should be new when the first test is performed. The stirrer should also be clean. DO NOT USE THE STIRRER FOR THE CHLORINE SOLUTION MIXING AND HOLDING TANK. During the test the bucket should be kept cool so that the chlorine gas does not escape from the water and give false chlorine residual values.

The chlorine demand for groundwater changes slowly, or not at all; therefore, the "initial or target" chlorine residual does not have to be checked more frequently than once a month. Always be sure to measure the chlorine residual in the distribution system on a daily basis. This is also a good check that the chlorination equipment is working properly and that the chlorine stock solution is the correct concentration.

The chlorine demand for surface water can change continuously, especially during storms and the snow melt season. Experience has proven that the required "initial or target" chlorine residual at time zero is directly tied to the turbidity of the fin-

ished (treated) water. The higher the finished water turbidity, the higher the "initial" chlorine residual value will have to be to ensure the desired chlorine residual in the distribution system. Careful documentation of this information in your records will greatly reduce the lag time in chlorine addition changes to maintain the desired residual in the distribution system and the delivery of safe drinking water to your consumers. Experience and a review of your records will indicate that for a given turbidity value, you can estimate the desired "initial" chlorine residual, which will require a given chlorinator output level for a given water flow rate.

Acknowledgment

The information in Sections 6.270 and 6.271 was developed by Bill Stokes. His suggestions and procedures are greatly appreciated.

6.272 Critical Factors

Both *CHLORINE RESIDUAL* and *CONTACT TIME* are essential for effective killing or inactivation of pathogenic microorganisms. Complete initial mixing is very important. Changes in pH affect the disinfection ability of chlorine and you must reexamine the best combination of contact time and chlorine residual when the pH fluctuates. Critical factors influencing disinfection are summarized as follows:

1. Effectiveness of upstream treatment processes. The lower the turbidity (suspended solids, organic content, reducing agents) of the water, the better the disinfection.

2. Injection point and method of mixing to get disinfectant in contact with water being disinfected. Depends on whether using prechlorination or postchlorination.

3. Temperature. The higher the temperature, the more rapid the rate of disinfection.

4. Dosage and type of chemical. Usually the higher the dosage, the faster the disinfection rate. The form (chloramines or free chlorine residual) and type of chemical also influence the disinfection rate.

5. pH. The lower the pH, the better the disinfection.

6. Contact time. With good initial mixing, the longer the contact time, the better the disinfection.

7. Chlorine residual.

6.3 POINTS OF APPLICATION

6.30 Wells and Pumps[21]

6.300 Disinfecting New and Existing Wells

When a new well is completed, it is necessary to disinfect the well, pump, and screen. A 50-mg/L residual of free chlorine in contact with all surfaces of the well, screen, pump, and piping is recommended. Consideration must be given to the fact that the water aquifer could have been contaminated during the drilling process. Disinfection is usually accomplished by using a chlorine solution applied into the well and the aquifer around the well.

To accomplish this process the use of sodium hypochlorite is recommended. Inject the chlorine solution down the pump column pipe rather than through the vent pipe. As previously mentioned, 50 mg/L is needed and 24 hours of contact time is

---

[21] Also see Section 3.5, "Disinfection of Wells and Pumps," in Chapter 3, SMALL WATER SYSTEM OPERATION AND MAINTENANCE.

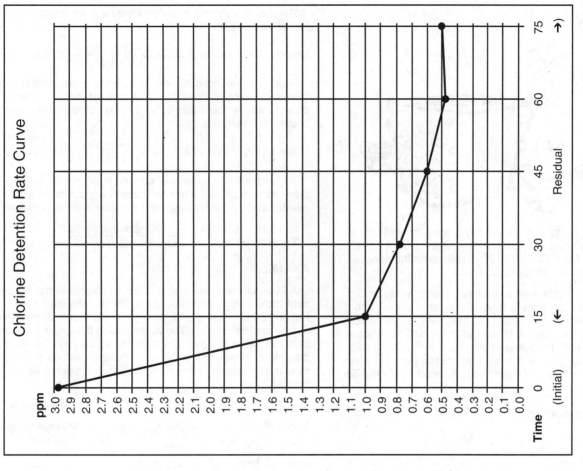

Fig. 6.5   Chlorine detention rate curves

recommended. At the end of this period, pump the well until all evidence of a chlorine residual is gone. Then take a sample and test for *TOTAL COLIFORMS*[22] to determine the effectiveness of the chlorine dosage.

Since organic matter such as oil may be used during drilling, the chlorine solution should have a concentration of about 1,000 mg/L before being injected into the well through the vent pipe. If the chlorine solution is applied to the well by injecting the chlorine through the pump column pipe, a chlorine dosage of 100 mg/L is acceptable. By using the column pipe, the oil on the water surface is avoided and the large amount of chlorine is not needed.

6.301   Sample Problem

FORMULAS

To disinfect a well, the first step is to determine the volume of water in the well. There are three approaches to calculating the volume of water in a well.

1. Calculate the volume of water using the diameter of the casing and the depth of water.

$$\text{Volume, gal} = \frac{(0.785)(\text{Diameter, in})^2(\text{Depth, ft})(7.48 \text{ gal/cu ft})}{144 \text{ sq in/sq ft}}$$

2. When the depth of water in a well changes from time to time, many operators calculate a constant for their well in terms of gallons of water per foot of water in the well.

$$\text{Volume, gal/ft} = \frac{(0.785)(\text{Diameter, in})^2(7.48 \text{ gal/cu ft})}{144 \text{ sq in/sq ft}}$$

3. Sometimes tables are available that allow you to find the volume in gallons per foot from the table if you know the diameter of the casing.

To find the pounds of chlorine needed to disinfect a well, we need to know the volume of water in gallons and the desired chlorine dose in milligrams per liter. Convert the volume in gallons to million gallons by dividing by 1,000,000.[23]

$$\text{Volume, M Gal} = \frac{\text{Volume, Gal}}{1,000,000/\text{Million}}$$

Chlorine, lbs = (Volume, M Gal)(Dose, mg/L)(8.34 lbs/gal)

We can "prove" the previous formula if we know that one liter of water weighs 1,000,000 mg.

$$\text{Chlorine, lbs} = (\text{Volume, M Gal})\left(\frac{\text{Dose, mg}}{1,000,000 \text{ mg}}\right)(8.34 \text{ lbs/gal})$$

Note that the dose is now in milligrams per million milligrams, or parts per million. Now change the milligrams to pounds so we will have dose, pounds chlorine per million pounds of water.

$$\text{Chlorine, lbs} = (\text{Volume, M Gal})\left(\frac{\text{Dose, lbs}}{\text{M lbs}}\right)(8.34 \text{ lbs/gal})$$

To calculate the gallons of sodium hypochlorite needed to disinfect a well, we need to know the pounds of chlorine needed and the percent available chlorine in the hypochlorite.

$$\frac{\text{Sodium Hypochlorite}}{\text{Solution, gallons}} = \frac{(\text{Chlorine, lbs})(100\%)}{(8.34 \text{ lbs/gal})(\text{Hypochlorite, \%})}$$

EXAMPLE 1

A 200-foot deep new well is to be disinfected with a five percent sodium hypochlorite solution. The top 150 feet of the well has a 12-inch casing and the bottom 50 feet has an 8-inch casing and well screen. The water level in the well is at 80 feet from the top of the well. Since this is a new well, the desired chlorine concentration in the initial dose should be 100 mg/L. How many gallons of five percent sodium hypochlorite will be needed?

	Known	Unknown
Depth of Well, ft	= 200 ft	5% Hypochlorite, gallons
Hypochlorite, %	= 5%	
Chlorine Dose, mg/L	= 100 mg/L	
Top 150 ft	= 12-in casing	
Bottom 50 ft	= 8-in casing and screen	
Water Level, ft	= 80 ft from top	
	or = 120 ft from bottom	

1. Find the volume of water in the well. Use Table 6.4.

TABLE 6.4   VOLUME OF WATER IN WELL PER FOOT OF DEPTH

Casing Size, in [a]	Volume, gallons per foot of depth
4	0.65
5	1.02
6	1.47
8	2.61
10	4.08
12	5.87
14	7.99
16	10.44
18	13.21
20	16.31

[a] Casing sizes are nominal values, not actual pipe diameters.

---

[22] *Total Coliforms.   See procedures for Coliform in Chapter 11, "Laboratory Procedures," WATER TREATMENT PLANT OPERATION, Volume I.*
[23] *For a more detailed explanation of how to convert units in a formula by using 1,000,000/1 Million, see Section A.130, "Flows," in the Arithmetic Appendix.*

a. From 80 feet from top to 150 feet from top we have 70 feet (150 ft − 80 ft) of 12-inch casing.

From Table 6.4, 12-inch casing has 5.87 gallons of water per foot.

b. From 150 feet from top to 200 feet from top we have 50 feet (200 ft − 150 ft) of 8-inch casing.

From Table 6.4, 8-inch casing has 2.61 gallons of water per foot.

$$\text{Total Volume of Water, gallons} = (\text{Length, ft})(\text{Volume, gal/ft})$$

$$= (70 \text{ ft})(5.87 \text{ gal/ft}) + (50 \text{ ft})(2.61 \text{ gal/ft})$$

$$= 411 \text{ gal} + 131 \text{ gal}$$

$$= 542 \text{ gallons of Water}$$

If Table 6.4 was not available, we could calculate the approximate volume in gallons per foot of depth and the total volume of water in gallons.

a. For the 12-inch casing, calculate the volume in gallons per foot of depth.

$$\text{Volume, gal} = (\text{Area, sq ft})(\text{Depth, ft})(7.48 \text{ gal/cu ft})$$

$$= (\text{Area, sq ft})(7.48 \text{ gal/cu ft})(\text{Depth, ft})$$

$$\text{Volume, gal/ft} = (\text{Area, sq ft})(7.48 \text{ gal/cu ft})$$

$$= \frac{(0.785)(12 \text{ in})^2(7.48 \text{ gal/cu ft})}{144 \text{ sq in/sq ft}}$$

$$= 5.87 \text{ gal/ft}$$

b. For the 8-inch casing, calculate the volume in gallons per foot of depth.

$$\text{Volume, gal/ft} = (\text{Area, sq ft})(7.48 \text{ gal/cu ft})$$

$$= \frac{(0.785)(8 \text{ in})^2(7.48 \text{ gal/cu ft})}{144 \text{ sq in/sq ft}}$$

$$= 2.61 \text{ gal/ft}$$

$$\text{Total Volume of Water, gallons} = (\text{Length, ft})(\text{Volume, gal/ft})$$

$$= (70 \text{ ft})(5.87 \text{ gal/ft}) + (50 \text{ ft})(2.61 \text{ gal/ft})$$

$$= 411 \text{ gal} + 131 \text{ gal}$$

$$= 542 \text{ gallons of Water}$$

This answer is the same as the 542 gallons we obtained using the values in Table 6.4.

2. Find the pounds of chlorine needed.

$$\text{Chlorine, lbs} = (\text{Vol Water, M Gal})(\text{Chlorine Dose, mg/}L)(8.34 \text{ lbs/gal})$$

$$= (0.000542 \text{ M Gal})(100 \text{ mg/}L)(8.34 \text{ lbs/gal})$$

$$= 0.45 \text{ lb Chlorine}$$

3. Calculate the gallons of 5 percent sodium hypochlorite solution needed.

$$\text{Sodium Hypochlorite Solution, gallons} = \frac{(\text{Chlorine, lbs})(100\%)}{(8.34 \text{ lbs/gal})(\text{Hypochlorite, }\%)}$$

$$= \frac{(0.45 \text{ lb})(100\%)}{(8.34 \text{ lbs/gal})(5\%)}$$

$$= 1.08 \text{ gallons}$$

A little over one gallon of 5 percent solution sodium hypochlorite should do the job.

The gravel in a gravel envelope well must be disinfected as the gravel is added to the well. This is accomplished by adding half a pound (227 gm) of five-gram *HTH*[24] granules per ton (900 kg) of gravel.

When disinfecting a well that draws from more than one aquifer, remember that there is nearly always flow from one aquifer to another. Also if water is flowing through a well, the procedures described in this chapter may not work very well because the flowing water will carry away the chlorine.

Once water from a well starts producing positive coliform test results, it is almost impossible to correct the problem with chlorine treatment. However, some wells will clear up after long periods of pumping (sometimes two years are required).

Disinfection of the well may not be accomplished the first time so this procedure may have to be repeated a second and possibly even a third time.

6.302 Procedures

Procedures described in this section are one method for disinfecting a well; however, other methods are available and may be used.

a. Calculate volume of water in well using guidelines in Example 1.

b. Wash the pump column or drop pipe (pipe attached to the pump) with chlorine solution as it is lowered into the well. The chlorine solution can be fed through a hose to wash the pipe.

---

[24] *HTH (pronounce as separate letters).* **H**igh **T**est **H**ypochlorite. *Calcium hypochlorite or Ca(OCl)$_2$.*

c. After the chlorine solution has been fed into the well, operate the pump so as to thoroughly mix the disinfectant with the water in the well. Pump until the water discharged has the odor of chlorine. Then shut down the pump and let water surge back down into the well. Repeat this procedure several times over an hour.

d. Allow the well to stand for 24 hours.

e. Pump well to waste until all traces of chlorine are gone.

f. Take a bacteriological sample for analysis. Use the 24-hour membrane filter method for the quickest results.

g. If results of bacteriological analysis for total coliforms indicate unsafe conditions, repeat disinfection procedure. Conditions are considered unsafe when the test results are positive (there are coliforms present).

6.303 Continuous Disinfecting of Wells

Normally wells are treated by a single application of disinfecting solution, but since more and more groundwater sources are becoming contaminated, it is becoming a standard practice to disinfect wells drawing groundwater on a continuous basis.

The type of disinfection installation to be used will depend on which type of chemical is to be used. From a safety standpoint the use of hypochlorite is the safest. Normally there are fewer hazards associated with hypochlorite compounds than with liquid or gaseous chlorine. However, liquid or gaseous chlorine is cheaper to use than hypochlorite, especially if large volumes of chlorine are required.

QUESTIONS

Write your answers in a notebook and then compare your answers with those on page 356.

6.3A   When disinfecting a well, why should the well, pump, screen, and aquifer around the well all be disinfected?

6.3B   Where should the chlorine be applied to a well being disinfected?

6.3C   How would you determine if a well has been successfully disinfected?

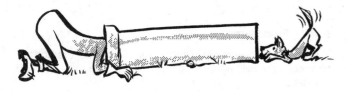

6.31 Mains

6.310 Procedures

A. Preventive Measures

One of the most effective steps in disinfecting water mains is to do everything possible to prevent the mains from becoming contaminated. Keep outside material such as dirt, construction materials, animals, rodents, and dirty water out of mains being installed or repaired. Inspect the interior of all pipes for cleanliness as the main is laid in the trench. Keep the trenches dry or dewatered. Install watertight plugs in all open-end joints whenever the trench is unattended for any length of time. An exception to the use of watertight plugs in the open ends of joints is when groundwater could cause the pipe to float. When a long pipeline is to be laid, a dry trench may not always be possible during the night when crews are not on duty.

B. Preliminary Measures

Flush the mains before attempting to disinfect them. Flushing is not an alternative to cleanliness when laying or repairing mains. Flushing cannot be expected to remove debris caked on joints, crevices, and other parts of the system. Flush the mains using water flowing with a velocity of at least 2.5 ft/sec (0.75 m/sec). See Table 6.5 for flows required in gallons per minute (GPM) to flush various diameter mains with 2.5 ft/sec (0.75 m/sec) and 5.0 ft/sec (1.5 m/sec) flushing velocities.

TABLE 6.5   FLOWS REQUIRED FOR VARIOUS FLUSHING VELOCITIES

Line Size, inches[c]	Flow Required, GPM[a,b]	
	Velocity, 2.5 ft/sec[d]	Velocity, 5 ft/sec[d]
4	100	200
6	220	440
8	390	780
10	610	1,220
12	880	1,760
14	1,200	2,400
16	1,570	3,140

[a] Line should be flushed for at least 30 minutes.
[b] Multiply GPM x 3.785 to obtain liters/min.
[c] Multiply inches x 2.5 to obtain cm.
[d] Multiply ft/sec x 0.3 to obtain m/sec.

Special care must be exercised to be sure fittings and valves are clean before disinfecting a main. Also, all air pockets or other conditions that would prevent proper disinfection should be eliminated. Many water mains contain mechanical joints. These joints have spaces that are difficult to chlorinate once they become filled with water. This is one reason why initial flushing may be ineffective. Some operators crush calcium hypochlorite tablets and place the crushed tablets in joints to improve the disinfection process.

If you wish to calculate the flow required in gallons per minute for any pipe diameter and flushing velocity not shown in Table 6.5, or if you wish to verify any number in Table 6.5, use the following procedure.

1. Calculate the cross-sectional area of the pipe in square feet. Use a 10-inch diameter pipe.

$$\text{Area, sq ft} = \frac{(0.785)(\text{Diameter, in})^2}{144 \text{ sq in/sq ft}}$$

$$= \frac{(0.785)(10 \text{ in})^2}{144 \text{ sq in/sq ft}}$$

$$= 0.545 \text{ sq ft}$$

2. Calculate the flow rate in cubic feet per second (CFS). Use a velocity of five feet per second and Q = AV.

$$\text{Flow Rate, CFS} = (\text{Area, sq ft})(\text{Velocity, ft/sec})$$

$$= (0.545 \text{ sq ft})(5 \text{ ft/sec})$$

$$= 2.725 \text{ CFS}$$

3. Convert the flow from cubic feet per second to gallons per minute.

$$\text{Flow Rate,}\atop\text{GPM} = (\text{Flow, cu ft/sec})(7.48 \text{ gal/cu ft})(60 \text{ sec/min})$$

$$= (2.725 \text{ cu ft/sec})(7.48 \text{ gal/cu ft})(60 \text{ sec/min})$$

$$= 1,223 \text{ GPM}$$

NOTE: Table 6.5 shows a flow of 1,220 GPM because this is as close as you can read most flowmeters, as well as regulate the flow.

C. Disinfection Methods

Next a method of disinfection must be selected. Chlorine gas may be used. A trained operator is required to operate the solution-feed chlorinator in combination with a booster pump. Applying gas directly from a cylinder is dangerous and if proper mixing of chlorine and water is not obtained, a highly corrosive condition could develop.

Another danger is that water could be drawn back into the chlorine cylinder. A mixture of water and liquid chlorine will form a very concentrated hydrochloric acid solution. This acid solution could "eat" a hole in the wall of the cylinder and allow liquid or gaseous chlorine to escape. For these reasons, *NEVER* use water on a chlorine leak. The corrosive action of chlorine and water will *ALWAYS* make a leak worse.

Calcium hypochlorite is available in powder, granular, or tablet form at 65 percent available chlorine. Calcium hypochlorite is relatively soluble in water and can be applied by the use of a solution feeder. When using calcium hypochlorite (chlorine) tablets, the tablets will not dissolve readily if the water temperature is below 41°F (5°C), which will reduce the chlorine concentration in the water being disinfected. Temperature control of water is difficult, but you can control contact time. Therefore, if you are disinfecting when the water temperature is low, increase the contact time to achieve effective disinfection.

Calcium hypochlorite requires special storage to avoid contact with organic material. When organic material and calcium hypochlorite come in contact, the resulting chemical reactions can generate enough heat and oxygen to start and support a fire. When calcium hypochlorite is mixed with water, heat is given off. To adequately disperse the heat generated, the dry calcium hypochlorite should be added to the correct volume of water, rather than adding water to the calcium hypochlorite.

Sodium hypochlorite is available in liquid form at five to fifteen percent available chlorine and can be fed by the use of a hypochlorinator. Sodium hypochlorite can lose from two to four percent of its available chlorine content per month at room temperatures. Therefore, manufacturers recommend a maximum storage period of from 60 to 90 days.

See Section 6.4, "Operation of Chlorination Equipment," for procedures on how to operate equipment.

Three common methods of disinfecting mains are summarized in Table 6.6, "Chlorination Methods for Disinfecting Water Mains," and two methods are illustrated in Figures 6.6 and 6.7.

TABLE 6.6 CHLORINATION METHODS FOR DISINFECTING WATER MAINS

Chlorination Method Used	Maximum Chlorine Dose, mg/L [a]	Minimum Contact Time, hr	Minimum Chlorine Residual, mg/L
Continuous	50	24	25
Slug	500	3	300
Tablet [b]	50	24	25

[a] NOTE: AWWA Standard C651-92 recommends the following doses: Continuous, 25 mg/L; Slug, 100 mg/L; and Tablet, 25 mg/L. The minimum chlorine dose depends on whether you are disinfecting an existing main (high dose, 500 mg/L, and short contact time (15 minutes)), or a new main (use a continuous minimum residual of 25 mg/L for 24 hours). Use whatever dose you need that will produce no positive coliform test results.

[b] Tablets must be placed at inside top of pipe when the pipe is being laid. Also, two tablets must be placed at all joints on both sides of the pipe at the half-full location. Place one ounce (28 gm) of HTH powder per inch (25 mm) of pipe diameter in the first length of pipe and again after each 500 feet (150 m) of pipe. This ensures that the first water entering the spaces at joints will have a high chlorine residual. Fill the pipe with water at velocities of less than one ft/sec.

> **WARNING**
>
> Do not use HTH powder in pipes with solvent-welded plastic or screwed-joint steel pipe because reaction between the joint compounds and the calcium hypochlorite could cause a fire or an explosion.

1. *NEW MAINS*

All pipes, fittings, valves, and other items that will not be disinfected by the filled line must be precleaned and disinfected. Valve bonnets and other high spots where cleaning may not be effective due to air pockets should be precleaned and chlorinated. Calcium hypochlorite tablets may be crushed and placed in joints and hydrant branches to assist in the total disinfection process.

When applying chlorine by either the continuous or slug method, be sure that the chlorine fed is well mixed with the water used to fill the pipe. Solution feeders can be used to inject chlorine into the water (continuous or slug method) used to disinfect the main. If possible, recycle the flows from the continuous method to minimize the use of chlorine and any problems that might be encountered from disposal of water with a high chlorine residual.

Care must be exercised when disposing of all water with a high chlorine residual (greater than 1 mg/L). Possible means of disposal include sanitary sewers, storm or on land. If sanitary sewers are used, there sho_ equate dilution and travel time so there will be residual when the water reaches the wastew_ plant. Be sure to notify the plant operator _ storm sewer is used, be sure there is n_ remaining when the water reaches th_ (creek, river, or lake). Chlorine is toxi_

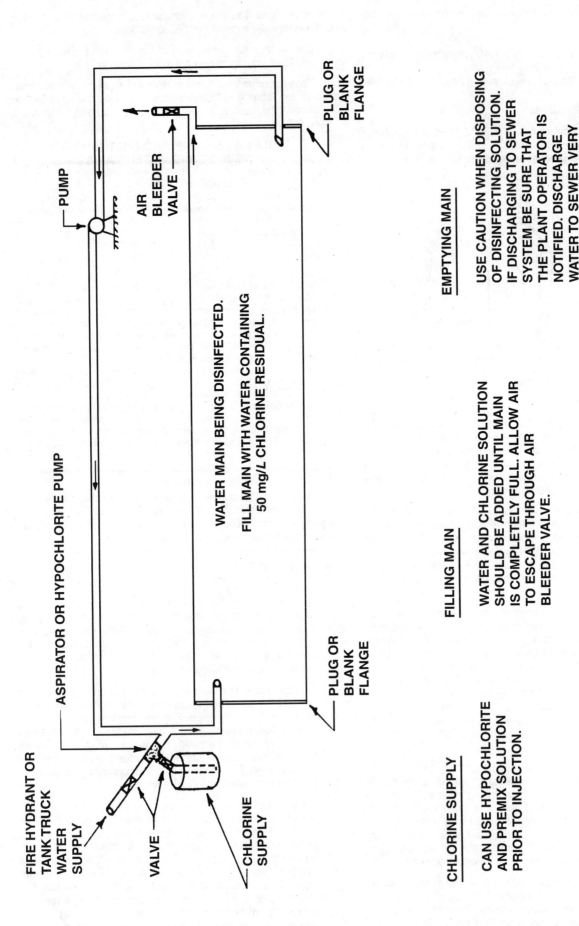

FIRE HYDRANT OR
TANK TRUCK
WATER
SUPPLY

ASPIRATOR OR HYPOCHLORITE PUMP

PUMP

AIR
BLEEDER
VALVE

VALVE

CHLORINE
SUPPLY

WATER MAIN BEING DISINFECTED.

FILL MAIN WITH WATER CONTAINING
50 mg/L CHLORINE RESIDUAL.

PLUG OR
BLANK
FLANGE

PLUG OR
BLANK
FLANGE

CHLORINE SUPPLY

CAN USE HYPOCHLORITE
AND PREMIX SOLUTION
PRIOR TO INJECTION.

FILLING MAIN

WATER AND CHLORINE SOLUTION
SHOULD BE ADDED UNTIL MAIN
IS COMPLETELY FULL. ALLOW AIR
TO ESCAPE THROUGH AIR
BLEEDER VALVE.

EMPTYING MAIN

USE CAUTION WHEN DISPOSING
OF DISINFECTING SOLUTION.
IF DISCHARGING TO SEWER
SYSTEM BE SURE THAT
THE PLANT OPERATOR IS
NOTIFIED. DISCHARGE
WATER TO SEWER VERY
SLOWLY OVER A PERIOD OF
TIME RATHER THAN IN
ONE LARGE SLUG.

Fig. 6.6   Disinfection of a water main by the continuous method

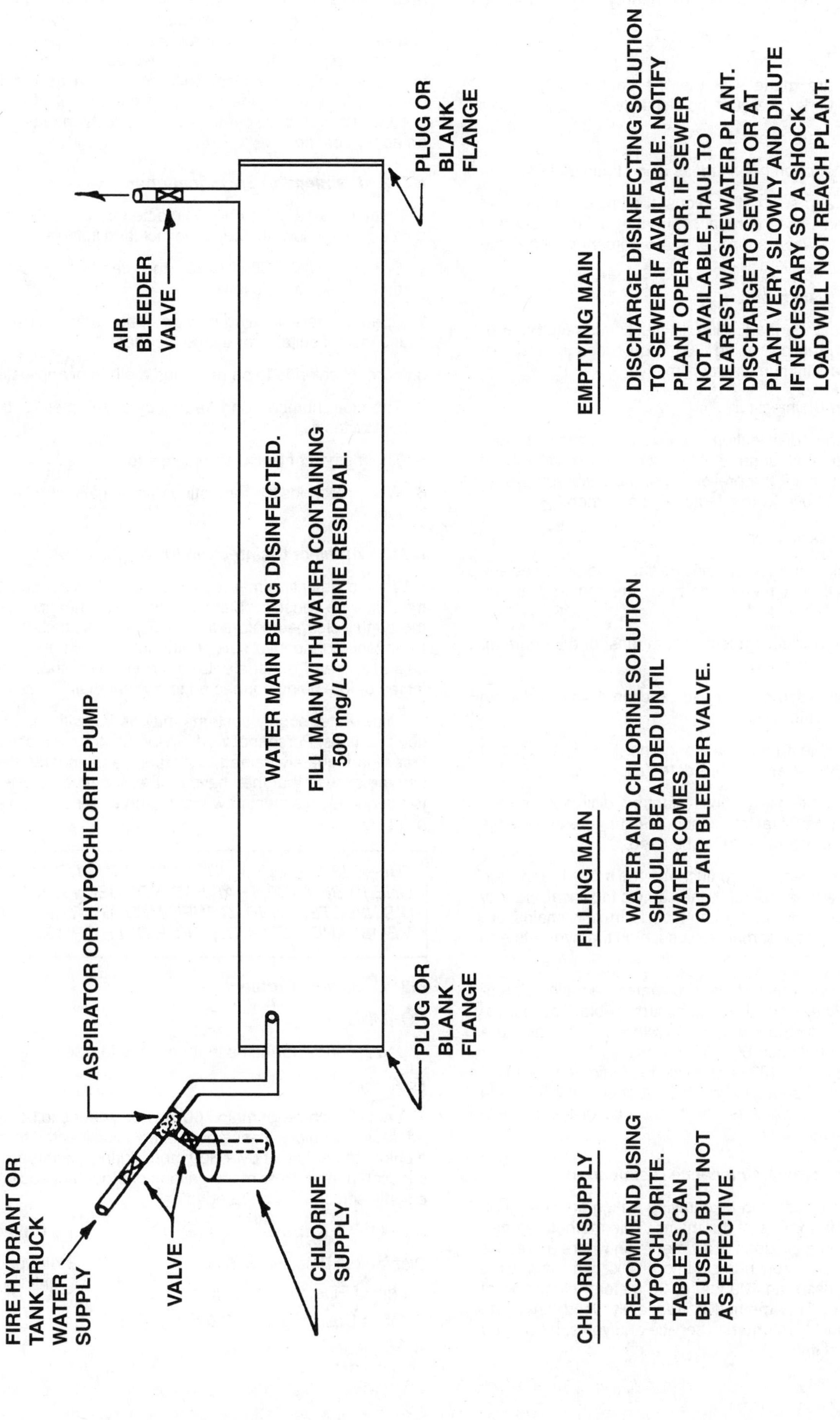

FIRE HYDRANT OR
TANK TRUCK
WATER
SUPPLY

ASPIRATOR OR HYPOCHLORITE PUMP

VALVE

CHLORINE
SUPPLY

PLUG OR
BLANK
FLANGE

WATER MAIN BEING DISINFECTED.

FILL MAIN WITH WATER CONTAINING
500 mg/L CHLORINE RESIDUAL.

AIR
BLEEDER
VALVE

PLUG OR
BLANK
FLANGE

CHLORINE SUPPLY

RECOMMEND USING
HYPOCHLORITE.
TABLETS CAN
BE USED, BUT NOT
AS EFFECTIVE.

FILLING MAIN

WATER AND CHLORINE SOLUTION
SHOULD BE ADDED UNTIL
WATER COMES
OUT AIR BLEEDER VALVE.

EMPTYING MAIN

DISCHARGE DISINFECTING SOLUTION
TO SEWER IF AVAILABLE. NOTIFY
PLANT OPERATOR. IF SEWER
NOT AVAILABLE, HAUL TO
NEAREST WASTEWATER PLANT.
DISCHARGE TO SEWER OR AT
PLANT VERY SLOWLY AND DILUTE
IF NECESSARY SO A SHOCK
LOAD WILL NOT REACH PLANT.

Fig. 6.7   Disinfection of a water main by slug method

aquatic life. Land disposal may be acceptable if percolation rates are high and there are no nearby wells pumping groundwater.

2. *MAIN REPAIRS*

a. If repairs are made with the line continuously full of water and under pressure, no disinfection is required.

b. Where lines are opened:

(1) Dewater open trench areas using trash pumps,

(2) Flush and swab all portions of all pipe, fittings, and materials used in repairs, and that will be in contact with the water supply, with a 5 percent hypochlorite solution,

(3) Flush system, and

(4) Disinfect the main. Slug disinfection is recommended if practical using 500 mg/L dosage and 30-minute minimum contact time.

D. Flushing After Disinfection

Flush lines after disinfection under all alternative procedures until residual chlorine is less than 1 mg/L. Velocity of flushing is not critical if the preventive and preliminary procedures described above were adequately performed.

E. Testing After Disinfection

After disinfection and prior to placing line in service, bacteriological tests (24-hour membrane filter) are required as follows:

1. In a chlorinated water system, test at least one sample for each section disinfected.

2. Test at least two samples for each section disinfected in an unchlorinated water system.

3. For long lines, test samples along line. Line in excess of 2,500 feet (750 meters) is considered a long line.

If bacteriological tests are unsatisfactory, disinfection must be repeated. Repeat of tablet method, of course, is impossible and an alternative procedure must be used.

A suggested procedure is to flush the main again and then take additional samples. Be sure the sample tap is satisfactory because many unsatisfactory samples from new mains are the result of using poor sample taps such as fire hydrants and blowoff valves.

If samples are still unsatisfactory, dewater the main as completely as possible. Use air compressors to blow out all remaining water. Refill the empty main with water containing a chlorine residual between 50 and 100 mg/L. Allow this water to stand in the main for 48 hours. Flush the main again and resample. Also collect samples from the water entering the main to be sure that this water is not the source of the positive coliform test results.

6.311 Emergency or Maintenance Disinfection

Mains may be disinfected by spraying a high concentration of chlorine on the insides of the mains. This method is often used for disinfecting sections of a water main where a break is repaired or where a crew has cut into an existing main, or in the very short (less than 100 feet or 30 meters) extension of an existing main. This method is frequently used when the main must be quickly returned to service and you can't wait for a 24-hour contact time.

Spray with a solution of one to five percent (10,000 to 50,000 mg/L) chlorine at a pressure of 100 psi (690 kPa or 7 kg/sq cm). An ordinary pressure-type, stainless-steel fire extinguisher will do the job. Spray the complete interior of each section of pipe or fitting as it is lowered into the trench. The stainless-steel extinguisher will not corrode as rapidly from contact with the chlorine solution if the interior of the tank is lined with a protective coating, such as a chlorinated rubber or an epoxy coating.

6.312 Disinfection Specifications

If water mains are to be disinfected by others as part of a contract, the following should be included in the specifications:

1. Specify AWWA C651-92 as the standard for disinfection procedures to be followed,

2. Specify where flushing may be done, rates of flushing, and location of suitable drainage facilities,

3. Form of chlorine to be used and method of application,

4. The type, number, and frequency of samples for bacteriological tests,

5. The method of taking samples, and

6. Who is responsible for testing and use of a certified laboratory.

6.313 Areas of Disinfection Problems

When disinfecting mains, there are certain aspects that are most likely to cause problems. Joints and connections must be thoroughly cleaned before the mains are disinfected. Whenever anyone connects a service line into a main, they must use clean tools and materials. Each service line should be disinfected before being connected to a water main.

Some water supply systems may be lax with their disinfection practices. Personnel working for these systems are sometimes inadequately trained, underpaid, and understaffed. They use the excuse that they have neither the time nor the budget to adequately disinfect new mains and existing mains after repairs.

> *THERE IS RARELY A JUSTIFICATION FOR PLACING A LINE IN SERVICE THAT HAS NOT BEEN ADEQUATELY DISINFECTED AND CONFIRMED BY THE 24-HOUR MEMBRANE FILTER TEST FOR COLIFORMS.*

6.314 Sample Problem

FORMULAS

Use the same formulas that are used to disinfect a well.

EXAMPLE 2

A new 6-inch water main 500 feet long needs to be disinfected. An initial chlorine dose of 400 mg/L is expected to maintain a chlorine residual of over 300 mg/L during the three-hour disinfection period. How many gallons of 5 percent sodium hypochlorite solution will be needed?

Known		**Unknown**
Diameter of Pipe, in	= 6 in	5% Hypochlorite, gallons
Length of Pipe, ft	= 500 ft	
Chlorine Dose, mg/L	= 400 mg/L	
Hypochlorite, %	= 5%	

1. Calculate the volume of water in the pipe in gallons.

$$\text{Pipe Volume, gallons} = (0.785)(\text{Diameter, ft})^2(\text{Length, ft})(7.48 \text{ gal/cu ft})$$

$$= \frac{(0.785)(6 \text{ in})^2(500 \text{ ft})(7.48 \text{ gal/cu ft})}{144 \text{ sq in/sq ft}}$$

$$= 734 \text{ gallons of water}$$

2. Determine the pounds of chlorine needed.

$$\text{Chlorine, lbs} = (\text{Volume, M Gal})(\text{Dose, mg/}L)(8.34 \text{ lbs/gal})$$

$$= (0.000734 \text{ M Gal})(400 \text{ mg/}L)(8.34 \text{ lbs/gal})$$

$$= 2.45 \text{ lbs Chlorine}$$

3. Calculate the gallons of 5 percent sodium hypochlorite solution needed.

$$\text{Sodium Hypochlorite Solution, gallons} = \frac{(\text{Chlorine, lbs})(100\%)}{(8.34 \text{ lbs/gal})(\text{Hypochlorite, \%})}$$

$$= \frac{(2.45 \text{ lbs})(100\%)}{(8.34 \text{ lbs/gal})(5\%)}$$

$$= 5.9 \text{ gallons}$$

Six gallons of 5 percent sodium hypochlorite solution should do the job.

6.32 Tanks

6.320 Procedures

Procedures for disinfecting tanks are similar to those used for mains. Thoroughly clean the tank after construction, maintenance, or repairs. Add chlorine to the water used to fill the tank during the disinfection process and mix thoroughly. Maintain a chlorine residual of at least 50 mg/L for at least six hours and preferably for 24 hours. When the disinfection procedure is completed, carefully dispose of the disinfection water using the same procedures as for disposal of water used to disinfect mains.

When disinfecting large tanks that hold more than one million gallons, it may not be practical or economical to fill and drain the tank with a disinfecting solution with a high chlorine residual. A solution to this problem is to fill the tank, increase the chlorine residual slightly, collect samples, and run bacteriological tests while keeping the tank full of water (see SPECIAL NOTE, Section 6.34). If results from the 24-hour membrane filter tests are acceptable, the tank may be placed in service. If necessary, dilute the tank contents as the water flows into the distribution system.

Another approach to disinfecting large tanks is to spray the walls with a jet of water containing a chlorine residual of 200 mg/L. Thoroughly spraying the walls with water with a high chlorine residual is an effective means of disinfecting large tanks. After the tank has been sprayed, allow the tank to stand unused for 30 minutes before filling. Fill the tank with distribution system water that has been treated with chlorine to provide a residual of 3 mg/L. Let the water in the tank stand for 3 to 6 hours. Operators doing the spraying should wear a self-contained breathing apparatus because the chlorine fumes from the spray water are very unpleasant and could be hazardous to your health. Be sure to provide plenty of ventilation.

See Chapter 5, Section 5.82, "Disinfection of Storage Facilities," page 264, for a more detailed discussion of three different methods of disinfecting water storage tanks. Typically, only one method is used for a particular storage facility disinfection, but combinations of the three methods' chlorine concentrations and contact times may be used.

6.321 Sample Problem

FORMULAS

Use the same formulas used to disinfect wells.

EXAMPLE 3

A new service storage reservoir needs to be disinfected before being placed in service. The tank is 8 feet high and 20 feet in diameter. An initial chlorine dose of 100 mg/L is expected to maintain a chlorine residual of over 50 mg/L during the 24-hour disinfection period. How many gallons of five percent sodium hypochlorite solution will be needed?

Known		Unknown
Diameter of Tank, ft	= 20 ft	5% Hypochlorite, gallons
Height of Tank, ft	= 8 ft	
Chlorine Dose, mg/L	= 100 mg/L	
Hypochlorite, %	= 5%	

1. Calculate the volume of water in the tank in gallons.

$$\text{Tank Volume, Gallons} = (0.785)(\text{Diameter, ft})^2(\text{Height, ft})(7.48 \text{ gal/cu ft})$$

$$= (0.785)(20 \text{ ft})^2(8 \text{ ft})(7.48 \text{ gal/cu ft})$$

$$= 18,800 \text{ gallons}$$

2. Determine the pounds of chlorine needed.

$$\text{Chlorine, lbs} = (\text{Vol Water, M Gal})(\text{Chlorine Dose, mg/}L)(8.34 \text{ lbs/gal})$$

$$= (0.0188 \text{ M Gal})(100 \text{ mg/}L)(8.34 \text{ lbs/gal})$$

$$= 15.68 \text{ lbs chlorine}$$

3. Calculate the gallons of 5 percent sodium hypochlorite solution needed.

$$\text{Sodium Hypochlorite Solution, gallons} = \frac{(\text{Chlorine, lbs})(100\%)}{(8.34 \text{ lbs/gal})(\text{Hypochlorite, \%})}$$

$$= \frac{(15.68 \text{ lbs})(100\%)}{(8.34 \text{ lbs/gal})(5\%)}$$

$$= 37.6 \text{ gallons}$$

Thirty-eight gallons of five percent sodium hypochlorite solution should do the job.

6.33 Water Treatment Plants

For additional information on disinfection at water treatment plants, see Chapter 4, "Small Water Treatment Plants," Section 4.5, "Disinfection," in *SMALL WATER SYSTEM OPERATION AND MAINTENANCE*, and Chapter 7, "Disinfection," in *WATER TREATMENT PLANT OPERATION*, Volume I, in this series of manuals.

6.34 Sampling

SPECIAL NOTE

Whenever you collect a sample for a bacteriological test (coliforms), be sure to use a sterile plastic or glass bottle. If the sample contains any chlorine residual, sufficient sodium thiosulfate should be added to neutralize all of the chlorine residual. Usually 0.1 mL of a 10 percent solution of sodium thiosulfate added to the 120-mL (4-oz) bottle before sterilization is sufficient, unless you are disinfecting mains or storage tanks. If the chlorine residual in the sample is greater than 15 mg/L, more "thio" is required to neutralize the chlorine.

QUESTIONS

Write your answers in a notebook and then compare your answers with those on pages 356 and 357.

6.3D   How can water mains be kept clean during construction or repair?

6.3E   Before a water main is disinfected, what flushing velocity should be used and for how long?

6.3F   What areas in a water main require extra effort for successful disinfection?

6.3G   List three forms of chlorine used for disinfection.

6.3H   What precautions should be taken when disposing of water with a high chlorine residual?

End of Lesson 1 of 2 Lessons on DISINFECTION

Please answer the discussion and review questions next.

DISCUSSION AND REVIEW QUESTIONS

Chapter 6. DISINFECTION

(Lesson 1 of 2 Lessons)

At the end of each lesson in this chapter you will find some discussion and review questions. The purpose of these questions is to indicate to you how well you understand the material in the lesson. Write the answers to these questions in your notebook before continuing.

1. Why is drinking water disinfected?

2. If a water is disinfected, will it be safe to drink?

3. How would you determine the chlorine dose for water?

4. What is the chlorine demand of a water?

5. Why should a chlorine residual be maintained in a water distribution system?

6. Why would you consider recycling water used to disinfect a water main when using the continuous chlorination method?

7. How would you dispose of the water used to disinfect a water main when using the slug chlorination method?

CHAPTER 6. DISINFECTION

(Lesson 2 of 2 Lessons)

6.4 OPERATION OF CHLORINATION EQUIPMENT

6.40 Description of Various Units

6.400 *Field Equipment*

1. *HYPOCHLORINATORS* (equipment that feeds liquid chlorine (bleach) solutions)

Hypochlorinators used on small water systems are very simple and relatively easy to install. Typical installations are shown in Figures 6.8 and 6.9. Hypochlorinator systems usually consist of a chemical solution tank for the hypochlorite, diaphragm-type pump (Figure 6.10), power supply, water pump, pressure switch, and water storage tank.

There are two methods of feeding the hypochlorite solution into the water being disinfected. The hypochlorite solution may be pumped directly into the water (Figure 6.11). In the other method, the hypochlorite solution is pumped through an *EJECTOR*[25] (also called an eductor or injector) which draws in additional water for dilution of the hypochlorite solution (Figure 6.12).

2. *CHLORINATORS* (equipment that feeds gaseous chlorine)

Disinfection by means of gaseous chlorine is typically accomplished in small systems with the equipment shown in Figure 6.13. For small water treatment systems, small chlorinators that are mounted directly on a chlorine container (as shown in Figures 6.13, 6.14, and 6.15) have proven to be

TYPICAL INSTALLATION

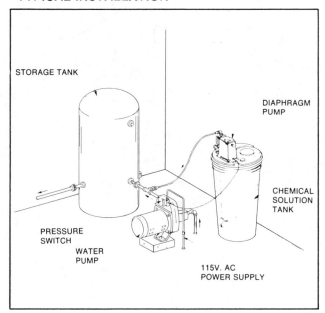

STORAGE TANK

DIAPHRAGM PUMP

CHEMICAL SOLUTION TANK

PRESSURE SWITCH

WATER PUMP

115V. AC POWER SUPPLY

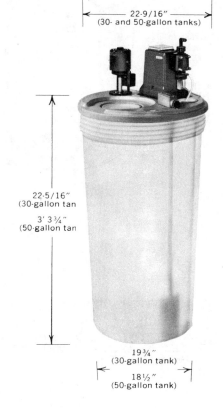

22-9/16" (30- and 50-gallon tanks)

22-5/16" (30-gallon tan

3' 3¾" (50-gallon tan

19¾" (30-gallon tank)

18½" (50-gallon tank)

Pump-tank system for chemical mixing and metering. Cover supports pump, impeller-type mixer, and liquid-level switch.

Fig. 6.8 Typical hypochlorinator installation
(Permission of Wallace & Tiernan Division, Pennwalt Corporation)

---

[25] *Ejector. A device used to disperse a chemical solution into water being treated.*

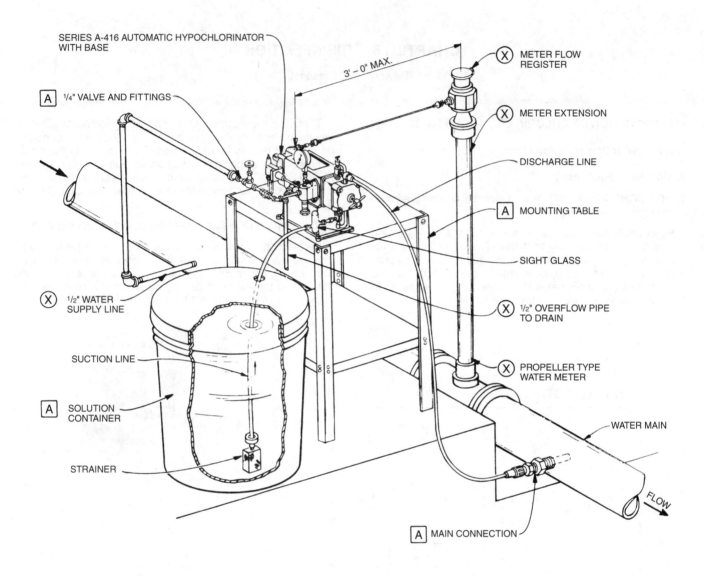

SERIES A-416 AUTOMATIC HYPOCHLORINATOR
WITH BASE

3' – 0" MAX.

Ⓧ METER FLOW
REGISTER

Ⓐ ¼" VALVE AND FITTINGS

Ⓧ METER EXTENSION

DISCHARGE LINE

Ⓐ MOUNTING TABLE

SIGHT GLASS

Ⓧ ½" WATER
SUPPLY LINE

Ⓧ ½" OVERFLOW PIPE
TO DRAIN

SUCTION LINE

Ⓧ PROPELLER TYPE
WATER METER

Ⓐ SOLUTION
CONTAINER

WATER MAIN

STRAINER

FLOW

Ⓐ MAIN CONNECTION

Ⓧ *NOT FURNISHED BY W & T.*

Ⓐ *ACCESSORY ITEM FURNISHED ONLY IF
SPECIFICALLY LISTED IN QUOTATION
AND AS CHECKED ON THIS DRAWING.*

NOTE: Hypochlorinator paced by a propeller-type water meter.

Fig. 6.9   Typical hypochlorinator installation
(Permission of Wallace & Tiernan Division, Pennwalt Corporation)

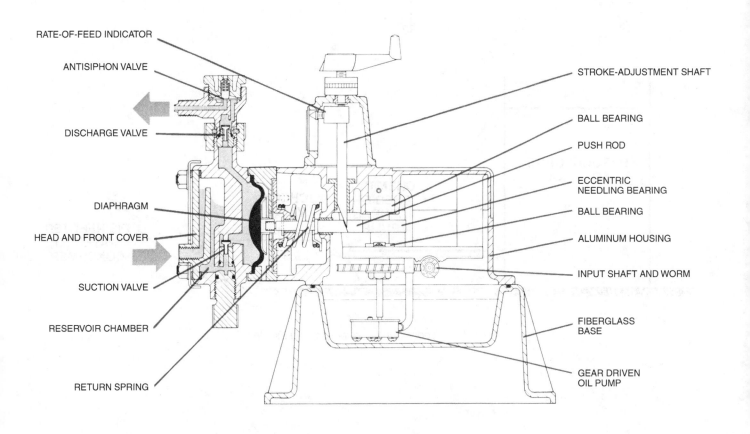

RATE-OF-FEED INDICATOR

ANTISIPHON VALVE

DISCHARGE VALVE

DIAPHRAGM

HEAD AND FRONT COVER

SUCTION VALVE

RESERVOIR CHAMBER

RETURN SPRING

STROKE-ADJUSTMENT SHAFT

BALL BEARING

PUSH ROD

ECCENTRIC NEEDLING BEARING

BALL BEARING

ALUMINUM HOUSING

INPUT SHAFT AND WORM

FIBERGLASS BASE

GEAR DRIVEN OIL PUMP

Belt guard removed to show step pulley

Fig. 6.10   Diaphragm-type pump
(Permission of Wallace & Tiernan Division, Pennwalt Corporation)

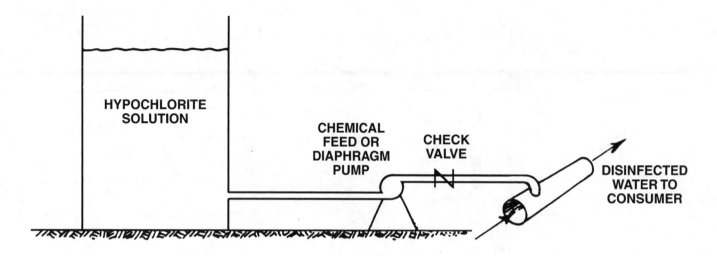

Fig. 6.11   Hypochlorinator direct pumping system

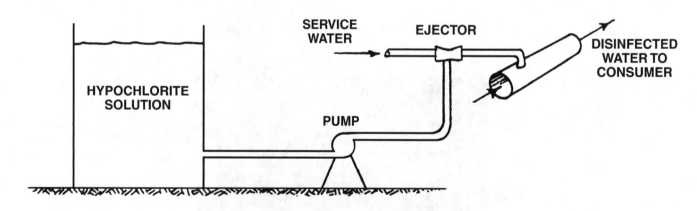

NOTE: Pump is chemical feed or diaphragm pump.

Fig. 6.12   Hypochlorinator injector feed system

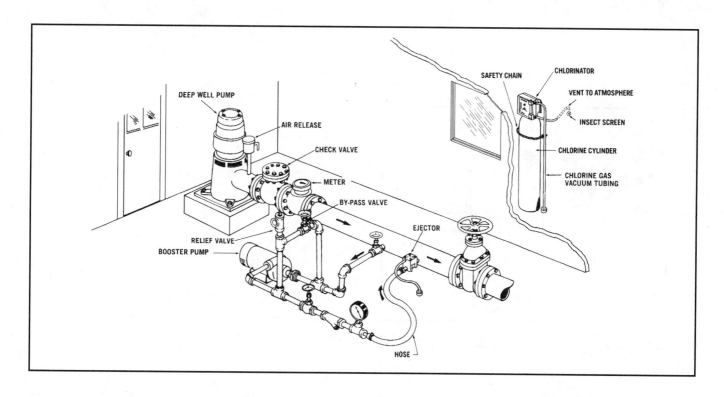

Fig. 6.13 Typical deep well chlorination system
(Permission of Capital Controls Company, Colmar, PA)

safer, easier to operate and maintain, and less expensive to install than larger in-place chlorination systems, yet they provide the same reliable service.

A direct mounted chlorinator meters prescribed (preset, or selected) doses of chlorine gas from a chlorine cylinder, conveys it under a vacuum, and injects it into the water supply. Direct cylinder mounting is the safest and simplest way to connect the chlorinator to the chlorine cylinder. The valves on the cylinder and chlorinator inlet are connected by a positive metallic yoke, which is sealed by a single lead or fiber gasket.

CHLORINATOR PARTS AND THEIR PURPOSE

THE EJECTOR: The ejector, fitted with a Venturi nozzle, creates the vacuum that moves the chlorine gas. Water supplied by a pump moves across the Venturi nozzle creating a differential pressure which establishes the vacuum. The gas chlorinator is able to transport the chlorine gas to the water supply by reducing the gas pressure from the chlorine cylinder to less than the atmospheric pressure (vacuum). Figure 6.13 illustrates such an arrangement. The flow diagrams in Figure 6.14 are cutaway views of the ejector and check valve assembly.

In the past it was not uncommon to find the ejector and the vacuum regulator mounted inside some type of cabinet. However, it makes better sense to locate the ejector at the site where the chlorine is to be applied, eliminating the necessity of pumping the chlorine over long distances and the associated problems inherent with gas pressure lines. Also, by placing the ejector at the application point, any tubing break will cause

the chlorinator to shut down. This halting of operation stops the flow of gas and any damage that could result from a chlorine solution leak.

CHECK VALVE ASSEMBLY: The vacuum created by the ejector moves through the check valve assembly. This assembly prevents water from back-feeding, that is, entering the vacuum-regulator portion of the chlorinator (Figure 6.14).

RATE VALVE: The rate valve controls the flow rate at which chlorine gas enters the chlorinator. The rate valve controls the vacuum level and thus directly affects the action of the diaphragm assembly in the vacuum regulator. A reduction in vacuum lets the diaphragm close, causing the needle valve to reduce the inlet opening which restricts chlorine gas flow to the chlorinator. An increase in the rate valve setting applies more vacuum to the diaphragm assembly, pulling the needle valve back away from the inlet opening and permitting an increased chlorine gas flow rate.

DIAPHRAGM ASSEMBLY: This assembly connects directly to the inlet valve of the vacuum regulator, as described above. A vacuum (of at least 20 inches (508 mm) of water column) exists on one side of the diaphragm; the other side is open to atmospheric pressure through the vent. This differential in pressure causes the diaphragm to open the chlorine inlet valve allowing the gas to move (under vacuum) through the *ROTAMETER,*[26] past the rate valve and through the tubing to the check valve assembly, into the ejector nozzle area, and then to the point of application. If for some reason the vacuum is lost, the diaphragm will seat the needle valve on the inlet, stopping chlorine gas flow to the chlorinator.

---

[26] *Rotameter (RODE-uh-ME-ter).   A device used to measure the flow rate of gases and liquids. The gas or liquid being measured flows vertically up a tapered, calibrated tube. Inside the tube is a small ball or bullet-shaped float (it may rotate) that rises or falls depending on the flow rate. The flow rate may be read on a scale behind or on the tube by looking at the middle of the ball or at the widest part or top of the float.*

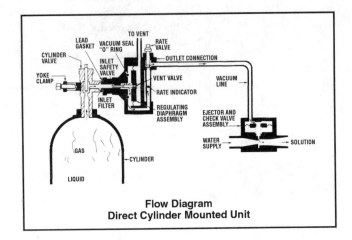

**Flow Diagram
Direct Cylinder Mounted Unit**

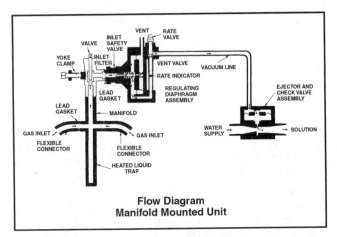

**Flow Diagram
Manifold Mounted Unit**

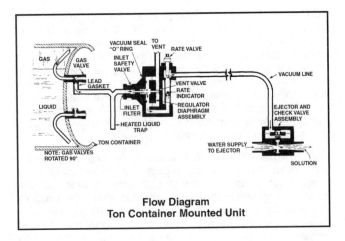

**Flow Diagram
Ton Container Mounted Unit**

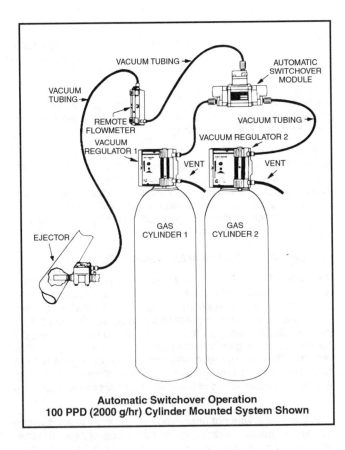

**Automatic Switchover Operation
100 PPD (2000 g/hr) Cylinder Mounted System Shown**

Fig. 6.14   Typical chlorinator flow diagrams
(Permission of Capital Controls Company, Colmar, PA)

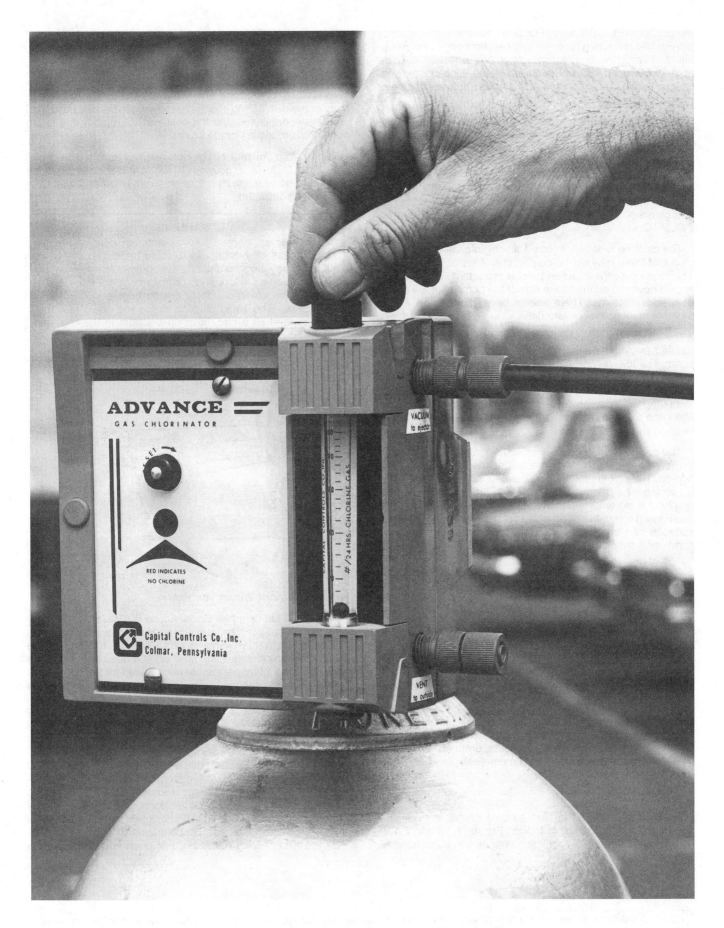

Fig. 6.15   Chlorinator with rotameter showing feed rate in lbs/24 hr chlorine gas

(Permission of Capital Controls Company, Colmar, PA)

INTERCONNECTION MANIFOLD: If several gas cylinders provide the chlorine gas, direct cylinder mounting is not possible. An interconnection manifold made of seamless steel pipe and flexible connectors of cadmium-plated copper fitted with isolation valves must be used as the bridge between the chlorinator and the various cylinders.

The steel gas manifold with chlorine valve is mounted to the chlorinator. The flexible connector links the rigid manifold and the chlorine cylinder. The isolation valve between the flexible connector and the cylinder valve provides a way to close off the flexible connector when a new gas cylinder must be attached. This limits the amount of moisture that enters the system. Moisture in the system will combine with the chlorine gas and cause corrosion. *CORROSION CAN CAUSE THE MANIFOLD TO FAIL.*

The chlorine is usually injected directly into the water supply pipe and there may not be contact chambers or mixing units. The location of the injection point is important. The injection should never be on the intake side of the pump as it will cause corrosion problems. There should be a check valve and a meter to monitor the chlorine dose.

On most well applications a chlorine booster pump is needed to overcome the higher water pump discharge pressures. The low-volume, high-pressure booster pump shown in Figure 6.13 must have extremely small clearance between the impeller and the casing. If the well produces sand in the water, this pump will wear rapidly and become unreliable. In this situation, the chlorine solution should be introduced down the well through a polyethylene tube.

The polyethylene tube (1/2 inch or 12 mm) must be installed in the well so as to discharge a few inches below the suction screen. The chlorinator should operate *ONLY* when the pump is running. The chlorine solution flowing through the polyethylene tube is extremely corrosive. If the tube does not discharge into flowing water, the effect of the solution impinging on metal surfaces can be disastrous. Wells have been destroyed by corrosion from chlorine.

6.401 Chlorine Containers

1. HYPOCHLORINATORS

Plastic containers make suitable chlorine solution tanks (Figures 6.8 and 6.9). The size needed will depend on usage, but containers should be large enough to hold a two or three days' supply of hypochlorite solution. A fresh solution should be prepared every two or three days because the solution may lose its strength over time and this will affect the actual chlorine feed rate. Normally a week's supply of hypochlorite should be in storage and available for preparing hypochlorite solutions. Store the hypochlorite in a cool, dark place. Sodium hypochlorite can lose from two to four percent of its available chlorine content per month at room temperatures. Therefore, manufacturers recommend a maximum shelf life of 60 to 90 days.

2. GAS CHLORINATORS

Chlorine is delivered for use by chlorinators in 100- and 150-pound (45- to 68-kg) cylinders (Figures 6.16 and 6.17), one-ton (900-kg) tanks or chlorine tank cars in sizes from 16 to 90 tons (14,500 to 81,800 kg). The 100- and 150-pound (45- to 68-kg) cylinders will be discussed in this section.

These cylinders are usually made of seamless carbon steel. A fusible plug is placed in the valve below the valve seat (Figure 6.18). This plug is a safety device. The fusible metal softens or melts at 158 to 165°F (70 to 74°C) to prevent buildup of excessive pressures and the possibility of rupture due to fire or high surrounding temperatures.

The maximum rate of chlorine removal from a 150-pound (68-kg) cylinder is 40 pounds (18 kg) of chlorine per day. If the rate of removal is greater, "freezing" can occur and less chlorine will be delivered.

WARNING

When frost appears on valves and flex connectors, the chlorine gas may condense to liquid (reliquify). The liquid chlorine may plug the chlorine supply lines (sometimes this is referred to as chlorine ice or frozen chlorine). If you disconnect the chlorine supply line to unplug it, *BE VERY CAREFUL.* The liquid chlorine in the line could reevaporate, expand as a gas, build up pressure in the line, and cause liquid chlorine to come shooting out the open end of a disconnected chlorine supply line.

6.402 Chlorine Room Ventilation

The 1991 Uniform Fire Code, Article 80, Hazardous Materials, states that exhaust ventilation for chlorine shall be taken from a point within twelve inches (30.5 centimeters) of the floor. Mechanical ventilation shall be at a rate of not less than one cubic foot of air per minute per square foot (0.00508 cubic meter of air per second per square meter) of floor area in the storage area. (The system should not draw air through the fan itself because chlorine gas can damage the fan motor.) Normally ventilated air from chlorine storage rooms is discharged to the atmosphere. When a chlorine leak occurs, the ventilated air containing the chlorine shall be routed to a treatment system. Treatment systems shall be used to process all exhaust ventilation containing chlorine from a leak that will be discharged from chlorine storage rooms. A caustic scrubbing system can be used to treat air containing chlorine from a leak. Treatment systems shall be designed to reduce the maximum allowable discharge concentration of chlorine to one-half the *IDLH*[27] level at the point of discharge to the atmos-

---

[27] *IDLH. **I**mmediately **D**angerous to **L**ife or **H**ealth. The atmospheric concentration of any toxic, corrosive, or asphyxiant substance that poses an immediate threat to life or would cause irreversible or delayed adverse health effects or would interfere with an individual's ability to escape from a dangerous atmosphere.*

NOTES:

1. Scale for weighing chlorine cylinders and chlorine.
2. Flexible tubing (pigtail).
3. Cylinders chained to wall.

Fig. 6.17   Typical chlorine cylinder station for water treatment
(Courtesy of PPG Industries)

Chlorine Cylinder

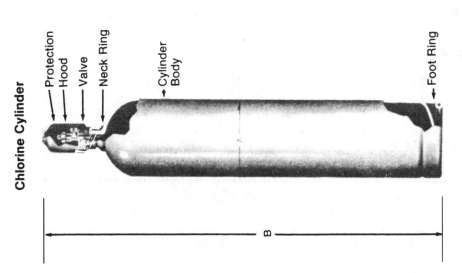

Net Cylinder Contents	Approx. Tare, Lbs.*	Dimensions, Inches	
		A	B
100 Lbs.	73	8 ¹/₄	54 ¹/₂
150 Lbs.	92	10 ¹/₄	54 ¹/₂

*Stamped tare weight on cylinder shoulder does not include valve protection hood.

Fig. 6.16   Chlorine cylinder
(Courtesy of PPG Industries)

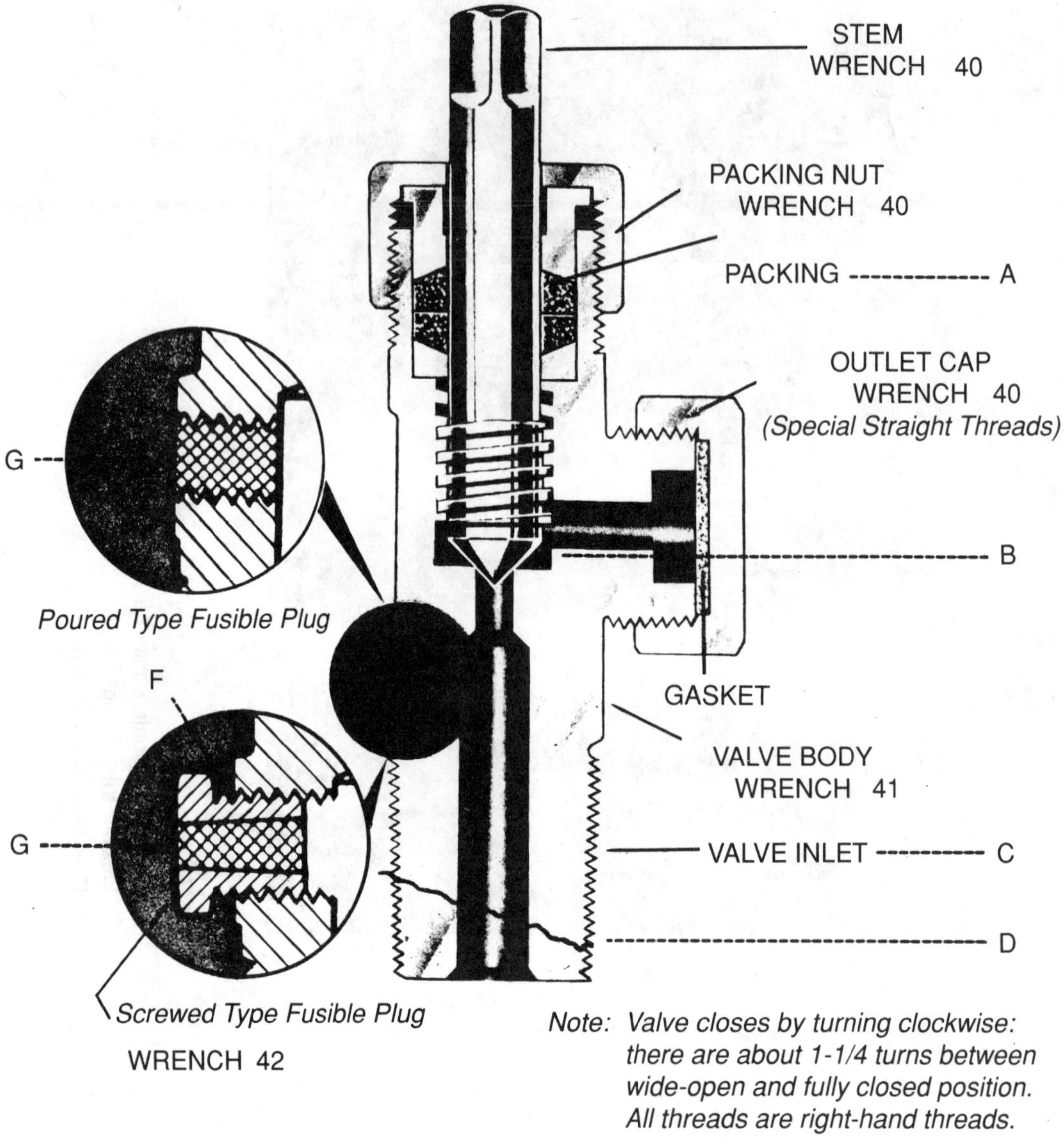

STEM
WRENCH   40

PACKING NUT
WRENCH   40

PACKING ------------- A

OUTLET CAP
WRENCH   40
(Special Straight Threads)

B

GASKET

VALVE BODY
WRENCH   41

VALVE INLET --------- C

D

G --- *Poured Type Fusible Plug*

F

G --- *Screwed Type Fusible Plug*
WRENCH   42

Note: *Valve closes by turning clockwise: there are about 1-1/4 turns between wide-open and fully closed position. All threads are right-hand threads.*

TYPICAL VALVE LEAKS OCCUR THROUGH ...

A - VALVE PACKING GLAND

B - VALVE SEAT

C - VALVE INLET THREADS

D - BROKEN VALVE

E - VALVE BLOWN OUT

F - FUSIBLE PLUG THREADS

G - FUSIBLE METAL OF PLUG

H - VALVE STEM BLOWN OUT
      (not shown)

Fig. 6.18   *Standard chlorine cylinder valve*
(Permission of Chlorine Specialties, Inc.)

phere. The IDLH for chlorine is 10 ppm. A secondary standby source of power is required for the chlorine detection, alarm, ventilation, and treatment systems.

6.41 Chlorine Handling

Cylinders containing 100 to 150 pounds (45 to 68 kg) of chlorine are convenient for very small plants with capacities less than 0.5 MGD (1,890 cu m/day or 1.89 ML/day).[28]

The following are procedures for safely handling chlorine cylinders.

1. Move cylinders with a properly balanced hand truck with clamp supports that fasten at least two-thirds of the way up the cylinder (Figure 6.19).

Fig. 6.19   Hand truck for chlorine cylinder
(Courtesy of PPG Industries)

2. 100- and 150-pound (45- to 68-kg) cylinders can be rolled in a vertical position. Avoid lifting these cylinders except with approved equipment. Never lift with chains, rope slings, or magnetic hoists.

3. Always replace the protective cap when moving a cylinder.

4. Keep cylinders away from direct heat (steam pipes or radiators).

5. Store cylinders in an upright position. All empty chlorine cylinders (depressurized) must be tagged as empty.

6. Remove the outlet cap from the cylinder and inspect the threads on the outlet. Cylinders having outlet threads that are corroded, worn, cross-threaded, broken, or missing should be rejected and returned to the supplier.

7. The specifications and regulations of the U.S. Interstate Commerce Commission require that chlorine cylinders be tested at 800 psi (5,516 kPa or 56.24 kg/sq cm) every five

years. The date of testing is stamped on the dome of the cylinder. Cylinders that have not been tested within that period of time should be rejected and returned to the supplier.

QUESTIONS

Write your answers in a notebook and then compare your answers with those on page 357.

6.4A   List the two major types of chlorine feeders.

6.4B   Why are chlorine booster pumps needed on most well applications?

6.4C   What is the maximum rate of chlorine removal from a 150-pound cylinder?

6.42 Performance of Chlorination Units

6.420 Hypochlorinators

1. *HYPOCHLORINATOR START-UP*

   a. Solution. Chemical solutions have to be made up. Most agencies buy commercial or industrial hypochlorite at around 12 to 15 percent chlorine. This solution is usually diluted down to a two percent solution. If using commercially prepared solutions, dosage rates will have to be calculated.

   b. Electrical. Lock out the circuit while making an inspection of an electric circuit. Normally no adjustments are needed. Look for frayed wires. Turn the power back on. Leave the solution switch off.

   c. Turn the chemical pump on. Make any necessary adjustments while the pump is running. Never adjust the pump while it is off because damage to the pump will occur.

   d. Make sure the solution is being fed into the system.

   e. Check the chlorine residual in the system. Residual should be at least 0.2 mg/L throughout the distribution system. Adjust the chemical feed as required.

2. *HYPOCHLORINATOR SHUTDOWN*

   a. Short Duration

      (1) Turn the water supply pump off. You do not want to pump any unchlorinated water and possibly contaminate the rest of the system.

---

[28] ML/day.   Megaliters or million liters per day.

(2) Turn the hypochlorinator off.

(3) When making any repairs, lock out the circuit or pull the plug from an electric socket.

b. Long Duration

(1) Obtain another hypochlorinator as a replacement if your existing hypochlorinator will be out of service.

3. *NORMAL OPERATION OF HYPOCHLORINATOR*

Normal operation of the hypochlorination process requires routine observation and preventive maintenance.

DAILY

a. Inspect the building to make sure only authorized personnel have been there.

b. Read and record the level of the solution tank at the same time every day. When preparing hypochlorite solutions, prepare only enough for a two- or three-day supply.

c. Read the meters and record the amount of water pumped.

d. Check the chlorine residual (0.2 mg/L) in the system and adjust the chlorine feed rate as necessary. Try to maintain a chlorine residual of 0.2 mg/L at the most remote point in the distribution system. The suggested free chlorine residual for treated water or well water is 0.5 mg/L at the point of chlorine application provided the 0.2 mg/L is maintained throughout the distribution system and coliform test results are negative.

e. Check the chemical feed pump operation. Most hypochlorinators have a dial with a range from 0 to 10 which indicates the chlorine feed rate. Initially set the pointer on the dial to approximately 6 or 7 on the dial and use a two percent hypochlorite solution. The pump should be operated in the upper ranges of the dial. This will require the frequency of the strokes or pulses from the pump to be frequent enough so that the chlorine will be fed continuously to the water being treated. Adjust feed rate after testing chlorine residual levels.

WEEKLY

a. Clean the building.

b. Replace the chemicals and wash the chemical storage tank. Try to have a 15- to 30-day supply of chlorine in storage for future needs. When preparing hypochlorite solutions, prepare only enough for a two- or three-day supply.

MONTHLY

a. Check the operation of the check valve.

b. Perform any preventive maintenance suggested by the manufacturer.

c. Cleaning

Commercial sodium hypochlorite solutions (such as Clorox) contain an excess of caustic (sodium hydroxide or NaOH). When this solution is diluted with water containing calcium and also carbonate alkalinity, the resulting solution becomes supersaturated with calcium carbonate. This calcium carbonate tends to form a coating on the poppet valves in the solution feeder. The coated valves will not seal properly and the feeder will fail to feed properly.

Use the following procedure to remove the carbonate scale:

(1) Fill a one-quart (one-liter) Mason jar half full of tap water.

(2) Place one fluid ounce (20 mL) of 30 to 37 percent hydrochloric acid (swimming pool acid) in the jar. *ALWAYS ADD ACID TO WATER, NEVER THE REVERSE.*

(3) Fill the jar with tap water.

(4) Place the suction hose of the hypochlorinator in the jar and pump the entire contents of the jar through the system.

(5) Return the suction hose to the hypochlorite solution tank and resume normal operation.

You can prevent the formation of the calcium carbonate coatings by obtaining the dilution water from an ordinary home water softener.

NORMAL OPERATION CHECKLIST

a. Check chemical usage. Record solution level and the water pump meter reading or number of hours of pump operation. Calculate the amount of chemical solution used and compare with the desired feed rate. See Example 4.

b. Determine if every piece of equipment is operating.

c. Inspect the lubrication of the equipment.

d. Check the building for any possible problems.

e. Clean up the area.

FORMULAS

To determine the actual chlorine dose of water being treated by either a chlorinator or a hypochlorinator, we need to know the gallons of water treated and the pounds of chlorine used to disinfect the water.

1. Calculate the pounds of water disinfected.

Water, lbs = (Water Pumped, gallons)(8.34 lbs/gal)

2. To calculate the volume of hypochlorite used from a container, we need to know the dimensions of the container.

$$\text{Volume, gallons} = \frac{(0.785)(\text{Diameter, in})^2(\text{Depth, ft})(7.48 \text{ gal/cu ft})}{144 \text{ sq in/sq ft}}$$

3. To calculate the pounds of chlorine used to disinfect water, we need to know the gallons of hypochlorite used and the percent available chlorine in the hypochlorite solution.

$$\text{Chlorine, lbs} = (\text{Hypochlorite, gal})(8.34 \text{ lbs/gal})\left(\frac{\text{Hypochlorite, \%}}{100\%}\right)$$

4. To determine the actual chlorine dose in milligrams per liter, divide the pounds of chlorine used by the millions of pounds of water treated. Pounds of chlorine per million pounds of water is the same as parts per million or milligrams per liter.

$$\text{Chlorine Dose, mg}/L = \frac{\text{Chlorine Used, lbs}}{\text{Water Treated, Million lbs}}$$

EXAMPLE 4

Water pumped from a well is disinfected by a hypochlorinator. A chlorine dosage of 1.2 mg/L is necessary to maintain an adequate chlorine residual throughout the system. During a one-week time period, the water meter indicated that 2,289,000 gallons of water were pumped. A two-percent sodium hypochlorite solution is stored in a three-foot diameter plastic tank. During this one-week period, the level of hypochlorite in the tank dropped 2 feet, 8 inches (2.67 feet). Does the chlorine feed rate appear to be too high, too low, or about right?

Known		**Unknown**
Desired Chlorine Dose, mg/L	= 1.2 mg/L	1. Actual Chlorine Dose, mg/L 2. Is Actual Dose OK?
Water Pumped, gal	= 2,289,000 gal	
Hypochlorite, %	= 2%	
Chemical Tank Diameter, ft	= 3 ft	
Chemical Drop in Tank, ft	= 2.67 ft	

1. Calculate the pounds of water disinfected.

   Water, lbs = (Water Pumped, gallons)(8.34 lbs/gal)

   = (2,289,000 gal)(8.34 lbs/gal)

   = 19,090,000 lbs

   or = 19.09 Million lbs

2. Calculate the volume of 2 percent sodium hypochlorite used in gallons.

   $$\text{Hypochlorite, gallons} = (0.785)(\text{Diameter, ft})^2(\text{Depth, ft})(7.48 \text{ gal/cu ft})$$

   $$= (0.785)(3 \text{ ft})^2(2.67 \text{ ft})(7.48 \text{ gal/cu ft})$$

   $$= 141.1 \text{ gallons}$$

3. Determine the pounds of chlorine used to disinfect the water.

   $$\text{Chlorine, lbs} = (\text{Hypochlorite, gal})(8.34 \text{ lbs/gal})\left(\frac{\text{Hypochlorite, \%}}{100\%}\right)$$

   $$= (141.1 \text{ gal})(8.34 \text{ lbs/gal})\left(\frac{2\%}{100\%}\right)$$

   $$= 23.5 \text{ lbs Chlorine}$$

4. Calculate the chlorine dosage in mg/L.

   $$\text{Chlorine Dose, mg}/L = \frac{\text{Chlorine Used, lbs}}{\text{Water Treated, Million lbs}}$$

   $$= \frac{23.5 \text{ lbs Chlorine}}{19.09 \text{ Million lbs Water}}$$

   $$= 1.23 \text{ mg}/L$$

Since actual chlorine dose (1.23 mg/L) was slightly greater than the desired dose of 1.2 mg/L, the chlorine feed rate appears OK.

4. *ABNORMAL OPERATION OF HYPOCHLORINATOR*

   a. Inform your supervisor of the problem.

   b. If the hypochlorinator malfunctions, it should be repaired immediately. See the shutdown operation (Step 2 in this section).

   c. Solution tank level.

      (1) If Too Low— Check the adjustment of the pump. Check the hour meter of the water pump.

      (2) If Too High—Check the chemical pump. Check the hour meter of the water pump.

   d. Determine if the chemical pump is not operating.

TROUBLESHOOTING GUIDELINES

(1) Check the electrical connection.
(2) Check the circuit breaker.
(3) Check for stoppages in the flow lines.

CORRECTIVE MEASURES

(1) Shut off the water pump so that no contaminated water is pumped into the system.
(2) Check for a blockage in the solution tank.
(3) Check the operation of the check valve.
(4) Check the electrical circuits.
(5) Replace the chemical feed pump with another pump while repairing the defective unit.

e. The solution is not being pumped into the water line.

TROUBLESHOOTING GUIDELINES

(1) Check the solution level.
(2) Check for blockages in the solution line.

5. *MAINTENANCE OF HYPOCHLORINATORS*

Hypochlorinators on small systems are normally small, sealed systems that cannot be repaired so replacement of the entire unit is the only solution. Maintenance requirements are normally minor such as changing the oil and lubricating the moving parts. Review the manufacturer's specifications for maintenance procedures.

QUESTIONS

Write your answers in a notebook and then compare your answers with those on page 357.

6.4D   When should the level of the hypochlorite solution tank be read?

6.4E   What is the basis for adjusting the chemical feed of a hypochlorinator?

6.4F   What is the basis for determining the strength of the hypochlorite solution in the solution tank?

6.4G   What maintenance is usually required on hypochlorinators?

6.421 *Chlorinators*

1. *SAFETY EQUIPMENT REQUIRED AND AVAILABLE OUTSIDE THE CHLORINATOR ROOM*

   a. Protective clothing

      (1) Gloves
      (2) Rubber suit

   b. Self-contained positive pressure/demand air supply system (Figure 6.20)

   c. Chlorine leak detector/warning device should be located outside the room storing chlorine and should have a battery backup in case of a power failure. The chlorine sensor, however, should be in the chlorine room and connected to the leak detector/warning device located outside the room.

2. *START-UP OF CHLORINATORS*

   Work in pairs. Never work alone when hooking up a chlorine system.

   a. Inspect the chlorine container for leaks. Position the container or chlorine cylinder in its location for connection. Install safety chains or locking devices to prevent cylinder or container movement. Remove the valve protective hood on cylinders, or the valve bonnet on one-ton containers.

   b. Inspect the chlorination equipment.

   c. Use new gaskets for connection of the chlorinator to the chlorine cylinder. The chlorine supplier usually attaches two fiber gaskets to the cylinder valve for connection. Most agencies prefer to use lead gaskets because they seat better, permitting fewer connection leaks. Fiber gaskets tend to leave a deposit of fiber material on the faces of both the cylinder valve outlet and the chlorinator inlet. This deposited material must be scraped or wire brushed off both faces to obtain a leakproof connection.

   d. Prior to the hookup, inspect for moisture and foreign substances in the lines.

   e. Have an ammonia bottle readily available to detect any leaks in the system. Dip a rag on the end of a stick in the ammonia[29] bottle and place the rag near the location of suspected leaks. A white cloud will reveal the location of a chlorine leak.

   f. Have safety equipment available which may be used in case of a leak.

   g. Hook up the chlorinator to the chlorine container with the chlorine valve turned off. Use the gas side (not liquid side) if using a ton tank. Remove the chlorine cylinder valve outlet cap and check the valve outlet face for burrs, deep scratches, or debris left from former connections; clean or wire brush if necessary. If the valve face is smooth, clean, and free of deep cuts or corrosion, proceed with hooking up the cylinder. If there is any evidence of damage to the valve face, replace the outlet cap and protection hood, and ask the supplier to pick up the cylinder and replace the damaged valve. Check the inlet face of the chlorinator and clean if necessary. Place a *NEW LEAD GASKET* on the chlorinator inlet, place the chlorinator on the cylinder valve, install the yoke clamp, and slowly tighten the yoke until the two faces are against the lead gasket. Continue to slowly tighten the yoke, compressing the gasket connection one-half to three-quarters of a turn. *DO NOT OVERTIGHTEN;* this can damage the yoke on the chlorinator resulting in a leak.

---

[29] *Use a concentrated ammonia solution containing 28 to 30 percent ammonia as NH_3 (this is the same as 58 percent ammonium hydroxide, NH_4OH, or commercial 26° Baumé).*

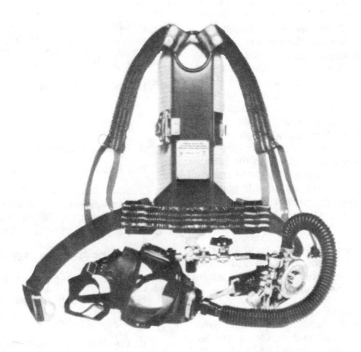

Fig. 6.20   Self-contained breathing apparatus
(Courtesy of PPG Industries)

h. Open the cylinder valve one-quarter of a turn and check for leaks. *IF THE VALVE IS DIFFICULT TO OPEN, RETURN THE CYLINDER TO YOUR SUPPLIER.* Some agencies require suppliers to torque the valve stems to no more than 35 psi (2.45 kg/sq cm); this permits operation of the valve stem with the correct valve wrench, without having to strike the wrench with the palm of your hand to unseat the valve stem. Striking the wrench usually results in the cylinder rotating because the safety chain usually does not securely hold the cylinder. Jarring the cylinder in this way may result in a gas leak, particularly if two cylinders are used with a manifold and pigtails. Remember—don't leave the valve in this position for a long time because it will plug up. Commercial chlorine usually contains small amounts of chlorinated organic compounds which are not volatile. Trace amounts of these compounds can be carried over in the chlorine vapor. A slightly opened valve, with the resulting change in velocity, is an ideal place for these materials to collect, build up, and cause the valve to plug up. Open the valve *ONE TURN*—this provides sufficient gas flow to prevent plugging and allows a quick shutoff in case of a leak.

i. Adjust the chlorinator to the proper setting to maintain the desired chlorine residual throughout the distribution system. Coliform test results should be negative when the chlorination system is operating properly.

j. Check the system for leaks by applying a concentrated ammonia solution (28 to 30 percent ammonia as NH_3) *VAPOR* from a "squeeze bottle" to the chlorine cylinder valve and the chlorinator. Any leaks will be detected by the presence of a white cloud. *IF A LEAK IS PRESENT, CLOSE THE GAS CYLINDER VALVE.*

- Make sure the packing nut on the chlorine cylinder valve stem is tight.

- Check the lead gasket. If the lead gasket is distorted, clean the connection and refit with a new gasket. Then repeat the above procedure for checking the system for leaks.

3. *CHLORINATOR SHUTDOWN*

Work in pairs. A plan should be used where both people are not exposed to the chlorine at the same time.

a. Have safety equipment available in the event of a chlorine leak.

b. Shut off the chlorine valve from the supply source.

c. Allow sufficient time for the chlorine to purge out of the line.

d. Turn the chlorinator off.

e. Leave the discharge line open. If chlorine is trapped between two valves, the chlorine gas can expand when heated by sunlight and develop high pressures in the line. Leave the discharge line open to prevent this hazard from developing.

4. *NORMAL OPERATION OF CHLORINATORS*

Normal operation of a chlorinator requires routine observation and preventive maintenance.

DAILY

a. Inspect the building to make sure that only authorized personnel have been there.

b. Read the chlorinator rotameter.

c. Record the reading, time, and date, and initial the entries.

d. Read the meters and record the number of gallons of water pumped.

e. Check the chlorine residual. If the residual is below 0.2 mg/L in the distribution system, increase the feed rate by adjusting the rotameter. If the residual is too high, lower the feed rate by adjusting the rotameter.

f. Calculate the chlorine usage. Refer to Examples 6 and 7.

WEEKLY

a. Clean the equipment and the building.

b. Perform preventive maintenance on the equipment.

Chlorine usage should be calculated so that replacement supply containers can be ordered and constant chlorination can be maintained. Refer to Example 7. Try to have a 15- to 30-day supply of chlorine in storage.

NORMAL OPERATION CHECKLIST

a. Chemical usage in pounds.

b. Meter readings for water usage in gallons.

c. Equipment log.

d. Lubrication inspection log.

e. Building inspection.

5. *ABNORMAL OPERATION*

a. Inform supervisor of the problem.

b. If the chlorinator malfunctions, repair the unit immediately.

c. If repairs cannot be completed quickly, shut off the water supply so that unchlorinated or contaminated water will not be delivered to the consumers.

ABNORMAL OPERATION, TROUBLESHOOTING

TABLE 6.7   DIRECT-MOUNT CHLORINATOR TROUBLESHOOTING GUIDE

Operating Symptoms	Probable Cause	Remedy
1. Water in the chlorine metering tube.	Check valve failure, deposits on seat of check valve, or check valve seat distorted by high pressure.	Clean deposits from check ball and seat with dilute muriatic acid. Badly distorted check valve may have to be replaced.
2. Water venting to atmosphere.	Excess water pressure in the vacuum regulator.	Remove vacuum regulator from chlorine cylinder and allow chlorinator to pull air until dry.
3. No indication on flowmeter when vacuum is present.	Vacuum leak due to bad or brittle vacuum tubing, connections, rate valve o-rings, or gasket on top of flowmeter.	Check the vacuum tubing, rate valve o-rings, and flowmeter gasket for vacuum leaks. Replace bad tubing connectors, o-rings, or gaskets.
4. Indication on flowmeter but air present, not chlorine gas.	Connection below meter tube gasket leaks.	Check connections and replace damaged elements.

6. *MAINTENANCE*

Most chlorinators are simple units and are more easily replaced than repaired on line. Remove and repair in shop or have repaired by others who do not have to operate the system.

FORMULAS

To determine the setting on a chemical feeder in pounds per day, multiply the flow in MGD times the dose in mg/L times 8.34 lbs/gal.

Feeder Setting, lbs/day = (Flow, MGD)(Dose, mg/L)(8.34 lbs/gal)

To calculate the number of chlorine cylinders used per month, determine the pounds of chlorine used per month and divide by the pounds of chlorine per cylinder.

$$\frac{\text{Cylinders Used,}}{\text{number/month}} = \frac{\text{Chlorine Used, lbs/mo}}{\text{Chlorine Cylinders, lbs/cylinder}}$$

EXAMPLE 5

A deep well turbine pump delivers approximately 200 GPM against typical operating heads. If the desired chlorine dose is 2 mg/L, what should be the setting on the rotameter for the chlorinator (lbs chlorine per 24 hours)?

Known	Unknown
Pump Flow, GPM = 200 GPM	Rotameter Setting,
Chlorine Dose, mg/L = 2 mg/L	lbs Chlorine/24 hours

1. Convert pump flow to million gallons per day (MGD).

$$\text{Flow, MGD} = \frac{(200 \text{ GPM})(60 \text{ min/hr})(24 \text{ hr/day})}{1{,}000{,}000/\text{M}}$$

$$= 0.288 \text{ MGD}$$

2. Calculate the rotameter setting in pounds of chlorine per 24 hours.

$$\frac{\text{Rotameter Setting,}}{\text{lbs/day}} = (\text{Flow, MGD})(\text{Dose, mg/}L)(8.34 \text{ lbs/gal})$$

$$= (0.288 \text{ M Gal/day})(2 \text{ mg/}L)(8.34 \text{ lbs/gal})$$

$$= 4.8 \text{ lbs/day}$$

$$= 4.8 \text{ lbs/24 hrs}$$

EXAMPLE 6

Using the results from Example 5 (a chlorinator setting of 4.8 lbs/day), how many pounds of chlorine would be used during one week if the pump hour meter showed 100 hours of pump operation? If the chlorine cylinder contained 78 pounds of chlorine at the start of the week, how many pounds of chlorine should be remaining at the end of the week?

Known	Unknown
Chlorinator Setting, lbs/day = 4.8 lbs/day	1. Chlorine Used, lbs/week
Time, hr/week = 100 hr/week	2. Chlorine Remaining, lbs
Chlorine Cylinder, lbs = 78 lbs	

1. Calculate the chlorine used in pounds per week.

$$\frac{\text{Chlorine Used,}}{\text{lbs/week}} = (\text{Chlorinator Setting, lbs/day})(\text{Time, hr/week})$$

$$= (4.8 \text{ lbs/day})\left(\frac{100 \text{ hr/week}}{24 \text{ hr/day}}\right)$$

$$= 20 \text{ lbs Chlorine/week}$$

2. Determine the amount of chlorine that should be remaining in the cylinder at the end of the week.

$$\frac{\text{Chlorine Remaining,}}{\text{lbs}} = \text{Chlorine at Start, lbs} - \text{Chlorine Used, lbs}$$

$$= 78 \text{ lbs} - 20 \text{ lbs}$$

$$= 58 \text{ lbs Chlorine remaining at end of week}$$

EXAMPLE 7

Given the pumping rate and chlorination system in Examples 5 and 6, if 20 pounds of chlorine are used during an average week, how many 150-pound chlorine cylinders will be used per month (assume 30 days per month)?

Known	Unknown
Chlorine Use, lbs/week = 20 lbs/week	1. Amount of Chlorine Used per Month, lbs
	2. Number of 150-lb Cylinders Used per Month

1. Calculate the amount of chlorine used in pounds of chlorine per month.

$$\frac{\text{Chlorine Used,}}{\text{lbs/month}} = (\text{Chlorine Use, lbs/week})(\text{Number Weeks/mo})$$

$$= (20 \text{ lbs/week})\left(\frac{(1 \text{ week})(30 \text{ days})}{(7 \text{ days})(1 \text{ month})}\right)$$

$$= 85.7 \text{ lbs/month}$$

2. Determine the number of 150-pound chlorine cylinders used per month.

$$\frac{\text{Cylinders Used,}}{\text{number/month}} = \frac{\text{Chlorine Used, lbs/mo}}{\text{Chlorine Cylinders, lbs/cylinder}}$$

$$= \frac{85.7 \text{ lbs/mo}}{150 \text{ lbs/cylinder}}$$

$$= 0.57 \text{ Cylinders/month}$$

This installation requires less than one 150-pound chlorine cylinder per month.

6.43 Laboratory Tests

The two most common water quality tests run on samples of water from a water supply system are (1) chlorine residual, and (2) coliform tests.

1. Chlorine Residual in System

Daily chlorine residual tests using the *DPD*[30] *METHOD*[31] should be taken at various locations in the system. A remote tap is ideal for one sampling location. Take the test sample from a tap as close to the main as possible. Allow the water to

---

[30] *DPD (pronounce as separate letters). A method of measuring the chlorine residual in water. The residual may be determined by either titrating or comparing a developed color with color standards. DPD stands for N,N-diethyl-p-phenylene-diamine.*
[31] *See Section 6.5, "Measurement of Chlorine Residual," for details on how to perform the DPD test for measuring chlorine residual.*

run at least 5 minutes before sampling to ensure a representative sample from the main. Chlorine residual test kits are available for small systems.

2. Bacteriological Analysis (Coliform Tests)

Samples should be taken routinely in accordance with EPA and health department requirements. Take samples according to approved procedures.[32] Be sure to use a sterile plastic or glass bottle. If the sample contains any chlorine residual, sufficient sodium thiosulfate should be added to neutralize all of the chlorine residual. Usually 0.1 milliliter of 10 percent sodium thiosulfate in a 120-mL (4-oz) bottle is sufficient for distribution systems. The "thio" should be added to the sample bottle before sterilization.

6.44   Troubleshooting

TABLE 6.8   DISINFECTION TROUBLESHOOTING GUIDE

Operating Symptom	Probable Cause	Remedy
1. Increase in coliform level	Low chlorine residual	Raise chlorine dose
2. Drop in chlorine level	a. Increase in chlorine demand	Raise chlorine dose and find out why chlorine demand increased or chlorine feed rate dropped
	b. Drop in chlorine feed rate	

6.45   Chlorination System Failure

IF YOUR CHLORINATION SYSTEM FAILS, DO NOT ALLOW UNCHLORINATED WATER TO ENTER THE DISTRIBUTION SYSTEM. Never allow unchlorinated water to be delivered to your consumers. If your chlorination system fails and cannot be repaired within a reasonable time period, notify your supervisor and officials of the health department. To prevent this problem from occurring, your plant should have backup or standby chlorination facilities.

6.46   Acknowledgment

Some of the material in this section on gas chlorinators was prepared by Joe Habraken, Treatment Supervisor, City of Tampa, Florida. His contribution is greatly appreciated.

QUESTIONS

Write your answers in a notebook and then compare your answers with those on page 357.

6.4H   What personal safety equipment should be available before attempting to locate and repair a chlorine gas leak?

6.4I   How is ammonia used to detect a chlorine leak?

6.4J   What would you do if you could not repair a broken chlorinator quickly?

6.4K   What two water quality tests are most commonly run on samples of water from a water supply system?

6.5   MEASUREMENT OF CHLORINE RESIDUAL

6.50   Methods of Measuring Chlorine Residual

AMPEROMETRIC TITRATION[33] provides for the most convenient and most repeatable chlorine residual results. However, amperometric titration equipment is more expensive than equipment for other methods. DPD tests can be used and are less expensive than other methods, but this method requires the operator to match the color of a sample with the colors on a comparator. See Chapter 11, "Laboratory Procedures," in WATER TREATMENT PLANT OPERATION, Volume I, for detailed information on these tests.

Residual chlorine measurements of treated water should be taken at least three times per day on small systems and once every two hours on large systems. Residuals are measured to ensure that the treated water is being adequately disinfected. A free chlorine residual of at least 0.5 mg/L in the treated water at the point of application is usually recommended.

6.51   Amperometric Titration for Free Residual Chlorine

1. Place a 200-mL sample of water in the titrator.

2. Start the agitator.

3. Add 1 mL of pH 7 buffer.

4. Titrate with 0.00564 N phenylarsene oxide solution.

5. End point is reached when one drop will cause a deflection on the microammeter and the deflection will remain.

6. mL of phenylarsene oxide used in titration is equal to mg/L of free chlorine residual.

6.52   DPD Colorimetric Method for Free Residual Chlorine (Figures 6.21 and 6.22)

This procedure is for the use of prepared powder pillows.

1. Collect a 100-mL sample.

2. Add color reagent.

3. Match color sample with a color on the comparator to obtain the chlorine residual in mg/L.

---

[32] See Chapter 11, "Laboratory Procedures," in WATER TREATMENT PLANT OPERATION, Volume I, for proper procedures for collecting and analyzing samples for chlorine residuals and coliform tests.

[33] Amperometric (am-PURR-o-MET-rick) Titration.   A means of measuring concentrations of certain substances in water (such as strong oxidizers) based on the electric current that flows during a chemical reaction.

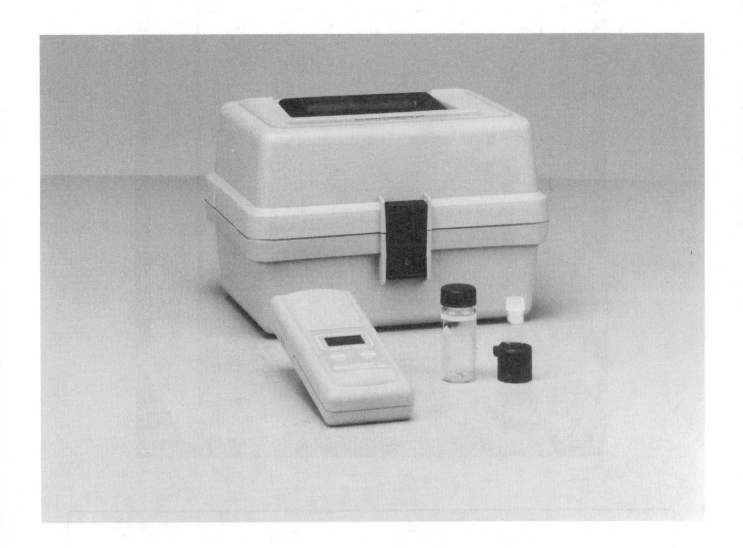

Fig. 6.21   Direct-reading colorimeter for free chlorine residuals
(Courtesy of the HACH Company)

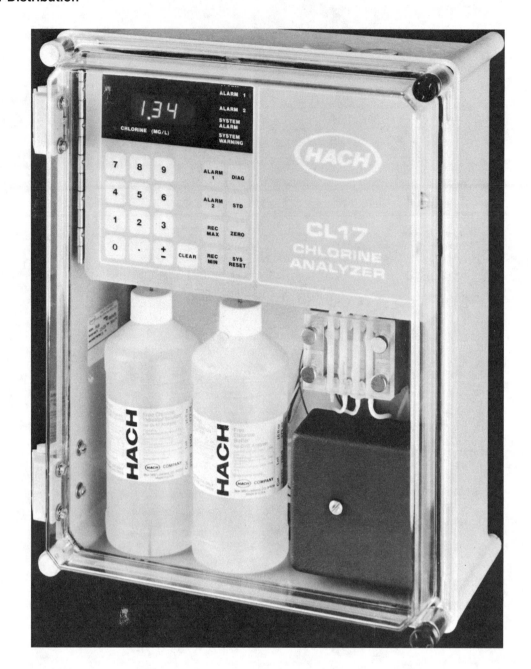

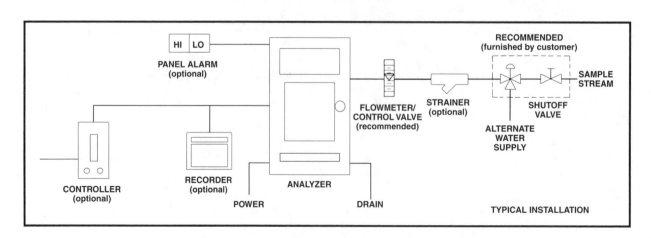

Fig. 6.22 Continuous on-line free chlorine residual analyzer
(Courtesy of the HACH Company)

Operators using the DPD colorimetric method to test water for a free chlorine residual need to be aware of a potential error that may occur. If the DPD test is run on water containing a combined chlorine residual, a precipitate may form during the test. The particles of precipitated material will give the sample a turbid appearance or the appearance of having color. This turbidity can produce a positive test result for free chlorine residual when there is actually no chlorine present. Operators call this error a "false positive" chlorine residual reading.

QUESTIONS

Write your answers in a notebook and then compare your answers with those on page 357.

6.5A   What two methods are commonly used to measure chlorine residual in treated water?

6.5B   How often should treated water residual chlorine measurements be made?

6.6 CHLORINE SAFETY PROGRAM

Every good safety program begins with cooperation between the employee and the employer. The employee must take an active part in the overall program. The employee must be responsible and should take all necessary steps to prevent accidents. This begins with the attitude that as good an effort as possible must be made by everyone. Safety is everyone's concern. The employer also must take an active part by supporting safety programs. There must be funding to purchase equipment and to enforce safety regulations required by OSHA and state industrial safety programs. The following items should be included in all safety programs.

1. Establishment of a formal safety program.

2. Written rules.

3. Periodic hands-on training using safety equipment.

    a. Leak-detection equipment
    b. Self-contained breathing apparatus (Figure 6.20)
    c. Atmospheric monitoring devices

4. Establishment of emergency procedures for chlorine leaks and first aid.

5. Establishment of a maintenance and calibration program for safety devices and equipment.

6. Provide police and fire departments with tours of facilities to locate hazardous areas and provide chlorine safety information.

All persons handling chlorine should be thoroughly aware of its hazardous properties. Personnel should know the location and use of the various pieces of protective equipment and be instructed in safety procedures. In addition, an emergency procedure should be established and each individual should be instructed how to follow the procedures. An emergency checklist also should be developed and available. For additional information on this topic, see the Chlorine Institute's *CHLORINE MANUAL*.[34] Also see Chapter 7, "Safety."

6.60 Chlorine Hazards

Chlorine is a gas that is heavier than air, extremely toxic, and corrosive in moist atmospheres. Dry chlorine gas can be safely handled in steel containers and piping, but with moisture must be handled in corrosion-resistant materials such as silver, glass, Teflon, and certain other plastics. Chlorine gas at container pressure should never be piped in silver, glass, Teflon, or any other material that cannot handle the pressure. Even in dry atmospheres chlorine combines with the moisture in the mucous membranes of the nose and throat, and with the fluids in the eyes and lungs; a very small percentage in the air can be very irritating and can cause severe coughing. Heavy exposure can be fatal (see Table 6.9).

TABLE 6.9 PHYSIOLOGICAL RESPONSE TO CONCENTRATIONS OF CHLORINE GAS[a, b]

Effect	Parts of Chlorine Gas Per Million Parts of Air by Volume (ppm)
Slight symptoms after several hours' exposure	1
Detectable odor	0.3 to 3.5
Noxiousness	5
Throat irritation	15
Coughing	30
Dangerous from one-half to one hour	40
Death after a few deep breaths	1,000

[a] Adapted from data in U.S. Bureau of Mines *TECHNICAL PAPER 248* (1955).

[b] The maximum **P**ermissible **E**xposure **L**imit (PEL) is 0.5 ppm (8-hour weighted average). The IDLH level is 10 ppm. IDLH is the **I**mmediately **D**angerous to **L**ife or **H**ealth concentration. The atmospheric concentration of any toxic, corrosive, or asphyxiant substance that poses an immediate threat to life or would cause irreversible or delayed adverse health effects or would interfere with an individual's ability to escape from a dangerous atmosphere.

WARNING

WHEN ENTERING A ROOM THAT MAY CONTAIN CHLORINE GAS, OPEN THE DOOR SLIGHTLY AND CHECK FOR THE SMELL OF CHLORINE. **NEVER** GO INTO A ROOM CONTAINING CHLORINE GAS WITH HARMFUL CONCENTRATIONS IN THE AIR WITHOUT A SELF-CONTAINED AIR SUPPLY, PROTECTIVE CLOTHING, AND HELP STANDING BY. HELP MAY BE OBTAINED FROM YOUR CHLORINE SUPPLIER AND YOUR LOCAL FIRE DEPARTMENT.

Most people can usually detect concentrations of chlorine gas above 0.3 ppm and you should not be exposed to concentrations greater than 1 ppm. However, chlorine gas can deaden your sense of smell and cause a false sense of security. *NEVER* rely on your sense of smell to protect you from chlorine because *YOUR* sense of smell might not be able to detect harmful levels of chlorine.

---

[34] *Write to: The Chlorine Institute, Inc., 1300 Wilson Boulevard, Arlington, VA 22209. Pamphlet 1. Price to members, $28.00; nonmembers, $70.00; plus $6.95 shipping and handling. The Chlorine Institute has also developed a wide variety of other training materials and video tapes that may be of interest to operators working with chlorine.*

6.61 Why Chlorine Must Be Handled With Care

You must always remember that chlorine is a hazardous chemical and must be handled with respect. Concentrations of chlorine gas in excess of 1,000 ppm (0.1% by volume in air) may be fatal after a few breaths.

Because the characteristic sharp odor of chlorine is noticeable even when the amount in the air is small, it is usually possible to get out of the gas area before serious harm is suffered. This feature makes chlorine less hazardous than gases such as carbon monoxide, which is odorless, and hydrogen sulfide, which impairs your sense of smell in a short time.

Inhaling chlorine causes general restlessness, panic, severe irritation of the throat, sneezing, and production of much saliva. These symptoms are followed by coughing, retching and vomiting, and difficulty in breathing. Chlorine is particularly irritating to persons suffering from asthma and certain types of chronic bronchitis. Liquid chlorine causes severe irritation and blistering on contact with the skin. Regular exposure to chlorine can produce *CHRONIC*[35] effects such as permanent damage to lung tissue.

6.62 Protect Yourself From Chlorine

Every person working with chlorine should know the proper ways to handle it, should be trained in the use of self-contained breathing apparatus (SCBA), and should know what to do in case of emergencies. Wear a SCBA that protects your face, eyes, and nose. The clothing of persons exposed to chlorine will be saturated with chlorine which will irritate the skin if exposed to moisture or sweat. These people should not enter confined spaces before their clothing is purged of chlorine (stand out in the open air for a while). This is particularly applicable to police and fire department personnel who leave the scene of a chlorine leak and ride back to their stations in closed vehicles. Suitable protective clothing for working in an atmosphere containing chlorine includes disposable rainsuits with hoods to protect your body, head, and limbs, and rubber boots to protect your feet.

> **WARNING**
>
> CANISTER TYPE 'GAS MASKS' ARE USUALLY **INADEQUATE** AND **INEFFECTIVE** IN SITUATIONS WHERE CHLORINE LEAKS OCCUR AND ARE THEREFORE NOT RECOMMENDED FOR USE UNDER ANY CIRCUMSTANCES. **SELF-CONTAINED AIR OR OXYGEN SUPPLY TYPE BREATHING APPARATUS ARE RECOMMENDED.** OPERATORS SERVING ON "EMERGENCY CHLORINE TEAMS" MUST BE CAREFULLY SELECTED AND RECEIVE REGULAR APPROVED TRAINING. THEY MUST BE PROVIDED THE PROPER EQUIPMENT WHICH RECEIVES REGULAR MAINTENANCE AND IS READY FOR USE AT ALL TIMES.

Self-contained air supply and positive pressure/demand breathing equipment must fit properly and be used properly. Pressure/demand units and rebreather kits may be safer. Pressure/demand units use more air from the air bottle which reduces the time a person may work on a leak. There are certain physical constraints when using respiratory protection. Contact your local safety regulatory agency to determine these requirements.

The 1991 Uniform Fire Code requires proper ventilation of chlorine storage rooms and rooms where chlorine is used. Mechanical exhaust systems must draw air from the room at a point no higher than 12 inches (30.5 cm) above the floor at a rate of not less than one cubic foot of air per minute per square foot (0.00508 cu m/sec/sq m) of floor area in the storage area. (The system should not draw air through the fan itself because chlorine gas can damage the fan motor.) Normally ventilated air from chlorine storage rooms is discharged to the atmosphere. When a chlorine leak occurs, the ventilated air containing the chlorine must be treated to reduce the chlorine concentration. A caustic scrubbing system can be used. The treatment must reduce the chlorine concentration to one-half the IDLH (**I**mmediately **D**angerous to **L**ife or **H**ealth) level at the point of discharge to the atmosphere. The IDLH level for chlorine is 10 ppm. A secondary standby source of power is also required for the chlorine detection, alarm, ventilation, and treatment systems.

Before entering an area with a chlorine leak, wear protective clothing. Gloves and a chemical suit will prevent chlorine from contacting the sweat on your body and forming hydrochloric acid. Chemical suits are very cumbersome, but should be worn when the chlorine concentration is high. A great deal of practice is required to perform effectively while wearing a chemical suit.

The best protection that one can have when dealing with chlorine is to respect it. Each individual should practice rules of safe handling and good PREVENTIVE MAINTENANCE.

PREVENTION IS THE BEST EMERGENCY TOOL YOU HAVE.

---

[35] *Chronic Health Effect. An adverse effect on a human or animal body with symptoms that develop slowly over a long period of time or that recur frequently.*

PLAN AHEAD.

1. Have your fire department and other available emergency response agencies tour the area so that they know where the facilities are located. Give them a clearly marked map indicating the location of the chlorine storage area, chlorinators, and emergency equipment.

2. Have regularly scheduled practice sessions in the use of respiratory protective devices, chemical suits, and chlorine repair kits. Involve all personnel who may respond to a chlorine leak.

3. Have a supply of ammonia available to detect chlorine leaks.

4. Write emergency procedures:

    Prepare a CHLORINE EMERGENCY LIST of names of companies and phone numbers of persons to call during an emergency and ensure that all involved personnel are trained in notification procedures. This list should be posted at plant telephones and should include:

    a. Fire department,

    b. Chlorine emergency personnel,

    c. Chlorine supplier, and

    d. Police department.

5. Follow established procedures during all emergencies.

    a. Never work alone during chlorine emergencies.

    b. Obtain help immediately and quickly repair the problem. *PROBLEMS DO NOT GET BETTER.*

    c. Only authorized and properly trained persons with adequate equipment should be allowed in the danger area to correct the problem.

    d. If you are caught in a chlorine atmosphere without appropriate respiratory protection, shallow breathing is safer than breathing deeply. Recovery depends upon the duration and amount of chlorine inhaled, so it is important to keep that amount as small as possible.

    e. If you discover a chlorine leak, leave the area immediately unless it is a very minor leak. Small leaks can be found by using a rag soaked with ammonia. A white cloud of gas will form near the leak so it can be located and corrected.

    f. Use approved respiratory protection and wear disposable clothing when repairing a chlorine leak.

    g. Notify your police department that you need help if it becomes necessary to stop traffic on roads and to evacuate persons in the vicinity of the chlorine leak.

6. Develop emergency evacuation procedures for use during a serious chlorine leak. Coordinate these procedures with your police department and other officials. Ensure that all facility personnel are thoroughly trained in any evacuation procedure developed.

7. Post emergency procedures in all operating areas.

8. Inspect equipment and routinely make any necessary repairs.

9. At least twice weekly, inspect area where chlorine is stored and where chlorinators are located. Remove all obstructions from the area.

10. Schedule routine maintenance on *ALL* chlorine equipment at least once every six months or more frequently.

11. Have health appraisal for employees on chlorine emergency duty. All those who have heart and/or respiratory problems should not be allowed on emergency teams. There may be other physical constraints. Contact your local safety regulatory agency for details.

REMEMBER:

Small amounts of chlorine cause large problems. Leaks never get better.

6.63 First-Aid Measures

MILD CASES

Whenever you have a mild case of chlorine exposure (which does happen from time to time around chlorination equipment), you should first leave the contaminated area. Move slowly, breathe lightly without exertion, remain calm, keep warm, and resist coughing. Notify other operators and have them repair the leak immediately.

If clothing has been contaminated, remove as soon as possible. Otherwise, the clothing will continue to give off chlorine gas, which will irritate the body even after leaving the contaminated area. Immediately wash off area affected by chlorine. Shower and put on clean clothes.

If victim has slight throat irritation, immediate relief can be accomplished by drinking milk. A mild stimulant such as hot coffee or hot tea is often used for coughing. Drinking spirits of peppermint also will help reduce throat irritation. See a physician.

EXTREME CASES

1. Follow established emergency procedures.

2. Always use proper safety equipment. Do not enter area without a self-contained breathing apparatus.

3. Remove patient from affected area immediately. Call a physician and begin appropriate treatment immediately.

4. First aid:

    a. Remove contaminated clothes to prevent clothing giving off chlorine gas, which will irritate the body.

    b. Keep patient warm and cover with blankets, if necessary.

    c. Place patient in a comfortable position on back.

    d. If breathing is difficult, administer oxygen if equipment and trained personnel are available.

    e. If breathing seems to have stopped, begin artificial respiration immediately. Mouth-to-mouth resuscitation or any of the approved methods may be used.

f. EYES!

If even a small amount of chlorine gets into the eyes, they should be flushed immediately with large amounts of lukewarm water so that all traces of chlorine are flushed from the eyes (at least 15 minutes). Hold the eyelids apart forcibly to ensure complete washing of all eye and lid tissues.

5. See a physician.

6.64 Hypochlorite Safety

Hypochlorite does not present the hazards that gaseous chlorine does and therefore is safer to handle. When spills occur, wash with large volumes of water. The solution is messy to handle. Hypochlorite causes damage to your eyes and skin upon contact. Immediately wash affected areas thoroughly with water. Consult a physician if the area appears burned. Hypochlorite solutions are very corrosive. Hypochlorite compounds are nonflammable; however, they can cause fires when they come in contact with organics or other easily oxidizable substances.

6.65 Chlorine Dioxide Safety

Chlorine dioxide is generated in much the same manner as chlorine and should be handled with the same care. Of special concern is the use of sodium chlorite to generate chlorine dioxide. Sodium chlorite is very combustible around organic compounds. Whenever spills occur, sodium chlorite must be neutralized with anhydrous sodium sulfite. Combustible materials (including gloves) should not be worn when handling sodium chlorite. If sodium chlorite comes in contact with clothing, the clothes should be removed immediately and soaked in water to remove all traces of sodium chlorite or the clothes should be burned immediately.

6.66 Operator Safety Training

Training is a concern to everyone, especially when your safety and perhaps your life is involved. Every utility agency should have an operator chlorine safety training program that introduces new operators to the program and updates previously trained operators. As soon as a training session ends, obsolescence begins. People will forget what they have learned if they don't use and practice their knowledge and skills. Operator turnover can dilute a well-trained staff. New equipment and also new techniques and procedures can dilute the readiness of trained operators. An ongoing training program can consist of a monthly luncheon seminar, a monthly safety bulletin that is to be read by every operator, and outside speakers can be brought in to reinforce and refresh specific elements of a safety training program.

6.67 CHEMTREC (800) 424-9300

Safely handling chemicals used in daily water treatment is an operator's responsibility. However, if the situation ever gets out of hand, there are emergency teams that will respond with help anywhere there is an emergency. If an emergency does develop in your plant and you need assistance, call CHEMTREC (Chemical Transportation Emergency Center) for assistance. CHEMTREC will provide immediate advice for those at the scene of an emergency and then quickly alert experts whose products are involved for more detailed assistance and appropriate follow-up.

CHEMTREC'S EMERGENCY TOLL-FREE PHONE NUMBER IS (800) 424-9300.

QUESTIONS

Write your answers in a notebook and then compare your answers with those on page 357.

6.6A   What properties make chlorine gas so hazardous?

6.6B   What type of breathing apparatus is recommended when repairing a chlorine leak?

6.6C   What first-aid measures should be taken if a person comes in contact with chlorine gas?

6.6D   What would you do if hypochlorite came in contact with your hand?

6.7   CHLORINATION ARITHMETIC

All calculations in this section can be performed by addition, subtraction, multiplication and division on a pocket electronic calculator.

FORMULAS

There are two approaches to calculating chlorine doses in milligrams per liter. They both give the same results, but have a slightly different form. From the basic equation on page 318,

Chlorine, lbs = (Volume, M Gal)(Dose, mg/L)(8.34 lbs/gal),

we can rearrange the equation and solve for the dose in milligrams per liter.

$$\text{Chlorine Dose, mg}/L = \frac{\text{Chlorine, lbs}}{(\text{Volume, M Gal})(8.34 \text{ lbs/day})}$$

If the basic equation is expressed as a chemical feeder setting in pounds per day, then the flow would be in million gallons per day (MGD).

$$\text{Chlorine Dose, mg}/L = \frac{\text{Chlorine, lbs/day}}{(\text{Flow, MGD})(8.34 \text{ lbs/gal})}$$

Both of the above equations are also expressed in terms of pounds or pounds per day of chlorine per million pounds or million pounds per day of water.

$$\text{Chlorine Dose, mg}/L = \frac{\text{Chlorine, lbs/day}}{\text{Water, Million lbs/day}}$$

6.70   Disinfection of Facilities

6.700   Wells and Pumps

EXAMPLE 8

How many gallons of 5.25 percent sodium hypochlorite will be needed to disinfect a well with an 18-inch diameter casing and well screen? The well is 300 feet deep and there is 200 feet of water in the well. Use an initial chlorine dose of 100 mg/L.

Known		Unknown
Hypochlorite, %	= 5.25%	5.25% Hypochlorite, gal
Chlorine Dose, mg/L	= 100 mg/L	
Well Casing, in	= 18 in	
Well Depth, ft	= 300 ft	
Water Depth, ft	= 200 ft	

1. Find the volume of water in the well in gallons.

$$\text{Well Vol, gal} = (0.785)(\text{Diameter, ft})^2(\text{Water Depth, ft})(7.48 \text{ gal/cu ft})$$

$$= \frac{(0.785)(18 \text{ in})^2(200 \text{ ft})(7.48 \text{ gal/cu ft})}{144 \text{ sq in/sq ft}}$$

$$= 2,642 \text{ gal}$$

2. Determine the pounds of chlorine needed.

$$\text{Chlorine, lbs} = (\text{Volume, M Gal})(\text{Dose, mg/L})(8.34 \text{ lbs/gal})$$

$$= (0.002642 \text{ M Gal})(100 \text{ mg/L})(8.34 \text{ lbs/gal})$$

$$= 2.2 \text{ lbs Chlorine}$$

3. Calculate the gallons of 5.25 percent sodium hypochlorite solution needed.

$$\text{Sodium Hypochlorite Solution, gallons} = \frac{(\text{Chlorine, lbs})(100\%)}{(8.34 \text{ lbs/gal})(\text{Hypochlorite, }\%)}$$

$$= \frac{(2.2 \text{ lbs})(100\%)}{(8.34 \text{ lbs/gal})(5.25\%)}$$

$$= 5.0 \text{ gallons}$$

Five gallons of 5.25 percent sodium hypochlorite should do the job.

6.701 Mains

EXAMPLE 9

A section of an old 8-inch water main has been replaced and a 350-foot section of pipe needs to be disinfected. An initial chlorine dose of 400 mg/L is expected to maintain a chlorine residual of over 300 mg/L during the three-hour disinfection period. How many gallons of 5.25 percent sodium hypochlorite solution will be needed?

Known		Unknown
Diameter of Pipe, in	= 8 in	5.25% Hypochlorite, gallons
or 8 in/12 in/ft	= 0.67 ft	
Length of Pipe, ft	= 350 ft	
Chlorine Dose, mg/L	= 400 mg/L	
Hypochlorite, %	= 5.25%	

1. Calculate the volume of water in the pipe in gallons.

$$\text{Pipe Volume, gallons} = (0.785)(\text{Diameter, ft})^2(\text{Length, ft})(7.48 \text{ gal/cu ft})$$

$$= (0.785)(0.67 \text{ ft})^2(350 \text{ ft})(7.48 \text{ gal/cu ft})$$

$$= 923 \text{ gallons of Water}$$

2. Determine the pounds of chlorine needed.

$$\text{Chlorine, lbs} = (\text{Volume, M Gal})(\text{Dose, mg/L})(8.34 \text{ lbs/gal})$$

$$= (0.000923 \text{ M Gal})(400 \text{ mg/L})(8.34 \text{ lbs/gal})$$

$$= 3.08 \text{ lbs Chlorine}$$

3. Calculate the gallons of 5.25 percent sodium hypochlorite solution needed.

$$\text{Sodium Hypochlorite Solution, gallons} = \frac{(\text{Chlorine, lbs})(100\%)}{(8.34 \text{ lbs/gal})(\text{Hypochlorite, }\%)}$$

$$= \frac{(3.08 \text{ lbs})(100\%)}{(8.34 \text{ lbs/gal})(5.25\%)}$$

$$= 7.0 \text{ gallons}$$

Seven gallons of 5.25 percent solution of sodium hypochlorite should do the job.

6.702 Tanks

EXAMPLE 10

An existing service storage reservoir has been taken out of service for inspection, maintenance, and repairs. The reservoir needs to be disinfected before being placed back on line. The reservoir is 6 feet deep, 10 feet wide, and 25 feet long. An initial chlorine dose of 100 mg/L is expected to maintain a chlorine residual of over 50 mg/L during the 24-hour disinfection period. How many gallons of 5.25 percent sodium hypochlorite solution will be needed?

Known		Unknown
Tank Depth, ft	= 6 ft	5.25% Hypochlorite, gal
Tank Width, ft	= 10 ft	
Tank Length, ft	= 25 ft	
Chlorine Dose, mg/L	= 100 mg/L	
Hypochlorite, %	= 5.25%	

1. Calculate the volume of water in the tank in gallons.

$$\text{Tank Volume, gallons} = (\text{Length, ft})(\text{Width, ft})(\text{Depth, ft})(7.48 \text{ gal/cu ft})$$

$$= (25 \text{ ft})(10 \text{ ft})(6 \text{ ft})(7.48 \text{ gal/cu ft})$$

$$= 11,220 \text{ gallons}$$

2. Determine the pounds of chlorine needed.

$$\text{Chlorine, lbs} = (\text{Volume, M Gal})(\text{Dose, mg/L})(8.34 \text{ lbs/gal})$$

$$= (0.01122 \text{ M Gal})(100 \text{ mg/L})(8.34 \text{ lbs/gal})$$

$$= 9.36 \text{ lbs Chlorine}$$

3. Calculate the gallons of 5.25 percent sodium hypochlorite solution needed.

$$\text{Sodium Hypochlorite Solution, gallons} = \frac{(\text{Chlorine, lbs})(100\%)}{(8.34 \text{ lbs/gal})(\text{Hypochlorite, \%})}$$

$$= \frac{(9.36 \text{ lbs})(100\%)}{(8.34 \text{ lbs/gal})(5.25\%)}$$

$$= 21.4 \text{ gallons}$$

Twenty-two gallons of 5.25 percent sodium hypochlorite solution should do the job.

6.71 Disinfection of Water From Wells

6.710 Chlorine Dose

EXAMPLE 11

A chlorine demand test from a well water sample produced a result of 1.2 mg/L. The water supplier would like to maintain a chlorine residual of 0.2 mg/L throughout the system. What should be the chlorine dose in mg/L from either a chlorinator or hypochlorinator?

Known	Unknown
Chlorine Demand, mg/L $= 1.2$ mg/L	Chlorine Dose, mg/L
Chlorine Residual, mg/L $= 0.2$ mg/L	

Calculate the chlorine dose in mg/L.

$$\text{Chlorine Dose, mg/L} = \text{Chlorine Demand, mg/L} + \text{Chlorine Residual, mg/L}$$

$$= 1.2 \text{ mg/L} + 0.2 \text{ mg/L}$$

$$= 1.4 \text{ mg/L}$$

NOTE: Be sure to check the chlorine residual regularly throughout the system. If the residual is low or there are coliforms present in the test results, then the residual should be increased.

6.711 Chlorinator

EXAMPLE 12

A deep well turbine pump is connected to a hydropneumatic tank. Under normal operating heads, the pump delivers 500 GPM. If the desired chlorine dosage is 3.5 mg/L, what should be the setting on the rotameter for the chlorinator (lbs chlorine per 24 hours)?

Known	Unknown
Pump Flow, GPM = 500 GPM	Rotameter Setting, lbs Chlorine/24 hours
Chlorine Dose, mg/L = 3.5 mg/L	

1. Convert pump flow to million gallons per day (MGD).

$$\text{Flow, MGD} = \frac{(500 \text{ GPM})(60 \text{ min/hr})(24 \text{ hr/day})}{1,000,000/M}$$

$$= 0.72 \text{ MGD}$$

2. Calculate the rotameter setting in pounds of chlorine per 24 hours.

$$\text{Rotameter Setting, lbs/day} = (\text{Flow, MGD})(\text{Dose, mg/L})(8.34 \text{ lbs/gal})$$

$$= (0.72 \text{ MGD})(3.5 \text{ mg/L})(8.34 \text{ lbs/gal})$$

$$= 21.0 \text{ lbs Chlorine/day}$$

$$= 21.0 \text{ lbs Chlorine/24 hours}$$

EXAMPLE 13

Using the results from Example 12 (a chlorinator setting of 21 lbs per 24 hours), how many pounds of chlorine would be used in one month if the pump hour meter shows the pump operates an average of 20 hours per day? The chlorinator operates only when the pump operates. How many 150-pound cylinders will be needed per month?

Known	Unknown
Chlorinator Setting, lbs/day $= 21$ lbs/day	1. Chlorine Used, lbs/mo
Pump Operation, hr/day $= 20$ hr/day	2. Cylinders Needed, no/mo
Chlorine Cylinders, lbs/cyl $= 150$ lbs/cyl	

1. Calculate the chlorine used in pounds per month.

$$\text{Chlorine Used, lbs/mo} = \frac{(\text{Cl Setting, lbs/day})(\text{Operation, hr/day})(30 \text{ days/mo})}{24 \text{ hr/day}}$$

$$= \frac{(21 \text{ lbs/day})(20 \text{ hr/day})(30 \text{ days/mo})}{24 \text{ hr/day}}$$

$$= 525 \text{ lbs/mo}$$

2. Determine the number of 150-pound cylinders used per month.

$$\text{Cylinders Needed, no/mo} = \frac{\text{Chlorine Used, lbs/mo}}{150 \text{ lbs Chlorine/cylinder}}$$

$$= \frac{525 \text{ lbs Cl/mo}}{150 \text{ lbs Cl/cylinder}}$$

$$= 3.5 \text{ Cylinders/month}$$

EXAMPLE 14

A deep well turbine pump delivers 400 GPM throughout a 24-hour period. The weight of chlorine in a 150-pound cylinder was 123 pounds at the start of the time period and 109 pounds at the end of the 24 hours. What was the chlorine dose rate in mg/L?

Known	Unknown
Pump Flow, GPM = 400 GPM	Chlorine Dose, mg/L
Time Period, hr = 24 hr	
Chlorine Wt at Start, lbs = 123 lbs	
Chlorine Wt at End, lbs = 109 lbs	

1. Convert flow of 400 GPM to MGD.

$$\text{Flow, MGD} = \frac{(400 \text{ GPM})(60 \text{ min/hr})(24 \text{ hr/day})}{1{,}000{,}000/M}$$

$$= 0.576 \text{ MGD}$$

2. Calculate the chlorine dose rate in mg/L.

$$\text{Chlorine Dose, mg/L} = \frac{\text{Chlorine Used, lbs/day}}{(\text{Flow, MGD})(8.34 \text{ lbs/gal})}$$

$$= \frac{(123 \text{ lbs} - 109 \text{ lbs})/1 \text{ day}}{(0.576 \text{ MGD})(8.34 \text{ lbs/gal})}$$

$$= \frac{14 \text{ lbs Chlorine/day}}{(0.576 \text{ MGD})(8.34 \text{ lbs/gal})}$$

$$= \frac{2.9 \text{ lbs Chlorine}}{1 \text{ M lbs Water}}$$

$$= 2.9 \text{ mg/L}$$

6.712 Hypochlorinator

EXAMPLE 15

Water from a well is being treated by a hypochlorinator. If the hypochlorinator is set at a pumping rate of 50 gallons per day (GPD) and uses a 3 percent available hypochlorite solution, what is the chlorine dose rate in mg/L if the pump delivers 350 GPM?

Known	Unknown
Hypochlorinator, GPD = 50 GPD	Chlorine Dose, mg/L
Hypochlorite, % = 3%	
Pump, GPM = 350 GPM	

1. Convert the pumping rate to MGD.

$$\text{Pumping Rate, MGD} = \frac{(350 \text{ GPM})(60 \text{ min/hr})(24 \text{ hr/day})}{1{,}000{,}000/M}$$

$$= 0.50 \text{ MGD}$$

2. Calculate the chlorine dose rate in pounds per day.

$$\text{Chlorine Dose, lbs/day} = \frac{(\text{Flow, gal/day})(\text{Hypochlorite, \%})(8.34 \text{ lbs/gal})}{100\%}$$

$$= \frac{(50 \text{ gal/day})(3\%)(8.34 \text{ lbs/gal})}{100\%}$$

$$= 12.5 \text{ lbs/day}$$

3. Calculate the chlorine dose in mg/L.

$$\text{Chlorine Dose, mg/L} = \frac{\text{Chlorine Dose, lbs/day}}{(\text{Flow, MGD})(8.34 \text{ lbs/gal})}$$

$$= \frac{12.5 \text{ lbs Chlorine/day}}{(0.50 \text{ M Gal/day})(8.34 \text{ lbs/gal})}$$

$$= 3 \text{ lbs Chlorine/M lbs Water}$$

$$= 3 \text{ mg/L}$$

EXAMPLE 16

Water pumped from a well is disinfected by a hypochlorinator. During a one-week time period, the water meter indicated that 1,098,000 gallons of water were pumped. A 2.0 percent sodium hypochlorite solution is stored in a 2.5-foot diameter plastic tank. During this one-week time period, the level of hypochlorite in the tank dropped 18 inches (1.50 ft). What was the chlorine dose in mg/L?

Known	Unknown
Water Treated, M Gal = 1.098 M Gal	Chlorine Dose, mg/L
Hypochlorite, % = 2.0%	
Hypochlorite Tank D, ft = 2.5 ft	
Hypochlorite Used, ft = 1.5 ft	

1. Calculate the pounds of water disinfected.

$$\text{Water, lbs} = (\text{Water Treated, M Gal})(8.34 \text{ lbs/gal})$$

$$= (1.098 \text{ M Gal})(8.34 \text{ lbs/gal})$$

$$= 9.16 \text{ M lbs Water}$$

2. Calculate the volume of hypochlorite solution used in gallons.

$$\text{Hypochlorite, gal} = (0.785)(\text{Diameter, ft})^2(\text{Drop, ft})(7.48 \text{ gal/cu ft})$$

$$= (0.785)(2.5 \text{ ft})^2(1.5 \text{ ft})(7.48 \text{ gal/cu ft})$$

$$= 55.0 \text{ gallons}$$

3. Determine the pounds of chlorine used to treat the water.

$$\text{Chlorine, lbs} = (\text{Hypochlorite, gal})\left(\frac{\text{Hypochlorite, \%}}{100\%}\right)(8.34 \text{ lbs/gal})$$

$$= (55.0 \text{ gal})\left(\frac{2.0\%}{100\%}\right)(8.34 \text{ lbs/gal})$$

$$= 9.17 \text{ lbs Chlorine}$$

4. Calculate the chlorine dose in mg/L.

$$\text{Chlorine Dose, mg/L} = \frac{\text{Chlorine Used, lbs}}{\text{Water Treated, Million lbs}}$$

$$= \frac{9.17 \text{ lbs Chlorine}}{9.16 \text{ M lbs Water}}$$

$$= \frac{1.0 \text{ lbs Chlorine}}{1 \text{ M lbs Water}}$$

$$= 1.0 \text{ mg/L}$$

EXAMPLE 17

Estimate the required concentration of a hypochlorite solution (%) if a pump delivers 600 GPM from a well. The hypochlorinator can deliver a maximum of 120 GPD and the desired chlorine dose is 1.8 mg/*L*.

Known	Unknown
Pump Flow, GPM = 600 GPM	Hypochlorite Strength, %
Hypochl Flow, GPD = 120 GPD	
Chlorine Dose, mg/*L* = 1.8 mg/*L*	

1. Calculate the flow of water treated in million gallons per day.

$$\text{Water Treated, M Gal/day} = \frac{(600 \text{ GPM})(60 \text{ min/hr})(24 \text{ hr/day})}{1,000,000/M}$$

$$= 0.864 \text{ MGD}$$

2. Determine the pounds of chlorine required per day.

$$\text{Chlorine Required, lbs/day} = (\text{Flow, MGD})(\text{Dose, mg/}L)(8.34 \text{ lbs/gal})$$

$$= (0.864 \text{ MGD})(1.8 \text{ mg/}L)(8.34 \text{ lbs/gal})$$

$$= 13.0 \text{ lbs Chlorine/day}$$

3. Calculate the hypochlorite solution strength as a percent.

$$\text{Hypochlorite Strength, \%} = \frac{(\text{Chlorine Required, lbs/day})(100\%)}{(\text{Hypochlorinator Flow, GPD})(8.34 \text{ lbs/gal})}$$

$$= \frac{(13.0 \text{ lbs/day})(100\%)}{(120 \text{ GPD})(8.34 \text{ lbs/gal})}$$

$$= 1.3\%$$

EXAMPLE 18

A hypochlorite solution for a hypochlorinator is being prepared in a 55-gallon drum. If 10 gallons of 5 percent hypochlorite is added to the drum, how much water should be added to the drum to produce a 1.3 percent hypochlorite solution?

Known	Unknown
Drum Capacity, gal = 55 gal	Water Added, gal
Hypochlorite, gal = 10 gal	
Actual Hypo, % = 5%	
Desired Hypo, % = 1.3%	

or

$$\text{Desired Hypo, \%} = \frac{(\text{Hypo, gal})(\text{Hypo, \%})}{\text{Hypo, gal} + \text{Water Added, gal}}$$

Rearrange the terms in the equation.

$(\text{Desired Hypo, \%})(\text{Hypo, gal} + \text{Water Added, gal}) = (\text{Hypo, gal})(\text{Hypo, \%})$

$(\text{Desired Hypo, \%})(\text{Hypo, gal}) + (\text{Desired Hypo, \%})(\text{Water Added, gal}) = (\text{Hypo, gal})(\text{Hypo, \%})$

$(\text{Desired Hypo, \%})(\text{Water Added, gal}) = (\text{Hypo, gal})(\text{Hypo, \%}) - (\text{Desired Hypo, \%})(\text{Hypo, gal})$

Calculate the volume of water to be added in gallons.

$$\text{Water Added, gal} = \frac{(\text{Hypo, gal})(\text{Actual Hypo, \%}) - (\text{Desired Hypo, \%})(\text{Hypo, gal})}{\text{Desired Hypo, \%}}$$

$$= \frac{(10 \text{ gal})(5\%) - (1.3\%)(10 \text{ gal})}{1.3\%}$$

$$= \frac{50 - 13}{1.3}$$

$$= 28.5 \text{ gallons of Water}$$

Add 28.5 gallons of water to the 10 gallons of 5 percent hypochlorite in the drum.

QUESTIONS

Write your answers in a notebook and then compare your answers with those on pages 357 and 358.

6.7A   A section of 12-inch water main has been repaired and a 400-foot section of pipe needs to be disinfected. An initial chlorine dose of 450 mg/*L* is expected to maintain a chlorine residual of over 300 mg/*L* during the three-hour disinfection period. How many gallons of 5 percent sodium hypochlorite solution will be needed?

6.7B   Estimate the chlorine demand of a water that is dosed at 2.0 mg/*L*. The chlorine residual is 0.2 mg/*L* after a 30-minute contact period.

6.7C   What should be the setting on a chlorinator (lbs chlorine per 24 hours) if a pump usually delivers 600 GPM and the desired chlorine dosage is 4.0 mg/*L*?

6.7D   Water from a well is being disinfected by a hypochlorinator. If the hypochlorinator is set at a pumping rate of 60 gallons per day (GPD) and uses a 2 percent available chlorine solution, what is the chlorine dose rate in mg/*L*? The pump delivers 400 GPM.

6.8 ARITHMETIC ASSIGNMENT

Turn to the Appendix, "How to Solve Water Distribution System Arithmetic Problems," at the back of this manual and read Section A.9, *STEPS IN SOLVING PROBLEMS*. Also work the example problems and check the arithmetic using your calculator.

In Section A.13, *TYPICAL WATER DISTRIBUTION SYSTEM PROBLEMS (ENGLISH SYSTEM)*, read and work the problems in Section A.134, Disinfection.

6.9 ADDITIONAL READING

1. *NEW YORK MANUAL*, Chapter 10,* "Chlorination."

2. *TEXAS MANUAL*, Chapter 10,* "Disinfection of Water."

3. *CHLORINE MANUAL*. Obtain from the Chlorine Institute, Inc., 1300 Wilson Boulevard, Arlington, VA 22209. Pamphlet 1. Price to members, $28.00; nonmembers, $70.00; plus $6.95 shipping and handling.

4. *CHLORINE MANUAL*. Obtain from PPG Industries, Inc., Chemicals Group, 440 College Park Drive, Monroeville, PA 15146. Available with the purchase of chlorine.

5. *CHLORINE SAFE HANDLING BOOKLET*. Obtain from PPG Industries, Inc., Chemicals Group, 440 College Park Drive, Monroeville, PA 15146. No charge.

6. *AWWA STANDARD FOR DISINFECTING WATER MAINS*, C651-99. Obtain from American Water Works Association (AWWA), Bookstore, 6666 West Quincy Avenue, Denver, CO 80235. Order No. 43651. Price to members, $42.00; nonmembers, $61.00; price includes cost of shipping and handling.

7. *AWWA STANDARD FOR DISINFECTION OF WATER-STORAGE FACILITIES*, C652-02. Obtain from American Water Works Association (AWWA), Bookstore, 6666 West Quincy Avenue, Denver, CO 80235. Order No. 43652. Price to members, $42.00; nonmembers, $61.00; price includes cost of shipping and handling.

* Depends on edition.

6.10 ACKNOWLEDGMENT

Malcolm P. Dalton, General Manager, and James F. Brace, Staff Assistant, Navajo Tribal Utility Authority, provided many helpful contributions to all aspects of safety throughout this entire manual.

End of Lesson 2 of 2 Lessons on DISINFECTION

Please answer the discussion and review questions next.

DISCUSSION AND REVIEW QUESTIONS

Chapter 6. DISINFECTION

(Lesson 2 of 2 Lessons)

Write the answers to these questions in your notebook. The question numbering continues from Lesson 1.

8. What procedures would you follow to safely handle chlorine cylinders?

9. What precautions would you take when shutting down a hypochlorinator to make repairs for a short duration?

10. How would you determine whether or not the chlorine feed rate of a chlorinator needs adjustment?

11. How would you determine the desired strength of a hypochlorite solution in the solution tank for effective operation of a hypochlorinator?

12. What would you do if you could not repair a broken chlorinator quickly?

13. Why is safety equipment needed when repairing a chlorine leak?

14. How would you detect a chlorine leak?

15. Why should clothing be removed from a person who has been in an area contaminated with liquid or gaseous chlorine?

SUGGESTED ANSWERS

Chapter 6.   DISINFECTION

ANSWERS TO QUESTIONS IN LESSON 1

Answers to questions on page 303.

6.0A   Pathogenic organisms are disease-producing organisms.

6.0B   Disinfection is the selective destruction or inactivation of pathogenic organisms.

6.0C   The U.S. Environmental Protection Agency establishes drinking water standards.

6.0D   MCL stands for **M**aximum **C**ontaminant **L**evel.

Answers to questions on page 305.

6.1A   Chlorine is a more effective disinfectant in water with a pH around 7.0 than in water above a pH of 8.0.

6.1B   Relatively cold water requires longer disinfection time or greater quantities of disinfectants.

6.1C   The number and type of organisms present in water influence the effectiveness of disinfection on microorganisms.

6.1D   Possible sources of drinking water contamination in distribution systems include new main installations, cross connections, and main breaks.

Answers to questions on page 307.

6.2A   Physical agents that have been used for disinfection include (1) ultraviolet rays, (2) heat, and (3) ultrasonic waves.

6.2B   Chemical agents that have been used for disinfection other than chlorine include (1) iodine, (2) bromine, (3) bases (sodium hydroxide and lime), and (4) ozone.

6.2C   A major limitation to the use of ozone is the inability of ozone to provide a residual in the distribution system.

Answers to questions on page 309.

6.2D $\dfrac{\text{Chlorine}}{\text{Dose, mg/}L} = \dfrac{\text{Chlorine}}{\text{Demand, mg/}L} + \dfrac{\text{Chlorine}}{\text{Residual, mg/}L}$

6.2E $\dfrac{\text{Chlorine}}{\text{Demand, mg/}L} = \dfrac{\text{Chlorine}}{\text{Dose, mg/}L} - \dfrac{\text{Chlorine}}{\text{Residual, mg/}L}$

6.2F   Hydrogen sulfide and ammonia are two inorganic reducing chemicals with which chlorine reacts rapidly.

Answers to questions on page 310.

6.2G   Chlorine gas tends to lower the pH while hypochlorite tends to increase the pH.

6.2H   The higher the pH level, the greater the percent of OCl^-.

Answers to questions on page 315.

6.2I   Breakpoint chlorination is the addition of chlorine to water until the chlorine demand has been satisfied and further additions of chlorine result in a free chlorine residual that is directly proportional to the amount of chlorine added beyond the breakpoint.

6.2J   An operator's decision to use chloramines depends on the ability to meet various regulations, the quality of the raw water, operational practices, and distribution system characteristics.

6.2K   The three primary methods by which chloramines are produced include: (1) preammoniation followed by chlorination, (2) concurrent addition of chlorine and ammonia, and (3) prechlorination/postammoniation.

6.2L   The *applied* chlorine to ammonia-nitrogen ratio is usually greater than the *actual* chlorine to ammonia-nitrogen ratio leaving the plant because of the chlorine demand of the water.

6.2M   Production of nitrite rapidly reduces free chlorine and can interfere with the measurement of free chlorine. The end result of incomplete nitrification may be a loss of total chlorine and ammonia and an increase in the concentration of heterotrophic plate count bacteria.

Answers to questions on page 320.

6.3A   When disinfecting a well, the well, pump, screen, and aquifer around the well all should be disinfected because they all could have been contaminated during construction and are potential sources of contamination.

6.3B   The chlorine solution should be applied to the well by injecting the chlorine down the pump column pipe rather than injecting if through the vent pipe.

6.3C   A well has been successfully disinfected if the results of a bacteriological analysis for total coliforms are negative (no coliforms).

Answers to questions on page 326.

6.3D   Water mains can be kept clean during construction and repair by keeping material such as dirt, construction materials, animals, rodents, and dirty water out of the mains. Inspecting the pipes when laid, keeping trenches dry, and installing watertight plugs all help to keep pipes clean.

6.3E   Before a water main is disinfected, the main should be flushed for at least 30 minutes with a flushing velocity of 2.5 ft/sec.

6.3F   Areas in a water main that require an extra effort for successful disinfection include fittings, valves, and air pockets where chlorine might not come in contact with the surface.

6.3G   Three forms of chlorine used for disinfection include:

1. Chlorine gas (liquid chlorine in cylinders),
2. Calcium hypochlorite, and
3. Sodium hypochlorite.

6.3H   Water used for disinfection that has a high chlorine residual may be disposed of in sanitary sewers, storm sewers, or on land, but should not be disposed of in a manner that will cause an adverse impact on the environment.

ANSWERS TO QUESTIONS IN LESSON 2

Answers to questions on page 337.

6.4A   The two major types of chlorine feeders are (1) hypochlorinators, and (2) chlorinators.

6.4B   Chlorine booster pumps are needed on most well applications to overcome the higher water pump discharge pressures.

6.4C   The maximum rate of chlorine removal from a 150-pound (68-kg) cylinder is 40 pounds (18 kg) of chlorine per day.

Answers to questions on page 340.

6.4D   The level of the hypochlorite solution tank should be read at the same time every day.

6.4E   The chemical feed of a hypochlorinator is adjusted until an adequate chlorine residual exists throughout the system and coliform test results are negative.

6.4F   The strength of the hypochlorite solution in the solution tank is adjusted so that the frequency of the strokes or pulses from the solution feed pump will be close together. This ensures that chlorine will be fed continuously to the water treated.

6.4G   Maintenance usually required of hypochlorinators includes oil changes and lubrication.

Answers to questions on page 344.

6.4H   Before attempting to locate and repair a chlorine gas leak, you should have protective clothing (gloves and rubber suit) and a self-contained pressure-demand air supply.

6.4I   A chlorine leak can be detected by holding an ammonia-soaked rag near suspected leaks. A white cloud will reveal the location of the leak.

6.4J   If a chlorinator cannot be repaired quickly, shut off the water supply so that unchlorinated or contaminated water will not be delivered to consumers.

6.4K   The two most common water quality tests run on samples of water from a water supply system are (1) chlorine residual, and (2) coliform tests.

Answers to questions on page 347.

6.5A   Chlorine residual is measured in treated water by the use of (1) amperometric titration, and (2) DPD colorimetric method.

6.5B   Residual chlorine measurements of treated water should be taken three times per day on small systems and once every two hours on large systems.

Answers to questions on page 350.

6.6A   Chlorine gas is extremely toxic and corrosive in moist atmospheres.

6.6B   A properly fitting self-contained air or oxygen supply type of breathing apparatus, positive pressure/demand, or a rebreather kit is recommended when repairing a chlorine leak.

6.6C   First-aid measures depend on the severity of the contact. Remove the victim from the gas area, remove clothing exposed to chlorine, and keep the victim warm and quiet. Call a doctor immediately. Keep the patient breathing.

6.6D   Whenever hypochlorite comes in contact with your hand, immediately wash the hypochlorite off and thoroughly wash your hand. Consult a physician if the area appears burned.

Answers to questions on page 354.

6.7A

Known		Unknown
Diameter of Pipe, in	= 12 in	5% Hypochlorite, gallons
or ft	= 1.0 ft	
Length of Pipe, ft	= 400 ft	
Chlorine Dose, mg/L	= 450 mg/L	
Hypochlorite, %	= 5%	

1. Calculate the volume of water in the pipe in gallons.

$$\text{Pipe Volume, gallons} = (0.785)(\text{Diameter, ft})^2(\text{Length, ft})(7.48 \text{ gal/cu ft})$$

$$= (0.785)(1.0 \text{ ft})^2(400 \text{ ft})(7.48 \text{ gal/cu ft})$$

$$= 2,349 \text{ gallons of Water}$$

2. Determine the pounds of chlorine needed.

$$\text{Chlorine, lbs} = (\text{Volume, M Gal})(\text{Dose, mg/L})(8.34 \text{ lbs/gal})$$

$$= (0.002349 \text{ M Gal})(450 \text{ mg/L})(8.34 \text{ lbs/gal})$$

$$= 8.82 \text{ lbs Chlorine}$$

3. Calculate the gallons of 5 percent sodium hypochlorite solution needed.

$$\text{Sodium Hypochlorite Solution, gallons} = \frac{(\text{Chlorine, lbs})(100\%)}{(8.34 \text{ lbs/gal})(\text{Hypochlorite, \%})}$$

$$= \frac{(8.82 \text{ lbs Chlorine})(100\%)}{(8.34 \text{ lbs/gal})(5\%)}$$

$$= 21.2 \text{ gal}$$

A little over 21 gallons of 5 percent solution of sodium hypochlorite should do the job.

6.7B

Known		Unknown
Chlorine Dose, mg/L	= 2.0 mg/L	Chlorine Demand, mg/L
Chlorine Residual, mg/L	= 0.2 mg/L	

Calculate the chlorine demand in mg/L.

$$\text{Chlorine Demand, mg/L} = \text{Chlorine Dose, mg/L} - \text{Chlorine Residual, mg/L}$$

$$= 2.0 \text{ mg/L} - 0.2 \text{ mg/L}$$

$$= 1.8 \text{ mg/L}$$

6.7C

	Known	**Unknown**
Pump Flow, GPM	= 600 GPM	Chlorinator Setting,
Chlorine Dose, mg/L	= 4.0 mg/L	lbs Chlorine/24 hours

1. Convert pump flow to million gallons per day (MGD).

$$\text{Flow, MGD} = \frac{(600 \text{ GPM})(60 \text{ min/hr})(24 \text{ hr/day})}{1,000,000/M}$$

$$= 0.864 \text{ MGD}$$

2. Calculate the chlorinator setting in pounds of chlorine per 24 hours.

$$\begin{aligned}\text{Chlorinator Setting, lbs/24 hr} &= (\text{Flow, MGD})(\text{Dose, mg/}L)(8.34 \text{ lbs/gal})\\[6pt]&= (0.864 \text{ MGD})(4.0 \text{ mg/}L)(8.34 \text{ lbs/gal})\\[6pt]&= 28.8 \text{ lbs Chlorine/day}\\[6pt]&= 28.8 \text{ lbs Chlorine/24 hours}\end{aligned}$$

6.7D

	Known	**Unknown**
Hypochlorinator, GPD	= 60 GPD	Chlorine Dose, mg/L
Hypochlorite, %	= 2%	
Pump, GPM	= 400 GPM	

1. Convert the pumping rate to MGD.

$$\text{Pumping Rate, MGD} = \frac{(400 \text{ GPM})(60 \text{ min/hr})(24 \text{ hr/day})}{1,000,000/M}$$

$$= 0.58 \text{ MGD}$$

2. Calculate the chlorine dose rate in pounds per day.

$$\begin{aligned}\text{Chlorine Dose, lbs/day} &= \frac{(\text{Flow, gal/day})(\text{Hypochlorite, \%})(8.34 \text{ lbs/gal})}{100\%}\\[6pt]&= \frac{(60 \text{ GPD})(2\%)(8.34 \text{ lbs/gal})}{100\%}\\[6pt]&= 10.0 \text{ lbs Chlorine/day}\end{aligned}$$

3. Calculate the chlorine dose in mg/L.

$$\begin{aligned}\text{Chlorine Dose, mg/}L &= \frac{\text{Chlorine Dose, lbs/day}}{(\text{Flow, MGD})(8.34 \text{ lbs/gal})}\\[6pt]&= \frac{10.0 \text{ lbs Chlorine/day}}{(0.58 \text{ MGD})(8.34 \text{ lbs/gal})}\\[6pt]&= 2.1 \text{ lbs Chlorine/M lbs Water}\\[6pt]&= 2.1 \text{ mg/}L\end{aligned}$$

CHAPTER 7

SAFETY

by

Dan Saenz

TABLE OF CONTENTS

Chapter 7.   SAFETY

OBJECTIVES

Chapter 7.   SAFETY

Following completion of Chapter 7, you should be able to:

1. *THINK SAFETY,*

2. Develop a safety program for a water utility agency,

3. Prepare and conduct tailgate safety sessions,

4. Safely operate and maintain pumps and wells, with attention to the safety of operators and consumers,

5. Inspect safety features of vehicles and equipment,

6. Drive vehicles defensively and safely,

7. Route traffic around a job site,

8. Work safely in streets,

9. Protect the motoring public and pedestrians from work areas in streets and sidewalks, and

10. Conduct a safety inspection of waterworks facilities.

WORDS

Chapter 7.   SAFETY

ACUTE HEALTH EFFECT

ACUTE HEALTH EFFECT

An adverse effect on a human or animal body, with symptoms developing rapidly.

CAUTION

CAUTION

This word warns against potential hazards or cautions against unsafe practices. Also see DANGER, NOTICE, and WARNING.

CHRONIC HEALTH EFFECT

CHRONIC HEALTH EFFECT

An adverse effect on a human or animal body with symptoms that develop slowly over a long period of time or that recur frequently.

COMPETENT PERSON

COMPETENT PERSON

A competent person is defined by OSHA as a person capable of identifying existing and predictable hazards in the surroundings, or working conditions which are unsanitary, hazardous or dangerous to employees, and who has authorization to take prompt corrective measures to eliminate the hazards.

CONFINED SPACE

CONFINED SPACE

Confined space means a space that:

A.   Is large enough and so configured that an employee can bodily enter and perform assigned work; and

B.   Has limited or restricted means for entry or exit (for example, manholes, tanks, vessels, silos, storage bins, hoppers, vaults, and pits are spaces that may have limited means of entry); and

C.   Is not designed for continuous employee occupancy.

(Definition from the Code of Federal Regulations (CFR) Title 29 Part 1910.146.)

CONFINED SPACE, NON-PERMIT

CONFINED SPACE, NON-PERMIT

A non-permit confined space is a confined space that does not contain or, with respect to atmospheric hazards, have the potential to contain any hazard capable of causing death or serious physical harm.

CONFINED SPACE, PERMIT-REQUIRED
(PERMIT SPACE)

CONFINED SPACE, PERMIT-REQUIRED
(PERMIT SPACE)

A confined space that has one or more of the following characteristics:

- Contains or has a potential to contain a hazardous atmosphere,
- Contains a material that has the potential for engulfing an entrant,
- Has an internal configuration such that an entrant could be trapped or asphyxiated by inwardly converging walls or by a floor which slopes downward and tapers to a smaller cross section, or
- Contains any other recognized serious safety or health hazard.

(Definition from the Code of Federal Regulations (CFR) Title 29 Part 1910.146.)

DANGER

DANGER

The word *DANGER* is used where an immediate hazard presents a threat of death or serious injury to employees. Also see CAUTION, NOTICE, and WARNING.

DANGEROUS AIR CONTAMINATION DANGEROUS AIR CONTAMINATION

An atmosphere presenting a threat of causing death, injury, acute illness, or disablement due to the presence of flammable and/or explosive, toxic or otherwise injurious or incapacitating substances.

A. Dangerous air contamination due to the flammability of a gas or vapor is defined as an atmosphere containing the gas or vapor at a concentration greater than 10 percent of its lower explosive (lower flammable) limit.

B. Dangerous air contamination due to a combustible particulate is defined as a concentration greater than 10 percent of the minimum explosive concentration of the particulate.

C. Dangerous air contamination due to the toxicity of a substance is defined as the atmospheric concentration immediately hazardous to life or health.

DECIBEL (DES-uh-bull) DECIBEL

A unit for expressing the relative intensity of sounds on a scale from zero for the average least perceptible sound to about 130 for the average level at which sound causes pain to humans. Abbreviated dB.

MATERIAL SAFETY DATA SHEET (MSDS) MATERIAL SAFETY DATA SHEET (MSDS)

A document which provides pertinent information and a profile of a particular hazardous substance or mixture. An MSDS is normally developed by the manufacturer or formulator of the hazardous substance or mixture. The MSDS is required to be made available to employees and operators whenever there is the likelihood of the hazardous substance or mixture being introduced into the workplace. Some manufacturers are preparing MSDSs for products that are not considered to be hazardous to show that the product or substance is *NOT* hazardous.

NOTICE NOTICE

This word calls attention to information that is especially significant in understanding and operating equipment or processes safely. Also see CAUTION, DANGER, and WARNING.

OXYGEN DEFICIENCY OXYGEN DEFICIENCY

An atmosphere containing oxygen at a concentration of less than 19.5 percent by volume.

OXYGEN ENRICHMENT OXYGEN ENRICHMENT

An atmosphere containing oxygen at a concentration of more than 23.5 percent by volume.

SPOIL SPOIL

Excavated material such as soil from the trench of a water main.

TAILGATE SAFETY MEETING TAILGATE SAFETY MEETING

Brief (10 to 20 minutes) safety meetings held every 7 to 10 working days. The term *TAILGATE* comes from the safety meetings regularly held by the construction industry around the tailgate of a truck.

WARNING WARNING

The word *WARNING* is used to indicate a hazard level between *CAUTION* and *DANGER.* Also see CAUTION, DANGER, and NOTICE.

CHAPTER 7.   SAFETY[1]

(Lesson 1 of 2 Lessons)

7.0 IMPORTANCE OF SAFETY

7.00   Think Safety

LET'S START THINKING SAFETY NOW! In this section we are going to provide you with some ideas for your safety program. The remaining sections of this chapter will discuss how to perform specific jobs safely. Two very important aspects of your safety program are:

1. Making people aware of unsafe acts, and

2. Conducting and/or participating in regular safety training programs.

7.01   What Is Safety?

Webster's dictionary states that safety is the following:

"The condition of being safe; freedom from exposure to danger; exemption from hurt, injury, or loss; to protect against failure, breakage, or other accidents; knowledge of or skill in methods of avoiding accidents or disease."

Safety is more than words. Safety is the action of Webster's definition. Safety is using one's knowledge or skill to avoid accidents or to protect oneself and others from accidents. Safety is a form of preventive maintenance which includes equipment and machinery and its proper handling. Who is responsible for safety? Everyone should be responsible from top management to all employees.

Safety is a program for everybody, not only on the job but also at home. Unfortunately, people miss more time at work from off-the-job accidents than from on-the-job accidents!

Management has its responsibilities for safety; its main function is to set the tone and provide the training and monies for an effective safety program. Management should:

1. Establish a safety policy;

2. Assign responsibility for accident prevention, which includes description of the duties and responsibilities of a safety officer, department heads, line supervisors, safety committees, and operators;

3. Appoint a safety officer or coordinator;

4. Establish realistic goals and periodically revise them to ensure continuous and maximum effort; and

5. Evaluate the results of the program.

For additional information about management's responsibilities for Safety, see Chapter 8, Section 8.11, "Safety Program."

While management has its responsibilities, the operators also have their particular responsibilities. Operators should:

1. Perform their jobs in accordance with established safe procedures,

2. Recognize the responsibility for their own safety and that of fellow operators,

3. Report all injuries,

4. Report all observed hazards, and

5. Actively participate in the safety program.

QUESTIONS

Write your answers in a notebook and then compare your answers with those on page 431.

7.0A   List two very important aspects of a safety program.

7.0B   What are management's responsibilities for safety?

7.0C   What are the operators' responsibilities for safety?

7.1 SAFETY PROGRAM

7.10   Objective of Safety Program

A SAFETY PROGRAM HAS ONE OBJECTIVE: TO PREVENT ACCIDENTS. Accidents do not happen, they are caused. They may be caused by unsafe acts of operators or result from hazardous conditions or may be a combination of

---

[1] *Portions of this chapter include safety material developed and distributed by the Department of Water and Power, City of Los Angeles.*

both. An analysis of accident statistics reveals that operator negligence and carelessness are the causes of most accidents. We know how to do our job safely, but we just don't do it safely.

Accidents reduce efficiency and effectiveness. An accident is that occurrence in a sequence of events that usually produces unintended injury, death, or property damage. Accidents affect the lives and morale of operators. They raise the costs not only to management but also to operators.

7.11 Unsafe Acts

Safety authorities tell us that nine out of ten injuries are the result of unsafe acts of either the person injured or someone else. Here are some of the principal reasons for unsafe acts.

IGNORANCE. This may be due to lack of experience or training, or to a temporary condition that prevents the recognition of a hazard.

INDIFFERENCE. Some know better but don't care. They take unnecessary risks and disregard the rules or instructions.

POOR WORK HABITS. Some people either don't learn the right way of doing things, or they develop a wrong way. Supervisors or fellow operators who see things done unsafely must speak up.

LAZINESS. Laziness affects the speed and quality of work. Laziness also affects safety because safety requires an effort. In most jobs you cannot reduce your "safety effort" and still maintain the same level of safety, even if you slow down or lower your quality standards.

HASTE. When we rush, we work too fast to think about what we are doing; we take dangerous shortcuts, and we are more likely to be injured.

POOR PHYSICAL CONDITION. Some of us just won't take reasonable care of ourselves. We ignore our bodily needs in regard to exercise, rest, and diet, lessening our endurance and alertness.

TEMPER. Impatience and anger cause many accidents. Again, our thinking is interfered with and the way prepared for an accident.

Therefore, every one of these unsafe acts could be considered due to operator negligence or carelessness.

SAFETY QUOTE: THE MORE YOU TALK ABOUT SAFETY, THE LESS YOU HEAR ABOUT ACCIDENTS.

QUESTIONS

Write your answers in a notebook and then compare your answers with those on page 431.

7.1A   What is an accident?

7.1B   List the principal reasons for unsafe acts.

7.1C   What problems are caused by laziness?

7.12 Vehicle and Equipment Inspection Procedures

Most of the tasks that we do in our everyday life require preparation, regardless of how routine or mundane the task. For example, going to the grocery store requires that your car will start, that it will have enough gas, that you have the shopping list and the checkbook, and that you allocate a certain amount of time to accomplish the task of going to the store. Human nature being what it is, we typically will try to substitute a modified plan where we have used poor planning.

In our water distribution business, the modifications and shortcuts that result from not following established procedures can have serious consequences. Because of the hazardous nature of our jobs, poor planning increases the risk of an injury and/or fatality to ourselves and co-workers. Therefore, one of the most critical aspects of your ability to perform your job in a professional manner occurs before you even leave the maintenance yard or shop. You and your agency should develop and follow established procedures for every water distribution system operation or activity.

Let us assume that today you are going to be working in a valve box and that access to the valve box is in the street. The water agency's standard procedures require completion of the following activities before the crew leaves the maintenance yard:

1. Discuss the work assignments for the day with your colleagues, co-workers, and supervisors.

2. Determine equipment needed; for example, atmospheric testing devices, traffic control devices, material handling devices such as slings, and hand tools not normally carried on utility vehicles. Do you need any specialized equipment?

3. Inspect each piece of equipment you will be using to ensure it is working. It doesn't make sense to have atmospheric testing devices available if they are not calibrated, or they are inoperative. Your inclination will be to short-cut procedures once in the field and jeopardize yourself and your co-workers' ability to survive. If there is any doubt about the functional capacity of an item, use replacement equipment while the defective or suspected equipment is repaired or further tested.

4. Inspect vehicles and towed equipment:

    a. Remember that most of our equipment is large/oversized/specialized equipment; therefore, procedures must be established to ensure safety and driveability once on the road. Driving equipment day in and day out may cause us to lose sight of the fact that this equipment has the following characteristics:

        1. Limited visibility,

        2. Reduced braking effectiveness,

        3. Top heavy (tankers),

        4. It may not be suitable for high-speed highway driving,

        5. Reduced maneuverability, such as the inability to perform quick maneuvers to avoid an accident,

        6. May have special requirements to prevent or accommodate load shifting,

7. Difficult to handle under unpredictable events such as a tire blowout, and

8. Susceptible to reduced control under icy, wet, or high-wind conditions.

b. Frequently it is necessary to tow trailer-mounted equipment such as generators, compressors, and pumps. Towing a trailer requires experience, knowledge, and a great amount of common sense. The job of towing a trailer seems to be a simple task; however, there are many pitfalls if good judgment is not exercised, or if the safety of others is not considered.

Many accidents related to trailer towing occur while simply coupling the trailer to the vehicle. Back strains and cuts and bruises to the hands and fingers are the injuries occurring before starting to tow a trailer.

As part of the procedure before leaving the yard, the vehicle and any towed equipment should undergo a thorough mechanical/safety item inspection including:

1. Mirrors and windows,

2. Lighting system, including backup lights, turn signals, and brake lights,

3. Brakes on vehicle and trailer if so equipped,

4. Tire tread and inflation and wheel attachment (vandals may have loosened lug nuts),

5. Trailer hitch/safety chain,

6. Towed vehicle lighting system, and

7. Auxiliary equipment such as winches or hoists.

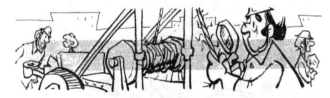

Once your equipment and vehicle have been checked out, you and your crew are ready to depart the maintenance area. Ironically, many vehicle accidents occur before the crew is even out of the yard, due to the unusual nature of the equipment they are driving. Vehicles should be equipped with a backup alarm that sounds when the vehicle is placed in reverse to alert personnel in the area that the vehicle is backing up, usually under a limited visibility situation. Congestion in the yard, as well as the amount of activity with other vehicles and crews, increases the hazard of an accident occurring at this time. An even better system, since most crews are at least two-person crews, is to have one person behind the vehicle to guide and warn the driver.

NOTE: It is a good idea to weigh your loaded vehicle at least annually to ensure that you are not exceeding its gross vehicle rating (GVR). Depending on how vehicles are loaded and equipped, they will have varying payload allowances. Weighing is easily accomplished for a very nominal fee at a large commercial truck scale. In the event that you are involved in an accident with an overweight vehicle, you could be subject to criminal or civil charges.

7.13 Defensive Driving

Having made it safely out of the maintenance yard, the next hazard we face is getting to the job site safely, usually during rush hour and with heavy vehicle traffic conditions. Don't forget to *FASTEN YOUR SEAT BELTS*. Many states now have seat belt laws, and most agencies have seat belt policies. (See *OPERATION AND MAINTENANCE OF WASTEWATER COLLECTION SYSTEMS*, Volume II, Chapter 11, Section 11.8, "Safety/Survival Program Policies/Standards," for a typical Seat Belt Policy.)

Because water distribution systems are spread out, operators drive more miles annually on the job than the average person's total miles driven during a year. Therefore, defensive driving is an important factor in our daily routine.

Defensive driving is not only a mature attitude toward driving, but a strategy as well. You must always be on the alert for the large number of new drivers, inattentive or sleepy drivers, people driving under the influence of alcohol and/or drugs, and those drivers who are just plain incompetent. In addition, weather, construction, and the type of equipment you are driving can also present hazards. Some of the elements of defensive driving are:

1. Always be aware of what is going on ahead of you, behind you, and on both sides of you;

2. Always have your vehicle under control;

3. Be willing to surrender your legal right-of-way if it might prevent an accident;

4. Know the limitations of the vehicle and towed equipment you are operating (stopping distance, impaired road vision, distance to change lanes safely, acceleration rate);

5. Take into consideration weather or other unusual conditions;

6. Develop a defensive driving attitude, that is, be aware of the situations that you are driving in and develop options to avoid accidents. For example, in residential areas you should assume that a child could run out into the street at any time, so always be thinking of what type of evasive action you could take. Be especially alert driving through construction areas during rush hour;

7. Rear end collisions, either due to driver inattentiveness or tailgating, are one of the most frequent types of vehicle accidents. Figure 7.1 illustrates the distance required for a vehicle to stop with ideal conditions and good brakes at various speeds, including the amount of time it takes for your brain to react. These figures are for passenger cars and, obviously, increase significantly with most of the equipment that we operate on the road; and

8. Keep in mind that you are driving a vehicle that is highly visible to the public. It may be necessary to swallow your pride when confronted by discourteous or unsafe drivers. Even when in the wrong, those types of drivers are quick to alert your agency, and thus your supervisor, to real or imagined violations that you committed while driving an agency vehicle.

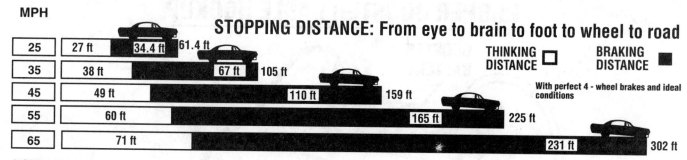

MPH

STOPPING DISTANCE: From eye to brain to foot to wheel to road

THINKING DISTANCE □   BRAKING DISTANCE ■

With perfect 4 - wheel brakes and ideal conditions

MPH	Thinking	Braking	Total
25	27 ft	34.4 ft	61.4 ft
35	38 ft	67 ft	105 ft
45	49 ft	110 ft	159 ft
55	60 ft	165 ft	225 ft
65	71 ft	231 ft	302 ft

NOTE: Multiply MPH x 1.6 to obtain km/hr; multiply ft x 0.3 to obtain meters.

Fig. 7.1   *Always maintain a safe stopping distance behind the vehicle in front of you*
(Courtesy of California Highway Patrol)

Formal defensive driving programs are offered by many public and private agencies and are one of the special programs that should be incorporated into your Distribution System Safety/Survival Program. These programs range from formal classroom training to actual defensive driving on a closed course using the vehicle you drive on your job every day. These courses are not only interesting and informative, but can be fun as well. Suggested contacts in your area for defensive driving opportunities include:

1. State highway patrol,

2. Fire departments and local public safety (police) agencies,

3. Private trucking companies,

4. Private bus companies, and

5. Public transportation companies.

For a nominal fee, these organizations frequently will allow you and your co-workers to participate in their defensive driving programs.

QUESTIONS

Write your answers in a notebook and then compare your answers with those on pages 431 and 432.

7.1D   Why must water distribution system operators follow established procedures?

7.1E   What tasks should be performed *BEFORE* leaving the maintenance yard or shop?

7.1F   What should an operator look for when checking out an atmospheric testing device before taking it out to a job site?

7.1G   List the items that should be inspected before towing a trailer.

7.1H   What is defensive driving?

7.14   How To Charge a Battery

At one time or another you may have given a battery a boost or may have seen it done. There is a correct procedure to follow to eliminate damage to electrical components and to prevent a battery explosion (Figure 7.2).

To boost the battery of a disabled vehicle from that of another vehicle, follow this procedure. First put out all cigarettes and flames. A spark can ignite hydrogen gas from the battery fluid.

Next take off the battery caps, if removable, and add distilled water if it is needed. Check for ice in the battery fluid. Never jump-start a frozen battery! Replace the caps.

Next, park the auto with the "live" battery close enough so the cables will reach between the batteries of the two autos. The cars can be parked close, but do not allow them to touch. If they touch, a dangerous "arcing" situation can occur. Now set each car's parking brake. Be sure that an automatic transmission is set in park; put a manual-shift transmission in neutral. Make sure your headlights, heater, and all other electrical accessories are off (you don't want to sap electricity away from your dead battery while you are trying to start your car). If the two batteries have vent caps, remove them. This will reduce the risk of explosion (relieves pressure within the battery). Then lay a cloth over the open holes.

Attach one end of the jumper cable to the positive terminal of the booster battery (A) (that's the good battery in the other car) and the other end to the positive terminal of your battery (D). The positive terminal is identified by a + sign, a red color, or a "P" on the battery in your car. Each of the two booster cables has an alligator clip at each end. To attach, you simply squeeze the clip, place it over the terminal, then let it shut. Now attach one end of the remaining booster cable to the negative terminal of the booster battery (B). The negative terminal is marked with a – sign, a black color, or the letter "N." Attach the other end of the cable to a metal part on the engine of your car (C). Many mechanics simply attach it to the negative post of the battery. This is not recommended because a resulting arc could ignite hydrogen gas present at the battery surface and cause an explosion. Be sure the cables do not interfere with the fan blades or belts. The engine in the other car should be running, although it is not an absolute necessity.

Get in the disabled car and start the engine in a normal manner. After it starts, remove the booster cables. Removal is the exact reverse of installation. Remove the black cable attached to your engine, then remove it from the negative terminal of the booster battery. Remove the remaining cable from the positive terminal of the "dead" battery and from the booster battery. Replace the vent caps and you are done. Have the battery and/or charging system of the car checked by a mechanic to correct any problems.

For maximum eye safety, everyone working with car batteries or standing nearby should wear protective goggles to keep flying battery fragments and chemicals out of the eyes.

Should battery acid get into the eyes, immediately flush them with water continuously for 15 minutes. Then see a doctor.

PROPER BOOSTER CABLE HOOKUP

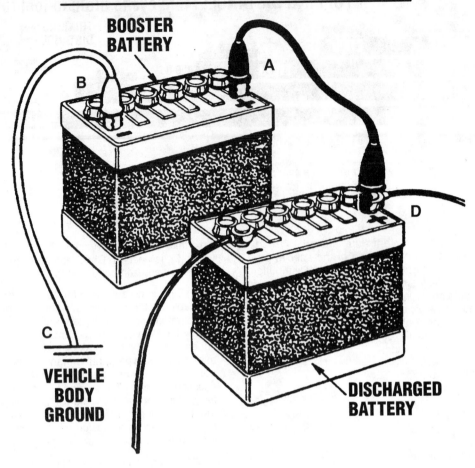

Fig. 7.2   Proper booster cable hookup

The recommended procedure for jump-starting is listed, step by step, on a 4-inch by 8-inch yellow vinyl sticker which has a permanent adhesive. The sticker can be affixed to any clean, dry surface under the hood or kept inside the car's glove compartment. Stickers are available—no charge for the first five, $27.00 per 100 after that—by writing to Prevent Blindness America, 211 West Wacker Drive, Suite 1700, Chicago, IL 60606.

7.15   Boat Safety

Boats are commonly used in the operation of a water utility. They are used for sampling, surveying, or inspecting reservoirs. Boats may be either motor operated (inboard or outboard) or rowboats. All boats should be inspected annually. The U.S. Coast Guard is usually the inspecting agency. They also recommend the safety practices to be used for boat safety. All boats should be equipped with a safety vest for each occupant or some type of approved flotation cushion. There should be oarlocks for the oars and oars should be of a suitable length for the respective boat and maintained in good operable condition. No boat should be loaded in excess of its loading capacity.

Motorized boats should meet the same standards as rowboats. In addition the motors (inboard and outboard) should be maintained in good operating condition. Periodic tune-ups and safety checks should be made following your utility's maintenance procedures or the manufacturer's recommendations. All motorized boats should have a set of oars to be used in an emergency and a fire extinguisher. At least two people should go out on a boat. If it is necessary that only one person use a boat, someone on shore should be aware that the person is out on the boat alone. Following safety practices when using boats could save someone from a potential drowning accident.

7.16   Corrosive Chemicals

A corrosive chemical is any chemical that may weaken, burn, and/or destroy a person's skin or eyes. A corrosive chemical may be either acid or base (alkali). The pH scale from 0 to 14 is an indication of the strength of a solution. The "acids" with the low pH values are the most corrosive. Hydrochloric (HCl), nitric (HNO_3), and sulfuric (H_2SO_4) are among the strongest acids. A "base" with the highest pH values of around 13 is the most corrosive. Sodium hydroxide (NaOH) is a strong base. Some common chemicals that are corrosive are:

1. Ammonium hydroxide,
2. Calcium hypochlorite,
3. Chlorine,
4. Ferric chloride,
5. Hydrochloric acid,
6. Potassium permanganate,

7 Sodium chlorite,
8. Sodium hydroxide,
9. Sodium hypochlorite,
10. Sodium nitrate,
11. Sulfuric acid, and
12. Zinc orthophosphate.

This list is by no means complete.

Safety procedures must be followed in the handling and use of all corrosive chemicals. Use safety protection equipment to protect your eyes and skin (see Section 7.21, "Protective Clothing"). Safety showers and eye/face wash stations should be installed at a location close to where any corrosive chemical is being used. Test the shower and eye/face wash station at least once a week. Run the water using the emergency levers (the pull chains, hand-actuated valves, or treadle foot-actuated valves). Let the water run for approximately three to five minutes or until the water runs clear. Rust may be in the water lines and may increase the problem of washing chemicals from your eyes.

7.17 Tailgate Safety Sessions

Tailgate safety sessions are so called because the session consists of a small group of operators gathered around the tailgate of a pickup or truck to discuss safety. These 10 to 20 minute sessions are usually held near the work site and new and old safety hazards and safer approaches or techniques to deal with the problems of the day or week are discussed. The amount or degree of organization or formality of the sessions depends on the types of safety problems and what you think is the most effective way of educating your crew. Table 7.1 is a typical topic for a tailgate safety session.

7.18 Employee "Right-To-Know" Laws

Employee "Right-To-Know" legislation has been implemented to require employers to inform employees (operators) of the possible health effects resulting from contact with hazardous substances. At locations where this legislation is in force, employers must meet certain requirements. Employers must provide operators with information regarding any hazardous substances that the operators might be exposed to under either normal work conditions or reasonably foreseeable emergency conditions resulting from workplace conditions. Information regarding hazardous substances is available from manufacturers in the form of Material Safety Data Sheets (MSDSs) for hazardous substances. Employers must provide operators with a copy of the MSDSs upon request and also must train operators to work safely with the hazardous substances that are encountered in the workplace.

As an operator you have the right to ask your employer if you are working with any hazardous substances. If you work with any hazardous substances, your employer must provide you with information on the health implications resulting from contact with the substances, the Material Safety Data Sheets, and training for working safely with the hazardous substances. For additional information, see Section 4.12, "Hazard Communication (Worker Right-To-Know Laws)," *OPERATION AND MAINTENANCE OF WASTEWATER COLLECTION SYSTEMS*, Volume I, in this series of operator training manuals.

TABLE 7.1   TAILGATE SAFETY TOPIC

Most injuries that happen on water utility jobs are caused directly by the injured person. Only a small percent are caused by defective equipment or devices. Because of this, you must be primarily responsible for your own safety.

Managers and supervisors are usually regarded as being the ones responsible for safety. Without proper interest on the part of management and supervisory staff, a total safety program can't be effective. But you must realize that you, more than anyone else, are responsible for not only your own safety but for the safety of your fellow operators. In other words, you must be your "brother's keeper."

Here's an example of what we're trying to say. Management can purchase new trucks and equip them with all the known safety devices and maintain them in perfect operating condition. But a truck has to be operated by a driver (you); you alone are responsible for its safe operation so that neither you nor any of your fellow operators will get hurt.

Here's another example. Take the simple wooden ladder. The ladder may be built to the best safety specifications; it may have been properly stored and frequently inspected for defects. Somebody, however, must place the ladder in position and somebody must use it. If it's not properly placed, if the footing is insecure, if it hasn't been secured to the building by some individual, it is entirely likely that the person using it or some other operator will get hurt. And how can either management or a supervisor be held responsible for such accidents? You must realize that you are the most vital factor in the control of accidents.

Some operators seem to think that the safety officer is responsible for preventing accidents. Even though the safety officer frequently makes inspections and counsels the operators and supervisors, the safety officer can't be in all places at all times. And the safety officer cannot, therefore, be blamed or held responsible whenever an accident occurs. So let's bear in mind that we ourselves, as individuals, must constantly be alert to the hazards around us. If we can't remove a hazard ourselves, we should call it to the attention of those who have the authority to do so.

QUESTIONS

Write your answers in a notebook and then compare your answers with those on page 432.

7.1I   Why should all cigarettes and flames be put out before attaching cables to charge a battery?

7.1J   List the safety practices for using boats.

7.1K   What is a corrosive chemical?

7.1L   What is a tailgate safety session?

7.2 PERSONAL SAFETY

Personal safety goes hand-in-hand with a good, effective safety program. You are the key individual to an effective safety program.

Don't you, as an operator, become a statistic

Table 7.2 is a tabulation of the types of accidents encountered in the water utility field. This chapter will provide you with safe procedures that will help you avoid becoming an accident statistic.

TABLE 7.2 ANALYSIS OF ACCIDENT TYPES[a]

Type of Accident	Percent
1. Sprains/strains in lifting, pulling, or pushing objects	28
2. Sprains/strains due to awkward position or sudden twist or slip	16
3. Struck by falling or flying objects	10
4. Struck against stationary or moving objects	9
5. Falls to different level from platform, ladder, stairs	7
6. Caught in, under or between objects	6
7. Falls on same level to working surface	5
8. Contact with radiation, caustics, toxic, and noxious substances	3
9. Animal or insect bites	1
10. Contact with temperature extremes	1
11. Rubbed or abraded	1
12. Contact with electric current	0
13. Miscellaneous	11

[a] AWWA Safety Bulletin

7.20 Monitoring Equipment

There are many locations in a water utility where it is necessary to monitor the atmosphere before entering. Some of these specific locations may be underground regulator vaults, solution vaults, manholes, tanks, trenches, and other *CONFINED SPACES*[2] (also see Section 7.63, "Confined Spaces"). There are a variety of devices that are used to monitor (check) the available oxygen and also combustible and toxic gases. Regardless of what type of monitoring device or devices are used, you should always test for *DANGEROUS AIR CONTAMINATION*[3] and/or *OXYGEN DEFICIENCY*[4] / *OXYGEN ENRICHMENT*[5] with an approved device immediately prior to an operator entering a confined space and at intervals frequent enough to ensure a safe atmosphere during the time an operator is in the structure. Dangerous air contamination includes explosive conditions and toxic gases. Toxic gases include hydrogen sulfide and carbon monoxide. *A RECORD OF THE TEST MUST BE KEPT AT THE JOB SITE* for the duration of the work. *YOUR LIFE* and the lives of *YOUR FELLOW OPERATORS* are in jeopardy if the air in a confined space is not tested before entering. The following five steps should be performed before entering a confined space:

1. *CALIBRATE* the gas detection device,

2. *BARELY OPEN* the confined space,

3. *TEST* for oxygen deficiency/enrichment, combustible gases, and toxic gases using a probe or tube to collect the sample,

4. *RECORD* the results of these tests in your gas log, and

5. *VENTILATE* the confined space.

If a hazardous atmospheric condition is discovered, you must repeat the tests for oxygen deficiency/enrichment and combustible and toxic gases while ventilating and again *RECORD* the results.

On rare occasions, unusual conditions may exist or be created even though the required testing procedures have been followed. Air currents through duct lines can easily change if other vaults in the same system are opened. Toxic or explosive gases from broken gas lines or decayed vegetation may then flow through the ducts into a previously gas-free vault. Therefore it is important that *ADEQUATE VENTILATION BE MAINTAINED AND THAT THE SPACE BE RECHECKED FOR HAZARDOUS ATMOSPHERES PERIODICALLY WHILE OPERATORS ARE WORKING IN SUCH LOCATIONS.* Use portable fans or ventilators to provide fresh air.

At least one person must stay outside the confined work area with another person standing by. This backup person should check continuously on the status of personnel working in the confined space. Should a person working in the confined space be rendered unconscious due to asphyxiation or lack of oxygen, the backup person at the surface could descend into the space by:

1. Securing the help of an additional backup rescue person, and

2. Putting on the correct type of respiratory protective equipment.

Self-contained breathing equipment should be nearby and used as necessary.

---

[2] *Confined Space.   Confined space means a space that:*
   A. Is large enough and so configured that an employee can bodily enter and perform assigned work; and
   B. Has limited or restricted means for entry or exit (for example, manholes, tanks, vessels, silos, storage bins, hoppers, vaults, and pits are spaces that may have limited means of entry); and
   C. Is not designed for continuous employee occupancy.
   (Definition from the Code of Federal Regulations (CFR) Title 29 Part 1910.146.)

[3] *Dangerous Air Contamination.   An atmosphere presenting a threat of causing death, injury, acute illness, or disablement due to the presence of flammable and/or explosive, toxic or otherwise injurious or incapacitating substances.*
   A. Dangerous air contamination due to the flammability of a gas or vapor is defined as an atmosphere containing the gas or vapor at a concentration greater than 10 percent of its lower explosive (lower flammable) limit.
   B. Dangerous air contamination due to a combustible particulate is defined as a concentration greater than 10 percent of the minimum explosive concentration of the particulate.
   C. Dangerous air contamination due to the toxicity of a substance is defined as the atmospheric concentration immediately hazardous to life or health.

[4] *Oxygen Deficiency.   An atmosphere containing oxygen at a concentration of less than 19.5 percent by volume.*

[5] *Oxygen Enrichment.   An atmosphere containing oxygen at a concentration of more than 23.5 percent by volume.*

Follow the procedures outlined in the *SAFETY NOTICE* whenever anyone must enter a confined space.

SAFETY NOTICE

Before anyone ever enters a tank for any reason, these safety procedures must be followed:

1. Test atmosphere in the tank for toxic and explosive gases and for oxygen deficiency/enrichment. Contact your local safety equipment supplier for the proper types of atmospheric testing devices. These devices should have alarms that are activated whenever an unsafe atmosphere is encountered;

2. Provide adequate ventilation, especially when painting. A self-contained, positive-pressure breathing apparatus may be necessary when painting;

3. All persons entering a tank must wear a safety harness; and

4. One person must be at the tank entrance to observe the actions of all people in the tank. An additional person must be readily available to help the person at the tank entrance with any rescue operation.

Table 7.3 lists some common dangerous gases that may be encountered by the operators of water supply systems and water treatment plants.

Not every mixture of gas with air will explode or burn. The "proper" mixture of oxygen with a gas is necessary to create explosive or flammable conditions. If there is not enough gas present (too much oxygen), the mixture is too lean and will not burn or explode. Also, if there is too much gas (not enough oxygen), the mixture is too rich and will not burn or explode. The range of explosive or flammable gas mixtures is defined by the Lower Explosive Limit (LEL) and Upper Explosive Limit (UEL). The objective of gas detection equipment is to warn us when we encounter explosive conditions within 10 percent of the Lower Explosive Limit. Any time a gas mixture is within 10 percent of the Lower Explosive Limit, this is considered a hazardous condition.

Figure 7.3 shows the relationship between the Lower Explosive Limit (LEL) and the Upper Explosive Limit (UEL) of a mixture of air and gas. Note in Table 7.3 that the percent by volume of a gas in air for the Lower Explosive Limit to occur is different for each gas. Also note in Figure 7.3 that if you have all gas (100% gas and 0% oxygen), the mixture is too rich and will not explode. If a mixture of gas and oxygen from the air is greater than 10 percent of the Lower Explosive Limit, the mixture is considered hazardous.

Monitoring equipment is essential to protect you from hazardous atmospheres when entering and working in enclosed spaces. This type of monitoring equipment is available in three different types:

1. Portable and attached to your clothing or your body,

2. Portable and carried by you or placed near your work site, and

3. Permanently installed in locations where hazardous atmospheres could develop.

You may encounter three types of hazardous atmospheres:

1. Toxic gases such as chlorine, carbon monoxide, or hydrogen sulfide,

2. Explosive or flammable conditions, and

3. Oxygen deficiency/enrichment.

Regardless of the type of detection devices that you are using to monitor hazardous atmospheres, regularly calibrate the detection devices according to the manufacturer's recommendations. These calibration procedures must be scheduled and performed as part of your preventive maintenance program. Remember, the detection devices are sensitive instruments. Handle them gently and keep them accurate if you wish to stay alive.

The recommended warning devices should be an audible noise and/or a flashing light that should start whenever a hazardous condition is detected. An operator should not be expected to read a meter while working and decide whether or not a hazard exists.

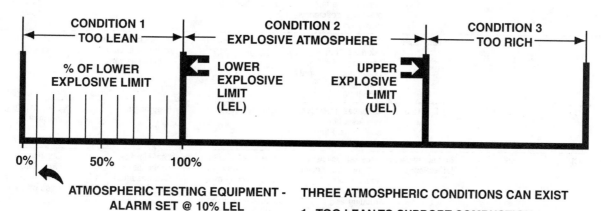

THREE ATMOSPHERIC CONDITIONS CAN EXIST
1. TOO LEAN TO SUPPORT COMBUSTION
2. MIXTURE JUST RIGHT, EXPLOSION OCCURS
3. MIXTURE TOO RICH TO SUPPORT COMBUSTION

Fig. 7.3 Relationship between the lower explosive limit or level (LEL) and the upper explosive limit (UEL) of a mixture of air and gas

TABLE 7.3   COMMON DANGEROUS GASES ENCOUNTERED IN WATER SUPPLY SYSTEMS AND AT WATER TREATMENT PLANTS [a]

Name of Gas	Chemical Formula	Specific Gravity or Vapor Density [b] (Air = 1)	Explosive Range (% by volume in air)		Common Properties (Percentages below are percent in air by volume)	Physiological Effects (Percentages below are percent in air by volume)	Most Common Sources in Sewers	Simplest and Cheapest Safe Method of Testing [c]
			Lower Limit	Upper Limit				
Oxygen (in Air)	O_2	1.11	Not flammable		Colorless, odorless, tasteless, non-poisonous gas. Supports combustion.	Normal air contains 20.93% of O_2. If O_2 is less than 19.5%, do not enter space without respiratory protection.	Oxygen depletion from poor ventilation and absorption or chemical consumption of available O_2.	Oxygen deficiency indicator.
Gasoline Vapor	C_5H_{12} to C_9H_{20}	3.0 to 4.0	1.3	7.0	Colorless, odor noticeable in 0.03%. Flammable. Explosive.	Anesthetic effects when inhaled. 2.43% rapidly fatal. 1.1% to 2.2% dangerous for even short exposure.	Leaking storage tanks, discharges from garages, and commercial or home dry-cleaning operations.	1. Combustible gas indicator. 2. Oxygen deficiency indicator for concentrations over 0.3%.
Carbon Monoxide	CO	0.97	12.5	74.2	Colorless, odorless, nonirritating, tasteless. Flammable. Explosive.	Hemoglobin of blood has strong affinity for gas, causing oxygen starvation. 0.2 to 0.25% causes unconsciousness in 30 minutes.	Manufactured fuel gas.	CO ampoules.
Hydrogen	H_2	0.07	4.0	74.2	Colorless, odorless, tasteless, non-poisonous, flammable. Explosive. Propagates flame rapidly; very dangerous.	Acts mechanically to deprive tissues of oxygen. Does not support life. A simple asphyxiant.	Manufactured fuel gas.	Combustible gas indicator.
Methane	CH_4	0.55	5.0	15.0	Colorless, tasteless, odorless, non-poisonous. Flammable. Explosive.	See hydrogen.	Natural gas, marsh gas, manufactured fuel gas, gas found in sewers.	1. Combustible gas indicator. 2. Oxygen deficiency indicator.
Hydrogen Sulfide	H_2S	1.19	4.3	46.0	Rotten egg odor in small concentrations, but sense of smell rapidly impaired. Odor not evident at high concentrations. Colorless. Flammable. Explosive. Poisonous.	Death in a few minutes at 0.2%. Paralyzes respiratory center.	Petroleum fumes, from blasting, gas found in sewers.	1. H_2S analyzer. 2. H_2S ampoules.
Carbon Dioxide	CO_2	1.53	Not flammable		Colorless, odorless, nonflammable. Not generally present in dangerous amounts unless there is already a deficiency of oxygen.	10% cannot be endured for more than a few minutes. Acts on nerves of respiration.	Issues from carbonaceous strata. Gas found in sewers.	Oxygen deficiency indicator.
Nitrogen	N_2	0.97	Not flammable		Colorless, tasteless, odorless. Non-flammable. Non-poisonous. Principal constituent of air (about 79%).	See hydrogen.	Issues from some rock strata. Gas found in sewers.	Oxygen deficiency indicator.
Ethane	C_2H_4	1.05	3.1	15.0	Colorless, tasteless, odorless, non-poisonous. Flammable. Explosive.	See hydrogen.	Natural gas.	Combustible gas indicator.
Chlorine	Cl_2	2.5	Not flammable Not explosive		Greenish yellow gas, or amber color liquid under pressure. Highly irritating and penetrating odor. Highly corrosive in presence of moisture.	Respiratory irritant, irritating to eyes and mucous membranes. 30 ppm causes coughing. 40-60 ppm dangerous in 30 minutes. 1,000 ppm apt to be fatal in a few breaths.	Leaking pipe connections. Overdosage.	Chlorine detector. Odor. Strong ammonia on a swab gives off white fumes.

[a] Originally printed in Water and Sewage Works, August 1953. Adapted from "Manual of Instruction for Sewage Treatment Plant Operators," State of New York.
[b] Gases with a specific gravity less than 1.0 are lighter than air; those more than 1.0 heavier than air.
[c] The first method given is the preferable testing procedure.

Sometimes operators working in small communities or for small agencies experience problems convincing their supervisors of the need for gas detection devices, portable ventilators, safety harnesses, and qualified people standing by when they must enter tanks, vaults, and manholes. OSHA regulations allow operators to refuse to perform work under unsafe conditions. This is a very difficult situation. We recommend that operators contact their local or state OSHA officials or other industrial safety officials. Obtain copies of the rules and regulations for entering tanks and confined spaces (see Section 7.63, "Confined Spaces"). These regulations usually specify the conditions that must be monitored. More important, however, is that these regulations will specify the penalties imposed on supervisors and employers for failing to provide operators with the proper safety equipment, procedures, and training. These penalties include stiff fines and jail sentences. Responsible employers and supervisors would rather comply with safety regulations than pay fines and spend time in jail.

OPERATORS HAVE THE RIGHT TO REFUSE TO DO ANYTHING THAT ENDANGERS THEIR LIVES.

QUESTIONS

Write your answers in a notebook and then compare your answers with those on page 432.

7.2A   List the typical locations in a water utility where it is necessary to monitor the atmosphere before entering.

7.2B   Why should a backup person stand outside the confined work area?

7.2C   At what level is a mixture of gas in air considered hazardous?

7.21   Protective Clothing (Figure 7.4)

Protective clothing is required for the protection of operators. Some clothing may be the responsibility of each operator and other protective equipment may be supplied by the water utility. Typical protective equipment supplied by the operators may be:

1. Safety toe (hard toe) shoes, and

2. Safety prescription glasses.

Protective equipment that may be required or supplied includes:

1. Safety (hard) hats,

2. Earplugs,

3. Safety glasses (nonprescription),

4. Safety goggles,

5. Face shields,

6. Respirators and masks,

7. Canvas gloves,

8. Rubber gloves for use with corrosive chemicals,

9. Safety aprons,

10. Leggings or knee pads,

11. Safety toe caps,

12. Protective clothing (lab coats or rubber suits), and

13. Safety belts or harnesses.

Protective clothing must be worn and equipment used (as appropriate) whenever *YOU* are exposed to safety hazards as you do your job.

7.22   Slips and Falls

Slips and falls are common accidents which can be a major problem in a water utility. All floors should be level and kept as slip resistant as possible. Floors should be free of all debris. Material that drips or spills can be collected in drip pans, appropriate gutters, or splash guards may be used to deflect drips. If liquids do get on floors, nonflammable absorbent materials should be available for cleaning up.

CAUTION: Sawdust should never be used as an absorbent because it is combustible.

When cracks, splinters, ruts, or breaks occur in floors they should be repaired as soon as possible. Good housekeeping and maintenance go hand in hand to prevent slips and falls. If water and chemicals spill on the floor, they should be wiped up as soon as possible. Catwalks or safety tread may be necessary on floors where water or chemicals are commonly spilled. The catwalk or safety tread will help prevent unnecessary slips and falls.

All stairways and elevated catwalks should have safety railings. Approved *CAUTION*[6] signs should be used where necessary to call attention to potential hazards or remind operators to use caution or to use safety equipment.

SAFETY QUOTE: Falls cause more accidents than summers, winters, or springs.

7.23   Handling and Lifting

Injuries caused by handling and lifting are many and varied. Approximately 28 percent of operator injuries can be directly attributed to the handling of objects. A large portion of strains and sprains, fractures, bruises, back injuries, and hernias are the result of common handling injuries. These injuries are caused primarily by unsafe work practices such as:

1. Improper lifting,

2. Carrying too heavy a load,

3. Incorrect gripping,

4. Failing to observe proper foot or hand clearance, and

5. Failing to use or wear proper equipment.

---

[6] *Caution.   This word warns against potential hazards or cautions against unsafe practices.*

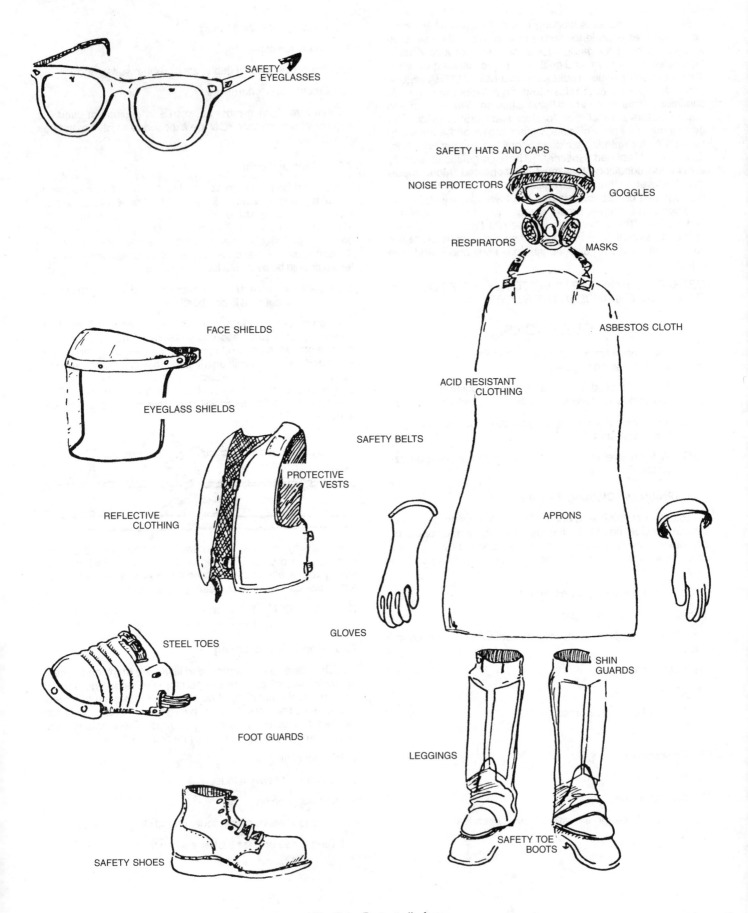

SAFETY EYEGLASSES

SAFETY HATS AND CAPS

NOISE PROTECTORS

GOGGLES

RESPIRATORS

MASKS

FACE SHIELDS

ASBESTOS CLOTH

ACID RESISTANT CLOTHING

EYEGLASS SHIELDS

SAFETY BELTS

PROTECTIVE VESTS

REFLECTIVE CLOTHING

APRONS

STEEL TOES

GLOVES

SHIN GUARDS

FOOT GUARDS

LEGGINGS

SAFETY SHOES

SAFETY TOE BOOTS

Fig. 7.4   Protect all of you

Many of us do not think much of lifting but if we are lifting an object incorrectly, we may suffer bad results such as pulled muscles, disc lesions, or painful hernias. Remember Figure 7.5 whenever you have a lifting job.

Here are seven steps to safe lifting:

1. Keep feet parted—one along the side of and one behind the object,

2. Keep back straight—nearly vertical,

3. Tuck your chin in,

4. Grip the object with the whole hand,

5. Tuck elbows and arms in,

6. Keep body weight directly over feet, and

7. Lift slowly.

Safety equipment should be used to help protect your body. To protect your hands, use gloves; to protect your feet, use safety shoes, instep protectors, and ankle guards. Estimate the weight of objects you must lift. Never lift more than you can safely carry. Use handles or holders that attach to the objects such as handles for moving auto batteries or baskets for carrying laboratory samples. If the object to be moved is awkward or too large or heavy for one operator, have two operators move the object in question. When more than one operator is used, the object to be lifted must be lifted in unison (all at once) and everyone must know how the object is to be lifted and handled. Only one operator is assigned the task of giving orders as necessary such as when to start, lift, carry, and set down the object. All parties concerned should lift correctly with the legs bearing the bulk of the physical effort.

If the object to be lifted and handled is too heavy for a person or persons to move safely, then use hand trucks, dollies, or powered hand trucks to move and/or lift objects.

Remember, 28 percent of injuries are caused by improper lifting and handling. Your health and safety depend on you using correct lifting and handling procedures.

7.24 Electrical

Electrical safety is very important to your life. Whenever you must work on electrical equipment or equipment that is run by electric motors, be sure to lock out and tag all of the electric switches. Only you should ever be allowed to remove a lockout device and tag that you have installed. Also you should never remove a tag or lockout device installed by another operator.

Water is a good conductor of electricity. Do not stand in water when working on electrical equipment.

Many water supply and water treatment monitoring, measuring, and control systems are electrical. Always be very careful when working around or troubleshooting these instrumentation systems.

Throughout this manual we have outlined safe procedures to follow whenever you must operate or maintain electrical equipment or equipment run by electric motors. However, two sections presented later in this chapter are of particular importance to all operators. Section 7.37, "Working Around Electrical Units," describes lockout/tagout procedures operators should use whenever it is necessary to shut down electrically operated equipment for repairs or adjustment. In addition, Section 7.42, "Maintenance and Repair," explains the importance of not only de-energizing electrical equipment but also physically blocking equipment to prevent unexpected movement that could injure someone working on the equipment. Read both of these sections carefully and make the suggested safety procedures part of your standard work routine.

BE CAREFUL . . .
THE BACK YOU SAVE MAY BE YOUR OWN!
1. Size up the load.
2. Bend your knees.
3. Get a firm grip.
4. Lift with your legs . . . gradually.
5. Keep the load close to you.

IT's HOW YOU LIFT as well as WHAT YOU LIFT.

Fig. 7.5 Lifting guidelines

QUESTIONS

Write your answers in a notebook and then compare your answers with those on page 432.

7.2D   List as many items of protective clothing for operators as you can recall.

7.2E   How can liquids spilled on floors be cleaned up?

7.2F   List the steps to safe lifting.

7.2G   What precautions should be taken when working on electric motors?

7.3   SAFETY AROUND WELLS

7.30   Importance of Well Safety

Safety around wells is important for two reasons:

1. Safety control around the well site is necessary to prevent contamination or pollution of the well that could affect the health and safety of the consumer, and

2. Safety housekeeping of the site will help prevent an accident to operators, or employees working near or around the well.

The well operator has the responsibility to preserve the quality of the well through preventive maintenance of the area where the well is located.

7.31   Location of Well Site

Selection of a well site requires serious consideration of several factors. The well should be located a suitable distance from any potential pollution source that may affect the groundwater. Some potential sources of pollution may be septic tanks, subsurface leaching systems, mining operations, underground storage tanks for fuel and gasoline, solid and hazardous waste disposal sites, and wastewater treatment facilities.

Some wells have been located in areas where initially the environment was correct for the site. The following are examples of well sites that were contaminated after a well had been in operation for some time.

EXAMPLE 1

A Girl Scout Camp leased a camp in the Angeles National Forest. This camp had two wells which supplied all the drinking water. The two wells were located in a small, narrow, secluded valley and for several years the bacteriological samples were negative. Camp officials decided to purchase horses to provide a special camping session where the camper could specialize in equestrian events. After a few years the well water bacteriological samples began to turn up positive. This water even had positive coliform samples with chlorination. After an investigation it was determined that the well was being contaminated because the horse stables were located above the well. During the rainy season the rain washed the horse manure down into the wells located below the stables.

EXAMPLE 2

A utility began to cut personnel because of fiscal problems. Some of the first cuts affected gardening personnel who cleaned up around the well site. After a few months the area around the well site became a dumping site. Old lumber, furniture, sofas, chairs, and garden clippings began to pile up

around the well site. This debris attracted rodents, rats, ground squirrels, and snakes close to the well site. This caused a potential problem to the employees and to the utility's customers.

7.32   New Wells

No sewers should be permitted within 50 feet (15 m) of any well. If the well must be located within 100 feet (30 m) of a sanitary or storm sewer under gravity flow, special pipe and fittings must be used (see Chapter 3 Appendix, Section 3.12, "Well Site Selection," in the *SMALL WATER SYSTEM OPERATION AND MAINTENANCE* manual and check with your local health department for distances that apply to your well). This is necessary to prevent intrusion of leaking wastewater into the well.

7.33   Sanitary Seal

The sanitary seal on a well is the most important factor to prevent contamination of any type of well. This seal will prevent contaminated water or debris from entering around the well casing and into the water (review Chapter 3 Appendix, Section 3.14, "Well Structure and Components," in *SMALL WATER SYSTEM OPERATION AND MAINTENANCE).*

When applying disinfectant or using well cleaning agents, the operator must protect the well against the entrance of surface water or foreign matter.

7.34   Surface Portion of Well

All surface openings installed on top of a well must be protected against the entrance of surface water or foreign matter. Well casing vents should be constructed so that openings are in a vertical, downward position. They should be a minimum of 36 inches (90 centimeters) in height above the finished surface of the well lot (ground surface) and should be covered with a fine mesh screen or device to exclude small insects from the inside of the well.

Gravel tubes must be capped at the top end, kept tightly sealed, and elevated above ground level to prevent flood waters from entering the well.

The sounding tube must be kept tightly sealed. All pipe vents, sounding lines, and gravel fill pipes must be continuous conduit through the concrete pedestal. Also, all conduits that penetrate the casing must be provided with a continuous watertight weld at the point of entry into the well interior. A pump motor base seal must be watertight. All these safety requirements are necessary to ensure that contaminants do not get into the well's interior. When these safety features are incorporated in a well's construction, the probability is lessened that the well will be contaminated by surface debris or waters. Review Chapter 3, "Wells," Section 3.1, "Surface Features of a Well," in *SMALL WATER SYSTEM OPERATION AND MAINTENANCE.*

7.35   Tank Coatings

Hydropneumatic pressure tanks incorporated with wells are normally painted on the interior surface with a protective coating to extend the life of the tank. Special consideration must be used in selecting the proper type of coating. The coating should be approved by the AWWA or the Additives Evaluation Branch of EPA and not be a cause of pollution or contamination of the drinking water. The application procedures are most important and the curing time is essential for proper results.

CAUTION

When applying a coating, personnel should use forced ventilation and/or a protective mask with an air line to provide an adequate air supply that is safe to work in. Coatings used in tanks are *VOLATILE* and can affect the respiratory system.

When applying coating, at least two operators should work together—one inside the tank and the other close by. The operator outside checks on the status of the person working inside and checks periodically on the forced ventilation system and air pressure for the air mask, if used. A third person must be readily available to assist the person outside with any rescue operations. See *SAFETY NOTICE.*

SAFETY NOTICE

Before anyone ever enters a tank for any reason, these safety procedures must be followed:

1. Test atmosphere in the tank for toxic and explosive gases and for oxygen deficiency/enrichment. Contact your local safety equipment supplier for the proper types of atmospheric testing devices. These devices should have alarms that are activated whenever an unsafe atmosphere is encountered;

2. Provide adequate ventilation, especially when painting. A self-contained, positive-pressure breathing apparatus may be necessary when painting;

3. All persons entering a tank must wear a safety harness; and

4. One person must be at the tank entrance to observe the actions of all people in the tank. An additional person must be readily available to help the person at the tank entrance with any rescue operation.

QUESTIONS

Write your answers in a notebook and then compare your answers with those on page 432.

7.3A   What are the safety responsibilities of the well operator?

7.3B   How is contaminated water or debris kept from entering around the well casing and reaching the groundwater?

7.3C   List the surface openings on top of a well that must be protected against the entrance of surface water or debris into the well.

7.36   Well Chemicals

7.360   *Acid Cleaning*

An acid treatment may be necessary to loosen encrustations from a well casing and well. Normally hydrochloric or sulfamic acids are used. Hydrochloric (HCl) acid should be used at full strength. Hydrochloric acid can be introduced into a well by means of a wide-mouthed funnel and a $3/4$- or 1-inch (16- or 25-mm) place pipe. Extend the place pipe into the well and attach the funnel to the end outside of the well. Pour the acid into the funnel and allow it to run into the well. If hydrochloric acid (muriatic acid—industrial name) is used, be careful! The pH of hydrochloric acid is approximately 1.0. This acid is very corrosive.

Inhalation of hydrochloric acid fumes can cause coughing, choking, and inflammation of the entire respiratory tract.

Swallowing even a small amount of this acid can cause corrosion of mucous membranes, esophagus, and stomach; difficulty swallowing; nausea; vomiting; intense thirst; and diarrhea. Circulatory collapse and death may occur.

Hydrochloric acid causes severe skin burns. Spills, splashes or fumes may affect the eyes and could cause permanent eye damage.

Safety apparatus must be used when working with hydrochloric acid. Protective clothing includes rubber gloves for hands—preferably 15 to 18 inches (38 to 45 cm) long, rubber apron for clothes protection, boots for foot protection in case of spillage, and face shield for protection against splashes and fumes.

Extreme care should be used in transporting acid. Acid should be carried in cases (gallons) or in carboys for protection and must be placed securely in a vehicle whenever transported.

Sulfamic acid has advantages over hydrochloric acid. Sulfamic acid may be used in granular form or as an acid solution mixed on site. Granular sulfamic acid is nonirritating to dry skin and its solution gives off no fumes except when reacting with encrusting materials. Spillage, therefore, presents no hazard and handling is easier, cheaper, and safer.

Although sulfamic acid is nonirritating to dry skin and it gives off no fumes, rubber gloves should still be used because hands may become wet from perspiration and the acid will cause burns on the skin. Safety goggles or glasses, preferably a face shield, should be used in case granules are blown into the face and/or the eyes. If sulfamic acid is mixed on site, care should be used in mixing.

CAUTION

Acid should always be added slowly to the water. Use an acid-proof crock for mixing. Mix with a wooden paddle or electric mixer. Again, always use rubber gloves, rubber boots, a rubber apron, and a face shield when mixing sulfamic acid.

NEVER ADD WATER TO ACID. ACID COULD SPLASH ALL OVER YOU.

When using any *CORROSIVE* or *CAUSTIC* chemical, an eye/face wash station must be available in case of an emergency. There are a variety of portable eye/face wash units commercially available. The unit should be located at a site that has easy access in case it is needed, preferably within 20 feet (6 m) of where the acid is being used. All personnel concerned should know the location of the eye/face wash station and be instructed on how to use it.

If acid is spilled on a person's skin, wash the exposed area with water as soon as possible. A shower and change of clothing may also be necessary. If acid is splashed on the face or in the eyes, immediately flush them at the eye/face wash station for at least 15 minutes. Make sure the eyes are thoroughly flushed with water. If irritation continues, resume washing eyes with water. Keep eyes moist with a wet towel and have a physician examine them for further treatment.

QUESTIONS

Write your answers in a notebook and then compare your answers with those on page 432.

7.3D   Why must hydrochloric acid be handled carefully?

7.3E   What are the advantages of using sulfamic acid over hydrochloric acid to acid clean a well?

7.3F   What safety precautions should be used when working with sulfamic acid in the granular form?

7.361   Chlorine Treatment

When a well requires treatment with chlorine, either granular calcium hypochlorite or liquid sodium hypochlorite is commonly used. Regardless of the type of hypochlorite used, safety equipment should be used. Again, use rubber gloves, preferably 15 to 18 inches (38 to 45 cm) long, a rubber apron, rubber boots, and a face shield to protect against splashes. A portable eye/face wash station should be available in case of splashes. Wear a "nuisance mask" when working with granular or powdered chemicals because these dusts can be harmful.

Use caution in the transfer of hypochlorite compounds. They should be transferred in their original containers and handled with care. Operators must be trained in the use of calcium and sodium hypochlorite products. Follow manufacturers' recommendations or your utility's safety policy.

CAUTION

When using acid or chlorine products, be mindful of the dangerous potential that fumes may have. Therefore, whenever using an acid or a chlorine product, use it where there is adequate ventilation. If necessary, use a forced ventilation system in pump houses or confined spaces.

When working with chemicals around a well, keep all personnel out of pits or depressions around the well. During treatment, some of the toxic gases, such as hydrogen sulfide, may rise from the well and settle in the lowest areas nearby because these gases are usually heavier than air.

7.362   Polyphosphate

Sodium polymetaphosphate is a chemical sometimes added to wells. This chemical may be sold commercially as Graham's salt, "Sodium hexametaphosphate," glassy sodium metaphosphate, or Hy-Phos $(NaPO_3)x$. The chemical is clear, hygroscopic (attracts moisture from the air), and soluble in water, but it dissolves slowly in water. The chemical is supplied in the form of a powder, flakes, or as small particles resembling broken glass; they are very sharp and possess dispersing and deflocculating properties.

CAUTION

Glassy phosphate consists of broken glass particles which are very sharp and can cut your skin. Use heavy-duty gloves when handling. Respiration equipment or a face shield should be used.

An eye/face wash station should be available if solution is splashed into the face or eyes. Wash face continuously for 15 minutes. Get medical attention.

7.363   Disinfection of Wells and Pumps

Care should be taken during the construction of new wells. All pipes used in the construction of a well should be as clean as possible. Regardless of how clean the pipe for the well casing is kept, a new well pump should be disinfected. This disinfection is necessary because contamination could be introduced into the well from the drilling tools and mud, makeup water, topsoil falling into or sticking to tools, and from the gravel itself.

Disinfect a well or well pump with a chlorine solution strong enough to produce a chlorine concentration of 50 mg/L in the well casing. Use caution when disinfecting wells and pumps. Follow safety procedures listed in Chapter 6, "Disinfection," and under chlorine treatment in this chapter. Also refer to Chapter 3, Section 3.5, "Disinfection of Wells and Pumps," in *SMALL WATER SYSTEM OPERATION AND MAINTENANCE.*

7.37   Working Around Electrical Units

Electricity is supplied as an alternating current (A.C.) at 120, 240, or 480 volts. Most wells or large pumps will be energized by 240 or 480 volts. Extremely large pumps may be energized by even higher voltages. Place nonconducting rubber mats on the floor in front of all power panels and motor control centers.

Care and caution should be used when working around equipment that is hooked up to any electric current. Water distribution system operators often work in damp and wet places using electric hand tools for routine maintenance work. The risk of shock or electrocution in such environments can be reduced substantially by the use of a device called a ground fault interrupter (GFI).

A GFI consists of three essential elements: a sensing element known as a differential transformer; an electronic amplifying section; and a circuit interrupting section. The differential transformer continuously monitors the current in the two conductors of a circuit, which are the phase or "hot" line, and the "neutral" line. Under normal conditions these two currents are always equal. If the differential transformer senses a difference between them of as little as 5 milliamps, it assumes the difference is being lost to ground through a fault of some sort. A partial shredded winding in your electric drill, for example, may cause some of the current to flow into the body of the drill and then into your hand. The electronic amplifying section of the GFI boosts the signal and trips the circuit in 1/40th of a second or less.

Ground fault receptacles should be installed in convenience receptacles where it is wet or damp, and a portable GFI, available for less than $50, should be carried on your work vehicle and used whenever you are using portable electric tools in the field.

When maintenance is to be done on wells or pumps, lock out the electric current to the equipment. Open the breakers so that electric current is turned off.[7] Place a tag (Figure 7.6) with your name on it on opened breakers and locks to indicate that you turned off the breakers and locks. No one should remove the tag or close the breakers or locks but you. This will allow the equipment to be pulled and worked on without the fear of being shocked or burned. Personnel working around high-voltage equipment should be trained and have respect for electric current.

After work is completed on the equipment, personnel should stand clear when the equipment is energized (electric current turned on). Make sure that the equipment is properly grounded. Personnel working with electrical equipment should have a good knowledge of electric circuits and circuit testing and should be qualified to do this work. They should get clearance (approval) to work on the equipment before starting any repairs. If no one is knowledgeable about electric circuits, the water utility should contact an electrician from a local electrical firm.

Safe procedures that must be used when working around electrical equipment include:

1. Only permit qualified persons to work on electrical equipment,

2. Maintain electrical installations in a safe condition,

3. Protect electrical equipment and wiring from mechanical damage and environmental deterioration,

4. Install covers or barriers on boxes, fittings, and enclosures to prevent accidental contact with live parts,

DANGER

OPERATOR WORKING ON LINE

DO NOT CLOSE THIS SWITCH WHILE THIS TAG IS DISPLAYED

TIME OFF: _____

DATE: _____

SIGNATURE: _____

This is the ONLY person authorized to remove this tag.

INDUSTRIAL INDEMNITY/INDUSTRIAL UNDERWRITERS/
INSURANCE COMPANIES

Fig. 7.6 Typical lockout warning tag

(Source: Industrial Indemnity/Industrial Underwriters Insurance Companies.)

---

[7] *When an electrician "closes" an electrical circuit, the circuit is connected together and electricity will flow. Closing a circuit is similar to opening a valve in a pipeline. The reverse is also true. Opening an electrical circuit is similar to closing a valve on a pipeline.*

5. Use an acceptable service pole,

6. Ground all electrical equipment,

7. Provide suitable overcurrent protection,

8. Lock out machinery during cleaning, servicing, or adjusting,

9. De-energize, lock out, and/or block all machinery to prevent movement if exposed parts are dangerous to personnel, and

10. If a switch or circuit breaker is tagged and locked out, do not remove the tag unless you are the person who locked out the device.

Whenever major replacement, repair, renovation, or modification of equipment is performed, OSHA regulations require that all equipment that could unexpectedly start up or release stored energy must be locked or tagged out to protect against accidental injury to personnel. Some of the most common forms of stored energy are electrical and hydraulic energy.

The energy isolating devices (switches, valves) for the equipment must be designed to accept a lockout device. A lockout device uses a positive means such as a lock, either key or combination type, to hold the switch in the safe position and prevent the equipment from becoming energized. In addition, prominent warnings, such as the tag illustrated in Figure 7.6, should be securely fastened to the energy isolating device and the equipment (in accordance with an established procedure) to indicate that both it and the equipment being controlled may not be operated until the tag and lockout device are removed by the person who installed them.

For the safety of all personnel, each plant should develop a standard operating procedure that must be followed whenever equipment must be shut down or turned off for repairs. If every operator follows the same procedures, the chances of an accidental start-up injuring someone will be greatly reduced. The following procedures, prepared by the State of Oklahoma, can be used as a model for developing your own standard operating procedure for lockouts.

BASIC LOCKOUT/TAGOUT PROCEDURES

1. Notify all affected employees that a lockout or tagged system is going to be utilized and the reason why. The authorized employee shall know the type and magnitude of energy that the equipment utilizes and shall understand the hazard thereof.

2. If the equipment is operating, shut it down by the normal stopping procedure.

3. Operate the switch, valve, or other energy isolating device(s) so that the equipment is isolated from its energy source(s). Stored energy such as that in springs; elevated machine members; rotating flywheels; hydraulic systems; and air, gas, steam, or water pressure must be dissipated or restrained by methods such as repositioning, blocking, or bleeding down.

4. Lock out and/or tag out the energy isolating device with assigned individual lock or tag.

5. After ensuring that no personnel are exposed, and as a check that the energy source is disconnected, operate the pushbutton or other normal operating controls to make certain the equipment will not operate. *CAUTION! RETURN OPERATING CONTROLS TO THE NEUTRAL OR OFF POSITION AFTER THE TEST.*

6. The equipment is now locked out or tagged out and work on the equipment may begin.

7. After the work on the equipment is complete, all tools have been removed, guards have been reinstalled, and employees are in the clear, remove all lockout or tagout devices. Operate the energy isolating devices to restore energy to the equipment.

8. Notify affected employees that the lockout or tagout device(s) has been removed before starting the equipment.

EMERGENCY PROCEDURES

In the event of electric shock, the following steps should be taken:

1. Survey the scene and see if it is safe to enter.

2. If necessary, free the victim from a live power source by shutting power off at a nearby disconnect, or by using a dry stick or some other nonconducting object to move the victim.

3. Send for help, calling 911 or other appropriate emergency number in your community. Check for breathing and pulse. Begin CPR (cardiopulmonary resuscitation) immediately if needed.

REMEMBER, only trained and qualified individuals working in pairs should be allowed to service, repair, or troubleshoot electrical equipment and systems.

7.38 Abandoning and Plugging Wells

Wells that are no longer useful should be abandoned and plugged for the following reasons:

1. To ensure that groundwater is protected and preserved for further use,

2. To eliminate the potential physical hazard to people, and

3. To protect nearby wells from contamination.

Obtain the appropriate permits from the proper agency. Depending on the locality, this may be a local, state, or federal agency. Follow the guidelines issued by the permitting agency and use experts (such as a geologist) where necessary.

Review Chapter 3 Appendix, Section 3.18, "Abandoning and Plugging Wells," in *SMALL WATER SYSTEM OPERATION AND MAINTENANCE.*

7.39 Safety Inspection

Safety around wells is the responsibility of the well operator. A periodic inspection should be performed at each well site whether the well is in service or out of service. Keep a record of each monthly inspection and use the remarks column to make special notations. Some things that you should inspect for are:

1. Cleanliness of site,

2. Locks on electrical panels,

3. Condition of fencing around site, and

4. Caution/warning signs in place where necessary.

Any problems or hazards, safety or otherwise, should be corrected as soon as possible. If you cannot take care of the problem, notify your supervisor about the nature of the problem and what type of support you will need to correct the problem.

QUESTIONS

Write your answers in a notebook and then compare your answers with those on pages 432 and 433.

7.3G   What protective clothing should be worn when handling hypochlorite?

7.3H   How should glassy phosphate be handled?

7.3I   If no one working for a water utility is knowledgeable about electric circuits, how can a water utility have a malfunctioning electrical system repaired?

7.3J   List the items that should be included in the safety inspection of a well site.

End of Lesson 1 of 2 Lessons on SAFETY

Please answer the discussion and review questions next.

DISCUSSION AND REVIEW QUESTIONS

Chapter 7. SAFETY

(Lesson 1 of 2 Lessons)

At the end of each lesson in this chapter you will find some discussion and review questions. The purpose of these questions is to indicate to you how well you understand the material in the lesson. Write the answers to these questions in your notebook before continuing.

1. What is safety?

2. Who is responsible for safety?

3. What is the one objective of a safety program?

4. What are the causes of accidents?

5. Which items on a vehicle and towed equipment should be inspected before leaving the yard?

6. How could you obtain information on defensive driving programs in your area?

7. How would you test a safety shower and eye/face wash station?

8. What should be done before entering a confined space?

9. Why should hazardous gas detection devices be calibrated regularly?

10. How can an object be moved if it is too heavy for operators to lift?

11. Why is safety around wells important?

12. How should hydrochloric acid be transported?

13. What would you do if acid splashed in your eyes?

CHAPTER 7.   SAFETY

(Lesson 2 of 2 Lessons)

7.4   PUMP SAFETY

7.40   Uses of a Pump

When we think of pumps, we commonly think of pumping water. Actually, pumps can do many things as is illustrated by the following definitions of a pump:

1. An apparatus or machine that forces liquids, air, or gas into or out of things.

2. A machine that increases the static pressure of fluids (air and water).

3. Pumping is the addition of energy to a fluid to move the fluid from one point to another.

7.41   Guards Over Moving Parts

A pump has moving parts. Where there are moving parts there is potential danger. All mechanical action or motion is hazardous, but in varying degrees. *ROTATION MEMBERS, RECIPROCATING ARMS, MOVING PARTS,* and *MESHING GEARS* are some examples of action and motion requiring protection on pumps.

Rotating, reciprocating, and transverse (crosswise) motions (Figure 7.7) create hazards in two general areas:

1. At the point of operation where work is done, and

2. At the points where power or motion is being transmitted from one part of a mechanical linkage to another.

Any rotating device (pump shaft) is dangerous. Whatever is rotating can grip clothing or hair and possibly force an arm or hand into a dangerous position. While accidents due to contact with rotating objects are not frequent, the severity of injury is very high.

Whenever hazardous machine actions or motions are identified, a means for providing protection for operators is essential. Enclosed guards are used to protect operators from rotating parts on pumps. The enclosed guard prevents access to dangerous parts at all times by enclosing the hazardous operation completely. The basic requirement for an enclosed guard is that it must prevent hands, arms, or any other part of an operator's body from making contact with dangerous moving parts. A good guarding system eliminates the possibility of the operator or another worker placing their hands near hazardous moving parts. Figures 7.8 and 7.9 are examples of good, effective guarding systems.

QUESTIONS

Write your answers in a notebook and then compare your answers with those on page 433.

7.4A   What are the uses of pumps?

7.4B   List the moving parts on a pump that require guards.

7.4C   How can rotating devices injure you?

7.4D   What is the basic requirement for an enclosed guard?

7.42   Maintenance and Repair

Pumps, like any other equipment, need preventive maintenance and periodic repair. A proper maintenance program will allow you to obtain satisfactory service from your pumps and minimize any potential safety hazards.

Remember your routine inspections will help you prolong the life of your pump and other equipment. Figure 7.10 identifies "danger" areas on pumps. These "danger" areas could be safety hazards to operators or areas where the pump could fail.

CAUTION
Do not operate a pump against a closed discharge valve, otherwise overheating will occur.

Whenever maintenance is to be performed on pumps or any other equipment, the work must be carefully planned and co-ordinated with other operators and the operation of the facilities. Good planning includes figuring out how to do the job from start to finish. Planning includes identifying potential safety hazards *BEFORE* starting a job and deciding how they can be avoided. To do a job safely, use the proper tools and equipment and *ISOLATE THE PUMP AND/OR EQUIPMENT FROM THE POWER SOURCE.*

Isolate power by throwing the appropriate switches and circuit breakers, locking them out, and tagging the locks with your name on the tag. The tag must be clearly visible. Only the person who signed the tag may turn the power back on. Power can also be isolated by unplugging the pump or equipment. *ALWAYS ISOLATE THE POWER BEFORE BEGINNING ANY MAINTENANCE.* Follow the basic lockout/tagout procedures described in Section 7.37, "Working Around Electrical Units." Even properly locked out equipment may be hazardous to

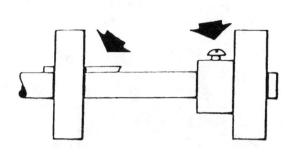

Rotating shaft and pulleys with
projecting key and set screw

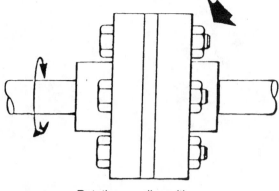

Rotating coupling with
projecting bolt heads

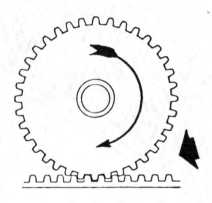

Rack and gear

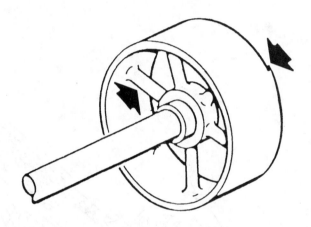

Rotating pulley with spokes and
projecting burr on face of pulley

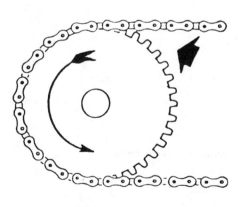

Chain and sprocket

Belt and pulley

NOTE: Short, thick arrows indicate danger points

Fig. 7.7   *Examples of typical rotating, reciprocating, and transverse mechanisms*

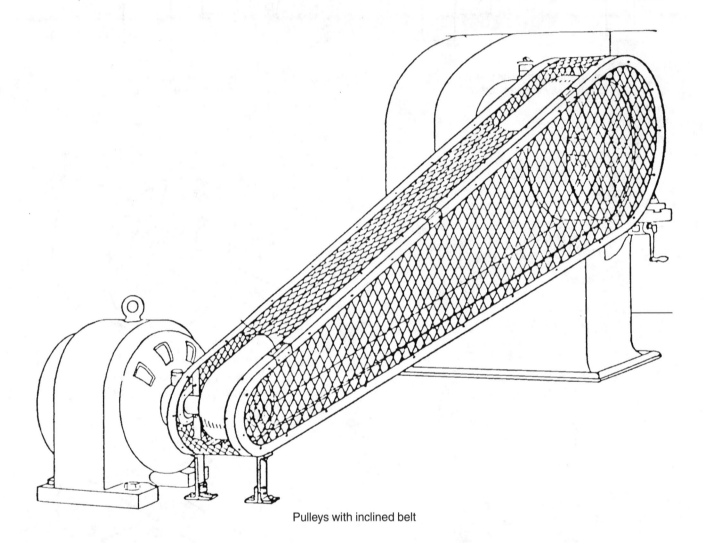

Pulleys with inclined belt

Fig. 7.8 Enclosed guard protecting operators from moving pulleys and belt

Guard over belt on gas engine

Guard over coupling

Guard on gas engine and pump

Guard on hypochlorinator

Fig. 7.9   Effective guarding systems

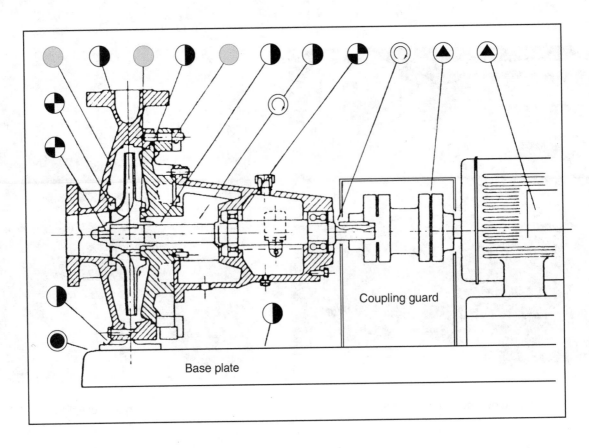

Coupling guard

Base plate

Fracture can cause leakage
and mechanical damage and
injury to personnel

Leakage and possibility
of collection of leakage
(vaporizing)

Seizure of parts or other
mechanical failure

Rotating parts area needs
protection and care

Lifting facility

Electric protection and/or
sparks possibility

Fig. 7.10 "Danger" areas on pumps

operators unless movement of the equipment is also physical-ly blocked. Energy may be stored in the form of hydraulic pres-sure in pipelines, electric energy in capacitors, compressed air, or springs under tension. Also, potential energy that may need to be blocked can come from suspended parts which may slip loose and swing or fall. Any source of energy that might cause unexpected movement must be disengaged and physically blocked to prevent movement and possible injury to operators.

Safety coordination and planning requires that supervisory personnel and all other potentially involved or affected person-nel be notified and necessary approvals obtained *BEFORE* starting any maintenance work. This means that any other people who operate the equipment must be notified. If the electricity must be shut off to do any work or maintenance, all persons affected by the lack of power also must be notified. A good safety practice is to have other persons leave the work area until the job is completed so no one will accidentally turn

on the power or equipment. When maintenance personnel must leave a job site for any reason, on returning they must check and be sure that the circuit is still locked out *BEFORE* resuming maintenance on the equipment.

If, or when, the electric circuit must be turned back on, all persons in the area must be notified that the current is going to be turned back on again.

When equipment may be operated by a standby or backup power source, this power source must also be isolated so the equipment cannot be energized. When emergency standby engines and pumps that are started by a battery must be maintained, the battery must be disconnected and the fuel source shut off. Appropriate warning signs and tags must be posted where they are visible indicating that the power source is disconnected.

ALWAYS ISOLATE EQUIPMENT FROM THE DRIVING POWER SOURCE BEFORE STARTING ANY WORK.

If a stationary pump or motor must be moved, use a lifting sling. Attach the sling properly to the equipment and lift the equipment up off the base. Set the pump or motor down on a cart or pallet and transfer the pump or motor to a workshop for repair or to another location as necessary for repair.

> *NEVER LEAVE ANY EQUIPMENT HANGING FROM A SLING IN MIDAIR. NEVER WORK ON EQUIPMENT WHEN IT IS HANGING FROM A SLING BECAUSE THE EQUIPMENT COULD SLIP FROM THE SLING AND FALL ON YOU. ALWAYS LOWER EQUIPMENT TO THE FLOOR, SET IT ON A CART, OR PLACE IT ON A WORKBENCH BEFORE STARTING MAINTENANCE.*

When transferring pumps or equipment from one location to another, use a sturdy cart or truck, or transfer the equipment on a pallet by using a forklift. Many operators have seriously injured their backs attempting to lift and carry heavy pumps, motors and other equipment.

7.43 Storage of Lubricants and Fuel

Lubricants and fuels must be stored in a separate facility from other storage areas because of the potential fire hazard. Clean up all spills immediately and keep the area neat and clean. A fire extinguisher must be readily available near the storage facility, but never inside the facility.

QUESTIONS

Write your answers in a notebook and then compare your answers with those on page 433.

7.4E   What is safety planning and coordination?

7.4F   What should be done before beginning maintenance on a pump?

7.4G   Why should you never work on equipment that is hanging from a sling in midair?

7.4H   Why should lubricants and fuels be stored in a separate facility from other storage areas?

7.5 WORKING IN STREETS

7.50 Need for Traffic Control

The primary function of streets and highways is to provide for the movement of traffic. A common secondary use within the right-of-way of streets or highways is for the placement of public and private utilities such as water distribution systems. While the movement of traffic is very important, streets need to be constructed, reconstructed, or maintained, and utility facilities need to be repaired, modified, or expanded. Consequently, traffic movements and street or utility repair work must be regulated to provide optimum safety and convenience for all.

Working in a roadway represents a significant hazard to a distribution system operator as well as pedestrians and drivers. Motor vehicle drivers can be observed reading books, files, and newspapers, shaving and applying makeup, talking on cellular phones, and changing tapes, CDs, or radio stations (and using headsets to listen to them), rather than concentrating on driving. At any given time of the night or day, a certain percentage of drivers can be expected to be driving while under the influence of drugs or alcohol. Given the amount of time distribution system operators work in traffic while performing inspection, cleaning, rehabilitation, and repairs, the control of traffic is necessary if we want to reduce the risk of injury or death while working in this hazardous area. The purpose of traffic control is to provide safe and effective work areas and to warn, control, protect, and expedite vehicular and pedestrian traffic. This can be accomplished by appropriate and prudent use of traffic control devices.

Most states, counties, and cities have adopted standards and guidelines to control traffic and reduce the risk under different circumstances. This section illustrates examples of traffic control which may or may not meet the specific requirements of the laws in your geographical area, but should serve to make you aware of various aspects of traffic control.

Any time traffic may be affected, appropriate authorities in your area must be notified before leaving for the job site. These could be state, county, or local depending on whether it is a state, county, or local street. Frequently a permit must be issued by the authority that has jurisdiction before traffic can be diverted or disrupted. In some cases, traffic diversion or disruption may have an impact on the emergency response system in your area, such as access by fire or police, and so these agencies may be involved as well. In most cases, you will need to plan ahead to secure permits and notify authorities. This may mean only a phone call or two or it could mean several days' or weeks' advance planning if you need to make extensive traffic control arrangements.

Upon arrival at the job site, look for a safe place to park vehicles. If they must be parked in the street to do the job, route traffic around the job site *BEFORE* parking vehicles in the street. If practical, park vehicles between oncoming traffic and the job site to serve as a warning barricade and to discourage reckless drivers from plowing into operators. Remember—you need protection from the drivers as well! Always try to park work equipment and workers' private vehicles and also store materials and debris in such a manner to reduce the chances of being impacted by vehicles that run off the designated traffic lanes.

Traffic must be warned of your presence in the street. BE PREPARED TO STOP, SHOULDER WORK, and UTILITY WORK AHEAD are signs that are effective. Signs with flashing warning lights and vehicles with yellow rotating beacons are used to warn other motorists (Figure 7.11). Vehicle-mounted traffic guides are also helpful. Use trained flaggers to alert drivers and to direct traffic around the work site. Warning signs and flaggers must be located far enough in advance of the work area to allow motorists time to realize they must slow

Fig. 7.11   Signs warning traffic (advance warning area)

down, be alert for activity, and safely change lanes or follow a detour. Exact distances and the nature of the advance warning depend on traffic speed, congestion, roadway conditions, and local regulations.

Once motorists have been warned, they must be safely routed around the job site. Traffic routing includes the use of flaggers and directional signs. Properly placed channelizing devices, such as barricades, drums, or cones, can effectively channel traffic from one lane to another around the job site. Retain a traffic officer to direct traffic if streets are narrow or if there is considerable traffic or congestion.

When work is not in progress or the hazard no longer exists, promptly remove, fold, or cover the traffic control devices. Drivers quickly begin to ignore traffic control signs when it becomes obvious no work is in progress, and they assume construction hazards exist only when crews are on the job. Therefore, all temporary traffic control devices should be removed as soon as practical when they are no longer needed. When work is suspended for short periods of time, devices that are no longer appropriate should be removed or covered.

Answers to several questions will determine how traffic control should be accomplished.

1. Is traffic moving at a low speed (0-35 MPH or 0-56 km/hr) or a high speed (40-55 MPH or 64-88 km/hr)?

2. Is the street two-lane, one-way, or two-way?

3. Is it undivided four-lane?

4. Is it multilane one-way?

5. Are pedestrian walkways affected?

6. Is it in a residential area?

7. Will a lane closure be required?

8. Will more than one lane be closed?

9. Will traffic control be required during peak traffic periods or at night?

Because of the need for consistency when traveling from one city, town, or state to another, most states have developed work zone traffic controls based on the U.S. Department of Transportation's *MANUAL ON UNIFORM TRAFFIC CONTROL DEVICES FOR STREET AND HIGHWAY* (see Section 7.584, "Additional Reading," reference 1). The information that follows in Section 7.51 through Section 7.55 on work zone traffic control was adapted from the above-mentioned manual and a state handbook. Check your local and state standards and guidelines for traffic control guidelines in specific circumstances and incorporate these control requirements into your agency's procedures.

7.51 Definitions of Terms

The following definitions apply to work zone traffic control:

Short-Term Work Any work on a street or highway where it is anticipated the activity will take one work shift (typically 6 to 12 hours or less) to complete. This time period may be modified by written approval of the governing road authority.

Intermediate-Term Work Any work on a street or highway that occupies a location from overnight to three days.

Street or Highway The entire width between boundary lines of any way or place that is open to the use of the public, as a matter of right, for the purpose of vehicular traffic.

Low-Speed Street or Highway Any street or highway where the speed limit is 35 miles per hour (56 km/hr) or less.

High-Speed Street or Highway Any street or highway where the speed limit is 40 miles per hour (64 km/hr) or greater.

Low-Speed Residential Street or Highway Any low-speed street or highway that serves only residential areas where through traffic is deliberately discouraged.

Low-Speed Collector or Arterial Street or Highway Any low-speed street or highway that serves as residential and business access and is primarily used to collect traffic from residential streets and channel it into the arterial system, or any low-speed street or highway that permits travel between communities and serves as a through facility.

Traffic Control Plan (TCP) A TCP describes traffic controls to be used for facilitating all traffic through a temporary traffic control zone. The degree of detail in the TCP depends entirely on the complexity of the job site. Persons knowledgeable about the principles of temporary traffic controls should prepare TCPs.

Traffic Control Zone A work zone in an area of a highway with construction, maintenance, or utility work activities. A work zone is typically marked by signs, channelizing devices, barriers, pavement markings, or work vehicles. It extends from the first warning sign or high-intensity rotating, flashing, oscillating, or strobe lights on a vehicle to the END ROAD WORK sign or the last traffic control device.

Stationary Traffic Control Zone Any traffic control zone that remains in one place for longer than 15 minutes.

Mobile Traffic Control Zone Any traffic control zone where the traffic control zone remains in one place for less than 15 minutes.

Moving Traffic Control Zone Any traffic control zone that is continuously moving.

Special Traffic Control Zone Any traffic control zone where the workers are performing tasks with little or no interference to traffic. Generally the presence of the vehicle and worker should not "surprise" the driver or cause any erratic maneuvers.

Low-Volume Street or Highway Any street or highway where the ADT (Average Daily Traffic) is less than 1,500 vehicles.

Decision Sight Distance The distance required by a driver to properly react to hazardous, unusual, or unexpected situations even where an evasive maneuver is more desirable than a hurried stop.

Good Visibility Location Any location where the sight distance to the work area is sufficient to meet decision sight distance.

QUESTIONS

Write your answers in a notebook and then compare your answers with those on page 433.

7.5A   What is the purpose of traffic control around work areas in streets and highways?

7.5B   Who should be contacted before setting up a work site in a street?

7.5C   How can traffic be warned of your presence in a street?

7.5D   How can motorists be safely routed around a job site?

7.5E   What area at a work site is considered a "traffic control zone"?

7.5F   What is the meaning of the term "decision sight distance"?

7.52   Individuals Qualified To Control Traffic

Each person whose actions affect temporary traffic control zone safety, from the upper-level management through the field workers, should receive training appropriate to the job decisions each individual is required to make. Only those individuals who are trained in proper temporary traffic control practices and have a basic understanding of the principles (established by applicable standards and guidelines, including those of the *MANUAL ON UNIFORM TRAFFIC CONTROL DEVICES FOR STREET AND HIGHWAY*) should supervise the selection, placement, and maintenance of traffic control devices used for temporary traffic control zones and for incident management. Individuals become qualified to control traffic by gaining the following knowledge and experience:

- A basic understanding of the principles of traffic control in work zones,

- Knowledge of the standards and guidelines governing traffic control,

- Adequate training in safe traffic control practices, and

- Experience in applying traffic control in work zones.

Only qualified individuals should supervise the selection, placement, and maintenance of traffic control devices in work and incident management areas.

7.53   Permission To Work Within the Right-of-Way of Streets or Highways

Any work in public streets or highways must be regulated to ensure proper coordination of the work, thus protecting the public's interest. To accomplish this, any person, firm, corporation, or agency must obtain permission from the governing road authority before starting work within the right-of-way of any street or highway. Regulatory traffic control devices or signs must be approved by the governing road authority before they are installed on any street or highway.

When working in or near an intersection where traffic is controlled by a traffic signal light, the owner of the signal must be notified before work begins. This is necessary to ensure the proper operation of the signal while the work is in progress. Also, the signals may be used to control traffic in the vicinity of the work area.

7.54   General Responsibilities

In most areas of the country, any agency performing work within the right-of-way of streets or highways is responsible for:

1. Supplying, installing, and maintaining all necessary traffic control devices as required by the governing road authority to protect the work area and safely direct traffic around the work zone,

2. Supplying their own flagger(s) when required,

3. Informing occupants of abutting properties, either orally or by written notice, of parking prohibitions or access limitations,

4. Notifying the governing road authority when existing traffic signs need to be removed or relocated or any regulatory sign must be installed for construction or maintenance work,

5. Replacing or reimbursing the governing road authority for any damage to or loss of existing traffic signs,

6. Keeping all traffic control devices clean and in proper position to ensure optimum effectiveness,

7. Removing traffic control equipment when it is no longer required or appropriate, and

8. Keeping proper records of traffic control that contain starting and ending times, location, names of personnel involved, and the traffic controls used.

7.55   Regulations Concerning All Street or Highway Work

7.550   Time of Work

The governing road authority may determine or define the times when work may be performed. During peak traffic periods, construction work may be restricted or not permitted. Peak periods of traffic movement may vary in different areas.

The governing road authority may require work to be performed at night or on weekends when the location of the work within the roadway is considered critical from a traffic standpoint. Some examples of these locations include the vicinity of major signalized intersections, shopping centers, and downtown areas.

When work is planned for nighttime hours, the traffic controls should be modified to adequately control traffic through the work zone. Some items that should be considered for nighttime activities are the addition of warning lights to traffic control devices and signs, use of retroreflective signs, use of larger, more reflective channelizing devices such as drums, use of floodlighting for any flagger stations and for the work area, and addition of reflective devices to work vehicles that are typically used for nighttime work.

7.551 Specific Situations

Situations may arise where the typical traffic control guidelines do not apply and a special traffic control plan is needed. This special control plan should be developed by people with traffic engineering expertise.

Although it is usually desirable to set up traffic controls as shown in the example layouts on pages 401 through 406, situations sometimes arise where this becomes impractical and modifications to the typical layouts are needed. When modifications are made, factors such as traffic volume, speed, sight distance, and type of work must be considered.

Situations may occur where consideration should be given to using a stationary zone where a mobile zone normally would be indicated. If the tasks to be performed require that the work area remain in one place less than 15 minutes, but several individual work areas are concentrated within a relatively short section of roadway, it would be desirable to handle this situation as a stationary zone. An example of this type of work would be the repair of many potholes on a long section of roadway.

7.552 Partial Street or Highway Closures

Unless a section of street is to be completely closed to vehicular traffic, the work needs to be done so that as few traffic lanes as practical are blocked. Where a traffic lane is blocked, parking prohibition(s) may be necessary to permit continued traffic movement. Closing of lanes requires permission from the governing road authority and must be accomplished using standard procedures and traffic control devices.

The governing road authority may require that excavations and cuts be properly backfilled, bridged, or plated to allow the passage of vehicles. The governing road authority may allow an exception to these requirements when traffic volumes or other special conditions warrant such action. All *SPOIL*[8] material from the excavation is to be removed from the pavement surface when bridges or plates are used. All bridging or plating needs to be of sufficient strength to accommodate all legal traffic loads. The plates need to be secured in place so they will not slip out of position during use.

7.553 Complete Street or Highway Closures and Detours

Permission needs to be obtained from the governing road authority for all street and highway closures. When detours are necessary and authorized by the governing road authority, they must be installed in accordance with the local standards and guidelines.

7.554 Emergency Situations

An emergency situation may arise where immediate action to protect the safety of the general public requires work to be done on a street or highway even though full compliance with the local standards and guidelines cannot be immediately provided. In this case, proper traffic control must be provided as soon as possible. Orange flags (24 inch by 24 inch (60 cm by 60 cm)) may be used to control traffic during emergencies only.

QUESTIONS

Write your answers in a notebook and then compare your answers with those on page 433.

7.5G How could an operator become qualified to control traffic?

7.5H Records of traffic control activities should contain what types of information?

7.5I When distribution system work is scheduled for nighttime hours, what additional traffic control safety measures should be considered?

7.56 Traffic Control Zones

When traffic is affected by construction, maintenance, or utility activities, traffic control is needed to safely guide and protect motorists, pedestrians, and operators in a traffic control zone. The traffic control zone is the area between the first advance warning sign and the point beyond the work area where traffic is no longer affected.

Most traffic control zones can be divided into these specific areas:

1. Advance warning area,

2. Transition area,

3. Buffer space,

4. Work area, and

5. Termination area.

Figure 7.12 illustrates these five parts of a traffic control zone. The following paragraphs discuss each of the five parts for one direction of travel. If the work activity affects more than one direction of travel, the same principles apply to traffic in all directions.

7.560 Advance Warning Area

An advance warning area is necessary for all traffic control zones because drivers need to know what to expect. Before reaching the work area, drivers should have enough time to alter their driving patterns. The advance warning area may vary from a series of signs starting several hundred feet in advance of the work area to a single sign or flashing lights on a vehicle.

When the work area, including access to the work area, is entirely off the shoulder and the work does not interfere with traffic, an advance warning sign may not be needed. However, an advance warning sign should be used when any problems or conflicts with the flow of traffic are anticipated. The ROAD WORK AHEAD sign may be replaced with other appropriate signs, such as the SHOULDER WORK sign. The SHOULDER WORK sign may be used for work adjacent to the shoulder. The ROAD WORK AHEAD sign may be omitted where the work space is behind a barrier, more than 24 inches (60 cm) behind the curb, or 15 feet (5 meters) or more from the edge of any roadway.

The advance warning area, from the first sign to the start of the next area, should be long enough to give the motorists adequate time to respond to the conditions. For most activities the length can be those listed in Table 7.4.

---

[8] *Spoil. Excavated material such as soil from the trench of a water main.*

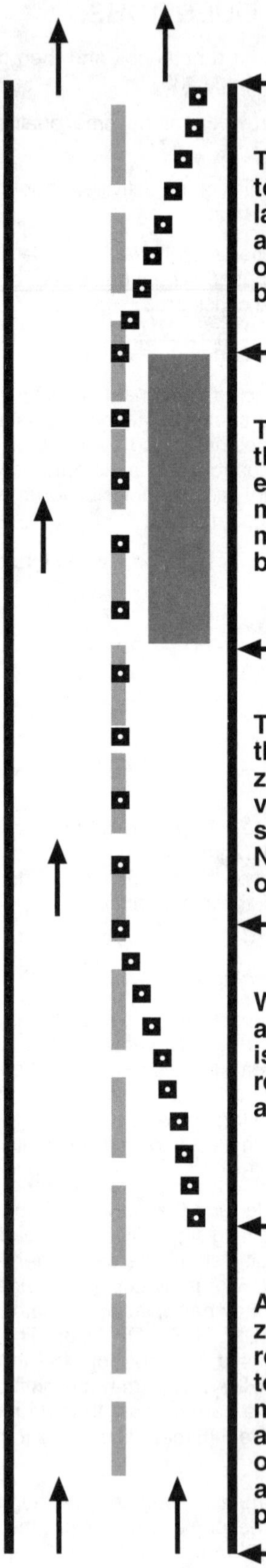

TERMINATION AREA

The termination area provides a short distance for traffic to clear the work area and to return to the normal traffic lanes. A downstream taper may be placed in the termination area. A downstream taper is used at the downstream end of the work area to indicate to drivers that they can move back into the lane that was closed.

WORK AREA

The work area is that portion of the roadway that contains the work activity and is closed to traffic and set aside for exclusive use by workers, equipment, and construction materials. Work areas may remain in fixed locations or may move as work progresses. The work area is usually delineated by channelizing devices or shielded by barriers.

BUFFER SPACE

The buffer space is the open or unoccupied space between the transition and work areas. With a mobile traffic control zone, the buffer space is the space between the shadow vehicle, if one is used, and the work vehicle. The buffer space provides recovery space for an out-of-control vehicle. Neither work activity nor storage of equipment, vehicles, or material should occur in this space.

TRANSITION AREA

When work is performed within one or more traveled lanes, a lane closure(s) is required. In the transition area, traffic is channelized from the normal highway lanes to the path required to move traffic around the work area. The transition area contains the tapers that are used to close lanes.

ADVANCE WARNING AREA

An advance warning area is necessary for all traffic control zones because drivers need to know what to expect. Before reaching the work area, drivers should have enough time to alter their driving patterns. The advance warning area may vary from a series of signs starting a mile (1.6 km) in advance of the work area to a single sign or flashing lights on a vehicle. The true test of whether sign spacing is adequate is to evaluate how much time the driver has to perceive and to react to the conditions ahead.

(not to scale)

Fig. 7.12   Parts of a traffic control zone

TABLE 7.4   DISTANCES BETWEEN SIGNS[a]

Road Type	Distance Between Signs[b]		
	A	B	C
Urban (low speed)[c]	30 (100)	30 (100)	30 (100)
Urban (high speed)[c]	100 (350)	100 (350)	100 (350)
Rural	150 (500)	150 (500)	150 (500)
Expressway/Freeway	300 (1,000)	450 (1,500)	800 (2,640)

[a]   Source: Manual on Uniform Traffic Control Devices, 2003.
[b]   Speed category to be determined by highway agency.
[c]   Distances are shown in meters (feet). The column headings A, B, and C are the dimensions shown in Figure 7.13. The A dimension is the distance from the transition or point of restriction to the first sign. The B dimension is the distance between the first and second signs. The C dimension is the distance between the second and third signs. (The third sign is the first one in a three-sign series encountered by a driver approaching a temporary traffic control (TTC) zone.

Available decision time should allow a driver time to detect a hazard in a cluttered environment, recognize it and its potential hazard, and select the appropriate speed and path. If there is only enough time to stop the vehicle after seeing a hazard, the driver usually will not have time to perform an evasive maneuver, which is often a safer action. Therefore, the recommended decision sight distance should be provided when the following conditions exist:

- When the driver needs to process relatively complex information,

- When the hazard is difficult to perceive,

- When unexpected or unusual maneuvers are required, and

- Where an evasive maneuver is more desirable than a hurried stop.

This decision sight distance is the sum of perception time, reaction time, and vehicle maneuver time. In other words, it is the time it takes a driver to recognize the hazard plus the time it takes to react plus the time for the vehicle to actually respond to the driver's reaction. Typical stopping distances are shown in Table 7.5 and typical decision sight distances for use in short-term work zones are shown in Table 7.6.

TABLE 7.6   SUGGESTED DECISION SIGHT DISTANCE

Posted Speed Limit (MPH)[a]	Suggested Decision Sight Distance (feet)[b]
0 – 35	750
40 – 45	950
50 – 55	1,200
60 – 65	1,400

[a]   Multiply MPH x 1.6 to obtain km/hr.
[b]   Multiply ft x 0.3 to obtain meters.

7.561   Transition Area (Figure 7.14)

When work is performed within one or more traveled lanes, a lane closure(s) is required. (If no lane or shoulder closure is involved, the transition area is not used.) The transition area is that section of highway where road users are redirected out of their normal path. Transition areas usually involve strategic use of tapers. Tapers are created by using a series of channelizing devices or pavement markings to move traffic out of or into the normal path.

TABLE 7.5   STOPPING DISTANCES[a]

Speed (MPH)[b]	Driver Reaction Distance (feet)[c]	Braking Distance (feet)[c]	Total Stopping Distance (feet)[c]
20	22 (20 + 2)	18 – 22	40 – 44
40	44 (40 + 4)	64 – 80	108 – 124
55	60 (55 + 5)	132 – 165	192 – 225
65	71 (65 + 6)	160 – 224	231 – 295

[a]   Source: State of California, Department of Motor Vehicles Hand Book.
[b]   Multiply MPH x 1.6 to obtain km/hr.
[c]   Multiply ft x 0.3 to obtain meters.

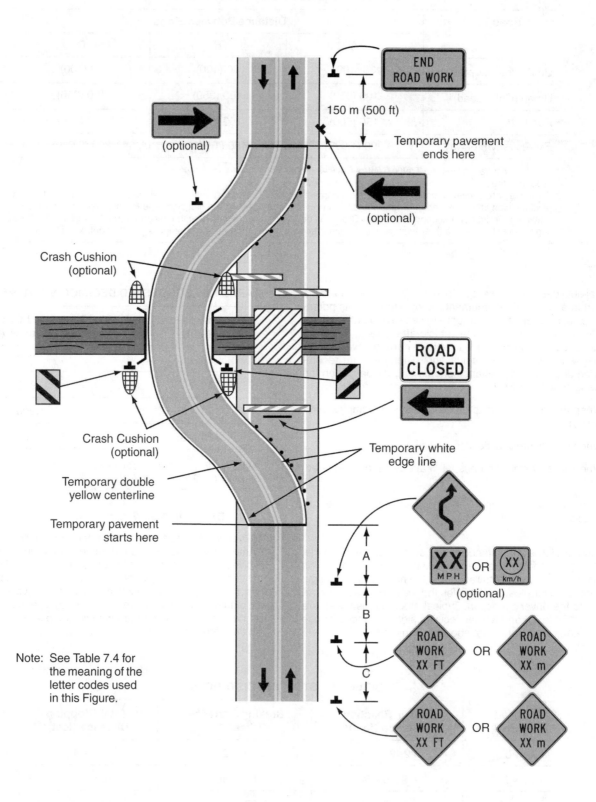

Fig. 7.13   Road closure with diversion

(Source: Manual on Uniform Traffic Control Devices, 2003)

Fig. 7.14   Portable arrows reduce driver confusion in construction area

(Permission of Safety Tech, Inc.)

Four general types of tapers used in traffic control zones are:

1. Lane closure tapers necessary for closing lanes of moving traffic (sometimes referred to as channelizing or merging tapers);

2. Two-way traffic tapers needed to control two-way traffic where traffic is required to alternately use a single lane (commonly used when flaggers are present);

3. Shoulder closure tapers needed to close shoulder areas; and

4. Downstream tapers installed to direct traffic back into its normal path.

1. Lane Closure Taper (L). The length of taper used to close a lane is determined by the speed of traffic and, in some locations, the width of the lane to be closed. Table 7.7 lists typical tapering guidelines. These tapers require the longest distance because drivers are required to merge into common road space. A merging taper should be long enough to enable merging drivers to have adequate advance warning and suffi-cient length to adjust their speeds and merge into a single lane before the end of the transition (Figure 7.15).

If sight distance is restricted by obstacles such as sharp vertical or horizontal curves, the taper should begin well in advance of the view obstruction. The beginning of tapers should not be hidden behind curves.

Generally, merging tapers should be lengthened, not short-ened, to increase their effectiveness. Observe traffic to see if the taper is working correctly. Frequent use of brakes and evidence of skid marks may be indications that either the taper is too short or the advance warning is inadequate.

2. Two-Way Traffic Taper. The two-way traffic taper is used in advance of a work area that occupies part of a two-way, two-lane road. The taper is placed in such a way that the remainder of the road can be used alternately by vehicles traveling in either direction. A short taper is used to cause traffic to slow down by giving the appearance of restricted alignment and a flagger is usually stationed at the taper to assign the right of way. Two-way tapers are 50 feet (15 m) to 100 feet (30 m) maximum (Figure 7.16).

TABLE 7.7   LANE CLOSURE TAPER SPECIFICATIONS AND MAXIMUM DEVICE SPACING[a]

Approach Speed (MPH)	Lane Closure Taper (feet)	Maximum Device Spacing[b] (feet)	Two-Way Traffic Taper (feet)	Shoulder Taper (feet)	Buffer Length (feet)
25	125	25	50–100	42	155
30	180	30	50–100	60	200
35	245	35	50–100	82	250
40	320	40	50–100	107	305
45	540	45	50–100	180	360
50	600	50	50–100	200	425
55	660	55	50–100	220	495
60	720	60	50–100	240	570
65	780	65	50–100	260	645
70	840	70	50–100	280	730

[a] Source: Manual on Uniform Traffic Control Devices, 2003.
[b] Device spacing applies to Lane Closure Taper. Use 10–20 feet (3–6 m) spacing for Two-Way Traffic Taper.

3. Shoulder Closure Taper. When an improved shoulder is closed on a high-speed roadway, it should be treated as a closure of a portion of the roadway that the motorist expects to use in an emergency. The work area on the shoulder should be preceded by a taper that is one-third of the length of a lane closure taper, provided the shoulder is not used as a travel lane (Figure 7.17).

4. Downstream Taper. The downstream taper is used in terminating areas to provide a visual cue to drivers that access is available to the original lane. The taper should have a maximum length of about 100 feet (30 meters) per lane, with channelizing devices spaced about 20 feet (6 meters) apart.

7.562   Buffer Space

The buffer space is a lateral or longitudinal area that separates road user flow from the work space or an unsafe area, and might provide some recovery space for an errant vehicle. With a mobile traffic control zone, the buffer space is the space between the shadow vehicle, if one is used, and the work vehicle.

The buffer space provides a margin of safety for both traffic and operators. If a driver does not see the advance warning or fails to negotiate the transition, a buffer space provides room to stop before entering the work area. It is important that the buffer space be free of equipment, operators, materials, and operators' vehicles.

Channelizing devices should be placed along the edge of the buffer space. In general, the spacing of these devices should not exceed two times the posted speed. For example, if the posted speed is 50 mph (80 km/hr), the spacing of devices along the buffer space should not exceed 100 feet (30 m).

Situations occur where opposing streams of traffic on multilane streets or highways are transitioned so one lane of traffic uses a lane that normally flows in the opposite direction. To help prevent head-on collisions, a buffer space should be used to separate the two tapers for opposing directions of traffic.

7.563   Work Area

The work area is that portion of the roadway that contains the work activity and is closed to traffic and set aside for exclusive use by operators, equipment, and construction materials. Work areas may remain in fixed locations or they may move as work progresses. An empty buffer space may be included at the upstream end. The work area is usually delineated by channelizing devices or shielded by barriers to exclude traffic and pedestrians.

Conflicts between traffic and the work activity or potential hazards increase when the following conditions occur:

● The work area is close to the traveled lanes,

● Physical deterrents to normal operation exist, such as uneven pavements or vehicle loading and unloading,

● Speed and volume of traffic increase, and

● The change in travel path gets more complex, shifting traffic across the median and into lanes normally used by opposing traffic rather than shifting traffic a few feet or meters.

Work areas that remain set up overnight have a greater need for delineation than daytime activities. Consideration should be given to selecting larger, more retroreflective devices and use of additional lighting devices for short-term work zones that are in place during nighttime hours.

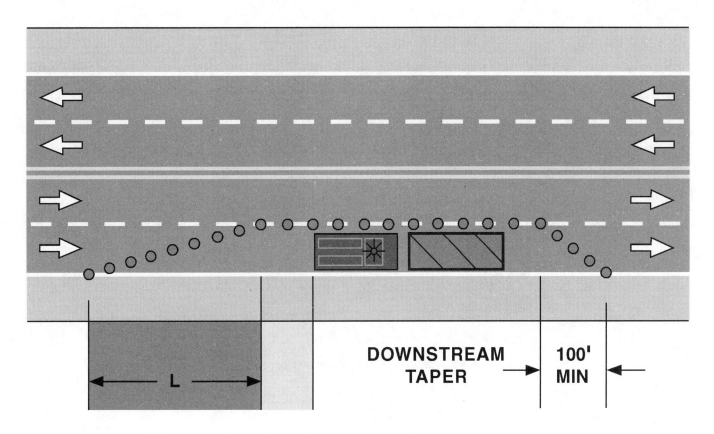

Fig. 7.15   Length of taper

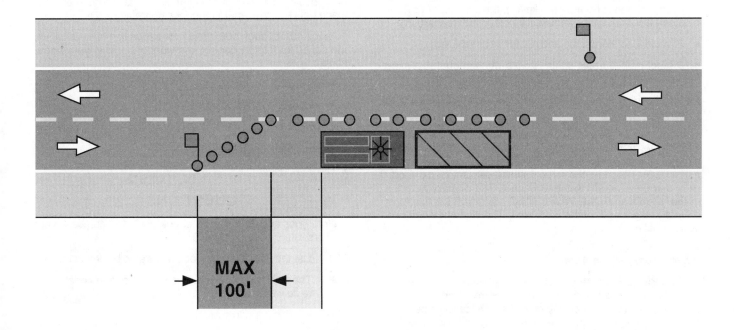

Fig. 7.16   Two-way traffic taper

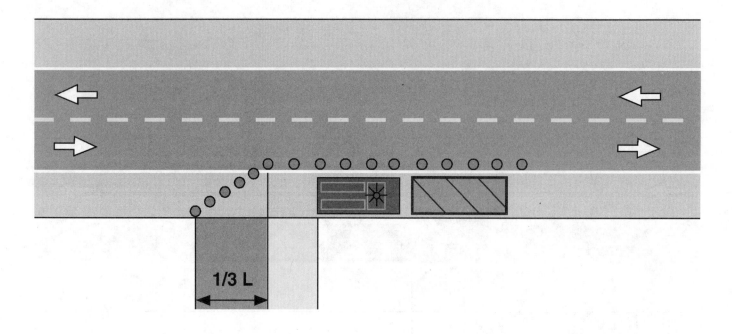

Fig. 7.17   Shoulder closure taper

Traffic control measures for work areas should meet the following guidelines:

• Use traffic control devices to make the work area clearly visible to traffic;

• Channelizing devices should be placed along the work space. In general, the spacing of these devices should not exceed two times the posted speed. For example, if the posted speed is 50 mph (80 km/hr), the spacing of devices along the work space should not exceed 100 feet (30 m).

• Provide a safe entrance and exit for work vehicles;

• Protect mobile traffic control zones with adequate warning devices on the work and/or shadow vehicles; and

• Use flashing lights or flags on work vehicles exposed to traffic.

Avoid gaps in the traffic control that may falsely lead drivers to think they have passed through the work area. For example, if the work area includes intermittent activity throughout a one-mile (1.6-km) section, the drivers should be reminded periodically that they are still in the work area. The primary purpose of the guide sign ROAD WORK NEXT____MILES is to inform drivers of the length of the work area. The sign should not be set up until work begins.

7.564   Termination Area

The termination area provides a short distance for traffic to clear the work area and return to the normal traffic lanes. A downstream taper may be placed in the termination area to indicate to drivers that they can move back into the lane that was closed. Closing tapers are optional and may not be advisable when material trucks move into the work area by backing up from the downstream end of the work area. Closing tapers

are similar in length and spacing to two-way traffic tapers with a maximum length of 100 feet (30 m).

There are occasions where the termination area could include a transition. For example, if a taper was used to shift traffic into opposing lanes around the work area, then the termination area should have a taper to shift traffic back to its normal path. This taper would then be in the transition area for the opposing direction of traffic. It is advisable to use a buffer space between the tapers for opposing traffic.

Pages 401 through 406 show how to position hi flags, signs, and cones for flagger control and work under five typical conditions:

1. Work area in center of road,

2. Work beyond intersection,

3. Closing left lane,

4. Closing half of roadway, and

5. Closing right lane.

QUESTIONS

Write your answers in a notebook and then compare your answers with those on page 433.

7.5J   List the five different areas of a traffic control zone.

7.5K   The decision sight distance should give motorists time to do what?

7.5L   What is a traffic control taper?

7.5M   What should be done to prevent head-on collisions on a multilane street when one lane of traffic uses a lane that normally flows in the opposite direction?

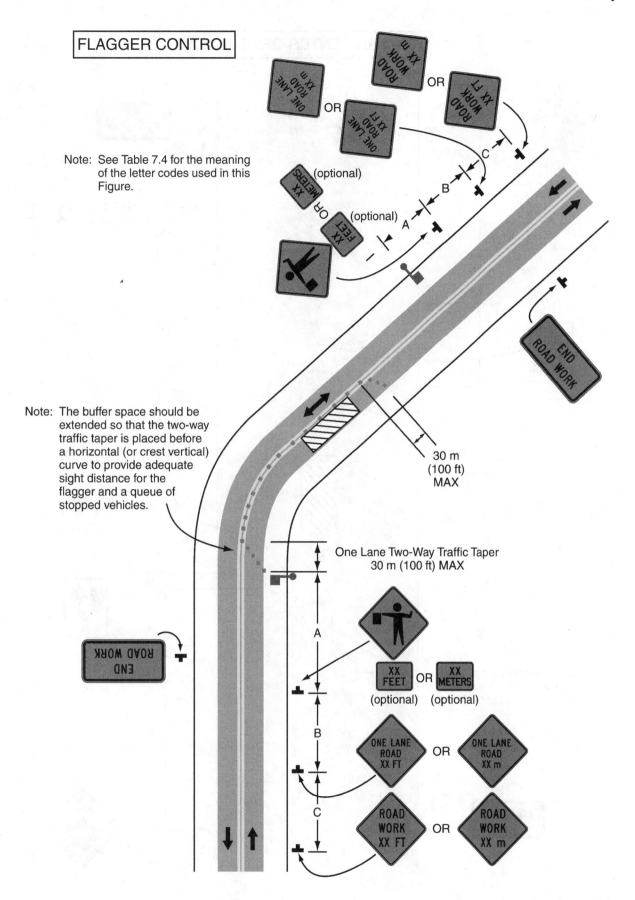

FLAGGER CONTROL

Note: See Table 7.4 for the meaning of the letter codes used in this Figure.

Note: The buffer space should be extended so that the two-way traffic taper is placed before a horizontal (or crest vertical) curve to provide adequate sight distance for the flagger and a queue of stopped vehicles.

30 m (100 ft) MAX

One Lane Two-Way Traffic Taper 30 m (100 ft) MAX

LANE CLOSURE ON TWO-LANE ROAD USING FLAGGERS

(Source: Manual on Uniform Traffic Control Devices, 2003)

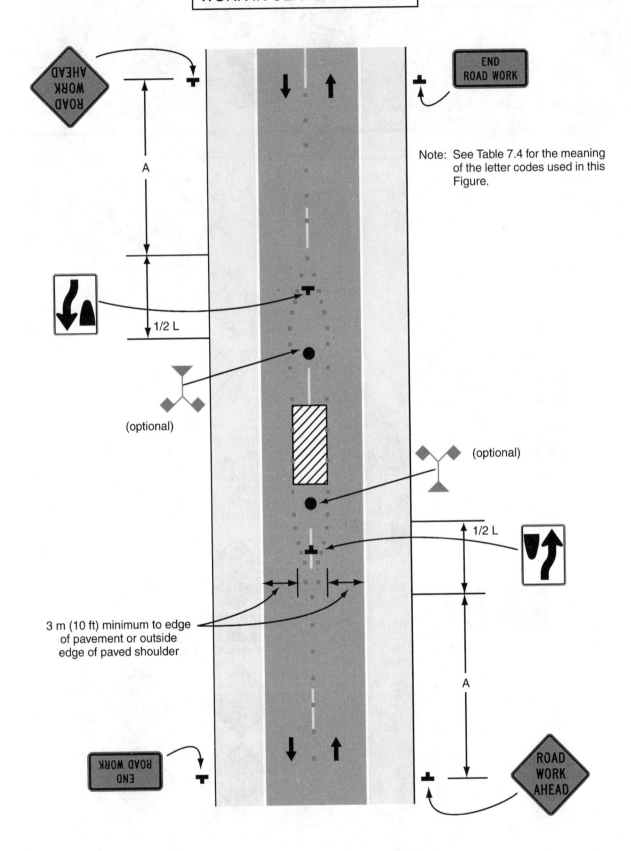

WORK IN CENTER OF ROAD

Note: See Table 7.4 for the meaning of the letter codes used in this Figure.

(optional)

(optional)

3 m (10 ft) minimum to edge of pavement or outside edge of paved shoulder

WORK IN CENTER OF ROAD WITH LOW TRAFFIC VOLUMES

(Source: Manual on Uniform Traffic Control Devices, 2003)

WORK BEYOND INTERSECTION

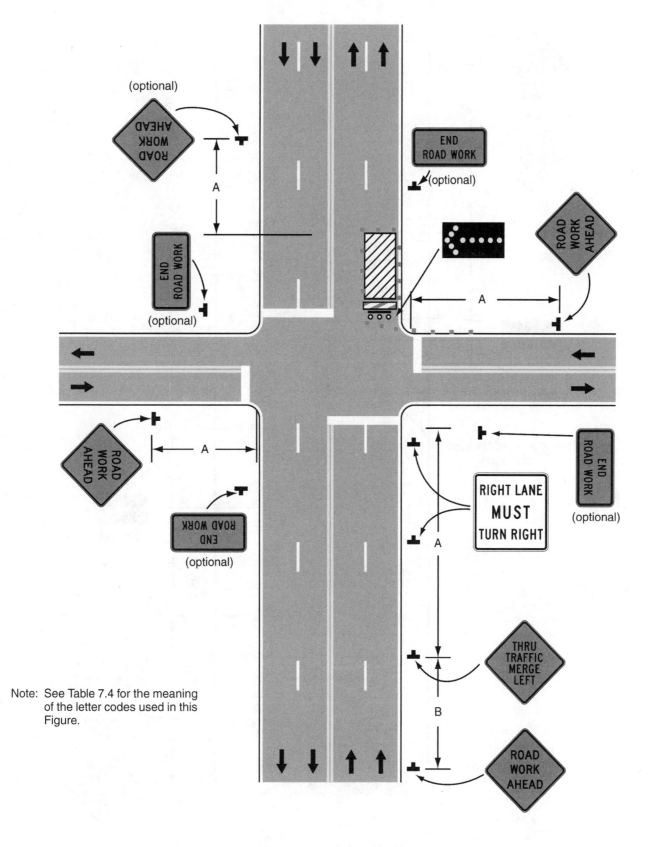

Note: See Table 7.4 for the meaning of the letter codes used in this Figure.

RIGHT LANE CLOSURE ON FAR SIDE OF INTERSECTION

(Source: Manual on Uniform Traffic Control Devices, 2003)

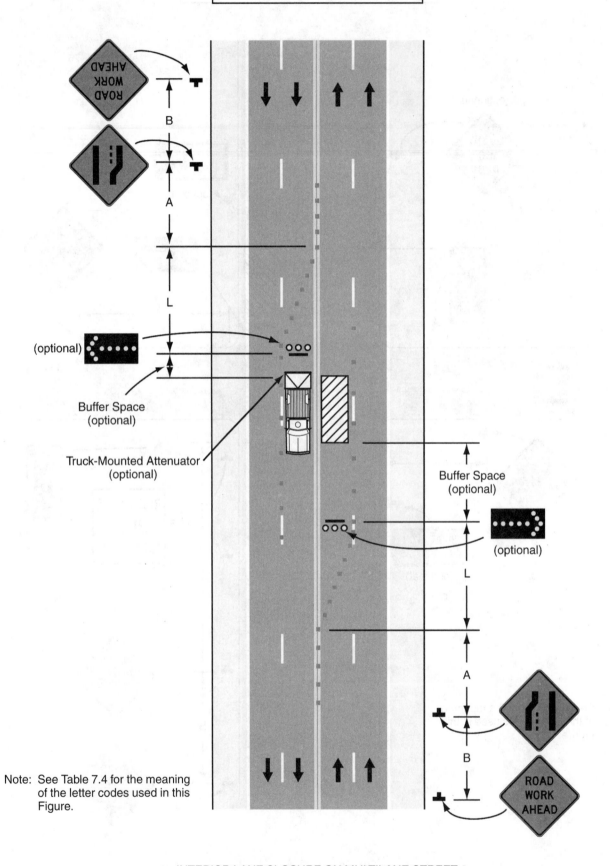

CLOSING INTERIOR LANE

ROAD WORK AHEAD

B

A

L

(optional)

Buffer Space
(optional)

Truck-Mounted Attenuator
(optional)

Buffer Space
(optional)

(optional)

L

A

B

ROAD
WORK
AHEAD

Note: See Table 7.4 for the meaning
of the letter codes used in this
Figure.

INTERIOR LANE CLOSURE ON MULTILANE STREET

(Source: Manual on Uniform Traffic Control Devices, 2003)

CLOSING HALF-ROADWAY

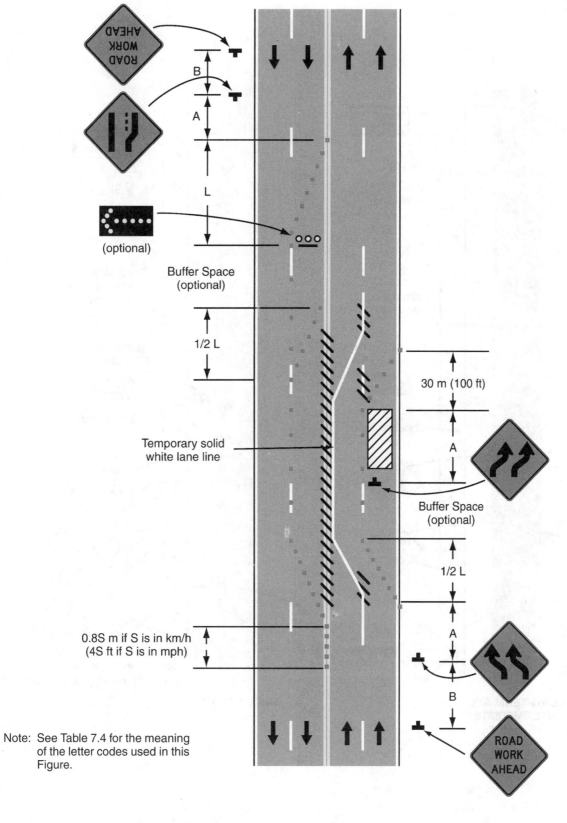

LANE CLOSURES ON STREET WITH UNEVEN DIRECTIONAL VOLUMES

(Source: Manual on Uniform Traffic Control Devices, 2003)

Note: See Table 7.4 for the meaning of the letter codes used in this Figure.

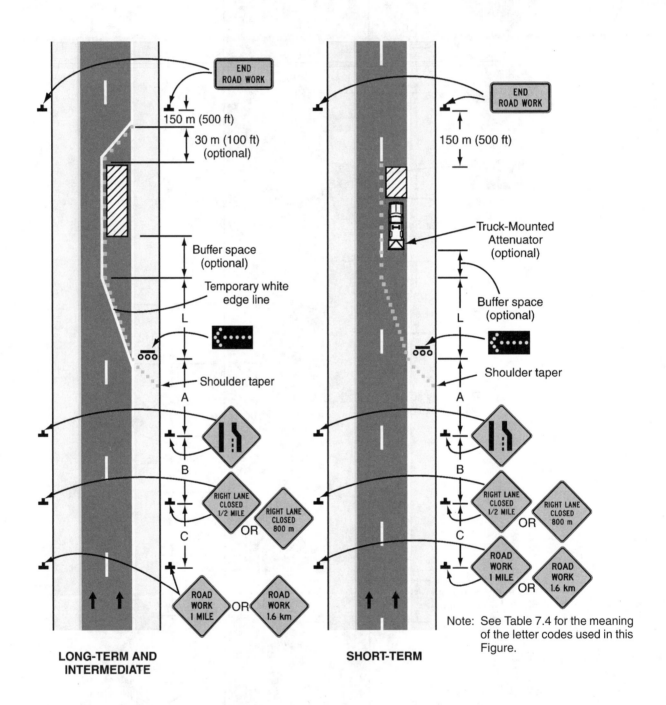

CLOSING RIGHT LANE

LONG-TERM AND INTERMEDIATE

SHORT-TERM

Note: See Table 7.4 for the meaning of the letter codes used in this Figure.

STATIONARY LANE CLOSURE ON DIVIDED HIGHWAY

(Source: Manual on Uniform Traffic Control Devices, 2003)

7.57   How To Use Traffic Control Devices

7.570   General Requirements

One of the best ways to increase the safety of both drivers and operators at construction sites is by using traffic control devices in a consistent, predictable manner. Drivers learn by experience what to expect when they see traffic control devices. If the devices themselves and their placement in the roadway are consistent over time, drivers can more quickly understand and respond to the changed road conditions. Traffic control devices should always be placed where they will convey their messages most effectively and give drivers adequate time to react. All signs, barricades, drums, and vertical panels must be retroreflectorized. Cones and tubes need to have two retroreflective white bands when used at night. All traffic control devices must be kept clean to ensure proper effectiveness and retroreflectivity.

All traffic control devices should be constructed to yield upon impact. The purpose of this precaution is to minimize damage to a vehicle that strikes them and to minimize hazards to motorists and workers. No traffic control device (signs, channelizing devices, sign trailers, or arrow panels) should be weighted so heavily that it becomes hazardous to motorists or operators. The approved ballast system for devices mounted on temporary portable supports is sandbags. The sandbags should be constructed so they do not readily rot or allow the sand to leach when exposed to the highway environment. Also, the sandbag should be constructed of a material that will allow the bag to break if struck by a vehicle. During cold weather, sandbags should be filled with a mixture of sand and a deicer. They should not be too heavy to be readily moved when a traffic control device is relocated. The number and size of sandbags used as traffic control device ballast should be kept to the minimum needed to provide stability for the device. Sandbags should not be suspended from the traffic control device and should be as close to the ground as possible. Never place sandbags on top of drums or ant-striped rail of a barricade.

7.571   Signs

The sizes of construction and maintenance signs also need to be consistent and are usually specified by local regulations. For example, warning signs should be at least 36 inches by 36 inches (90 cm by 90 cm) for low-speed applications and 48 inches by 48 inches (120 cm by 120 cm) for high-speed applications. Smaller signs may be used, if approved by the local regulations, on low-volume streets or highways or where larger signs become an additional hazard to motorists and pedestrians.

Construction and maintenance signs must not be mounted on existing traffic signs, posts, or other utility structures with-out permission from the proper authority. The minimum mounting height on fixed supports should be seven feet (2.1 m) from the ground to the bottom of the sign in urban districts and five feet (1.5 m) from pavement elevation to the bottom of the sign in rural districts. Signs mounted on barricades or other temporary supports may be installed at lower heights, but the bottom of the sign should not be less than one foot (0.3 m) above the pavement elevation. See Figure 7.18.

For maximum mobility, which may be needed on certain types of maintenance activities, a large sign may be mounted on a vehicle stationed in advance of the work. These mobile sign displays may be mounted on a trailer with self-contained electric power units for flashers and lights, or they may be mounted on a regular maintenance vehicle.

Regulatory signs, such as STOP, YIELD, NO PARKING, NO TURNS, and SPEED LIMIT, impose legal obligations and restrictions on all traffic. It is essential, therefore, that their use be authorized by the governing road authority having jurisdiction over traffic in the work area.

Construction and maintenance signs need to be placed where they will convey their messages effectively. Signs mounted on temporary supports should not be placed in the open, traveled lane where they pose a hazard to traffic. Generally these signs are placed on the right-hand side of the roadway or in the parking lane of the street or highway. Care should be taken to ensure that the motorists' view of the signs is not blocked by parked vehicles, trees, or other sight obstructions on or near the roadway. When special emphasis is needed, signs may be placed on both the left and right sides of the roadway.

All signs used at night should be either retroreflective or illuminated to show similar shape and color both day and night. Roadway lighting doesn't meet the requirements for illumination.

7.572   Channelizing Devices (Figure 7.19)

Channelizing devices are used to warn drivers and alert them to conditions created by work activities in or near the roadway, to protect workers in the temporary traffic control (TTC) zone, and to guide drivers and pedestrians. Consideration should be given to equipping the devices with warning lights in fog or rain areas, along severely curved roadways, and in unusually cluttered environments.

Channelizing devices include barricades, traffic cones and tubular markers, drums, and vertical panels. These devices are not interchangeable because they have different effects on traffic. Therefore, judgment must be used in selecting the appropriate channelizing device(s) for a specific work zone. Many factors, such as traffic volumes, speeds, and sight distances, should be considered when selecting the appropriate channelizing device(s). Channelizing devices should be constructed and ballasted to perform in a predictable manner when inadvertently struck by a vehicle. Channelizing devices should be crashworthy. Fragments or other debris from the device or the ballast should not pose a significant hazard to road users or workers.

7.573   Barricades

Barricades are commonly used to outline excavation or construction areas, close or restrict the right-of-way, channelize traffic, mark hazards, or mount signs. Barricades need to be placed so that the diagonal stripes slope down toward traffic. Barricades are classified as Type I, Type II, or Type III (Figure 7.19). Type I and Type II barricades are used in situations where traffic continues to move through the area, and Type III

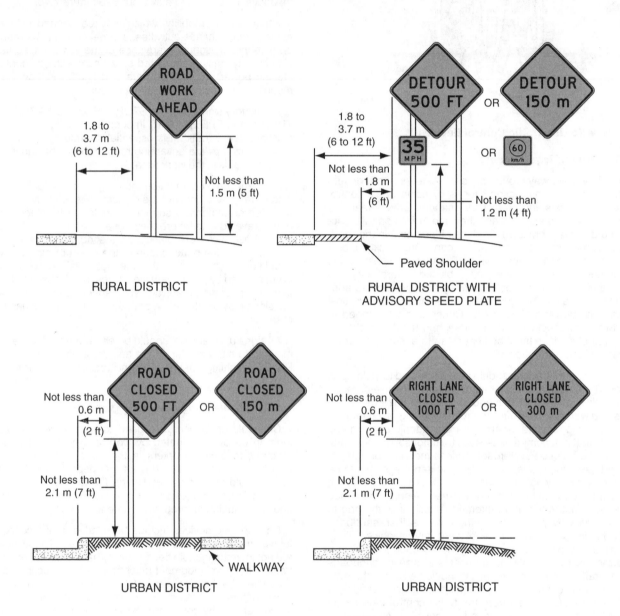

Fig. 7.18   Height and lateral location of signs—typical installations

(Source: Manual on Uniform Traffic Control Devices, 2003)

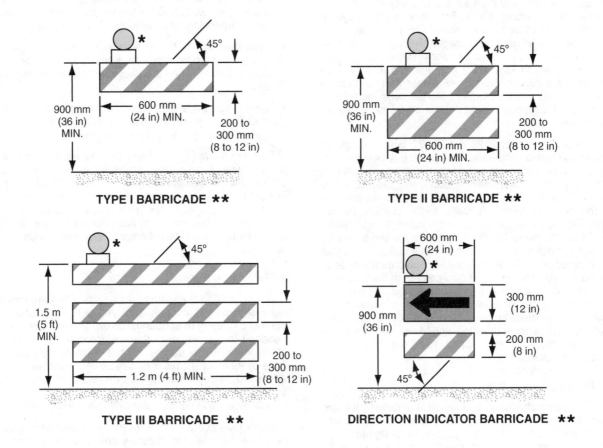

TYPE I BARRICADE ✱✱

TYPE II BARRICADE ✱✱

TYPE III BARRICADE ✱✱

DIRECTION INDICATOR BARRICADE ✱✱

✱ Warning lights (optional)
✱✱ Rail stripe widths shall be 150 mm (6 in), except that 100 mm (4 in) wide stripes
   may be used if rail lengths are less than 900 mm (36 in). The sides of barricades
   facing traffic shall have retroreflective rail faces.

Note: If barricades are used to channelize pedestrians, there shall be continuous detectable bottom and top rails with
      no gaps between individual barricades to be detectable to users of long canes. The bottom of the bottom rail shall
      be no higher than 150 mm (6 in) above the ground surface. The top of the top rail shall be no lower than 900 mm
      (36 in) above the ground surface.

Fig. 7.19   Channelizing devices and barricades
(Source: Manual on Uniform Traffic Control Devices, 2003)

barricades are used where the street or highway is partially or totally closed. Type III barricades should be set up at the point of closure. Where provision is made for access of authorized equipment and vehicles, the responsibility for Type III barricades should be assigned to a person to ensure proper closure at the end of each work period.

7.574 Traffic Cones and Tubular Markers (Figure 7.20)

Traffic cones and tubular markers are an effective method of channelizing traffic along a specified route during daylight hours. Use of cones or tubular markers in transition areas during nighttime activities is discouraged. Traffic cones and tubular markers are easily moved by passing vehicles so monitoring is necessary to keep them in place. Steps should be taken to minimize cone displacement. Cones can be doubled up to increase their weight. Some cones are constructed with bases that can be filled with ballast while others have special weighted bases.

Cones should be at least 28 inches (70 cm) in height in high-speed roadways. Some states even require 36-inch (90-cm) cones. Some cones are constructed with bases that can be filled with ballast. Others have specially weighted bases or weights, such as sandbag rings that can be dropped over the cones and onto the base to provide added stability. Ballast should be kept to the minimum amount needed. Tubular markers should be stabilized by affixing them to the pavement by using weighted bases or weights, such as sandbag rings that can be dropped over the tubular markers and onto the base to provide added stability. Ballast should be kept to the minimum amount needed. Tubular markers may be used effectively to divide opposing lanes of road users, divide vehicular traffic lanes when two or more lanes of moving motor vehicle traffic are kept open in the same direction, and to delineate the edge of a pavement drop-off where space limitations do not allow the use of larger devices.

7.575 Drums (Figure 7.20)

Drums are commonly used to channelize or delineate traffic flow. Drums should not be weighted to the extent that they become hazardous to motorists. They can be designed with a base that will separate from the drum if struck by a vehicle. Drums should not be weighted with sand, water, or any material to the extent that would make them hazardous to road users or workers if struck. Follow the manufacturer's instructions for ballasting. Drums used in regions susceptible to freezing should have drain holes in the bottom so that water will not accumulate and freeze causing a hazard if struck by a road user. Ballast should not be placed on the top of a drum.

7.576 Vertical Panels (Figure 7.20)

Vertical panels are most commonly used for traffic separation or shoulder barricading where only limited space is available. Vertical panels should be 8 to 12 inches (20 to 30 cm) in width and at least 24 inches (60 cm) in height. They should have orange and white diagonal stripes and be retroreflectorized. Vertical panels should be mounted with the top a minimum of 36 inches (90 cm) above the roadway. Where the height of the vertical panel itself is 36 inches (90 cm) or greater, a panel stripe width of 6 inches (15 cm) should be used. Where the height of the vertical panel itself is less than 36 inches (90 cm), a panel stripe width of 4 inches (10 cm) may be used. Markings for vertical panels should be alternating orange and white retroreflective stripes, sloping downward at an angle of 45 degrees in the direction vehicular traffic is to pass.

7.577 Lighting Devices

Construction and maintenance activities often create conditions that become particularly hazardous at night. Therefore, lighting devices should be used in addition to the traffic control devices. Warning lights are classified as Type A, B, or C as follows:

- Type A lights are low-intensity, flashing yellow lights most commonly mounted on barricades, drums, or advance warning signs, and are intended to warn drivers that they are approaching, or in, an area of potential danger.

- Type B lights are high-intensity, flashing yellow lights normally mounted on advance warning signs or on independent supports. They are designed to operate 24 hours per day.

- Type C lights are steady-burn lights used to delineate (define) the edge of the desired path, such as on detour curves, on lane changes, and on lane closures.

Flashing lights (Types A and B) must not be used for delineation because a series of flashers may obscure the desired path and would be confusing to the drivers. When used, flashing warning lights are normally installed above the first advance warning sign. Flashing warning lights also may be used to supplement barricades at hazardous locations, particularly where there is a sudden change in alignment, such as T-intersections.

7.578 Arrow Panels

Arrow panels (or flashing arrow signs, FAS) are rectangular black sign panels covered with yellow lights capable of displaying a flashing or sequential arrow. These arrow panels are intended to supplement other traffic control devices, not replace them. Arrow panels will not solve difficult traffic problems by themselves, but they can be very effective when properly used to reinforce signs, barricades, cones, and other traffic control devices. All required traffic control devices should be used in combination with the arrow panel.

Arrow panels are vehicle- or trailer-mounted for easy transport to a job site. The displays are operated from a remote control panel and powered by a self-contained power source (batteries or an electric generator) mounted on the vehicle or trailer.

An arrow panel in the arrow or chevron mode should be used to advise approaching traffic of a lane closure along major multilane roadways in situations involving heavy traffic volumes, high speeds, and/or limited sight distances, or at other locations and under other conditions where road users are less likely to expect such lane closures. If used, an arrow panel should be used in combination with appropriate signs, channelizing devices, or other traffic control devices. An arrow panel should be placed on the shoulder of the roadway or, if practical, further from the traveled lane. It should be delineated with retroreflective traffic control devices. When an arrow panel is not being used, it should be removed; if not removed, it should be shielded; or if the previous two options are not feasible, it should be delineated with retroreflective traffic control devices.

For shoulder work, blocking the shoulder, for roadside work near the shoulder, or for temporarily closing one lane on a two-lane, two-way roadway, an arrow panel should be used only in the caution mode (four corners flashing or flashing bar) if allowed.

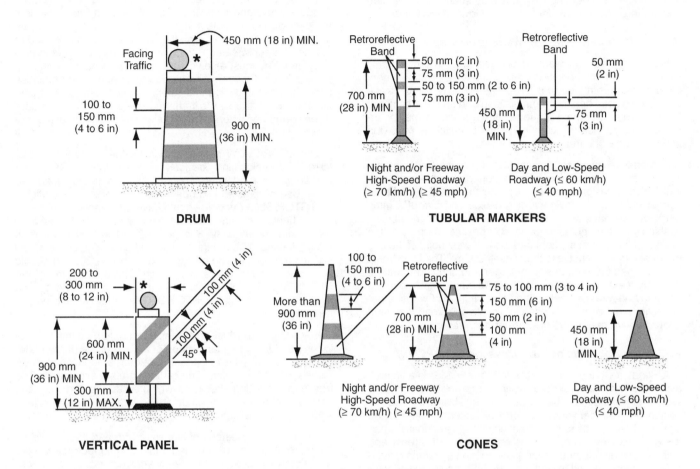

DRUM

TUBULAR MARKERS

VERTICAL PANEL

CONES

* Warning lights (optional)

Note: If drums, cones, or tubular markers are used to channelize pedestrians, they shall be located such that there are no gaps between the bases of the devices, in order to create a continuous bottom, and the height of each individual drum, cone, or tubular marker shall be no less than 900 mm (36 in) to be detectable to users of long canes.

Fig. 7.20   Delineating and channelizing devices

(Source: Manual on Uniform Traffic Control Devices, 2003)

Arrow panel elements should be capable of at least a 50 percent dimming from full brilliance. The dimmed mode should be used for nighttime operation of arrow panels.

7.579   *Flashing Yellow Vehicle Lights*

Work vehicles in or near the traffic areas should be equipped with flashing yellow lights. The vehicle warning lights may be emergency flashers, flashing strobe lights, or rotating beacons. High-intensity lights are effective for both day and nighttime work. The governing road authority should be contacted concerning requirements for flashing yellow vehicle lights.

7.5710   *High-Level Flag Tree Warning Devices* (Figure 7.21)

High-level flag tree warning devices are most commonly used in urban high-density traffic situations to supplement other traffic control devices. The high-level warning device should be placed behind channelization and near the center of the blocked traffic lane. High-level warning devices may be attached to a service vehicle when such a vehicle is placed in advance of a work area.

A high-level warning device (flag tree) may supplement other traffic control devices in work zones. A high-level warning device is designed to be seen over the top of typical passenger cars. A high-level warning device consists of a minimum of two flags with or without a Type B high-intensity flashing warning light. The distance from the roadway to the bottom of the lens of the light and to the lowest point of the flag material should be not less than 8 feet (2.5 m). The flag should be 16 inches (40 cm) square or larger and should be orange or fluorescent red-orange in color. An appropriate warning sign may be mounted below the flags. High-level warning devices are most commonly used in high-density road user situations to warn road users of short-term operations.

7.5711   *Portable Changeable Message Sign (PCMS)*

A Portable Changeable Message Sign (PCMS) is a lighted, electronic traffic control device with the flexibility to display a variety of messages. The PCMS should always be used in conjunction with conventional signs, pavement markings, and lighting. They are used most frequently on high-density urban freeways, but have applications on all types of streets and highways where alignment and traffic routing problems require advance warning and information. PCMSs serve a wide variety of functions in work zones including advising motorists of road closures, accident management instructions, narrow lanes, construction schedules, traffic management and diversion activities, and adverse conditions.

PCMSs should be placed in advance of a temporary traffic control zone and should not replace any required sign. Each PCMS display should convey a single thought in as brief a message as possible. The entire message cycle should be readable at least twice at the posted speed.

7.5712   *Flagger Control* (Figure 7.22)

Flagging procedures, when used, can provide positive guidance to motorists driving through the work area. Specific methods, procedures, and specifications for flagging must be used. In general, flaggers may be needed when the following conditions exist:

1. Operators or equipment intermittently block a traffic lane,

2. One lane must be used for two directions of traffic (a flagger is required for each direction of traffic), and

3. The safety of the public and/or operators requires it.

Use of flaggers is a last resort. For lane closures, use tapers and channelizing devices instead of flaggers.

The flaggers should be positioned at least 100 feet (30 m) in front of the work space and a sign reading FLAGGER AHEAD should be placed as far ahead of the flaggers as practical (500 feet or 150 meters minimum). (Be sure to check with your governing road authority for local regulations concerning the use of flaggers to control traffic.) Flagger stations should be located such that approaching road users will have sufficient distance to stop at an intended stopping point. Figure 7.22 shows flaggers and hand signaling motions and devices. Flaggers are expected to regulate and control the flow of traffic in a safe and orderly fashion. Because flaggers are responsible for employee and public safety, they must have a sense of responsibility and they must receive appropriate training in traffic control practices. Flaggers and utilities have been sued because improper actions by flaggers caused an accident. Once again, consistency is the key to traffic control safety. Use standard devices and consistent signal motions to avoid confusing motorists.

The STOP/SLOW paddle should be the primary and preferred hand-signaling device because the STOP/SLOW paddle gives road users more positive guidance than red flags. Use of flags should be limited to emergency situations. The STOP/SLOW paddle should have an octagonal shape on a rigid handle, at least 5 feet (1.5 m) high. STOP/SLOW paddles should be at least 18 inches (45 cm) wide with letters at least 6 inches (15 cm) high and should be fabricated from light semirigid material. The background of the STOP face should be red with white letters and border. The background of the SLOW face should be orange with black letters and border. When used at night, the STOP/SLOW paddle should be retroreflectorized.

Flags, when used for emergencies, should be a minimum of 24 inches (60 cm) square, made of a good grade of red material, and securely fastened to a staff that is approximately 36 inches (90 cm) in length. Orange flags are not acceptable. The free edge of a flag should be weighted so the flag will hang vertically, even in heavy winds. When used at nighttime, flags should be retroreflectorized red.

The following methods of signaling with paddles should be used: To stop road users, the flagger should face road users and aim the STOP paddle face toward road users in a stationary position with the arm extended horizontally away from the body. The free arm should be held with the palm of the hand above the shoulder level toward approaching traffic. To direct stopped road users to proceed, the flagger should face road users with the SLOW paddle face aimed toward road users in a stationary position with the arm extended horizontally away from the body. The flagger should motion with the free hand for road users to proceed. To alert or slow traffic, the flagger should face road users with the SLOW paddle face aimed toward road users in a stationary position with the arm extended horizontally away from the body. To further alert or slow traffic, the flagger holding the SLOW paddle face toward road users may motion up and down with the free hand, palm down.

The following methods of signaling with a flag should be used: To stop road users, the flagger should face road users and extend the flag staff horizontally across the road users' lane in a stationary position so that the full area of the flag is visibly hanging below the staff. The free arm should be held with the palm of the hand above the shoulder level toward approaching traffic. To direct stopped road users to proceed, the flagger should stand parallel to the road user movement and with flag and arm lowered from the view of the road users, and

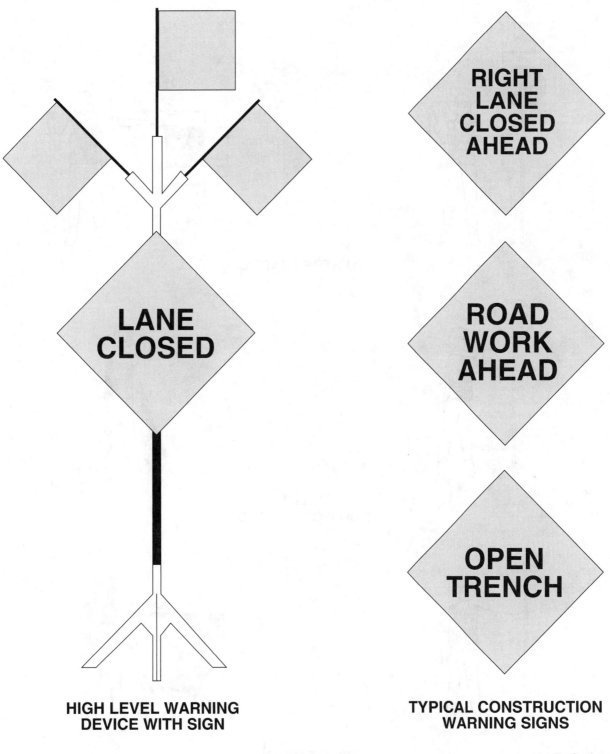

**HIGH LEVEL WARNING
DEVICE WITH SIGN**

**TYPICAL CONSTRUCTION
WARNING SIGNS**

Not To Scale

Fig. 7.21   Warning devices and signs

PREFERRED METHOD **STOP/SLOW Paddle**	**EMERGENCY SITUATIONS ONLY** **Red Flag**

450 mm (18 in)
MIN.

900 mm
(36 in)

600 mm
(24 in)

600 mm
(24 in)

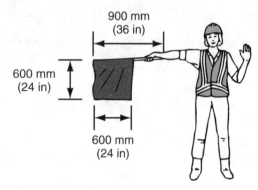

TO STOP TRAFFIC

TO LET
TRAFFIC PROCEED

TO ALERT AND
SLOW TRAFFIC

Fig. 7.22   Use of hand-signaling devices by flaggers

(Source: Manual on Uniform Traffic Control Devices, 2003)

should motion with the free hand for road users to proceed. Flags should not be used to signal road users to proceed. To alert or slow traffic, the flagger should face road users and slowly wave the flag in a sweeping motion of the extended arm from shoulder level to straight down without raising the arm above a horizontal position. The flagger should keep the free hand down.

For daytime and nighttime activity, flaggers should wear safety apparel meeting the requirements of ISEA "American National Standard for High-Visibility Apparel" and labeled as meeting the ANSI 107-1999 standard performance for Class 2 risk exposure. The apparel background (outer) material should be either fluorescent orange-red or fluorescent yellow-green as defined in the standard. The retroreflective material color should be either orange, yellow, white, silver, yellow-green, or a fluorescent version of these colors, and should be visible at a minimum distance of 1,000 ft (300 m). The retrore-flective safety apparel should be designed to clearly identify the wearer as a person. For nighttime activity, safety apparel meeting the requirements of ISEA "American National Standard for High-Visibility Apparel" and labeled as meeting the ANSI 107-1999 standard performance for Class 3 risk exposure should be considered for flagger wear (instead of the Class 2 safety apparel described earlier.)

REMEMBER, YOU ARE SPEAKING A SIGN LANGUAGE... MAKE YOUR SIGNS SPEAK LOUD AND CLEAR!

7.58 Other Traffic Control Concerns

7.580 Pedestrians, Bicyclists, and Persons With Disabilities

If the work will block pedestrian walkways, bicycle lanes, or bicycle pathways, the governing road authority may require you to provide alternative walkways or bikeways. Alternative walkways or bikeways must provide protection from hazardous work areas and vehicular traffic in conformance with the Americans With Disabilities Act of 1990. Temporary facilities, including reasonably safe pedestrian routes around work sites, are also covered by the accessibility requirements of the Americans With Disabilities Act of 1990 (ADA) (Public Law 101-336, 104 Stat. 327, July 26, 1990. 42 USC 12101–12213 (as amended)). The most desirable way to provide information to pedestrians with visual disabilities that is equivalent to visual signage for notification of sidewalk closures is a speech message provided by an audible information device. Devices that provide speech messages in response to passive pedestrian actuation are the most desirable. Other devices that continuously emit a message, or that emit a message in response to use of a pushbutton, are also acceptable. Signage information can also be transmitted to personal receivers, but currently such receivers are not likely to be carried or used by pedestrians with visual disabilities in temporary traffic control (TTC) zones. Audible information devices might not be needed if detectable channelizing devices make an alternate route of travel evident to pedestrians with visual disabilities. Tape, rope, or plastic chain strung between devices are not detectable, do not comply with the design standards in the Americans With Disabilities Act, and should not be used as a control for pedestrian movements.

Consideration should be made to separate pedestrian movements from both work site activity and vehicular traffic. Unless a reasonably safe route that does not involve crossing the roadway can be provided, pedestrians should be appropriately directed with advance signing that encourages them to cross to the opposite side of the roadway. In urban and suburban areas with high vehicular traffic volumes, these signs should be placed at intersections (rather than midblock locations) so that pedestrians are not confronted with midblock work sites that will induce them to attempt skirting the work site or making a midblock crossing. Temporary traffic barriers or longitudinal channelizing devices may be used to discourage pedestrians from unauthorized movements into the work space. They may also be used to inhibit conflicts with vehicular traffic by minimizing the possibility of midblock crossings.

Every effort must be made to separate pedestrians from the work area. Protective barricades, fencing, handrails, and bridges must be used with warning devices and signs to direct foot traffic to areas where pedestrians may pass safely. Walkways in construction areas should be at least four feet (1.2 meters) wide and free of abrupt changes in grade. Where traffic must change grade, an incline should be used. Special consideration should be given to the blind and handicapped. There should be a minimum vertical clearance of seven feet (2.1 meters) and illumination should be provided during hours of darkness. Occasionally, a walkway must be closed completely. When this is necessary, foot traffic should be diverted at the closest crosswalks with the appropriate signs and barricades (Figure 7.23).

When welding is necessary on the job, suitable screens must be placed to protect the public from arc welding flash or sparks from cutting operations.

When equipment such as pipe is stored on the job site, it should be securely blocked so it cannot be displaced.

When work is to be done in a residential area, it is good public relations to advise the public what is to be done, how long it is expected to take, that all safety regulations will be observed, and that every effort will be made not to disrupt the routine of the neighborhood.

7.581 Worker Visibility

All workers not separated from traffic by a positive barrier must be required to wear high-visibility clothing such as a vest, shirt, or jacket as approved by the governing road authority. In addition, any time it is raining, snowing, sleeting, or hailing, and any other time visibility is reduced by weather, smoke, fog, or other conditions, reflectorized garments should be worn. The reflective material may be orange, yellow, white, silver, or yellow-green and fluorescent versions and should be visible from at least 1,000 feet (300 m). The shape of the reflective garment should be visible through the full range of body motions. Details on the approved types of high-visibility garments should be obtained from the governing road authority.

Worker Safety Apparel—All workers exposed to the risks of moving roadway traffic or construction equipment should wear high-visibility safety apparel meeting the requirements of ISEA "American National Standard for High-Visibility Safety Apparel" (see Section 1A.11), or equivalent revisions, and labeled as ANSI 107-1999 standard performance for Class 1, 2, or 3 risk exposure. A competent person designated by the employer to be responsible for the worker safety plan within the activity area of the job site should make the selection of the appropriate class of garment.

7.582 Speed Limits in Work Zones

Proper uniform application of speed limits in work zones will help to improve the safety of operators working in and near roadways as well as protecting the motoring public. In high-traffic, high-hazard areas, consideration should be given to re-

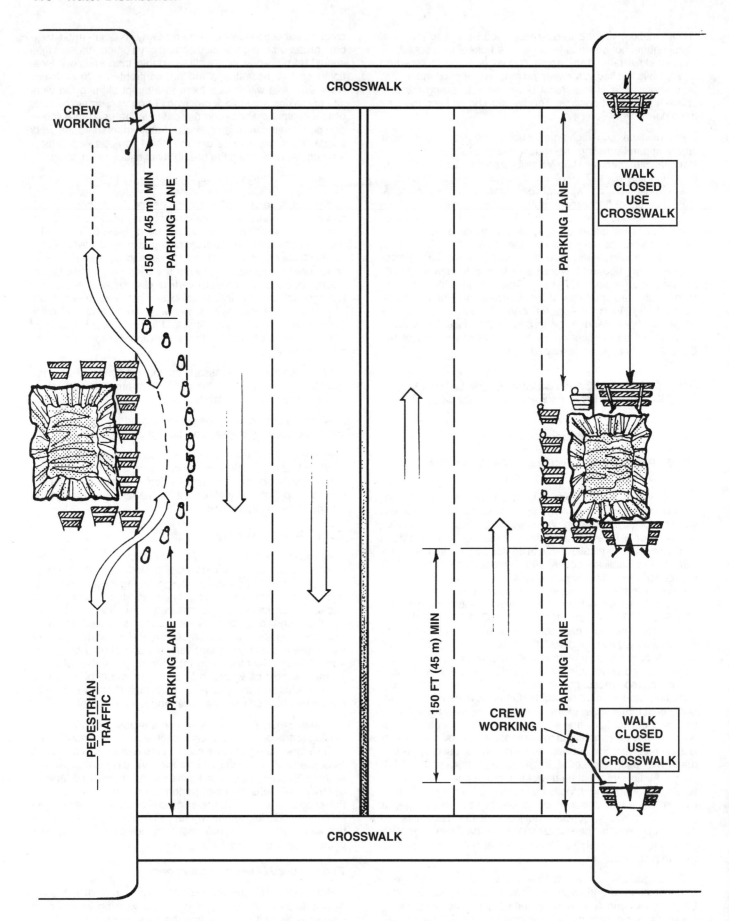

Fig. 7.23   Pedestrian control

ducing the speed of traffic through regulatory speed zoning, use of police, lane reduction, or flaggers. The use of flaggers should be discouraged in high-speed areas.

Reduced speed zoning (lowering the regulatory speed limit) should be avoided as much as practical because drivers will reduce their speeds only if they clearly perceive a need to do so. If warranted, lowering the posted speed limit requires the use of regulatory signs, which must be approved by the proper highway authority. Reduced speed limits should be used only in the specific portion of the temporary traffic control (TTC) zone where conditions or restrictive features are present. However, frequent changes in the speed limit should be avoided. A TTC plan should be designed so that vehicles can reasonably safely travel through the TTC zone with a speed limit reduction of no more than 10 mph (16 km/hr).

7.583   Worker Considerations

The safety of workers in a temporary traffic control (TTC) zone is as important as the safety of the public. In addition to worker visibility and speed controls mentioned earlier, the following elements should also be included to ensure worker safety:

- Training—all workers must be trained in how to safely work next to traffic. Workers with traffic control responsibilities must be trained in traffic control techniques and device usage and placement.

- Barriers—barriers along the work area should be considered if the work area is close to traffic, if traffic speed is excessive, or if traffic volume is high.

- Lighting—if work is required at night, lighting the area and approaches will provide drivers with a better understanding of the requirements being imposed.

- Public information—a public relations effort that tells what the nature of the work is, its duration, anticipated effects on traffic, and alternate routes may improve drivers' performance and reduce traffic exposure.

- Road closure—if an alternate route is available to handle detoured traffic, a temporary road closure would reduce traffic hazards greatly and would probably allow the job to be completed much faster.

Equally as important as the safety of road users traveling through the temporary traffic control (TTC) zone is the safety of workers. TTC zones present temporary and constantly changing conditions that are unexpected by the road user. This creates an even higher degree of vulnerability for workers on or near the roadway. Maintaining TTC zones with road user flow inhibited as little as possible, and using TTC devices that get the road users' attention and provide positive direction are of particular importance. Likewise, equipment and vehicles moving within the activity area create a risk to workers on foot. When possible, the separation of moving equipment and construction vehicles from workers on foot provides the operators of these vehicles with a greater separation clearance and improved sight lines to minimize exposure to the hazards of moving vehicles and equipment.

The following are the key elements of worker safety and TTC management that should be considered to improve worker safety:

- Training—All workers should be trained on how to work next to motor vehicle traffic in a way that minimizes their vulnerability. Workers having specific traffic control responsibilities should be trained in TTC techniques, device usage, and placement.

- Worker Safety Apparel—All workers exposed to the risks of moving roadway traffic or construction equipment should wear high-visibility safety apparel meeting the requirements of ISEA "American National Standard for High-Visibility Safety Apparel" (see Section 1A.11) or equivalent revisions, and labeled as ANSI 107-1999 standard performance for Class 1, 2, or 3 risk exposure. A competent person designated by the employer to be responsible for the worker safety plan within the activity area of the job site should make the selection of the appropriate class of garment.

- Temporary Traffic Barriers—Temporary traffic barriers should be placed along the work space depending on factors such as lateral clearance of workers from adjacent traffic, speed of traffic, duration and type of operations, time of day, and volume of traffic.

- Speed Reduction—Reducing the speed of vehicular traffic, mainly through regulatory speed zoning, funneling, lane reduction, or the use of uniformed law enforcement officers, or flaggers, should be considered.

- Activity Area—Planning the internal work activity area to minimize backing-up maneuvers of construction vehicles should be considered to minimize the exposure to risk.

- Worker Safety Planning—A competent person designated by the employer should conduct a basic hazard assessment for the work site and job classifications required in the activity area. This safety professional should determine whether engineering, administrative, or personal protection measures should be implemented. This plan should be in accordance with the Occupational Safety and Health Act of 1970, as amended, "General Duty Clause" Section 5(a)(1) - Public Law 91-596, 84 Stat. 1590, December 29, 1970, as amended, and with the requirement to assess worker risk exposures for each job site and job classification, as per 29 CFR 1926.20 (b)(2) of "Occupational Safety and Health Administration Regulations, General Safety and Health Provisions."

7.584   Additional Reading

1. *MANUAL ON UNIFORM TRAFFIC CONTROL DEVICES*, Federal Highway Administration (FHWA). The 2003 Edition is currently available as an electronic version on the FHWA website at www.fhwa.dot.gov. Hard copies are available from the American Association of State Highway Traffic Officials (AASHTO), Publications Order Department, PO Box 96716, Washington, DC 20090-6716. Price, $75.00.

2. *WORK AREA TRAFFIC CONTROL HANDBOOK*, Building News, Inc., 1612 South Clementine Street, Anaheim, CA 92802. Price, $7.95.

7.585   Acknowledgment

Material in this section was reviewed by Juan M. Morales on behalf of the American Traffic Safety Services Association. His comments and suggestions were excellent and are greatly appreciated.

QUESTIONS

Write your answers in a notebook and then compare your answers with those on page 434.

7.5N   Why is it important to use traffic control devices in a consistent, predictable manner?

7.5O   List four types of channelizing devices.

7.5P   What is a Portable Changeable Message Sign (PCMS) and when is it used?

7.5Q   Describe the motions a flagger would use to alert and slow daytime traffic.

7.5R   What would you tell the public when excavation work is planned in a residential neighborhood?

7.59   Excavations in Streets[9]

7.590   Safety Rules for Excavations

Excavations are necessary to install or repair water lines in streets or roadways. Regulations and requirements governing excavations may be found in your state construction safety orders or other similar documents. A permit may be required for any excavations four feet (1.2 meters) or more in depth in which personnel must work. Shoring is also required.

General safety rules include:

1. Employer must inspect any excavation for hazards from possible moving ground before employees may work in or adjacent to the excavation.

2. The location of underground installations (such as utility lines) must be determined before excavation begins.

3. Excavations greater than four feet deep (1.2 m) must be tested as often as necessary for oxygen deficiency/enrichment and hazardous atmospheres, and appropriate precautions must be taken to protect workers.

4. Excavations must be inspected daily by a *COMPETENT PERSON*[10] and if a problem exists, workers must be removed until precautions have been taken to ensure their safety.

5. Properly qualified supervisors must be on site at all times during excavations.

6. Safety provisions must be taken to protect workers installing or removing shoring systems.

7. Spoil (material removed from the excavation) must be kept back at least two feet (0.6 m) from the edge of all excavations.

8. Safe and convenient access to an excavation must be provided.

9. Effective barriers must be installed at excavations adjacent to mobile equipment.

10. Water must not be allowed to accumulate in any excavation.

7.591   Trenches

A trench is an excavation in which the average depth exceeds the width and the width is 15 feet (4.5 meters) or less at the bottom. The same general requirements that apply to excavations apply to trenches. Trenching is more common than excavations in water utility work. Exits must be provided at 25-foot (7.4-meter) intervals for all occupied trenches four or more feet (1.2 m) deep.

Covers placed over trenches in roadways must be secured against displacement. Covers are usually placed over a trench at the end of the day. This provides for opening of the street for peak traffic flow. Occasionally excavations or trenches must cross sidewalks or other pedestrian rights-of-way. When this is necessary, a safe, temporary right-of-way must be provided or foot traffic should be rerouted. Figure 7.24 illustrates various types of trench shoring conditions.

7.592   Cave-Ins

Improper shoring can contribute to cave-ins. Many cave-ins can be avoided by using the appropriate shoring for the particular soil type and existing conditions. Also, correct procedures should be followed when installing shoring. The safe way of installing screw jacks and braces is to start at the top of the trench and work downward.

Another hazard is the excavated spoil bank. Usually the spoil bank is set back far enough to prevent the material from falling back into the trench and also far enough to provide a walkway or tool-rest for shovels and crowbars. Of great significance, however, is the fact that the proper setback of the spoil bank will keep the weight away from the edge of the trench. If the spoil is too heavy and too close to the edge, the wall of the trench could cave in on top of you.

The following procedures are recommended in order to prevent violations of the safety rules, to prevent cave-ins, and to help keep us from getting hurt.

1. In excavations 5 feet (1.5 meters) or deeper (4 feet (1.2 meters) in some regulations), shoring is required and the spoil bank is to be kept back a minimum of 4 feet (1.2 meters) from the edge of the trench. Increase this distance consistent with the type and condition of the soil. If the depth is less than 5 feet (1.5 meters), the spoil should be placed 2 feet (0.6 meter) or more back from the edge of the trench. (*NOTE:* OSHA requires a minimum spoil setback of 2 feet (0.6 meter) from the edge of a trench. Some states require greater spoil setbacks depending on local conditions.)

---

[9] *Also see Chapter 3, "Distribution System Facilities," Section 3.653, "Excavation and Shoring."*

[10] *Competent Person.   A competent person is defined by OSHA as a person capable of identifying existing and predictable hazards in the surroundings, or working conditions which are unsanitary, hazardous or dangerous to employees, and who has authorization to take prompt corrective measures to eliminate the hazards.*

SATURATED, FILLED OR UNSTABLE GROUND

Sheeting must be provided, and must be sufficient to hold the material in place.

Longitudinal-stringer dimensions depend upon the strut and stringer spacing and upon the degree of instability encountered.

HARD COMPACT GROUND

Trenches 5 feet or more deep and over 8 feet long must be braced at intervals of 8 feet or less.

RUNNING MATERIAL

Sheet Piling or equivalent solid sheeting is required for trenches four feet or more deep.

Longitudinal-stringer dimensions depend upon the stringer spacing, and the depth of stringer below the ground surface.

Greater loads are encountered as the depth increases, so more or stronger stringers and struts are required near the trench bottom.

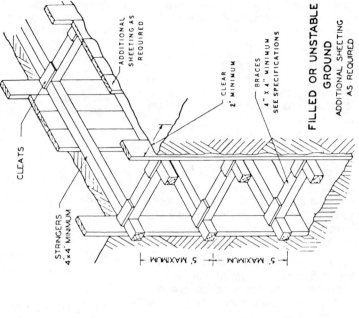

ADDITIONAL SHEETING AS REQUIRED

CLEATS

STRINGERS 4 x 4 MINIMUM

CLEAR 2' MINIMUM

BRACES 4" X 4" MINIMUM SEE SPECIFICATIONS

5' MAXIMUM — 5' MAXIMUM

FILLED OR UNSTABLE GROUND
ADDITIONAL SHEETING AS REQUIRED

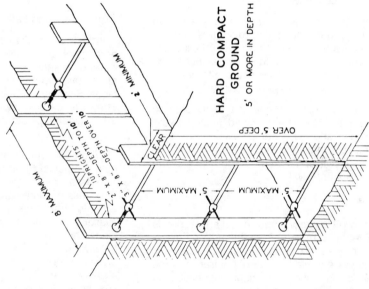

2' MINIMUM

UPRIGHTS DEPTH TO 10'
2" X 8"—DEPTH OVER 10'
3" X 8"

8' MAXIMUM

8' MAXIMUM

CLEAR

5' MAXIMUM

5' MAXIMUM

OVER 5' DEEP

HARD COMPACT GROUND
5' OR MORE IN DEPTH

CLEATED

STRINGERS

SHEET PILINGS TRENCH DEPTH
4' TO 8'—2" MIN. THICK.
OVER 8'—3" MIN. THICK.

CLEAR 2' MINIMUM

BRACES 4" X 4" MINIMUM SEE SPECIFICATIONS

5' MAXIMUM — 5' MAXIMUM

RUNNING MATERIAL

NOTE: Multiply inches x 2.5 to obtain centimeters

Fig. 7.24   Various types of trench shoring conditions

2. When installing braces or screw jacks, start with the top one first and work downward. Removal is in reverse order, starting with the bottom brace and working upward.

3. Eight feet (2.4 meters) is the maximum spacing of uprights under the ideal conditions of hard, compact soil excavated to a depth not exceeding 10 feet (3 meters). Spacing of the uprights shall be decreased and solid sheathing may be required when the depth exceeds 10 feet (3 meters) or if unstable soil is encountered.

4. When sources of vibration (trucks, railroads) are nearby, special safety precautions may be necessary. Additional bracing or other effective means are recommended.

5. In trenches four feet (1.2 meters) or deeper, ladders or ramps should be provided for access and should be located so that workers need not move farther than 25 feet (7.5 meters). The ladders should extend a minimum of 30 inches (75 cm) above the top of the excavations.

Field inspections and injury reports have indicated that the items listed above are frequently overlooked. Attention to these items is necessary if we are to reduce accidents in this area. If we continue to ignore safety regulations, the pain and suffering of accidents will continue.

The economic impact of shoring requirements is a known factor. Provision of the proper materials, equipment and the time to install and remove the shoring is a required expense. The installation of a substandard shoring system is just as expensive as one properly installed. Given the alternative, it makes sense to install the shoring system properly!

These requirements are minimal and, on occasion, the shoring and methods may have to be improved. In large excavations or where complex shoring systems deviate from regular shoring procedures, alternate shoring systems must be designed by a registered civil engineer and submitted to the appropriate authorities for approval.

7.593 Ladders

Ladders must be placed in excavations so that any worker is never more than 25 feet (7.5 meters) from the ladder. For example, one ladder is required for each 50 feet (15 meters) of trench if the ladder is placed in the center of the trench. If the ladder is placed near one end of the trench, two ladders would be required. The ladders must extend 3 feet (90 centimeters) above the surface of the excavation. Ladders are required in all excavations where a problem exists in entering or exiting the excavation. All ladders must be kept clean and in good repair.

7.594 WARNING: Look out for Other Underground Utilities

Before you do any excavating, determine the location of underground utilities and advise the utility owners of the proposed work. Some underground installations that could be damaged by excavations include underground cables, electrical services, gas mains, and occasionally storm water and sewer lines that may be adjacent to proposed excavations. *NEVER* disturb a thrust block supporting an underground utility. Many phone books list an "Underground Service Alert" or "One-Call" phone number (in the "white" pages) which you should call *BEFORE* you do any underground digging or drilling. Figures 7.25 and 7.26 give some typical telephone company warnings and indicators of underground cable lines.

QUESTIONS

Write your answers in a notebook and then compare your answers with those on page 434.

7.5S   List the general safety rules for an excavation.

7.5T   What type of shoring material should be used with trenches in running material?

7.5U   List the underground utilities that could be damaged by water utility personnel searching for a leaking underground water main.

7.6 SAFETY AROUND WATER STORAGE FACILITIES

7.60 Slips and Falls

Industrial accident reports indicate that approximately 25 percent of all accidents are caused by slips and falls. The water utility industry is no different than other industries. Operators suffer from too many accidents due to slippery, wet surfaces. Slippery surfaces are common around storage facilities. They are caused by water or chemical spills on the floor. When cleaning a storage facility, one cannot avoid working on a wet surface.

When you must work inside a water tank, wear boots that have a special tread to help prevent a person from slipping. The area around a storage facility may be wet and could be the cause of slimes and growths that create slippery surfaces. Since rubber boots with special anti-slip tread are not always available, it may be necessary to treat the wet, slippery surface. There are many anti-slip coatings commercially available. Many of these anti-slip coatings may be applied with a brush, rollers, or sprayer. These coatings usually have an abrasive (carbide crystals or ground walnut shells) substance suspended in a resin and use of them can change a slippery surface to a safe, nonslip surface.

In addition to coatings, there are other products you might consider. These are specially treated strips of tape-like materials that can be affixed to areas where foot traffic is heavy. This type of tape material is often affixed every 6 to 12 inches (15 to 30 cm) along a slippery walkway.

7.61 Ladders

7.610 Fixed Ladders (Figure 7.27)

Most storage facilities are equipped with a fixed ladder that provides the only means of access to the top of an elevated facility (tank). A fixed ladder usually consists of individual rungs that are attached to the tank. A recommended type of fixed ladder should have provisions for use of a safety belt equipped with hooks and collar. The major hazard in the use of a fixed ladder is free fall. Other hazards include falls from carrying loads, running up or down, jumping from a ladder, and reaching too far out to the side while working from the ladders. Electric shocks can cause people to fall from ladders. The major hazard may be eliminated by the installation and proper use of available safety climbing devices.

WARNING
UNDERGROUND CABLES
BEFORE DIGGING
FOR FREE CABLE LOCATING SERVICE
call collect
AREA CODE 213
621·3III
24 HOUR SERVICE
PACIFIC TELEPHONE

WARNING
UNDERGROUND CABLE

BELL SYSTEM

BEFORE DIGGING
PLEASE CALL COLLECT
AREA CODE 213
621-3111
24 HOUR SERVICE

Fig. 7.25   Typical identification and warning signs used by a phone company to identify facilities

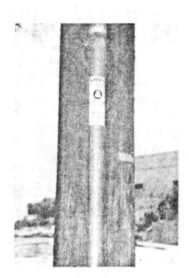

Fig. 7.26   Typical identification that can be found near areas where the telephone company has installations

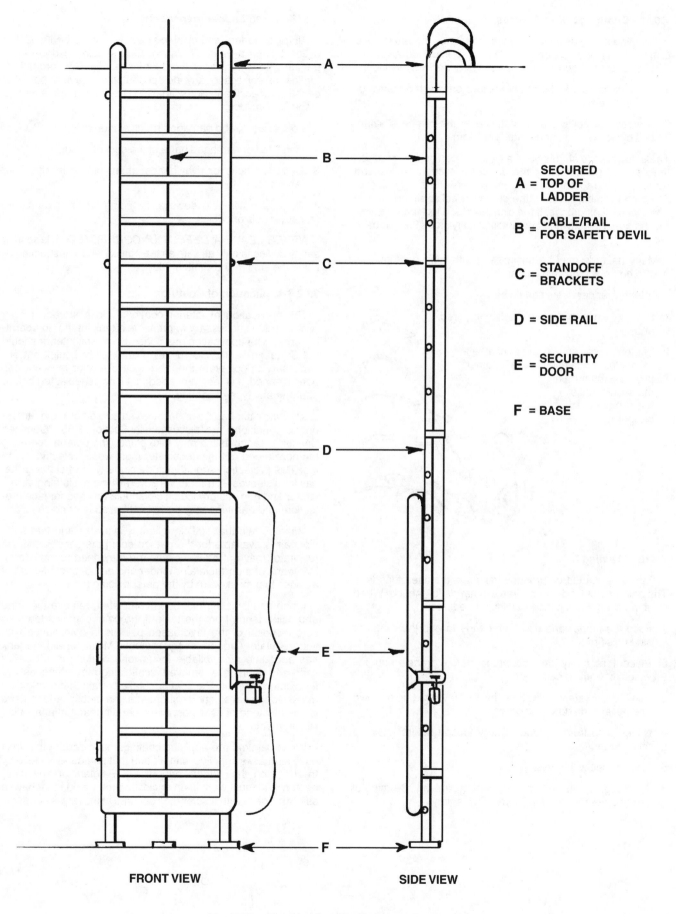

A = SECURED TOP OF LADDER

B = CABLE/RAIL FOR SAFETY DEVIL

C = STANDOFF BRACKETS

D = SIDE RAIL

E = SECURITY DOOR

F = BASE

FRONT VIEW

SIDE VIEW

Fig. 7.27   Fixed ladder with climbing safety device

7.611   *Climbing Safety Devices* (Figure 7.27)

Climbing safety devices are intended to prevent free fall of a climber if, for any reason, the climber should lose a grip or footing. There are two basic types of safety devices available:

1. A rail or cable attached to the ladder on which a sleeve or collar travels, and

2. A sleeve or collar that is fastened to the climber's safety belt by hooks and short lengths of chain.

The safety sleeve should be of a type that can be operated entirely by the person using the device to ascend or descend without having to continuously manipulate the safety sleeve. At normal climbing speed, the sleeve or collar slides up and down without hindrance. If the climber falls, a locking trigger or friction brake is activated and automatically stops the climber's fall.

Many hazards may be minimized by training personnel in how to go up and down a ladder.

Potential ladder users should be:

1. Physically capable of the exertion required, and

2. Without a previous history of heart ailment, dizzy or fainting spells, or other physical impairment that would make this climbing dangerous.

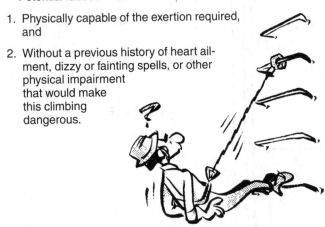

7.612   *Training*

Personnel should be properly trained in the use of ladders. The safe use of ladders requires two hands for the job. When climbing a ladder, use the following procedures:

1. Face the ladder and use both hands to grip the rungs or side rails firmly.

2. Place feet firmly on each rung before transferring full weight of body to each foot.

3. Climb deliberately, but without haste. Never run up or down and never slide down a ladder.

4. Never jump from a ladder. Check footing before stepping off a ladder.

5. Wear shoes with heels.

6. Always give your full attention when climbing. Be mindful that you are responsible for your own safety.

7.613   *Fixed Ladder Inspection*

All fixed ladders and climbing devices should be inspected at least once a year. This inspection should be done by a person who is familiar with ladder inspection. Give prompt attention to correcting any defect that may be the potential cause of an accident. Some items to check for during a routine inspection include:

1. Loose, worn, and damaged rungs or side rails,

2. Corroded or damaged parts of the security door,

3. Corroded bolts or rivet heads where ladder is affixed to tank, and

4. Defects in climbing devices, including loose or damaged carrier rails or cables.

NEVER LEAVE A DEFECT UNCORRECTED. A ladder or safety device is only as safe as the person who is inspecting or using the ladder and climbing device.

7.62   Application of Coatings

The application of interior coatings[11] to a storage facility (water tank) is necessary to preserve a tank interior in operating order. The application of a coating to a tank interior should not be taken lightly. Coordination and safety equipment are necessary to complete the job effectively. After the tank has been cleaned, it is recommended that it be sandblasted before applying any type of coating.

After the tank has been declared safe to enter (safe atmospheric conditions are discussed in Section 7.63, "Confined Spaces"), sandblasting may take place. The person operating the nozzle end of the sandblaster must wear protective clothing. This protection should include some type of clothing that has long sleeves—preferably a coverall type of clothing; a respirator and goggles or face shield should be used for face protection; and earplugs may be helpful because of noise.

Make certain that sufficient light is available and that sufficient air is available for the person doing the sandblasting. A respirator may not be sufficient. You may need to provide air by means of a portable air cylinder and an adequate length of air hose to a mask worn by the person doing the work.

When sandblasting has been completed, remove the sand and paint debris from the tank. If the paint contains lead, be sure disposal of the lead-based paint is in compliance with your regulatory agency's requirements. Make certain the tank has adequate air available for anyone working inside. After sand and debris are removed, scaffolding may be necessary for painters to apply the coating. Make certain that scaffolding meets your local safety requirements. If a mobile ladder stand is used instead of scaffolds, make certain that it meets safety requirements.

While painters are applying coating, make certain that they have air masks supplied with air and that someone is close by to check on their status periodically. Painters or any other person who must work in an enclosed area (tank) must have a safety harness with a safety line secured at the tank's entrance.

---

[11] *For information about types of coatings, surface preparation, and application procedures, see Chapter 2, Section 2.41.*

7.63 Confined Spaces
by Russ Armstrong

Water distribution system operators are sometimes required to perform work in spaces such as tanks, channels, manholes, or valve vaults. Written, understandable operating procedures must be developed and provided to all persons whose duties may involve work in confined spaces. Training in the use of procedures must also be provided. The procedures presented here are intended as guidelines. Exact procedures for work in confined spaces may vary with different agencies and geographical locations and must be confirmed with the appropriate regulatory safety agency.

A confined space is a space that requires a permit (Figure 7.28) for entry because it:

- Is large enough and so configured that an employee can bodily enter and perform assigned work; and

- Has limited or restricted means for entry or exit (for example, tanks, vessels, silos, storage bins, hoppers, vaults, and pits are spaces that may have limited means of entry); and

- Is not designed for continuous employee occupancy; and

- It has one or more of the following characteristics:

  - Contains or has the potential to contain a hazardous atmosphere;

  - Contains a material that has the potential for engulfing an entrant;

  - Has an internal configuration such that an entrant could be *trapped or asphyxiated* by inwardly converging walls or by a floor which slopes downward and tapers to a smaller cross section; or

  - Contains any other recognized serious safety or health hazard.

One easy way to identify a confined space is by whether or not you can enter it by simply walking while standing fully upright. In general, if you must duck, crawl, climb, or squeeze into the space, it is considered a confined space.

Covered water storage facilities are considered a confined space. This includes elevated tanks, facilities with roofs or covers, and underground facilities. Hazardous atmospheric conditions in confined spaces are a threat to the health of anyone who attempts to enter the facility. Other potential confined spaces encountered by water distribution system operators include manholes and valve vaults.

The federal regulations regarding confined spaces identify both permit-required confined spaces and non-permit-required confined spaces. All spaces must be considered permit-required confined spaces until a competent person following prescribed pre-entry procedures determines otherwise. The permit must be renewed each time the space is left and re-entered, even if only for a break or lunch or to go get a tool.

A non-permit confined space means a confined space that does not contain or, with respect to atmospheric hazards, have the potential to contain any hazard capable of causing death or serious physical harm.

The competent person is a person designated in writing as capable (by education and/or specialized training) of anticipating, recognizing, and evaluating employee exposure to hazardous substances or other unsafe conditions in a confined space. This person must be capable of specifying the control procedures and protective actions necessary to ensure worker safety.

A written copy of Operating and Rescue Procedures that are required by the confined space entry procedure must be at the work site during any work. The checklist includes the control of atmospheric and engulfment hazards, surveillance of the surrounding area to avoid hazards such as drifting vapors, and atmospheric testing for oxygen deficiency/enrichment and lower explosive limit.

The potential for encountering dangerous air contamination (buildup of toxic or explosive gas mixtures and/or oxygen deficiency/enrichment) exists in all confined spaces. These conditions present a serious threat of causing death, injury, acute illness, or disablement, and precautions must be taken to ensure the safety of anyone working in a confined space.

In water distribution systems, we are concerned primarily with *OXYGEN DEFICIENCY* (less than 19.5 percent oxygen by volume in air). The atmosphere must be checked with reliable, calibrated instruments (see Section 7.20, "Monitoring Equipment") prior to every entry. The oxygen concentration in normal breathing air is 20.9 percent. The atmosphere in the confined space must not fall below 19.5 percent oxygen. In atmospheres where the oxygen content is less than 19.5 percent, a self-contained breathing apparatus (SCBA) is required. SCBAs are sometimes referred to as scuba gear because they look and work much like the oxygen tanks used by divers, but they are not waterproof.

Entry into confined spaces is never permitted until the space has been properly ventilated using specially designed forced-air ventilators. These blowers force all the existing air out of the space, replacing it with fresh air from outside. This crucial step must *ALWAYS* be taken even if gas detection and oxygen deficiency/enrichment instruments show the atmosphere to be safe. Because some of the gases likely to be encountered in a confined space are combustible or explosive, the blowers must be specially designed so that the blower itself will not create a spark which could cause an explosion.

The following steps are recommended *PRIOR* to entry into a confined space.

1. Identify and close off or reroute any lines that may convey harmful substance(s) to, or through, the work area.

2. Empty, flush, or purge the space of any harmful substance(s) to the extent possible.

3. Monitor the atmosphere at the work site and within the space to determine if oxygen deficiency/enrichment or dangerous air contamination exists.

Confined Space Pre-Entry Checklist/Confined Space Entry Permit

Date and Time Issued: _____ Date and Time Expires: _____ Job Site/Space I.D.: _____

Job Supervisor: _____ Equipment to be worked on: _____ Work to be performed: _____

Standby personnel: _____   _____   _____

1. Atmospheric Checks: Time _____ Oxygen _____ %   Toxic _____ ppm

   Explosive _____ % LEL   Carbon Monoxide _____ ppm

2. Tester's signature: _____

3. Source isolation: (No Entry)   N/A   Yes   No

   Pumps or lines blinded,
   disconnected, or blocked   ( )   ( )   ( )

4. Ventilation Modification: N/A   Yes   No

   Mechanical   ( )   ( )   ( )

   Natural ventilation only   ( )   ( )   ( )

5. Atmospheric check after isolation and ventilation: Time _____

   Oxygen _____ % > 19.5% < 23.5%   Toxic _____ ppm   < 10 ppm H$_2$S

   Explosive _____ % LEL   < 10%   Carbon Monoxide _____ ppm   < 35 ppm CO

Tester's signature: _____

6. Communication procedures: _____

7. Rescue procedures: _____

   _____

8. Entry, standby, and backup persons   Yes   No

   Successfully completed required training?   ( )   ( )

   Is training current?   ( )   ( )

9. Equipment: N/A   Yes   No

   Direct reading gas monitor tested   ( )   ( )   ( )

   Safety harnesses and lifelines for entry and standby persons   ( )   ( )   ( )

   Hoisting equipment   ( )   ( )   ( )

   Powered communications   ( )   ( )   ( )

   SCBAs for entry and standby persons   ( )   ( )   ( )

   Protective clothing   ( )   ( )   ( )

   All electric equipment listed for Class I, Division I,
   Groups A, B, C, and D, and nonsparking tools.   ( )   ( )   ( )

10. Periodic atmospheric tests:

   Oxygen: ____% Time ___;   ____% Time ___;   ____% Time ___;   ____% Time ___;

   Explosive: ____% Time ___;   ____% Time ___;   ____% Time ___;   ____% Time ___;

   Toxic: ____ppm Time ___;   ____ppm Time ___;   ____ppm Time ___;   ____ppm Time ___;

   Carbon Monoxide: ____ppm Time ___;   ____ppm Time ___;   ____ppm Time ___;   ____ppm Time ___;

We have reviewed the work authorized by this permit and the information contained herein. Written instructions and safety procedures have been received and are understood. Entry cannot be approved if any brackets ( ) are marked in the "No" column. This permit is not valid unless all appropriate items are completed.

Permit Prepared By: (Supervisor) _____   Approved By: (Unit Supervisor) _____

Reviewed By: (CS Operations Personnel) _____

(Entrant)   (Attendant)   (Entry Supervisor)

This permit to be kept at job site. Return job site copy to Safety Office following job completion.

Fig. 7.28   Confined space pre-entry checklist/confined space entry permit

4. Record the atmospheric test results and keep them at the site throughout the work period.

5. If the space is interconnected with another space, each space should be tested and the most hazardous conditions found should govern subsequent steps for entry into the space.

6. If an atmospheric hazard is noted, use portable blowers to further ventilate the area; retest the atmosphere after a suitable period of time. Do not place the blowers inside the confined space.

7. Provide appropriate, approved respiratory protective equipment for the standby person and place it outside the confined space where it will be readily available for immediate use in case of emergency.

8. If oxygen deficiency/enrichment and/or dangerous air contamination do not exist prior to or following ventilation, entry into the area may proceed. *NOTE:* The space must not contain any other recognized serious safety or health hazard. *IF IT DOES, OBSERVE THE FOLLOWING PROCEDURES:*

1. If the confined space has both side and top openings, enter through the side opening whenever possible.

2. Wear appropriate, approved respiratory protective equipment.

3. Wear an approved safety belt with an attached line. The free end of the line should be secured outside the entry point.

4. Station at least one person to stand by on the outside of the confined space and at least one additional person within sight or call of the standby person.

5. Maintain frequent, regular communication between the standby person and the entry person.

6. The standby person, equipped with appropriate respiratory protection, should only enter the confined space in case of emergency.

7. If the entry is made through a top opening, use a hoisting device with a harness that suspends a person in an upright position.

8. If the space contains, or is likely to develop, flammable or explosive atmospheric conditions, do not use any tools or equipment (including electrical) that may provide a source of ignition.

9. Wear appropriate protective clothing when entering a confined space that contains corrosive substances or other substances harmful to the skin.

10. At least one person trained in first aid and cardiopulmonary resuscitation (CPR) should be immediately available during any confined space job.

Confined space work can present serious hazards if you are uninformed and/or untrained. The procedures presented are only guidelines and exact requirements for confined space work for your locale may vary. *CONTACT YOUR LOCAL REGULATORY SAFETY AGENCY FOR SPECIFIC REQUIREMENTS.*

Safety is a must—never cut corners no matter how cumbersome the safety equipment may be—its use may save your life. Always review the *SAFETY NOTICE* before entering a confined space.

SAFETY NOTICE

Before anyone ever enters a tank for any reason, these safety procedures must be followed:

1. Test atmosphere in the tank for toxic and explosive gases and for oxygen deficiency/enrichment. Contact your local safety equipment supplier for the proper types of atmospheric testing devices. These devices should have alarms that are activated whenever an unsafe atmosphere is encountered;

2. Provide adequate ventilation, especially when painting. A self-contained, positive-pressure breathing apparatus may be necessary when painting;

3. All persons entering a tank must wear a safety harness; and

4. One person must be at the tank entrance to observe the actions of all people in the tank. An additional person must be readily available to help the person at the tank entrance with any rescue operation.

QUESTIONS

Write your answers in a notebook and then compare your answers with those on page 434.

7.6A How do anti-slip coatings prevent slippery surfaces?

7.6B What are the two basic types of safety devices used by operators when climbing fixed ladders?

7.6C What are the physical requirements for operators who must climb fixed ladders?

7.6D A safety inspection of a fixed ladder should include what items?

7.6E Why do storage tanks need interior coatings?

7.6F What atmospheric hazards may be encountered in a confined space?

7.7 WORKING NEAR NOISE

Operators must be protected against the effects of exposure to noise when the sound levels exceed the limits listed in Table 7.8. Engineering methods of controlling noise must be initiated

TABLE 7.8 PERMISSIBLE NOISE EXPOSURES [a]

1. When employees are exposed to loud or extended noise, earplugs and/or protective devices shall be provided and employees shall wear them accordingly.

2. Protection against the effects of noise exposure shall be provided when sound levels exceed those shown in rule number 7. These measurements shall be made on the A-scale of a standard sound level meter at a slow response.

3. When employees are subjected to sound levels exceeding those listed in rule number 7, feasible administrative or engineering controls shall be utilized. If such controls fail to reduce employee exposure to within the permissible noise levels listed in this table, personal protective equipment shall be provided and used.

4. Plain cotton shall not be considered acceptable as an earplug.

5. Machinery creating excessive noise shall be equipped with mufflers.

6. Exposure to impulsive or impact noise should not exceed 140 dB peak sound pressure level.

7. Employees shall not be exposed to noise which exceeds the levels listed below for a time period not to exceed those listed below.

DURATION (Hours Per Day)	SOUND LEVEL (dB [b] Slow Response)
8	90
6	92
4	95
3	97
2	100
1½	102
1	105
½ (thirty minutes)	110
one-fourth (¼) hour, or less	115

[a] Occupational Safety and Health Standards, Subpart G, Paragraph 1910.94, Noise Exposure

[b] Decibel (DES-uh-bull). A unit for expressing the relative intensity of sounds on a scale from zero for the average least perceptible sound to about 130 for the average level at which sound causes pain to humans. Abbreviated dB.

whenever possible. If noise levels cannot be controlled within acceptable limits, operators must be provided and use approved earplugs, muffs, and/or personal protective equipment.

7.8 SAFETY INSPECTIONS

Regular safety inspections are a good way to be sure that equipment and facilities receive a regular safety review. To ensure that important items are not overlooked or forgotten, inspection forms or reports are very helpful. These forms also provide a record of who inspected what items and when. Figures 7.29 and 7.30 are typical "Safety Inspection Reports."

QUESTIONS

Write your answers in a notebook and then compare your answers with those on page 434.

7.7A How can operators protect themselves from excessive noise?

7.8A Why are safety inspection forms helpful?

7.9 ARITHMETIC ASSIGNMENT

There is no arithmetic assignment for this chapter.

7.10 ADDITIONAL READING

1. *NEW YORK MANUAL*, Chapter 19,* "Treatment Plant Maintenance and Accident Prevention."

2. *TEXAS MANUAL*, Chapter 21,* "Safety."

3. Chapter 20, "Safety," in *WATER TREATMENT PLANT OPERATION*, Volume II. Obtain from the Office of Water Programs, California State University, Sacramento, 6000 J Street, Sacramento, CA 95819-6025. Price, $45.00.

4. *SAFETY BASICS FOR WATER UTILITIES*. DVD, 27 minutes. Obtain from American Water Works Association (AWWA), Bookstore, 6666 West Quincy Avenue, Denver, CO 80235. Order No. 64119. Price to members, $208.00; nonmembers, $313.00; price includes cost of shipping and handling.

5. *LET'S TALK SAFETY—2005 SAFETY TALKS*. A series of 52 lectures on common water utility safety practices. Obtain from American Water Works Association (AWWA), Bookstore, 6666 West Quincy Avenue, Denver, CO 80235. Order No. 10123. ISBN 1-58321-354-6. Price to members, $42.00; nonmembers, $61.00; price includes cost of shipping and handling.

6. The American Red Cross is another source of up-to-date information on safety and first aid. Contact your local Area Chapter for a catalog of materials.

7. Read the *SAFETY* chapters in the other operator manuals in this series.

* Depends on edition.

End of Lesson 2 of 2
Lessons on SAFETY

Please answer the discussion and review questions next.

SAFETY INSPECTION REPORT
SANITARY ENGINEERING DIVISION

Location _____

Inspec. By – Date						
Housekeeping						
Material Handling						
Material Storage						
Aisles and Walkways						
Ladders and Stairs						
Floors and Railings						
Exits						
Lighting						
Ventilation						
Hand Tools						
Electric Equipment						
Machinery						
Safety Guards						
Safety Devices						
Clothing and Equipment						
Dusts and Fumes						
Fire Hazards						
Explosion Hazards						
Chemical Hazards						
Fire Equipment						
Unsafe Practices						
Other						

Reviewed By – Date						

Remarks:

Fig. 7.29   *Safety inspection report used by operators to inspect facilities*

Location Inspected _Palos Verdes Chlorination Station_   Date _April 28, 2004_
Division _Sanitary Engineering_   Section _Water Treatment_
Supervisor _Danny Saenz_   Title _____

Reason For Call				X		Routing	X			
	Routine	Request	Special	Annual	Calif. D.I.S.		Supervisor	Section	Division	System

Inspected By _Clarence Butrum_   Discussed With _Danny Saenz_

CHECKLIST

Numeric Rating Value: 0 Poor or Deficient, 2 Fair or Average, 3 Good, 4 Excellent

Items marked X were not inspected on this visit

1. **4** Housekeeping
2. **4** Material Handling
3. **4** Material Storage
4. **4** Aisles and Walkways
5. **2** Ladders and Stairs
6. **4** Floors, Platforms, and Railings
7. **4** Exits
8. **4** Lighting
9. **4** Ventilation
10. **−** Hand Tools
11. **4** Electric Tools and Equipment
12. **−** Machinery and Equipment
13. **−** Guards and Safety Devices

14. **−** Welding Equipment
15. **4** Protective Clothing and Equipment
16. **−** Personal Tools and Equipment
17. **4** Dusts, Fumes, Mists, Gases, and Vapors
18. **−** Fire Hazards
19. **−** Explosion Hazards
20. **4** Chemical Hazards
21. **−** Hand and Power Trucks
22. **4** Firefighting Equipment
23. **−** Vehicles
24. **−** Unsafe Practices
25. **−** Horseplay
26. **−** Other _____

TO SUPERVISOR: Items found to be deficient are described in detail below.

#5 Ladder needs 2004 inspection.

MAY 0 4 2004

J K Jacobs

MAY 0 4 2004

D. Saenz

Signed _Clarence Butrum - Safety_

Fig. 7.30   Safety inspection report used by safety section personnel

DISCUSSION AND REVIEW QUESTIONS

Chapter 7.   SAFETY

(Lesson 2 of 2 Lessons)

Write the answers to these questions in your notebook. The question numbering continues from Lesson 1.

14. Where do rotating, reciprocating, and transverse (crosswise) motions on pumps create hazards?

15. How can traffic be warned of your presence in the street or highway?

16. What factors should be considered when deciding where to station flaggers?

17. What are the definitions of a low-speed and high-speed street or highway?

18. What is the purpose of the advance warning area?

19. Why should traffic control devices be constructed to yield upon impact?

20. List six types of traffic control devices.

21. How can the public be protected from excavation hazards?

22. What types of clothing should operators wear to increase their visibility to motorists?

23. How can cave-ins be prevented?

24. How can you reduce the chances of operators slipping and falling on slippery surfaces?

25. In water distribution systems, what type of atmospheric hazard is of greatest concern with regard to confined spaces?

SUGGESTED ANSWERS

Chapter 7.   SAFETY

ANSWERS TO QUESTIONS IN LESSON 1

Answers to questions on page 366.

7.0A   Two very important aspects of a safety program are:

1. Making people aware of unsafe acts, and
2. Conducting and/or participating in regular safety training programs.

7.0B   Management's responsibilities for safety include:

1. Establishing a safety policy,
2. Assigning responsibility for accident prevention,
3. Appointing a safety officer,
4. Establishing realistic goals and periodically revising them, and
5. Evaluating the results of the program.

7.0C   Operators' responsibilities for safety include:

1. Performing their jobs in accordance with established safe procedures,
2. Recognizing the responsibility for their own safety and that of fellow operators,
3. Reporting all injuries,
4. Reporting all observed hazards, and
5. Actively participating in the safety program.

Answers to questions on page 367.

7.1A   An accident is that occurrence in a sequence of events that usually produces unintended injury, death, or property damage.

7.1B   The principal reasons for unsafe acts include ignorance, indifference, poor work habits, laziness, haste, poor physical condition, and temper.

7.1C   Laziness affects the speed and quality of work. Laziness also affects safety because safety requires an effort. In most jobs you cannot reduce your "safety effort" and still maintain the same level of safety, even if you slow down or lower your quality standards.

Answers to questions on page 369.

7.1D   Water distribution system operators must follow established procedures because of the serious consequences that could result due to the nature of the job and the increased risk of injury and/or fatality to the operator and co-workers.

7.1E   Before leaving the maintenance yard or shop, discuss work assignments, determine equipment needed, inspect equipment that will be used, and inspect vehicles and towed equipment.

7.1F   Before taking an atmospheric testing device out to a job site it should be calibrated and working properly.

7.1G   Before towing a trailer, inspect:

1. Mirrors and windows,
2. Backup lights, turn signals, and brake lights,
3. Trailer brake operation if equipped with brakes,
4. Tire tread and inflation,
5. Wheel attachment, and
6. Trailer hitch and safety chain.

7.1H   Defensive driving is not only a mature attitude toward driving, but a strategy as well. You must always be on the alert for the large number of new drivers, inattentive or sleepy drivers, people driving under the influence of alcohol and/or drugs, and those drivers who are just plain incompetent. In addition, weather, construction, and the type of equipment you are driving can also present hazards.

Answers to questions on page 371.

7.1I   All cigarettes and flames should be put out before attaching cables to charge a battery because a spark can ignite the hydrogen gas from the battery fluid.

7.1J   Good safety practices for using boats include:

1. An annual safety inspection,
2. A safety vest or flotation device for each occupant,
3. Oarlocks and oars of suitable length,
4. Not overloading a boat, and
5. At least two people should go out in a boat.

7.1K   A corrosive chemical is any chemical that may weaken, burn, and/or destroy a person's skin or eyes.

7.1L   A tailgate safety session is usually held near the work site and new and old safety hazards and safer approaches and techniques to deal with the problems of the day or week are discussed.

Answers to questions on page 375.

7.2A   Some of the typical locations in a water utility where it is necessary to monitor the atmosphere before entering include underground regulator vaults, solution vaults, manholes, tanks, trenches, and other confined spaces.

7.2B   The backup person outside the confined area should check continuously on the status of personnel working in the confined space. Should a person working in the confined space be rendered unconscious due to asphyxiation or lack of oxygen, the backup person at the surface could enter into the space by:

1. Securing the help of an additional backup rescue person, and
2. Putting on the correct type of respiratory protective equipment.

7.2C   A mixture of gas in air is considered hazardous when the mixture exceeds 10 percent of the Lower Explosive Limit (LEL).

Answers to questions on page 378.

7.2D   Items of protective clothing for operators include:

1. Safety toe (hard toe) shoes,
2. Safety prescription glasses,
3. Safety (hard) hats,
4. Earplugs,
5. Safety glasses (nonprescription),
6. Safety goggles,
7. Face shields,
8. Respirators and mask,
9. Canvas gloves,
10. Rubber gloves for use with corrosive chemicals,
11. Safety aprons,
12. Leggings or knee pads,
13. Safety toe caps,
14. Protective clothing (lab coats or rubber suits), and
15. Safety belts or harnesses.

7.2E   Liquids spilled on floors can be cleaned up by using nonflammable absorbent materials.

7.2F   The steps to safe lifting include:

1. Keep feet parted—one along the side of and one behind the object,
2. Keep back straight—nearly vertical,
3. Tuck in your chin,
4. Grip the object with the whole hand,
5. Tuck elbows and arms in,
6. Keep body weight directly over feet, and
7. Lift slowly.

7.2G   Before attempting to work on electric motors, be sure to lock out and tag all of the electric switches.

Answers to questions on page 379.

7.3A   The well operator has the responsibility to preserve the quality of the well through preventive maintenance of the area where the well is located. This responsibility also extends to good housekeeping to prevent the development of safety hazards to operators working around the well site.

7.3B   A sanitary seal on a well is used to keep contaminated water and debris from entering around the well casing and reaching the groundwater.

7.3C   Surface openings on top of a well that must be protected against the entrance of surface water or debris into the well include:

1. Well casing vents,
2. Gravel tubes,
3. Sounding tube,
4. Conduits, and
5. Pump motor base seal.

Answers to questions on page 380.

7.3D   Hydrochloric acid must be handled carefully because it is extremely corrosive. Hydrochloric acid can burn your eyes and skin, if inhaled it will cause coughing and choking, and if swallowed will cause corrosion of your insides (including your circulatory system).

7.3E   Sulfamic acid has advantages over hydrochloric acid for acid cleaning a well. Granular sulfamic acid is non-irritating to dry skin and its solution gives off no fumes except when reacting with encrusting materials. Spillage, therefore, presents no hazard and handling is easier, cheaper, and safer.

7.3F   Although the granular form of sulfamic acid is nonirritating to dry skin and it gives off no fumes, rubber gloves should still be used because hands may become wet from perspiration and the acid will cause burns on the skin. Safety goggles or glasses, preferably a face shield, should be used in case granules are blown into the face and/or eyes. Acid should always be added slowly to the water.

Answers to questions on page 383.

7.3G   When handling hypochlorite, use rubber gloves, a rubber apron, rubber boots, and a face shield.

7.3H   Glassy phosphate should be handled very carefully because the glass-like particles are very sharp and can cut your skin. Use heavy-duty gloves when handling. Respiration equipment or a face shield should be worn.

7.3I If no one working for a water utility is knowledgeable about electric circuits, the utility should contact an electrician from a local electrical firm.

7.3J Items that should be included in a safety inspection of a well site include:

1. Cleanliness of site,
2. Locks on electrical panels,
3. Fencing around site, and
4. Caution/warning signs in place.

ANSWERS TO QUESTIONS IN LESSON 2

Answers to questions on page 384.

7.4A Pumps are used to move liquids, air, or gas. They are also used to increase pressures and energy of fluids.

7.4B Moving parts on a pump that require guards include rotation members, reciprocating arms, moving parts, and meshing gears. This includes rotating shafts, couplings, belts, pulleys, chains, and sprockets.

7.4C Rotating devices can injure you by gripping your clothing or hair and possibly forcing an arm or hand into a dangerous position.

7.4D The basic requirement for an enclosed guard is that it must prevent hands, arms, or any other part of an operator's body from making contact with dangerous moving parts. A good guarding system eliminates the possibility of the operator or another worker placing their hands near hazardous moving parts.

Answers to questions on page 389.

7.4E Safety planning includes identifying potential safety hazards *BEFORE* starting a job and deciding how they can be avoided. Safety coordination requires that supervisory personnel be notified and necessary approvals obtained *BEFORE* starting any maintenance work.

7.4F *ALWAYS* isolate the power *BEFORE* beginning any maintenance on a pump.

7.4G Never work on any equipment when it is hanging from a sling in midair because the equipment could slip from the sling and fall on you.

7.4H Lubricants and fuels should be stored in a separate facility from other storage areas because of the potential fire hazard.

Answers to questions on page 392.

7.5A The purpose of traffic control around work areas in streets and highways is to provide safe and effective work areas and to warn, control, protect, and expedite vehicular and pedestrian traffic.

7.5B Any time traffic may be affected, appropriate authorities must be notified, which could be state, county, or local. Frequently a permit must be issued by the authority that has jurisdiction before traffic can be diverted or disrupted.

7.5C Traffic can be warned of your presence in a street by BE PREPARED TO STOP, SHOULDER WORK, and UTILITY WORK AHEAD signs. Signs with flashing warning lights and vehicles with yellow rotating beacons are used to warn other motorists. Vehicle-mounted traffic guides are also helpful. Use trained flaggers to alert drivers and to direct traffic around the work site.

7.5D Once motorists have been warned, they must be safely routed around the job site. Traffic routing includes the use of flaggers and directional signs. Properly placed channelizing devices, such as barricades, drums, or cones, can effectively channel traffic from one lane to another around the job site. Retain a traffic officer to direct traffic if streets are narrow or if there is considerable traffic or congestion.

7.5E The traffic control zone is a work zone in an area of a highway with construction, maintenance, or utility work activities. A work zone is typically marked by signs, channelizing devices, barriers, pavement markings, or work vehicles. It extends from the first warning sign or high-intensity rotating, flashing, oscillating, or strobe lights on a vehicle to the END ROAD WORK sign or the last traffic control device.

7.5F Decision sight distance is the distance required by a driver to properly react to hazardous, unusual, or unexpected situations even where an evasive maneuver is more desirable than a hurried stop.

Answers to questions on page 393.

7.5G An operator can become qualified to control traffic by gaining the following knowledge and experience:

1. A basic understanding of the principles of traffic control in work zones,
2. Knowledge of the standards and guidelines governing traffic control,
3. Adequate training in safe traffic control practices, and
4. Experience in applying traffic control in work zones.

7.5H Records of traffic control activities should contain the starting and ending times, location, names of personnel involved, and the traffic controls used.

7.5I When work is planned for nighttime hours, consider adding warning lights to traffic control devices and signs, using retroreflective signs, using larger, more reflective channelizing devices such as drums, using floodlights for any flagger stations and for the work area, and adding reflective devices to work vehicles that are typically used for nighttime work.

Answers to questions on page 400.

7.5J The five different areas of a traffic control zone are: (1) advance warning, (2) transition, (3) buffer, (4) work, and (5) termination areas.

7.5K The decision sight distance should give a motorist time to recognize a hazard, plus time to react to the hazard, and time for the vehicle to respond to the driver's reaction.

7.5L A traffic control taper is created by using a series of channelizing devices or pavement markings to move traffic out of or into the normal path.

7.5M A buffer space should be used to separate the two tapers for opposing directions of traffic to prevent head-on collisions on multilane streets when one lane of traffic must use a lane that normally flows in the opposite direction.

Answers to questions on page 418.

7.5N   Using traffic control devices in a consistent, predictable manner is one of the best ways to increase the safety of both drivers and operators at construction sites. Drivers learn by experience what to expect when they see traffic control devices. If the devices themselves and their placement in the roadway are consistent over time, drivers can more quickly understand and respond to the changed road conditions.

7.5O   Types of channelizing devices include barricades, traffic cones and tubular markers, drums, and vertical panels.

7.5P   A Portable Changeable Message Sign (PCMS) is a lighted, electronic traffic control device with the flexibility to display a variety of messages. PCMSs are used to give motorists advance warning of road closures, accident management instructions, narrow lanes, construction schedules, traffic management and diversion activities, and adverse conditions.

7.5Q   To alert and slow daytime traffic using a SLOW paddle, the flagger should face the road users with the paddle face aimed toward the road users in a stationary position with the arm extended horizontally away from the body. To further alert or slow traffic, the flagger may motion up and down with the free hand, palm down. To alert and slow traffic using a flag, the flagger should face the road users and slowly wave the flag in a sweeping motion of the extended arm from shoulder level to straight down without raising the arm above a horizontal position and keeping the free hand down.

7.5R   When work is to be done in a residential neighborhood, it is good public relations to advise the public what is to be done, how long it is expected to take, that all safety regulations will be observed, and that every effort will be made not to disrupt the routine of the neighborhood.

Answers to questions on page 420.

7.5S   General safety rules for an excavation include:

1. Employer must inspect excavations for hazards before employees may work in or adjacent to the excavation.
2. The location of underground installations (such as utility lines) must be determined before excavation begins.
3. Excavations greater than four feet deep (1.2 m) must be tested as often as necessary for oxygen deficiency/enrichment and hazardous atmospheres, and appropriate precautions must be taken to protect workers.
4. Excavations must be inspected daily by a competent person and if a problem exists, workers must be removed until precautions have been taken to ensure their safety.
5. Properly qualified supervisors shall be on site at all times during excavations.
6. Safety provisions must be taken to protect workers installing or removing shoring systems.
7. Spoil must be kept back at least two feet (0.6 m) from the edge of all excavations.
8. Safe and convenient access to an excavation must be provided.
9. Effective barriers must be installed at excavations adjacent to mobile equipment.
10. Water must not be allowed to accumulate in any excavation.

7.5T   Trenches in running material should have sheet piling or equivalent solid sheeting, longitudinal stringers, and cross braces.

7.5U   Underground utilities that could be damaged by water utility personnel searching for a leaking underground water main include telephone cables, television cables, electrical services, gas mains, and occasionally storm water and sewer lines.

Answers to questions on page 427.

7.6A   Anti-slip coatings usually have an abrasive (carbide crystals or ground walnut shells) substance suspended in a resin which can change a slippery surface to a safe, nonslip surface.

7.6B   The two basic types of safety devices used by operators when climbing fixed ladders are:

1. A rail or cable attached to the ladder on which a sleeve or collar travels, and
2. A sleeve or collar that is fastened to the climber's safety belt by hooks and short lengths of chain.

7.6C   Physical requirements for operators who must climb fixed ladders include:

1. Physically capable of the exertion required, and
2. Without a previous history of heart ailment, dizzy or fainting spells, or other physical impairment that would make this climbing dangerous.

7.6D   A safety inspection of a fixed ladder should include checking for:

1. Loose, worn, and damaged rungs or side rails,
2. Corroded or damaged parts of the security door,
3. Corroded bolts or rivet heads where ladder is affixed to tank, and
4. Defects in climbing devices, including loose or damaged carrier rails or cables.

7.6E   Storage tanks need interior coatings to preserve a tank interior in operating order.

7.6F   Atmospheric hazards that may be encountered in a confined space include explosive conditions, toxic gases, and oxygen deficiency/enrichment.

Answers to questions on page 428.

7.7A   Operators can protect themselves from excessive noise by using approved earplugs, muffs, and/or personal protective equipment.

7.8A   Safety inspection forms are helpful because (1) they help to ensure that important safety items are not overlooked or forgotten during regular safety inspections, and (2) they provide a record of who inspected what items and when.

CHAPTER 8

DISTRIBUTION SYSTEM ADMINISTRATION

by

Lorene Lindsay

With Portions by

Tim Gannon

and

Jim Sequeira

TABLE OF CONTENTS

Chapter 8.   DISTRIBUTION SYSTEM ADMINISTRATION

OBJECTIVES

Chapter 8.   DISTRIBUTION SYSTEM ADMINISTRATION

Following completion of Chapter 8, you should be able to:

1. Identify the functions of a manager,

2. Describe the benefits of short-term, long-term, and emergency planning,

3. Define the following terms:

   a. Authority,   d. Accountability, and
   b. Responsibility,   e. Unity of command,
   c. Delegation,

4. Read and construct an organizational chart identifying lines of authority and responsibility,

5. Write a job description for a specific position within the utility,

6. Write good interview questions,

7. Conduct employee evaluations,

8. Describe the steps necessary to provide equal and fair treatment to all employees,

9. Prepare a written or oral report on the distribution system's operations,

10. Communicate effectively within the organization, with media representatives, and with the community,

11. Describe the financial strength of your distribution system,

12. Calculate your utility's operating ratio, coverage ratio, and simple payback,

13. Prepare a contingency plan for emergencies,

14. Prepare a plan to strengthen the security of your utility's facilities,

15. Set up a safety program for your utility, and

16. Collect, organize, file, retrieve, use, and dispose of distribution system records.

WORDS

Chapter 8. DISTRIBUTION SYSTEM ADMINISTRATION

ACCOUNTABILITY ACCOUNTABILITY

When a manager gives power/responsibility to an employee, the employee ensures that the manager is informed of results or events.

AUTHORITY AUTHORITY

The power and resources to do a specific job or to get that job done.

BACK PRESSURE BACK PRESSURE

A pressure that can cause water to backflow into the water supply when a user's water system is at a higher pressure than the public water system.

BACKFLOW BACKFLOW

A reverse flow condition, created by a difference in water pressures, which causes water to flow back into the distribution pipes of a potable water supply from any source or sources other than an intended source. Also see BACKSIPHONAGE.

BACKSIPHONAGE BACKSIPHONAGE

A form of backflow caused by a negative or below atmospheric pressure within a water system. Also see BACKFLOW.

BOND BOND

(1) A written promise to pay a specified sum of money (called the face value) at a fixed time in the future (called the date of maturity). A bond also carries interest at a fixed rate, payable periodically. The difference between a note and a bond is that a bond usually runs for a longer period of time and requires greater formality. Utility agencies use bonds as a means of obtaining large amounts of money for capital improvements.

(2) A warranty by an underwriting organization, such as an insurance company, guaranteeing honesty, performance, or payment by a contractor.

CALL DATE CALL DATE

First date a bond can be paid off.

CERTIFICATION EXAMINATION CERTIFICATION EXAMINATION

An examination administered by a state agency that water distribution system operators take to indicate a level of professional competence. In the United States, certification of water distribution system operators is mandatory.

CODE OF FEDERAL REGULATIONS (CFR) CODE OF FEDERAL REGULATIONS (CFR)

A publication of the United States Government which contains all of the proposed and finalized federal regulations, including environmental regulations.

CONFINED SPACE CONFINED SPACE

Confined space means a space that:

A. Is large enough and so configured that an employee can bodily enter and perform assigned work; and

B. Has limited or restricted means for entry or exit (for example, manholes, tanks, vessels, silos, storage bins, hoppers, vaults, and pits are spaces that may have limited means of entry); and

C. Is not designed for continuous employee occupancy.

(Definition from the Code of Federal Regulations (CFR) Title 29 Part 1910.146.)

CONFINED SPACE, PERMIT REQUIRED (PERMIT SPACE) CONFINED SPACE, PERMIT REQUIRED (PERMIT SPACE)

A confined space that has one or more of the following characteristics:

- Contains or has a potential to contain a hazardous atmosphere,

- Contains a material that has the potential for engulfing an entrant,

- Has an internal configuration such that an entrant could be trapped or asphyxiated by inwardly converging walls or by a floor which slopes downward and tapers to a smaller cross section, or

- Contains any other recognized serious safety or health hazard.

(Definition from the Code of Federal Regulations (CFR) Title 29 Part 1910.146.)

COVERAGE RATIO

COVERAGE RATIO

The coverage ratio is a measure of the ability of the utility to pay the principal and interest on loans and bonds (this is known as "debt service") in addition to any unexpected expenses.

DEBT SERVICE

DEBT SERVICE

The amount of money required annually to pay the (1) interest on outstanding debts; or (2) funds due on a maturing bonded debt or the redemption of bonds.

DELEGATION

DELEGATION

The act in which power is given to another person in the organization to accomplish a specific job.

MATERIAL SAFETY DATA SHEET (MSDS)

MATERIAL SAFETY DATA SHEET (MSDS)

A document which provides pertinent information and a profile of a particular hazardous substance or mixture. An MSDS is normally developed by the manufacturer or formulator of the hazardous substance or mixture. The MSDS is required to be made available to employees and operators whenever there is the likelihood of the hazardous substance or mixture being introduced into the workplace. Some manufacturers are preparing MSDSs for products that are not considered to be hazardous to show that the product or substance is *NOT* hazardous.

OSHA (O-shuh)

OSHA

The Williams-Steiger **O**ccupational **S**afety and **H**ealth **A**ct of 1970 (OSHA) is a federal law designed to protect the health and safety of industrial workers and also the operators of water supply systems and treatment plants. The Act regulates the design, construction, operation and maintenance of water supply systems and water treatment plants. OSHA also refers to the federal and state agencies which administer the OSHA regulations.

OPERATING RATIO

OPERATING RATIO

The operating ratio is a measure of the total revenues divided by the total operating expenses.

ORGANIZING

ORGANIZING

Deciding who does what work and delegating authority to the appropriate persons. A utility should have a written organizational plan and written policies.

OUCH PRINCIPLE

OUCH PRINCIPLE

This principle says that as a manager when you delegate job tasks you must be **O**bjective, **U**niform in your treatment of employees, **C**onsistent with utility policies, and **H**ave job relatedness.

PLANNING

PLANNING

Management of utilities to build the resources and financial capability to provide for future needs.

PRESENT WORTH

PRESENT WORTH

The value of a long-term project expressed in today's dollars. Present worth is calculated by converting (discounting) all future benefits and costs over the life of the project to a single economic value at the start of the project. Calculating the present worth of alternative projects makes it possible to compare them and select the one with the largest positive (beneficial) present worth or minimum present cost.

RESPONSIBILITY

RESPONSIBILITY

Answering to those above in the chain of command to explain how and why you have used your authority.

SCADA (ss-KAY-dah) SYSTEM

SCADA SYSTEM

Supervisory **C**ontrol **A**nd **D**ata **A**cquisition system. A computer-monitored alarm, response, control and data acquisition system used by drinking water facilities to monitor their operations.

TAILGATE SAFETY MEETING

TAILGATE SAFETY MEETING

Brief (10 to 20 minutes) safety meetings held every 7 to 10 working days. The term *TAILGATE* comes from the safety meetings regularly held by the construction industry around the tailgate of a truck.

CHAPTER 8.   DISTRIBUTION SYSTEM ADMINISTRATION

(Lesson 1 of 3 Lessons)

8.0   NEED FOR DISTRIBUTION SYSTEM MANAGEMENT

The management of a public or private distribution system, large or small, is a complex and challenging job. Communities are concerned about their drinking water and their wastewater. They are aware of past environmental disasters and they want to protect their communities, but they want this protection with a minimum investment of money. In addition to the local community demands, the distribution system manager must also keep up with increasingly stringent regulations and monitoring from regulatory agencies. While meeting these "external" (outside the water distribution system) concerns, the manager faces the normal challenges from within the organization: personnel, resources, equipment, and preparing for the future. For the successful manager, all of these responsibilities combine to create an exciting and rewarding job.

A brief quiz is given in Table 8.1 that asks some basic management questions. This quiz can be used as a guide to management areas that may need some attention in your water distribution system. You should be able to answer yes to most of the questions; however, all water distribution systems have areas which can be improved.

In the environmental field, as well as other fields, the workforce itself is changing. Minorities, women, and people with disabilities provide new opportunities for growth in the water distribution system. For the employee, however, overcoming employment barriers can be difficult, especially when the workload is demanding and physically challenging. The water distribution system manager must provide adequate support services for these operators and learn to deal with organized operator groups.

Changes in the environmental workplace also are created by advances in technology. The environmental field has exploded with new technologies, such as computer-controlled water treatment processes and distribution systems. The water distribution system manager must keep up with these changes and provide the leadership to keep everyone at the water distribution system up to speed on new ways of doing things. In addition, the water distribution system manager must provide a safer, cleaner work environment while constantly training and retraining operators to understand new technologies.

QUESTIONS

Write your answers in a notebook and then compare your answers with those on page 511.

8.0A   What are the local community demands on a water distribution system manager?

8.0B   What has created changes in the environmental workplace?

8.1   FUNCTIONS OF A MANAGER

The functions of a water distribution system manager are the same as for the CEO (Chief Executive Officer) of any big company: planning, organizing, staffing, directing, and controlling. In small communities the water distribution system manager may be the only one who has these responsibilities and the community depends on the manager to handle everything.

Planning (see Section 8.2) consists of determining the goals, policies, procedures, and other elements to achieve the goals and objectives of the agency. Planning requires the manager to collect and analyze data, consider alternatives, and then make decisions. Planning must be done before the other managing functions. Planning may be the most difficult in smaller communities, where the future may involve a decline in population instead of growth.

Organizing (see Section 8.3) means that the manager decides who does what work and delegates authority to the appropriate operators. The organizational function in some distribution system agencies may be fairly loose while some communities are very tightly controlled.

Staffing (see Section 8.4) is the recruiting of new operators and staff and determining if there are enough qualified operators and staff to fill available positions. The distribution system manager's staffing responsibilities include selecting and training employees, evaluating their performance, and providing opportunities for advancement for operators and staff in the agency.

Directing includes guiding, teaching, motivating, and supervising operators and water distribution system staff members. Direction includes issuing orders and instructions so that activities at the facilities or in the field are performed safely and are properly completed.

TABLE 8.1   HOW WELL DOES YOUR SYSTEM MANAGE?

The following self-test is designed for small water treatment facilities to provide a guide for identifying areas of concern and for improving water distribution system management.

1. Is the treatment system budget separate from other accounts so that the true cost of treatment can be determined?

2. Are the funds adequate to cover operating costs, debt service, and future capital improvements?

3. Do operational personnel have input into the budget process?

4. Is there a monthly or quarterly review of the actual operating costs compared to the budgeted costs?

5. Does the user charge system adequately reflect the cost of treatment?

6. Are all users properly metered and does the unaccounted for water not exceed 20 percent of the total flow?

7. Are finished water quality tests representative of plant performance?

8. Are operational control decisions based on process control testing within the plant?

9. Are provisions made for continued training for plant personnel?

10. Are qualified personnel available to fill job vacancies and is job turnover relatively low?

11. Are the energy costs for the system not more than 20 to 30 percent of the total operating costs?

12. Is the ratio of corrective (reactive) maintenance to preventive (proactive) maintenance remaining stable and is it less than 1.0?

13. Are maintenance records available for review?

14. Is the spare parts inventory adequate to prevent long delays in equipment repairs?

15. Are old or outdated pieces of equipment replaced as necessary to prevent excessive equipment downtime, inefficient process performance, or unreliability?

16. Are technical resources and tools available for repairing, maintaining, and installing equipment?

17. Is the utility's pump station equipment providing the expected design performance?

18. Are standby units for key equipment available to maintain process performance during breakdowns or during preventive maintenance activities?

19. Are the plant processes adequate to meet the demand for treatment?

20. Does the facility have an adequate emergency response plan including an alternate water source?

Controlling involves taking the steps necessary to ensure that essential activities are performed so that objectives will be achieved as planned. Controlling means being sure that progress is being made toward objectives and taking corrective action as necessary. The water distribution system manager is directly involved in controlling the distribution system to ensure that safe water is being delivered to consumers' taps and to make sure that the utility is meeting its short- and long-term goals.

QUESTIONS

Write your answers in a notebook and then compare your answers with those on page 511.

8.1A   What are the functions of a distribution system manager?

8.1B   In small communities, what does the community depend on the utility manager to do?

8.2 PLANNING[1]

A very large portion of any manager's typical work day will be spent on activities that can be described as planning activities since nearly every area of a manager's responsibilities require some type of planning.

Planning is one of the most important functions of utility management and one of the most difficult. Communities must have good, safe drinking water and the management of water utilities must include building the resources and financial capability to provide for future needs. The utility must plan for future growth, including industrial development, and be ready to provide the water that will be needed as the community grows. The most difficult problem for some small communities is recognizing and planning for a decline in population. The utility manager must develop reliable information to plan for growth or decline. Decisions must be made about goals, both short- and long-term. The manager must prepare plans for the next two years and the next 10 to 20 years. Remember that water distribution system planning should include operational personnel, local officials (decision makers), and the public. Everyone must understand the importance of planning and be willing to contribute to the process.

Operation and maintenance of a water distribution system also involves planning by the distribution system manager. A preventive maintenance program should be established to keep the system performing as intended and to protect the community's investment in water supply and distribution facilities. (Section 8.9 describes the various types of maintenance and the benefits of establishing maintenance programs.)

The water distribution system also must have an Emergency Response Plan to deal with natural or human disasters. Without adequate planning your water distribution system will be facing system failures, inability to meet compliance regulations, and inadequate service capacity to meet community needs. Plan today and avoid disaster tomorrow. (Section 8.10, "Emergency Response," describes the basic elements of an Emergency Operations Plan.)

8.3 ORGANIZING[2]

A water distribution system should have a written organizational plan and written policies. In some communities the organizational plan and policies are part of the overall community plan. In either case, the water distribution system manager and all plant personnel should have a copy of the organizational plan and written policies of the water distribution system.

The purpose of the organizational plan is to show who reports to whom and to identify the lines of authority. The organizational plan should show each person or job position in the organization with a direct line showing to whom each person reports in the organization. Remember, an employee can serve only one supervisor (unity of command) and each supervisor should ideally manage only six or seven employees. The organizational plan should include a job description for each of the positions on the organizational chart. When the organizational plan is in place, employees know who is their immediate boss and confusion about job tasks is eliminated. A sample organizational plan for a water/wastewater utility is shown in Figure 8.1. The basic job duties for some typical utility positions are described in Table 8.2.

To understand organization and its role in management, we need to understand some other terms including authority, responsibility, delegation, and accountability. AUTHORITY means the power and resources to do a specific job or to get that job done. Authority may be given to an employee due to their position in the organization (this is formal authority) or authority may be given to the employee informally by their co-workers when the employee has earned their respect. RESPONSIBILITY may be described as answering to those above in the chain of command to explain how and why you have used your authority. DELEGATION is the act in which power is given to another person in the organization to accomplish a specific job. Finally when a manager gives power/responsibility to an employee, then the employee is ACCOUNT-ABLE[3] for the results.

Organization and effective delegation are very important to keep any utility operating efficiently. Effective delegation is uncomfortable for many managers since it requires giving up power and responsibility. Many managers believe that they can do the job better than others, they believe that other employees are not well trained or experienced, and they are afraid of mistakes. The water distribution system manager retains some responsibility even after delegating to another employee and, therefore, the manager is often reluctant to delegate or may delegate the responsibility but not the authority to get the job done. For the water distribution system manager, good organization means that employees are ready to accept responsibility and have the power and resources to make sure that the job gets done.

Employees should not be asked to accept responsibilities for job tasks that are beyond their level of authority in the organizational structure. For example, an operator or lead distribution system operator should not be asked to accept responsibility for additional lab testing. The responsibility for additional lab testing must be delegated to the lab supervisor. Authority and responsibility must be delegated properly to be effective.

---

[1] Planning.   Management of utilities to build the resources and financial capability to provide for future needs.

[2] Organizing.   Deciding who does what work and delegating authority to the appropriate persons. A utility should have a written organizational plan and written policies.

[3] Accountability.   When a manager gives power/responsibility to an employee, the employee ensures that the manager is informed of results or events.

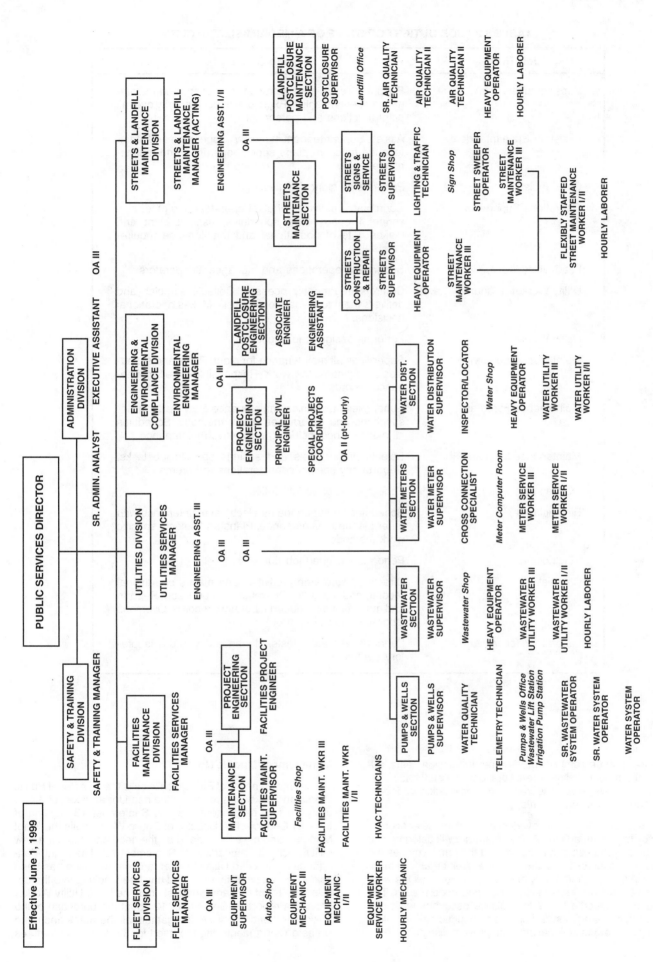

Fig. 8.1   Organizational chart for medium-sized utility
(Courtesy of City of Mountain View, California)

TABLE 8.2 JOB DUTIES FOR STAFF OF A MEDIUM-SIZED UTILITY

Job Title	Job Duties
Superintendent	Responsible for administration, operation, and maintenance of entire facility. Exercises direct authority over all plant functions and personnel.
Assistant Superintendent	Assists Superintendent in review of operation and maintenance function, plans special operation and maintenance tasks.
Clerk/Typist	Performs all clerical duties.
Operations Supervisor	Coordinates activities of plant operators and other personnel. Prepares work schedules, inspects plant, and makes note of operational and maintenance requirements.
Lead Utility Worker	Supervises operations and manages all operators.
Utility Worker II (Journey Level)	Controls treatment processes. Collects samples and delivers them to the lab for analysis. Makes operational decisions.
Utility Worker I	Performs assigned job duties.
Maintenance Supervisor	Supervises all maintenance for plant. Plans and schedules all maintenance work. Responsible for all maintenance records.
Maintenance Foreman	Supervises mechanical maintenance crew. Performs inspections and determines repair methods. Schedules all maintenance including preventive maintenance.
Maintenance Mechanic II	Selects proper tools and assigns specific job tasks. Reports any special considerations to Foreman.
Maintenance Mechanic I	Performs assigned job duties.
Electrician II	Schedules and coordinates electrical maintenance with other planned maintenance. Plans and selects specific work methods.
Electrician I	Performs assigned job duties.
Chemist	Directs all laboratory activities and makes operational recommendations to Operations Supervisor. Reports and maintains all required laboratory records. Oversees laboratory quality control.
Laboratory Technician	Performs laboratory tests. Manages day-to-day laboratory operations.

When these three components—proper job assignments, authority, and responsibility—are all present, the supervisor has successfully delegated. The success of delegation is dependent upon all three components.

An important and often overlooked part of delegation is *follow-up* by the supervisor. A good manager will delegate and follow up on progress to make sure that the employee has the necessary resources to get the job done. Well-organized managers can delegate effectively and do not try to do all the work themselves, but are responsible for getting good results. The Management Muddle No. 1 that follows describes what can happen when delegation is improperly conducted, and illustrates how an organizational plan can prevent disaster.

Management Muddle No. 1

The City Manager of Pleasantville calls the Director of Public Works and asks for a report on the need for and cost of a new utility truck to be presented at the September 13 meeting of the City Council. The Director of Public Works calls the Plant Manager and asks for a report on the need and cost for a new utility truck with a deadline of September 12. The Plant Manager calls the Lead Utility Worker, an operator, who has been asking for a new utility truck and has been looking into the details. The Plant Manager requests that the Lead Utility Worker provide a report on September 12 about the purchase of the truck. The Lead Utility Worker gathers all the notes and hand writes a report identifying the need for the truck, the features required, and the cost. The Lead Utility Worker takes the report to City Hall to be typed and leaves it with a secretary on September 12. On September 13, the City Manager is preparing for the City Council meeting and does not have the report. Who is responsible? Who is accountable? How could this situation have been avoided?

Responsibility: The Lead Utility Worker's responsibility has been carried out with the authority and resources made available. Both the Director of Public Works and the Plant Manager failed to follow up on the report on September 12. No one informed the Lead Utility Worker that the report must be presented to the City Council on September 13, nor was the Lead Utility Worker supplied with the resources for getting the report in final form. However, the City Manager is ultimately responsible for reporting to the City Council.

Accountability: Starting with the Lead Utility Worker and working upward, each employee is accountable to his or her supervisor and should have communicated the status of the report.

How to avoid this situation: Good communication and follow-up by each of these supervisors could have prevented this situation completely. The City Manager should have asked to see the report on September 12; the Director of Public Works should have asked the Plant Manager to deliver the report no later than September 11; and the Plant Manager should have asked the Lead Utility Worker to submit the typed report (to the Plant Manager) no later than September 10. When delegating this task, the Plant Manager should have arranged for a secretary or clerk to assist the Lead Utility Worker in typing the report. Providing clerical support enables the Lead Utility Manager to complete the assigned task in a timely manner.

At each step in this chain of delegation, setting an early deadline gives the individual receiving the report an opportunity to review the document and make revisions, if necessary, before forwarding it up the chain of authority and ensures that the report reaches the City Manager no later than September 12.

QUESTIONS

Write your answers in a notebook and then compare your answers with those on page 511.

8.2A   Who must be included in distribution system planning?

8.3A   What is the purpose of an organizational plan?

8.3B   Why is it sometimes difficult or uncomfortable for supervisors or managers to delegate effectively?

8.3C   What is an important and often overlooked part of delegation?

8.4   STAFFING

8.40   The Distribution System Manager's Responsibilities

The distribution system manager is also responsible for staffing, which includes hiring new employees, training employees, and evaluating job performance. The distribution system should have established procedures for job hiring which include requirements for advertising the position, application procedures, and the procedures for conducting interviews.

NOTICE

The information provided in this section on staffing should *not* be viewed as *legal advice*. The purpose of this section is simply to identify and describe in general terms the major components of a distribution system manager's responsibilities in the area of staffing. One issue, harassment, is discussed in somewhat greater detail to illustrate the broad scope of a manager's responsibilities within a single policy area. Personnel administration is affected by many federal and state regulations. Legal requirements of legislation such as the Americans With Disabilities Act (ADA), Equal Employment Opportunity (EEO) Act, Family and Medical Leave Act, and wages and hours laws are complex and beyond the scope of this manual. If your distribution system agency does not have established personnel policies and procedures, consider getting help from a labor law attorney to develop appropriate policies. At the very least, you should get help from a recruitment specialist to develop and document hiring procedures which meet the federal guidelines for Equal Employment Opportunity.

In the area of staffing, more than any other area of responsibility, a manager must be extremely cautious and consider the consequences before taking action. Personnel management practices have changed dramatically in the past few years and continue to be redefined almost daily by the courts. A manager who violates an employee's or job applicant's rights can be held both personally and professionally liable in court. Throughout this section on staffing you will repeatedly find references to two terms: **job-related** and **documentation**. These are key concepts in personnel management today. Any personnel action you take must be job-related, from the questions you ask during interviews to disciplinary actions or promotions. And while almost no one wants more paperwork, documentation of personnel actions detailing what you did, when you did it, and why you did it (the reasons will be job-related, of course) is absolutely essential. There is no way to predict when you might be called upon to defend your actions in court. Good records not only serve to refresh your memory about past events but can also be used to demonstrate your pattern of lawful behavior over time.

8.41   How Many Employees Are Needed?

There is a common tendency for organizations to add personnel in response to changing conditions without first examining how the existing workforce might be reorganized to achieve greater efficiency and meet the new work demands. In water supply and distribution utilities, aging of the system, changes in use, and expansion of the system often mean changes in the operation and maintenance tasks being performed. The manager of a distribution system should periodically review the agency's work requirements and staffing to ensure that the distribution system is operating as efficiently as possible. A good time to conduct such a review is during the annual budgeting process or when you are considering hiring a new employee because the workload seems to be greater than the current staff can adequately handle.

The staffing analysis procedure outlined in this section illustrates how to conduct a comprehensive analysis of the type needed for a complete reorganization of the agency. In practice, however, a complete reorganization may not be desirable or even possible. Frequent organizational changes can make employees anxious about their jobs and may interfere with their work performance. Some employees show strong resistance to any change in job responsibilities. Nonetheless, by thoroughly examining the functions and staffing of the utility on a periodic basis, the manager may spot trends (such as an increase in the amount of time spent maintaining certain equipment or portions of the system) or discover inefficiencies that could be corrected over a period of time.

The first step in analyzing the distribution system's staffing needs is to prepare a detailed list of all the tasks to be performed to operate and maintain the distribution system. Next, estimate the number of staff hours per year required to perform each task. Be sure to include the time required for supervision and training.

When you have completed the task analysis, prepare a list of the distribution system's current employees. Assign tasks to each employee based on the person's skills and abilities. To the extent possible, try to minimize the number of different work activities assigned to each person but also keep in mind the need to provide opportunities for career advancement. One full-time staff year equals 260 days, including vacation and holiday time: (52 wk/yr)(5 days/wk) = 260 days/yr.

You can expect to find that this "ideal" staffing arrangement does not exactly match up with your current employees' job assignments. Most likely, you will also find that the number of staff hours required does not exactly equal the number of staff

hours available. Your responsibility as a manager is to create the best possible fit between the work to be done and the personnel/skills available to do it. In addition to shifting work assignments between employees, other options you might consider are contracting out some types of work, hiring part-time or seasonal staff, or setting up a second shift (to make fuller use of existing equipment). Of course, you may find that it is time to hire another full- or part-time operator.

8.42   Qualifications Profile

Hiring new employees requires careful planning before the personal interview process. In an effort to limit discriminatory hiring practices, the law and administrative policy have carefully defined the hiring methods and guidelines employers may use. The selection method and examination process used to evaluate applicants must be limited to the applicant's knowledge, skills, and abilities to perform relevant job-related activities. In all but rare cases, factors such as age and level of education may not be used to screen candidates in place of performance testing. A description of the duties and qualifications for the job must be clearly defined in writing. The job description may be used to develop a qualifications profile. This qualifications profile clearly and precisely identifies the required job qualifications. All job qualifications must be relevant to the actual job duties that will be performed in that position. The following list of typical job qualifications may be used to help you develop your own qualifications profiles with advice from a recruitment specialist.

1. General Requirements:

    a. Knowledge of methods, tools, equipment, and materials used in water distribution systems,

    b. Knowledge of work hazards and applicable safety precautions,

    c. Ability to establish and maintain effective working relations with employees and the general public, and

    d. Possession of a valid state driver's license for the class of equipment the employee is expected to drive.

2. General Educational Development:

    a. Reasoning: Apply common sense understanding to carry out instructions furnished in oral, written, or diagrammatical form,

    b. Mathematical: Use a pocket calculator to make arithmetic calculations relevant to the water distribution system's operation and maintenance processes, and

    c. Language: Communicate with fellow employees and train subordinates in work methods. Fill out maintenance report forms.

3. Specific Vocational Preparation: Three years of experience in water distribution system operation and maintenance.

4. Interests: May or may not be relevant to knowledge, skills, and ability; for example, an interest in activities concerned with objects and machines, ecology, or business management.

5. Temperament: Must adjust to a variety of tasks requiring frequent change and must routinely use established standards and procedures.

6. Physical Demands: Medium to heavy work involving lifting, climbing, kneeling, crouching, crawling, reaching, hearing, and seeing. Must be able to lift and carry _____ number of pounds for a distance of _____ feet.

7. Working Conditions: The work involves wet conditions, cramped and awkward spaces, noise, risks of bodily injury, and exposure to weather.

QUESTIONS

Write your answers in a notebook and then compare your answers with those on pages 511 and 512.

8.4A   What do staffing responsibilities include?

8.4B   What are two key personnel management concepts a manager should always keep in mind?

8.4C   List the steps involved in a staffing analysis.

8.4D   What is a qualifications profile?

8.43   Applications and the Selection Process

8.430   Advertising the Position

To advertise a job opening, first prepare a written description of the required job qualifications, compensation, job duties, selection process, and application procedures (with a closing date). The distribution system agency should have established procedures about how to advertise the position and conduct the application process. The application procedure may require that the job be posted first within the utility to allow existing personnel first chance at the job opportunity.

8.431   Paper Screening

The next step in the selection process is known as paper screening. The personnel department and the distribution system manager review each application and eliminate those who are not qualified. The qualified applicants may be given examinations to verify their qualifications. Usually the top three to twelve applicants are selected for an interview, depending on the agency's preference.

8.432   Interviewing Applicants

The purpose of the job interview is to gain additional information about the applicants so that the most qualified person can be selected. The distribution system manager should prepare for the interview in advance. Review the background information on each applicant. Draw up a list of job-related questions that will be asked of each applicant. During the interviews, briefly note the answers each applicant gives.

It used to be thought that the best way to learn about applicants was to give them plenty of time to talk about themselves because the content and type of information applicants volunteer might provide a deeper insight into the person and what type of employee they will become. Be very careful about open-ended, unstructured conversations with job applicants, even the friendly remarks you make initially to put the applicant at ease during the interview. If the applicant begins to volunteer information which you could not otherwise legally ask for (such as marital status, number of children, religious affiliation, or age), be polite but firm in promptly redirecting the conversation. Even if this information was provided to you voluntarily, an applicant who did not get the job could later allege that you discriminated against them based on age or religion.

The only type of information you may legally request is information about the applicant's job skills, abilities, and experience relating directly to the job for which the person is applying. You must always be sensitive to the civil rights of the applicant and the affirmative action policies of the distribution system agency, which is another good reason to prepare a list of questions before the interview process begins. Structure the questions so that you avoid simple yes-and-no answers. Table 8.3 summarizes acceptable and unacceptable pre-employment inquiries to guide you in developing a good list of questions.

If other distribution system staff members are participating in the interviews, their participation should be confined to the pre-selected questions. Under no circumstances should front line employees conduct interviews in the absence of the manager or another person knowledgeable about personnel policies and practices.

The interview should be conducted in a quiet room without interruptions. Most applicants will be nervous, so start the interview on a positive note with introductions and some general remarks to put the applicant at ease. Explain the details of the job, working conditions, wages, benefits, and potential for advancement. Allow the applicant a chance to ask questions about the job. Ask each applicant the questions you have prepared and jot down brief notes on their responses. Taking notes while interviewing is awkward for some people but it becomes easier with practice. Notes are important because after interviewing several candidates you may not be able to remember what each one said. Also, as mentioned earlier, notes taken at the time of an interview can be valuable evidence in court if an unsuccessful applicant files a lawsuit for unfair hiring practices. At the end of the interview, tell the applicant when a decision will be made and how the applicant will be informed of the decision.

If an applicant's responses during the interview indicate that the person clearly is not qualified for this job but may be qualified for another job, briefly describe the other opportunity and how the person can apply for that position. The applicant may ask to be interviewed immediately for the second position. However, do not violate the distribution system agency's hiring procedures for the convenience of a job applicant. The same sequence of hiring procedures should be followed each time a position is filled. Tell this applicant that it will still be necessary to apply for the other position and that another interview may or may not follow, depending on the qualifications of the other applicants for the position.

8.433   Selecting the Most Qualified Candidate

Once the interviews are over, the job of evaluating and selecting the successful candidate begins. Review your interview notes and check the candidates' references. Checking references will verify the job experience of the applicant and may

TABLE 8.3 ACCEPTABLE AND UNACCEPTABLE PRE-EMPLOYMENT INQUIRIES[a]

Acceptable Pre-Employment Inquiries	Subject	Unacceptable Pre-Employment Inquiries
"Have you ever worked for this agency under a different name?"	NAME	Former name of applicant whose name has been changed by court order or otherwise.
Applicant's place of residence. How long applicant has been resident of this state or city.	ADDRESS OR DURATION OF RESIDENCE	
"If hired can you submit a birth certificate or other proof of U.S. citizenship or age?"	BIRTHPLACE	Birthplace of applicant. Birthplace of applicant's parents, spouse, or other relatives. Requirement that applicant submit a birth certificate, naturalization, or baptismal record.
"If hired, can you furnish proof of age?" /or/ Statement that hire is subject to verification that applicant's age meets legal requirements.	AGE	Questions which tend to identify applicants 40 to 64 years of age.
Statement by employer of regular days, hours, or shift to be worked.	RELIGIOUS	Applicant's religious denomination or affiliation, church, parish, pastor, or religious holidays observed. "Do you attend religious services /or/ a house of worship?" Applicant may not be told, "This is a Catholic/Protestant/Jewish/atheist organization."
	RACE OR COLOR	Complexion, color of skin, or other questions directly or indirectly indicating race or color.
Statement that photograph may be required after employment.	PHOTOGRAPH	Requirement that applicant affix a photograph to his/her application form. Request applicant, at his/her option, to submit photograph. Requirement of photograph after interview but before hiring.
Statement by employer that if hired applicant may be required to submit proof of eligibility to work in the United States.	CITIZENSHIP	"Are you a United States citizen?" Whether applicant or applicant's parents or spouse are naturalized or native-born U.S. citizens. Date when applicant or parents or spouse acquired U.S. citizenship. Requirement that applicant produce naturalization papers or first papers. Whether applicant's parents or spouse are citizens of the U.S.
Applicant's work experience. Applicant's military experience in armed forces of United States, in a state militia (U.S.), or in a particular branch of the U.S. armed forces.	EXPERIENCE	"Are you currently employed?" Applicant's military experience (general). Type of military discharge.
Applicant's academic, vocational, or professional education; schools attended.	EDUCATION	Date last attended high school.
Language applicant reads, speaks, or writes fluently.	NATIONAL ORIGIN OR ANCESTRY	Applicant's nationality, lineage, ancestry, national origin, descent, or parentage. Date of arrival in United States or port of entry; how long a resident. Nationality of applicant's parents or spouse; maiden name of applicant's wife or mother. Language commonly used by applicant. "What is your mother tongue?" How applicant acquired ability to read, write, or speak a foreign language.
	CHARACTER	"Have you ever been arrested?"

TABLE 8.3   ACCEPTABLE AND UNACCEPTABLE PRE-EMPLOYMENT INQUIRIES (continued)

Acceptable Pre-Employment Inquiries	Subject	Unacceptable Pre-Employment Inquiries
Names of applicant's relatives already employed by the agency.	RELATIVES	Marital status or number of dependents.
		Name or address of relative, spouse, or children of adult applicant.
		"With whom do you reside?"
		"Do you live with your parents?"
Organizations, clubs, professional societies, or other associations of which applicant is a member, excluding any names the character of which indicate the race, religious creed, color, national origin, or ancestry of its members.	ORGANIZATIONS	"List all organizations, clubs, societies, and lodges to which you belong."
"By whom were you referred for a position here?"	REFERENCES	Requirement of submission of a religious reference.
"Do you have any physical condition which may limit your ability to perform the job applied for?"	PHYSICAL CONDITION	"Do you have any physical disabilities?"
		Questions on general medical condition.
Statement by employer that offer may be made contingent on passing a physical examination.		Inquiries as to receipt of Workers' Compensation.
Notice to applicant that any misstatements or omissions of material facts in his/her application may be cause for dismissal.	MISCELLANEOUS	Any inquiry that is not job-related or necessary for determining an applicant's eligibility for employment.

[a]   Courtesy of Marion B. McCamey, Affirmative Action Officer, California State University, Sacramento, CA.

provide insight into the applicant's work habits. Questions you might ask previous employers include: Was the employee reliable and punctual? How well did the employee relate to co-workers? Did the employee consistently practice safe work procedures? Would you rehire this employee?

The rights of certain "protected groups" in the workforce today, such as minorities, women, disabled persons, persons over 40 years of age, and union members, are protected by law. A manager's responsibilities regarding protected groups begins with the hiring process and continues for as long as the employer/employee relationship lasts. The best principle to deal with protected groups (and all other employees, for that matter) is the "OUCH" principle. The OUCH principle says that when you hire new employees or delegate job tasks to current employees, you must be OBJECTIVE, UNIFORM in your treatment of applicants or employees, CONSISTENT with utility policies, and HAVE JOB RELATEDNESS. If you don't manage with all of these characteristics, you may find yourself in a "hurting" position with regard to protected workers.

Objectivity is the first hurdle. Often the physical characteristics of a person, such as large or small size, may make a person seem more or less job capable. However, many distribution system agency jobs are done with power tools or other technology which allows all persons, regardless of size, to manage most tasks. Try to remain objective but reasonable in assessing job applicants and making job assignments.

Uniform treatment of job applicants and employees is necessary to protect yourself and other employees. Nothing will destroy morale more quickly than unequal treatment of employees. Your role as a manager is to consistently apply the policies and procedures that have been adopted by the distribution system agency. Often, policies and procedures exist that are not popular and may not even be appropriate. However, the job of the distribution system manager is to consistently uphold and apply the policies of the distribution system agency.

The last part of the OUCH principle is having job relatedness. Any hiring decision must be based on the applicant's qualifications to meet the specific job requirements and any job assignment given to an employee must be related to that employee's job description. Extra assignments, such as buying personal gifts for the boss's family or washing the boss's car, are not appropriate. These types of job assignments will eventually catch up to the manager and can be particularly embarrassing if the public gets involved. So to protect yourself and your utility, remember the OUCH principle as you hire and manage your employees.

Once you have made your selection, the applicant is usually required to pass a medical examination. When this has been successfully completed and the applicant has accepted the position, notify the other applicants that the position has been filled.

8.44   New Employee Orientation

During the first day of work, a new employee should be given all the information available in written and verbal form on the policies and practices of the distribution system agency including compensation, benefits, attendance expectations, alcohol and drug testing (if the distribution system agency does this), and employer/employee relations. Answer any questions from the new employee at this time and try to explain the overall structure of the distribution system agency as well as identify who can answer employee questions when they arise. Introduce the new employee to co-workers and tour the work area. Every distribution system agency should have a safety training session for all new employees and specific safety training for some job categories. Provide safety training (see Section 8.11, "Safety Program") for new employees on the first day of employment or as soon thereafter as possible. Establishing safe work practices is a very important function of management.

QUESTIONS

Write your answers in a notebook and then compare your answers with those on page 512.

8.4E   What is the purpose of a job interview?

8.4F   List four "protected groups."

8.4G   What does the "OUCH" principle stand for?

8.4H   When should a new employee's safety training begin?

8.45   Employment Policies and Procedures

8.450   Probationary Period

Many employers now use a probationary period for all new employees. The probationary period is typically three to six months but may be as long as a year. This period begins on the first day of work. Management may reserve the right to terminate employment of the person with or without cause during this probationary period. The employee must be informed of this probationary period and must understand that successful completion of the probationary period is required in order to move into regular employment status.

The probationary period provides a time during which both the employer and employee can assess the "fit" between the job and the person. Normally a performance evaluation is completed near the end of the probationary period. A satisfactory performance evaluation is the mechanism used to move an employee from probationary status into regular employment.

8.451   Compensation

The compensation an employee receives for the work performed includes satisfaction, recognition, security, appropriate pay, and benefits. All are important to keep good employees satisfied. Salaries should be a function of supply and demand. Pay should be high enough to attract and retain qualified employees. Salaries are usually determined by the governing body of the water distribution system agency in negotiation with employee groups, when appropriate. The salary structure should meet all state and federal regulations and accurately reflect the level of service given by the employee. A survey of salaries from other distribution system agencies in the area may provide valuable information in the development of a salary structure.

The benefits supplied by the employer are an important part of the compensation package. Benefits generally include the following: retirement, health insurance, life insurance, employer's portion of social security, holiday and vacation pay, sick leave, personal leave, parental leave, worker's compensation, and protective clothing. Many employers now provide dental and/or vision insurance, long-term disability insurance, educational bonus or costs and leave, bereavement leave, and release time for jury duty. Some employers also include in their benefit package cash bonus programs and longevity pay. The value of an employee's entire benefit package is often computed and printed on the pay stub as a reminder that salary alone is not the only compensation being provided.

8.452   Training and Certification [4]

Training has become an ongoing process in the workplace. The distribution system manager must provide new employee

---

[4] Certification Examination.   An examination administered by a state agency that water distribution system operators take to indicate a level of professional competence. In the United States, certification of water distribution system operators is mandatory.

training as well as ongoing training for all employees. Safety training is particularly important for all utility operators and staff members and is discussed in detail later in Section 8.11. Certified water distribution system and treatment plant operators earn their certificates by knowing how to do their jobs safely. Preparing for certification examinations is one means by which operators learn to identify safety hazards and to follow safe procedures at all times under all circumstances.

Although it is extremely important, safety is not the sole benefit to be derived from a certification program. Other benefits include protection of the public's investment in water distribution and treatment facilities and employee pride and recognition.

Vast sums of public funds have been invested in the construction of water distribution and treatment facilities. Certification of operators assures utilities that these facilities will be operated and maintained by qualified operators who possess a certain level of competence. These operators should have the knowledge and skills not only to prevent unnecessary deterioration and failure of the facilities, but also to improve operation and maintenance techniques.

Achievement of a level of certification is a public acknowledgment of a water distribution system or treatment plant operator's skills and knowledge. Presentation of certificates at an official meeting of the governing body will place the operators in a position to receive recognition for their efforts and may even get press coverage and public opinion that is favorable. An improved public image will give the certified operator more credibility in discussions with property owners.

Recognition for their personal efforts will raise the self-esteem of all certified operators. Certification will also give water distribution system and treatment plant operators an upgraded image that has been too long denied them. If properly publicized, certification ceremonies will give the public a more accurate image of the many dedicated, well-qualified operators working for them. Certification provides a measurable goal that operators can strive for by preparing themselves to do a better job. Passing a certification exam should be recognized by an increase in salary and other employee benefits.

Most states and Canadian provinces now require that water distribution system operators be certified. To maintain current certification, these operators must complete additional training classes every one to five years. In the environmental field, new technologies and regulations require operators to attend training to keep up with their field. The distribution system manager has the responsibility to provide employees with high-quality training opportunities. Many types of training are available to meet the different training needs of distribution system operators, for example, in-house training, training conducted by training centers, professional organizations, engineering firms, or regulatory agencies, and correspondence courses such as this one by the Office of Water Programs.

ABC stands for the Association of Boards of Certification for Operating Personnel in Water Utilities and Pollution Control Systems. If you wish to find out how operators can become certified in your state or province, contact:

Executive Director, ABC
208 Fifth Street
Ames, IA 50010-6259
Phone: (515) 232-3623

ABC will provide you with the name and address of the appropriate contact person.

One area of training that is frequently overlooked is training for supervisors. Managing people requires a different set of skills than performing the day-to-day work of operating and maintaining a water distribution system facility. Supervisors need to know how to communicate effectively and how to motivate others, as well as how to delegate responsibility and hold people accountable for their performance. Supervisors share management's responsibility for fair and equitable treatment of all workers and are required to act in accordance with applicable state and federal personnel regulations. Making the transition from operator to supervisor also requires a change in attitude. A supervisor is part of the management team and is therefore obliged to promote the best interests of the distribution system agency at all times. When the interests of the agency conflict with the desires of one or more employees, the supervisor must support management's decisions and policies regardless of the supervisor's own personal opinion about the issue. It is the responsibility of the distribution system manager to ensure that supervisors receive appropriate training in all of these areas.

Training on how to motivate people, deal with co-workers, and supervise or manage people working for you has become a very highly specialized field of training. These are complex topics that are beyond the scope of this manual. If you have a need for or wish to learn more about how to deal with people, consider enrolling in courses or reading books on supervision or personnel management.

QUESTIONS

Write your answers in a notebook and then compare your answers with those on page 512.

8.4I   What is the purpose of a probationary period for new employees?

8.4J   What kind of compensation does an employee receive for work performed?

8.4K   Why should distribution system managers provide training opportunities for employees?

8.453 Performance Evaluation

Most organizations conduct some type of performance evaluation, usually on an annual basis. The evaluation may be written and/or oral; however, a written evaluation is strongly recommended because it will provide a record of the employee's performance. Documentation of this type may be needed in the future to support taking disciplinary action if the employee's performance consistently fails to meet expectations. The evaluation of employee performance can be a challenging task, especially when performance has not been acceptable. However, evaluations are also an opportunity to provide employees with positive feedback and let them know their contributions to the organization have been noticed and appreciated.

A formal performance evaluation typically begins with an employee's immediate supervisor filling out the performance evaluation form (a sample evaluation form is shown in Figure 8.2). Complete the entire form and be specific about the employee's achievements as well as areas needing improvement. Next, schedule a private meeting with the employee to discuss the evaluation. Give the employee frequent opportunities to be heard and listen carefully. If some of the employee's accomplishments were overlooked, note them on the evaluation form and consider whether this new information changes your overall rating of performance in one or more categories.

After reviewing the employee's performance for the past year, set performance goals for the next year. Be sure to document the goals you have agreed upon. Setting performance goals is particularly important if an employee's performance has been poor and improvement is needed. If appropriate, develop a written performance improvement plan which includes specific dates when you will again review the employee's progress in meeting the performance goals. Some supervisors find it helpful to schedule an informal mid-year meeting with each employee to review their progress and to avoid surprises during the next performance evaluation.

Many employees and managers dread even the thought of a performance evaluation and see it as an ordeal to be endured. When properly conducted, however, a performance review can strengthen the lines of communication and increase trust between the employee and the manager. Use this opportunity to acknowledge the employee's unique contributions and to seek solutions to any problems the employee may be having in completing work assignments. Ask the employee how you can be of assistance in removing any obstacles to getting the job done. If necessary, provide coaching to help the employee understand both how and why certain tasks are performed. Be generous (but sincere) with praise for the good work the employee does well every day and try to keep the employee's shortcomings in perspective. If the person is doing a good job 95 percent of the time, don't let the entire discussion consist of criticism about the remaining 5 percent of the person's job assignments.

At the end of the meeting, ask the employee to sign the evaluation form to acknowledge having seen and discussed it. Give the employee a copy of the evaluation. If the employee disagrees with any part of the evaluation, invite the person to submit a written statement describing the reasons for their disagreement. The written statement should be filed with the completed performance evaluation form.

8.454   Dealing With Disciplinary Problems

Handling employee discipline problems is difficult, even for an experienced manager. But remember, **no discipline problem ever solves itself and the sooner you deal with the problem, the better the outcome will be**. If problem behavior is not corrected, then other employees will become dissatisfied and the problems will increase.

Every water distribution system agency, no matter how small, should have written employment policies enabling the manager to deal effectively with employee problems. It should also provide a formal complaint or grievance procedure by which employees can have their complaint heard and resolved without fear of retaliation by the supervisor.

Dealing with employee discipline requires tact and skill. You will have to find your own style and then try to stay FLEXIBLE, CALM, and OPEN-MINDED when the situation gets really tough. If you repeatedly find yourself unable to deal successfully with disciplinary problems, consult with the distribution system agency's personnel office (if available) or consider enrolling in a management training course designed specifically around strategies and techniques for disciplining employees.

A commonly accepted method for dealing with job-related employee problems is to first discuss the problem with the employee in private. Most employers will give a person two or three verbal warnings; then the warnings should be written with copies given to the employee. Finally, if the written warnings do not produce positive results, the employee may have to be suspended or dismissed. Your job is to make sure that all warnings are documented with specific descriptions of unsatisfactory behaviors and to make sure that all employees are treated fairly.

Start the disciplinary discussion with a positive comment about the employee. Then identify the problem but keep emotion and blame out of the discussion. The best approach is to state the problem and then ask the employee to suggest a solution. If they respond inappropriately, you must restate the problem and explain that you are trying to find a positive solution that is acceptable to everyone.

Try to keep the discussion focused on solving the problems and do not permit the employee to heap on general complaints, report on what other employees do, or wander from the topic. The following is an example of how you might start the discussion for an employee who is tardy every day. "Joe, you have done a good job in keeping that north side pump station running. You are an asset to this operation. Your tardiness every morning, however, is causing problems. Is there some reason for you to be tardy? We need to find a solution to this problem because your being late creates a bad situation for the night shift. What do you suggest?"

Always remain calm and do not allow yourself to become angry when dealing with an employee about performance issues. If you begin to feel angry and are about to lose control, or if the employee becomes combative or abusive, suggest that the meeting is not producing positive results and schedule an alternative meeting time. Do not let the emotions of the moment carry you into a rage in front of employees. If either you or one of your employees expresses extreme emotions, then the discussion should be postponed until everyone cools down. The following steps may serve as a guide to dealing with confrontation; they apply equally to the employee and the supervisor.

EMPLOYEE EVALUATION FORM

Employee Name _____ Date _____

Job Title _____ Department _____

Evaluate the employee on the job now being performed. Circle the number which most nearly expresses your overall judgment. In the space for comments, consider the employee's performance since their last evaluation and make notes about the progress or specific concerns in that area. The care and accuracy of this appraisal will determine its value to you, the employee, and your employer.

JOB KNOWLEDGE: (Consider knowledge of the job gained through experience, education, and special training)

5. Well informed on all phases of work
4. Knowledge thorough enough to perform well without assistance
3. Adequate grasp of essentials, some assistance required
2. Requires considerable assistance to perform
1. Inadequate knowledge

Comments: _____

_____

_____

QUALITY OF WORK: (Consider accuracy and dependability of the results)

5. Exceptionally accurate, practically no mistakes
4. Usually accurate, seldom necessary to check results
3. Acceptable, occasional errors
2. Often unacceptable, frequent errors, needs supervision
1. Unacceptable, too many errors

Comments: _____

_____

_____

INITIATIVE: (Consider the speed with which the employee grasps new job skills)

5. Excellent, grasps new ideas and suggests improvements, is a leader with others
4. Very resourceful, can work unsupervised, manages time well, is reliable
3. Shows initiative on occasion, is reliable
2. Lacks initiative, must be reminded to complete tasks
1. Needs constant prodding to complete job tasks, is unreliable

Comments: _____

_____

_____

Fig. 8.2   Employee evaluation form

COOPERATION AND RELATIONSHIPS: (Consider manner of handling relationships with co-workers, superiors, and the public)

5. Excellent cooperation and communication with co-workers, supervisors, and others, takes and gives instructions well
4. Gets along well with co-workers
3. Acceptable, usually gets along well, occasionally complains
2. Shows a reluctance to cooperate, complains
1. Very poor cooperation, does not follow instruction, dislikes fellow employees

Comments: _____

_____

_____

ATTENDANCE: (Consider frequency of absences, reasons for absences or tardiness, and promptness in giving notice about absences)

5. Excellent, absent only for emergencies, illness, civic duties, always on time, gives notice when absent
4. Rarely absent or late, always gives notice and good reason
3. Occasionally absent, less important reasons, usually gives notice, but not always in time
2. Often absent, lack of adequate notice or reasons for absenteeism
1. Unexcusable absenteeism, does not give notice, reasons are unacceptable, cannot be depended upon

Comments: _____

_____

_____

OVERALL EVALUATION: Superior _____ Good _____ Satisfactory _____ Unsatisfactory _____

Comments: _____

_____

_____

I hereby certify that this appraisal is my best judgment of the service value of this employee and is based on personal observation and knowledge of the employee's work.

Supervisor's Signature _____ Date _____

I hereby certify that I have personally reviewed this report.

Employee's Signature _____ Date _____

Fig. 8.2   Employee evaluation form (continued)

- Maintain an adult approach—positive criticism should be taken/given to improve job skills.

- Create a private environment—job performance issues should be discussed in private between the employee and supervisor.

- Listen very carefully—be sure that both you and the employee understand the situation in the same way. If not, you need to keep talking until both parties are in agreement about the problem and the solution.

- Keep your language appropriate—anger and bad language will cause the situation to escalate. Keep your cool and hold your tongue.

- Stay focused in the present—let go of all the past slights, misunderstandings, and dissatisfaction. Problems must be solved one at a time.

- Aim for a permanent solution—changes in job performance need to be permanent to be effective.

Reports of violence in the workplace appear regularly in newspapers and on television. Managers and supervisors should be alert for signs that an employee might become violent and should take any threat of violence seriously.

Violence in the workplace may take the form of physical harm, psychological harm, and/or property damage. Common warning signs include abusive language, threatening or confrontational behavior, and brandishing a weapon. Some examples of physical harm include pushing, hitting, shoving, or any other form of physical assault. Threats and harassment are forms of psychological harm, and property damage can range from theft or destruction of equipment to sabotage of the employer's computer systems.

The distribution system agency's safe workplace policy should be a "zero tolerance" policy. Any employee who is the target of violent behavior or who witnesses such behavior should be encouraged to immediately report the behavior to a supervisor and the incident should be investigated promptly. If necessary for the immediate safety of other employees, the offending employee should be placed on administrative leave, escorted from the work environment, and permitted to return to work only after the investigation has been completed.

Management Muddle No. 2

Sue has been working for five years as a laboratory technician and was recently passed over for promotion. A lab director who has more college experience than Sue was hired from another plant. Since then Sue's work has not been very good, she has come to work late, and she does not always get all of the lab tests done during her workday. What should be done about Sue? If you were the supervisor, how would you handle this problem with Sue?

Actions: As the supervisor, you should ask Sue to come by your office. In private, you should discuss with Sue why the new lab director was hired. Discuss with her the good work record she has maintained over the past five years, and explain the changes you have seen in her work recently. At this point you might ask her to evaluate her own performance or what she would do if she were in your situation. She might need to express her resentment about the new lab director. If she does, let her ramble and rave just for a few minutes, then stop her. You might say, "OK, you're unhappy, but what are we going to do to change this situation? How can I help you to regain your motivation and improve your work habits?" If possible you might help her figure out a way to continue her college education, go to additional training classes, or reorganize the lab so that her job duties change somewhat. There are many other possibilities for helping Sue to become motivated again but she should be part of the process. It is important to communicate clearly that her job performance must improve.

QUESTIONS

Write your answers in a notebook and then compare your answers with those on page 512.

8.4L Who is the appropriate person to conduct an employee's performance evaluation?

8.4M What should be the attitude of a supervisor or manager when dealing with disciplinary problems?

8.4N What are some common warning signs that an employee could become violent?

8.455 *Example Policy: Harassment*

Harassment is any behavior that is offensive, annoying, or humiliating to an individual and that interferes with a person's ability to do a job. This behavior is uninvited, often repeated, and creates an uncomfortable or even hostile environment in the workplace. Harassment is not limited to physical behavior but also may be verbal or involve the display of offensive pictures or other images. Sexual harassment is legally defined as unwanted sexual advances, or visual, verbal, or physical conduct of a sexual nature. Any type is harassment is inappropriate in the workplace. A manager's responsibilities with regard to harassment include:

- Establish a written policy (such as the one shown in Figure 8.3) that clearly defines and prohibits harassment of any type,

- Distribute copies of the harassment policy to all employees and take whatever steps are necessary (small group discussions, general staff meetings, training sessions) to ensure that all employees understand the policy,

| SUBJECT: HARASSMENT POLICY AND | NO: _____ |
| COMPLAINT PROCEDURE | |

PURPOSE:

To establish a strong commitment to prohibit harassment in employment, to define discrimination harassment and to set forth a procedure for investigating and resolving internal complaints of harassment.

POLICY:

Harassment of an applicant or employee by a supervisor, management employee, or co-worker on the basis of race, religion, color, national origin, ancestry, handicap, disability, medical condition, marital status, familial status, sex, sexual orientation, or age will not be tolerated. This policy applies to all terms and conditions of employment, including, but not limited to, hiring, placement, promotion, disciplinary action, layoff, recall, transfer, leave of absence, compensation, and training.

Disciplinary action up to and including termination will be instituted for behavior described in the definition of harassment set forth below:

* Any retaliation against a person for filing a harassment charge or making a harassment complaint is prohibited. Employees found to be retaliating against another employee shall be subject to disciplinary action up to and including termination.

DEFINITION:

Harassment includes, but is not limited to:

A. Verbal Harassment—For example, epithets, derogatory comments or slurs on the basis of race, religious creed, color, national origin, ancestry, handicap, disability, medical condition, marital status, familial status, sex, sexual orientation, or age. This might include inappropriate sex-oriented comments on appearance, including dress or physical features or race-oriented stories.

B. Physical Harassment—For example, assault, impeding or blocking movement, with a physical interference with normal work or movement when directed at an individual on the basis of race, religion, color, national origin, ancestry, handicap, disability, medical condition, marital status, familial status, age, sex, or sexual orientation. This could be conduct in the form of pinching, grabbing, patting, propositioning, leering, or making explicit or implied job threats or promises in return for submission to physical acts.

C. Visual Forms of Harassment—For example, derogatory posters, notices, bulletins, cartoons, or drawings on the basis of race, religious creed, color, national origin, ancestry, handicap, disability, medical conditions, marital status, familial status, sex, sexual orientation, or age.

D. Sexual Favors—Unwelcome sexual advances, requests for sexual favors, and other verbal or physical conduct of a sexual nature which is conditioned upon an employment benefit, unreasonably interferes with an individual's work performance, or creates an offensive work environment.

Fig. 8.3 Harassment policy
(Courtesy of City of Mountain View, California)

SUBJECT: HARASSMENT POLICY AND COMPLAINT PROCEDURE (continued)

COMPLAINT PROCEDURE:

A. Filing:

An employee who believes he or she has been harassed may make a complaint orally or in writing with any of the following:

1. Immediate supervisor.
2. Any supervisor or manager within or outside of the department.
3. Department head.
4. Employee Services Director (or his/her designee).

Any supervisor or department head who receives a harassment complaint should notify the Employee Services Director immediately.

B. Upon notification of the harassment complaint, the Employee Services Director shall:

1. Authorize the investigation of the complaint and supervise and/or investigate the complaint. The investigation will include interviews with:

    (a) The complainant;
    (b) The accused harasser; and
    (c) Any other persons the Employee Services Director has reasons to believe has relevant knowledge concerning the complaint. This may include victims of similar conduct.

2. Review factual information gathered through the investigation to determine whether the alleged conduct constitutes harassment; giving consideration to all factual information, the totality of the circumstances, including the nature of the verbal, physical, visual, or sexual conduct and the context in which the alleged incidents occurred.

3. Report the results of the investigation and the determination as to whether harassment occurred to appropriate persons, including to the complainant, the alleged harasser, the supervisor, and the department head. If discipline is imposed, the discipline may or may not be communicated to the complainant.

4. If harassment occurred, take and/or recommend to the appropriate department head or other appropriate authority prompt and effective remedial action against the harasser. The action will be commensurate with the severity of the offense.

5. Take reasonable steps to protect the victim and other potential victims from further harassment.

6. Take reasonable steps to protect the victim from any retaliation as a result of communicating the complaint.

DISSEMINATION OF POLICY:

All employees, supervisors, and managers shall be sent copies of this policy.

Effective Date: May, 1992
Revision Date: March 1, 1993   _____
                                                                     City Manager

Fig. 8.3   Harassment policy (continued)
(Courtesy of City of Mountain View, California)

- Encourage employees to report incidents of harassment to their immediate supervisor, a manager, or the personnel department,

- Investigate every reported case of harassment, and

- Document all aspects of the complaint investigation, including the procedures followed, statements by witnesses, the complainant, and the accused person, the conclusions reached in the case, and the actions taken (if any).

How do you know when offensive behavior could be considered sexual harassment? Unwelcome sexual advances or other verbal or physical conduct of a sexual nature could be interpreted by an employee as sexual harassment under the following conditions:

- A person is required, or feels they are required, to accept unwelcome sexual conduct in order to get a job or keep a job,

- Decisions about an employee's job or work status are made based upon either the employee's acceptance or rejection of unwelcome sexual conduct, or

- The conduct interferes with the employee's work performance or creates an intimidating, hostile, or offensive working environment.

The following is a list of examples of the kind of behavior that is unacceptable and illegal. It is only a partial list to give you an idea of the scope of the requirements.

- Unwanted hugging, patting, kissing, brushing up against someone's body, or other inappropriate sexual touching,

- Subtle or open pressure for sexual activity,

- Persistent sexually explicit or sexist statements, jokes, or stories,

- Repeated leering or staring at a person's body,

- Suggestive or obscene notes or phone calls, and/or

- Display of sexually explicit pictures or cartoons.

The best way to prevent harassment is to set an example by your own behavior and to keep communication open between employees. In most cases an open discussion with employees about harassment can help everyone understand that innuendo and slurs about a person's race, religion, sex, appearance, or any other personal belief or characteristic are humiliating. The most productive way to control such behavior is by enlisting the help of all employees to feel that they have the right and the responsibility to stop harassment. When you get employees to think about their behavior and how their behavior makes others feel, they will usually realize they should speak up to prevent harassment.

Here is an example of how employees handled a problem of harassment. A group of operators often had coffee in the office of the utility during their morning break. One of the operators often used foul language, which was embarrassing and offensive to one of the operator's co-workers. The manager sent a memo to all employees that mentioned respect for fellow workers and included a reminder about inappropriate language. The next day all of the other operators had taped their copy of the memo to the mailbox of the one operator who was most vocal. These operators found a way to send their message loud and clear—no one wants to work in an environment that is unpleasant to others. As a manager, you must establish an atmosphere that is open, congenial, and harmonious for all employees.

Occasional flirting, innuendo, or jokes may not meet the legal definition of sexual harassment. Nevertheless, they may be offensive or intimidating to others. Every employee has a right to a workplace free of discrimination and harassment, and every employee has a responsibility to respect the rights of others.

Managers need to **be aware of** and **take action to prevent** any type of harassment in the workplace. It is not enough to simply distribute copies of the utility's harassment policy. If legal action is taken against the utility due to harassment or the existence of a hostile work environment, the utility manager may face both personal and professional liability if it can be shown that the manager *should have known* harassment was occurring or that the manager permitted a hostile environment to continue to exist.

Management Muddle No. 3

The maintenance crew is a group of five men who have been with the utility for many years and are well respected for their work habits. However, in the maintenance shed the walls are covered by calendars of scantily clad women and the language used out in the shed is sometimes pretty rough. One of your operators is a woman; she is well respected by her co-workers and is a very good operator. She comes to your office to complain about the situation in the maintenance shed and demands that you remove the pictures from the walls in the shed. She goes on to report that when she went out to the shed and requested assistance to check on a pump, which was noisy and running hot, she was told "Kiss my _____, toots!" As the manager, what should you do? Should you immediately go to the maintenance shed and rip down the pictures? What should you say to the female operator in your office?

Actions: Your first response should be to reassure your operator that you understand her anger and frustration. Let her know that you will investigate the matter immediately and take action to correct any problems you find. Ask her to write out a complete statement of the facts, including her concerns about the pump, when and how her request for assistance was made, to whom the request was made, who made the offensive remark, the names of any witnesses, and what responses she has gotten from the maintenance crew in the past. Try to establish if this is a one-time response or if this problem has been going on for some time.

Begin your investigation with a trip to the maintenance shed to observe and evaluate what is hanging on the walls. Discuss the situation with the crew. Try to make this an open discussion so that everybody understands how the pictures affect the atmosphere and the image of professionalism of the utility. The best solution is to get the crew to understand how this type of behavior looks to persons outside their own small group and then let them take down the pictures. (Be sure to follow up later to confirm that the calendars or other offensive material has not reappeared.)

Next, set up private interviews with the person accused of making the offensive remark and each of the witnesses. Ask each person to describe the encounter in the maintenance shed and make detailed notes of their responses. (Depending on the complexity of the situation, it may sometimes be appropriate to have each person involved submit a written statement describing what occurred.)

After you have thoroughly investigated the incident, discuss with the crew the use of acceptable language in response to other employees. An open discussion and increased awareness of sexual harassment should be all that is necessary to change this situation. If it is not, arrange for a training program in sexual harassment awareness, if one is available. Be sure to establish a policy on the consequences of inappropriate behavior and be prepared to enforce the policy when needed.

Retaliation against an employee for filing a complaint about harassment or a hostile work environment is also illegal. Some examples of retaliation are demotion, suspension, failure to hire or consider for hire, failure to make impartial employment recommendations or decisions, adversely changing working conditions, spreading rumors, or denying any employment benefit. Retaliation could be the basis for a lawsuit involving not only the person who is accused of retaliation, but also the immediate supervisor, the manager, and the utility.

Most areas of personnel management, including harassment and retaliation issues, are complex and have significant legal consequences for everyone involved. The discussion in this section is *NOT* a complete explanation of harassment or retaliation. If your utility is large enough to employ a personnel specialist or a labor law attorney, ask them to review your staffing policies and procedures and consult with them whenever you have questions about personnel matters. If you manage a small utility and have no in-house sources for technical or legal advice, enroll in appropriate training courses or consider working with an attorney on a contract basis.

8.456 Laws Governing Employer/Employee Relations

Many employers take pride in advertising that they are an "equal opportunity employer," and one often sees this claim in newspaper help wanted ads and other forms of job postings. It means an employer's staffing policies and procedures do not discriminate against anyone based on race, religion, national origin, color, citizenship, marital status, gender, age, Vietnam era or disabled veteran status, or the presence of a physical, mental, or sensory disability. An employer must meet specific requirements of the federal Equal Employment Opportunity Act to be eligible to advertise as an equal opportunity employer. These requirements include adoption of nondiscriminatory personnel policies and procedures and periodic submission of reports of personnel actions for review by the Equal Employment Opportunity Commission.

The Family and Medical Leave Act of 1993 (FMLA) is a federal law that requires all public agencies as well as companies with 50 or more employees to permit eligible employees to take up to 12 weeks of time off in a 12-month period for the following purposes: (1) the employee's own serious health condition, (2) to care for a child following birth or placement for adoption or foster care, or (3) to care for the employee's spouse, child, or parent with a serious health condition. To be eligible to receive this benefit, an employee must have been employed for at least one year prior to the leave. The employer is not required to pay the employee's salary during the time off work, but many employers permit (or require) employees to use accrued sick leave and vacation time during the period of unpaid FMLA leave.

The Americans With Disabilities Act of 1990 (ADA) prohibits employment discrimination based on a person's mental or physical disability. The law applies to employers engaged in an industry affecting commerce who have 15 or more employees.

In general, the ADA defines disability as a physical or mental impairment that substantially limits one or more of the major life activities of an individual. The exact meaning of this definition is evolving as the courts settle lawsuits in which individuals allege they were discriminated against because of a physical or mental disability. The original ADA legislation listed more than 40 specific types of impairments and the courts continue to expand the list.

Under the ADA, employers must make reasonable accommodations to enable a disabled person to function successfully in the work environment, for example, installing a ramp to make facilities accessible to someone in a wheelchair, or restructuring an individual's job or work schedule, or providing an interpreter. The requirements of each situation are unique. In each case, the nature and extent of the disability and the "reasonableness" (including cost factors) of the requested accommodation by the employer must be weighed. Employers are not automatically required to do everything possible to accommodate disabled persons, but rather to take whatever reasonable steps they can to do so.

All of these personnel laws are very complex and managers of any utility that may be covered by them are strongly urged to seek the assistance of an experienced labor law attorney or personnel specialist.

8.457   Personnel Records

A personnel file should be maintained for each utility employee. This file should contain all documents related to the employee's hiring, performance reviews, promotions, disciplinary actions, and any other records of employment-related matters. Since these records often contain sensitive, confidential information, access to personnel records should be closely controlled. (Also see Section 8.12, "Recordkeeping," for more information about what records should be kept and how long they should be kept.)

QUESTIONS

Write your answers in a notebook and then compare your answers with those on page 512.

8.4O   What is harassment?

8.4P   List three types of behavior that could be considered sexual harassment.

8.4Q   What is the best way to prevent harassment?

8.4R   What is the meaning of "disability" under the Americans With Disabilities Act?

8.46   Unions

Whether your utility operators belong to a union now or may join one in the future, a good employee-management relationship is crucial to keeping an agency functioning properly. Managers, supervisors, crew leaders, and operators all have to work together to develop this relationship.

Most of a manager's union contacts are with a shop steward. The shop steward is elected by the union employees and is their official representative to management and the local union. The steward is in an awkward position because the steward is an employee who is expected to do a full-time job like other employees, while also representing all of the employees. The steward must create an effective link between the utility manager and/or supervisors and the employees.

During contract negotiations between management and the employees' union representatives, management should be in constant consultation with the supervisors. Many employee demands regarding working conditions originate from the supervisor's daily dealings with the employees. An effective supervisor can minimize unreasonable demands. Also any demands that are agreed upon must be implemented and carried out by a supervisor. A supervisor can help both sides reach an acceptable contractual agreement.

Once a contract has been agreed upon by both the union and the utility, the utility manager and the other supervisors must manage the organization within the framework of the contract. Do not attempt to ignore or "get around" the contract even if you disagree with some aspects of it. If you do not understand certain contract provisions, ask for clarification before you begin implementation of those provisions.

Contracts do not change the supervisor's delegated authority or responsibility. Operators must carry out the supervisor's orders and get the work done properly, safely, and within a reasonable amount of time. As a supervisor, you have the right and even the duty to make decisions. However, a contract gives a union the right to protest or challenge your decision. When an operator requires discipline, disciplinary action is a management responsibility.

Handling employee grievances within the framework of a union contract can be a very time-consuming job for the supervisor and the steward. Union contracts usually spell out in great detail the steps and procedures the steward and the supervisor must follow to settle differences. Grievances can develop over disciplinary action, distribution of overtime, transfers, promotions, demotions, and interpretation of labor contracts. The shop steward must communicate complaints and grievances from operators to the supervisor. Then the supervisor and the steward must work together to settle complaints and adjust grievances. When a shop steward and supervisor can work together, the steward can help the supervisor to be an effective manager.

An effective manager is available to discuss problems. **Dealing with grievances as quickly as possible** often prevents small problems from growing into large problems. When a shop steward presents a grievance to you, listen carefully and sympathetically to the steward. Discuss the problem with the employee directly with the help of the shop steward. Try to identify the facts and cause of the problem and keep a written record of your findings. Focus on the problem and do not get caught up in irrelevant issues. Make every effort to settle the grievance quickly and to everyone's satisfaction.

The consequences of any solution to a grievance must be considered and solutions must be consistent and fair to other operators. The solution or settlement should be clear and understandable to everyone involved. Once a solution has been agreed upon, prepare a written summary of the agreement. Review this final report with the shop steward to be sure the intent of the solution is understood and properly documented. The entire grievance procedure must be documented and properly filed, from initial presentation to final solution and settlement.

Union activities are governed by the National Labor Relations Act. When a union attempts to organize the utility's employees, your rights and actions as a manager are also governed by this Act. Be sure to seek competent legal assistance if your experience in dealing with a union is limited or if you have no such experience.

QUESTIONS

Write your answers in a notebook and then compare your answers with those on page 512.

8.4S   What is the role of a shop steward?

8.4T   How does a union contract affect a supervisor's authority?

END OF LESSON 1 OF 3 LESSONS

ON

DISTRIBUTION SYSTEM ADMINISTRATION

Please answer the discussion and review questions next.

DISCUSSION AND REVIEW QUESTIONS

Chapter 8.   DISTRIBUTION SYSTEM ADMINISTRATION

(Lesson 1 of 3 Lessons)

At the end of each lesson in this chapter you will find some discussion and review questions. The purpose of these questions is to indicate to you how well you understand the material in the lesson. Write the answers to these questions in your notebook before continuing.

1. What are the different types of demands on a distribution system manager?

2. List the basic functions of a manager.

3. What can happen without adequate distribution system agency planning?

4. What is the purpose of an organizational plan?

5. Define the following terms:

   1. Authority,

   2. Responsibility,

   3. Delegation, and

   4. Accountability.

6. When has a supervisor successfully delegated?

7. What two concepts should a manager keep in mind to avoid violating the rights of an employee or job applicant?

8. Why should a manager thoroughly examine the functions and staffing of the distribution system agency on a periodic basis?

9. When hiring new employees, the selection method and examination process used to evaluate applicants must be based on what criteria?

10. Why should you make notes of applicants' responses during job interviews?

11. What information should be provided to a new employee during orientation?

12. What type of training should be provided for supervisors?

13. Why is it important to formally document each employee's performance on a regular basis?

14. How should discipline problems be solved?

15. What steps can you take to help reach a successful resolution to a confrontation with an employee?

16. What is a manager's responsibility for preventing harassment in the workplace?

17. How are employee grievances usually handled under a union contract?

CHAPTER 8.   DISTRIBUTION SYSTEM ADMINISTRATION

(Lesson 2 of 3 Lessons)

8.5   COMMUNICATION

Good communication is an essential part of good management skills. Both written and oral communication skills are needed to effectively organize and direct the operation of a water distribution system. Remember that communication is a two-part process; information must be given and it must be understood. Good listening skills are as important in communication as the information you need to communicate. As the manager of a water utility, you will need to communicate with employees, with your governing body, and with the public. Your communication style will be slightly different with each of these groups but you should be able to adjust easily to your audience.

8.50   Oral Communication

Oral communication may be informal, such as talking with employees, or it may be formal, such as giving a technical presentation. In both cases your words should be appropriate to the audience, for example, avoid technical jargon when talking with nontechnical audiences. As you talk, you should be observing your audience to be sure that what you are saying is getting across. If you are talking with an employee, it is a good idea to ask for feedback from the employee, especially if you are giving instructions. When the employee is talking, watch and listen carefully and clarify areas that seem unclear. Likewise, in a more formal presentation, watching your audience will give you feedback about how well your message is being received. Some tips for preparing a formal speech are given in Table 8.4.

8.51   Written Communication

Written communication is more demanding than oral communication and requires more careful preparation. Again, keep your audience in mind and use language that will be understood. Written communication requires more organization since you cannot clarify and explain ideas in response to your audience. Before you begin you should have a clear idea of exactly what you wish to communicate, then keep your language as concise as possible. Extra words and phrases tend to confuse and clutter your message. Good writing skills develop slowly, but you should be able to find good writing class-

es in your community if you need help improving your skills. In addition, many publications and computer software programs are available which assist you in writing the most commonly needed documents such as memos, letters, press releases, resumes, monitoring reports, and the annual report.

Before you can write a report you must first organize your thoughts. Ask yourself, what is the objective of this report? Am I trying to persuade someone of something? What information is important to communicate in this report? For whom is the report being written? How can I make it interesting? What does the reader want to learn from this report?

After you have answered the above questions, the next step is to prepare a general outline of how you intend to proceed with the preparation of the report. List not only key topics, but try to list all of the related topics. Then arrange the key topics in sequence so there is a workable, smooth flow from one topic to the next. Do not attempt to make your outline perfect. It is just a guide. It should be flexible. As you write you will find that you need to remove nonessential points and expand on more important points.

You might, for example, outline the following points in preparation for writing a report on a polymer testing program.

- A problem condition of high turbidity was discovered

- Polymer testing offered the best means of reducing turbidity

- Funds, equipment, and material were acquired

- Operators were trained

- Tests were conducted

- Results and conclusions were reached

- Corrective actions were planned and taken

- Conclusion, the tests did or did not produce the anticipated results or correct the problem

Once you are fairly sure you have included all the major topics you will want to discuss, go through the outline and write down facts you want to include on each topic. As you work through it, you may decide to move material from one topic to another. The outline will help you organize your ideas and facts.

TABLE 8.4   TIPS FOR GIVING AN ORAL PRESENTATION

1. Arrive early. Give yourself plenty of time to become familiar with the room, practice using your audiovisuals, and make any necessary changes in room setup.

2. Be ready for mistakes. Number the pages in your presentation and your audiovisuals. Check the order of the pages before the meeting begins.

3. Pace yourself. Don't speak too quickly, speak slowly and carefully. Keep a careful eye on audience reaction to be sure that you are speaking at a pace that can be understood.

4. Project yourself. Speak loudly and look at the audience. Do not talk with your back to the audience. Check that those in the back can hear you.

5. Be natural. Try not to read your presentation. Practice ahead of time so that you can speak normally and keep eye contact with your audience.

6. Connect with the audience. Try to smile and make eye contact with audience.

7. Involve the audience. Allow for audience questions and invite their comments.

8. Repeat audience questions. Always repeat the question so everyone can hear and to be sure that you hear the question correctly.

9. Know when to stop. Keep your remarks within the time allocated for your presentation and be aware that long, rambling speeches create a negative impression on the audience.

10. Use readable audiovisuals. Audiovisuals should enhance and reinforce your words. Be sure that all members of the audience can see and read your audiovisuals. Normal typewritten text is not readable on overheads; use large type so everyone can see. Use no more than five to seven key ideas per overhead.

11. Organize your presentation. Prepare an introduction, body of the speech, and conclusion. "Tell them what you're going to say, say it, and then tell them what you said." The presentation should have three to five main points presented in some logical order, for example, chronologically or from simple to complex.

When your outline is complete, you will have the essentials of your report. Now you need to tailor it to the audience that will be reading it. Take a few minutes to think about your audience. What information do they want? What aspects of the topics will they be most interested in reading? Each of the following groups may be interested in specific topics in the report. Consider these interests as you write.

1. Management

Management will have specific interests that relate to the cost effectiveness of the program. A report to management should include a summary that presents the essential information, procedures used, an analysis of the data (including trends), and conclusions. Be sure to include complete cost information. Did the benefits warrant the costs? As a result of the tests, can future expenses be reduced? Backup information and field data can be included in an appendix for those who want more information.

2. Other Utilities

Other utilities will be interested in costs but will also want more detailed information about how the program was performed. They will also be interested in the results and benefits of the program. Explain how the tests were done, the procedures, size of the crew, equipment used, source and availability of materials, difficulties encountered, and how they were overcome.

3. Citizen Groups

Citizens' interest will be more general. What is a polymer test and why is it needed? Is the polymer harmless? Will it injure fish or birds? How does the polymer test work? Who pays for the test? How much will it cost to implement results and will they be effective?

Your report may be written to include all of these groups. Adjust the outline to include the topics of interest to each group and identify the topics so readers can find the information most interesting to them. Keep the following information in mind as you write your report:

- Drafts. Good reports are not perfect the first time. Re-read and improve your report several times.

- Facts. Confine your writing to the facts and events that occurred. Include figures and statistics only when they make the report more effective. Include only the relevant facts. Large amounts of data should be put in the appendix. Do not clutter the report with unimportant data.

- Continuity. To be interesting and understood, a report must have continuity. It must make sense to the reader and be organized logically. In the report on the polymer testing, the report should be organized to show you had a problem, you had to find a way to identify where the problem existed, you did the testing, you identified the problems and the corrective actions.

- Effective. To be effective, a report should achieve the objective for which it was written. In this example we wanted to justify the costs for the program to management, help other utilities in conducting a similar program, and help citizens to understand what we were doing and why.

- Candid. A good report should be frank and straightforward. Keep the language appropriate for the audience. Do not try to impress your readers with technical terms they do not understand. Your purpose is to communicate information. Keep the information accurate and easy to understand.

The annual report is an important part of the management of the water distribution system. It is one of the most involved writing projects that the distribution system agency must put together. The annual report should be a review of what and how the agency operated during the past year and it should also include the goals for the next year. In many small communities the annual report may be presented orally to the city council rather than written. If this report is well written, it can be used to highlight accomplishments and provide support for future planning.

The first step to organizing the report is to make a list of three or four major accomplishments of the last year, then make a list of the top three goals for next year. These accomplishments and goals should be the focus of the report. The annual report should be a summary of the expenses, distribution system services provided, and revenues generated over the last year. As you organize this information, keep those accomplishments in mind and let the data tell the story of how the agency accomplished last year's tasks. The data by itself may seem boring but as you organize the data it becomes a meaningful description of the year's accomplishments. Conclude with projections for next year. The facts and figures should tell the audience how you plan to accomplish your goals for the next year. The annual report may be simple or complex depending on your community needs. A sample Table of Contents for a medium-sized agency is given in Table 8.5. When you are finished the annual report will be a valuable planning tool for the agency and can be used to build support for new projects.

TABLE 8.5   EXAMPLE OF TABLE OF CONTENTS FOR THE ANNUAL REPORT OF A DISTRIBUTION SYSTEM AGENCY

TABLE OF CONTENTS

Executive Summary

Summary of the Distribution System Services Including Flows and Costs

Review of Goals and Objectives for the Year

Special Projects Completed

Professional Awards or Recognition for the Agency or Its Staff

General Operating Conditions Including Regulatory Requirements

Expectations for the Next Year—Goals and Objectives

Recommended Changes for the Agency in Organizing, Staffing, Equipment, of Resources Summary

Appendixes: Operating Data

Budget

Information on Special Projects

QUESTIONS

Write your answers in a notebook and then compare your answers with those on page 512.

8.5A   What kinds of communication skills are needed by a manager?

8.5B   What are the most common written documents that a distribution system manager must write?

8.5C   What should be included in the annual report?

8.6   CONDUCTING MEETINGS

As a distribution system manager you will be asked to conduct meetings. These meetings may be with employees, your governing board, the public, or with other professionals in your field. Many new managers fail to prepare for these meetings and the meetings end up as a terrible waste of time. As a manager you need to learn to conduct meetings in a way that is productive and guides the participants into an active role. The following steps should be taken to conduct a productive meeting.

Before the meeting:

- Prepare an agenda and distribute it to all participants.

- Find an adequate meeting room.

- Set a beginning and ending time for the meeting.

During the meeting:

- Start the meeting on time.

- Clearly state the purpose and objectives of the meeting.

- Involve all the participants.

- Do not let one or two individuals dominate the meeting.

- Keep the discussion on track and on time with the agenda.

- When the group makes a decision or reaches consensus, restate your understanding of the results.

- Make clear assignments for participants and review them with everyone during the meeting.

After the meeting:

- Send out minutes of the meeting.

- Send out reminders, when appropriate, about any assignments made for participants, and the next meeting time.

QUESTIONS

Write your answers in a notebook and then compare your answers with those on page 512.

8.6A   With whom may a distribution system manager be asked to conduct meetings?

8.6B   What should be done before a meeting?

8.7   PUBLIC RELATIONS

8.70   Establish Objectives

The first step in organizing an effective public relations campaign is to establish objectives. The only way to know whether your program is a success is to have a clear idea of what you expect to achieve—for example, better customer relations, greater water conservation, and enhanced organizational credibility. Each objective must be specific, achievable, and measurable. It is also important to know your audience and tailor various elements of your public relations effort to specific groups you wish to reach, such as community leaders, school children, or the average customer. Your objective may be the same in each case, but what you say and how you say it will depend upon your target audience.

8.71   Distribution System Operations

Good public relations begin at home. Dedicated, service-oriented employees provide for better public relations than paid advertising or complicated public relations campaigns. For most people, contact with an agency employee establishes their first impression of the competence of the organization, and those initial opinions are difficult to change.

In addition to ensuring that employees are adequately trained to do their jobs and knowledgeable about the distribution system agency's operations, management has the responsibility to keep employees informed about the organization's plans, practices, and goals. Newsletters, bulletin boards, and regular, open communication between supervisors and subordinates will help build understanding and contribute to a team spirit.

Despite the old adage to the contrary, the customer is not always right. Management should try to instill among its employees the attitude that while the customer may be confused or unclear about the situation, everyone is entitled to courteous treatment and a factual explanation. Whenever possible, employees should phrase responses as positively, or neutrally, as possible, avoiding negative language. For example, "Your complaint" is better stated as "Your question." "You should have ..." is likely to make the customer defensive, while "Will you please ..." is courteous and respectful. "You made a mis-take" emphasizes the negative, "What we'll do..." is a positive, problem-solving approach.

8.72   The Mass Media

We live in the age of communications, and one of the most effective and least expensive ways to reach people is through the mass media—radio, television, and newspapers. Each medium has different needs and deadlines, and obtaining coverage for your issue or event is easier if you are aware of these constraints. Television must have strong visuals, for example. When scheduling a press conference, provide an interesting setting and be prepared to suggest good shots to the reporter. Radio's main advantage over television and newspapers is immediacy, so have a spokesperson available and prepared to give the interview over the telephone if necessary. Newspapers give more thorough, in-depth coverage to stories than do the broadcast media, so be prepared to spend extra time with print reporters and provide written backup information and additional contacts.

It is not difficult to get press coverage for your event or press conference if a few simple guidelines are followed:

1. Demonstrate that your story is newsworthy, that it involves something unusual or interesting.

2. Make sure your story will fit the targeted format (television, radio, or newspaper).

3. Provide a spokesperson who is interesting, articulate, and well prepared.

8.73   Being Interviewed

Whether you are preparing for a scheduled interview or are simply contacted by the press on a breaking news story, here are some key hints to keep in mind when being interviewed.

1. Speak in personal terms, free of institutional jargon.

2. Do not argue or show anger if the reporter appears to be rude or overly aggressive.

3. If you don't know an answer, say so and offer to find out. Don't bluff.

4. If you say you will call back by a certain time, do so. Reporters face tight deadlines.

5. State your key points early in the interview, concisely and clearly. If the reporter wants more information, he or she will ask for it.

6. If a question contains language or concepts with which you disagree, don't repeat them, even to deny them.

7. Know your facts.

8. Never ask to see a story before it is printed or broadcast. Doing so indicates that you doubt the reporter's ability and professionalism.

8.74 Public Speaking

Direct contact with people in your community is another effective tool in promoting your distribution system agency. Though the audiences tend to be small, a personal, face-to-face presentation generally leaves a strong and long-lasting impact on the listener.

Depending upon the size of the organization, your agency may wish to establish a speaker's bureau and send a list of topics to service clubs in the area. Visits to high schools and college campuses can also be beneficial, and educators are often looking for new and interesting topics to supplement their curriculum.

Effective public speaking takes practice. It is important to be well prepared while retaining a personal, informal style. Find out how long your talk is expected to be, and don't exceed that time frame. Have a definite beginning, middle, and end to your presentation. Visual aids such as charts, slides, or models can assist in conveying your message. The use of humor and anecdotes can help to warm up the audience and build rapport between the speaker and the listener. Just be sure the humor is natural, not forced, and that the point of your story is accessible to the particular audience. Try to keep in mind that audiences only expect you to do your best. They are interested in learning about their water distribution system, the quality of the safe drinking water you deliver, and they will appreciate that you are making a sincere effort to inform them about an important subject.

8.75 Telephone Contacts

First impressions are extremely important, and frequently a person's first contact with your water distribution system agency is over the telephone. A person who answers the phone in a courteous, pleasant, and helpful manner goes a long way toward establishing a friendly, cooperative atmosphere. Be sure anyone answering telephone inquiries receives appropriate training and conveys a positive image for the distribution system agency.

Following a few simple guidelines will help to start your agency off on the right note with your customers:

1. ANSWER CALLS PROMPTLY. Your conversation will get off to a better start if the phone is answered by the third or fourth ring.

2. IDENTIFY YOURSELF. This adds a personal note and lets the caller know whom he or she is talking to.

3. PAY ATTENTION. Don't conduct side conversations. Minimize distractions so you can give the caller your full atten-

tion, avoiding repetitions of names, addresses, and other pertinent facts.

4. MINIMIZE TRANSFERS. Nobody likes to get the runaround. Few things are more frustrating to a caller than being transferred from office to office, repeating the situation, problem, or concern over and over again. Transfer only those calls that must be transferred, and make certain you are referring the caller to the right person. Then, explain why you are transferring the call. This lets the caller know you are referring him or her to a co-worker for a reason and reassures the customer that the problem or question will be dealt with. In some cases, it may be better to take a message and have someone return the call than to keep transferring the customer's call.

8.76 Consumer Inquiries

No single set of rules can possibly apply to all types of consumer questions or complaints about water quality and service. There are, however, basic principles to follow in responding to inquiries and concerns.

1. BE PREPARED. Your employees should be familiar enough with your agency's organization, services, and policies to either respond to the question or complaint or locate the person who can.

2. LISTEN. Ask the customer to describe the problem and listen carefully to the explanation. Take written notes of the facts and addresses.

3. DON'T ARGUE. Callers often express a great deal of pent-up frustration in their contacts with an agency. Give the caller your full attention. Once you've heard them out, most people will calm down and state their problems in more reasonable terms.

4. AVOID JARGON. The average consumer lacks the technical knowledge to understand the complexities of water quality. Use plain, nontechnical language and avoid telling the consumer more than he or she needs to know.

5. SUMMARIZE THE PROBLEM. Repeat your understanding of the situation back to the caller. This will assure the customer that you understand the problem and offer the opportunity to clear up any confusion or missed communication.

6. PROMISE SPECIFIC ACTION. Make an effort to give the customer an immediate, clear, and accurate answer to the problem. Be as specific as possible without promising something you can't deliver.

In some cases, you may wish to have a representative of the agency visit the customer and observe the problem first hand. If the complaint involves water quality, take samples if necessary and report back to the customer to be sure the problem has been resolved.

Complaints can be a valuable asset in determining consumer acceptance and pinpointing water quality problems. Customer calls are frequently your first indication that something may be wrong. Responding to complaints and inquiries promptly can save the distribution system agency money and staff resources, and minimize the number of customers who are inconvenienced. Still, education can greatly reduce complaints about water quality. Information brochures, utility bill inserts, and other educational tools help to inform customers and avoid future complaints.

8.77 Distribution System Facility Tours

Tours of water distribution systems and treatment plants can be an excellent way to inform the public about your agency's efforts to provide a safe, high-quality water supply. Political leaders, such as the City Council and members of the Board of Supervisors, should be invited and encouraged to tour the facilities, as should school groups and service clubs.

A brochure describing your agency's goals, accomplishments, operations, and processes can be a good supplement to the tour and should be handed out at the end of the visit. The more visually interesting the brochure is, the more likely that it will be read, and the use of color, photographs, graphics, or other design features is encouraged. If you have access to the necessary equipment, production of a videotape program about the agency can also add interest to the facility tour.

The tour itself should be conducted by an employee who is very familiar with the distribution system and plant operations and can answer the types of questions that are likely to arise. Consider including:

1. A description of the sources of water supply,

2. History of the system, plant, the years of operation, modifications, and innovations over the years,

3. Major system and plant design features, including system and plant capacity and safety features,

4. Observation of the system, storage facilities, and treatment processes, including filtration, sedimentation, flocculation, and disinfection,

5. A visit to the laboratory, including information on the quality of water distributed to consumers, and

6. Anticipated improvements, expansions, and long-range plans for meeting future service needs.

System and plant tours can contribute to a water agency's overall program to gain financing for capital improvements. If the City Council or other governing board has seen the system and treatment processes first hand, it is more likely to understand the need for enhancement and support future funding.

QUESTIONS

Write your answers in a notebook and then compare your answers with those on pages 512 and 513.

8.7A What is the first step in organizing a public relations campaign?

8.7B How can employees be kept informed of the distribution system agency's plans, practices, and goals?

8.7C Which news medium is more likely to give a story thorough, in-depth coverage?

8.7D What is the key to effective public speaking?

8.7E How do customer complaints help a distribution system agency?

8.8 FINANCIAL MANAGEMENT

Financial management for a distribution system agency should include providing financial stability for the utility, careful budgeting, and providing capital improvement funds for future distribution system agency expansion. These three areas must be examined on a routine basis to ensure the continued operation of the utility. They may be formally reviewed on an annual basis or more frequently when the utility is changing rapidly. The utility manager should understand what is required for each of the three areas and be able to develop record systems that keep the utility on track and financially prepared for the future.

8.80 Financial Stability

How do you measure financial stability for a distribution system agency? Two very simple calculations can be used to help you determine how healthy and stable the finances are for the utility. These two calculations are the OPERATING RATIO and the COVERAGE RATIO. The operating ratio is a measure of the total revenues divided by the total operating expenses. The coverage ratio is a measure of the ability of the utility to pay the principal and interest on loans and bonds (this is known as *DEBT SERVICE*[5]) in addition to any unexpected expenses. A utility that is in good financial shape will have an operating ratio and coverage ratio above 1.0. In fact, most bonds and loans require the utility to have a coverage ratio of at least 1.25. As state and federal funds for utility improvements have become much more difficult to obtain, these financial indicators have become more important for utilities. Being able to show and document the financial stability of the utility is an important part of getting funding for more capital improvements.

The operating ratio is perhaps the simplest measure of a utility's financial stability. In essence, the utility must be generating enough revenue to pay its operating expenses. The actual ratio is usually computed on a yearly basis, since many utilities may have monthly variations that do not reflect the overall performance. The total revenue is calculated by adding up all revenue generated by user fees, hook-up charges, taxes or assessments, interest income, and special income. Next determine the total operating expenses by adding up the expenses of the utility, including administrative costs, salaries, benefits, energy costs, chemicals, supplies, fuel, equipment costs, equipment replacement fund, principal and interest payments, and other miscellaneous expenses.

---

[5] *Debt Service.   The amount of money required annually to pay the (1) interest on outstanding debts; or (2) funds due on a maturing bonded debt or the redemption of bonds.*

EXAMPLE 1

The total revenues for a utility are $1,686,000 and the operating expenses for the utility are $1,278,899. The debt service expenses are $560,000. What is the operating ratio? What is the coverage ratio?

Known		Unknown
Total Revenue, $	= $1,686,000	Operating Ratio
Operating Expenses, $	= $1,278,899	Coverage Ratio
Debt Service Expenses, $	= $560,000	

1. Calculate operating ratio.

$$\text{Operating Ratio} = \frac{\text{Total Revenue, \$}}{\text{Operating Expenses, \$}}$$

$$= \frac{\$1,686,000}{\$1,278,899}$$

$$= 1.32$$

2. Calculate non-debt expenses.

$$\text{Non-Debt Expenses, \$} = \text{Operating Exp, \$} - \text{Debt Service Exp, \$}$$
$$= \$1,278,899 - \$560,000$$
$$= \$718,899$$

3. Calculate coverage ratio.

$$\text{Coverage Ratio} = \frac{\text{Total Revenue, \$} - \text{Non-Debt Expenses, \$}}{\text{Debt Service Expenses, \$}}$$

$$= \frac{\$1,686,000 - \$718,899}{\$560,000}$$

$$= 1.73$$

These calculations provide a good starting point for looking at the financial strength of the utility. Both of these calculations use the total revenue for the utility, which is an important component for any utility budgeting. As managers we often focus on the expense side and forget to look carefully at the revenue side of utility management. The fees collected by the utility, including hook-up fees and user fees, must accurately reflect the cost of providing service. These fees must be reviewed annually and they must be increased as expenses rise to maintain financial stability. Some other areas to examine on the revenue side include how often and how well user fees are collected, the number of delinquent accounts, and the accuracy of meters. Some small communities have found they can cut their administrative costs significantly by switching to a quarterly billing cycle. The utility must have the support of the community to determine and collect user fees, and the utility must keep track of revenue generation as carefully as resource spending.

8.81 Budgeting

Budgeting for the utility is perhaps the most challenging task of the year for many managers. The list of needs usually is much larger than the possible revenue for the utility. The only way for the manager to prepare a good budget is to have good records from the year before. A system of recording or filing purchase orders (see Section f, "Procurement Records") or a requisition records system must be in place to keep track of expenses and prevent spending money that is not in the budget.

To budget effectively, a manager needs to understand how the money has been spent over the last year, the needs of the utility, and how the needs should be prioritized. The manager

also must take into account cost increases that cannot be controlled while trying to minimize the expenses as much as possible. The following problem is an example of the types of decisions a manager must make to keep the budget in line while also improving service from the utility.

EXAMPLE 2

A pump that has been in operation for 25 years pumps a constant 600 GPM through 47 feet of dynamic head. The pump uses 6,071 kilowatt-hours of electricity per month, at a cost of $0.085 per kilowatt-hr. The old pump efficiency has dropped to 63 percent. Assuming a new pump that operates at 86 percent efficiency is available for $9,730.00, how long would it take to pay for replacing the old pump?

Known		Unknown
Electricity, kW-hr/mo	= 6,071 kW-hr/mo	New Pump Payback Time, yr
Electricity Cost, $/kW-hr	= $0.085/kW-hr	
Old Pump Efficiency, %	= 63%	
New Pump Efficiency, %	= 86%	
New Pump Cost, $	= $9,730	

1. Calculate old pump operating costs in dollars per month.

$$\text{Old Pump Operating Costs, \$/mo} = (\text{Electricity, kW-hr/mo})(\text{Electricity Cost, \$/kW-hr})$$

$$= (6,071 \text{ kW-hr/mo})(\$0.085/\text{kW-hr})$$

$$= \$516.04/\text{mo}$$

2. Calculate new pump operating electricity requirements.

$$\text{New Pump Electricity, kW-hr/mo} = (\text{Old Pump Electricity, kW-hr/mo})\frac{(\text{Old Pump Eff, \%})}{(\text{New Pump Eff, \%})}$$

$$= (6,071 \text{ kW-hr/mo})\frac{(63\%)}{(86\%)}$$

$$= 4,447 \text{ kW-hr/mo}$$

3. Calculate new pump operating costs in dollars per month.

$$\text{New Pump Operating Costs, \$/mo} = (\text{Electricity, kW-hr/mo})(\text{Electricity Cost, \$/kW-hr})$$

$$= (4,447 \text{ kW-hr/mo})(\$0.085/\text{kW-hr})$$

$$= \$378.03/\text{mo}$$

4. Calculate annual cost savings of new pump.

$$\text{Cost Savings, \$/yr} = (\text{Old Costs, \$/mo} - \text{New Costs, \$/mo})(12 \text{ mo/yr})$$

$$= (\$516.04/\text{mo} - \$378.03/\text{mo})(12 \text{ mo/yr})$$

$$= \$1,656.12/\text{yr}$$

5. Calculate the new pump payback time in years.

$$\text{Payback Time, yr} = \frac{\text{Initial Cost, \$}}{\text{Savings, \$/yr}}$$

$$= \frac{\$9,730.00}{\$1,656.12/\text{yr}}$$

$$= 5.9 \text{ years}$$

In this example a payback time of 5.9 years is acceptable and would probably justify the expense for a new pump. This calculation was a simple payback calculation which did not take into account the maintenance on each pump, depreciation, and inflation. Many excellent references are available from EPA to help utility managers make more complex decisions about purchasing new equipment.

The annual report should be used to help develop the budget so that long-term planning will have its place in the budgeting process. The utility manager must track revenue generation and expenses with adequate records to budget effectively. The manager must also get input from other personnel in the utility as well as community leaders as the budgeting process proceeds. This input from others is invaluable to gain support for the budget and to keep the budget on track once adopted.

8.82   Equipment Repair/Replacement Fund

To adequately plan for the future, every utility must have a repair/replacement fund. The purpose of this fund is to generate additional revenue to pay for the repair and replacement of capital equipment as the equipment wears out. To prepare adequately for this repair/replacement, the manager should make a list of all capital equipment (this is called an asset inventory) and estimate the replacement cost for each item. The expected life span of the equipment must be used to determine how much money should be collected over time. When a treatment plant is new, the balance in repair/replacement fund should be increasing each year. As the plant gets older, the funds will have to be used and the balance may get dangerously low as equipment breakdowns occur. Perhaps the hardest job for the utility manager is to maintain a positive balance in this account with the understanding that this account is not meant to generate a "profit" for the utility but rather to plan for future equipment needs. In water treatment and distribution system facilities construction, providing an adequate repair/replacement fund is very important, but if this repair/replacement fund hasn't been reviewed annually, it must be updated.

To set up a repair/replacement fund for your utility, you should first put together a list of the equipment required for each treatment process and distribution system in your utility. Once you have this list, you need to estimate the life expectancy of the equipment and the replacement cost. From this list you can predict the amount of money you should set aside each year so that when each piece of equipment wears out, you will have enough money to replace that piece of equipment. Several EPA publications listed in Section 8.14, "Additional Reading," at the end of this chapter are excellent references for utility planning.

8.83   Water Rates

The process of determining the cost of water and establishing a water rate schedule for customers is a subject of much controversy. There is no single set of rules for determining water rates. The establishment of a rate schedule involves many factors including the form of ownership (investor or publicly owned), differences in regulatory control over the water utility (state commission or local authority), and differences in individual viewpoints and preferences concerning the appropriate philosophy to be followed to meet local conditions and requirements.

Generally, the development of water rate schedules involves the following procedures:

- A determination of the total revenue requirements for the period that the rates are to be effective (usually one year).

- A determination of all the cost components of system operations. That is, how much does it cost to treat the water? How much does it cost to distribute? How much does it cost to install a water service to a customer? How much are administrative costs?

- Distribution of the various component costs to the various customer classes in accordance with their requirements for service.

- The design of water rates that will recover from each class of customers, within practical limits, the cost to serve that class of customers.

Sales of water to customers may be metered or unmetered. In the case of metered sales, the charge to the customer is based on a rate schedule applied to the amount of water used through each water meter. If meters are not used, the charge per customer is based on a flat rate per period of time per fixture, foot of frontage, number of rooms, or other measurable unit. Although the flat rate basis still is fairly common, meter-based rates are more widely used.

See *SMALL WATER SYSTEM OPERATION AND MAINTENANCE*, Chapter 8, "Setting Water Rates for Small Water Utilities," for an explanation and examples of how to determine and set water rates. This publication is available from the Office of Water Programs, California State University, Sacramento, 6000 J Street, Sacramento, CA 95819-6025. Price, $45.00.

8.84   Capital Improvements and Funding in the Future

A capital improvements fund must be a part of the utility budget and included in the operating ratio. Your responsibility as the utility manager is to be sure that everyone, your governing body and the public, understands the capital improvement fund is not a profit for the utility but a replacement fund to keep the utility operating in the future.

Capital planning starts with a look at changes in the community. Where are the areas of growth in the community, where are the areas of decline, and what are the anticipated changes in industry within the community? After identifying the changing needs in the community, you should examine the existing utility structure. Identify your weak spots (in the distribution system or with in-plant processes). Make a list of the areas that will be experiencing growth, weak spots in the system, and anticipated new regulatory requirements. The list should include expected capital improvements that will need to be made over the next year, two years, five years, and ten years. You can use the information in your annual reports and other operational logs to help compile the list.

Once you have compiled this information, prioritize the list and make a timetable for improving each of the areas. Starting at the top of the priority list, estimate the costs for improvements and incorporate these costs into your capital improvement budget. The calculations you have made previously, including corrective to preventive maintenance ratios, operating ratio, coverage ratio, and payback time, will all be useful in prioritizing and streamlining your list of needs. Another useful ratio is the corrective to preventive maintenance ratio. `

You may find that some of your capital improvement needs could be met in more than one way. How do you decide which of several options is most cost-effective? How do you compare fundamentally different solutions? For example, assume your community's population is growing rapidly and you will need to increase treated water production by 20 percent by the end of the next ten years. Your existing wells cannot provide that amount of additional water and your filtration equipment is operating at 90 percent of design capacity. Possible solutions might include a combination of the following options, some of which might be implemented immediately while others might be brought on line in five or ten years:

- Rehabilitate some declining wells,

- Drill additional new wells,

- Develop an available surface water source,

- Install another filtration unit, or

- Install additional distribution system storage reservoirs.

To compare alternative plans you will need to calculate the present value (or present worth) of each plan; that is, the costs and benefits of each plan in today's dollars. This is done by identifying all the costs and benefits of each alternative plan over the same time period or time horizon. Costs should include not only the initial purchase price or construction costs, but also financing costs over the life of the loans or bonds and all operation and maintenance costs. Benefits include all of the revenue that would be produced by this facility or equipment, including connection and user fees. With the help of an experienced accountant, apply standard inflation, depreciation, and other economic discount factors to calculate the present value of all the benefits and costs of each plan during the same planning period. This will give you the cost of each plan in the equivalent of today's dollars.

Remember to involve all of your local officials and the public in the development of this capital improvement budget so they understand what will be needed.

Long-term capital improvements such as a new distribution system, a new plant, or a new treatment process are usually anticipated in your 10-year or 20-year projection. These long-term capital improvements usually require some additional financing. The basic ways for a utility to finance capital improvements are through general obligation bonds, revenue bonds, or loan funding programs.

General obligation bonds or *ad valorem* (based on value) taxes are assessed based on property taxes. These bonds usually have a lower interest rate and longer payback time, but the total bond limit is determined for the entire community. This means that the water utility will have available only a portion of the total bond capacity of the community. These bonds are not often used for funding water utility improvements today.

The second type of bond, the revenue bond, is commonly used to fund water distribution system improvements. This bond has no limit on the amount of funds available and the user charges provide repayment on the bond. To qualify for these bonds, the utility must show sound financial management and the ability to repay the bond. As the utility manager you should be aware of the provisions of the bond. Be sure the bond has a call date, which is the first date when you can pay off the bond. The common practice is for a 20-year bond to have a 10-year call date and for a 15-year bond to have an 8-year call date. The bond will also have a call premium, which is the amount of extra funds needed to pay off the debt on the call date. You should try to get your bonds a call premium of no more than 102 percent par. This means that for a debt of $200,000 on the call date, the total payoff would be $204,000, which includes the extra two percent for the call premium. You will need to get help from a financial advisor to prepare for and issue the bonds. These advisors will help you negotiate the best bond structure for your community.

Special assessment bonds may be used to extend services into specific areas. The direct users pay the capital costs and the assessment is usually based on frontage or area of real estate. These special assessments carry a greater risk to investors but may be the best way to extend service to some areas.

The most common way to finance water supply system improvements in the past has been federal and state grant programs. The Block Grants from HUD are still available for some projects and Rural Utilities Service (RUS) loans may also be used as a funding source. In addition, state revolving fund (SRF) programs provide loans (but not direct grants) for improvements. The SRF program has already been implemented with wastewater improvements and the new Safe Drinking Water Regulations include an SRF program for funding water treatment improvements. These SRF programs are very competitive and utilities must provide evidence of sound financial management to qualify for these loans. You should contact your state regulatory agency to find out more about the SRF program in your state.

8.85   Financial Assistance

Many small water treatment and distribution systems need additional funds to repair and upgrade their systems. Potential funding sources include loans and grants from federal and state agencies, banks, foundations, and other sources. Some of the federal funding programs for small public utility systems include:

- Appalachian Regional Commission (ARC),

- Department of Housing and Urban Development (HUD) (provides Community Development Block Grants),

- Economic Development Administration (EDA),

- Indian Health Service (IHS), and

- Rural Utilities Service (RUS)(formerly Farmer's Home Administration (FmHA) and Rural Development Administration (RDA)).

For additional information see, "Financing Assistance Available for Small Public Water Systems," by Susan Campbell, Benjamin W. Lykins, Jr., and James A. Goodrich, *Journal American Water Works Association*, June 1993, pages 47-53.

Another valuable contact is the Environmental Financing Information Network which provides information on financing alternatives for state and local environmental programs and projects in the form of abstracts of publications, case studies, and contacts. Contact Environmental Financing Information Network, U.S. Environmental Protection Agency (EPA), EFIN (mail code 2731R), Ariel Rios Building, 1200 Pennsylvania Avenue, NW, Washington, DC 20460. Phone (202) 564-4994 and FAX (202) 565-2587.

Also many states have one or more special financing mechanisms for small public utility systems. These funds may be in the form of grants, loans, bonds, or revolving loan funds. Contact your state drinking water agency for more information.

QUESTIONS

Write your answers in a notebook and then compare your answers with those on page 513.

8.8A   List the three main areas of financial management for a distribution system agency.

8.8B   How is a distribution system agency's operating ratio calculated?

8.8C   Why is it important for a manager to consult with other distribution system agency personnel and with community leaders during the budgeting process?

8.8D   How can long-term capital improvements be financed?

8.8E   What is a revenue bond?

END OF LESSON 2 OF 3 LESSONS
ON
DISTRIBUTION SYSTEM ADMINISTRATION

Please answer the discussion and review questions next.

DISCUSSION AND REVIEW QUESTIONS

Chapter 8.   DISTRIBUTION SYSTEM ADMINISTRATION

(Lesson 2 of 3 Lessons)

Write the answers to these questions in your notebook before continuing. The question numbering continues from Lesson 1.

18. With whom do managers need to communicate?

19. What information should be included in the distribution system agency's annual report?

20. List four steps that can be taken during a meeting to make sure it is a productive meeting.

21. What happens any time you or a member of your distribution system agency comes in contact with the public?

22. What attitude should management try to develop among its employees regarding the consumer?

23. What is the value of consumer complaints?

24. How do you measure financial stability for a utility?

25. How can a manager prepare a good budget?

CHAPTER 8.   DISTRIBUTION SYSTEM ADMINISTRATION

(Lesson 3 of 3 Lessons)

8.9   OPERATIONS AND MAINTENANCE

8.90   The Manager's Responsibilities

A distribution system agency manager's specific operation and maintenance (O & M) responsibilities vary depending on the size of the agency. At a small agency, the manager may oversee all agency operations while also serving as chief operator and supervising a small staff of operations and maintenance personnel. In larger agencies, the manager may have no direct, day-to-day responsibility for operations and maintenance but is ultimately responsible for efficient, cost-effective operation of the entire agency. Whether large or small, every agency needs an effective operations and maintenance program.

8.91   Purpose of O & M Programs

The purpose of O & M programs is to maintain design functionality (capacity) and/or to restore the system components to their original condition and thus functionality. Stated another way, does the system perform as designed and intended? The ability to effectively operate and maintain a water distribution system so it performs as intended depends greatly on proper design (including selection of appropriate materials and equipment), construction and inspection, acceptance, and system start-up. Permanent system deficiencies that affect O & M of the system are frequently the result of these phases. O & M staff should be involved at the beginning of each project, including planning, design, construction, acceptance, and start-up. When a water distribution system is designed with future O & M considerations in mind, the result is a more effective O & M program in terms of O & M cost and performance.

Effective O & M programs are based on knowing what components make up the system, where they are located, and the condition of the components. With that information, proactive maintenance can be planned and scheduled, rehabilitation needs identified, and long-term Capital Improvement Programs (CIP) planned and budgeted. High-performing agencies have all developed performance measurements of their O & M program and track the information necessary to evaluate performance.

8.92   Types of Maintenance

Water supply and distribution system maintenance can be either a proactive or a reactive activity. In general, maintenance activities can be classified as corrective maintenance, preventive maintenance, and predictive maintenance.

Corrective maintenance, including emergency maintenance, is reactive. For example, a piece of equipment or a system is allowed to operate until it fails, with little or no scheduled maintenance occurring prior to the failure. Only when the equipment or system fails is maintenance performed. Reli-

ance on reactive maintenance will always result in poor system performance, especially as the system ages. Distribution system agencies taking a corrective maintenance approach are characterized by:

● The inability to plan and schedule work,

● The inability to budget adequately,

● Poor use of resources, and

● A high incidence of equipment and system failures.

Emergency maintenance involves two types of emergencies: normal emergencies and extraordinary emergencies. Public distribution system agencies are faced with normal emergencies such as water main breaks on a daily basis. Normal emergencies can be reduced by an effective maintenance program. Extraordinary emergencies, such as high-intensity rainstorms, hurricanes, floods, and earthquakes, will always be unpredictable occurrences. However, the effects of extraordinary emergencies on the agency's performance can be minimized by implementation of a planned maintenance program and development of a comprehensive emergency response plan (see Section 8.10).

Preventive maintenance is proactive and is defined as a programmed, systematic approach to maintenance activities. This type of maintenance will always result in improved system performance except in the case where major chronic problems are the result of design and/or construction flaws that cannot be corrected by O & M activities. Proactive maintenance is performed on a periodic (preventive) basis or an as-needed (predictive) basis. Preventive maintenance can be scheduled on the basis of specific criteria such as equipment operating time since the last maintenance was performed, or passage of a certain amount of time (calendar period). Lubrication of motors, for example, is frequently based on running time.

The major elements of a good preventive maintenance program include the following:

1. Planning and scheduling,

2. Records management,

3. Spare parts management,

4. Cost and budget control,

5. Emergency repair procedures, and

6. Training program.

Some benefits of taking a preventive maintenance approach are:

1. Maintenance can be planned and scheduled,

2. Work backlog can be identified,

3. Adequate resources necessary to support the maintenance program can be budgeted,

4. Capital Improvement Program (CIP) items can be identified and budgeted for, and

5. Human and material resources can be used effectively.

Predictive maintenance, which is also proactive, is a method of establishing baseline performance data, monitoring performance criteria over a period of time, and observing changes in performance so that failure can be predicted and maintenance can be performed on a planned, scheduled basis. Knowing the condition of the system makes it possible to plan and schedule maintenance on an "as required" basis and thus avoid unnecessary maintenance.

In reality, every agency operates their system with corrective and emergency maintenance, preventive maintenance, and predictive maintenance methods. The goal, however, is to reduce the corrective and emergency maintenance efforts by performing preventive maintenance which will minimize system failures that result in leakage and water main breaks.

System performance is frequently a reliable indicator of how the system is operated and maintained. Agencies that rely primarily on corrective maintenance as their method of operating and maintaining the system are never able to focus on preventive and predictive maintenance since most of their resources are directed at corrective maintenance activities and it is difficult to free up these resources to begin developing preventive maintenance programs. For an agency to develop an effective proactive maintenance program, they must add initial resources over and above those currently existing in order to establish preventive and predictive maintenance programs.

8.93 Benefits of Managing Maintenance

The goal of managing maintenance is to minimize investments of labor, materials, money, and equipment. In other words, we want to manage our human and material resources as effectively as possible, while delivering a high level of service to our customers. The benefits of an effective operation and maintenance program are as follows:

- Ensuring the availability of facilities and equipment as intended.

- Maintaining the reliability of the equipment and facilities as designed. Water distribution systems are required to oper-

ate 24 hours per day, 7 days per week, 365 days per year. Reliability is a critical component of the operation and maintenance program. If equipment and facilities are not reliable, then the ability of the system to perform as designed is impaired.

- Maintaining the value of the investment. Water treatment and distribution systems represent major capital investments for communities and are major capital assets of the community. If maintenance of the system is not managed, equipment and facilities will deteriorate through normal use and age. Maintaining the value of the capital asset is one of the water distribution system manager's major responsibilities. Accomplishing this goal requires ongoing investment to maintain existing facilities and equipment, extend the life of the system, and establish a comprehensive O & M program.

- Obtaining full use of the system throughout its design life.

- Collecting accurate information and data on which to base the operation and maintenance of the system and justify requests for the financial resources necessary to support it.

QUESTIONS

Write your answers in a notebook and then compare your answers with those on page 513.

8.9A What is the purpose of an operation and maintenance (O & M) program?

8.9B What are the three common types of maintenance?

8.9C List the major elements of a good preventive maintenance program.

8.94 SCADA Systems

8.940 Description of SCADA (ss-KAY-dah) Systems

SCADA stands for Supervisory Control And Data Acquisition system. This is a computer-monitored alarm, response, control, and data acquisition system used by operators to monitor and adjust their distribution system facilities and treatment processes.

A SCADA system collects, stores, and analyzes information about all aspects of operation and maintenance, transmits alarm signals when necessary, and allows fingertip control of alarms, equipment, and processes. SCADA provides the information that operators need to solve minor problems before they become major incidents. As the nerve center of a water distribution system, the SCADA system allows operators to enhance the efficiency of their water distribution system by keeping the operators fully informed and fully in control.

A typical SCADA system is made up of basically four groups of components:

1. Field-mounted sensors, instrumentation, and controlled or monitored equipment. These devices sense system variables and generate input signals to the SCADA system for monitoring. These devices also receive the command output signals from the SCADA system.

2. Remote Terminal Units (RTUs). These devices gather the data from the field-mounted instruments and provide the control signals to the field equipment.

3. Communications medium. This is the link between the SCADA RTUs and the main control location. There are many communications mediums available for transmitting signals between the RTUs and the supervisory control station. Some of the options presently available include:

- FM (VHF/UHF) radio,
- Dedicated leased telephone circuits,
- Privately owned metallic signal lines,
- Conventional dial telephone lines (pulse or tone/DTMF),
- Coaxial cable networks,
- Spread spectrum radio,
- Fiber-optic cable,
- Microwave,
- 900 MHz radio,
- Cellular telephone, and
- Ground station satellites.

4. Supervisory control and monitoring equipment. There are three basic categories of supervisory equipment:

a. Hardware-based systems include graphic displays, annunciator lamp boxes, chart recorders, and similar equipment.

b. Software-based systems include all computer-based systems such as microcomputers, workstations, mini and mainframes. Software-based systems sometimes offer more flexibility and capabilities than hardware-based systems and are often less expensive.

c. Hybrid systems are a combination of hardware and software systems; for example, a PC (microcomputer) with a graphic display panel.

Applications for SCADA systems include water distribution system control and monitoring, water treatment plant control monitoring, and other related applications. SCADA systems can vary from merely data collection and storage to total data analysis, interpretation, and process control.

In water applications, SCADA systems monitor levels, pressures, and flows and also operate pumps, valves, and alarms. They monitor temperatures, speeds, motor currents, pH, chlorine residuals, and other operating guidelines, and provide control as necessary. SCADA also logs event and analog signal trends and monitors equipment operating time for maintenance purposes.

A SCADA system might include liquid level, pressure, and flow sensors. The measured (sensed) information could be transmitted by one of the communications systems listed earlier to a computer system which stores, analyzes, and presents the information. The information may be read by an operator on dials or as digital readouts or analyzed and plotted by the computer as trend charts.

Most SCADA systems present a graphical picture of the overall system on the screen of a computer monitor. In addition, detailed pictures of specific portions of the system can be examined by the operator following a request and instructions to the computer. The graphical displays on a TV or computer screen can include current operating information. The operator can observe this information, analyze it for trends, or determine if it is within acceptable operating ranges, and then decide if any adjustments or changes are necessary.

SCADA systems are capable of analyzing data and providing operating, maintenance, regulatory, and annual reports. In some plants operators rely on a SCADA system to help them prepare daily, weekly, and monthly maintenance schedules, monitor the spare parts inventory status, order additional spare parts when necessary, print out work orders and record completed work assignments.

SCADA systems can also be used to enhance energy conservation programs. For example, operators can develop energy management routines that allow for both maximum energy savings and maximum water storage prior to entering "on-peak" periods. In this type of system, power meters are used to accurately measure and record power consumption and the information can then be reviewed by operators to watch for changes that may indicate equipment problems.

Emergency response procedures can be programmed into a SCADA system. Operator responses can be provided for different operational scenarios that could be encountered as a result of adverse weather changes, fires, or earthquakes.

QUESTIONS

Write your answers in a notebook and then compare your answers with those on page 513.

8.9D　What does SCADA stand for?

8.9E　What does a SCADA system do?

8.9F　How could measured (sensed) information be transmitted?

8.941　Typical Water Treatment and Distribution SCADA Systems

Water treatment and water distribution SCADA systems are usually linked together with the controls located at the water treatment plant. Information that historically was recorded on paper strip charts is now being recorded and stored (archived) by computers. This information can be easily retrieved and reviewed by a SCADA system, rather than examining years of strip chart records to find needed information. Therefore,

SCADA systems quickly provide operators with the information they need to make informed decisions.

SCADA systems can be used at water treatment plants to provide a constant flow for treatment for long periods of time. The raw water can be monitored continuously for turbidity, pH, and temperature. If these water quality indicators change significantly or exceed predetermined levels, the SCADA system will alert the operator to the change. The operator must then determine if a change in chemical doses or operating procedures is necessary.

Filtration systems are commonly monitored by SCADA systems. Influent and effluent turbidity levels, head losses through the filter, filter flows, and filter valve settings (70 percent open, for example) are all measured, recorded, and stored for future reference. If the flow needs to be increased, the operator has "fingertip control" of each filter valve and can open each filter valve as much as desired from the control room. This procedure is much easier, faster, and more accurate than manually turning each valve.

Filter backwashing is another function that a SCADA system can be programmed to control using any of three basic approaches: (1) SCADA monitors filter head loss and automatically starts backwashing after an elapsed time; (2) the SCADA system initiates backwashing when a target head loss is exceeded; or (3) the operator can start the backwash on the basis of an analysis of a head loss and/or filtered water turbidity trend chart drawn by the SCADA system over a time period since the last backwash. Also, by analyzing filtered water turbidity, the operator can decide to reduce the flow to the filter or make adjustments to the coagulant chemical and/or the prefilter polymer.

Clear well water levels and fluctuations, as well as expected consumer demands, can be studied by operators to determine if the raw water flows to the treatment plant need to be increased or decreased.

Historical operating data stored in a SCADA system is readily available at any time. The SCADA system can be asked to identify all days during the last two years when the temperature exceeded 10°F (38°C) or the demand was greater than 50 MGD (190 ML/day). The system demands and plant performance under these conditions can be recalled by the SCADA system, analyzed by the operators, and the results used to operate the plant under these conditions (100°F or 38°C or demand greater than 50 MGD (190 ML/day)).

Distribution system service storage tank levels, system demands, and system pressures are all recorded and plotted by SCADA systems. Operators study this information and decide

if adjustments are necessary in booster pump operation and/or target levels in service storage tanks. SCADA systems can plot system pressures over time against flow demands. A steady pressure in spite of fluctuating flows shows that the operators are in control of the system. If a customer phones and complains about a loss of water pressure, a review of system pressures can indicate if a problem exists or if the problem might be in the customer's home.

Hand-held computers are being used to assign work orders to field crews. Crews could be instructed to perform routine preventive maintenance on hydrants and valves or to investigate a drinking water taste or odor complaint. The field crew uses the computer to record standard tasks performed and also to record special comments. The standardized comments allow for field information to be retrieved and analyzed (queried).

Electrical energy consumption can be optimized by the use of SCADA systems. Time-power management describes procedures used to minimize power costs and meet consumer water and pressure demands. Most power companies are anxious to work with operators to try to increase water treatment and distribution power consumption during periods when electrical system power demands are low and also to decrease water power consumption during periods of peak demands on electrical power supplies. SCADA systems also monitor power consumption and conduct a diagnostic performance of pumps. Operators can review this information and then identify potential pump problems before they become serious.

SCADA systems are continually improving and helping operators do a better job. Today operators can create their own screens, their own graphics, and show whatever operating characteristics they wish to display on the screen. The main screen could be a flow diagram of the supply, treatment, or distribution system. Critical operating information could be displayed for each pumping station or each treatment process and detailed screens could be easily reached for each piece of equipment.

Information on the screen should be color coded with the colors of red, yellow, green, blue, white, and any other necessary colors. Colors could be used to indicate if a pump is running, ready, unavailable, or failed and/or if a valve is open, closed, moving, unavailable, or failed. A "failed" signal is used by the computer to inform the operator that something is wrong with the information or signal the computer is receiving or is being instructed to display. The signal is not logical with the rest of the information available to the computer. For example, if there is no power to a motor, then the motor can't be running even though the computer is receiving a signal that indicates it is running. Therefore the computer would send a "failed" signal.

The operator can request a computer to display a summary of all alarm conditions in a plant, a particular plant area, or a segment of the distribution system. A blinking alarm signal indicates that the alarm condition has not yet been acknowledged by the operator. On the other hand, a steady alarm signal, one that is not blinking, indicates that the alarm has been acknowledged but the alarm indicator will stay on until the condition causing it is fixed. Also the screen could be set up to automatically designate certain alarm conditions as *PRIORITY* alarms, which means they require immediate operator attention.

Current laptop computers allow operators to plug into a telephone at home or on vacation, access their distribution system or plant SCADA system, and help operators on duty solve operational problems. Computer networking systems allow operators at terminals in offices, in plants, and in the field to work together and use the same information or whatever information they need from one central file service (computer database).

A drawback of some SCADA systems is that when the system fails ("goes down") due to a power failure, the system will often display the numbers that were registered immediately before the failure and not display the current numbers. The operator may therefore experience a period of time when accurate, current information about the system is not immediately available.

Customer satisfaction with the performance of a water utility can be enhanced by the use of an effective SCADA system. Historically when a booster pump failed or "tripped out," the first a utility learned of this problem was when an irate consumer phoned the water agency and complained about being out of water. The utility then had to contact the operators and send a crew into the field to correct the problem. Today, SCADA systems often alert operators to a booster pump failure or "trip out" immediately. The operator may be able to correct or "override" the failure or "trip out" from the office without ever having to travel to the problem booster pump in the field. Thus, the problem is corrected without the customers ever being aware that a problem occurred and was corrected almost immediately.

When operators decide to initiate or expand the SCADA system for their plant or distribution system, the first step is to decide what the SCADA system should do to make the operators' jobs easier, more efficient, and safer and to make their facilities' performance more reliable and cost effective. Cost savings associated with the use of a SCADA system frequently include reduced labor costs for operation, maintenance, and monitoring functions that were formerly performed manually. Precise control of chemical feed rates by a SCADA system eliminates wasteful overdosing. Preventive maintenance monitoring can save on equipment and repair costs and, as previously noted, energy savings may result from use of off-peak electrical power rates. Operators should visit facilities with SCADA systems and talk to the operators about what they find beneficial and also detrimental with regard to SCADA systems and how the systems contribute to their performance as operators.

The greatest challenge for operators using SCADA systems is to realize that just because a computer says something (a pump is operating as expected), *THIS DOES NOT*

MEAN THAT THE COMPUTER IS ALWAYS CORRECT. Also when the system fails due to a power failure or for any other reason (natural disaster), operators will be required to operate the plant manually and without critical information. Could you do this?

Operators will always be needed to question and analyze the results from SCADA systems. They will be needed to *SEE* if the floc looks OK, to *LISTEN* to a pump to be sure it sounds proper, and to *SMELL* the water and the equipment to determine if unexpected or unidentified changes are occurring. Water treatment plants and distribution systems will always need alert, knowledgeable, and experienced operators who have a "feel" for their plant and their distribution system.

QUESTIONS

Write your answers in a notebook and then compare your answers with those on page 513.

8.9G   What do SCADA systems do with information that was formerly recorded on paper strip charts?

8.9H   What water quality indicators can be easily monitored by a SCADA system?

8.9I   What can an operator do when a customer phones and complains about a loss of water pressure?

8.9J   What are the greatest challenges for operators using SCADA systems?

8.95   Cross-Connection Control Program

8.950   Importance of Cross-Connection Control

BACKFLOW[6] of contaminated water through cross connections into community water systems is not just a theoretical problem. Contamination through cross connections has consistently caused more waterborne disease outbreaks in the United States than any other reported factor. Inspections have often disclosed numerous unprotected cross connections between public water systems and other piped systems on consumers' premises which might contain wastewater, stormwater, processed waters (containing a wide variety of chemicals), and untreated supplies from private wells, streams, and ocean waters. Therefore, an effective cross-connection control program is essential.

Backflow results from either *BACK PRESSURE*[7] or *BACK-SIPHONAGE*[8] situations in the distribution system. Back pressure occurs when the user's water supply is at a higher pressure than the public water supply system (Figure 8.4). Typical locations where back pressure problems could develop include services to premises where wastewater or toxic chemicals are handled under pressure or where there are unapproved auxiliary water supplies such as a private well or the use of surface water or seawater for firefighting. Backsiphonage is caused by the development of negative or below atmospheric pressures in the water supply piping (Figure 8.5). This condition can occur when there are extremely high water demands (firefighting), water main breaks, or the use of on-line booster pumps.

---

[6] *Backflow.   A reverse flow condition, created by a difference in water pressures, which causes water to flow back into the distribution pipes of a potable water supply from any source or sources other than an intended source. Also see BACKSIPHONAGE.*

[7] *Back Pressure.   A pressure that can cause water to backflow into the water supply when a user's water system is at a higher pressure than the public water system.*

[8] *Backsiphonage.   A form of backflow caused by a negative or below atmospheric pressure within a water system. Also see BACKFLOW.*

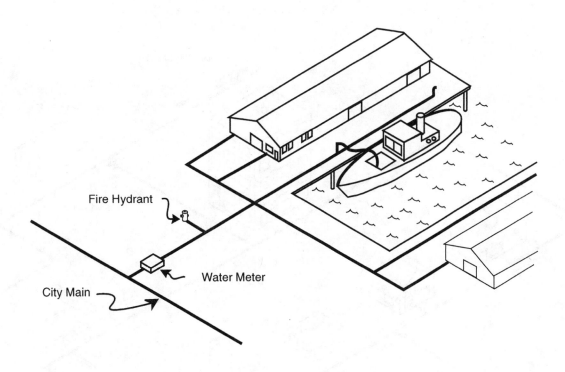

HYDRAULIC GRADIENT

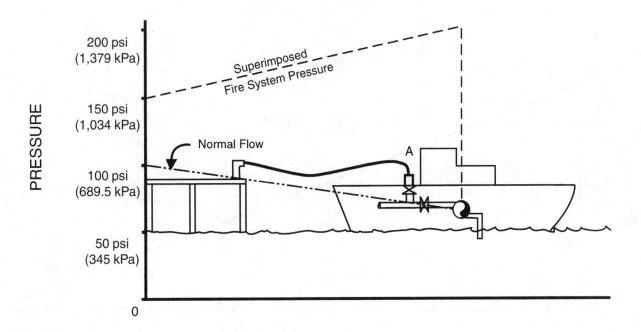

Fig. 8.4   Backflow due to back pressure

(Source: *MANUAL OF CROSS-CONNECTION CONTROL PRACTICES AND PROCEDURES*,
Sanitary Engineering Branch, California Department of Health Services, Berkeley, CA)

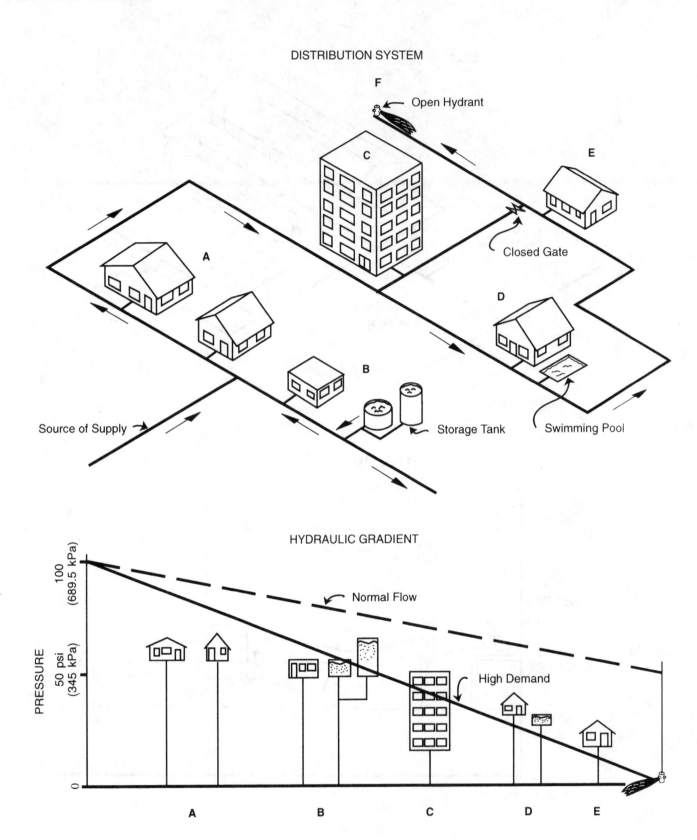

DISTRIBUTION SYSTEM

F
← Open Hydrant

C

E

Closed Gate

A

D

Source of Supply →

B

Storage Tank

Swimming Pool

HYDRAULIC GRADIENT

← Normal Flow

High Demand

PRESSURE

100 (689.5 kPa)

50 psi (345 kPa)

0

A   B   C   D   E

Fig. 8.5   Backsiphonage due to extremely high water demand
(Source: *MANUAL OF CROSS-CONNECTION CONTROL PROCEDURES AND PRACTICES*,
Sanitary Engineering Branch, California Department of Health Services, Berkeley, CA)

The best way to prevent backflow is to permanently eliminate the hazard. Back pressure hazards can be eliminated by severing (eliminating) any direct connection at the pump causing the back pressure with the domestic water supply system. Another solution to the problem is to require an air gap separation device (Figure 8.6) where the water supply service line connects to the private system under pressure from the pump. To eliminate or minimize backsiphonage problems, proper enforcement of plumbing codes and improved water distribution and storage facilities will be helpful. As an additional safety factor for certain selected conditions, a double check valve (Figure 8.7) may be required at the meter.

8.951 Program Responsibilities

Responsibilities in the implementation of cross-connection control programs are shared by water suppliers, water users (businesses and industries), health agencies, and plumbing officials. The water supplier is responsible for preventing contamination of the public water system by backflow. This responsibility begins at the source, includes the entire distribution system, and ends at the user's connection. To meet this responsibility, the water supplier must issue (promulgate) and enforce needed laws, rules, regulations, and policies. Water service should not be provided to premises where the strong possibility of an unprotected cross connection exists. The essential elements of a water supplier cross-connection control program are discussed in the next section.

The water user is responsible for keeping contaminants out of the potable water system on the user's premises. When backflow prevention devices are required by the health agency or water supplier, the water user must pay for the installation, testing, and maintenance of the approved devices. The user is also responsible for preventing the creation of cross connections through modifications of the plumbing system on the premises. The health agency or water supplier may, when necessary, require a water user to designate a water supervisor or foreman to be responsible for the cross-connection control program within the water user's premises.

The local or state health agency is responsible for issuing and enforcing laws, rules, regulations, and policies needed to control cross connections. Also this agency must have a program that ensures maintenance of an adequate cross-connection control program. Protection of the system on the user's premises is provided where needed by the water utilities.

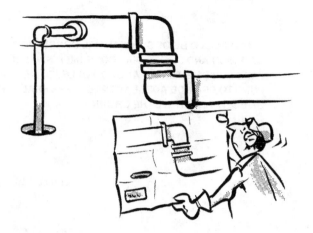

The plumbing agency (building inspectors) is responsible for the enforcement of building regulations relating to prevention of cross connections on the user's premises.

QUESTIONS

Write your answers in a notebook and then compare your answers with those on page 513.

8.9K   What has caused more waterborne disease outbreaks in the United States than any other reported factor?

8.9L   Who is usually responsible for the implementation of cross-connection control programs?

8.9M   What is the water user's cross-connection control responsibility?

8.952 Water Supplier Program

The following elements should be included in each water supplier's cross-connection control program:

1. Enactment of an ordinance providing enforcement authority if the supplier is a government agency, or enactment of appropriate rules of service if the system is investor-owned.[9]

2. Training of personnel on the causes of and hazards from cross connections and procedures to follow for effective cross-connection control.

3. Listing and inspection or reinspection on a priority basis of all existing facilities where cross connections are of concern.

4. Review and screening of all applications for new services or modification of existing services for cross-connection hazards to determine if backflow protection is needed.

5. Obtaining and maintaining a list of approved backflow prevention devices and a list of certified testers, if available.

6. Acceptable installation of the proper type of device needed for the specific hazard on the premises.

---

[9] A typical ordinance is available in CROSS-CONNECTION CONTROL MANUAL, available from National Technical Information Service (NTIS), 5285 Port Royal Road, Springfield, VA 22161. Order No. PB91-145490. EPA No. 570-9-89-007. Price, $33.50, plus $5.00 shipping and handling per order.

TANK SHOULD BE OF SUBSTANTIAL CONSTRUCTION AND
OF A KIND AND SIZE TO SUIT CONSUMER'S NEEDS.
TANK MAY BE SITUATED AT GROUND LEVEL (WITH A
PUMP TO PROVIDE ADEQUATE PRESSURE HEAD) OR
BE ELEVATED ABOVE THE GROUND.

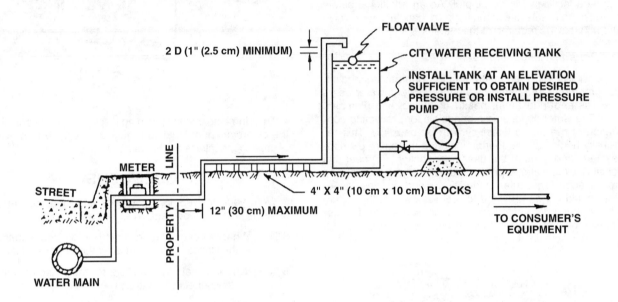

Fig. 8.6   *Typical air gap separation*

(From *MANUAL OF CROSS-CONNECTION CONTROL PRACTICES AND PROCEDURES*, Sanitary Engineering Branch, California Department of Health Services, Berkeley, CA)

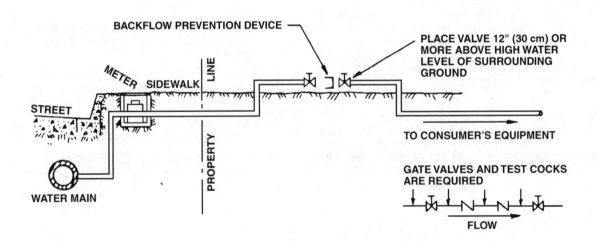

Fig. 8.7   *Typical double check valve backflow prevention device*

(From *MANUAL OF CROSS-CONNECTION CONTROL PRACTICES AND PROCEDURES*, Sanitary Engineering Branch, California Department of Health Services, Berkeley, CA)

7. Routine testing of installed backflow prevention devices as required by the health agency or the water supplier. Contact the health agency for approved procedures.

8. Maintenance of adequate records for each backflow prevention device installed, including records of inspection and testing.

9. Notification of each water user when a backflow prevention device has to be tested. This should be done at least once a year after installation/repair of the device.

10. Maintenance of adequate pressures throughout the distribution system at all times to minimize the hazards from any undetected cross connections that may exist.

All field personnel should be constantly alert for situations where cross connections are likely to exist, whether protection has been installed or not. An example is a contractor using a fire hose from a hydrant to fill a tank truck for dust control or the jetting (for compaction) of pipe trenches. Operators should especially be on the lookout for illegal bypassing of installed backflow prevention devices. For additional information about the potential for cross connections occurring at various type of industries, see Chapter 5, Section 5.3, "Cross-Connection Control."

8.953 Types of Backflow Prevention Devices

Different types of backflow prevention devices are available. The particular type of device most suitable for a given situation depends on the degree of health hazard, the probability of backflow occurring, the complexity of the piping on the premises, and the probability of the piping being modified. The higher the assessed risk due to these factors, the more reliable and positive the type of device needed. The types of devices normally approved are listed below according to the degree of assessed risk, with the type of device providing the greatest protection listed first. Only the first three devices are approved for use at service connections.

1. Air gap separation.

2. Reduced-pressure principle (RPP) device.

3. Double check valve.

4. Pressure vacuum breaker (only used for internal protection on the premises).

5. Atmospheric (non-pressure) vacuum breaker.

Figure 8.6 shows a typical air gap separation device and its recommended location. Figure 8.7 shows the installation of a typical double check valve backflow prevention device. These devices are normally installed on the water user's side of the connection to the utility's system and as close to the connection as practical. Figure 8.8 shows a typical installation of pressure vacuum breakers.

Only backflow prevention devices that have passed both laboratory and field evaluations by a recognized testing agency and that have been accepted by the health agency and the water supplier should be used.

8.954 Devices Required for Various Types of Situations

The state or local health agency should be contacted to determine the actual types of devices acceptable for various situations inside the consumer's premises. However, the types of devices generally acceptable for particular situations can be mentioned.

An air gap or a reduced-pressure principle (RPP) device is normally required at services to wastewater treatment plants, wastewater pumping stations, reclaimed water reuse areas, areas where toxic substances in toxic concentrations are handled under pressure, and premises having an auxiliary water supply that is or may be contaminated. The ultimate degree of protection is also needed in cases where fertilizer, herbicides, or pesticides are injected into a sprinkler system.

A double check valve device should be required when a moderate hazard exists on the premises or where an auxiliary supply exists, but adequate protection on the premises is provided.

Atmospheric and pressure vacuum breakers are usually required for irrigation systems; however, they are not adequate in situations where they may be subject to back pressure. If there is a possibility of back pressure, a reduced-pressure principle device is needed.

QUESTIONS

Write your answers in a notebook and then compare your answers with those on page 513.

8.9N What factors should be considered when selecting a backflow prevention device?

8.9O Air gap or reduced-pressure principle backflow prevention devices are installed under what conditions?

8.10 EMERGENCY RESPONSE

8.100 Planning

Contingency planning is an essential facet of utility management and one that is often overlooked. Although utilities in various locations will be vulnerable to somewhat different kinds of natural disasters, the effects of these disasters in many cases will be quite similar. As a first step toward an effective contingency plan, each utility should make an assessment of its own vulnerability and then develop and implement a comprehensive plan of action.

All utilities suffer from common problems such as equipment breakdowns and leaking pipes. During the past few years there has also been an increasing amount of vandalism, civil disorder, toxic spills, and employee strikes which have threatened to disrupt utility operations. In observing today's international tension and the potential for nuclear war or the effects of terrorist-induced chemical or biological warfare, water utilities must seriously consider how to respond. Natural disasters such as floods, earthquakes, hurricanes, forest fires, avalanches, and blizzards are a more or less routine occurrence for some utilities. When such catastrophic emergencies occur, the utility must be prepared to minimize the effects of the event

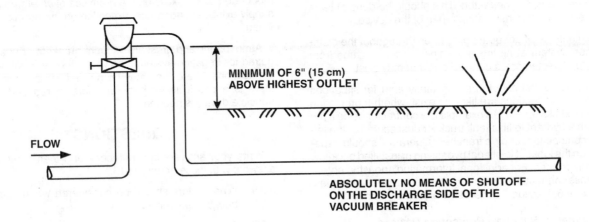

MINIMUM OF 6" (15 cm)
ABOVE HIGHEST OUTLET

FLOW

ABSOLUTELY NO MEANS OF SHUTOFF
ON THE DISCHARGE SIDE OF THE
VACUUM BREAKER

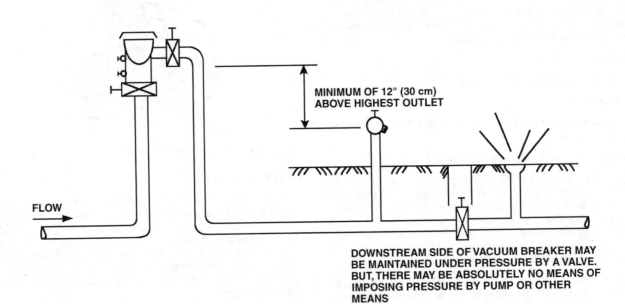

MINIMUM OF 12" (30 cm)
ABOVE HIGHEST OUTLET

FLOW

DOWNSTREAM SIDE OF VACUUM BREAKER MAY
BE MAINTAINED UNDER PRESSURE BY A VALVE.
BUT, THERE MAY BE ABSOLUTELY NO MEANS OF
IMPOSING PRESSURE BY PUMP OR OTHER
MEANS

Fig. 8.8   Typical installations of atmospheric (top) and pressure (bottom) vacuum breakers

(From *MANUAL OF CROSS-CONNECTION CONTROL PRACTICES AND PROCEDURES*, Sanitary Engineering Branch, California Department of Health Services, Berkeley, CA)

and have a plan for rapid recovery. Such preparation should be a specific obligation of every utility manager.

Start by assessing the vulnerability of the utility during various types of emergency situations. If the extent of damage can be estimated for a series of most probable events, the weak elements can be studied and protection and recovery operations can center on these elements. Although all elements are important for the utility to function, experience with disasters points out elements that are most subject to disruption. These elements are:

1. The absence of trained personnel to make critical decisions and carry out orders,

2. The loss of power to the utility's facilities,

3. An inadequate amount of supplies and materials, and

4. Inadequate communication equipment.

The following steps should be taken in assessing the vulnerability of a system:

1. Identify and describe the system components,

2. Assign assumed disaster characteristics,

3. Estimate disaster effects on system components,

4. Estimate customer demand for service following a potential disaster, and

5. Identify key system components that would be primarily responsible for system failure.

If the assessment shows a system is unable to meet estimated requirements because of the failure of one or more critical components, the vulnerable elements have been identified. Repeating this procedure using several "typical" disasters will usually point out system weaknesses. Frequently the same vulnerable element appears for a variety of assumed disaster events.

You might consider, for example, the case of the addition of toxic pollutants to water supplies. The list of toxic agents that may have a harmful effect on humans is almost endless. However, it is recognized that there is a relationship between the quantity of toxic agents added to the treatment provided for the supply. Adequate chlorination is effective against most biological agents. Other considerations are the amount of dilution water and the solubility of the chemical agents. There is the possibility that during normal detention times many of the biological agents will die off with adequate chlorination.

Although the drafting of an emergency plan for a water system may be a difficult job, the existence of such a plan can be of critical importance during an emergency situation.

An Emergency Operations Plan need not be too detailed, since all types of emergencies cannot be anticipated and a complex response program can be more confusing than helpful. Supervisory personnel must have a detailed description of their responsibilities during emergencies. They will need information, supplies, equipment, and the assistance of trained personnel. All these can be provided through a properly constructed emergency operations plan that is not extremely detailed.

The following outline can be used as the basis for developing an Emergency Operations Plan:

1. Make a vulnerability assessment,

2. Inventory organizational personnel,

3. Provide for a recovery operation (plan),

4. Provide training programs for operators in carrying out the plan,

5. Coordinate with local and regional agencies such as the health, police, and fire departments to develop procedures for carrying out the plan,

6. Establish a communications procedure, and

7. Provide protection for personnel, plant equipment, records, and maps.

By following these steps, an emergency plan can be developed and maintained even though changes in personnel may occur. "Emergency Simulation" training sessions, including the use of standby power, equipment, and field test equipment will ensure that equipment and personnel are ready at times of emergency.

A list of phone numbers for operators to call in an emergency should be prepared and posted by a phone for emergency use. The list should include:

1. Plant supervisor,

2. Director of public works or head of utility agency,

3. Police,

4. Fire,

5. Doctor (2 or more),

6. Ambulance (2 or more),

7. Hospital (2 or more),

If appropriate for your utility, also include the following phone numbers on the emergency list:

8. Chlorine supplier and manufacturer,

9. CHEMTREC (800) 424-9300 for the hazardous chemical spills; sponsored by the Manufacturing Chemists Association,

10. U.S. Coast Guard's National Response Center (800) 424-8802,

11. Local and state poison control centers, and

12. Local hazardous materials spill response team.

You should prepare a list for your plant *NOW*, if you have not already done so. Be sure to review and update the list at least once a year, for example, during the annual review of the entire Emergency Response Plan.

Other sections of this manual also contain some useful information on emergencies: Chapter 5, Section 5.12, "Emergency Planning"; Chapter 6, Section 6.311, "Emergency or Maintenance Disinfection"; and Chapter 6, Section 6.63, "First-Aid Measures" (for chlorine exposure).

For additional information on emergencies, see *WATER TREATMENT PLANT OPERATION*, Volume I, Chapter 7, "Disinfection," Section 7.52, "Chlorine Leaks," and Chapter 10, "Plant Operation," Section 10.9, "Emergency Conditions and Procedures."

QUESTIONS

Write your answers in a notebook and then compare your answers with those on pages 513 and 514.

8.10A   What is the first step toward an effective contingency plan for emergencies?

8.10B   Why is too detailed an Emergency Operations Plan not needed or even desirable?

8.10C   An Emergency Operations Plan should include what specific information?

8.101   Homeland Defense

World events in recent years have heightened concern in the United States over the security of one of America's most valuable resources, the critical drinking water treatment, storage, and distribution infrastructure. Water distribution pipelines form an extensive network that runs near or beneath key buildings and roads and is physically close to many communication and transportation networks. Significant damage to the nation's water supply or distribution facilities could result in: loss of life, contamination of drinking water supplies, long-term public health impacts, catastrophic environmental damage to rivers, lakes, and wetlands, and disruption to commerce, the economy, and our normal way of life.

Water treatment and distribution facilities have been identified as a target for international and domestic terrorism. This knowledge, coupled with the responsibility of the facility to provide a safe and healthful workplace, requires that management establish rules to protect the workers as well as the facilities. Emergency action and fire prevention plans must identify what steps need to be taken when the threat analysis indicates a potential for attack. These plans must be in writing and be practiced periodically so that all workers know what actions to take.

Some actions that should be taken at all times to reduce the possibility of a terrorist attack are:

- Ensure that all visitors sign in and out of the facilities with a positive ID check,

- Reduce the number of visitors to a minimum,

- Discourage parking by the public near critical buildings to eliminate the chances of car bombs,

- Be cautious with suspicious packages that arrive,

- Be aware of the hazardous chemicals used and how to defend against spills,

- Keep emergency numbers posted near telephones and radios,

- Patrol the facilities frequently, looking for suspicious activity or behavior, and

- Maintain, inspect, and use your PPE (hard hats, respirators).

The following recommendations by the EPA[10] include many straightforward, common-sense actions a utility can take to increase security and reduce threats from terrorism.

Guarding Against Unplanned Physical Intrusion

- Lock all doors and set alarms at your office, booster pump stations, treatment plants, and vaults, and make it a rule that doors are locked and alarms are set.

- Limit access to facilities and control access to booster pump stations, chemical and fuel storage areas, giving close scrutiny to visitors and contractors.

- Post guards at treatment plants and post "Employee Only" signs in restricted areas.

- Increase lighting in parking lots, treatment bays, and other areas with limited staffing.

- Control access to computer networks and control systems and change the passwords frequently.

- Do not leave keys in equipment or vehicles at any time.

Making Security a Priority for Employees

- Conduct background security checks on employees at hiring and periodically thereafter.

- Develop a security program with written plans and train employees frequently.

- Ensure all employees are aware of established procedures for communicating with law enforcement, public health, environmental protection, and emergency response organization.

- Ensure that employees are fully aware of the importance of vigilance and the seriousness of breaches in security.

- Make note of unaccompanied strangers on the site and immediately notify designated security officers or local law enforcement agencies.

- If possible, consider varying the timing of operational procedures so that, to anyone watching for patterns, the pattern changes.

- Upon the dismissal of an employee, change pass codes and make sure keys and access cards are returned.

- Provide customer service staff with training and checklists of how to handle a threat if it is called in.

---

[10] Adapted from *"What Wastewater Utilities Can Do Now to Guard Against Terrorist and Security Threats,"* U.S. Environmental Protection Agency, Office of Wastewater Management, October 2001.

Coordinating Actions for Effective Emergency Response

- Review existing emergency response plans and ensure that they are current and relevant.

- Make sure employees have the necessary training in emergency operating procedures.

- Develop clear procedures and chains-of-command for reporting and responding to threats and for coordinating with emergency management agencies, law enforcement personnel, environmental and public health officials, consumers, and the media. Practice the emergency procedures regularly.

- Ensure that key utility personnel (both on and off duty) have access to critical telephone numbers and contact information at all times. Keep the call list up to date.

- Develop close relationships with local law enforcement agencies and make sure they know where critical assets are located. Ask them to add your facilities to their routine rounds.

- Report to county or state health officials any illness among the employees that might be associated with water contamination.

- Immediately report criminal threats, suspicious behavior, or attacks on water utilities to law enforcement officials and the nearest field office of the Federal Bureau of Investigation.

Investing in Security and Infrastructure Improvements

- Assess the vulnerability of the distribution system, water storage facilities, major pumping stations, water treatment plants, chemical and fuel storage areas, and other key infrastructure elements.

- Improve computer system and remote operational security.

- Use local citizen watches.

- Seek financing for more expensive and comprehensive system improvements.

The U.S. Terrorism Alert System (Figure 8.9) is a color-coded system that identifies the potential for terrorist activity and suggests specific actions to be taken. Your safety plan should identify the actions that your facility will take when the threat level changes. Tables 8.6 and 8.7 show examples of security measures that should be taken to improve safety at a water treatment and/or distribution facility when the threat level is YELLOW and when it is ORANGE. (The utility's safety plan should include similar lists of actions for the RED, BLUE, and GREEN levels as well.)

For additional information on homeland defense, see the U.S. Environmental Protection Agency's Security Product Guide available online at:

www.epa.gov/safewater/watersecurity/guide/index.html

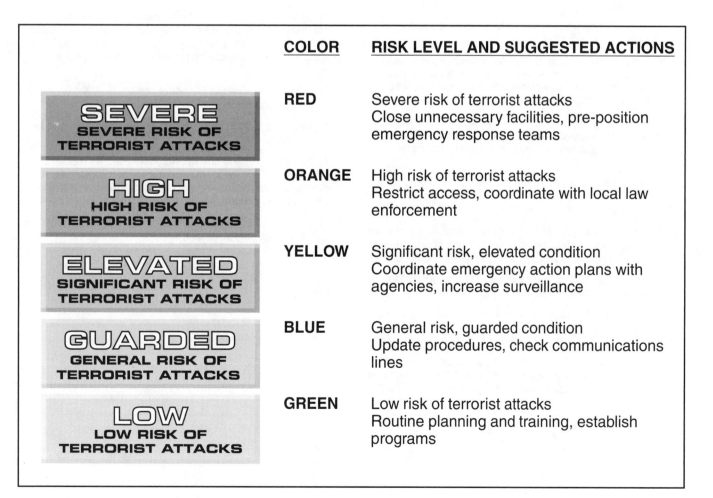

COLOR	RISK LEVEL AND SUGGESTED ACTIONS
SEVERE — SEVERE RISK OF TERRORIST ATTACKS	**RED** — Severe risk of terrorist attacks. Close unnecessary facilities, pre-position emergency response teams
HIGH — HIGH RISK OF TERRORIST ATTACKS	**ORANGE** — High risk of terrorist attacks. Restrict access, coordinate with local law enforcement
ELEVATED — SIGNIFICANT RISK OF TERRORIST ATTACKS	**YELLOW** — Significant risk, elevated condition. Coordinate emergency action plans with agencies, increase surveillance
GUARDED — GENERAL RISK OF TERRORIST ATTACKS	**BLUE** — General risk, guarded condition. Update procedures, check communications lines
LOW — LOW RISK OF TERRORIST ATTACKS	**GREEN** — Low risk of terrorist attacks. Routine planning and training, establish programs

Fig. 8.9   Threat level categories established by the U.S. Department of Homeland Defense

TABLE 8.6   SECURITY MEASURES FOR THREAT LEVEL YELLOW (CONDITION ELEVATED)

Continue to introduce all measures listed in BLUE: Condition Guarded.	
Detection	**Prevention**
• To the extent possible, increase the frequency and extent of monitoring the flow coming into and leaving the treatment facility and review results against baseline quantities. Increase review of operational and analytical data (including customer complaints) with an eye toward detecting unusual variability (as an indicator of unexpected changes in the system). Variations due to natural or routine operational variability should be considered first. • Increase surveillance activities in water supply, treatment, and distribution facilities.	• Carefully review all facility tour requests before approving. If allowed, implement security measures to include list of names prior to tour, request identification of each attendee prior to tour, prohibit backpacks, duffle bags, and cameras, and identify parking restrictions. • On a daily basis, inspect the interior and exterior of buildings in regular use for suspicious activity or packages, signs of tampering, or indications of unauthorized entry. • Implement mail room security procedures. Follow guidance provided by the United States Postal Service.
Preparedness	**Protection**
• Continue to review, update and test emergency response procedures and communication protocols. • Establish unannounced security spot checks (such as verification of personal identification and door security) at access control points for critical facilities. • Increase frequency for posting employee reminders of the threat situation and about events that constitute security violations. • Ensure employees understand notification procedures in the event of a security breach. • Conduct security audit of physical security assets, such as fencing and lights, and repair or replace missing/broken assets. Remove debris from along fence lines that could be stacked to facilitate scaling. • Maximize physical control of all equipment and vehicles; make them inoperable when not in use (for example, lock steering wheels, secure keys, chain, and padlock on front-end loaders). • Review draft communications on potential incidents; brief media relations personnel of potential for press contact and/or issuance of press releases. • Ensure that list of sensitive customers (such as government agencies and medical facilities) within the service area is accurate and shared with appropriate public health officials. • Contact neighboring water utilities to review coordinated response plans and mutual aid during emergencies. • Review whether critical replacement parts are available and accessible. • Identify any work/project taking place in proximity to events where large attendance is anticipated. Consult with the event organizers and local law enforcement regarding contingency plans, security awareness, and site accessibility and control.	• Verify the identity of all persons entering the water utility. Mandate visible use of identification badges. Randomly check identification badges and cards of those on the premises. • At the discretion of the facility manager or security director, remove all vehicles and objects (such as trash containers) located near mission critical facility security perimeters and other sensitive areas. • Verify the security of critical information systems (for example, Supervisory Control and Data Acquisition (SCADA), Internet, e-mail) and review safe computer and Internet access procedures with employees to prevent cyber intrusion. • Consider steps needed to control access to all areas under the jurisdiction of the utility. • Implement critical infrastructure facility surveillance and security plans. • At the beginning and end of each work shift, as well as at other regular and frequent intervals, inspect the interior and exterior of buildings in regular use for suspicious packages, persons, and circumstances. • Lock and regularly inspect all buildings, rooms, and storage areas not in regular use.

TABLE 8.7 SECURITY MEASURES FOR THREAT LEVEL ORANGE (CONDITION HIGH)

Continue to introduce all measures listed in YELLOW: Condition Elevated.	
Detection	**Prevention**
• Increase the frequency and extent of monitoring activities. Review results against baseline measurements. • Confirm that county and state health officials are on high alert and will inform the utility of any potential waterborne illnesses. • If a neighborhood watch-type program is in place, notify the community and request increased awareness.	• Discontinue tours and prohibit public access to all operational facilities. • Consider requesting increased law enforcement surveillance, particularly of critical assets and otherwise unprotected areas. • Limit access to computer facilities. No outside visitors. • Increase monitoring of computer and network intrusion detection systems and security monitoring systems.
Preparedness	**Protection**
• Confirm that emergency response and laboratory analytical support network are ready for deployment 24 hours per day, 7 days a week. • Reaffirm liaison with local police, intelligence, and security agencies to determine likelihood of an attack on the water supply, treatment, or distribution utility personnel and facilities and consider appropriate protective measures (such as road closing and extra surveillance). • Practice communications procedures with local authorities and others cited in the facility's emergency response plan. • Post frequent reminders for staff and contractors of the threat level, along with a reminder of what events constitute security violations. • Ensure employees are fully aware of emergency response communication procedures and have access to contact information for relevant law enforcement, public health, environmental protection, and emergency response organizations. • Have alternative water supply plan ready to implement (for example, bottled water delivery for employees and other critical business uses). • Place all emergency management and specialized response teams on full alert status. • Ensure personal protective equipment (PPE) and specialized response equipment is checked, issued, and readily available for deployment. • Review all plans, procedures, guidelines, personnel details, and logistical requirements related to the introduction of a higher threat condition level.	• Evaluate the need to staff the water treatment facility at all times. • Increase security patrol activity to the maximum level sustainable and ensure tight security in the vicinity of mission critical facilities. Vary the timing of security patrols. • Request employees change their passwords on critical information management systems. • Limit building access points to the absolute minimum, strictly enforce entry control procedures. Identify and protect all designated vulnerable points. Give special attention to vulnerable points outside of the critical facility. • Lock all exterior doors except the main facility entrance(s). Check all visitors' purpose, intent, and identification. Ensure that contractors have valid work orders. Require visitor's to sign in upon arrival; verify and record their identifying information. Escort visitors at all times when they are in the facility.

8.102 Handling the Threat of Contaminated Water Supplies[11]

8.1020 Importance

More than 50 water utilities in southern Louisiana were threatened with cyanide poisoning in their water supplies in one year. Such threats can occur anywhere, and every water utility should be prepared to handle this type of emergency.

8.1021 Toxicity

The term toxicity is often used when discussing contamination of a water supply with the intention of creating a serious health hazard. Toxicity is the ability of a contaminant (chemical or biological) to cause injury when introduced into the body. The degree of toxicity varies with the concentration of contaminant required to cause injury, the speed with which the injury takes place, and the severity of the injury.

The effect of a toxic contaminant, once added to a water supply, depends on several things. First the amount of contaminant added can vary, as can the size of the water supply. In general, it takes larger quantities of a contaminant to be toxic in a larger water supply. Second, the solubility of the contaminant can vary. The more soluble the substance is in water, the more likely it is to cause problems. Finally the detention time of the contaminant in the water can vary. For example, many biological agents will die before they can cause a problem in the water supply.

Generally, the terms acute and chronic are used to describe toxic agents and their effects. An acute toxic agent causes injury quickly. When the contaminant causes illness in seconds, minutes, or hours after a single exposure or a single dose, it is considered an acute toxic agent. A chronic agent causes injury to occur over an extended period of exposure. Generally, the contaminant is ingested in repeated doses over a period of days, months, or years.

8.1022 Effective Dosages (See Table 8.8)

When determining the effective dosage of a contaminant (the amount of that contaminant necessary to cause injury), the following facts must be considered:

1. Quantity or concentration of the contaminant,

2. Duration of exposure to the contaminant,

3. Physical form of the contaminant (size of particle; physical state—solid, liquid, gas),

4. Attraction of the contaminant to the organism being contaminated,

5. Solubility of the contaminant in the organism, and

6. Sensitivity of the organism to the contaminant.

Concentration of a contaminant can be expressed in two ways. The maximum allowable concentration (MAC) is the maximum concentration of the contaminant allowed in drinking water. Table 8.8 lists several contaminants and their MACs, specifically for short-term emergencies ranging up to three days. The MACs should not be confused with concentration required to have an acute effect on the population. Lethal dose 50 (LD 50) is used to express the concentration of a contaminant that will produce 50 percent fatalities from an average exposure.

TABLE 8.8 EMERGENCY LIMITS OF SOME CHEMICAL POLLUTANTS IN DRINKING WATER[a]

Chemical	Concentration Limits, mg/L	
	Emergency Short Term (Three days)	Long Term
Cyanide (CN)	5.0	0.01
Aldrin	0.05	0.032
Chlordane	0.06	0.003
DDT	1.4	0.042
Dieldrin	0.05	0.017
Endrin	0.01	0.001
Heptachlor	0.1	0.018
Heptachlor epoxide	0.05	0.018
Lindane	2.0	0.056
Methoxychlor	2.8	0.035
Toxaphene	1.4	0.005
Beryllium	0.1	0.000
Boron	25.0	1.000
2,4-D	2.0	0.1
Ethylene chlorohydrin	2.0	
Organophosphorus and carbamate pesticides	2.0	0.100
Trinitrotoluene $(NO_2)(C_6H_2CH_3)$	0.75	0.005

[a] These limits, based on current knowledge and informed judgment, have been recommended by knowledgeable persons in the field of toxicology. They are subject to change should new information indicate the need. Additional information on some of the chemicals listed can be found in "Report of the Secretary's Commission on Pesticides and Their Relationship to Environmental Health," Parts I and II, USDHEW, Washington, DC, Dec. 1969.

8.1023 Protective Measures

A utility can take three approaches to protect its water supply from contamination. First, the utility can isolate those reservoirs that offer easy access to the general public. These reservoirs can be fenced off and patrolled, or they can be cov-

[11] This section was reprinted from OPFLOW, Vol. No. 3, March 1983, by permission. Copyright 1983, the American Water Works Association.

ered. If access to on-line reservoirs is limited, persons attempting to contaminate the water supply will generally be forced to look to larger bodies of water. Contamination of these large water bodies requires larger quantities of contaminant, increases the detention time of the contaminant, and increases the likelihood of its detection.

As a second means of protection, the water utility can develop an extensive detection and monitoring program. Detecting any contaminant that might be added to a water supply is difficult and expensive. However, because most contaminants cause secondary effects in a water supply, such as taste, color, odor, or chlorine demand, detection is easier.

Because utility operators "know" their water supply (they know its characteristics), any subtle changes in taste, odor, color, and chlorine demand are instantly recognized. Once it has been determined that the water supply may be contaminated, water samples can be tested. Tests can either be done at the utility's laboratory, if it is a large utility, or the samples can be sent to the state health department.

Finally, the utility can maintain a high chlorine residual. Generally, chlorine residuals of one mg/L or higher effectively oxidize or destroy most contaminants. For example, infectious hepatitis virus will not survive a free residual chlorine level of 0.7 mg/L.

8.1024 Emergency Countermeasures

Following is a list of emergency countermeasures that, when used over a short time period, can increase protection of a water supply:

1. Maintaining a high chlorine residual in the system,

2. Having engineers, chemists, and medical personnel on 24-hour alert,

3. Continuously monitoring key points in the distribution system (monitoring chlorine residual is mandatory),

4. Increasing security around exposed on-line reservoirs,

5. Sealing off access to manholes within a three- to six-block radius of highly populated areas,

6. Setting up emergency crews who can isolate sections of the distribution system, and

7. Staffing the treatment facility on a 24-hour basis.

8.1025 In Case of Contamination

If contamination of the water supply is discovered, the immediate concern must be the safety of the public. If the contaminated water has entered the distribution system, immediate public notification is the highest priority. The local police chief, sheriff, or other responsible governmental authority will help to spread the word. The contaminated portions of the distribution system should be thoroughly flushed and alternative sources of water may need to be provided.

If the contaminated water has not entered the distribution system, it may be possible to isolate the contaminated source and continue to supply water from other, unaffected sources. If the contaminated water is the only source for the community, treatment measures may be available that will remove the contaminant or reduce its toxicity.

Table 8.9 lists a series of emergency treatment steps that can be taken when identified chemicals are added to the system. These emergency treatment methods are effective only if the contaminant has been identified.

Write your answers in a notebook and then compare your answers with those on page 514.

8.10D   What are the color codes and risk levels for the U.S. Terrorism Alert System?

8.10E   What does the word toxicity mean?

8.10F   The degree of toxicity varies with what factors?

8.10G   List possible secondary effects in a water supply that may allow detection of a contaminant without specific testing.

8.1026 Cryptosporidium

Cryptosporidium is a parasite that has become a significant public health concern for drinking water utilities. Even when drinking water meets or exceeds all current state and federal standards, it may contain sufficient Cryptosporidium to cause serious illness in sensitive individuals. This section contains suggestions that operators should consider to protect consumers from "crypto."

Operators need to develop a strategy for protecting their source water supplies and optimizing the operation of their water treatment plant to protect the public health. Cryptosporidium oocyst contamination is widespread in the water environment. Potential sources of the parasite in the watersheds of your water supply need to be identified and controlled. The most serious threat to a public water supply occurs during periods of heavy rains or snow melts that flush areas that are sources of high concentrations of oocysts into waters upstream of a plant intake. In the watershed, these potentially contaminated areas include wastewater treatment plants, cattle feedlots, and pastures where livestock graze.

Conventional water treatment plants using coagulation, flocculation, sedimentation, filtration, and disinfection can provide effective treatment to protect drinking water from Cryptosporidium. Be especially alert during periods when the intake water has high turbidity levels resulting from storm water runoff, snow melt runoff, and/or lake overturns. During these periods all treatment processes must operate effectively. Try to achieve turbidity levels of 0.1 NTU or lower at all times. Run frequent jar tests to determine or simply to verify that you are using the optimum coagulant doses as intake turbidity and other water quality indicators change (pH, temperature, alkalinity). Also monitor your filtration processes continuously to avoid any increases in turbidity in the treated water. If the water used for backwashing filters is routinely returned to the headworks for conservation purposes, avoid recycling the backwash water during periods of high intake water turbidity. Recycling backwash water may tend to concentrate oocysts in the filter media. Instead, consider wasting the backwash water until the high turbidity levels drop back to normal levels.

TABLE 8.9 EMERGENCY TREATMENT FOR REDUCING CONCENTRATION OF SPECIFIC CHEMICALS IN COMMUNITY WATER SUPPLIES[a]

Concentration	Treatment	Concentration	Treatment
Arsenicals		*Nerve Agents*	
Unknown organic and inorganic arsenicals in groundwater at concentrations of 100 mg/L	Precipitation with ferric sulfate and liming to pH 6.8, followed by sedimentation and filtration.	(Organophosphorus compounds)	Superchlorination at pH 7 to provide at least 40 mg/L residual after 30-min chlorine contact time, followed by dechlorination and conventional clarification processes.
Cyanides			
Hydrogen cyanide	Prechlorination to free residual with pH 7, followed by coagulation, sedimentation, and filtration. *Caution:* housed facilities must be adequately ventilated.	*Pesticides*	
		2,4-DCP (2,4-Dichlorophenol) an impurity in commercial 2,4-D herbicides	Adsorption on activated carbon followed by coagulation, sedimentation, and filtration.
	Precipitation with ferrous or ferric salts to form Prussian blue (iron ferric cyanide) followed by coagulation, sedimentation, and clarification. As long as an excess of iron coagulant is applied, the filtered water should be nontoxic even though it is blue.	DDT (dichloro-diphenyltrichloroethane), concentrations of 10 gm/L	Chemical coagulation, sedimentation, and filtration.
		Dieldrin, concentrations of 10 gm/L	Chemical coagulation, sedimentation, and filtration. Supplemental treatment with activated carbon may be necessary.
Acetone cyanohydrin	Same as for hydrogen cyanide.	Endrin, concentrations of 10 gm/L	Chemical coagulation, sedimentation, and filtration. Supplemental treatment with activated carbon may be necessary.
Cyanogen chloride	Same as for hydrogen cyanide.		
Hydrocarbons		Lindane, concentrations of 10 gm/L	Application of activated carbon followed by chemical coagulation, sedimentation, and filtration.
Kerosene, peak concentrations of 140 mg/L	Preapplications of bleaching clay and activated carbon, plus some increase in normal dosages of alum, chlorine dioxide, lime, and carbon, to provide treatment enabling continued production of water.	Parathion, concentrations of 10 gm/L	Chemical coagulation, sedimentation, and filtration. Supplemental treatment with activated carbon may be necessary. Omit prechlorination as chlorine reacts with parathion to form paraoxon, which is more toxic than parathion.
Miscellaneous Organic Chemicals			
LSD (lysergic acid derivative)	Chlorination in alkaline water, or water made alkaline by addition of lime or soda ash, to provide a free chlorine residual. Two parts free chlorine are required to react with each part LSD.		

[a] Source: Graham Walton, Chief, Technical Services, National Water Supply Research Laboratory, USSR Program, Oct. 24, 1968.

Chemical inactivation of *Cryptosporidium* oocysts by disinfection is influenced by several factors. The effectiveness of chemical disinfectants such as chlorine, chlorine dioxide, and ozone is reduced by the presence of high levels of total organic carbon (TOC)(caused by color and turbidity), lower water temperatures, and shorter disinfection contact times. Therefore, effective disinfection can be difficult during periods of high turbidity caused by high storm water or snow melt runoff flows.

The best approach to evaluating the potential threat of the drinking water to the public is the analysis for *Cryptosporidium* oocysts in the treated drinking water. However, current sampling and analytical methods make it difficult for operators to use the detection of *Cryptosporidium* oocysts for determining the efficiency and effectiveness of filtration and chemical inactivation treatment processes. Today operators are using turbidity and particle counting measurements as a means of determining the treatment processes' ability to remove and/or inactivate oocysts. Test methods for identifying the presence or potential presence of *Cryptosporidium* are evolving and should be used to analyze both source and treated (finished) waters for "crypto."

Keeping the public informed is another good strategy for preventing outbreaks of disease due to *Cryptosporidium*. Operators need to educate the media and consumers regarding the sources of the parasite, possible health risks, monitoring efforts, and treatment processes. Sensitive populations must be informed that even properly treated municipal drinking water, bottled water, and water treated by a home water treatment device still may not be free of *Cryptosporidium*. The Centers for Disease Control and Prevention suggests, "Immunocompromised persons who wish to take independent action to reduce the risk of waterborne *Cryptosporidium* may choose to take precautions similar to those recommended during outbreaks (such as boiling tap water for one minute). Such decisions should be made in conjunction with their health care provider."

There are alternatives to boiling water that may be effective when used with proper precautions. Point-of-use filters that remove particles one micrometer (1 μm) in diameter or smaller are effective. One-micrometer filters for cyst removal and reverse osmosis units are also acceptable. Bottled water from protected springs and wells may be a safe drinking water, especially if treated by reverse osmosis or distillation to remove *Cryptosporidium* before bottling.

Operators must stay current with the efforts of our profession to prevent outbreaks of waterborne *Cryptosporidium*.

QUESTIONS

Write your answers in a notebook and then compare your answers with those on page 514.

8.10H   Why is *Cryptosporidium* a significant public health concern for drinking water utilities?

8.10I   How can a water treatment plant be operated to protect the public from *Cryptosporidium*?

8.10J   When does the most serious threat of *Cryptosporidium* occur to a public water supply?

8.11   SAFETY PROGRAM

8.110   Responsibilities

8.1100   Everyone Is Responsible for Safety

Waterworks utilities, regardless of size, must have a safety program if they are to realize a low frequency of accident occurrence. A safety program also provides a means of comparing frequency, disability, and severity of injuries with other utilities. The utility should identify the causes of accidents and injuries, provide safety training, implement and maintain an accident reporting system, and hold supervisors responsible for implementing the safety program. Each utility should have a safety officer or supervisor evaluate every accident, offer recommendations, and keep and apply statistics.

The effectiveness of any safety program will depend upon how the utility holds its supervisors responsible. If the utility holds only the safety officer or the employees responsible, the program will fail. The supervisors are key in any organization. If they disregard safety measures, essential parts of the program will not work. The results will be an overall poor safety record. After all, the first line supervisor is where the work is being performed, and some may take advantage of an unsafe situation in order to get the job completed. The organization must discipline such supervisors and make them aware of their responsibility for their own and their operators' safety.

Safety is good business for both the operator and the agency. For a good safety record to be accomplished, all individuals must be educated about safety and must believe in the safety program. All individuals involved must have the conviction that accidents can be prevented. The operations should be studied to determine the safe way of performing each job. Safety pays, both in monetary savings and in the health and well-being of the operating staff.

8.1101   Regulatory Agencies

Many state and federal agencies are involved in ensuring safe working conditions. The one law that has had the greatest impact has been the *Occupational Safety and Health Act of 1970 (OSHA)*,[12] Public Law 91-596, which took effect on December 29, 1970. This legislation affects more than 75,000,000 employees and has been the basis for most of the current state laws covering employees. Both state and federal regulatory agencies enforce the OSHA requirements.

The OSHA regulations provide for safety inspections, penalties, recordkeeping, and variances. Managers and supervisors must understand the OSHA Act and must furnish each operator with the rules of conduct in order to comply with occupational safety and health standards. The intent of the regulations is to create a place of employment that is free from recognized hazards that could cause serious physical harm or death to an operator.

Civil and criminal penalties are allowed under the OSHA Law, depending upon the size of the business and the seriousness of the violation. A routine violation could cost an employer or supervisor up to $1,000 for each violation. A serious, willful, or repeated violation could cause the employer or supervisor to be assessed a penalty of not more than $10,000 for each violation. Penalties are assessed against the supervisor responsible for the injured operator. Operators should become familiar with the OSHA regulations as they apply to their organizations. Managers and supervisors must correct violations and prevent others from occurring.

8.1102   Managers

The utility manager is responsible for the safety of the agency's personnel and the public exposed to the water utility's operations. Therefore, the manager must develop and administer an effective safety program and must provide new employee

---

[12] *OSHA (O-shuh). The Williams-Steiger **O**ccupational **S**afety and **H**ealth **A**ct of 1970 (OSHA) is a federal law designed to protect the health and safety of industrial workers and also the operators of water supply systems and treatment plants. The Act regulates the design, construction, operation and maintenance of water supply systems and water treatment plants. OSHA also refers to the federal and state agencies which administer the OSHA regulations.*

safety training as well as ongoing training for all employees. The basic elements of a safety program include a safety policy statement, safety training and promotion, and accident investigation and reporting.

A safety policy statement should be prepared by the top management of the utility. The purpose of the statement is to let employees know that the safety program has the full support of the agency and its management. The statement should:

1. Define the goals and objectives of the program,

2. Identify the persons responsible for each element of the program,

3. Affirm management's intent to enforce safety regulations, and

4. Describe the disciplinary actions that will be taken to enforce safe work practices.

Give a copy of the safety policy statement to every current employee and each new employee during orientation. Figure 8.10 is an example of a safety policy statement for a water supply utility.

The following list of responsibilities for safety is from the *PLANT MANAGER'S HANDBOOK.*[13] These responsibilities represent a typical list but may be incomplete if your agency is subject to stricter local, state, and/or federal regulations than what is shown here. Check with your safety professional.

Management has the responsibility to:

1. Formulate a written safety policy,

2. Provide a safe workplace,

3. Set achievable safety goals,

4. Provide adequate training, and

5. Delegate authority to ensure that the program is properly implemented.

The manager is the key to any safety program. Implementation and enforcement of the program is the responsibility of the manager. The manager also has the responsibility to:

1. Ensure that all employees are trained and periodically retrained in proper safe work practices,

2. Ensure that proper safety practices are implemented and continued as long as the policy is in effect,

3. Investigate all accidents and injuries to determine their cause,

4. Institute corrective measures where unsafe conditions or work methods exist, and

5. Ensure that equipment, tools, and the work are maintained to comply with established safety standards.

QUESTIONS

Write your answers in a notebook and then compare your answers with those on page 514.

8.11A   Who should be responsible for the implementation of a safety program?

8.11B   What should be the duties of a safety officer?

8.11C   Who enforces the OSHA requirements?

8.11D   What are the utility manager's responsibilities with regard to safety?

8.11E   What should be included in a utility's policy statement on safety?

SAFETY POLICY STATEMENT

It is the policy of the Las Vegas Valley Water District that every employee shall have a safe and healthy place to work. It is the District's responsibility; its greatest asset, the employees and their safety.

When a person enters the employ of the District, he or she has a right to expect to be provided a proper work environment, as well as proper equipment and tools, so that they will be able to devote their energies to the work without undue danger. Only under such circumstances can the association between employer and employee be mutually profitable and harmonious. It is the District's desire and intention to provide a safe workplace, safe equipment, proper materials, and to establish and insist on safe work methods and practices at all times. It is a basic responsibility of all District employees to make the SAFETY of human beings a matter for their daily and hourly concern. This responsibility must be accepted by everyone who works at the District, regardless of whether he or she functions in a management, supervisory, staff, or the operative capacity. Employees must use the SAFETY equipment provided; Rules of Conduct and SAFETY shall be observed; and, SAFETY equipment must not be destroyed or abused. Further, it is the policy of the Water District to be concerned with the safety of the general public. Accordingly, District employees have the responsibility of performing their duties in such a manner that the public's safety will not be jeopardized.

The joint cooperation of employees and management in the implementation and continuing observance of this policy will provide safe working conditions and relatively accident-free performance to the mutual benefit of all involved. The Water District considers the SAFETY of its personnel to be of primary importance, and asks each employee's full cooperation in making this policy effective.

Fig. 8.10   Safety policy statement

(Permission of Las Vegas Valley Water District)

---

[13] *PLANT MANAGER'S HANDBOOK (MOP SM-4), Water Environment Federation (WEF), no longer in print.*

8.1103 Supervisors

The success of any safety program will depend upon how the supervisors of the utility view their responsibility. The supervisor who has the responsibility for directing work activities must be safety conscious. This supervisor controls the operators' general environment and work habits and influences whether or not the operators comply with safety regulations. The supervisor is in the best position to counsel, instruct, and review the operators' working methods and thereby effectively ensure compliance with all aspects of the utility's safety program.

The problem, however, is one of the supervisor accepting this responsibility. The supervisor who wishes to complete the job and go on to the next one without taking time to be concerned about working conditions, the welfare of operators, or considering any aspects of safety is a poor supervisor. Only after an accident occurs will a careless supervisor question the need for a work program based on safety. At this point, however, it is too late, and the supervisor may be tempted to simply cover up past mistakes. As sometimes happens, the supervisor may even be partially or fully responsible for the accident by causing unsafe acts to take place, by requiring work to be performed in haste, by disregarding an unsafe work environment, or by overlooking or failing to consider any number of safety hazards. This negligent supervisor could be fined, sentenced to a jail term, or even be barred from working in the profession.

All utilities should make their supervisors bear the greatest responsibility for safety and hold them accountable for planning, implementing, and controlling the safety program. If most accidents are caused and do not just happen, then it is the supervisor who can help prevent most accidents.

Equally important are the officials above the supervisor. These officials include commissioners, managers, public works directors, chief engineers, superintendents, and chief operators. The person in responsible charge for the entire agency or operation must believe in the safety program. This person must budget, promote, support, and enforce the safety program by vocal and visible examples and actions. The top person's support is absolutely essential for an effective safety program.

8.1104 Operators

Each operator also shares in the responsibility for an effective safety program. After all, operators have the most to gain since they are the most likely victims of accidents. A review of accident causes shows that *THE ACCIDENT VICTIM OFTEN HAS NOT ACTED RESPONSIBLY.* In some way the victim has not complied with the safety regulations, has not been fully aware of the working conditions, has not been concerned about fellow employees, or just has not accepted any responsibility for the utility's safety program.

Each operator must accept, at least in part, responsibility for fellow operators, for the utility's equipment, for the operator's own welfare, and even for seeing that the supervisor complies with established safety regulations. As pointed out above, the operator has the most to gain. If the operator accepts and uses unsafe equipment, it is the operator who is in danger if something goes wrong. If the operator fails to protect the other operators, it is the operator who must make up the work lost because of injury. If operators fail to consider their own welfare, it is they who suffer the pain of any injury, the loss of income, and maybe even the loss of life.

The operator must accept responsibility for an active role in the safety program by becoming aware of the utility's safety policy and conforming to established regulations. *THE OPERATOR SHOULD ALWAYS CALL TO THE SUPERVISOR'S ATTENTION UNSAFE CONDITIONS,* environment, equipment, or other concerns operators may have about the work they are performing. Safety should be an essential part of the operator's responsibility.

8.111 First Aid

By definition, first aid means emergency treatment for injury or sudden illness, before regular medical treatment is available. Everyone in an organization should be able to give some degree of prompt treatment and attention to an injury.

First-aid training in the basic principles and practices of lifesaving steps that can be taken in the early stages of an injury are available through the local Red Cross, Heart Association, local fire departments, and other organizations. Such training should periodically be reinforced so that the operator has a complete understanding of water safety, cardiopulmonary resuscitation (CPR), and other life-saving techniques. All operators need training in first aid, but it is especially important for those who regularly work with electrical equipment or must handle chlorine and other dangerous chemicals. (*WATER TREATMENT PLANT OPERATION,* Volume II, Chapter 20, "Safety," lists specific first-aid procedures for exposure to a variety of water treatment chemicals.)

First aid has little to do with preventing accidents, but it has an important bearing upon the survival of the injured patient. A well-equipped first-aid chest or kit is essential for proper treatment. The kit should be inspected regularly by the safety officer to ensure that supplies are available when needed. First-aid kits should be prominently displayed throughout the treatment plant and in company vehicles. Special consideration must be given to the most hazardous areas of the plant such as shops, laboratories, and chemical handling facilities.

Regardless of size, each utility should establish standard operating procedures (SOPs) for first-aid treatment of injured personnel. All new operators should be instructed in the utility's first-aid program.

QUESTIONS

Write your answers in a notebook and then compare your answers with those on page 514.

8.11F How could a supervisor be responsible for an accident?

8.11G What types of safety-related responsibilities must each operator accept?

8.11H What is first aid?

8.11I First-aid training is most important for operators involved in what types of activities?

8.112 Hazard Communication Program and Worker Right-To-Know (RTK) Laws

In the past few years there has been an increased emphasis nationally on hazardous materials and wastes. Much of this attention has focused on hazardous and toxic waste dumps and the efforts to clean them up after the long-term effects on human health were recognized. Each year thousands of new chemical compounds are produced for industrial, commercial, and household use. Frequently, the long-term effects of these chemicals are unknown. As a result, federal and state laws have been enacted to control all aspects of hazardous materials handling and use. These laws are more commonly known as Worker Right-To-Know (RTK) laws which are enforced by OSHA.

As noted earlier, the intent of the OSHA regulations is to create a place of employment that is free from recognized hazards that could cause serious physical harm or death to an operator (or other employee). In many cases, the individual states have the authority under the OSHA Standard to develop their own state RTK laws and most states have adopted their own laws. The Federal OSHA Standard 29 CFR[14] 1910.1200—Hazard Communication forms the basis of most of these state RTK laws. Unfortunately, state laws vary significantly from state to state. The state laws that have been passed are at least as stringent as the federal standard and, in most cases, are even more stringent. State laws are also under continuous revision and, because a strong emphasis is being placed on hazardous materials and worker exposure, state laws can be expected to be amended in the future to apply to virtually everybody in the workplace. Managers should become familiar with both the state and federal OSHA regulations that apply to their organizations.

The basic elements of a hazard communication program are described in the following paragraphs.

1. Identify Hazardous Materials—While there are thousands of chemical compounds that could be considered hazardous, focus on the ones to which operators and other personnel in your utility are most likely to be exposed.

2. Obtain Chemical Information and Define Hazardous Conditions—Once the inventory of hazardous chemicals is complete, the next step is to obtain specific information on each of the chemicals. This information is generally incorporated into a standard format form called the *MATERIAL SAFETY DATA SHEET (MSDS)*.[15] This information is commonly available from manufacturers. Many agencies request an MSDS when the purchase order is generated and will refuse to accept delivery of the shipment if the MSDS is not included. Figure 8.11 shows OSHA's standard MSDS form, but other forms are also acceptable provided they contain all of the required information.

The purpose of the MSDS is to have a readily available reference document that includes complete information on common names, safe exposure level, effects of exposure, symptoms of exposure, flammability rating, type of first-aid procedures, and other information about each hazardous substance. Operators should be trained to read and understand the MSDS forms. The forms themselves should be stored in a convenient location where they are readily available for reference.

3. Properly Label Hazards—Once the physical, chemical, and health hazards have been identified and listed, a labeling and training program must be implemented. To meet labeling requirements on hazardous materials, specialized labeling is available from a number of sources, including commercial label manufacturers. Exemptions to labeling requirements do exist, so consult your local safety regulatory agency for specific details.

4. Train Operators—The last element in the hazard communication program is training and making information available to utility personnel. A common-sense approach eliminates the confusing issue of which of the thousands of substances operators should be trained for, and concentrates on those that they will be exposed to or use in everyday maintenance routines.

The Hazard Communication Standard and the individual state requirements are obviously a very complex set of regulations. Remember, however, that the ultimate goal of these regulations is to provide additional operator protection. These standards and regulations, once the intent is understood, are relatively easy to implement.

---

[14] *Code of Federal Regulations (CFR). A publication of the United States Government which contains all of the proposed and finalized federal regulations, including environmental regulations.*

[15] *Material Safety Data Sheet (MSDS). A document which provides pertinent information and a profile of a particular hazardous substance or mixture. An MSDS is normally developed by the manufacturer or formulator of the hazardous substance or mixture. The MSDS is required to be made available to employees and operators whenever there is the likelihood of the hazardous substance or mixture being introduced into the workplace. Some manufacturers are preparing MSDSs for products that are not considered to be hazardous to show that the product or substance is NOT hazardous.*

Material Safety Data Sheet

May be used to comply with
OSHA's Hazard Communication Standard,
29 CFR 1910.1200 Standard must be
consulted for specific requirements.

U.S. Department of Labor

Occupational Safety and Health Administration
(Non-Mandatory Form)
Form Approved
OMB No. 1218-0072

IDENTITY *(As Used on Label and List)*

Note: Blank spaces are not permitted. If any item is not applicable, or no information is available, the space must be marked to indicate that.

Section I

Manufacturer's Name	Emergency Telephone Number
Address *(Number, Street, City, State, and ZIP Code)*	Telephone Number for Information
	Date Prepared
	Signature of Preparer *(optional)*

Section II—Hazardous Ingredients/Identity Information

Hazardous Components (Specific Chemical Identity: Common Name(s))	OSHA PEL	ACGIH TLV	Other Limits Recommended	%(optional)

Section III—Physical/Chemical Characteristics

Boiling Point		Specific Gravity (H_2O = 1)	
Vapor Pressure (mm Hg.)		Melting Point	
Vapor Density (Air = 1)		Evaporation Rate (Butyl Acetate = 1)	

Solubility in Water

Appearance and Odor

Section IV—Fire and Explosion Hazard Data

Flash Point (Method Used)	Flammable Limits	LEL	UEL

Extinguishing Media

Special Fire Fighting Procedures

Unusual Fire and Explosion Hazards

(Reproduce locally)   OSHA 174, Sept. 1985

Fig. 8.11   Material Safety Data Sheet

Section V—Reactivity Data

Stability	Unstable		Conditions to Avoid
	Stable		
Incompatibility *(Materials to Avoid)*			
Hazardous Decomposition or Byproducts			
Hazardous Polymerization	May Occur		Condition to Avoid
	Will Not Occur		

Section VI—Health Hazard Data

Route(s) of Entry:	Inhalation?	Skin?	Ingestion?
Health Hazards *(Acute and Chronic)*			
Carcinogenicity	NTP?	IARC Monographs?	OSHA Regulated?
Signs and Symptoms of Exposure			
Medical Conditions Generally Aggravated by Exposure			
Emergency and First Aid Procedures			

Section VII—Precautions for Safe Handling and Use

Steps to Be Taken in Case Material is Released or Spilled

Waste Disposal Method

Precautions to be Taken in Handling and Storing

Other Precautions

Section VIII—Control Measures

Respiratory Protection (Specify Type)		
Ventilation	Local Exhaust	Special
	Mechanical *(General)*	Other
Protective Gloves		Eye Protection
Other Protective Clothing or Equipment		
Work/Hygienic Practices		

Page 2

* USGPO 1986-491-529/45775

Fig. 8.11 Material Safety Data Sheet (continued)

8.113 Confined Space Entry Procedures

CONFINED SPACES[16, 17] pose significant risks for a large number of workers, including many utility operators. OSHA has therefore defined very specific procedures to protect the health and safety of operators whose jobs require them to enter or work in a confined space. The regulations (which can be found in the Code of Federal Regulations at 29 CFR 1910.146) require conditions in the confined space to be tested and evaluated before anyone enters the space. If conditions exceed OSHA's limits for safe exposure, additional safety precautions must be taken and a confined space entry permit (Figure 8.12) must be approved by the appropriate authorities prior to anyone entering the space.

The managers of water utilities may or may not be involved in the day-to-day details of enforcing the agency's confined space policy and procedures. However, every utility manager should be aware of the current OSHA requirements and should ensure that the utility's policies not only comply with current regulations, but that the agency's policies are vigorously enforced for the safety of all operators.

QUESTIONS

Write your answers in a notebook and then compare your answers with those on page 514.

8.11J List the basic elements of a hazard communication program.

8.11K What are a manager's responsibilities for ensuring the safety of operators entering or working in confined spaces?

8.114 Reporting

The mainstay of a safety program is the method of reporting and keeping of statistics. These records are needed regardless of the size of the utility because they provide a means of identifying accident frequencies and causes as well as the personnel involved. The records can be looked upon as the operator's safety report card. Therefore, it becomes the responsibility of each injured operator to fill out the utility's accident report.

All injuries should be reported, even if they are minor in nature, so as to establish a record in case the injury develops into a serious injury. It may be difficult at a later date to prove whether the accident occurred on or off the job. This information may determine who is responsible for the costs. The responsibility for reporting accidents affects several levels of personnel. First, of course, is the injured person. Next, it is the responsibility of the supervisor, and finally, the

Responsibility of Management to review the causes and take steps to prevent such accidents from happening in the future.

Accident report forms may be very simple. However, they must record all details required by law and all data needed for statistical purposes. The forms shown in Figures 8.13 and 8.14 are examples for you to consider for use in your plant. The report must show the name of the injured, employee number, division, time of accident, nature of injury, cause of accident, first aid administered, and remarks for items not covered elsewhere. There should be a review process by foreman, supervisor, safety officer, and management. *RECOMMENDATIONS ARE NEEDED AS WELL AS A FOLLOW-UP REVIEW TO BE SURE THAT PROPER ACTION HAS BEEN TAKEN TO PREVENT RECURRENCE.*

In addition to reports needed by the utility, other reports may be required by state or federal agencies. For example, vehicle accident reports must be submitted to local police departments. If a member of the public is injured, additional forms are needed because of possible subsequent claims for damages. If the accident is one of occupational injury, causing lost time, other reports may be required. Follow-up investigations to identify causes and responsibility may require the development of other specific types of record forms.

---

[16] *Confined Space.   Confined space means a space that:*
  A. Is large enough and so configured that an employee can bodily enter and perform assigned work; and
  B. Has limited or restricted means for entry or exit (for example, manholes, tanks, vessels, silos, storage bins, hoppers, vaults, and pits are spaces that may have limited means of entry); and
  C. Is not designed for continuous employee occupancy.
  (Definition from the Code of Federal Regulations (CFR) Title 29 Part 1910.146.)

[17] *Confined Space, Permit-Required (Permit Space).   A confined space that has one or more of the following characteristics:*
  ● *Contains or has a potential to contain a hazardous atmosphere,*
  ● *Contains a material that has the potential for engulfing an entrant,*
  ● *Has an internal configuration such that an entrant could be trapped or asphyxiated by inwardly converging walls or by a floor which slopes downward and tapers to a smaller cross section, or*
  ● *Contains any other recognized serious safety or health hazard.*
  (Definition from the Code of Federal Regulations (CFR) Title 29 Part 1910.146.)

Confined Space Pre-Entry Checklist/Confined Space Entry Permit

Date and Time Issued: _____ Date and Time Expires: _____ Job Site/Space I.D.: _____

Job Supervisor: _____ Equipment to be worked on: _____ Work to be performed: _____

Standby personnel: _____ _____ _____

1. Atmospheric Checks: Time _____ Oxygen _____ % Toxic _____ ppm

   Explosive _____ % LEL Carbon Monoxide _____ ppm

2. Tester's signature: _____

3. Source isolation: (No Entry)   N/A   Yes   No

   Pumps or lines blinded,
   disconnected, or blocked   ( )   ( )   ( )

4. Ventilation Modification: N/A   Yes   No

   Mechanical   ( )   ( )   ( )

   Natural ventilation only   ( )   ( )   ( )

5. Atmospheric check after isolation and ventilation: Time _____

   Oxygen _____ % > 19.5%   < 23.5%   Toxic _____ ppm   < 10 ppm H_2S

   Explosive _____ % LEL   < 10%   Carbon Monoxide _____ ppm   < 35 ppm CO

Tester's signature: _____

6. Communication procedures: _____

7. Rescue procedures: _____

_____

8. Entry, standby, and backup persons   Yes   No

   Successfully completed required training?   ( )   ( )

   Is training current?   ( )   ( )

9. Equipment: N/A   Yes   No

   Direct reading gas monitor tested   ( )   ( )   ( )

   Safety harnesses and lifelines for entry and standby persons   ( )   ( )   ( )

   Hoisting equipment   ( )   ( )   ( )

   Powered communications   ( )   ( )   ( )

   SCBAs for entry and standby persons   ( )   ( )   ( )

   Protective clothing   ( )   ( )   ( )

   All electric equipment listed for Class I, Division I,
   Groups A, B, C, and D, and nonsparking tools   ( )   ( )   ( )

10. Periodic atmospheric tests:

    Oxygen: ____% Time ____; ____% Time ____; ____% Time ____; ____% Time ____;

    Explosive: ____% Time ____; ____% Time ____; ____% Time ____; ____% Time ____;

    Toxic: ____ppm Time ____; ____ppm Time ____; ____ppm Time ____; ____ppm Time ____;

    Carbon Monoxide: ____ppm Time ____; ____ppm Time ____; ____ppm Time ____; ____ppm Time ____;

We have reviewed the work authorized by this permit and the information contained herein. Written instructions and safety procedures have been received and are understood. Entry cannot be approved if any brackets ( ) are marked in the "No" column. This permit is not valid unless all appropriate items are completed.

Permit Prepared By: (Supervisor) _____ Approved By: (Unit Supervisor) _____

Reviewed By: (CS Operations Personnel) _____

         (Entrant)   (Attendant)   (Entry Supervisor)

This permit to be kept at job site. Return job site copy to Safety Office following job completion.

Fig. 8.12 Confined space pre-entry checklist/confined space entry permit

Date _____

Name of injured employee _____ Employee # _____ Area _____

Date of accident _____ Time _____ Employee's Occupation _____

Location of accident _____ Nature of injury _____

Name of doctor_____ Address _____

Name of hospital _____ Address _____

Witnesses (name & address)_____

PHYSICAL CAUSES

Indicate below by an "X" whether, in your opinion, the accident was caused by:

_____ Improper guarding

_____ Defective substances or equipment

_____ Hazardous arrangement

_____ Improper illumination

_____ Improper dress or apparel

_____ Not listed (describe briefly) _____

_____ No mechanical cause

_____ Working methods

_____ Lack of knowledge or skill

_____ Wrong attitude

_____ Physical defect

UNSAFE ACTS

Sometimes the injured person is not directly associated with the causes of an accident. Using an "X" to represent the injured worker and an "O" to represent any other person involved, indicate whether, in your opinion, the accident was caused by:

_____ Operating without authority

_____ Failure to secure or warn

_____ Working at unsafe speed

_____ Made safety device inoperative

_____ Unsafe equipment or hands instead of equipment

_____ No unsafe act

_____ Not listed (describe briefly) _____

_____ Unsafe loading, placement & etc.

_____ Took unsafe position

_____ Worked on moving equipment

_____ Teased, abused, distracted & etc.

_____ Did not use safe clothing or personal protective equipment

What job was the employee doing? _____

What specific action caused the accident? _____

What steps will be taken to prevent recurrence? _____

_____

_____

Date of Report _____ Immediate Supervisor _____

REVIEWING AUTHORITY

Comments:

Comments:

_____ _____

Safety Officer   Date   Department Director   Date

Fig. 8.13 Supervisor's accident report

INJURED: COMPLETE THIS SECTION

Name _____ Age _____ Sex _____

Address _____

Title _____ Dept. Assigned _____

Place of Accident _____

Street or Intersection _____

Date _____ Hour _____ A.M. _____ P.M. _____

Type of Job You Were Doing When Injured

_____

Object Which Directly Injured You   Part of Body Injured

_____

How Did Accident Happen? (Be specific and give details; use back of sheet if necessary.)

	Yes	No
Did You Report Accident or Exposure at Once? (Explain "No")	☐	☐
Did You Report Accident or Exposure to Supervisor? Give Name	☐	☐
Were There Witnesses to Accident or Exposure? Give Names	☐	☐
Did You See a Doctor? (If Yes, Give Name)	☐	☐
Are You Going to See a Doctor? (Give Name)	☐	☐

_____   _____
           Date   Signature

SUPERVISOR: COMPLETE THIS SECTION — (Return to Personnel as Soon as Possible)

	Yes	No
Was an Investigation of Unsafe Conditions and/or Unsafe Acts Made? If Yes, Please Submit Copy.	☐	☐
Was Injured Intoxicated or Behaving Inappropriately at Time of Accident? (Explain "Yes")	☐	☐

Date Disability   Last Day   Date Back
Commenced _____ Wages Earned _____ on Job _____

Date Report Completed _____ 19____ Signed By _____

                                           Title _____

Distribution: Canary - Department Head, Pink - Supervisor, White - Personnel

Fig. 8.14 Accident report

In the preparation of accident reports, it is the operator's responsibility to correctly fill out each form, giving complete details. The supervisor must be sure no information is overlooked that may be helpful in preventing recurrence.

The Safety Officer must review the reports and determine corrective actions and make recommendations.

In day-to-day actions, operators, supervisors, and management often overlook opportunities to counsel individual operators in safety matters. Then, when an accident occurs, they are not inclined to look too closely at accident reports. First, the accident is a series of embarrassments, to the injured person, to the supervisor, and to management. Therefore, there is a reluctance to give detailed consideration to accident reports. However, if a safety program is to function well, it will require a thorough effort on the part of the operator, supervisor, and management in accepting their responsibility for the accident and making a greater effort through good reporting to prevent future similar accidents. Accident reports must be analyzed, discussed, and the real cause of the accident identified and corrected.

Emphasis on the prevention of future accidents cannot be overstressed. We must identify the causes of accidents and implement whatever measures are necessary to protect operators from becoming injured.

8.115 Training

If a safety program is to work well, management will have to accept responsibility for the following three components of training:

1. Safety education of all employees,

2. Reinforced education in safety, and

3. Safety education in the use of tools and equipment.

Or, to put it another way, the three most important controlling factors in safety are education, education, and more education.

Responsibility for overall training must be that of upper management. A program that will educate operators and then reinforce this education in safety must be planned systematically and promoted on a continuous basis. There are many avenues to achieving this goal.

The safety education program should start with the new operator. Even before employment, verify the operator's past record and qualifications and review the pre-employment physical examination. In the new operator's orientation, include instruction in the importance of safety at your utility or plant. Also discuss the matter of proper reporting of accidents as well as the organization's policies and practices. Give new operators copies of all safety SOPs and direct their attention to parts that directly involve them. Ask the safety officer to give a talk about utility policy, safety reports, and past accidents, and to orient the new operator toward the importance of safety to operators and to the organization.

The next consideration must be one of training the new operator in how to perform assigned work. Most supervisors think in terms of On-the-Job Training (OJT). However, OJT is not a good way of preventing accidents with an inexperienced operator. The idea is all right if the operator comes to the organization trained in how to perform the work, such as a distribution system operator from another distribution system. Then you only need to explain your safety program and how your policies affect the new operator. For a new operator who is inexperienced in water distribution or in utility operation, the supervisor must give detailed consideration to the operator's welfare. In this instance, the training should include not only a safety talk, but the foreman (supervisor) must train the inexperienced operator in all aspects of distribution system safety. This training includes instruction in the handling of chemicals, the dangers of electrical apparatus, fire hazards, and proper maintenance of equipment to prevent accidents. Special instructions will also be needed for specific work environments such as manholes, gases (chlorine and hydrogen sulfide (H_2S)), water safety, and any specific hazards that are unique to your facility. The new operator must be checked out on any equipment personnel may operate such as vehicles, forklifts, valve operators, and radios. All new operators should be required to participate in a safety orientation program during the first few days of their employment, and an overall training program in the first few months.

The next step in safety education is reinforcement. Even if the operator is well trained, mistakes can occur; therefore, the education must be continual. Many organizations use the *TAILGATE SAFETY MEETING*[18] method as a means of maintaining the operator's interest in safety. The program should be conducted by the first line supervisor. Schedule the informal tailgate meeting for a suitable location, keep it short, avoid distractions, and be sure that everyone can hear. Hand out literature, if available. Tailgate talks should communicate to the operator specific considerations, new problems, and accident information. These topics should be published. One resource for such meetings can be those operators who have been involved in an accident. Although it is sometimes embarrassing to the injured, the victim is now the expert on how the accident occurred, what could have been done to prevent it, and how it felt to have the injury. Encourage all operators, new and old, to participate in tailgate safety sessions.

---

[18] *Tailgate Safety Meeting. Brief (10 to 20 minutes) safety meetings held every 7 to 10 working days. The term TAILGATE comes from the safety meetings regularly held by the construction industry around the tailgate of a truck.*

Use safety posters to reinforce safety training and to make operators aware of the location of dangerous areas or show the importance of good work habits. Such posters are available through the National Safety Council's catalog.[19]

Awards for good safety records are another means of keeping operators aware of the importance of safety. The awards could be given to individuals in recognition of a good safety record. Publicity about the awards may provide an incentive to the operators and demonstrates the organization's determination to maintain a good safety record. The awards may include: AWWA's water drop pins, certificates, and/or plaques showing number of years without an accident. Consider publishing a utility newsletter on safety tips or giving details concerning accidents that may be helpful to other operators in the organization. Awards may be given to the organization in recognition of its effort in preventing accidents or for its overall safety program. A suggestion program concerning safety will promote and reinforce the program and give recognition to the best suggestions. The goal of all these efforts is to reinforce concerns for the safety of all operators. If safety, as an idea, is present, then accidents can be prevented.

Education of the operator in the use of tools and equipment is necessary. As pointed out above, OJT is not the answer to a good safety record. A good safety record will be achieved only with good work habits and safe equipment. If the operator is trained in the proper use of equipment (hand tools or vehicles), the operator is less likely to misuse them. However, if the supervisor finds an operator misusing tools or equipment, then it is the supervisor's responsibility to reprimand the operator as a means of reinforcing utility policies. The careless operator who misuses equipment is a hazard to other operators. Careless operators will also be the cause of a poor safety record in the operator's division or department.

An important part of every job should be the consideration of its safety aspects by the supervisor. The supervisor should instruct the foreman or operators about any dangers involved in job assignments. If a job is particularly dangerous, then the supervisor must bring that fact to everyone's attention and clarify utility policy in regard to unsafe acts and conditions.

If the operator is unsure of how to perform a job, then it is the operator's responsibility to ask for the training needed. Each operator must think, act, and promote safety if the organization is to achieve a good safety record. Training is the key to achieving this objective and training is everyone's responsibility—management, the supervisors, foremen, and operators.

QUESTIONS

Write your answers in a notebook and then compare your answers with those on pages 514 and 515.

8.11L   What is the mainstay of a safety program?

8.11M   Why should you report even a minor injury?

8.11N   Why should a safety officer review an accident report form?

8.11O   A new, inexperienced operator must receive instruction on what aspects of treatment plant safety?

8.11P   What should an operator do if unsure of how to perform a job?

8.116 Measuring

To be complete, a safety program must also include some means of identifying, measuring, and analyzing the effects of the program. The systematic classification of accidents, injuries, and lost time is the responsibility of the safety officer. This person should use an analytical method which would refer to types and classes of accidents. Reports should be prepared using statistics showing lost time, costs, type of injuries, and other data, based on a specific time interval. Such data call at-

tention to the effectiveness of the program and promote awareness of the types of accidents that are happening. Management can use this information to decide where the emphasis should be placed to avoid accidents. However, statistical data are of little value if a report is prepared and then set on the bookshelf or placed in a supervisor's desk drawer. The data must be distributed and read by all operating and maintenance personnel.

As an example, injuries can be classified as fractures, burns, bites, eye injuries, cuts, and bruises. Causes can be classified as related to heat, machinery, falling, handling objects, chemicals, unsafe acts, and miscellaneous. Cost can be considered as lost time, lost dollars, lost production, contaminated water, or any other means of showing the effects of the accidents.

Good analytical reporting will provide a great deal of detail without a lot of paper to read and comprehend. Keep the method of reporting simple and easy to understand by all operators, so they can identify with the causes and be aware of how to prevent the accident happening to themselves or other operators. Table 8.10 gives one method of summarizing the causes of various types of injuries.

---

[19] *Write or call your local safety council or National Safety Council, 1121 Spring Lake Drive, Itasca, IL 60143-3201.*

TABLE 8.10   SUMMARY OF TYPES AND CAUSES OF INJURIES

Type of Injury	PRIMARY CAUSE OF INJURY										
	Unsafe Act	Chemical	Falls	Handling Objects	Heat	Machinery	Falling Objects	Stepping	Striking	Miscellaneous	TOTAL
Fractures											
Sprains											
Eye Injuries											
Bites											
Cuts											
Bruises											
Burns											
Miscellaneous											

There are many other methods of compiling data. Table 8.10 could reflect cost in dollars or in work hours lost. Not all accidents mean time lost, but there can be other cost factors. The data analysis should also indicate if the accidents involve vehicles, company personnel, the public, company equipment, loss of chemical, or other factors. Results also should show direct cost and indirect cost to the agency, operator, and the public.

Once the statistical data have been compiled, someone must be responsible for reviewing it in order to take preventive actions. Frequently such responsibility rests with the safety committee. In fact, safety committees may operate at several levels, for example management committee, working committee, or an accident review board. In any event, the committee must be active, be serious, and be fully supported by management.

Another means of measuring safety is by calculating the injury frequency rate for an indication of the effectiveness of your safety program. Multiply the number of disabling injuries by one million and divide by the total number of employee-hours worked. An average work year equals 2,080 work hours. The number of injuries per year is multiplied by one million in order to obtain injury frequency rate values or numbers that are easy to use. In our example problems, we obtained numbers between one and one thousand.

$$\text{Injury Frequency Rate} = \frac{(\text{Number of Disabling Injuries/year})(1,000,000)}{\text{Number of Hours Worked/year}}$$

These calculations indicate a frequency rate per year, which is the usual means of showing such data. Note that this calculation accounts only for disabling injuries. You may wish to show all injuries, but the calculations are much the same.

EXAMPLE 3

A rural water company employs 36 operators who work in many small towns throughout a three-state area. The operators suffered four injuries in one year while working 74,880 hours. Calculate the injury frequency rate.

Known	Unknown
Number of Operators = 36	Injury Frequency Rate
Number of Injuries/yr = 4 injuries/yr	
Total Hours Worked/yr = 74,880 hr/yr	

Calculate the injury frequency rate.

$$\text{Injury Frequency Rate} = \frac{(\text{Number of Injuries/year})(1,000,000)}{\text{Total Hours Worked/year}}$$

$$= \frac{(4 \text{ Injuries/yr})(1,000,000)}{74,880 \text{ hr/yr}}$$

$$= 53.4$$

EXAMPLE 4

Of the four injuries suffered by the operators in Example 3, one was a disabling injury. Calculate the injury frequency rate for the disabling injuries.

Known	Unknown
Number of Disabling Injuries/yr = 1 injury/yr	Injury Frequency Rate
Total Hours Worked/yr = 74,880 hr/yr	

Calculate the injury frequency rate.

$$\text{Injury Frequency Rate (Disabling Injuries)} = \frac{(\text{Number of Disabling Injuries/yr})(1,000,000)}{\text{Total Hours Worked/yr}}$$

$$= \frac{(1 \text{ injury/yr})(1,000,000)}{74,880 \text{ hr/yr}}$$

$$= 13.4$$

Yet another consideration may be lost-time accidents. The safety officer's analysis may take into account many other considerations, but in any event, the method given here will provide a means of recording and measuring injuries in the treatment plant. In measuring lost-time injuries, a severity rate can be considered.

A severity rate is based on one lost hour for every million operator hours worked. The rate is found by multiplying the number of hours lost by one million and dividing by the total of operator-hours worked.

$$\text{Injury Severity Rate} = \frac{(\text{Number of Hours Lost/yr})(1,000,000)}{\text{Total Hours Worked/yr}}$$

EXAMPLE 5

The water company described in Examples 3 and 4 experienced 40 operator-hours lost due to injuries while the operators worked 74,880 hours. Calculate the injury severity rate.

Known	Unknown
Number of Hours Lost/yr = 40 hr/yr	Injury Severity Rate
Total Hours Worked/yr = 74,880 hr/yr	

Calculate the injury severity rate.

$$\text{Injury Severity Rate} = \frac{(\text{Number of Hours Lost/yr})(1{,}000{,}000)}{\text{Total Hours Worked/yr}}$$

$$= \frac{(40 \text{ hr/yr})(1{,}000{,}000)}{74{,}880 \text{ hr/yr}}$$

$$= 534$$

Notice that all these data points are based on a one-year time interval which makes them suitable for use by the safety officer in preparing an annual report.

8.117   Human Factors

First, you may ask, what is a human factor? Well, it is not too often that a safety text considers human factors as part of the safety program. However, if these factors are understood and emphasis is given to their practical application, then many accidents can be prevented. Human Factors Engineering is the specialized study of technology relating to the design of operator-machine interface. That is to say, it examines ways in which machinery might be designed or altered to make it easier to use, safer, and more efficient for the operator. We hear a lot about making computers more user friendly, but human factors engineering is just as important to everyday operation of other machinery in the everyday plant.

Many accidents occur because the operator forgets the human factors. The ultimate responsibility for accidents due to human factors belongs to the management group. However, this does not relieve the operator of the responsibility to point out the human factors as they relate to safety. After all, it is the operator using the equipment who can best tell if it meets all the needs for an interrelationship between operator and machine.

The first step in the prevention of accidents takes place in the plant design. Even with excellent designs, accidents can and do happen. However, every step possible must be taken during design to ensure a maximum effort of providing a safe plant environment. Most often the operator has little to do with design, and therefore needs to understand human factors engineering so as to be able to evaluate these factors as the

plant is being operated. As newer plants become automated, this type of understanding may even be more important.

Other contributing human factors are the operator's mental and physical characteristics. The operator's decision-making abilities and general behavior (response time, sense of alarm, and perception of problems and danger) are all important factors. Ideally, tools and machines should function as intuitive extensions of the operator's natural senses and actions. Any factors disrupting this flow of action can cause an accident. Therefore, be on the lookout for such factors. When you find a system that cannot be operated in a smooth, logical sequence of steps, change it. You may prevent an accident. If the everyday behavior of an operator is inappropriate with regard to a specific job, reconsider the assignment to prevent an accident.

The human factor in safety is the responsibility of design engineers, supervisors, and operators. However, the operator who is doing the work will have a greater understanding of the operator-machine interface. For this reason, the operator is the appropriate person to evaluate the means of reducing the human factor's contribution to the cause of accidents, thereby improving the plant's safety record.

QUESTIONS

Write your answers in a notebook and then compare your answers with those on page 515.

8.11Q   Statistical accident reports should contain what types of accident data?

8.11R   How can injuries be classified?

8.11S   How can causes of injuries be classified?

8.11T   How can costs of accidents be classified?

8.12   RECORDKEEPING

8.120   Purpose of Records

Accurate records are a very important part of effective operation of water distribution system facilities. Records are a valuable source of information. They can save time when trouble develops and provide proof that problems were identified and solved. Pertinent and complete records should be used as a basis for distribution system operation, interpreting results of distribution system operation, preparing preventive maintenance programs, and preparing budget requests. When accurately kept, records provide an essential basis for design of future changes or expansions of the distribution system, and also can be used to aid in the design of other water distribution systems where similar water may be distributed and similar problems may develop.

If legal questions or problems occur in connection with the treatment of the water, the operation of the system, or the distribution of water, accurate and complete records will provide evidence of what actually occurred and what procedures were followed.

Records are essential for effective management of water distribution and treatment facilities and to satisfy legal requirements. Some of the important uses of records include:

1. Aiding operators in solving treatment, distribution, and water quality problems,

2. Providing a method of alerting operators to changes in source water quality,

3. Showing that the treated water is acceptable to the consumer,

4. Documenting that the final product meets plant performance standards, as well as the standards of the regulatory agencies,

5. Determining performance of treatment processes, equipment, and the plant,

6. Satisfying legal requirements,

7. Aiding in answering complaints,

8. Anticipating routine maintenance,

9. Providing data for cost analyses and preparation of budgets,

10. Providing data for future engineering designs, and

11. Providing information for monthly and annual reports.

8.121 Types of Records

There are many different types of records that are required for effective management and operation of water supply, treatment, and distribution system facilities. Below is a listing of some essential records:

1. Source of supply,

2. Operation,

3. Laboratory,

4. Maintenance,

5. Chemical inventory and usage,

6. Purchases,

7. Chlorination station,

8. Main disinfection,

9. Cross-connection control,

10. Personnel,

11. Accidents, and

12. Customer complaints.

8.122 Types of Plant Operations Data[20]

Plant operations logs can be as different as the plants and water systems whose information they record. The differences in amount, nature, and format of data are so significant that any attempt to prepare a "typical" log would be very difficult. This section will outline the kinds of data that are usually required to help you develop a useful log for your facilities.

Treatment plant data such as total flows, chemical use, chemical doses, filter performance, reservoir levels, quality control tests, and rainfall and runoff information represent the bulk of the data required for proper plant operation. Frequently, however, source and distribution system data such as reservoir storage and water quality data are included because of the impact of this information on plant operation and operator responsibilities. Typical plant operations data include:

1. Plant title, agency, and location;

2. Date;

3. Names of operators and supervisors on duty;

4. Source of supply,

   a. Reservoir elevation and volume of storage,
   b. Reservoir inflow and outflow,
   c. Evaporation and precipitation,
   d. Apparent runoff, seepage loss, or infiltration gain, and
   e. Production figures from wells;

5. Water treatment plant,

   a. Plant inflow,
   b. Treated water flow,
   c. Plant operating water (backwash), and
   d. Clear well level;

6. Distribution system,

   a. Flows to system (system demand),
   b. Distribution system reservoir levels and changes, and
   c. Comparison of production with deliveries (the difference is "unaccounted for water");

7. Chemical inventory and usage,

   a. Chemical inventory/storage (measured use and deliveries),
   b. Metered or estimated plant usages, and
   c. Calculated usage of chemicals (compare with actual use);

8. Quality control tests,

   a. Turbidity,
   b. Chlorine residual,
   c. Coliforms,
   d. Odor,
   e. Color, and
   f. Other;

9. Filter performance,

   a. Operation,

      (1) Total hours, all units,
      (2) Filtered water turbidities,
      (3) Head losses,
      (4) Levels, and
      (5) Flow rates;

   b. Backwash,

      (1) Total hours,
      (2) Head losses,
      (3) Total wash water used, and
      (4) Duration and rate of back/surface wash;

---

[20] Also see WATER TREATMENT PLANT OPERATION, Volume I, Chapter 10, "Plant Operation," Section 10.6, "Operating Records and Reports," for additional details and recordkeeping forms.

10. Meteorologic,

   a. Rainfall, evaporation, and temperature of both water and air, and
   b. Weather (clear, cloudy, windy); and

11. Remarks.

Space should be provided to describe or explain unusual data or events. Extensive notes should be entered on a daily worksheet or diary.

8.123   Maintenance Records

A good plant maintenance effort depends heavily upon good recordkeeping. There are several areas where proper records and documentation can definitely improve overall plant and distribution system performance.

8.124   Procurement Records

Ordering repair parts and supplies usually is done when the on-hand quantity of a stocked part or chemical falls below the reorder point, a new item is added to stock, or an item has been requested that is not stocked. Most organizations require employees to submit a requisition (similar to the one shown in Figure 8.15) when they need to purchase equipment or supplies. When the requisition has been approved by the authorized person (a supervisor or purchasing agent, in most cases) the items are ordered using a form called a purchase order. A purchase order contains a number of important items. These items include: (1) the date, (2) a complete description of each item and quantity needed, (3) prices, (4) the name of the vendor, and (5) a purchase order number.

A copy of the purchase order should be retained in a suspense file or on a clipboard until the ordered items arrive. This procedure helps keep track of the items that have been ordered but have not yet been received.

All supplies should be processed through the storeroom immediately upon arrival. When an item is received it should be so recorded on an inventory card. The inventory card will keep track of the numbers of an item in stock, when last ordered, cost, and other information. Furthermore, by always logging in supplies immediately upon receipt, you are in a position to reject defective or damaged shipments and control shortages or errors in billing. Some utilities use personal computers to keep track of orders and deliveries.

8.125   Inventory Records

An inventory consists of the supplies the treatment plant and distribution system needs to keep on hand to operate the facilities. These maintenance supplies may include repair parts, spare valves, electrical supplies, tools, and lubricants. The purpose of maintaining an inventory is to provide needed parts and supplies quickly, thereby reducing equipment downtime and work delays.

In deciding what supplies to stock, keep in mind the economics involved in buying and stocking an item as opposed to depending upon outside availability to provide needed supplies. Is the item critical to continued plant or process operation? Should certain frequently used repair parts be kept on hand? Does the item have a shelf life?

Inventory costs can be held to a minimum by keeping on hand only those parts and supplies for which a definite need exists or which would take too long to obtain from an outside vendor. A "definite need" for an item is usually demonstrated by a history of regular use. Some items may be infrequently used but may be vital in the event of an emergency; these items should also be stocked. Take care to exclude any parts and supplies that may become obsolete, and do not stock parts for equipment scheduled for replacement.

Tools should be inventoried. Tools that are used by operators on a daily basis should be permanently signed out to them. More expensive tools and tools that are only occasionally used, however, should be kept in a storeroom. These tools should be signed out only when needed and signed back in immediately after use.

8.126   Equipment Records

You will need to keep accurate records to monitor the operation and maintenance of plant and distribution system equipment. Equipment control cards and work orders (this information can be in a computer) can be used to:

● Record important equipment data such as make, model, serial number, and date purchased,

● Record maintenance and repair work performed to date,

● Anticipate preventive maintenance needs, and

● Schedule future maintenance work.

See *WATER TREATMENT PLANT OPERATION*, Volume II, Chapter 18, Section 18.00, "Preventive Maintenance Records," for additional information.

8.127   Disposition of Plant and System Records

Good recordkeeping is very important because records indicate potential problems, adequate operation, and are a good waterworks practice. Usually the only record required by the health agency is the summary of the daily turbidity of the treated surface water as it enters the distribution system. Chlorine residual and bacterial counts are often required. Other records that may also be required include:

1. Total trihalomethane (TTHM) data (frequency of this report is based on the number of people served),

2. The daily log and records of the analyses to control the treatment process may be required when there are chronic treatment problems,

3. Chlorination, constituent removal, and sequestering records may be required from small systems (especially those demonstrating little understanding of the processes), and

4. Records showing the quantity of water from each source in use may be required from systems with sources producing water not meeting state and/or local health department water quality standards.

An important question is how long records should be kept. Records should be kept as long as they may be useful. Some information will become useless after a short time, while other

P-1

CITY OF SACRAMENTO
REQUISITION

1 BID NO.

DUE DATE

2 PURCHASE ORDER NO.

DATE

3 DELIVER TO

4 FUND GP	5 FUND	6 ORGAN.	7 REV. COST CTR	8 OBJECT	

9 REQUISITION NO

DATE

10 REFER QUESTIONS TO:

11 ITEM NO.	12 COMMODITY CODE NO.	13 DESCRIPTION	14 QUANTITY UNIT	15 PRICE UNIT	VENDORS: USE THESE COLS. ONLY 16 BRAND OFFERED	17 LOT PRICE FOR EACH ITEM

18 DEPT

19 ORIG BY

TEL. NO.

THIS REQUISITION WILL BE REPRODUCED TO CREATE A PURCHASE ORDER QUICKLY AND PROVIDE FASTER DELIVERY OF THESE ITEMS TO YOUR DEPARTMENT. PLEASE FOLLOW INSTRUCTIONS CAREFULLY.

SUB TOTAL

SALES TAX

INVOICE AMOUNT

20 EST. OF COST

22 SPECIAL REQUIREMENTS OR INSTRUCTIONS - PREVIOUS P.O. NO. OR BID NO., OR RECOMMENDED VENDORS

21

NUMBER OF REQUISITION ATTACHMENT SHEETS

23 CERTIFICATION IS HEREBY MADE THAT THE ABOVE IS A LEGAL CHARGE AGAINST THE APPROPRIATION INDICATED

DEPT. HEAD ____________ BY ____________

____
OFFICIAL TITLE

24

VENDOR ____________ (A)

ADDRESS ____________ (B)

CITY, STATE AND ZIP ____________ (C)

TERMS ____________ TEL. NO. ____________ (D)

DELIVERY ____________ FOB DESTINATION (E)

BY ____________ TITLE ____________

Fig. 8.15 Requisition/purchase order form

data may be valuable for many years. Data that might be used for future design or expansion should be kept indefinitely. Laboratory data will always be useful and should be kept indefinitely. Regulatory agencies may require you to keep certain water quality analyses (bacteriological test results) and customer complaint records on file for specified time periods (10 years for chemical analyses and bacteriological tests).

Even if old records are not consulted every day, this does not lessen their potential value. For orderly records handling and storage, set up a schedule to periodically review old records and to dispose of those records that are no longer needed. A decision can be made when a record is established regarding the time period for which it must be retained.

QUESTIONS

Write your answers in a notebook and then compare your answers with those on page 515.

8.12A   List some of the important uses of records.

8.12B   What is "unaccounted for water"?

8.12C   What chemical inventory and usage records should be kept?

8.12D   List the important items usually contained on a purchase order.

8.13   ACKNOWLEDGMENTS

During the writing of this material on administration, Lynne Scarpa, Phil Scott, Chris Smith, and Rich von Langen, all members of California Water Environment Association (CWEA), provided many excellent materials and suggestions for improvement. Their generous contributions are greatly appreciated.

8.14   ADDITIONAL READING

1. *MANAGE FOR SUCCESS*. Obtain from the Office of Water Programs, California State University, Sacramento, 6000 J Street, Sacramento, CA 95819-6025. Price, $45.00.

2. *WATER UTILITY MANAGEMENT* (M5). Obtain from American Water Works Association (AWWA), Bookstore, 6666 West Quincy Avenue, Denver, CO 80235. Order No. 30005. ISBN 0-89867-063-2. Price to members, $69.00; nonmembers, $99.00; price includes cost of shipping and handling.

3. *A WATER AND WASTEWATER MANAGER'S GUIDE FOR STAYING FINANCIALLY HEALTHY*, July 1989. Obtain from National Technical Information Service (NTIS), 5285 Port Royal Road, Springfield, VA 22161. Order No. PB90-114455. EPA No. 430-9-89-004. Price, $33.50, plus $5.00 shipping and handling.

4. *WASTEWATER UTILITY RECORDKEEPING, REPORTING AND MANAGEMENT INFORMATION SYSTEMS*, July 1982. Obtain from National Technical Information Service (NTIS), 5285 Port Royal Road, Springfield, VA 22161. Order No. PB83-109348. EPA No. 430-9-82-006. Price, $39.50, plus $5.00 shipping and handling.

5. *SUPERVISION: CONCEPTS AND PRACTICES OF MANAGEMENT*, Ninth Edition, 2004, Hilgert, Raymond L., and Edwin Leonard, Jr. Obtain from Thomson Learning, Order Fulfillment, PO Box 6904, Florence, KY 41022-6904. ISBN 0-324-17881-6. Price, $102.95, plus shipping and handling.

6. *CUSTOMER SERVICE II: A TEAM EFFORT*. Video, 13 minutes. Obtain from American Water Works Association (AWWA), Bookstore, 6666 West Quincy Avenue, Denver, CO 80235. Order No. 65065V. Price to members, $208.00; nonmembers, $313.00; price includes cost of shipping and handling.

7. *TEXAS MANUAL*, Chapter 18,* "Effective Public Relations in Water Works Operations," and Chapter 19,* "Planning and Financing."

8. *PRINCIPLES OF WATER RATES, FEES, AND CHARGES* (M1). Obtain from American Water Works Association (AWWA), Bookstore, 6666 West Quincy Avenue, Denver, CO 80235. Order No. 30001. ISBN 1-58321-069-5. Price to members, $84.00; nonmembers, $126.00; price includes cost of shipping and handling.

9. *EMERGENCY PLANNING FOR WATER UTILITIES* (M19). Obtain from American Water Works Association (AWWA), Bookstore, 6666 West Quincy Avenue, Denver, CO 80235. Order No. 30019. ISBN 1-58321-135-7. Price to members, $69.00; nonmembers, $99.00; price includes cost of shipping and handling.

* Depends on edition.

END OF LESSON 3 OF 3 LESSONS

ON

DISTRIBUTION SYSTEM ADMINISTRATION

Please answer the discussion and review questions next.

DISCUSSION AND REVIEW QUESTIONS

Chapter 8.   DISTRIBUTION SYSTEM ADMINISTRATION

(Lesson 3 of 3 Lessons)

Write the answers to these questions in your notebook. The question numbering continues from Lesson 2.

26. What can happen when agencies rely primarily on corrective maintenance to keep the system running?

27. What does a SCADA system do?

28. What items should be included in a water supplier's cross-connection control program?

29. What factors should be considered when selecting a backflow prevention device for a particular situation?

30. How would you assess the vulnerability of a water supply system?

31. How can a utility protect its water supply from contamination?

32. What would you do if you discovered that contaminated water has entered your distribution system?

33. Why must waterworks utilities have a safety program?

34. How can a good safety record be accomplished?

35. What is the intent of the OSHA regulations?

36. What are the four main elements of a hazard communication program?

37. What topics should be included in a safety officer's talk to new operators?

38. Why are records important?

SUGGESTED ANSWERS

Chapter 8.   DISTRIBUTION SYSTEM ADMINISTRATION

ANSWERS TO QUESTIONS IN LESSON 1

Answers to questions on page 442.

8.0A   Local utility demands on a water distribution system manager include protection from environmental disasters with a minimum investment of money.

8.0B   Changes in the environmental workplace are created by changes in the workforce and advances in technology.

Answers to questions on page 444.

8.1A   The functions of a distribution system manager include planning, organizing, staffing, directing, and controlling.

8.1B   In small communities the community depends on the manager to handle everything.

Answers to questions on page 447.

8.2A   Distribution system planning must include operational personnel, local officials (decision makers), and the public.

8.3A   The purpose of an organizational plan is to show who reports to whom and to identify the lines of authority.

8.3B   Effective delegation is uncomfortable for many managers since it requires giving up power and responsibility. Many managers believe that they can do the job better than others, they believe that other employees are not well trained or experienced, and they are afraid of mistakes. The water distribution system manager retains some responsibility even after delegating to another employee and, therefore, the manager is often reluctant to delegate or may delegate the responsibility but not the authority to get the job done.

8.3C   An important and often overlooked part of delegation is follow-up by the supervisor.

Answers to questions on page 449.

8.4A   Staffing responsibilities include hiring new employees, training employees, and evaluating job performance.

8.4B   The two personnel management concepts a manager should always keep in mind are "job-related" and "documentation."

8.4C   The steps involved in a staffing analysis include:

1. List the tasks to be performed,
2. Estimate the number of staff hours per year required to perform each task,
3. List the utility's current employees,
4. Assign tasks based on each employee's skills and abilities, and
5. Adjust the work assignments as necessary to achieve the best possible fit between the work to be done and the personnel/skills available to do it.

8.4D   A qualifications profile is a clear statement of the knowledge, skills, and abilities a person must possess to perform the essential job duties of a particular position.

Answers to questions on page 452.

8.4E   The purpose of a job interview is to gain additional information about the applicants so that the most qualified person can be selected to fill a job opening.

8.4F   Protected groups include minorities, women, disabled persons, persons over 40 years of age, and union members.

8.4G   The "OUCH" principle stands for:
Objectivity,
Uniform treatment of employees,
Consistency, and
Having job relatedness.

8.4H   A new employee's safety training should begin on the first day of employment or as soon thereafter as possible.

Answers to questions on page 453.

8.4I   The purpose of a probationary period for new employees is to provide a time during which both the employer and employee can assess the "fit" between the job and the person.

8.4J   The compensation an employee receives for the work performed includes satisfaction, recognition, security, appropriate pay, and benefits.

8.4K   Distribution system managers should provide training opportunities for employees so they can keep informed of new technologies and regulations. Training for supervisors is also important to ensure that supervisors have the knowledge, skills, and attitude that will enable them to be effective supervisors.

Answers to questions on page 457.

8.4L   An employee's immediate supervisor should conduct the employee's performance evaluation.

8.4M   Dealing with employee discipline requires tact and skill. The manager or supervisor should stay flexible, calm, and open-minded.

8.4N   Common warning signs of potential violence include abusive language, threatening or confrontational behavior, and brandishing a weapon.

Answers to questions on page 462.

8.4O   Harassment is any behavior that is offensive, annoying, or humiliating to an individual and that interferes with a person's ability to do a job. This behavior is uninvited, often repeated, and creates an uncomfortable or even hostile environment in the workplace.

8.4P   Types of behavior that could be considered sexual harassment include:

- Unwanted hugging, patting, kissing, brushing up against someone's body, or other inappropriate sexual touching,
- Subtle or open pressure for sexual activity,
- Persistent sexually explicit or sexist statements, jokes, or stories,
- Repeated leering or staring at a person's body,
- Suggestive or obscene notes or phone calls, and/or
- Display of sexually explicit pictures or cartoons.

8.4Q   The best way to prevent harassment is to set an example by your own behavior and to keep communication open between employees. A manager must also be aware of and take action to prevent any type of harassment in the workplace.

8.4R   In general, the ADA defines disability as a physical or mental impairment that substantially limits one or more of the major life activities of an individual.

Answers to questions on page 462.

8.4S   The shop steward is elected by the union employees and is their official representative to management and the local union.

8.4T   Union contracts do not change the supervisor's delegated authority or responsibility. Operators must carry out the supervisor's orders and get the work done properly, safely, and within a reasonable amount of time. However, a contract gives a union the right to protest or challenge a supervisor's decision.

ANSWERS TO QUESTIONS IN LESSON 2

Answers to questions on page 466.

8.5A   A manager needs both written and oral communication skills.

8.5B   The most common written documents that a distribution system manager must write include memos, business letters, press releases, monitoring reports, and the annual report.

8.5C   The annual report should be a review of what and how the agency operated during the past year and also the goals for the next year.

Answers to questions on page 466.

8.6A   A distribution system manager may be asked to conduct meetings with employees, the governing board, the public, and with other professionals in your field.

8.6B   Before a meeting (1) prepare an agenda and distribute, (2) find an adequate meeting room, and (3) set a beginning and ending time.

Answers to questions on page 469.

8.7A   The first step in organizing a public relations campaign is to establish objectives so you will have a clear idea of what you expect to achieve.

8.7B   Employees can be informed about the distribution system agency's plans, practices, and goals through newsletters, bulletin boards, and regular, open communication between supervisors and subordinates.

8.7C Newspapers give more thorough, in-depth coverage to stories than do the broadcast media.

8.7D Practice is the key to effective public speaking.

8.7E Complaints can be a valuable asset in determining consumer acceptance and pinpointing water quality problems. Customer calls are frequently the first indication that something may be wrong. Responding to complaints and inquiries promptly can save the distribution system agency money and staff resources, and minimize customer inconvenience.

Answers to questions on page 473.

8.8A The three main areas of financial management for a distribution system agency include providing financial stability for the utility, careful budgeting, and providing capital improvement funds for future utility expansion.

8.8B The operating ratio for a distribution system agency is calculated by dividing total revenues by total operating expenses.

8.8C It is important for a manager to get input from other personnel in the distribution system agency as well as community leaders as the budgeting process proceeds in order to gain support for the budget and to keep the budget on track once adopted.

8.8D The basic ways for a distribution system agency to finance capital improvements are through general obligation bonds, revenue bonds, or loan funding programs.

8.8E A revenue bond is a debt incurred by the community, often to finance distribution system agency improvements. User charges provide repayment on the bond.

ANSWERS TO QUESTIONS IN LESSON 3

Answers to questions on page 475.

8.9A The purpose of an O & M program is to maintain design functionality and/or to restore the system components to their original condition and thus functionality. That is, to ensure that the system performs as designed and intended.

8.9B In general, maintenance activities can be classified as corrective maintenance, preventive maintenance, and predictive maintenance.

8.9C The major elements of a good preventive maintenance program include the following:

1. Planning and scheduling,
2. Records management,
3. Spare parts management,
4. Cost and budget control,
5. Emergency repair procedures, and
6. Training program.

Answers to questions on page 476.

8.9D SCADA stands for **S**upervisory **C**ontrol **A**nd **D**ata **A**cquisition system.

8.9E A SCADA system collects, stores, and analyzes information about all aspects of operation and maintenance, transmits alarm signals when necessary, and allows fingertip control of alarms, equipment, and processes.

8.9F Measured (sensed) information could be transmitted by FM radio, leased telephone circuits, private signal lines, dial telephone lines, coaxial cable networks, spread spectrum radio, fiber-optic cable, microwave, 900 MHz radio, cellular telephone, or satellite communications systems.

Answers to questions on page 478.

8.9G SCADA systems record and store (archive) data in computers for easy retrieval and analysis.

8.9H Water quality indicators that can be easily monitored by a SCADA system include turbidity, pH, and temperature.

8.9I When a customer phones and complains about a loss of water pressure, the operator can determine the recorded system pressures near the customer's residence. If there is a loss of water pressure the operator can try to determine the cause (if the water pressure was below an acceptable minimum (30 psi), the SCADA system should have alerted the operators immediately). If there is not a system loss of water pressure, the operator can suggest how the customer can try to locate the cause of the problem on the customer's property.

8.9J The greatest challenges for operators using SCADA systems are to realize that computers may not be correct and to have the ability to operate when the SCADA system fails.

Answers to questions on page 481.

8.9K Contamination through cross connections has consistently caused more waterborne disease outbreaks in the United States than any other reported factor.

8.9L Responsibilities in the implementation of cross-connection control programs are shared by water suppliers, water users, health agencies, and plumbing officials.

8.9M The water user's cross-connection control responsibility is to keep contaminants out of the potable water system on the user's premises. The user also has the responsibility to prevent the creation of cross connections through modifications of the plumbing system on the premises.

Answers to questions on page 483.

8.9N Factors that should be considered when selecting a backflow prevention device include the (1) degree of hazard, (2) probability of backflow occurring, (3) complexity of the piping on the premises, and (4) the probability of the piping being modified.

8.9O Air gap or reduced-pressure principle backflow prevention devices are installed at wastewater treatment plants, wastewater pumping stations, reclaimed water reuse areas, areas where toxic substances in toxic concentrations are handled under pressure, premises having an auxiliary water supply that is or may be contaminated, and in locations where fertilizer, herbicides, or pesticides are injected into a sprinkler system.

Answers to questions on page 486.

8.10A The first step toward an effective contingency plan for emergencies is to make an assessment of vulnerability. Then a comprehensive plan of action can be developed and implemented.

8.10B   A detailed Emergency Operations Plan is not needed since all types of emergencies cannot be anticipated and a complex response program can be more confusing than helpful.

8.10C   An Emergency Operations Plan should include:

1. Vulnerability assessment,
2. Inventory of personnel,
3. Provisions for a recovery operation,
4. Provisions for training programs for operators in carrying out the plan,
5. Inclusion of coordination plans with health, police, and fire departments,
6. Establishment of a communications procedure, and
7. Provisions for protection of personnel, plant equipment, records, and maps.

Answers to questions on page 491.

8.10D | COLOR | RISK LEVEL AND SUGGESTED ACTIONS |
|---|---|
| RED | Severe risk of terrorist attacks |
| ORANGE | High risk of terrorist attacks |
| YELLOW | Significant risk, elevated condition |
| BLUE | General risk, guarded condition |
| GREEN | Low risk of terrorist attacks |

8.10E   Toxicity is the ability of a contaminant (chemical or biological) to cause injury when introduced into the body.

8.10F   The degree of toxicity varies with the concentration of contaminant required to cause injury, the speed with which the injury takes place, and the severity of the injury.

8.10G   Possible secondary effects in a water supply that may allow detection of a contaminant without specific testing include taste, odor, color, and chlorine demand.

Answers to questions on page 493.

8.10H   *Cryptosporidium* is a significant public health concern because sensitive populations may become seriously ill even when the drinking water meets or exceeds all current state and federal standards.

8.10I   A water treatment plant can be operated to protect the public from *Cryptosporidium* by developing a strategy for protecting their source water supplies and optimizing the treatment processes. The operator must continually verify optimum coagulant doses, be sure filtered water turbidity levels are low, and provide adequate disinfection chemicals and contact times.

8.10J   The most serious threat of *Cryptosporidium* to a public water supply occurs during periods of heavy rains or snow melts that flush areas that are sources of high concentrations of oocysts into waters upstream of a plant intake.

Answers to questions on page 494.

8.11A   The supervisors should be responsible for the implementation of a safety program.

8.11B   A safety officer should evaluate every accident, offer recommendations, and keep and apply statistics.

8.11C   Both state and federal regulatory agencies enforce the OSHA requirements.

8.11D   The utility manager is responsible for the safety of the agency's personnel and the public exposed to the water utility's operations. Therefore, the manager must develop and administer an effective safety program and must provide new employee safety training as well as ongoing training for all employees.

8.11E   A safety policy statement should:

1. Define the goals and objectives of the program,
2. Identify the persons responsible for each element of the program,
3. Affirm management's intent to enforce safety regulations, and
4. Describe the disciplinary actions that will be taken to enforce safe work practices.

Answers to questions on page 495.

8.11F   A supervisor may be responsible, in part or completely, for an accident by causing unsafe acts to take place, by requiring that work be performed in haste, by disregarding an unsafe work environment, or by failing to consider any number of safety hazards.

8.11G   Each operator must accept, at least in part, responsibility for fellow operators, for the utility's equipment, for the operator's own welfare, and even for seeing that the supervisor complies with established safety regulations.

8.11H   First aid means emergency treatment for injury or sudden illness, before regular medical treatment is available.

8.11I   First-aid training is most important for operators who regularly work with electrical equipment and those who must handle chlorine and other dangerous chemicals.

Answers to questions on page 499.

8.11J   The basic elements of a hazard communication program include the following:

1. Identify hazardous materials,
2. Obtain chemical information and define hazardous conditions,
3. Properly label hazards, and
4. Train operators.

8.11K   A utility manager may or may not be involved in the day-to-day details of enforcing the agency's confined space policy and procedures. However, every utility manager should be aware of the current OSHA requirements and should ensure that the utility's policies not only comply with current regulations, but that the agency's policies are vigorously enforced for the safety of all operators.

Answers to questions on page 504.

8.11L   The mainstay of a safety program is the method of reporting and keeping statistics.

8.11M   All injuries should be reported, even if they are minor in nature, so as to establish a record in case the injury develops into a serious injury. It may be difficult at a later date to prove whether the accident occurred on or off the job. This information may determine who is responsible for the costs.

8.11N A safety officer should review an accident report form to (1) determine corrective actions, and (2) make recommendations.

8.11O A new, inexperienced operator must receive instruction on all aspects of plant safety. This training includes instruction in the handling of chemicals, the dangers of electrical apparatus, fire hazards, and proper maintenance of equipment to prevent accidents. Special instructions are required for specific work environments such as manholes, gases (chlorine and hydrogen sulfide (H_2S)), water safety, and any specific hazards that are unique to your facility. All new operators should be required to participate in a safety orientation program during the first few days of employment, and an overall training program in the first few months.

8.11P If an operator is unsure of how to perform a job, then it is the operator's responsibility to ask for the training needed.

Answers to questions on page 506.

8.11Q Statistical accident reports should contain accident statistics showing lost time, costs, type of injuries, and other data, based on a specific time interval.

8.11R Injuries can be classified as fractures, burns, bites, eye injuries, cuts, and bruises.

8.11S Causes of injuries can be classified as related to heat, machinery, falling, handling objects, chemicals, unsafe acts, and miscellaneous.

8.11T Costs of accidents can be classified as lost time, lost dollars, lost production, contaminated water, or any other means of showing the effects of the accidents.

Answers to questions on page 510.

8.12A Some of the important uses of records include:

1. Aiding operators in solving treatment, distribution, and water quality problems,
2. Providing a method of alerting operators to changes in source water quality,
3. Showing that the treated water is acceptable to the consumer,
4. Documenting that the final product meets plant performance standards, as well as the standards of the regulatory agencies,
5. Determining performance of treatment processes, equipment, and the plant,
6. Satisfying legal requirements,
7. Aiding in answering complaints,
8. Anticipating routine maintenance,
9. Providing data for cost analyses and preparation of budgets,
10. Providing data for future engineering designs, and
11. Providing information for monthly and annual reports.

8.12B "Unaccounted for water" is the difference between the amount of treated water that enters the distribution system and water that is delivered to consumers.

8.12C Chemical inventory and usage records that should be kept include:

1. Chemical inventory/storage (measured use and deliveries),
2. Metered or estimated plant usages, and
3. Calculated usage of chemicals (compare with actual use).

8.12D Important items usually contained on a purchase order include: (1) the date, (2) a complete description of each item and quantity needed, (3) prices, (4) the name of the vendor, and (5) a purchase order number.

APPENDIX

WATER DISTRIBUTION SYSTEM
OPERATION AND MAINTENANCE

Final Examination and
Suggested Answers

How to Solve Water Distribution
System Arithmetic Problems

Water Abbreviations

Water Words

Subject Index

FINAL EXAMINATION

This final examination was prepared *TO HELP YOU RE-VIEW* the material in this manual. *YOU DO NOT HAVE TO SEND YOUR ANSWERS TO CALIFORNIA STATE UNIVERSITY, SACRAMENTO.* The questions are divided into five types:

1. True-False,

2. Best Answer,

3. Multiple Choice,

4. Short Answer, and

5. Problems.

To work this examination:

1. Write the answers to each question in your notebook,

2. After you have worked a group of questions (you decide how many), check your answers with the suggested answers at the end of this exam, and

3. If you missed a question and don't understand why, reread the material in the manual.

You may wish to use this examination for review purposes when preparing for civil service and certification examinations.

Since you have already completed this course, please *DO NOT* send your answers to California State University, Sacramento.

True-False

1. Water distribution system operators have the responsibility for the health and well-being of the community they serve.

   1. True
   2. False

2. Frequently the water supplier tries to maintain specified minimum target water pressures throughout the distribution system.

   1. True
   2. False

3. Cathodic protection is regarded as a substitute for the use of a proper interior coating or the application of chemical inhibitors.

   1. True
   2. False

4. Positive coliform test results indicate that a water storage facility has been adequately disinfected.

   1. True
   2. False

5. When water is flowing into an elevated water storage reservoir, the hydraulic grade line will slope upward toward the reservoir.

   1. True
   2. False

6. All trenches and excavations are unsafe until properly shored, used, and backfilled.

   1. True
   2. False

7. The quality of water coming out of the consumer's tap can be affected by what happens to water as it travels through the consumer's own system.

   1. True
   2. False

8. Biofilms will grow more easily on materials that supply nutrients.

   1. True
   2. False

9. Operators should be required to follow a prepared preventive maintenance (PM) schedule and document the work they do.

   1. True
   2. False

10. A lack of chlorine residual in a distribution system could indicate the presence of heavy contamination.

    1. True
    2. False

11. Dry out the inside of a water meter after the meter is taken out of service in the field.

    1. True
    2. False

12. Longer contact times are required to disinfect water at higher temperatures.

    1. True
    2. False

13. When a new well is completed, it is necessary to disinfect the well, pump, and screen.

    1. True
    2. False

14. The chlorine injection should never be on the intake side of the pump because it will cause corrosion problems.

    1. True
    2. False

15. Only acids are corrosive.

    1. True
    2. False

16. Traffic cones and tubes are an effective method of channeling traffic along a specified route during nighttime hours.

    1. True
    2. False

17. The location of underground installations (such as utility lines) must be determined before excavation begins.

    1. True
    2. False

18. A confined space permit must be renewed each time an operator leaves and re-enters the confined space.

    1. True
    2. False

19. Nearly every area of a manager's responsibilities requires some type of planning.

    1. True
    2. False

20. The sooner you deal with a discipline problem, the better the outcome will be.

    1. True
    2. False

21. The water supplier is responsible for preventing contamination of the public water system by backflow.

    1. True
    2. False

22. A supervisor negligent of safety hazards could be fined, sentenced to a jail term, or even be barred from working in the profession.

    1. True
    2. False

Best Answer (Select only the closest or best answer.)

1. What is the operator's most important responsibility with regard to safety?

    1. Awareness of why safe procedures must be followed
    2. Detection of explosive conditions inside an elevated tank
    3. Following safe procedures
    4. Learning about the hazards of working in traffic

2. What is the main purpose of a water storage facility?

    1. To allow for a reserve storage
    2. To help maintain adequate pressures throughout the entire system
    3. To meet the needs for fire protection
    4. To provide a sufficient amount of water to average or equalize the daily demands

3. What is the purpose of surge tanks?

    1. To absorb energy
    2. To control water hammer
    3. To provide surges to clean mains
    4. To store water for surge demands

4. Why should all storage facilities be sampled regularly?

    1. To check on the operating levels of the storage tank
    2. To determine if the atmosphere above the water is hazardous
    3. To determine the need for pump maintenance
    4. To determine the quality of the water that enters and leaves the facility

5. Why is backflow a concern for water distribution system operators?

    1. Flow from back direction could drain storage tanks
    2. Flow from nonpotable source could contaminate distribution system
    3. Flow from reverse direction could exceed maximum allowable pipe pressure
    4. Flow from wrong direction could cause lead and copper compliance violation

6. Why can air become a problem in a water main?

    1. Air can cause water hammer
    2. Air can collect and stop the flow of water completely
    3. Air can dissolve in the water and cause supersaturation
    4. An air vacuum can cause the main to collapse

7. What is a cross connection?

    1. Connection between two drinking water systems, one across the street from the other
    2. Location where two water mains are connected to form a cross
    3. Protected connection between a drinking water supply and a system that is potable
    4. Unprotected connection between a drinking water supply and a system *NOT* approved as safe

8. What is back pressure?

    1. Backflow caused by a negative pressure in the water supply piping
    2. Backflow caused when the user's water system pressure is higher than the public system
    3. Pressure back up the distribution system in a water main
    4. Pressure back up the distribution system in a water storage reservoir

9. Which type of corrosion-control inhibitor may stimulate bacterial growth?

    1. Hydroxide
    2. Lime
    3. Phosphate-based product
    4. Sequestering agent

10. What is the main objective of distribution system maintenance?

    1. Implement and monitor an accurate distribution system recordkeeping program
    2. Keep all equipment clean and polished so it appears like new during system tours
    3. Maintain an up-to-date maintenance program schedule
    4. Provide maximum continuous equipment service at the lowest possible cost

11. What should be done immediately if there is any evidence that distribution system water has been tampered with by vandals or terrorists?

    1. Apprehend the vandals or terrorists
    2. Determine mechanism of tampering
    3. Investigate extent of tampering
    4. Take facility out of service

12. What is an air gap separation device?

    1. A check valve that prevents air from flowing back into the system
    2. A cross-connection device with an open gap or space
    3. A vertical empty space that separates a drinking water supply from an unapproved water system
    4. An air valve that isolates a section of the distribution system

13. What information is available from a geographic information system (GIS) computer program?

    1. Map detailing financial records of distribution system
    2. Map displaying USGS topographical maps
    3. Map showing distribution system hydraulic grade line contours
    4. Map with detailed information about the physical structures within geographic areas

14. What is disinfection?

    1. Complete destruction of all organisms
    2. Die-away of coliforms in drinking water
    3. Reinfection of organisms in distribution systems
    4. Selective destruction or inactivation of pathogenic organisms

15. What happens when chlorine is added to water and the reaction of chlorine with organic and inorganic materials stops?

    1. Chlorine demand has been satisfied
    2. Chlorine vapor pressure has dropped too low
    3. Formation of chloramines starts
    4. Total trihalomethanes have reached peak concentrations

16. What is the chlorine dose?

    1. Chlorine demand plus chlorine residual
    2. Chlorine demand plus free chlorine
    3. Combined chlorine forms plus free chlorine
    4. Free chlorine plus chlorine residual

17. What is the one objective of a safety program?

    1. To achieve and maintain OSHA compliance
    2. To keep accurate safety records
    3. To lower insurance rates
    4. To prevent accidents

18. What should an operator do if the operator cannot remove a safety hazard?

    1. Call hazard to attention of those who have authority to remove hazard
    2. Inform safety compliance inspector of hazard
    3. Try to avoid hazard
    4. Warn other operators of hazard

19. What is the basic requirement for an enclosed guard?

    1. Enclose moving parts to prevent contact with corrosive chemicals
    2. Guard moving parts from harmful pollutants
    3. Prevent hands and arms from making contact with dangerous moving parts
    4. Protect moving parts from damage by mechanical objects

20. In water distribution systems, what is the primary concern when entering a confined space?

    1. Explosive conditions
    2. Flammable conditions
    3. Oxygen deficiency
    4. Toxic gases

21. What is the meaning of accountability?

    1. When an operator answers to those above in the chain of command
    2. When an operator ensures that the manager is informed of results or events
    3. When an operator gives power to another person to accomplish a specific job
    4. When an operator has the power and resources to do a specific job

22. The selection method and examination process used to evaluate applicants must be limited to which areas?

    1. Applicant's ability to perform physical tasks in a job-related field
    2. Applicant's experience in a job-related field
    3. Applicant's formal education in a job-related field
    4. Applicant's knowledge, skills, and abilities to perform job-related activities

23. What is the purpose of a distribution system O & M program?

    1. To deliver safe drinking water
    2. To ensure that the system performs as designed and intended
    3. To keep operations and maintenance personnel communicating effectively
    4. To provide a program for maintaining equipment warranties

24. If contamination of a water supply is discovered, what is the most important immediate concern?

    1. Determining the cause of contamination
    2. Flushing out the system
    3. Neutralizing the contamination
    4. Safety of the public

Multiple Choice (Select all correct answers.)

1. What are the responsibilities of water distribution system operators?

    1. Collecting adequate revenue to operate and maintain the system
    2. Ensuring that safe and pleasant drinking water is delivered to everyone's tap
    3. Maintaining the financial integrity of the distribution system
    4. Making sure that adequate amounts of water and pressure are available during emergencies
    5. Waiting for failures to occur and then repairing them

2. Why do surface waters require treatment?

   1. To keep costs of treatment low
   2. To remove, kill, or inactivate disease-causing organisms
   3. To remove impurities such as iron and manganese
   4. To remove suspended and dissolved materials
   5. To treat excessive hardness

3. Which factors influence reserve storage requirements?

   1. Alternate sources of water supply
   2. Available capital improvement program (CIP) funds
   3. Fire insurance regulations
   4. Future growth and development demands
   5. Standby requirements

4. Which items are abnormal operating conditions for water storage tanks?

   1. Broken or out-of-service pumps
   2. Excessive water demands, such as fire demands
   3. Peak demands during middle of day
   4. Peak electrical rates during middle of day
   5. Stale water in tanks creating tastes and odors in system

5. Which of the following items are possible causes of tastes and odors in storage tanks and distribution systems?

   1. Algal growth
   2. Calcium carbonate
   3. Dead ends
   4. Floc carryover
   5. High chlorine residual

6. What causes water hammer?

   1. Filling of water storage reservoirs
   2. Improper location of thrust blocks
   3. Operating valves too quickly
   4. Rapid increase or decrease in water flow
   5. Turning pumps on or off

7. Which of the following factors influence distribution system performance?

   1. Demands for water throughout the entire system
   2. Firefighting demands
   3. Potential service interruptions resulting from power supply failures
   4. Protection against unauthorized entry and vandalism
   5. Quality of the water supply which may affect the type of materials used

8. Why are thrust blocks installed?

   1. To block the flow of water into dead ends
   2. To prevent leakage
   3. To prevent separation of joints
   4. To provide additional thrust to the flowing water
   5. To provide suitable bedding for water mains

9. What is the most positive type of backflow prevention available?

   1. Air gap devices
   2. Double check valve assembly
   3. Independently acting check valves
   4. Pressure and vacuum relief valves
   5. Reduced-pressure principle devices

10. When is a domestic water supply considered to be of good quality?

    1. Attractive in taste and appearance
    2. Desirable levels of nutrients
    3. Distributed without undue corrosive or scale-forming effects
    4. Free of disease-causing organisms
    5. Free of toxic chemicals

11. Which of the following inorganic substances can be found in water?

    1. Arsenic
    2. Herbicides
    3. Lead
    4. Mercury
    5. Nitrate

12. Which of the following factors may cause water quality degradation in water distribution systems?

    1. Biological growths
    2. Corrosion
    3. Cross connections
    4. Dead ends
    5. Operational procedures

13. Which items should be part of a daily inspection of a storage reservoir?

    1. Change in water appearance
    2. Evidence water has been tampered with
    3. Fence openings
    4. Roof damage
    5. Screen openings

14. Which of the following are purposes of water quality monitoring?

    1. Answer consumer questions
    2. Assist in determining the source of any contamination that may have reached the system
    3. Comply with requirements of regulatory agencies
    4. Provide a record of the quality of the water served
    5. Verify or refute claims

15. Which of the following backflow prevention devices are approved for use at service connections?

    1. Air gap separation
    2. Atmospheric vacuum breaker
    3. Double check valve
    4. Pressure vacuum breaker
    5. Reduced-pressure principle device

16. Which factors can cause a loss of water carrying capacity in a pipe?

    1. Corrosion
    2. Deposition of sediment
    3. Pitting
    4. Slime growth
    5. Tuberculation

17. When does field disinfection of distribution system mains and storage reservoirs become necessary?

    1. After rainstorms and floods
    2. After repairs are made
    3. Before spring turnover in raw surface water reservoirs
    4. When they are new
    5. Whenever there is any possibility of contamination

18. Which of the following factors influence disinfection?

    1. Organic matter
    2. pH
    3. Temperature
    4. Time of day
    5. Turbidity

19. Which types of disinfectants are used to disinfect water that is delivered to consumers through distribution systems?

    1. Free chlorine
    2. Hydrochloric acid
    3. Iodine
    4. Monochloramine
    5. Sodium chloride

20. Which of the following are critical factors influencing disinfection?

    1. Chlorine residual
    2. Contact time
    3. Injection point
    4. Method of mixing
    5. pH

21. Which of the following items are specific safety responsibilities for distribution system operators?

    1. Actively participate in the safety program
    2. Perform their jobs in accordance with safe procedures
    3. Recognize the responsibility for their own safety and that of fellow operators
    4. Report all injuries
    5. Report all observed hazards

22. Which of the following are elements of defensive driving?

    1. Always have your vehicle under control
    2. Be willing to surrender your legal right-of-way if it might prevent an accident
    3. Develop a defensive driving attitude
    4. Know the limitations of the vehicle and towed equipment you are operating
    5. Take into consideration weather or other unusual conditions

23. Safety around wells requires inspection of which items?

    1. Caution/warning signs in place where necessary
    2. Cleanliness of site
    3. Condition of fencing around site
    4. Height of groundwater table
    5. Locks on electrical panels

24. Which of the following elements should be included to ensure worker safety in a temporary traffic control zone?

    1. Barriers
    2. Lighting
    3. Public information
    4. Road closure
    5. Training

25. Which staffing tasks are performed by distribution system managers?

    1. Developing an employee benefit package
    2. Evaluating job performance
    3. Hiring new employees
    4. Organizing employee labor groups
    5. Training employees

26. It is illegal to ask a job applicant about which of the following items?

    1. Age
    2. Marital status
    3. Number of children
    4. Religious affiliation
    5. Years of experience

27. Which of the following items should be kept in mind when writing a report?

    1. Include a summary that presents the essential information
    2. Keep information accurate and easy to understand
    3. Organize material logically
    4. Try to achieve objective for which report is being written
    5. Use technical terms to impress the readers

28. Which of the following items are important for the financial management of a utility?

    1. Careful budgeting
    2. Effectively responding to consumer inquiries
    3. Implementing a public relations campaign
    4. Providing capital improvement funds for future utility expansion
    5. Providing financial stability

Short Answer

1. Why is the appearance of pump stations and elevated tanks and the grounds around them important?

2. Why should operators know how high the level in a storage tank should be each morning?

3. How can the depth of water in a storage facility be measured?

4. What is the difference between preventive and corrective maintenance?

5. Regulator stations are used for what purpose?

6. Why must a continuous positive water pressure be maintained in the distribution system at all times?

7. What causes friction losses in pipes?

8. List the forces water distribution system pipes must be capable of resisting.

9. Where should the excavated soil from a trench be piled?

10. What are the consequences of delivering a poor quality water to the consumer?

11. Why have water quality standards been prepared and used by the waterworks industry?

12. MCL stands for what three words?

13. How can corrosion cause health problems?

14. How frequently should a storage tank be cleaned?

15. How could you determine the condition (smoothness) of the inside of a pipe?

16. What problems may be created if distribution system lines are not regularly flushed?

17. What factors would you consider when selecting a method for using chlorine to disinfect a storage facility?

18. How should complaints be handled?

19. Define the following terms:

    a. Coupon
    b. SCADA
    c. Tuberculation

20. What are possible sources of drinking water contamination in distribution systems?

21. Why should a chlorine residual be maintained in a water distribution system?

22. What two water quality tests are most commonly run on samples of water from a water supply system?

23. How would you detect a chlorine leak?

24. Define the following terms:

    a. Breakpoint Chlorination
    b. Pathogenic Organisms

25. What tasks should be performed *BEFORE* leaving the maintenance yard or shop?

26. What is a tailgate safety session?

27. At what level is a mixture of gas in air considered hazardous?

28. How can liquids spilled on floors be cleaned up?

29. List the five different areas of a traffic control zone.

30. What is the purpose of an organizational plan?

31. Why should a manager thoroughly examine the functions and staffing of the distribution system agency on a periodic basis?

32. Why is it important to formally document each employee's performance on a regular basis?

33. How do you measure financial stability for a utility?

Problems

1. A water storage tank contains 2,400 cubic feet of water. How many gallons of water are in the tank?

    1. 17,950 gallons
    2. 18,700 gallons
    3. 20,000 gallons
    4. 20,850 gallons
    5. 22,950 gallons

2. How many gallons of water are in a circular water storage tank 25 feet in diameter when the water is 6 feet deep?

    1. 880 gallons
    2. 17,052 gallons
    3. 19,012 gallons
    4. 21,722 gallons
    5. 22,020 gallons

3. During a pressure test of a distribution system a pressure gage read 35 psi. What was the pressure head in feet?

    1. 15 ft
    2. 18 ft
    3. 20 ft
    4. 81 ft
    5. 104 ft

4. An elevated storage tank contains water 28 feet deep. What is the pressure in pounds per square inch (psi) on the bottom of the tank?

    1. 9.5 psi
    2. 12.1 psi
    3. 21.2 psi
    4. 50.8 psi
    5. 64.7 psi

5. Estimate the water velocity in feet per second in an open channel if a float travels 55 feet in 25 seconds.

    1. 2.2 ft/sec
    2. 2.4 ft/sec
    3. 2.6 ft/sec
    4. 2.8 ft/sec
    5. 3.0 ft/sec

6. Estimate the water flow rate in cubic feet per second (CFS) in a canal five feet wide when the water is two feet deep and the water velocity is three feet per second.

    1. 30 CFS
    2. 36 CFS
    3. 45 CFS
    4. 48 CFS
    5. 54 CFS

7. During a test of a totalizing flowmeter, the water depth rose 18 inches in a tank 30 inches in diameter. What was the total flow in gallons?

    1. 49 gallons
    2. 55 gallons
    3. 102 gallons
    4. 180 gallons
    5. 200 gallons

8. Estimate the flow from a pipe in gallons per minute if a 55-gallon drum fills in 6 minutes and 10 seconds.

    1. 7.5 GPM
    2. 7.6 GPM
    3. 8.0 GPM
    4. 8.9 GPM
    5. 13.3 GPM

9. How many kilowatt-hours per day are required by a pump with a motor input horsepower of 50 horsepower when the pump operates 8 hours per day?

    1. 248 kWh/day
    2. 298 kWh/day
    3. 358 kWh/day
    4. 448 kWh/day
    5. 537 kWh/day

10. What should be the chlorine dose of a water that has a chlorine demand of 2.6 mg/L if a residual of 0.4 mg/L is desired?

    1. 0.4 mg/L
    2. 2.2 mg/L
    3. 2.4 mg/L
    4. 2.6 mg/L
    5. 3.0 mg/L

11. A ten-inch water main 750 feet long needs to be disinfected. An initial chlorine dose of 400 mg/L is expected to maintain a chlorine residual of over 300 mg/L during the three-hour disinfection period. How many gallons of 5.25 percent sodium hypochlorite solution will be needed?

    1. 19.4 gallons
    2. 23.3 gallons
    3. 24.6 gallons
    4. 25.9 gallons
    5. 32.8 gallons

12. Estimate the gallons of hypochlorite pumped by a hypochlorinator if the hypochlorite solution is in a container with a diameter of 30 inches and the hypochlorite level drops 18 inches during a specific time period.

    1. 27.5 gallons
    2. 36.1 gallons
    3. 43.2 gallons
    4. 49.5 gallons
    5. 55.0 gallons

SUGGESTED ANSWERS
FOR FINAL EXAMINATION

True-False

1. **True** — Water distribution system operators have the responsibility for the health and well-being of the community they serve.

2. **True** — Frequently the water supplier tries to maintain specified minimum target water pressures throughout the distribution system.

3. **False** — Cathodic protection must *NOT* be regarded as a substitute for the use of a proper interior coating or the application of chemical inhibitors.

4. **False** — Negative coliform test results indicate that a water storage facility has been adequately disinfected.

5. **False** — When water is flowing into an elevated water storage reservoir, the hydraulic grade line will slope downward toward the reservoir.

6. **True** — All trenches and excavations are unsafe until properly shored, used, and backfilled.

7. **True** — The quality of water coming out of the consumer's tap can be affected by what happens to water as it travels through the consumer's own system.

8. **True** — Biofilms will grow more easily on materials that supply nutrients.

9. **True** — Operators should be required to follow a prepared preventive maintenance (PM) schedule and document the work they do.

10. **True** — A lack of chlorine residual in a distribution system could indicate the presence of heavy contamination.

11. **False** — Never allow the inside of a water meter to dry out after the meter is taken out of service in the field.

12. **False** — Longer contact times are required to disinfect water at lower temperatures.

13. **True** — When a new well is completed, it is necessary to disinfect the well, pump, and screen.

14. **True** — The chlorine injection should never be on the intake side of the pump because it will cause corrosion problems.

15. **False** — A corrosive chemical may be either acid or base (alkali).

16. **False** — The use of traffic cones and tubes in transition areas during nighttime activities is discouraged.

17. **True** — The location of underground installations (such as utility lines) must be determined before excavation begins.

18. **True** — A confined space permit must be renewed each time an operator leaves and re-enters the confined space.

19. **True** — Nearly every area of a manager's responsibilities requires some type of planning.

20. **True** — The sooner you deal with a discipline problem, the better the outcome will be.

21. True   The water supplier is responsible for preventing contamination of the public water system by backflow.

22. True   A supervisor negligent of safety hazards could be fined, sentenced to a jail term, or even be barred from working in the profession.

Best Answer

1. 3   The operator's most important responsibility is to follow safe procedures.

2. 4   The main purpose of a water storage facility is to provide sufficient water to equalize the daily demands.

3. 2   The purpose of surge tanks is to control water hammer.

4. 4   Sample storage facilities regularly to determine the quality of the water that enters and leaves the facility.

5. 2   Backflow from a nonpotable source could contaminate the distribution system.

6. 2   Air can become a problem in a water main by collecting and stopping the flow of water completely.

7. 4   A cross connection is an unprotected connection between a drinking water supply and a system *NOT* approved as safe.

8. 2   Back pressure is backflow caused when the user's water system pressure is higher than the public system.

9. 3   A phosphate-based corrosion-control inhibitor may stimulate bacterial growth.

10. 4   The main objective of distribution system maintenance is to provide maximum continuous service at the lowest possible cost.

11. 4   If there is any evidence that distribution system water has been tampered with by vandals or terrorists, immediately take the facility out of service.

12. 3   An air gap separation device is a vertical empty space that separates a drinking water supply from an unapproved water system.

13. 4   A geographic information system (GIS) shows detailed information about the physical structures within geographic areas.

14. 4   Disinfection is the selective destruction or inactivation of pathogenic organisms.

15. 1   When chlorine is added to water and the reaction of chlorine with organic and inorganic materials stops, the chlorine demand has been satisfied.

16. 1   The chlorine dose is chlorine demand plus chlorine residual.

17. 4   Preventing accidents is the one objective of a safety program.

18. 1   An operator should call a hazard to the attention of those who have the authority to remove the hazard.

19. 3   The basic requirement for an enclosed guard is to protect the body from making contact with dangerous moving parts.

20. 3   Oxygen deficiency is the primary concern when entering a confined space.

21. 2   Accountability is when an operator ensures that the manager is informed of results or events.

22. 4   The selection method and examination process used to evaluate applicants must be limited to the applicant's knowledge, skills, and abilities to perform job-related activities.

23. 2   The purpose of a distribution system O & M program is to ensure that the system performs as designed and intended.

24. 4   Safety of the public is the most important immediate concern when contamination of a water supply is discovered.

Multiple Choice

1. 2, 4   The responsibilities of water distribution system operators include ensuring that safe and pleasant drinking water is delivered to everyone's tap and making sure that adequate amounts of water and pressure are available during emergencies.

2. 2, 4   Surface waters require treatment to remove, kill, or inactivate disease-causing organisms and to remove suspended and dissolved materials.

3. 1, 3, 4, 5   Factors that influence reserve storage requirements include alternate sources of water supply, fire insurance regulations, future growth and development demands, and standby requirements.

4. 1, 2, 5   Operating conditions for water storage tanks are abnormal when pumps are broken or out-of-service, when water demands are excessive, such as fire demands, and when stale water in the tanks is creating tastes and odors.

5. 1, 3, 5   Possible causes of tastes and odors in storage tanks and distribution systems include algal growth, dead ends, and high chlorine residual.

6. 3, 4, 5   Water hammer can be caused by operating valves too quickly, rapid increases or decreases in water flow, and turning pumps on or off.

7. 1, 2, 3, 4, 5   Distribution system performance is influenced by demands for water throughout the entire system, firefighting demands, potential service interruptions resulting from power supply failures, protection against unauthorized entry and vandalism, and the quality of the water supply which may affect the type of materials used.

8. 2, 3   Thrust blocks are installed to prevent leakage and to prevent separation of joints.

9. 1   Air gap devices are the most positive type of backflow prevention available.

10. 1, 3, 4, 5   The domestic water supply is considered to be of good quality when it is attractive in taste and appearance, when it is not corrosive or scale forming, when it is free of disease-causing organisms, and when it is free of toxic chemicals.

11. 1, 3, 4, 5   Inorganic substances that can be found in water include arsenic, lead, mercury, and nitrate.

12. 1, 2, 3, 4, 5 Factors that may cause water quality degradation in water distribution systems include biological growths, corrosion, cross connections, dead ends, and operational procedures.

13. 1, 2, 3, 4, 5 Storage reservoirs should be inspected daily for changes in water appearance, evidence that the water has been tampered with, fence openings, roof damage, and screen openings.

14. 1, 2, 3, 4, 5 Water quality is monitored to answer consumer questions, to determine the source of any contamination that may have reached the system, to comply with requirements of regulatory agencies, to provide a record of the quality of the water served, and to verify or refute claims.

15. 1, 3, 5   Backflow prevention devices that are approved for use at service connections include air gap separation devices, double check valves, and reduced-pressure principle devices.

16. 1, 2, 3, 4, 5 A loss of water carrying capacity in a pipe can be caused by corrosion, deposition of sediment, pitting, slime growth, and tuberculation.

17. 2, 4, 5   Field disinfection of distribution system mains and storage reservoirs becomes necessary after repairs are made, when mains and storage reservoirs are new, and whenever there is any possibility of contamination.

18. 1, 2, 3, 5   Organic matter influences disinfection as well as the pH, temperature, and turbidity of the water.

19. 1, 4   Free chlorine and monochloramine are used to disinfect water that is delivered to consumers through distribution systems.

20. 1, 2, 3, 4, 5 Factors that critically influence disinfection include chlorine residual, contact time, injection point, method of mixing, and pH.

21. 1, 2, 3, 4, 5 Distribution system operators have the specific safety responsibilities to actively participate in the safety program, to perform their jobs in accordance with safe procedures, to recognize the responsibility for their own safety and that of fellow operators, to report all injuries, and to report all observed hazards.

22. 1, 2, 3, 4, 5 Defensive driving includes always having your vehicle under control, developing a defensive driving attitude, and taking into consideration weather and unusual conditions. It also means being willing to surrender your legal right-of-way if it might prevent an accident and knowing the limitations of the vehicle and towed equipment you are operating.

23. 1, 2, 3, 5   Safety inspections around wells include making sure caution/warning signs are in place where necessary, checking the cleanliness of the site, checking the condition of the fencing around the site, and checking the locks on electrical panels.

24. 1, 2, 3, 4, 5 Elements that should be included to ensure worker safety in a temporary traffic control zone include barriers, lighting, public information, road closure, and training.

25. 2, 3, 5   Staffing tasks performed by distribution system managers include evaluating job performance, hiring new employees, and training employees.

26. 1, 2, 3, 4   It is illegal to ask a job applicant about age, marital status, number of children, and religious affiliation.

27. 1, 2, 3, 4   When writing a report, include a summary that presents the essential information, keep the information accurate and easy to understand, organize the material logically, and try to achieve the objective for which the report is being written.

28. 1, 4, 5   Financial management of a utility includes careful budgeting, providing capital improvement funds for utility expansion, and providing financial stability.

Short Answer

1. The appearance of pump stations and elevated tanks and the grounds around them indicates to the public the type of operation you maintain.

2. Operators should know how high the level in a storage tank should be each morning so that the system's demands will be met during the rest of the day.

3. The depth of water in a storage facility can be measured by the use of floats, electrodes, ultrasonic signals, pressure switches, solid-state electronic sensors, and differential-pressure altitude valves.

4. Preventive maintenance is something that is done before some type of deterioration takes place or repairs become necessary. Corrective maintenance is maintenance or repair that is necessary because a problem exists and needs repair or fixing.

5. Regulator stations are used to reduce pressure and to maintain an even (low), acceptable pressure downstream from a high-pressure system.

6. A continuous positive water pressure must be maintained in the distribution system at all times to protect the distribution system from the entrance of toxic and other undesirable substances.

7. Friction losses in pipes are caused by the roughness of the inside of the pipe creating turbulence proportional to the velocity of the flowing water.

8. Water distribution system pipes must be capable of resisting forces from (1) backfill, (2) weight of passing traffic, and (3) internal water pressures such as delivery pressure, surges, and water hammer.

9. Excavated soil is piled on the side of the trench between the trench and the traffic and far enough away (2 to 5 feet (0.6 to 1.5 m)) so the crew can walk between the trench and the excavated material.

10. The consequences of delivering a poor quality water to the consumer range from the water not being acceptable to the consumer because of its appearance or taste, to illness or even death of some susceptible consumers.

11. Water quality standards have been prepared and used by the waterworks industry to provide the needed quality control and to ensure the acceptability of the product water.

12. MCL stands for **M**aximum **C**ontaminant **L**evel.

13. Corrosion can cause health problems by increasing the levels of toxic or suspected toxic substances in water such as lead, cadmium, copper, zinc, asbestos, and certain organic compounds. Increased incidences of cardiovascular (heart) disease have been associated with consumption of soft, corrosive water.

14. At least once a year, the interior of each storage tank should be inspected to determine the condition of the interior coating and whether the tank needs to be washed out, completely cleaned, or recoated.

15. The condition of the inside of a pipe can be determined by (1) looking during repairs or additions to the system, (2) collecting pipe cut-outs from main tapping operations, and (3) placing test pipe specimens in the distribution system and periodically checking them for deterioration.

16. Failure to regularly flush distribution system lines may cause numerous problems. Deposits that have settled out and accumulated in the lines may result in taste, odor, and turbidity problems. Encrustations may restrict the water flow. Sand, rust, and biological materials may cause water quality problems.

17. When selecting which chlorination method is most suitable for disinfecting a storage facility, consider the availability of materials and equipment, training status of the operators who will be doing the disinfecting, and safety considerations.

18. When handling a complaint, show concern, listen carefully, offer to look into the complaint, and help correct any problems found. Every effort should be made to give the customer an immediate, clear, and accurate answer to the problem in nontechnical language.

19. Define the following terms:

    a. Coupon. A steel specimen inserted into water to measure the corrosiveness of water. The rate of corrosion is measured as the loss of weight of the coupon (in milligrams) per surface area (in square decimeters) exposed to the water per day. 10 decimeters = 1 meter = 100 centimeters.

    b. SCADA. **S**upervisory **C**ontrol **A**nd **D**ata **A**cquisition system. A computer-monitored alarm, response, control and data acquisition system used by drinking water facilities to monitor their operations.

    c. Tuberculation. The development or formation of small mounds of corrosion products (rust) on the inside of iron pipe. These mounds (tubercles) increase the roughness of the inside of the pipe thus increasing resistance to water flow (decreases the C Factor).

20. Possible sources of drinking water contamination in distribution systems include new main installations, cross connections, and main breaks.

21. A chlorine residual should be maintained in a water distribution system to control any microorganisms that could produce slimes, tastes, or odors in the water in the distribution system. Also, if any pathogenic organisms reach the water in the distribution system due to cross connections or leakage into the system, the residual chlorine provides a means of disinfection.

22. The two most common water quality tests run on samples of water from a water supply system are (1) chlorine residual, and (2) coliform tests.

23. A chlorine leak may be detected by the use of a rag on the end of a stick dipped into a solution of ammonia water. Place the rag near the location of a suspected chlorine leak. A white cloud of gas will reveal the location of the leak.

24. Define the following terms:

    a. Breakpoint Chlorination. Addition of chlorine to water until the chlorine demand has been satisfied. At this point, further additions of chlorine will result in a free chlorine residual that is directly proportional to the amount of chlorine added beyond the breakpoint.

    b. Pathogenic Organisms. Organisms, including bacteria, viruses or cysts, capable of causing diseases (giardiasis, cryptosporidiosis, typhoid, cholera, dysentery) in a host (such as a person). There are many types of organisms which do *NOT* cause disease. These organisms are called non-pathogenic.

25. Before leaving the maintenance yard or shop, discuss work assignments, determine equipment needed, inspect equipment that will be used, and inspect vehicles and towed equipment.

26. A tailgate safety session is usually held near the work site and new and old safety hazards and safer approaches and techniques to deal with the problems of the day or week are discussed.

27. A mixture of gas in air is considered hazardous when the mixture exceeds 10 percent of the Lower Explosive Limit (LEL).

28. Liquids spilled on floors can be cleaned up by using non-flammable absorbent materials.

29. The five different areas of a traffic control zone are: (1) advance warning, (2) transition, (3) buffer, (4) work, and (5) termination areas.

30. The purpose of the organizational plan is to show who reports to whom and to identify the lines of authority within the organization.

31. By thoroughly examining the functions and staffing of the utility on a periodic basis, the manager may spot trends or discover inefficiencies that could be corrected over a period of time.

32. It is important to document each employee's performance on a regular basis because documentation of this type may be needed in the future to support taking disciplinary action if the employee's performance consistently fails to meet expectations.

33. Financial stability is measured by calculating (1) the operating ratio, and (2) the coverage ratio. Both of these ratios should be greater than 1.0.

Problems

1. A water storage tank contains 2,400 cubic feet of water. How many gallons of water are in the tank?

Known	Unknown
Tank Vol, cu ft = 2,400 cu ft	Tank Vol, gal

Convert the tank volume from cubic feet to gallons.

$$\text{Tank Volume, gal} = (\text{Tank Vol, cu ft})(7.48 \text{ gal/cu ft})$$
$$= (2,400 \text{ cu ft})(7.48 \text{ gal/cu ft})$$
$$= 17,950 \text{ gallons}$$

2. How many gallons of water are in a circular water storage tank 25 feet in diameter when the water is 6 feet deep?

Known	Unknown
Diameter, ft = 25 ft	Tank Vol, gal
Depth, ft = 6 ft	

Calculate the gallons of water stored in the tank.

$$\text{Tank Vol, gal} = (0.785)(\text{Diameter, ft})^2(\text{Depth, ft})(7.48 \text{ gal/cu ft})$$
$$= (0.785)(25 \text{ ft})^2(6 \text{ ft})(7.48 \text{ gal/cu ft})$$
$$= 22,020 \text{ gallons}$$

3. During a pressure test of a distribution system a pressure gage read 35 psi. What was the pressure head in feet?

Known	Unknown
Pressure, psi = 35 psi	Pressure Head, ft

Calculate the pressure head in feet.

$$\text{Pressure Head, ft} = (\text{Pressure, psi})(2.31 \text{ ft/psi})$$
$$= (35 \text{ psi})(2.31 \text{ ft/psi})$$
$$= 81 \text{ ft}$$

4. An elevated storage tank contains water 28 feet deep. What is the pressure in pounds per square inch (psi) on the bottom of the tank?

Known	Unknown
Depth, ft = 28 ft	Pressure, psi

Calculate the pressure in psi on the bottom of the tank.

$$\text{Pressure, psi} = \frac{\text{Depth, ft}}{2.31 \text{ ft/psi}}$$
$$= \frac{28 \text{ ft}}{2.31 \text{ ft/psi}}$$
$$= 12.1 \text{ psi}$$

5. Estimate the water velocity in feet per second in an open channel if a float travels 55 feet in 25 seconds.

Known	Unknown
Distance, ft = 55 ft	Velocity, ft/sec
Time, sec = 25 sec	

Calculate the velocity in feet per second.

$$\text{Velocity, ft/sec} = \frac{\text{Distance, ft}}{\text{Time, sec}}$$
$$= \frac{55 \text{ ft}}{25 \text{ sec}}$$
$$= 2.2 \text{ ft/sec}$$

6. Estimate the water flow rate in cubic feet per second (CFS) in a canal five feet wide when the water is two feet deep and the water velocity is three feet per second.

Known	Unknown
Width, ft $\quad$ = 5 ft	Flow Rate, CFS
Depth, ft $\quad$ = 2 ft	
Velocity, ft/sec = 3 ft/sec	

Calculate the flow rate in cubic feet per second (CFS).

Flow Rate, CFS = (Area, sq ft)(Velocity, ft/sec)

$$= (5 \text{ ft})(2 \text{ ft})(3 \text{ ft/sec})$$

$$= (10 \text{ sq ft})(3 \text{ ft/sec})$$

$$= 30 \text{ cu ft/sec}$$

7. During a test of a totalizing flowmeter, the water depth rose 18 inches in a tank 30 inches in diameter. What was the total flow in gallons?

Known	Unknown
Depth, in $\quad$ = 18 in	Volume, gal
Diameter, in = 30 in	
Diameter, ft = 2.5 ft	

Calculate the total flow in gallons.

$$\frac{\text{Volume,}}{\text{gal}} = \frac{(0.785)(\text{Diameter, ft})^2(\text{Depth, in})(7.48 \text{ gal/cu ft})}{12 \text{ in/ft}}$$

$$= \frac{(0.785)(2.5 \text{ ft})^2(18 \text{ in})(7.48 \text{ gal/cu ft})}{12 \text{ in/ft}}$$

$$= 55 \text{ gallons}$$

8. Estimate the flow from a pipe in gallons per minute if a 55-gallon drum fills in 6 minutes and 10 seconds.

Known	Unknown
Volume, gal $\quad$ = 55 gal	Flow, GPM
Time, min and sec = 6 min and 10 sec	

1. Convert the time from minutes and seconds to minutes.

$$\text{Time, min} = 6 \text{ min} + \frac{10 \text{ sec}}{60 \text{ sec/min}}$$

$$= 6 \text{ min} + 0.17 \text{ min}$$

$$= 6.17 \text{ min}$$

2. Calculate the flow in gallons per minute.

$$\text{Flow, GPM} = \frac{\text{Volume, gal}}{\text{Time, min}}$$

$$= \frac{55 \text{ gal}}{6.17 \text{ min}}$$

$$= 8.9 \text{ GPM}$$

9. How many kilowatt-hours per day are required by a pump with a motor input horsepower of 50 horsepower when the pump operates 8 hours per day?

Known	Unknown
Motor Horsepower, HP = 50 HP	kWh/day
Time, hr/day $\quad$ = 8 hr/day	

Calculate the required kilowatt-hours per day.

kWh/day = (Motor Horsepower, HP)(Time, hr/day)(0.746 kW/HP)

$$= (50 \text{ HP})(8 \text{ hr/day})(0.746 \text{ kW/HP})$$

$$= 298 \text{ kilowatt-hours/day}$$

$$= 298 \text{ kWh/day}$$

10. What should be the chlorine dose of a water that has a chlorine demand of 2.6 mg/L if a residual of 0.4 mg/L is desired?

Known	Unknown
Chlorine Demand, mg/L $\quad$ = 2.6 mg/L	Chlorine Dose, mg/L
Chlorine Residual, mg/L $\quad$ = 0.4 mg/L	

Calculate the chlorine dose in mg/L.

$$\frac{\text{Chlorine Dose,}}{\text{mg/L}} = \text{Chlorine Demand, mg/L} + \text{Chlorine Residual, mg/L}$$

$$= 2.6 \text{ mg/L} + 0.4 \text{ mg/L}$$

$$= 3.0 \text{ mg/L}$$

11. A ten-inch water main 750 feet long needs to be disinfected. An initial chlorine dose of 400 mg/L is expected to maintain a chlorine residual of over 300 mg/L during the three-hour disinfection period. How many gallons of 5.25 percent sodium hypochlorite solution will be needed?

Known	Unknown
Diameter, in $\quad$ = 10 in	5.25% Hypochlorite, gal
Length, ft $\quad$ = 750 ft	
Chlorine Dose, mg/L = 400 mg/L	
Hypochlorite, % $\quad$ = 5.25%	

1. Calculate the volume of water in the main in gallons.

$$\frac{\text{Main Volume,}}{\text{gallons}} = \frac{(0.785)(\text{Diameter, in})^2(\text{Length, ft})(7.48 \text{ gal/cu ft})}{144 \text{ sq in/sq ft}}$$

$$= \frac{(0.785)(10 \text{ in})^2(750 \text{ ft})(7.48 \text{ gal/cu ft})}{144 \text{ sq in/sq ft}}$$

$$= 3{,}058 \text{ gal}$$

2. Determine the pounds of chlorine needed.

$$\frac{\text{Chlorine,}}{\text{lbs}} = (\text{Vol Water, M Gal})(\text{Chlorine Dose, mg/L})(8.34 \text{ lbs/gal})$$

$$= (0.003058 \text{ M Gal})(400 \text{ mg/L})(8.34 \text{ lbs/gal})$$

$$= 10.2 \text{ lbs Chlorine}$$

3. Calculate the gallons of 5.25 percent sodium hypochlorite solution needed.

$$\text{Sodium Hypochlorite Solution, gallons} = \frac{(\text{Chlorine, lbs})(100\%)}{(8.34 \text{ lbs/gal})(\text{Hypochlorite, \%})}$$

$$= \frac{(10.2 \text{ lbs})(100\%)}{(8.34 \text{ lbs/gal})(5.25\%)}$$

$$= 23.3 \text{ gallons}$$

12. Estimate the gallons of hypochlorite pumped by a hypochlorinator if the hypochlorite solution is in a container with a diameter of 30 inches and the hypochlorite level drops 18 inches during a specific time period.

Known	Unknown
Diameter, in = 30 in	Hypochlorite Pumped, gal
Drop, in = 18 in	

Estimate the gallons of hypochlorite pumped.

$$\text{Hypochlorite, gal} = (\text{Container Area, sq ft})(\text{Drop, ft})(7.48 \text{ gal/cu ft})$$

$$= \frac{(0.785)(30 \text{ in})^2(18 \text{ in})(7.48 \text{ gal/cu ft})}{(144 \text{ sq in/sq ft})(12 \text{ in/ft})}$$

$$= 55.0 \text{ gal}$$

APPENDIX

HOW TO SOLVE WATER DISTRIBUTION SYSTEM ARITHMETIC PROBLEMS

by

Ken Kerri

TABLE OF CONTENTS

HOW TO SOLVE WATER DISTRIBUTION SYSTEM ARITHMETIC PROBLEMS

OBJECTIVES

HOW TO SOLVE WATER DISTRIBUTION SYSTEM ARITHMETIC PROBLEMS

Following completion of this Appendix, you should be able to:

1. Add, subtract, multiply, and divide;

2. List from memory basic conversion factors and formulas; and

3. Solve water distribution system arithmetic problems.

APPENDIX

HOW TO SOLVE WATER DISTRIBUTION SYSTEM ARITHMETIC PROBLEMS

A.0   HOW TO STUDY THIS APPENDIX

This appendix may be worked early in your training program to help you gain the greatest benefit from your efforts. Whether to start this appendix early or wait until later is your decision. The chapters in this manual were written in a manner requiring very little background in arithmetic. You may wish to concentrate your efforts on the chapters and refer to this appendix when you need help. Some operators prefer to complete this appendix early so they will not have to worry about how to do the arithmetic when they are studying the chapters. You may try to work this appendix early or refer to it while studying the other chapters.

The intent of this appendix is to provide you with a quick review of the addition, subtraction, multiplication, and division needed to work the arithmetic problems in this manual. This appendix is not intended to be a math textbook. There are no fractions because you don't need fractions to work the problems in this manual. Some operators will be able to skip over the review of addition, subtraction, multiplication, and division. Others may need more help in these areas. If you need help in solving problems, read Section A.9, "Steps in Solving Problems." Basic arithmetic textbooks are available at every local library or bookstore and should be referred to if needed. Most instructional or operating manuals for pocket electronic calculators contain sufficient information on how to add, subtract, multiply, and divide.

After you have worked a problem involving your job, you should check your calculations, examine your answer to see if it appears reasonable, and if possible have another operator check your work before making any decisions or changes.

A.1   BASIC ARITHMETIC

In this section we provide you with basic arithmetic problems involving addition, subtraction, multiplication, and division. You may work the problems "by hand" if you wish, but we recommend you use an electronic pocket calculator. The operating or instructional manual for your calculator should outline the step-by-step procedures to follow. All calculators use similar procedures, but most of them are slightly different from others.

We will start with very basic, simple problems. Try working the problems and then comparing your answers with the given answers. If you can work these problems, you should be able to work the more difficult problems in the text of this training manual by using the same procedures.

A.10   Addition

2	6.2	16.7	6.12	43
3	8.5	38.9	38.39	39
5	14.7	55.6	44.51	34
				38
				39
2.12	0.12	63	120	37
9.80	2.0	32	60	29
11.92	2.12	95	180	259
				70
4	23	16.2	45.98	50
7	79	43.5	28.09	40
2	31	67.8	114.00	80
13	133	127.5	188.07	240

A.11   Subtraction

7	12	25	78	83
−5	− 3	− 5	−30	−69
2	9	20	48	14
61	485	4.3	3.5	123
−37	−296	−0.8	−0.7	−109
24	189	3.5	2.8	14
8.6	11.92	27.32	3.574	75.132
−8.22	− 3.70	−12.96	−0.042	−49.876
0.38	8.22	14.36	3.532	25.256

A.12   Multiplication

(3)(2)*	= 6		(4)(7)	= 28
(10)(5)	= 50		(10)(1.3)	= 13
(2)(22.99)	= 45.98		(6)(19.5)	= 117
(16)(17.1)	= 273.6		(50)(20,000)	= 1,000,000
(40)(2.31)	= 92.4		(80)(0.433)	= 34.64

(40)(20)(6)	= 4,800
(4,800)(7.48)	= 35,904
(1.6)(2.3)(8.34)	= 30.6912
(0.001)(200)(8.34)	= 1.668
(0.785)(7.48)(60)	= 352.308
(12,000)(500)(60)(24)	= 8,640,000,000 or 8.64×10^9
(4)(1,000)(1,000)(454)	= 1,816,000,000 or 1.816×10^9

NOTE: The term, $\times 10^9$, means that the number is multiplied by 10^9 or 1,000,000,000. Therefore $8.64 \times 10^9 = 8.64 \times 1,000,000,000 = 8,640,000,000.$

---

* (3)(2) is the same as $3 \times 2 = 6$.

A.13 Division

$$\frac{6}{3} = 2 \qquad\qquad \frac{48}{12} = 4$$

$$\frac{50}{25} = 2 \qquad\qquad \frac{300}{20} = 15$$

$$\frac{20}{7.1} = 2.8 \qquad\qquad \frac{11,400}{188} = 60.6$$

$$\frac{1,000,000}{17.5} = 57,143 \qquad\qquad \frac{861,429}{30,000} = 28.7$$

$$\frac{4,000,000}{74,880} = 53.4 \qquad\qquad \frac{1.67}{8.34} = 0.20$$

$$\frac{80}{2.31} = 34.6 \qquad\qquad \frac{62}{454} = 0.137$$

$$\frac{250}{17.1} = 14.6 \qquad\qquad \frac{4,000,000}{14.6} = 273,973$$

NOTE: When we divide $1/3 = 0.3333$, we get a long row of 3s. Instead of the row of 3s, we "round off" our answer so $1/3 = 0.33$. For a discussion of rounding off numbers, see Section A.95, "Significant Figures."

A.14 Rules for Solving Equations

Most of the arithmetic problems we work in the water distribution field require us to plug numbers into formulas and calculate the answer. There are a few basic rules that apply to solving formulas. These rules are:

1. Work from left to right.

2. Do all the multiplication and division above the line (in the numerator) and below the line (in the denominator); then do the addition and subtraction above and below the line.

3. Perform the division (divide the numerator by the denominator).

Parentheses ( ) are used in formulas to identify separate parts of a problem. A fourth rule tells us how to handle numbers within parentheses.

4. Work the arithmetic within the parentheses before working outside the parentheses. Use the same order stated in rules 1, 2, and 3: work left to right, above and below the line, then divide the top number by the bottom number.

Let's look at an example problem to see how these rules apply. This year one of the responsibilities of the operators at our plant is to paint both sides of the wooden fence across the front of the facility. The fence is 145 feet long and 9 feet high. The steel access gate, which does not need painting, measures 14 feet wide by 9 feet high. Each gallon of paint will cover 150 square feet of surface area. How many gallons of paint should be purchased?

STEP 1: Identify the correct formula.

$$\text{Paint Req, gal} = \frac{\text{Total Area, sq ft}}{\text{Coverage, sq ft/gal}}$$

or

$$\text{Paint Req,}\atop\text{gal} = \frac{(\text{Fence L, ft} \times \text{H, ft} \times \text{No. Sides}) - (\text{Gate L, ft} \times \text{H, ft} \times \text{No. Sides})}{\text{Coverage, sq ft/gal}}$$

STEP 2: Plug numbers into the formula.

$$\text{Paint Req, gal} = \frac{(145\ \text{ft} \times 9\ \text{ft} \times 2) - (14\ \text{ft} \times 9\ \text{ft} \times 2)}{150\ \text{sq ft/gal}}$$

STEP 3: Work the multiplication within the parentheses.

$$\text{Paint Req, gal} = \frac{(2,610\ \text{sq ft}) - (252\ \text{sq ft})}{150\ \text{sq ft/gal}}$$

STEP 4: Work the subtraction above the line.

$$\text{Paint Req, gal} = \frac{2,358\ \text{sq ft}}{150\ \text{sq ft/gal}}$$

STEP 5: Divide the numerator by the denominator.

$$\text{Paint Req, gal} = 15.72\ \text{gal}$$
$$\text{or } 16 \text{ gallons of paint will be needed.}$$

Instructions for your electronic calculator can provide you with the detailed procedures for working the practice problems below.

$$\frac{(3)(4)}{2} = 6 \qquad\qquad \frac{64}{(8)(4)} = 2$$

$$\frac{(2+3)(4)}{5} = 4 \qquad\qquad \frac{54}{(4+2)(3)} = 3$$

$$\frac{(7-2)(8)}{4} = 10 \qquad\qquad \frac{48}{(8-3)(4)} = 2.4$$

$$\frac{(0.1)(60)(24)}{3} = 48$$

$$\frac{(12,000)(500)(60)(24)}{(4)(1,000)(1,000)(454)} = 4.76$$

$$\frac{12}{(0.432)(8.34)} = 3.3$$

$$\frac{(274,000)(24)}{200,000} = 32.88$$

A.15 Actual Problems

Let's look at the last four problems in the previous Section A.14, "Rules for Solving Equations," as they might be encountered by an operator.

1. To determine the actual chemical feed rate from an alum feeder, an operator collects the alum from the feeder in a bucket for three minutes. The alum in the bucket weighs 0.1 pound.

Known	Unknown
Weight of Alum, lbs = 0.1 lb	Actual Alum Feed, lbs/day
Time, min = 3 min	

Calculate the actual alum feed rate in pounds per day.

$$\text{Actual Alum}\atop\text{Feed Rate,}\atop\text{lbs/day} = \frac{(\text{Alum Wt, lbs})(60\ \text{min/hr})(24\ \text{hr/day})}{\text{Time Alum Collected, min}}$$

$$= \frac{(0.1\ \text{lb})(60\ \text{min/hr})(24\ \text{hr/day})}{3\ \text{min}}$$

$$= 48\ \text{lbs/day}$$

2. A solution chemical feeder is calibrated by measuring the time to feed 500 milliliters of chemical solution. The test calibration run required four minutes. The chemical concentration in the solution is 12,000 mg/L or 1.2%. Determine the chemical feed in pounds per day.

Known		Unknown
Volume Pumped, mL	= 500 mL	Chemical Feed, lbs/day
Time Pumped, min	= 4 min	
Chemical Conc, mg/L	= 12,000 mg/L	

Estimate the chemical feed rate in pounds per day.

$$\begin{aligned}\text{Chemical Feed, lbs/day} &= \frac{(\text{Chem Conc, mg/}L)(\text{Vol Pumped, m}L)(60 \text{ min/hr})(24 \text{ hr/day})}{(\text{Time Pumped, min})(1,000 \text{ m}L/L)(1,000 \text{ mg/gm})(454 \text{ gm/lb})} \\[2mm] &= \frac{(12,000 \text{ mg/}L)(500 \text{ m}L)(60 \text{ min/hr})(24 \text{ hr/day})}{(4 \text{ min})(1,000 \text{ m}L/L)(1,000 \text{ mg/gm})(454 \text{ gm/lb})} \\[2mm] &= 4.76 \text{ lbs/day}\end{aligned}$$

3. A chlorinator is set to feed 12 pounds of chlorine per day to a flow of 300 gallons per minute (0.432 million gallons per day). What is the chlorine dose in milligrams per liter?

Known		Unknown
Chlorinator Feed, lbs/day	= 12 lbs/day	Chlorine Dose, mg/L
Flow, MGD	= 0.432 MGD	

Determine the chlorine dose in milligrams per liter.

$$\begin{aligned}\text{Chlorine Dose, mg/}L &= \frac{\text{Chlorinator Feed Rate, lbs/day}}{(\text{Flow, MGD})(8.34 \text{ lbs/gal})} \\[2mm] &= \frac{12 \text{ lbs/day}}{(0.432 \text{ MGD})(8.34 \text{ lbs/gal})} \\[2mm] &= 3.3 \text{ mg/}L\end{aligned}$$

4. Estimate the operating time of a water softening ion exchange unit before the unit needs regeneration. The unit can treat 274,000 gallons of water before the exchange capacity is exhausted. The average daily flow is 200,000 gallons per day.

Known		Unknown
Water Treated, gal	= 274,000 gal	Operating Time, hr
Avg Daily Flow, gal/day	= 200,000 gal/day	

Estimate the operating time of the ion exchange unit in hours.

$$\begin{aligned}\text{Operating Time, hr} &= \frac{(\text{Water Treated, gal})(24 \text{ hr/day})}{\text{Avg Daily Flow, gal/day}} \\[2mm] &= \frac{(274,000 \text{ gal})(24 \text{ hr/day})}{200,000 \text{ gal/day}} \\[2mm] &= 32.9^* \text{ hours}\end{aligned}$$

---

* We rounded off 32.88 hours to 32.9 hours.

A.2   AREAS

A.20   Units

Areas are measured in two dimensions or in square units. In the English system of measurement the most common units are square inches, square feet, square yards, and square miles. In the metric system the units are square millimeters, square centimeters, square meters, and square kilometers.

A.21   Rectangle

The area of a rectangle is equal to its length (L) multiplied by its width (W).

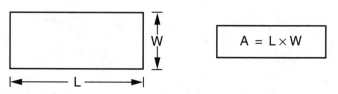

$$A = L \times W$$

EXAMPLE: Find the area of a rectangle if the length is 5 feet and the width is 3.5 feet.

$$\begin{aligned}\text{Area, sq ft} &= \text{Length, ft} \times \text{Width, ft} \\ &= 5 \text{ ft} \times 3.5 \text{ ft} \\ &= 17.5 \text{ ft}^2 \\ \text{or} &= 17.5 \text{ sq ft}\end{aligned}$$

EXAMPLE: The surface area of a settling basin is 330 square feet. One side measures 15 feet. How long is the other side?

$$A = L \times W$$

$$330 \text{ sq ft} = L, \text{ft} \times 15 \text{ ft}$$

$$\frac{L, \text{ft} \times 15 \text{ ft}}{15 \text{ ft}} = \frac{330 \text{ sq ft}}{15 \text{ ft}} \qquad \text{Divide both sides of equation by 15 ft.}$$

$$L, \text{ft} = \frac{330 \text{ sq ft}}{15 \text{ ft}}$$

$$= 22 \text{ ft}$$

A.22   Triangle

The area of a triangle is equal to one-half the base multiplied by the height. This is true for any triangle.

$$A = \tfrac{1}{2} B \times H$$

NOTE: The area of *ANY* triangle is equal to ½ the area of the rectangle that can be drawn around it. The area of the rectangle is B × H. The area of the triangle is ½ B × H.

EXAMPLE: Find the area of triangle ABC.

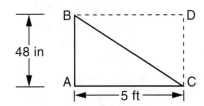

The first step in the solution is to make all the units the same. In this case, it is easier to change inches to feet.

$$48 \text{ in} = 48 \text{ in} \times \frac{1 \text{ ft}}{12 \text{ in}} = \frac{48}{12} \text{ ft} = 4 \text{ ft}$$

NOTE: All conversions should be calculated in the above manner. Since 1 ft/12 in is equal to unity, or one, multiplying by this factor changes the form of the answer but not its value.

Area, sq ft = ½(Base, ft)(Height, ft)

$$= \frac{1}{2} \times 5 \text{ ft} \times 4 \text{ ft}$$

$$= \frac{20}{2} \text{ sq ft}$$

$$= 10 \text{ sq ft}$$

NOTE: Triangle ABC is one-half the area of rectangle ABCD. The triangle is a special form called a *RIGHT TRIANGLE* since it contains a 90° angle at point A.

A.23   Circle

A square with sides of 2R can be drawn around a circle with a radius of R.

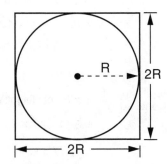

The area of the square is: $A = 2R \times 2R = 4R^2$.

It has been found that the area of any circle drawn within a square is slightly more than ¾ of the area of the square. More precisely, the area of the preceding circle is:

A circle = $3\frac{1}{7} R^2 = 3.14 R^2$

The formula for the area of a circle is usually written:

$$A = \pi R^2$$

The Greek letter π (pronounced pie) merely substitutes for the value 3.1416.

Since the diameter of any circle is equal to twice the radius, the formula for the area of a circle can be rewritten as follows:

$$A = \pi R^2 = \pi \times R \times R = \pi \times \frac{D}{2} \times \frac{D}{2} = \frac{\pi D^2}{4} = \frac{3.14}{4} D^2 = \boxed{0.785 \ D^2}$$

The type of problem and the magnitude (size) of the numbers in a problem will determine which of the two formulas will provide a simpler solution. All of these formulas will give the same results if you use the same number of digits to the right of the decimal point.

EXAMPLE: What is the area of a circle with a diameter of 20 centimeters?

In this case, the formula using a radius is more convenient since it takes advantage of multiplying by 10.

Area, sq cm = $\pi(R, \text{cm})^2$

$$= 3.14 \times 10 \text{ cm} \times 10 \text{ cm}$$

$$= 314 \text{ sq cm}$$

EXAMPLE: What is the area of a clarifier with a 50 foot radius?

In this case, the formula using diameter is more convenient.

Area, sq ft = $(0.785)(\text{Diameter, ft})^2$

$$= 0.785 \times 100 \text{ ft} \times 100 \text{ ft}$$

$$= 7,850 \text{ sq ft}$$

Occasionally the operator may be confronted with a problem giving the area and requesting the radius or diameter. This presents the special problem of finding the square root of the number.

EXAMPLE: The surface area of a circular clarifier is approximately 5,000 square feet. What is the diameter?

$A = 0.785 \ D^2$, or

Area, sq ft = $(0.785)(\text{Diameter, ft})^2$

5,000 sq ft = $0.785 \ D^2$ To solve, substitute given values in equation.

$\dfrac{0.785 \ D^2}{0.785} = \dfrac{5,000 \text{ sq ft}}{0.785}$ Divide both sides by 0.785 to find D^2.

$D^2 = \dfrac{5,000 \text{ sq ft}}{0.785}$

= 6,369 sq ft. Therefore,

D = square root of 6,369 sq ft, or

Diameter , ft = $\sqrt{6,369 \text{ sq ft}}$

Press the √ sign on your calculator and get D, ft = 79.8 ft.

Sometimes a trial-and-error method can be used to find square roots. Since 80 × 80 = 6,400, we know the answer is close to 80 feet.

Try 79 × 79   = 6,241

Try 79.5 × 79.5 = 6,320.25

Try 79.8 × 79.8 = 6,368.04

The diameter is 79.8 ft, or approximately 80 feet.

A.24 Cylinder

With the formulas presented thus far, it would be a simple matter to find the number of square feet in a room that was to be painted. The length of each wall would be added together and then multiplied by the height of the wall. This would give the surface area of the walls (minus any area for doors and windows). The ceiling area would be found by multiplying length times width and the result added to the wall area gives the total area.

The surface area of a circular cylinder, however, has not been discussed. If we wanted to know how many square feet of surface area are in a tank with a diameter of 60 feet and a height of 20 feet, we could start with the top and bottom.

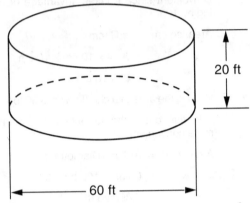

The area of the top and bottom ends are both $\pi \times R^2$.

Area, sq ft = (2 ends)(π)(Radius, ft)^2

   = $2 \times \pi \times (30 \text{ ft})^2$

   = 5,652 sq ft

The surface area of the wall must now be calculated. If we made a vertical cut in the wall and unrolled it, the straightened wall would be the same length as the circumference of the floor and ceiling.

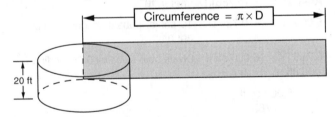

This length has been found to always be $\pi \times D$. In the case of the tank, the length of the wall would be:

Length, ft = (π)(Diameter, ft)

   = $3.14 \times 60 \text{ ft}$

   = 188.4 ft

Area would be:

A_W, sq ft = Length, ft × Height, ft

   = 188.4 ft × 20 ft

   = 3,768 sq ft

Outside Surface Area to Paint, sq ft = Area of Top and Bottom, sq ft + Area of Wall, sq ft

   = 5,652 sq ft + 3,768 sq ft

   = 9,420 sq ft

A container has inside and outside surfaces and you may need to paint both of them.

A.25 Cone

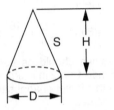

The lateral area of a cone is equal to $\frac{1}{2}$ of the slant height (S) multiplied by the circumference of the base.

$$A_L = \frac{1}{2} S \times \pi \times D = \pi \times S \times R$$

In this case the slant height is not given; it may be calculated by:

$$S = \sqrt{R^2 + H^2}$$

EXAMPLE: Find the entire outside area of a cone with a diameter of 30 inches and a height of 20 inches.

Slant Height, in = $\sqrt{(\text{Radius, in})^2 + (\text{Height, in})^2}$

   = $\sqrt{(15 \text{ in})^2 + (20 \text{ in})^2}$

   = $\sqrt{225 \text{ sq in} + 400 \text{ sq in}}$

   = $\sqrt{625 \text{ sq in}}$

   = 25 in

Lateral Area of Cone, sq in = π(Slant Height, in)(Radius, in)

   = $3.14 \times 25 \text{ in} \times 15 \text{ in}$

   = 1,177.5 sq in

Since the entire area was asked for, the area of the base must be added.

Area, sq in = (0.785)(Diameter, in)^2

   = $0.785 \times 30 \text{ in} \times 30 \text{ in}$

   = 706.5 sq in

Total Area, sq in = Area of Cone, sq in + Area of Bottom, sq in

   = 1,177.5 sq in + 706.5 sq in

   = 1,884 sq in

A.26   Sphere

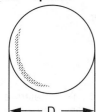

The surface area of a sphere or ball is equal to π multiplied by the diameter squared which is four times the cross-sectional area.

$$A_s = \pi D^2$$

If the radius is used, the formula becomes:

$$A_s = \pi D^2 = \pi \times 2R \times 2R = 4\pi R^2$$

EXAMPLE: What is the surface area of a sphere-shaped water tank 20 feet in diameter?

$$\text{Area, sq ft} = \pi (\text{Diameter, ft})^2$$
$$= 3.14 \times 20 \text{ ft} \times 20 \text{ ft}$$
$$= 1,256 \text{ sq ft}$$

A.3   VOLUMES

A.30   Rectangle

Volumes are measured in three dimensions or in cubic units. To calculate the volume of a rectangle, the area of the base is calculated in square units and then multiplied by the height. The formula then becomes:

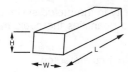

$$V = L \times W \times H$$

EXAMPLE: The length of a box is 2 feet, the width is 15 inches, and the height is 18 inches. Find its volume.

$$\text{Volume, cu ft} = \text{Length, ft} \times \text{Width, ft} \times \text{Height, ft}$$
$$= 2 \text{ ft} \times \frac{15 \text{ in}}{12 \text{ in/ft}} \times \frac{18 \text{ in}}{12 \text{ in/ft}}$$
$$= 2 \text{ ft} \times 1.25 \text{ ft} \times 1.5 \text{ ft}$$
$$= 3.75 \text{ cu ft}$$

A.31   Prism

The same general rule that applies to the volume of a rectangle also applies to a prism.

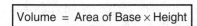
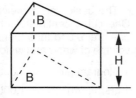

$$\boxed{\text{Volume} = \text{Area of Base} \times \text{Height}}$$

EXAMPLE: Find the volume of a prism with a base area of 10 square feet and a height of 5 feet. (Note that the base of a prism is triangular in shape.)

$$\text{Volume, cu ft} = \text{Area of Base, sq ft} \times \text{Height, ft}$$
$$= 10 \text{ sq ft} \times 5 \text{ ft}$$
$$= 50 \text{ cu ft}$$

A.32   Cylinder

The volume of a cylinder is equal to the area of the base multiplied by the height.

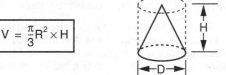

$$V = \pi R^2 \times H = 0.785\, D^2 \times H$$

EXAMPLE: A tank has a diameter of 100 feet and a depth of 12 feet. Find the volume.

$$\text{Volume, cu ft} = 0.785 \times (\text{Diameter, ft})^2 \times \text{Height, ft}$$
$$= 0.785 \times 100 \text{ ft} \times 100 \text{ ft} \times 12 \text{ ft}$$
$$= 94,200 \text{ cu ft}$$

A.33   Cone

The volume of a cone is equal to ⅓ the volume of a circular cylinder of the same height and diameter.

$$V = \frac{\pi}{3} R^2 \times H$$

EXAMPLE: Calculate the volume of a cone if the height at the center is 4 feet and the diameter is 100 feet (radius is 50 feet).

$$\text{Volume, cu ft} = \frac{\pi}{3} \times (\text{Radius})^2 \times \text{Height, ft}$$
$$= \frac{\pi}{3} \times 50 \text{ ft} \times 50 \text{ ft} \times 4 \text{ ft}$$
$$= 10,500 \text{ cu ft}$$

A.34   Sphere

The volume of a sphere is equal to π/6 times the diameter cubed.

$$\boxed{V = \frac{\pi}{6} \times D^3}$$

EXAMPLE: How much gas can be stored in a sphere with a diameter of 12 feet? (Assume atmospheric pressure.)

$$\text{Volume, cu ft} = \frac{\pi}{6} \times (\text{Diameter, ft})^3$$
$$= \frac{\pi}{6} \times \overset{2}{\cancel{12 \text{ ft}}} \times 12 \text{ ft} \times 12 \text{ ft}$$
$$= 904.32 \text{ cubic feet}$$

A.4   METRIC SYSTEM

The two most common systems of weights and measures are the English system and the metric system (*Le Système International d'Unités (SI)*). Of these two, the metric system is more popular with most of the nations of the world. The reason for this is that the metric system is based on a system of tens and is therefore easier to remember and easier to use than the English system. Even though the basic system in the United States is the English system, the scientific community uses the metric system almost exclusively. Many organizations have urged, for good reason, that the United States switch to the metric system. Today the metric system is gradually becoming the standard system of measurement in the United States.

As the United States changes from the English to the metric system, some confusion and controversy has developed. For example, which is the correct spelling of the following words:

1. Liter or litre?

2. Meter or metre?

The U.S. National Bureau of Standards, the Water Environment Federation, and the American Water Works Association use litre and metre. The U.S. Government uses liter and meter and accepts no deviations. Some people argue that METRE should be used to measure LENGTH and that METER should be used to measure FLOW RATES (like a water or electric meter). Liter and meter are used in this manual because this is most consistent with spelling in the United States.

One of the most frequent arguments heard against the U.S. switching to the metric system was that the costs of switching manufacturing processes would be excessive. Pipe manufacturers have agreed upon the use of a "soft" metric conversion system during the conversion to the metric system. Past practice in the U.S. has identified some types of pipe by external (outside) diameter while other types are classified by nominal (existing only in name, not real or actual) bore. This means that a six-inch pipe does not have a six-inch inside diameter. With the strict or "hard" metric system, a six-inch pipe would be a 152.4-mm (6 in x 25.4 mm/in) pipe. In the "soft" metric system a six-inch pipe is a 150-mm (6 in x 25 mm/in) pipe. Typical customary and "soft" metric pipe-size designations are shown below:

PIPE-SIZE DESIGNATIONS

Customary, in	2	4	6	8	10	12	15	18
"Soft" Metric, mm	50	100	150	200	250	300	375	450
Customary, in	24	30	36	42	48	60	72	84
"Soft" Metric, mm	600	750	900	1050	1200	1500	1800	2100

In order to study the metric system, you must know the meanings of the terminology used. Following is a list of Greek and Latin prefixes used in the metric system.

PREFIXES USED IN THE METRIC SYSTEM

Prefix	Symbol	Meaning
Micro	μ	1/1 000 000 or 0.000 001
Milli	m	1/1000 or 0.001
Centi	c	1/100 or 0.01
Deci	d	1/10 or 0.1
Unit		1
Deka	da	10
Hecto	h	100
Kilo	k	1000
Mega	M	1 000 000

A.40   Measures of Length

The basic measure of length is the meter.

1 kilometer (km)	= 1,000 meters (m)
1 meter (m)	= 100 centimeters (cm)
1 centimeter (cm)	= 10 millimeters (mm)

Kilometers are usually used in place of miles, meters are used in place of feet and yards, centimeters are used in place of inches, and millimeters are used for inches and fractions of an inch.

LENGTH EQUIVALENTS

1 kilometer	= 0.621 mile	1 mile	= 1.61 kilometers
1 meter	= 3.28 feet	1 foot	= 0.305 meter
1 meter	= 39.37 inches	1 inch	= 0.0254 meter
1 centimeter	= 0.3937 inch	1 inch	= 2.54 centimeters
1 millimeter	= 0.0394 inch	1 inch	= 25.4 millimeters

NOTE: The above equivalents are reciprocals. If one equivalent is given, the reverse can be obtained by division. For instance, if one meter equals 3.28 feet, one foot equals 1/3.28 meter, or 0.305 meter.

A.41   Measures of Capacity or Volume

The basic measure of capacity in the metric system is the liter. For measurement of large quantities the cubic meter is sometimes used.

1 kiloliter (k*L*) = 1,000 liters (*L*) = 1 cu meter (cu m)

1 liter (*L*) = 1,000 milliliters (m*L*)

Kiloliters, or cubic meters, are used to measure capacity of large storage tanks or reservoirs in place of cubic feet or gallons. Liters are used in place of gallons or quarts. Milliliters are used in place of quarts, pints, or ounces.

CAPACITY EQUIVALENTS

1 kiloliter	= 264.2 gallons	1 gallon	= 0.003785 kiloliter
1 liter	= 1.057 quarts	1 quart	= 0.946 liter
1 liter	= 0.2642 gallon	1 gallon	= 3.785 liters
1 milliliter	= 0.0353 ounce	1 ounce	= 29.57 milliliters

A.42   Measures of Weight

The basic unit of weight in the metric system is the gram. One cubic centimeter of water at maximum density weighs one gram, and thus there is a direct, simple relation between volume of water and weight in the metric system.

1 kilogram (kg)	= 1,000 grams (gm)
1 gram (gm)	= 1,000 milligrams (mg)
1 milligram (mg)	= 1,000 micrograms (μg)

Grams are usually used in place of ounces, and kilograms are used in place of pounds.

WEIGHT EQUIVALENTS

1 kilogram	= 2.205 pounds	1 pound	= 0.4536 kilogram
1 gram	= 0.0022 pound	1 pound	= 453.6 grams
1 gram	= 0.0353 ounce	1 ounce	= 28.35 grams
1 gram	= 15.43 grains	1 grain	= 0.0648 gram

A.43 Temperature

Just as you should become familiar with the metric system, it is also important to become familiar with the centigrade (Celsius) scale for measuring temperature. There is nothing magical about the centigrade scale—it is simply a different size than the Fahrenheit scale. The two scales compare as follows:

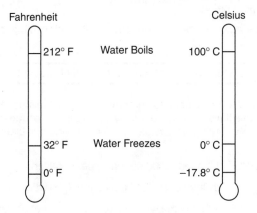

The two scales are related in the following manner:

$$\text{Fahrenheit} = (°C \times 9/5) + 32°$$
$$\text{Celsius} = (°F - 32°) \times 5/9$$

EXAMPLE: Convert 20° Celsius to degrees Fahrenheit.

$$°F = (°C \times 9/5) + 32°$$
$$°F = (20° \times 9/5) + 32°$$
$$°F = \frac{180°}{5} + 32°$$
$$= 36° + 32°$$
$$= 68°F$$

EXAMPLE: Convert −10°C to °F.

$$°F = (-10° \times 9/5) + 32°$$
$$°F = -90°/5 + 32°$$
$$= -18° + 32°$$
$$= 14°F$$

EXAMPLE: Convert −13°F to °C.

$$°C = (°F - 32°) \times \frac{5}{9}$$
$$°C = (-13° - 32°) \times \frac{5}{9}$$
$$= -45° \times \frac{5}{9}$$
$$= -5° \times 5$$
$$= -25°C$$

A.44 Milligrams Per Liter

Milligrams per liter (mg/L) is a unit of measurement used in laboratory and scientific work to indicate very small concentrations of dilutions. Since water contains small concentrations of dissolved substances and solids, and since small amounts of chemical compounds are sometimes used in water treatment processes, the term milligrams per liter is also common in treatment plants. It is a weight-volume relationship.

As previously discussed:

1,000 liters = 1 cubic meter = 1,000,000 cubic centimeters.

Therefore,

1 liter = 1,000 cubic centimeters.

Since one cubic centimeter of water weighs one gram,

1 liter of water = 1,000 grams or 1,000,000 milligrams.

$$\frac{1\text{ milligram}}{\text{liter}} = \frac{1\text{ milligram}}{1,000,000\text{ milligrams}} = \frac{1\text{ part}}{\text{million parts}} = \frac{1\text{ part per}}{\text{million (ppm)}}$$

Milligrams per liter and parts per million (parts) may be used interchangeably as long as the liquid density is 1.0 gm/cu cm or 62.43 lb/cu ft. A concentration of 1 milligram/liter (mg/L) or 1 ppm means that there is 1 part of substance by weight for every 1 million parts of water. A concentration of 10 mg/L would mean 10 parts of substance per million parts of water.

To get an idea of how small 1 mg/L is, divide the numerator and denominator of the fraction by 10,000. This, of course, does not change its value since 10,000 ÷ 10,000 is equal to one.

$$1\frac{mg}{L} = \frac{1\text{ mg}}{1,000,000\text{ mg}} = \frac{1/10,000\text{ mg}}{1,000,000/10,000\text{ mg}} = \frac{0.0001\text{ mg}}{100\text{ mg}} = 0.0001\%$$

Therefore, 1 mg/L is equal to one ten-thousandth of a percent, or

1% is equal to 10,000 mg/L.

To convert mg/L to %, move the decimal point four places or numbers to the left.

Working problems using milligrams per liter or parts per million is a part of everyday operation in most water treatment plants.

A.45 Example Problems

EXAMPLE: Raw water flowing into a plant at a rate of five million pounds per day is prechlorinated at 5 mg/L. How many pounds of chlorine are used per day?

$$5\text{ mg}/L = \frac{5\text{ lbs Chlorine}}{\text{million lbs Water}}$$

$$\begin{array}{l}\text{Chlorine}\\\text{Feed,}\\\text{lbs/day}\end{array} = \text{Conc, lbs/M lbs} \times \text{Flow, lbs/day}$$

$$= \frac{5\text{ lbs}}{\text{million lbs}} \times \frac{5\text{ million lbs}}{\text{day}}$$

$$= 25\text{ lbs/day}$$

There is one thing that is unusual about the previous problem and that is the flow is reported in pounds per day. In most treatment plants, flow is reported in terms of gallons per minute or gallons per day. To convert these flow figures to weight, an additional conversion factor is needed. One gallon of water weighs 8.34 pounds. Using this factor, it is possible to convert flow in gallons per day to flow in pounds per day.

EXAMPLE: A well pump with a flow of 3.5 million gallons per day (MGD) chlorinates the water with 2.0 mg/L chlorine. How many pounds of chlorine are used per day?

$$\text{Flow, lbs/day} = \text{Flow, } \frac{\text{M gal}}{\text{day}} \times \frac{8.34 \text{ lbs}}{\text{gal}}$$

$$= \frac{3.5 \text{ million } \text{gal}}{\text{day}} \times \frac{8.34 \text{ lbs}}{\text{gal}}$$

$$= 29.19 \text{ million lbs/day}$$

$$\begin{array}{l}\text{Chlorine} \\ \text{Feed,} \\ \text{lbs/day}\end{array} = \text{Level, mg/}L \times \text{Flow, M lbs/day}$$

$$= \frac{2.0 \text{ mg}^*}{\text{million mg}} \times \frac{29.19 \text{ million lbs}}{\text{day}}$$

$$= 58.38 \text{ lbs/day}$$

---

* Remember that $\dfrac{1 \text{ mg}}{\text{M mg}} = \dfrac{1 \text{ lb}}{\text{M lb}}$. They are identical ratios.

In solving the previous problem, a relation was used that is most important to understand and commit to memory.

$$\boxed{\text{Feed, lbs/day} = \text{Flow, MGD} \times \text{Dose, mg/}L \times 8.34 \text{ lbs/gal}}$$

EXAMPLE: A chlorinator is set to feed 50 pounds of chlorine per day to a flow of 0.8 MGD. What is the chlorine dose in mg/L?

$$\begin{array}{l}\text{Conc or Dose,} \\ \text{mg/}L\end{array} = \frac{\text{lbs/day}}{\text{MGD} \times 8.34 \text{ lbs/gal}}$$

$$= \frac{50 \text{ lbs/day}}{0.80 \text{ MG/day} \times 8.34 \text{ lbs/gal}}$$

$$= \frac{50 \text{ lbs}}{6.672 \text{ M lbs}}$$

$$= 7.5 \text{ mg/}L, \text{ or } 7.5 \text{ ppm}$$

EXAMPLE: A pump delivers 500 gallons per minute to a water treatment plant. Alum is added at 10 mg/L. How much alum is used in pounds per day?

$$\text{Flow, MGD} = \text{Flow, GPM} \times 60 \text{ min/hr} \times 24 \text{ hr/day}$$

$$= \frac{500 \text{ gal}}{\text{min}} \times \frac{60 \text{ min}}{\text{hr}} \times \frac{24 \text{ hr}}{\text{day}}$$

$$= 720,000 \text{ gal/day}$$

$$= 0.72 \text{ MGD}$$

$$\begin{array}{l}\text{Alum Feed,} \\ \text{lbs/day}\end{array} = \text{Flow, MGD} \times \text{Dose, mg/}L \times 8.34 \text{ lbs/gal}$$

$$= \frac{0.72 \text{ M gal}}{\text{day}} \times \frac{10 \text{ mg}}{\text{M mg}} \times \frac{8.34 \text{ lbs}}{\text{gal}}$$

$$= 60.048 \text{ lbs/day or about 60 lbs/day}$$

A.5 WEIGHT-VOLUME RELATIONS

Another factor for the operator to remember, in addition to the weight of a gallon of water, is the weight of a cubic foot of water. One cubic foot of water weighs 62.4 lbs. If these two weights are divided, it is possible to determine the number of gallons in a cubic foot.

$$\frac{62.4 \text{ pounds/cu ft}}{8.34 \text{ pounds/gal}} = 7.48 \text{ gal/cu ft}$$

Thus we have another very important relationship to commit to memory.

$$\boxed{8.34 \text{ lbs/gal} \times 7.48 \text{ gal/cu ft} = 62.4 \text{ lbs/cu ft}}$$

It is only necessary to remember two of the above items since the third may be found by calculation. For most problems, 8⅓ lbs/gal and 7½ gal/cu ft will provide sufficient accuracy.

EXAMPLE: Change 1,000 cu ft of water to gallons.

1,000 cu ft × 7.48 gal/cu ft = 7,480 gallons

EXAMPLE: What is the weight of three cubic feet of water?

62.4 lbs/cu ft × 3 cu ft = 187.2 lbs

EXAMPLE: The net weight of a tank of water is 750 lbs. How many gallons does it contain?

$$\frac{750 \text{ lbs}}{8.34 \text{ lbs/gal}} = 90 \text{ gal}$$

A.6 FORCE, PRESSURE, AND HEAD

In order to study the forces and pressures involved in fluid flow, it is first necessary to define the terms used.

FORCE: The push exerted by water on any surface being used to confine it. Force is usually expressed in pounds, tons, grams, or kilograms.

PRESSURE: The force per unit area. Pressure can be expressed in many ways, but the most common term is pounds per square inch (psi).

HEAD: Vertical distance from the water surface to a reference point below the surface. Usually expressed in feet or meters.

An *EXAMPLE* should serve to illustrate these terms.

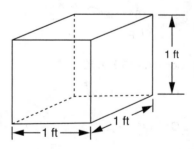

If water were poured into a one-foot cubical container, the *FORCE* acting on the bottom of the container would be 62.4 pounds.

The *PRESSURE* acting on the bottom would be 62.4 pounds per square foot. The area of the bottom is also 12 in × 12 in = 144 sq in. Therefore, the pressure may also be expressed as:

Pressure, psi $= \dfrac{62.4 \text{ lbs}}{\text{sq ft}} = \dfrac{62.4 \text{ lbs/sq ft}}{144 \text{ sq in/sq ft}}$

$= 0.433 \text{ lb/sq in}$

$= 0.433 \text{ psi}$

Since the height of the container is one foot, the *HEAD* would be one foot.

The pressure in any vessel at one foot of depth or one foot of head is 0.433 psi acting in any direction.

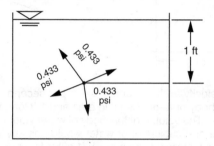

If the depth of water in the previous example were increased to two feet, the pressure would be:

$p = \dfrac{2(62.4 \text{ lbs})}{144 \text{ sq in}} = \dfrac{124.8 \text{ lbs}}{144 \text{ sq in}} = 0.866 \text{ psi}$

Therefore, we can see that for every foot of head, the pressure increases by 0.433 psi. Thus, the general formula for pressure becomes:

$\boxed{p, \text{ psi} = 0.433(H, \text{ ft})}$

$\boxed{P, \text{ lbs/sq ft} = 62.4(H, \text{ ft})}$

H = feet of head

p = pounds per square *INCH* of pressure

H = feet of head

P = pounds per square *FOOT* of pressure

We can now draw a diagram of the pressure acting on the side of a tank. Assume a 4-foot deep tank. The pressures shown on the tank are gage pressures. These pressures do not include the atmospheric pressure acting on the surface of the water.

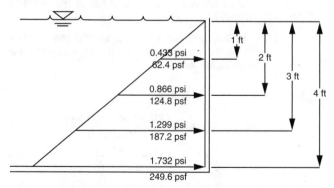

$p_0 = 0.433 \times 0 = 0.0 \text{ psi}$

$p_1 = 0.433 \times 1 = 0.433 \text{ psi}$

$p_2 = 0.433 \times 2 = 0.866 \text{ psi}$

$p_3 = 0.433 \times 3 = 1.299 \text{ psi}$

$p_4 = 0.433 \times 4 = 1.732 \text{ psi}$

$P_0 = 62.4 \times 0 = 0.0 \text{ lb/sq ft}$

$P_1 = 62.4 \times 1 = 62.4 \text{ lbs/sq ft}$

$P_2 = 62.4 \times 2 = 124.8 \text{lbs/sq ft}$

$P_3 = 62.4 \times 3 = 187.2 \text{lbs/sq ft}$

$P_4 = 62.4 \times 4 = 249.6 \text{lbs/sq ft}$

The average *PRESSURE* acting on the tank wall is 1.732 psi/2 = 0.866 psi, or 249.6 psf/2 = 124.8 psf. We divided by two to obtain the average pressure because there is zero pressure at the top and 1.732 psi pressure on the bottom of the wall.

If the wall were 5 feet long, the pressure would be acting over the entire 20-square-foot (5 ft × 4 ft) area of the wall. The total force acting to push the wall would be:

Force, lbs $= (\text{Pressure, lbs/sq ft})(\text{Area, sq ft})$

$= 124.8 \text{ lbs/sq ft} \times 20 \text{ sq ft}$

$= 2,496 \text{ lbs}$

If the pressure in psi were used, the problem would be similar:

Force, lbs $= (\text{Pressure, lbs/sq in})(\text{Area, sq in})$

$= 0.866 \text{ psi} \times 48 \text{ in} \times 60 \text{ in}$

$= 2,494 \text{ lbs*}$

---

* Difference in answer due to rounding off of decimal points.

The general formula, then, for finding the total force acting on a side wall of a tank is:

$\boxed{F = 31.2 \times H^2 \times L}$

F = force in pounds

H = head in feet

L = length of wall in feet

31.2 = constant with units of lbs/cu ft and considers the fact that the force results from H/2 or half the depth of the water, which is the average depth. The force is exerted at H/3 from the bottom.

EXAMPLE: Find the force acting on a five-foot long wall in a four-foot deep tank.

Force, lbs $= 31.2(\text{Head, ft})^2(\text{Length, ft})$

$= 31.2 \text{ lbs/cu ft} \times (4 \text{ ft})^2 \times 5 \text{ ft}$

$= 2,496 \text{ lbs}$

Occasionally an operator is warned: *NEVER EMPTY A TANK DURING PERIODS OF HIGH GROUNDWATER.* Why? The pressure on the bottom of the tank caused by the water surrounding the tank will tend to float the tank like a cork if the upward force of the water is greater than the weight of the tank.

F = upward force in pounds

H = head of water on tank bottom in feet

$\boxed{F = 62.4 \times H \times A}$

A = area of bottom of tank in square feet

62.4 = a constant with units of lbs/cu ft

This formula is approximately true if the tank doesn't crack, leak, or start to float.

EXAMPLE: Find the upward force on the bottom of an empty tank caused by a groundwater depth of 8 feet above the tank bottom. The tank is 20 ft wide and 40 ft long.

$$\text{Force, lbs} = 62.4(\text{Head, ft})(\text{Area, sq ft})$$

$$= 62.4 \text{ lbs/cu ft} \times 8 \text{ ft} \times 20 \text{ ft} \times 40 \text{ ft}$$

$$= 399,400 \text{ lbs}$$

A.7 VELOCITY AND FLOW RATE

A.70 Velocity

The velocity of a particle or substance is the speed at which it is moving. It is expressed by indicating the length of travel and how long it takes to cover the distance. Velocity can be expressed in almost any distance and time units. For instance, a car may be traveling at a rate of 280 miles per five hours. However, it is normal to express the distance traveled per unit time. The above example would then become:

$$\text{Velocity, mi/hr} = \frac{280 \text{ miles}}{5 \text{ hours}}$$

$$= 56 \text{ miles/hour}$$

The velocity of water in a channel, pipe, or other conduit can be expressed in the same way. If the particle of water travels 600 feet in five minutes, the velocity is:

$$\text{Velocity, ft/min} = \frac{\text{Distance, ft}}{\text{Time, minutes}}$$

$$= \frac{600 \text{ ft}}{5 \text{ min}}$$

$$= 120 \text{ ft/min}$$

If you wish to express the velocity in feet per second, multiply by 1 min/60 seconds.

NOTE: Multiplying by $\frac{1 \text{ minute}}{60 \text{ seconds}}$ is like multiplying by $\frac{1}{1}$; it does not change the relative value of the answer. It only changes the form of the answer.

$$\text{Velocity, ft/sec} = (\text{Velocity, ft/min})(1 \text{ min/60 sec})$$

$$= \frac{120 \text{ ft}}{\text{min}} \times \frac{1 \text{ min}}{60 \text{ sec}}$$

$$= \frac{120 \text{ ft}}{60 \text{ sec}}$$

$$= 2 \text{ ft/sec}$$

A.71 Flow Rate

If water in a one-foot wide channel is one foot deep, then the cross-sectional area of the channel is 1 ft × 1 ft = 1 sq ft.

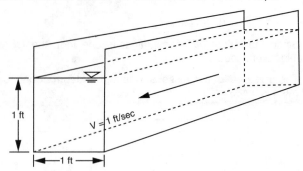

If the velocity in this channel is 1 ft per second, then each second a body of water 1 sq ft in area and 1 ft long will pass a given point. The volume of this body of water would be 1 cubic foot. Since one cubic foot of water would pass by every second, the flow rate would be equal to 1 cubic foot per second, or 1 CFS.

To obtain the flow rate in the above example the velocity was multiplied by the cross-sectional area. This is another important general formula.

$$\boxed{Q = V \times A}$$

Q = flow rate, CFS or cu ft/sec

V = velocity, ft/sec

A = area, sq ft

EXAMPLE: A rectangular channel 3 feet wide contains water 2 feet deep and flowing at a velocity of 1.5 feet per second. What is the flow rate in CFS?

$$Q = V \times A$$

$$\text{Flow Rate, CFS} = \text{Velocity, ft/sec} \times \text{Area, sq ft}$$

$$= 1.5 \text{ ft/sec} \times 3 \text{ ft} \times 2 \text{ ft}$$

$$= 9 \text{ cu ft/sec}$$

EXAMPLE: Flow in a 2.5-foot wide channel is 1.4 feet deep and measures 11.2 CFS. What is the average velocity?

In this problem we want to find the velocity. Therefore, we must rearrange the general formula to solve for velocity.

$$V = \frac{Q}{A}$$

$$\text{Velocity, ft/sec} = \frac{\text{Flow Rate, cu ft/sec}}{\text{Area, sq ft}}$$

$$= \frac{11.2 \text{ cu ft/sec}}{2.5 \text{ ft} \times 1.4 \text{ ft}}$$

$$= \frac{11.2 \text{ cu ft/sec}}{3.5 \text{ sq ft}}$$

$$= 3.2 \text{ ft/sec}$$

EXAMPLE: Flow in an 8-inch pipe is 500 GPM. What is the average velocity?

$$\text{Area, sq ft} = 0.785(\text{Diameter, ft})^2$$

$$= 0.785(8/12 \text{ ft})^2$$

$$= 0.785(0.67 \text{ ft})^2$$

$$= 0.785(0.67 \text{ ft})(0.67 \text{ ft})$$

$$= 0.785(0.45 \text{ sq ft})$$

$$= 0.35 \text{ sq ft}$$

$$\text{Flow, CFS} = \text{Flow, gal/min} \times \frac{\text{cu ft}}{7.48 \text{ gal}} \times \frac{1 \text{ min}}{60 \text{ sec}}$$

$$= \frac{500 \text{ gal}}{\text{min}} \times \frac{\text{cu ft}}{7.48 \text{ gal}} \times \frac{1 \text{ min}}{60 \text{ sec}}$$

$$= \frac{500 \text{ cu ft}}{448.8 \text{ sec}}$$

$$= 1.114 \text{ CFS}$$

$$\text{Velocity, ft/sec} = \frac{\text{Flow, cu ft/sec}}{\text{Area, sq ft}}$$

$$= \frac{1.114 \text{ cu ft/sec}}{0.35 \text{ sq ft}}$$

$$= 3.18 \text{ ft/sec}$$

A.8 PUMPS

A.80 Pressure

Atmospheric pressure at sea level is approximately 14.7 psi. This pressure acts in all directions and on all objects. If a tube is placed upside down in a basin of water and a 1 psi partial vacuum is drawn on the tube, the water in the tube will rise 2.31 feet.

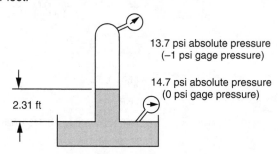

13.7 psi absolute pressure
(−1 psi gage pressure)

14.7 psi absolute pressure
(0 psi gage pressure)

2.31 ft

NOTE: 1 ft of water = 0.433 psi; therefore,

$$1 \text{ psi} = \frac{1}{0.433} \text{ ft} = 2.31 \text{ ft of water}$$

The action of the partial vacuum is what gets water out of a sump or well and up to a pump. It is not sucked up, but it is pushed up by atmospheric pressure on the water surface in the sump. If a complete vacuum could be drawn, the water would rise 2.31 × 14.7 = 33.9 feet; but this is impossible to achieve. The practical limit of the suction lift of a positive displacement pump is about 22 feet, and that of a centrifugal pump is 15 feet.

A.81 Work

Work can be expressed as lifting a weight a certain vertical distance. It is usually defined in terms of foot-pounds.

EXAMPLE: A 165-pound man runs up a flight of stairs 20 feet high. How much work did he do?

$$\text{Work, ft-lbs} = \text{Weight, lbs} \times \text{Height, ft}$$

$$= 165 \text{ lbs} \times 20 \text{ ft}$$

$$= 3,300 \text{ ft-lbs}$$

A.82 Power

Power is a rate of doing work and is usually expressed in foot-pounds per minute.

EXAMPLE: If the man in the above example runs up the stairs in three seconds, how much power has he exerted?

$$\text{Power, ft-lbs/sec} = \frac{\text{Work, ft-lbs}}{\text{Time, sec}}$$

$$= \frac{3,300 \text{ ft-lbs}}{3 \text{ sec}} \times \frac{60 \text{ sec}}{\text{minute}}$$

$$= 66,000 \text{ ft-lbs/min}$$

A.83 Horsepower

Horsepower is also a unit of power. One horsepower is defined as 33,000 ft-lbs per minute or 746 watts.

EXAMPLE: How much horsepower has the man in the previous example exerted as he climbs the stairs?

$$\text{Horsepower, HP} = (\text{Power, ft-lbs/min})\left(\frac{\text{HP}}{33,000 \text{ ft-lbs/min}}\right)$$

$$= 66,000 \text{ ft-lbs/min} \times \frac{\text{Horsepower}}{33,000 \text{ ft-lbs/min}}$$

$$= 2 \text{ HP}$$

Work is also done by lifting water. If the flow from a pump is converted to a weight of water and multiplied by the vertical distance it is lifted, the amount of work or power can be obtained.

$$\text{Horsepower, HP} = \frac{\text{Flow, gal}}{\text{min}} \times \text{Lift, ft} \times \frac{8.34 \text{ lbs}}{\text{gal}} \times \frac{\text{Horsepower}}{33,000 \text{ ft-lbs/min}}$$

Solving the above relation, the amount of horsepower necessary to lift the water is obtained. This is called water horsepower.

$$\text{Water, HP} = \frac{(\text{Flow, GPM})(\text{H, ft})}{3,960^*}$$

$$^* \frac{8.34 \text{ lbs}}{\text{gal}} \times \frac{\text{HP}}{33,000 \text{ ft-lbs/min}} = \frac{1}{3,960}$$

1 gallon weighs 8.34 pounds and 1 horsepower is the same as 33,000 ft-lbs/min.

H or Head in feet is the same as Lift in feet.

However, since pumps are not 100 percent efficient (they cannot transmit all the power put into them), the horsepower supplied to a pump is greater than the water horsepower. Horsepower supplied to the pump is called brake horsepower.

$$\text{Brake, HP} = \frac{\text{Flow, GPM} \times \text{H, ft}}{3,960 \times E_p}$$

E_p = Efficiency of Pump (Usual range 50-85 percent, depending on type and size of pump)

Motors are also not 100% efficient; therefore, the power supplied to the motor is greater than the motor transmits.

$$\text{Motor, HP} = \frac{\text{Flow, GPM} \times \text{H, ft}}{3,960 \times E_p \times E_m}$$

E_m = Efficiency of Motor (Usual range 80-95 percent, depending on type and size of pump)

The above formulas have been developed for the pumping of water and wastewater which have a specific gravity of 1.0. If other liquids are to be pumped, the formulas must be multiplied by the specific gravity of the liquid.

EXAMPLE: A flow of 500 GPM of water is to be pumped against a total head of 100 feet by a pump with an efficiency of 70 percent. What is the pump horsepower?

$$\text{Brake, HP} = \frac{\text{Flow, GPM} \times \text{H, ft}}{3,960 \times E_p}$$

$$= \frac{500 \times 100}{3,960 \times 0.70}$$

$$= 18 \text{ HP}$$

EXAMPLE: Find the horsepower required to pump gasoline (specific gravity = 0.75) in the above problem.

$$\text{Brake, HP} = \frac{500 \times 100 \times 0.75}{3,960 \times 0.70}$$

$$= 13.5 \text{ HP (gasoline is lighter and requires less horsepower)}$$

A.84   Head

Basically, the head that a pump must work against is determined by measuring the vertical distance between the two water surfaces, or the distance the water must be lifted. This is called the static head. Two typical conditions for lifting water are shown below.

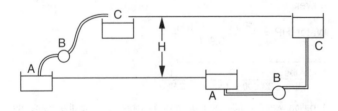

If a pump were designed in the above examples to pump only against head H, the water would never reach the intended point. The reason for this is that the water encounters friction in the pipelines. Friction depends on the roughness and length of pipe, the pipe diameter, and the flow velocity. The turbulence caused at the pipe entrance (point A); the pump (point B); the pipe exit (point C); and at each elbow, bend, or transition also adds to these friction losses. Tables and charts are available in Section A.88 for calculation of these friction losses so they may be added to the measured or static head to obtain the total head. For short runs of pipe that do not have high velocities, the friction losses are generally less than 10 percent of the static head.

EXAMPLE: A pump is to be located 8 feet above a wet well and must lift 1.8 MGD another 50 feet to a storage reservoir. If the pump has an efficiency of 75% and the motor an efficiency of 90%, what is the cost of the power consumed if one kilowatt-hour cost 4 cents?

Since we are not given the length or size of pipe and the number of elbows or bends, we will assume friction to be 10 percent of static head.

Static Head, ft = Suction Lift, ft + Discharge Head, ft

= 8 ft + 50 ft

= 58 ft

Friction Losses, ft = 0.1(Static Head, ft)

= 0.1(58 ft)

= 5.8 ft

Total Dynamic Head, ft = Static Head, ft + Friction Losses, ft

= 58 ft + 5.8 ft

= 63.8 ft

Flow, GPM $= \dfrac{1,800,000 \text{ gal}}{\text{day}} \times \dfrac{\text{day}}{24 \text{ hr}} \times \dfrac{1 \text{ hr}}{60 \text{ min}}$

= 1,250 GPM (assuming pump runs 24 hours per day)

Motor, HP $= \dfrac{\text{Flow, GPM} \times \text{H, ft}}{3,960 \times E_p \times E_m}$

$= \dfrac{1,250 \times 63.8}{3,960 \times 0.75 \times 0.9}$

= 30 HP

Kilowatt-hr = 30 ~~HP~~ × 24 hr/day × 0.746 kW/~~HP~~*

= 537 kilowatt-hr/day

Cost = kWh × $0.04/kWh

= 537 × 0.04

= $21.48/day

---

* See Section A.10, "Basic Conversion Factors (English System)," *POWER*, page 556.

A.85   Pump Characteristics

The discharge of a centrifugal pump, unlike a positive displacement pump, can be made to vary from zero to a maximum capacity which depends on the speed, head, power, and specific impeller design. The interrelation of capacity, efficiency, head, and power is known as the characteristics of the pump.

The first relation normally looked at when selecting a pump is the head vs. capacity. The head of a centrifugal pump normally rises as the capacity is reduced. If the values are plotted on a graph they appear as follows:

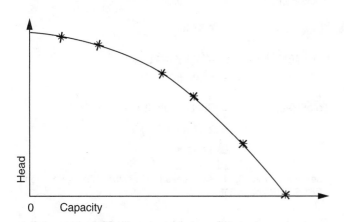

Another important characteristic is the pump efficiency. It begins from zero at no discharge, increases to a maximum, and then drops as the capacity is increased. Following is a graph of efficiency vs. capacity:

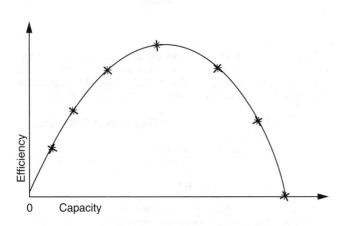

The last important characteristic is the brake horsepower or the power input to the pump. The brake horsepower usually increases with increasing capacity until it reaches a maximum, then it normally reduces slightly.

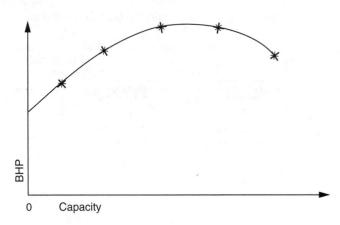

These pump characteristic curves are quite important. Pump sizes are normally picked from these curves rather than calculations. For ease of reading, the three characteristic curves are normally plotted together. A typical graph of pump characteristics is shown as follows:

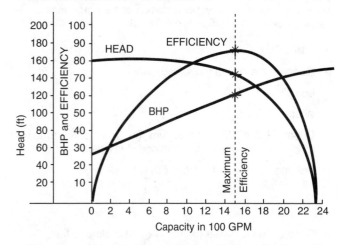

The curves show that the maximum efficiency for the particular pump in question occurs at approximately 1,475 GPM, a head of 132 feet, and a brake horsepower of 58. Operating at this point the pump has an efficiency of approximately 85%. This can be verified by calculation:

$$\text{BHP} = \frac{\text{Flow, GPM} \times \text{H, ft}}{3,960 \times \text{E}}$$

As previously explained, a number can be written over one without changing its value:

$$\frac{\text{BHP}}{1} = \frac{\text{GPM} \times \text{H}}{3,960 \times \text{E}}$$

Since the formula is now in ratio form, it can be cross multiplied.

$$\text{BHP} \times 3,960 \times \text{E} = \text{GPM} \times \text{H} \times 1$$

Solving for E,

$$\text{E} = \frac{\text{GPM} \times \text{H}}{3,960 \times \text{BHP}}$$

$$\text{E} = \frac{1,475 \text{ GPM} \times 132 \text{ ft}}{3,960 \times 58 \text{ HP}}$$

$$= 0.85 \text{ or } 85\% \text{ (Check)}$$

The preceding is only a brief description of pumps to familiarize the operator with their characteristics. The operator does not normally specify the type and size of pump needed at a plant. If a pump is needed, the operator should be able to supply the information necessary for a pump supplier to provide the best possible pump for the lowest cost. Some of the information needed includes:

1. Flow range desired;

2. Head conditions:

    a. Suction head or lift,
    b. Pipe and fitting friction head, and
    c. Discharge head;

3. Type of fluid pumped and temperature; and

4. Pump location.

A.86   Evaluation of Pump Performance

1. Capacity

Sometimes it is necessary to determine the capacity of a pump. This can be accomplished by determining the time it takes a pump to fill or empty a portion of a storage tank or diversion box when all inflow is blocked off.

EXAMPLE:

a. Measure the size of the storage tank.

Length = 10 ft

Width = 10 ft

Depth = 5 ft (We will measure the time it takes to lower the tank a distance of five feet.)

$$\text{Volume, cu ft} = L, ft \times W, ft \times D, ft$$

$$= 10\ ft \times 10\ ft \times 5\ ft$$

$$= 500\ cu\ ft$$

b. Record time for water to drop five feet in storage tank.

Time = 10 minutes 30 seconds

= 10.5 minutes

c. Calculate pumping rate or capacity.

$$\text{Pumping Rate, GPM} = \frac{\text{Volume, gallons}}{\text{Time, minutes}}$$

$$= \frac{(500\ cu\ ft)(7.5\ gal/cu\ ft)}{10.5\ min}$$

$$= \frac{3,750}{10.5}$$

$$= 357\ GPM$$

If you know the total dynamic head and have the pump's performance curves, you can determine if the pump is delivering at design capacity. If not, try to determine the cause (see Chapter 18, "Maintenance," in *WATER TREATMENT PLANT OPERATION*, Volume II). After a pump overhaul, the pump's actual performance (flow, head, power, and efficiency) should be compared with the pump manufacturer's performance curves. This procedure for calculating the rate of filling or emptying of a storage tank or diversion box can be used to calibrate flowmeters.

2. Efficiency

To estimate the efficiency of the pump in the previous example, the total head must be known. This head may be estimated by measuring the suction and discharge pressures. Assume these were measured as follows:

2 in Mercury
vacuum

20 psi

PUMP → flow

Suction side   Discharge side

No additional information is necessary if we assume the pressure gages are at the same height and the pipe diameters are the same. Both pressure readings must be converted to feet.

$$\text{Suction Lift, ft} = 2\ \text{in Mercury} \times \frac{1.133\ \text{ft water*}}{1\ \text{in Mercury}}$$

$$= 2.27\ ft$$

$$\text{Discharge Head, ft} = 20\ psi \times 2.31\ ft/psi*$$

$$= 46.20\ ft$$

$$\text{Total Head, ft} = \text{Suction Lift, ft} + \text{Discharge Head, ft}$$

$$= 2.27\ ft + 46.20\ ft$$

$$= 48.47\ ft$$

---

* See Section A.10, "Basic Conversion Factors (English System)," *PRESSURE*, page 556.

Calculate the power output of the pump or water horsepower:

$$\text{Water Horsepower, HP} = \frac{(\text{Flow, GPM})(\text{Head, ft})}{3,960}$$

$$= \frac{(357\ GPM)(48.47\ ft)}{3,960}$$

$$= 4.4\ HP$$

To estimate the efficiency of the pump, measure the kilowatts drawn by the pump motor. Assume the meter indicates 8,000 watts or 8 kilowatts. The manufacturer claims the electric motor is 80 percent efficient.

$$\text{Brake Horsepower, HP} = (\text{Power to Elec. Motor})(\text{Motor Eff.})$$

$$= \frac{(8\ kW)(0.80)}{0.746\ kW/HP}$$

$$= 8.6\ HP$$

$$\text{Pump Efficiency, \%} = \frac{\text{Water Horsepower, HP} \times 100\%}{\text{Brake Horsepower, HP}}$$

$$= \frac{4.4\ HP \times 100\%}{8.6\ HP}$$

$$= 51\%$$

The following diagram may clarify the previous problem:

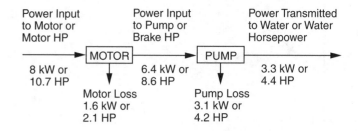

Power Input to Motor or Motor HP	Power Input to Pump or Brake HP	Power Transmitted to Water or Water Horsepower
8 kW or 10.7 HP	6.4 kW or 8.6 HP	3.3 kW or 4.4 HP
Motor Loss 1.6 kW or 2.1 HP	Pump Loss 3.1 kW or 4.2 HP	

The wire-to-water efficiency is the efficiency of the power input to produce water horsepower.

$$\text{Wire-to-Water Efficiency, \%} = \frac{\text{Water Horsepower, HP}}{\text{Power Input, HP}} \times 100\%$$

$$= \frac{4.4 \text{ HP}}{10.7 \text{ HP}} \times 100\%$$

$$= 41\%$$

The wire-to-water efficiency of a pumping system (pump and electric motor) can be calculated by using the following formula:

$$\text{Efficiency, \%} = \frac{(\text{Flow, GPM})(\text{TDH, ft})(100\%)}{(\text{Voltage, volts})(\text{Current, amps})(5.308)}$$

$$= \frac{(375 \text{ GPM})(48.47 \text{ ft})(100\%)}{(220 \text{ volts})(36 \text{ amps})(5.308)}$$

$$= 41\%$$

A.87 Pump Speed-Performance Relationships

Changing the velocity of a centrifugal pump will change its operating characteristics. If the speed of a pump is changed, the flow, head developed, and power requirements will change. The operating characteristics of the pump will change with speed approximately as follows:

$$\text{Flow, } Q_n = \left[\frac{N_n}{N_r}\right] Q_r$$

$$\text{Head, } H_n = \left[\frac{N_n}{N_r}\right]^2 H_r$$

$$\text{Power, } P_n = \left[\frac{N_n}{N_r}\right]^3 P_r$$

r = rated

n = now

N = pump speed

Actually, pump efficiency does vary with speed; therefore, these formulas are not quite correct. If speeds do not vary by more than a factor of two (if the speeds are not doubled or cut in half), the results are close enough. Other factors contributing to changes in pump characteristic curves include impeller wear and roughness in pipes.

EXAMPLE: To illustrate these relationships, assume a pump has a rated capacity of 600 GPM, develops 100 ft of head, and has a power requirement of 15 HP when operating at 1,500 RPM. If the efficiency remains constant, what will be the operating characteristics if the speed drops to 1,200 RPM?

Calculate new flow rate or capacity:

$$\text{Flow, } Q_n = \left[\frac{N_n}{N_r}\right] Q_r$$

$$= \left[\frac{1,200 \text{ RPM}}{1,500 \text{ RPM}}\right](600 \text{ GPM})$$

$$= \left(\frac{4}{5}\right)(600 \text{ GPM})$$

$$= (4)(120 \text{ GPM})$$

$$= 480 \text{ GPM}$$

Calculate new head:

$$\text{Head, } H_n = \left[\frac{N_n}{N_r}\right]^2 H_r$$

$$= \left[\frac{1,200 \text{ RPM}}{1,500 \text{ RPM}}\right]^2 (100 \text{ ft})$$

$$= \left(\frac{4}{5}\right)^2 (100 \text{ ft})$$

$$= \left(\frac{16}{25}\right)(100 \text{ ft})$$

$$= (16)(4 \text{ ft})$$

$$= 64 \text{ ft}$$

Calculate new power requirement:

$$\text{Power, } P_n = \left[\frac{N_n}{N_r}\right]^3 P_r$$

$$= \left[\frac{1,200 \text{ RPM}}{1,500 \text{ RPM}}\right]^3 (15 \text{ HP})$$

$$= \left(\frac{4}{5}\right)^3 (15 \text{ HP})$$

$$= \left(\frac{64}{125}\right)(15 \text{ HP})$$

$$= \left(\frac{64}{25}\right)(3 \text{ HP})$$

$$= 7.7 \text{ HP}$$

A.88 Friction or Energy Losses

Whenever water flows through pipes, valves, and fittings, energy is lost due to pipe friction (resistance), friction in valves and fittings, and the turbulence resulting from the flowing water changing its direction. Figure A.1 can be used to convert the friction losses through valves and fittings to lengths of straight pipe that would produce the same amount of friction losses. To estimate the friction or energy losses resulting from water flowing in a pipe system, we need to know:

1. Water flow rate,

2. Pipe size or diameter and length, and

3. Number, size, and type of valve fittings.

An easy way to estimate friction or energy losses is to follow these steps:

1. Determine the flow rate,

2. Determine the diameter and length of pipe,

3. Convert all valves and fittings to equivalent lengths of straight pipe (see Figure A.1),

4. Add up total length of equivalent straight pipe, and

5. Estimate friction or energy losses by using Figure A.2. With the flow in GPM and diameter of pipe, find the friction loss per 100 feet of pipe. Multiply this value by equivalent length of straight pipe.

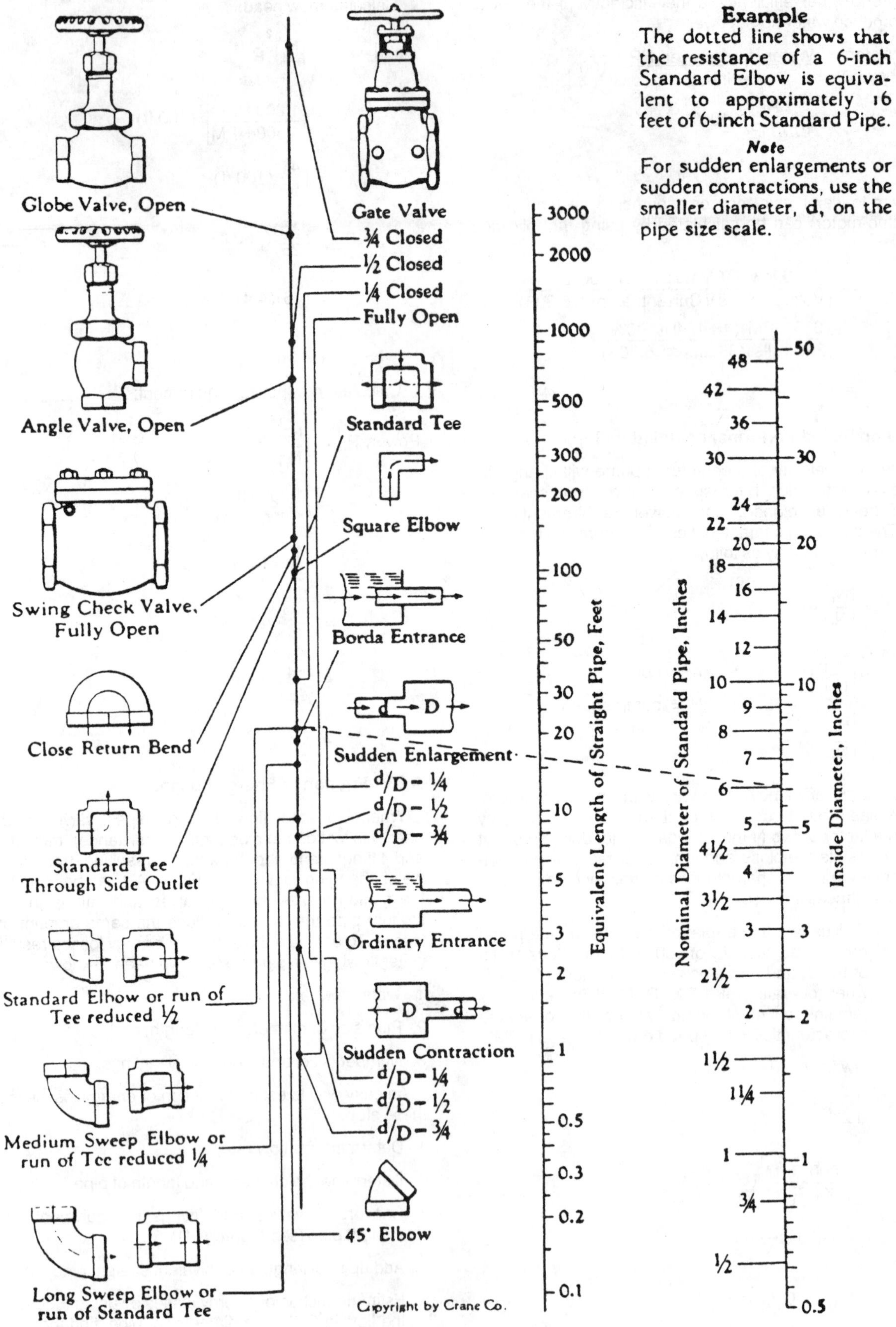

Globe Valve, Open

Angle Valve, Open

Swing Check Valve, Fully Open

Close Return Bend

Standard Tee Through Side Outlet

Standard Elbow or run of Tee reduced ½

Medium Sweep Elbow or run of Tee reduced ¼

Long Sweep Elbow or run of Standard Tee

Gate Valve
- ¾ Closed
- ½ Closed
- ¼ Closed
- Fully Open

Standard Tee

Square Elbow

Borda Entrance

Sudden Enlargement
- d/D − ¼
- d/D − ½
- d/D − ¾

Ordinary Entrance

Sudden Contraction
- d/D − ¼
- d/D − ½
- d/D − ¾

45° Elbow

Example
The dotted line shows that the resistance of a 6-inch Standard Elbow is equivalent to approximately 16 feet of 6-inch Standard Pipe.

Note
For sudden enlargements or sudden contractions, use the smaller diameter, d, on the pipe size scale.

Equivalent Length of Straight Pipe, Feet

3000
2000
1000
500
300
200
100
50
30
20
10
5
3
2
1
0.5
0.3
0.2
0.1

Nominal Diameter of Standard Pipe, Inches

48
42
36
30
24
22
20
18
16
14
12
10
9
8
7
6
5
4½
4
3½
3
2½
2
1½
1¼
1
¾
½

Inside Diameter, Inches

50
30
20
10
5
3
2
1
0.5

Copyright by Crane Co.

Fig. A.1   *Resistance of valves and fittings to flow of water*
(Reprinted by permission of Crane Co.)

U.S. GPM	0.5 in. Vel.	0.5 in. Frict.	0.75 in. Vel.	0.75 in. Frict.	1 in. Vel.	1 in. Frict.	1.25 in. Vel.	1.25 in. Frict.	1.5 in. Vel.	1.5 in. Frict.	2 in. Vel.	2 in. Frict.	2.5 in. Vel.	2.5 in. Frict.
10	10.56	95.9	6.02	23.0	3.71	6.86	2.15	1.77	1.58	.83	.96	.25	.67	.11
20	......	......	12.0	86.1	7.42	25.1	4.29	6.34	3.15	2.94	1.91	.87	1.34	.36
30					11.1	54.6	6.44	13.6	4.73	6.26	2.87	1.82	2.01	.75
40					14.8	95.0	8.58	23.5	6.30	10.79	3.82	3.10	2.68	1.28
50							10.7	36.0	7.88	16.4	4.78	4.67	3.35	1.94
60							12.9	51.0	9.46	23.2	5.74	6.59	4.02	2.72
70							15.0	68.8	11.03	31.3	6.69	8.86	4.69	3.63
80							17.2	89.2	12.6	40.5	7.65	11.4	5.36	4.66
90									14.2	51.0	8.60	14.2	6.03	5.82
100									15.8	62.2	9.56	17.4	6.70	7.11
120									18.9	88.3	11.5	24.7	8.04	10.0
140											13.4	33.2	9.38	13.5
160											15.3	43.0	10.7	17.4
180											17.2	54.1	12.1	21.9
200											19.1	66.3	13.4	26.7
220											21.0	80.0	14.7	32.2
240											22.9	95.0	16.1	38.1
260													17.4	44.5
280													18.8	51.3
300													20.1	58.5
350													23.5	79.2

U.S. GPM	3 in. Vel.	3 in. Frict.	4 in. Vel.	4 in. Frict.	5 in. Vel.	5 in. Frict.	6 in. Vel.	6 in. Frict.	8 in. Vel.	8 in. Frict.	10 in. Vel.	10 in. Frict.	12 in. Vel.	12 in. Frict.	14 in. Vel.	14 in. Frict.	16 in. Vel.	16 in. Frict.	18 in. Vel.	18 in. Frict.	20 in. Vel.	20 in. Frict.
20	.91	.15																				
40	1.82	.55	1.02	.13																		
50	2.72	1.17	1.53	.28	.96	.08																
80	3.63	2.02	2.04	.48	1.28	.14	.91	.06														
100	4.54	3.10	2.55	.73	1.60	.20	1.13	.10														
120	5.45	4.40	3.06	1.03	1.92	.29	1.36	.13														
140	6.35	5.93	3.57	1.38	2.25	.38	1.59	.18														
160	7.26	7.71	4.08	1.78	2.57	.49	1.82	.23														
180	8.17	9.73	4.60	2.24	2.89	.61	2.04	.28														
200	9.08	11.9	5.11	2.74	3.21	.74	2.27	.35														
220	9.98	14.3	5.62	3.28	3.53	.88	2.50	.42	1.40	.10												
240	10.9	17.0	6.13	3.88	3.85	1.04	2.72	.49	1.53	.12												
260	11.8	19.8	6.64	4.54	4.17	1.20	2.95	.57	1.66	.14												
280	12.7	22.8	7.15	5.25	4.49	1.38	3.18	.66	1.79	.16												
300	13.6	26.1	7.66	6.03	4.81	1.58	3.40	.75	1.91	.18												
350			8.94	8.22	5.61	2.11	3.97	1.01	2.24	.24												
400			10.20	10.7	6.41	2.72	4.54	1.30	2.55	.30												
460			11.45	13.4	7.22	3.41	5.11	1.64	2.87	.38	1.84	.12										
500			12.8	16.6	8.02	4.16	5.67	2.02	3.19	.46	2.04	.15	1.42	.06								
550			14.0	19.9	8.82	4.98	6.24	2.42	3.51	.56	2.25	.18	1.56	.07								
600					9.62	5.88	6.81	2.84	3.83	.66	2.45	.21	1.70	.08	1.25	.04						
700					11.2	7.93	7.94	3.87	4.47	.88	2.86	.29	1.99	.12	1.46	.05						
800					12.8	10.22	9.08	5.06	5.11	1.14	3.27	.37	2.27	.15	1.67	.07						
900					14.4	12.9	10.2	6.34	5.74	1.44	3.68	.46	2.55	.18	1.88	.09						
1000							11.3	7.73	6.38	1.76	4.09	.57	2.84	.22	2.08	.11						
1100							12.5	9.80	7.02	2.14	4.49	.68	3.12	.27	2.29	.13						
1200							13.6	11.2	7.66	2.53	4.90	.81	3.40	.32	2.50	.15	1.91	.08				
1300							14.7	13.0	8.30	2.94	5.31	.95	3.69	.37	2.71	.17	2.07	.09				
1400									8.93	3.40	5.72	1.09	3.97	.43	2.92	.20	2.23	.10				
1500									9.57	3.91	6.13	1.25	4.26	.49	3.13	.23	2.34	.12				
1600									10.2	4.45	6.54	1.42	4.54	.55	3.33	.25	2.55	.13	2.02	.07		
1700									10.8	5.00	6.94	1.60	4.87	.62	3.54	.29	2.71	.15	2.15	.08		
1800									11.5	5.58	7.35	1.78	5.11	.70	3.75	.32	2.87	.16	2.27	.09		
1900									12.1	6.19	7.76	1.97	5.39	.77	3.96	.35	3.03	.18	2.40	.10		
2000									12.8	6.84	8.17	2.17	5.67	.86	4.17	.39	3.19	.20	2.52	.11		
2500											10.2	3.38	7.10	1.33	5.21	.60	3.99	.31	3.15	.17		
3000											12.3	4.79	8.51	1.88	6.25	.86	4.79	.44	3.78	.24	3.06	.14
3500											14.3	6.55	9.93	2.56	7.29	1.16	5.58	.58	4.41	.32	3.57	.19
4000													11.3	3.31	8.34	1.50	6.38	.75	5.04	.42	4.08	.24
4500													12.8	4.18	9.38	1.88	7.18	.95	5.67	.53	4.59	.31
5000													14.7	5.13	10.4	2.30	7.98	1.17	6.30	.65	5.11	.38
6000															12.5	3.31	9.57	1.66	7.56	.92	6.13	.53
7000															14.6	4.50	11.2	2.26	8.83	1.24	7.15	.72
8000																	12.8	2.96	10.09	1.61	8.17	.94
9000																	14.4	3.73	11.3	2.02	9.19	1.18
10000																			12.6	2.48	10.2	1.45

> No allowance has been made for age, differences in diameter, or any other abnormal condition of interior surface. Any Factor of Safety must be estimated from the local conditions and the requirements of each particular installation. For general purposes, 15% is a responsible Factor of Safety.

Fig. A.2 Friction loss for water in feet per 100 feet of pipe

(Reprinted from the 10th Edition of the Standards of the Hydraulic Institute,
122 East 42nd Street, New York)

The procedure for using Figure A.1 is very easy. Locate the type of valve or fitting you wish to convert to an equivalent pipe length; find its diameter on the right-hand scale; and draw a straight line between these two points to locate the equivalent length of straight pipe.

EXAMPLE: Estimate the friction losses in the piping system of a pump station when the flow is 1,000 GPM. The 8-inch suction line is 10 feet long and contains a 90-degree bend (long sweep elbow), a gate valve and an 8-inch by 6-inch reducer at the inlet to the pump. The 6-inch discharge line is 30 feet long and contains a check valve, a gate valve, and three 90-degree bends (medium sweep elbows):

SUCTION LINE (8-inch diameter)

Item	Equivalent Length, ft
1. Length of pipe	10
2. 90-degree bend	14
3. Gate valve	4
4. 8-inch by 6-inch reducer	3
5. Ordinary entrance	12
Total equivalent length	43 feet

Friction loss (Figure A.2) = 1.76 ft/100 ft of pipe

DISCHARGE LINE (6-inch diameter)

Item	Equivalent Length, ft
1. Length of pipe	30
2. Check valve	38
3. Gate valve	4
4. Three 90-degree bends (3)(14)	42
Total equivalent length	114 feet

Friction loss (Figure A.2) = 7.73 ft/100 ft of pipe

Estimate the total friction losses in pumping system for a flow of 1,000 GPM.

SUCTION

Loss = (1.76 ft/100 ft)(43 ft) = 0.8 ft

DISCHARGE

Loss = (7.73 ft/100 ft)(114 ft) = 8.8 ft

Total Friction Losses, ft = 9.6 ft

A.9   STEPS IN SOLVING PROBLEMS

A.90   Identification of Problem

To solve any problem, you have to identify the problem, determine what kind of answer is needed, and collect the information needed to solve the problem. A good approach to this type of problem is to examine the problem and make a list of *KNOWN* and *UNKNOWN* information.

EXAMPLE: Find the theoretical detention time in a rectangular sedimentation tank 8 feet deep, 30 feet wide, and 60 feet long when the flow is 1.4 MGD.

Known	Unknown
Depth, ft = 8 ft	Detention Time, hours
Width, ft = 30 ft	
Length, ft = 60 ft	
Flow, MGD = 1.4 MGD	

Sometimes a drawing or sketch will help to illustrate a problem and indicate the knowns, unknowns, and possibly additional information needed.

A.91   Selection of Formula

Most problems involving mathematics in water distribution system operation can be solved by selecting the proper formula, inserting the known information, and calculating the unknown. In our example, we could look in a book containing a chapter on sedimentation, such as *WATER TREATMENT PLANT OPERATION*, Chapter 5, to find a formula for calculating detention time.

$$\text{Detention Time, hr} = \frac{(\text{Tank Volume, cu ft})(7.48 \text{ gal/cu ft})(24 \text{ hr/day})}{\text{Flow, gal/day}}$$

To convert the known information to fit the terms in a formula sometimes requires extra calculations. The next step is to find the values of any terms in the formula that are not in the list of known values.

Flow, gal/day = 1.4 MGD

= 1,400,000 gal/day

From Section A.30:

Tank Volume, cu ft = (Length, ft)(Width, ft)(Height, ft)

= 60 ft × 30 ft × 8 ft

= 14,400 cu ft

Solution of Problem:

$$\text{Detention Time, hr} = \frac{(\text{Tank Volume, cu ft})(7.48 \text{ gal/cu ft})(24 \text{ hr/day})}{\text{Flow, gal/day}}$$

$$= \frac{(14,400 \text{ cu ft})(7.48 \text{ gal/cu ft})(24 \text{ hr/day})}{1,400,000 \text{ gal/day}}$$

= 1.85 hr

The remainder of this section discusses the details that must be considered in solving this problem.

A.92   Arrangement of Formula

Once the proper formula is selected, you may have to rearrange the terms to solve for the unknown term. From Section A.71, Flow Rate," we can develop the formula:

$$\text{Velocity, ft/sec} = \frac{\text{Flow Rate, cu ft/sec}}{\text{Cross-Sectional Area, sq ft}}$$

or $V = \dfrac{Q}{A}$

In this equation if Q and A were given, the equation could be solved for V. If V and A were known, the equation would have to be rearranged to solve for Q. To move terms from one side of an equation to another, use the following rule:

When moving a term or number from one side of an equation to the other, move the numerator (top) of one side to the denominator (bottom) of the other; or from the denominator (bottom) of one side to the numerator (top) of the other.

$$V = \frac{Q}{A} \text{ or } Q = AV \text{ or } A = \frac{Q}{V}$$

If the volume of a sedimentation tank and the desired detention time were given, the detention time formula could be rearranged to calculate the design flow.

$$\frac{\text{Detention}}{\text{Time, hr}} = \frac{(\text{Tank Vol, cu ft})(7.48 \text{ gal/cu ft})(24 \text{ hr/day})}{\text{Flow, gal/day}}$$

By rearranging the terms,

$$\frac{\text{Flow,}}{\text{gal/day}} = \frac{(\text{Tank Vol, cu ft})(7.48 \text{ gal/cu ft})(24 \text{ hr/day})}{\text{Detention Time, hr}}$$

A.93 Unit Conversions

Each term in a formula or mathematical calculation must be of the correct units. The area of a rectangular clarifier (Area, sq ft = Length, ft × Width, ft) can't be calculated in square feet if the width is given as 246 inches or 20 feet 6 inches. The width must be converted to 20.5 feet. In the example problem, if the tank volume were given in gallons, the 7.48 gal/cu ft would not be needed. *THE UNITS IN A FORMULA MUST ALWAYS BE CHECKED BEFORE ANY CALCULATIONS ARE PERFORMED TO AVOID TIME-CONSUMING MISTAKES.*

$$\frac{\text{Detention}}{\text{Time, hr}} = \frac{(\text{Tank Volume, cu ft})(7.48 \text{ gal/cu ft})(24 \text{ hr/day})}{\text{Flow, gal/day}}$$

$$= \frac{\cancel{\text{cu ft}}}{} \times \frac{\text{gal}}{\cancel{\text{cu ft}}} \times \frac{\text{hr}}{\cancel{\text{day}}} \times \frac{\cancel{\text{day}}}{\cancel{\text{gal}}}$$

$$= \text{hr (all other units cancel)}$$

NOTE: We have hours = hr. One should note that the hour unit on both sides of the equation can be cancelled out and nothing would remain. This is one more check that we have the correct units. By rearranging the detention time formula, other unknowns could be determined.

If the design detention time and design flow were known, the required capacity of the tank could be calculated.

$$\frac{\text{Tank Volume,}}{\text{cu ft}} = \frac{(\text{Detention Time, hr})(\text{Flow, gal/day})}{(7.48 \text{ gal/cu ft})(24 \text{ hr/day})}$$

If the tank volume and design detention time were known, the design flow could be calculated.

$$\frac{\text{Flow,}}{\text{gal/day}} = \frac{(\text{Tank Volume, cu ft})(7.48 \text{ gal/cu ft})(24 \text{ hr/day})}{\text{Detention Time, hr}}$$

Rearrangement of the detention time formula to find other unknowns illustrates the need to always use the correct units.

A.94 Calculations

Sections A.12, "Multiplication," and A.13 "Division," outline the steps to follow in mathematical calculations. In general, do the calculations inside parentheses ( ) first and brackets [ ] next. Calculations should be done left to right above and below the division line before dividing.

$$\frac{\text{Detention}}{\text{Time, hr}} = \frac{[(\text{Tank Volume, cu ft})(7.48 \text{ gal/cu ft})(24 \text{ hr/day})]}{\text{Flow, gal/day}}$$

$$= \frac{[(14,400 \text{ cu ft})(7.48 \text{ gal/cu ft})(24 \text{ hr/day})]}{1,400,000 \text{ gal/day}}$$

$$= \frac{2,585,088 \text{ gal-hr/day}}{1,400,000 \text{ gal/day}}$$

$$= 1.85, \text{ or}$$

$$= 1.9 \text{ hr}$$

A.95 Significant Figures

In calculating the detention time in the previous section, the answer is given as 1.9 hr. The answer could have been calculated:

$$\frac{\text{Detention}}{\text{Time, hr}} = \frac{2,585,088 \text{ gal-hr/day}}{1,400,000 \text{ gal/day}}$$

$$= 1.846491429...\text{hours}$$

How does one know when to stop dividing? Common sense and significant figures both help.

First, consider the meaning of detention time and the measurements that were taken to determine the knowns in the formula. Detention time in a tank is a theoretical value and assumes that all particles of water throughout the tank move through the tank at the same velocity. This assumption is not correct; therefore, detention time can only be a representative time for some of the water particles.

Will the flow of 1.4 MGD be constant throughout the 1.9 hours, and is the flow exactly 1.4 MGD, or could it be 1.35 MGD or 1.428 MGD? A carefully calibrated flowmeter may give a reading within 2 percent of the actual flow rate. Flows into a tank fluctuate and flowmeters do not measure flows extremely accurately; so the detention time again appears to be a representative or typical detention time.

Tank dimensions are probably satisfactory within 0.1 ft. A flowmeter reading of 1.4 MGD is less precise and it could be 1.3 or 1.5 MGD. A 0.1 MGD flowmeter error when the flow is 1.4 MGD is (0.1/1.4) × 100% = 7 percent error. A detention time of 1.9 hours, based on a flowmeter reading error of plus or minus 7 percent, also could have the same error or more, even if the flow was constant. Therefore, the detention time error could be 1.9 hours × 0.07 = ±0.13 hour.

In most of the calculations in the operation of water distribution systems, the operator uses measurements determined in the lab or read from charts, scales, or meters. The accuracy of every measurement depends on the sample being measured, the equipment doing the measuring, and the operator reading or measuring the results. Your estimate is no better than the least precise measurement. Do not retain more than one doubtful number.

To determine how many figures or numbers mean anything in an answer, the approach called "significant figures" is used. In the example the flow was given in two significant figures (1.4 MGD), and the tank dimensions could be considered accurate to the nearest tenth of a foot (depth = 9.0 ft) or two significant figures. Since all measurements and the constants contained two significant figures, the results should be reported as two significant figures or 1.9 hours. The calculations are normally carried out to three significant figures (1.85 hours) and rounded off to two significant figures (1.9 hours).

Decimal points require special attention when determining the number of significant figures in a measurement.

Measurement	Significant Figures
0.00325	3
11.078	5
21,000.	2

EXAMPLE: The distance between two points was divided into three sections, and each section was measured by a different group. What is the distance between the two points if each group reported the distance it measured as follows?

Group	Distance, ft	Significant Figures
A	11,300.	3
B	2,438.9	5
C	87.62	4
Total Distance	13,826.52	

Group A reported the length of the section it measured to three significant figures; therefore, the distance between the two points should be reported as 13,800 feet (3 significant figures).

When adding, subtracting, multiplying, or dividing, the number of significant figures in the answer should not be more than the term in the calculations with the least number of significant figures.

A.96 Check Your Results

After completing your calculations, you should carefully examine your calculations and answer. Does the answer seem reasonable? If possible, have another operator check your calculations before making any operational changes.

A.10   BASIC CONVERSION FACTORS (ENGLISH SYSTEM)

UNITS
1,000,000	= 1 Million	1,000,000/1 Million

LENGTH
12 in	= 1 ft	12 in/ft
3 ft	= 1 yd	3 ft/yd
5,280 ft	= 1 mi	5,280 ft/mi

AREA
144 sq in	= 1 sq ft	144 sq in/sq ft
43,560 sq ft	= 1 acre	43,560 sq ft/ac

VOLUME
7.48 gal	= 1 cu ft	7.48 gal/cu ft
1,000 mL	= 1 liter	1,000 mL/L
3.785 L	= 1 gal	3.785 L/gal
231 cu in	= 1 gal	231 cu in/gal

WEIGHT
1,000 mg	= 1 gm	1,000 mg/gm
1,000 gm	= 1 kg	1,000 gm/kg
454 gm	= 1 lb	454 gm/lb
2.2 lbs	= 1 kg	2.2 lbs/kg

POWER
0.746 kW	= 1 HP	0.746 kW/HP

DENSITY
8.34 lbs	= 1 gal	8.34 lbs/gal
62.4 lbs	= 1 cu ft	62.4 lbs/cu ft

DOSAGE
17.1 mg/L	= 1 grain/gal	17.1 mg/L/gpg
64.7 mg	= 1 grain	64.7 mg/grain

PRESSURE
2.31 ft water	= 1 psi	2.31 ft water/psi
0.433 psi	= 1 ft water	0.433 psi/ft water
1.133 ft water	= 1 in Mercury	1.133 ft water/in Mercury

FLOW
694 GPM	= 1 MGD	694 GPM/MGD
1.55 CFS	= 1 MGD	1.55 CFS/MGD

TIME
60 sec	= 1 min	60 sec/min
60 min	= 1 hr	60 min/hr
24 hr	= 1 day	24 hr/day

NOTE: In our equations the values in the right-hand column may be written either as 24 hr/day or 1 day/24 hours depending on which units we wish to convert to obtain our desired results.

A.11   BASIC FORMULAS

FLOWS

1. $\text{Flow, MGD} = \dfrac{(\text{Flow, GPM})(60 \text{ min/hr})(24 \text{ hr/day})}{1,000,000/M}$

   or

   $\text{Flow, GPM} = \dfrac{(\text{Flow, MGD})(1,000,000/M)}{(60 \text{ min/hr})(24 \text{ hr/day})}$

CHEMICAL DOSES

2. Chemical Feed, lbs/day = (Flow, MGD)(Dose, mg/L)(8.34 lbs/gal)

Calibration of a Dry Chemical Feeder

3. Chemical Feed, lbs/day = $\dfrac{\text{Chemical Applied, lbs}}{\text{Length of Application, day}}$

Calibration of a Solution Chemical Feeder
(Chemical Feed Pump or a Hypochlorinator)

4. Chemical Feed, lbs/day = $\dfrac{(\text{Chem Conc, mg/L})(\text{Vol Pumped, m}L)(60\ \text{min/hr})(24\ \text{hr/day})}{(\text{Time Pumped, min})(1{,}000\ \text{m}L/L)(1{,}000\ \text{mg/gm})(454\ \text{gm/lb})}$

DISTRIBUTION SYSTEM FACILITIES

5. Pressure Head, ft = (Pressure, psi)(2.31 ft/psi)

6. Pressure, psi = $\dfrac{\text{Pressure Head, ft}}{2.31\ \text{ft/psi}}$

   or

   Pressure, psi = (Pressure Head, ft)(0.433 psi/ft)

7. Flow, cu ft/sec = (Area, sq ft)(Velocity, ft/sec)

8. Actual Leakage, GPD/mi-in = $\dfrac{\text{Leak Rate, GPD}}{(\text{Length, mi})(\text{Diameter, in})}$

9. Pressure, lbs/sq ft = (Density of Water, lbs/cu ft)(Height, ft)

9a. Uplift Force, lbs = (Area, sq ft)(Pressure, lbs/sq ft)

10. Slope = $\dfrac{\text{Energy Loss, ft}}{\text{Distance, ft}}$

11. C Factor = $\dfrac{\text{Flow, GPM}}{193.75(\text{Diameter, ft})^{2.63}(\text{Slope})^{0.54}}$

DISTRIBUTION SYSTEM OPERATION AND MAINTENANCE

12. Average = $\dfrac{\text{Sum of Measurements}}{\text{Number of Measurements}}$

13. Annual Running Average = $\dfrac{\text{Sum of All Averages}}{\text{Number of Averages}}$

14. Flow, CFS = (Area, sq ft)(Velocity, ft/sec)

14a. Flow, GPM = (Flow, cu ft/sec)(7.48 gal/cu ft)(60 sec/min)

15. Volume, gal = (Area, sq ft)(Depth, ft)(7.48 gal/cu ft)

15a. Flow, GPM = $\dfrac{\text{Volume, gal}}{\text{Time, min}}$

16. Meter Accuracy, % = $\dfrac{(\text{Meter Reading, GPM})(100\%)}{\text{Actual Volume, GPM}}$

17. Meter Accuracy, % = $\dfrac{(\text{Meter Reading, gal})(100\%)}{\text{Actual Volume, gal}}$

18. Volume, cu ft = (Length, ft)(Width, ft)(Depth, ft)

18a. Volume, cu yd = $\dfrac{\text{Volume, cu ft}}{27\ \text{cu ft/ cu yd}}$

19. Flow Volume, gal = (Flow Rate, gal/min)(Time, min)

20. Time to Fill, min = $\dfrac{\text{Tank Volume, gal}}{\text{Flow Rate, gal/min}}$

21. Volume, gal/ft = (Area, sq ft)(7.48 gal/cu ft)

21a. Volume, gal = (Volume, gal/ft)(Depth, ft)

22. Flow, GPM = $\dfrac{2.83(\text{Diameter, in})^{2}(\text{Distance, in})}{\sqrt{\text{Height, in}}}$

DISINFECTION

23. Chlorine, lbs = (Volume, M Gal)(Dose, mg/L)(8.34 lbs/gal)

23a. Sodium Hypochlorite Solution, gal = $\dfrac{(\text{Chlorine, lbs})(100\%)}{(8.34\ \text{lbs/gal})(\text{Hypochlorite, \%})}$

24. To disinfect a main use the same formulas as in 23 to disinfect a well.

25. To disinfect a storage tank use the same formulas as in 23 to disinfect a well.

26. Actual Dose, mg/L = $\dfrac{\text{Chlorine, lbs}}{(\text{Volume of Water, M Gal})(8.34\ \text{lbs/gal})}$

27. Chlorine Feed, lbs/day = (Flow, MGD)(Dose, mg/L)(8.34 lbs/gal)

28. Chlorine Demand, mg/L = Chlorine Dose, mg/L − Chlorine Residual, mg/L

29. Hypochlorite, gal = (Container Area, sq ft)(Drawdown, ft)(7.48 gal/cu ft)

30. Hypochlorite Strength, % = $\dfrac{(\text{Chlorine Required, lbs/day})(100\%)}{(\text{Hypochlorinator Flow, gal/day})(8.34\ \text{lbs/gal})}$

31. Water Added, gal (to Hypochlorite Solution) = $\dfrac{(\text{Hypo, gal})(\text{Hypo, \%}) - (\text{Hypo, gal})(\text{Des Hypo, \%})}{\text{Desired Hypo, \%}}$

A.12   HOW TO USE THE BASIC FORMULAS

One clever way of using the basic formulas is to use the Davidson* Pie Method. To apply this method to the basic formula for chemical doses,

1. Chemical
    Feed,   = (Flow, MGD)(Dose, mg/L)(8.34 lbs/gal)
    lbs/day

    (a) Draw a circle and draw a horizontal line through the middle of the circle;

    (b) Write the Chemical Feed, lbs/day in the top half;

    (c) Divide the bottom half into three parts; and

    (d) Write Flow, MGD; Dose, mg/L; and 8.34 lbs/gal in the other three parts.

    (e) The line across the middle of the circle represents the line of the equation. Items above the line stay above the line and those below the line stay below the line.

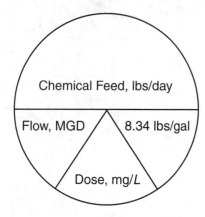

---

* Gerald Davidson, Manager, Clear Lake Oaks Water District, Clear Lake Oaks, California.

If you want to find the Chemical Feed, lbs/day, cover up the Chemical Feed, lbs/day, and what is left uncovered will give you the correct formula.

2. Chemical
    Feed,   = (Flow, MGD)(Dose, mg/L)(8.34 lbs/gal)
    lbs/day

If you know the chlorinator setting in pounds per day and the flow in MGD and would like to know the dose in mg/L, cover up the Dose, mg/L, and what is left uncovered will give you the correct formula.

3. Dose, mg/L $= \dfrac{\text{Chemical Feed, lbs/day}}{(\text{Flow, MGD})(8.34 \text{ lbs/gal})}$

Another approach to using the basic formulas is to memorize the basic formula, for example the detention time formula.

4. Detention Time, hr $= \dfrac{(\text{Tank Volume, gal})(24 \text{ hr/day})}{\text{Flow, gal/day}}$

This formula works fine to solve for the detention time when the Tank Volume, gal, and Flow, gal/day, are given.

If you wish to determine the Flow, gal/day, when the Detention Time, hr, and Tank Volume, gal, are given, you must change the basic formula. You want the Flow, gal/day, on the left of the equal sign and everything else on the right of the equal sign. This is done by moving the terms diagonally (from top to bottom or from bottom to top) past the equal sign.

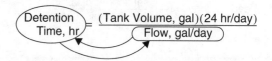

or

Flow, gal/day $= \dfrac{(\text{Tank Volume, gal})(24 \text{ hr/day})}{\text{Detention Time, hr}}$

This same approach can be used if the Tank Volume, gal, was unknown and the Detention Time, hr, and Flow, gal/day, were given. We want Tank Volume, gal, on one side of the equation and everything else on the other side.

Detention
Time, hr $= \dfrac{(\text{Tank Volume, gal})(24 \text{ hr/day})}{\text{Flow, gal/day}}$
      To Top
                                    To Bottom

or

$\dfrac{(\text{Detention Time, hr})(\text{Flow, gal/day})}{24 \text{ hr/day}} = \text{Tank Volume, gal}$

or

Tank Volume, gal $= \dfrac{(\text{Detention Time, hr})(\text{Flow, gal/day})}{24 \text{ hr/day}}$

One more check is to be sure the units in the rearranged formula cancel out correctly.

For additional information on the use of the basic formulas, refer to Sections:

A.91, "Selection of Formula,"

A.92, "Arrangement of Formula,"

A.93, "Unit Conversions," and

A.94, "Calculations."

A.13   TYPICAL WATER DISTRIBUTION SYSTEM PROBLEMS (ENGLISH SYSTEM)

A.130   Flows

EXAMPLE 1

Convert a flow of 450 gallons per minute to million gallons per day.

Known	Unknown
Flow, GPM = 450 GPM	Flow, MGD

Convert flow from 450 GPM to MGD.

$$\text{Flow, MGD} = \frac{(\text{Flow, GPM})(60 \text{ min/hr})(24 \text{ hr/day})}{1{,}000{,}000/M}$$

$$= \frac{(450 \text{ GPM})(60 \text{ min/hr})(24 \text{ hr/day})}{1{,}000{,}000/M}$$

$$= 0.648 \text{ MGD}$$

A.131   Chemical Doses

We use the "lbs/day formula" regularly in our work to calculate the setting on a chemical feeder (lbs chlorine per day) and the loading on a treatment process (lbs BOD per day). An explanation of how the units in the formula cancel helps us understand how to use and apply the formula.

How the formula works can be explained by an example where we want to calculate the setting on a chemical feeder to treat a given flow at a required chemical dose. For example, calculate the setting on a chlorinator in *POUNDS OF CHLORINE PER DAY* to the flow of two million gallons per day (2 MGD) at a chlorine dose of 5 milligrams of chlorine per liter of water (5 mg/L). To perform this calculation we need to realize that one liter of water weighs one million milligrams. Therefore,

$$\frac{\text{mg}}{L} = \frac{\text{mg}}{1 \text{ Million mg}} = \frac{\text{lbs}}{1 \text{ Million lbs}}$$

Calculate the chlorinator setting in pounds of chlorine per day.

$$\text{lbs Cl/day} = (\text{Flow, MGD})(\text{Conc, mg/}L)(8.34 \text{ lbs/gal})$$

$$\text{or} \qquad = (\text{Flow, MGD})(\text{Conc, lbs/M lbs})(8.34 \text{ lbs/gal})$$

$$= \frac{2 \text{ M gal } H_2O}{\text{day}} \times \frac{5 \text{ lbs Cl}}{1 \text{ M lbs } H_2O} \times \frac{8.34 \text{ lbs } H_2O}{\text{gal } H_2O}$$

$$= 83.4 \text{ lbs Cl/day}$$

In this formula the million (*M*) on top and bottom of the formula cancel, the gallons of water (*gal H₂O*) on top and bottom cancel and the pounds of water (*lbs H₂O*) on top and bottom cancel. This leaves us with pounds of chlorine (*lbs Cl*) on the top and day (*day*) on the bottom. The answer is the chlorinator setting in pounds of chlorine per day (*lbs Cl/day*).

EXAMPLE 2

Determine the chlorinator setting in pounds per 24 hours if a well pump delivers 300 GPM and the desired chlorine dose is 2.0 mg/L.

Known		Unknown
Flow, GPM	= 300 GPM	Chlorinator Setting, lbs/24 hours
Chlorine Dose, mg/L	= 2.0 mg/L	

1. Convert flow from gallons per minute to million gallons per day.

$$\text{Flow, MGD} = \frac{(\text{Flow, gal/min})(60 \text{ min/hr})(24 \text{ hr/day})}{1{,}000{,}000/M}$$

$$= \frac{(300 \text{ gal/min})(60 \text{ min/hr})(24 \text{ hr/day})}{1{,}000{,}000/M}$$

$$= 0.432 \text{ MGD}$$

NOTE: When we multiply in an equation by 1,000,000/M we do not change anything except the units. This is just like multiplying an equation by 12 inches/foot or 60 min/hr; all we are doing is changing units.

2. Determine the chlorinator setting in pounds per 24 hours or pounds per day.

$$\text{Chemical Feed, lbs/day} = (\text{Flow, MGD})(\text{Dose, mg/}L)(8.34 \text{ lbs/gal})$$

$$= (0.432 \text{ MGD})(2.0 \text{ mg/}L)(8.34 \text{ lbs/gal})$$

$$= 7.2 \text{ lbs/day}$$

EXAMPLE 3

Determine the actual chemical dose or chemical feed in pounds per day from a dry chemical feeder. A bucket placed under the chemical feeder weighed 0.2 pound empty and 2.6 pounds after 30 minutes.

Known		Unknown
Empty Bucket, lbs	= 0.2 lbs	Chemical Feed, lbs/day
Full Bucket, lbs	= 2.6 lbs	
Time to Fill, min	= 30 min	

Determine the chemical feed in pounds of chemical per day.

$$\text{Chemical Feed, lbs/day} = \frac{\text{Chemical Applied, lbs}}{\text{Length of Application, days}}$$

$$= \frac{(2.6 \text{ lbs} - 0.2 \text{ lbs})(60 \text{ min/hr})(24 \text{ hr/day})}{30 \text{ min}}$$

$$= 115 \text{ lbs/day}$$

EXAMPLE 4

Determine the chemical feed in pounds of chlorine per day from a hypochlorinator. The hypochlorite solution is 1.4 percent or 14,000 mg chlorine per liter. During a test run the hypochlorinator delivers 400 mL during 12 minutes.

Known		Unknown
Cl Solution, %	= 1.4%	Chemical Feed, lbs/day
Cl Conc, mg/L	= 14,000 mg/L	
Vol Pumped, mL	= 400 mL	
Time Pumped, min	= 12 min	

Calculate the chlorine fed by the hypochlorinator in pounds of chlorine per day.

$$\text{Chemical Feed, lbs/day} = \frac{(\text{Cl Conc, mg/}L)(\text{Vol Pumped, m}L)(60 \text{ min/hr})(24 \text{ hr/day})}{(\text{Time Pumped, min})(1{,}000 \text{ m}L/L)(1{,}000 \text{ mg/gm})(454 \text{ gm/lb})}$$

$$= \frac{(14{,}000 \text{ mg/}L)(400 \text{ m}L)(60 \text{ min/hr})(24 \text{ hr/day})}{(12 \text{ min})(1{,}000 \text{ m}L/L)(1{,}000 \text{ mg/gm})(454 \text{ gm/lb})}$$

$$= 1.48 \text{ lbs/day}$$

$$\text{or} \qquad = 1.5 \text{ lbs Chlorine/day}$$

A.132 Distribution System Facilities

EXAMPLE 5

A pressure gage at a fire hydrant reads 50 psi. What is the pressure head on the gage at the fire hydrant in feet?

Known	Unknown
Pressure, psi = 50 psi	Pressure Head, ft

Convert the pressure in psi to pressure head in feet.

Pressure Head, ft = (Pressure, psi)(2.31 ft/psi)

$$= (50 \text{ psi})(2.31 \text{ ft/psi})$$

$$= 115.5 \text{ ft}$$

EXAMPLE 6

A pressure gage on a fire hydrant reads 170 feet of pressure head. What is the pressure in psi?

Known	Unknown
Pressure Head, ft = 170 ft	Pressure, psi

Calculate the pressure head in pounds per square inch (psi).

$$\text{Pressure, psi} = \frac{\text{Pressure Head, ft}}{2.31 \text{ ft/psi}}$$

$$= \frac{170 \text{ ft}}{2.31 \text{ ft/psi}}$$

$$= 74 \text{ psi}$$

or

Pressure, psi = (Pressure Head, ft)(0.433 psi/ft)

$$= (170 \text{ ft})(0.433 \text{ psi/ft})$$

$$= 74 \text{ psi}$$

EXAMPLE 7

A 6-inch diameter water main needs to be flushed with a water velocity of five feet per second. What should be the flow through the main in cubic feet per second and gallons per minute?

Known	Unknown
Diameter, in = 6 in	1. Flow, cu ft/sec
Velocity, ft/sec = 5 ft/sec	2. Flow, GPM

1. Calculate the pipe area in square feet.

$$\text{Area, sq ft} = (0.785)(\text{Diameter, ft})^2$$

$$= \frac{0.785(6 \text{ in})^2}{(12 \text{ in/ft})^2}$$

$$= 0.196 \text{ sq ft}$$

2. Determine the flow in cubic feet per second.

Flow, cu ft/sec = (Area, sq ft)(Velocity, ft/sec)

$$= (0.196 \text{ sq ft})(5 \text{ ft/sec})$$

$$= 0.98 \text{ cu ft/sec}$$

3. Convert the flow from cubic feet per second to gallons per minute. To do this use the proper conversion factors to change cubic feet to gallons and also seconds to minutes.

Flow, GPM = (Flow, cu ft/sec)(7.48 gal/cu ft)(60 sec/min)

$$= (0.98 \text{ cu ft/sec})(7.48 \text{ gal/cu ft})(60 \text{ sec/min})$$

$$= 440 \text{ GPM}$$

EXAMPLE 8

Determine the actual leakage in GPD/mi-in from a new section of pipe if the leak rate during the test period was 60 GPD. The section tested is one-quarter of a mile (0.25 mi) long and the pipe diameter is 12 inches.

Known	Unknown
Leak Rate, GPD = 60 GPD	Actual Leakage, GPM/mi-in
Length, mi = 0.25 mi	
Diameter, in = 12 in	

Calculate the actual leakage from the test section in gallons per day per mile per inch of pipe diameter.

$$\frac{\text{Actual Leakage,}}{\text{GPD/mi-in}} = \frac{\text{Leak Rate, GPD}}{(\text{Length, mi})(\text{Diameter, in})}$$

$$= \frac{60 \text{ GPD}}{(0.25 \text{ mi})(12 \text{ in})}$$

$$= 20 \text{ GPD/mi-in}$$

EXAMPLE 9

An underground rectangular water storage tank must be drained and cleaned. The tank is 30 feet long and 10 feet wide. The groundwater is 5 feet above the bottom of the tank. What is the force of the water pressure lifting upward on the bottom of the tank?

Known	Unknown
Length, ft = 30 ft	Uplift Force, lbs
Width, ft = 10 ft	
Height, ft = 5 ft	

1. Calculate the area of the tank bottom.

Area, sq ft = (Length, ft)(Width, ft)

$$= (30 \text{ ft})(10 \text{ ft})$$

$$= 300 \text{ sq ft}$$

2. Calculate the pressure on the bottom of the tank in pounds per square foot (lbs/sq ft).

Pressure, lbs/sq ft = (Density of Water, lbs/cu ft)(Height, ft)

= (62.4 lbs/cu ft)(5 ft)

= 312 lbs/sq ft

3. Calculate the uplift force on the bottom of the tank.

Uplift Force, lbs = (Area, sq ft)(Pressure, lbs/sq ft)

= (300 sq ft)(312 lbs/sq ft)

= 93,600 lbs

If the tank weighs less than 93,600 lbs, the tank will float when emptied if the groundwater is five feet or higher above the bottom of the tank.

EXAMPLE 10 (Advanced Problem)

Estimate the C Factor for a 10-inch water main when the flow is 800 GPM and the drop in pressure head elevation between two pressure gages 500 feet apart is two feet.

Known		**Unknown**
Diameter, in	= 10 in	C Factor
Flow, GPM	= 800 GPM	
Energy Loss, ft	= 2 ft	
Distance, ft	= 500 ft	

1. Convert a diameter of 10 inches to feet.

$$\text{Diameter, ft} = \frac{\text{Diameter, in}}{12 \text{ in/ft}}$$

$$= \frac{10 \text{ in}}{12 \text{ in/ft}}$$

$$= 0.833 \text{ ft}$$

2. Calculate the slope.

$$\text{Slope} = \frac{\text{Energy Loss, ft}}{\text{Distance, ft}}$$

$$= \frac{2 \text{ ft}}{500 \text{ ft}}$$

$$= 0.004 \text{ ft/ft}$$

3. Calculate the C Factor.

$$\text{C Factor} = \frac{\text{Flow, GPM}}{193.75(\text{Diameter, ft})^{2.63}(\text{Slope})^{0.54}}$$

$$= \frac{800 \text{ GPM}}{193.75(0.833 \text{ ft})^{2.63}(0.004)^{0.54}}$$

$$= \frac{800 \text{ GPM}}{(193.75)(0.619)(0.0507)}$$

$$= 132$$

EXAMPLE 11 (Advanced Problem)

Estimate the C Factor for a 15-inch diameter pipe if a field test was conducted using a flow of 1,500 GPM. A pressure gage at elevation 51.0 feet at a fire hydrant read 40 psi. Another pressure gage at elevation 50.0 feet at a fire hydrant read 39.8 psi. The fire hydrants are 800 feet apart.

Known		**Unknown**
Diameter, in	= 15 in	C Factor
Flow, GPM	= 1,500 GPM	

GAGE NO. 1

Elev, ft	= 51.0 ft
Pres, psi	= 40 psi

GAGE NO. 2

Elev, ft	= 50.0 ft
Pres, psi	= 39.8 psi

Distance, ft	= 800 ft

1. Determine the elevations of the pressure heads at gages 1 and 2.

GAGE NO. 1

Elev Pres Head, ft = 51.0 ft + (40 psi)(2.31 ft/psi)

= 143.4 ft

GAGE NO. 2

Elev Pres Head, ft = 50.0 ft + (39.8 psi)(2.31 ft/psi)

= 141.9 ft

2. Determine the slope of the energy grade line.

$$\text{Slope} = \frac{\text{Difference in Elev Pres Head, ft}}{\text{Distance, ft}}$$

$$= \frac{143.4 \text{ ft} - 141.9 \text{ ft}}{800 \text{ ft}}$$

$$= \frac{1.5 \text{ ft}}{800 \text{ ft}}$$

$$= 0.001875 \text{ ft/ft}$$

3. Calculate the C Factor.

$$\text{C Factor} = \frac{\text{Flow, GPM}}{193.75(\text{Diameter, ft})^{2.63}(\text{Slope})^{0.54}}$$

$$= \frac{1,500 \text{ GPM}}{193.75(15 \text{ in}/12 \text{ in/ft})^{2.63}(0.001875)^{0.54}}$$

$$= \frac{1,500 \text{ GPM}}{(193.75)(1.798)(0.03368)}$$

$$= 128$$

A.133 Distribution System Operation and Maintenance

EXAMPLE 12

Calculate the quarterly average TTHM (Total Trihalomethane) for a water supply system in micrograms per liter based on the results of tests taken during a three-month period.

Test Number	1	2	3	4	5	6
TTHM, μg/L	140	100	80	160	150	90

Known	Unknown
Test Results Shown Above	Average TTHM, μg/L

$$\text{Avg TTHM, } \mu\text{g}/L = \frac{\text{Sum of Measurements, } \mu\text{g}/L}{\text{Number of Measurements}}$$

$$= \frac{140 + 100 + 80 + 160 + 150 + 90}{6}$$

$$= \frac{720 \ \mu\text{g}/L}{6}$$

$$= 120 \ \mu\text{g}/L$$

or

$$= 0.12 \ \text{mg}/L$$

EXAMPLE 13

Determine the annual running TTHM average using the quarterly results shown below.

Known	Unknown
Results of Quarterly TTHM Analyses	Annual Running TTHM Average, mg/L

YEAR 1

QUARTER	CONC, mg/L
1	0.07
2	0.09
3	0.12
4	0.08

YEAR 2

1	0.09
2	0.11
3	0.13
4	0.07

Calculate the annual running TTHM average for each quarter during Year 2. Use the quarterly TTHM average for the quarter being considered and the three quarters immediately before the one being considered.

YEAR 2, QUARTER 1

$$\text{Annual Running TTHM Average, mg}/L = \frac{\text{Sum of Measurements, mg}/L}{\text{Number of Measurements}}$$

$$= \frac{0.09 \text{ mg}/L + 0.08 \text{ mg}/L + 0.12 \text{ mg}/L + 0.09 \text{ mg}/L}{4}$$

$$= \frac{0.38 \text{ mg}/L}{4}$$

$$= 0.095 \text{ mg}/L$$

YEAR 2, QUARTER 2

$$\text{Annual Running TTHM Average, mg}/L = \frac{0.11 \text{ mg}/L + 0.09 \text{ mg}/L + 0.08 \text{ mg}/L + 0.12 \text{ mg}/L}{4}$$

$$= \frac{0.40 \text{ mg}/L}{4}$$

$$= 0.10 \text{ mg}/L$$

YEAR 2, QUARTER 3

$$\text{Annual Running TTHM Average, mg}/L = \frac{0.13 \text{ mg}/L + 0.11 \text{ mg}/L + 0.09 \text{ mg}/L + 0.08 \text{ mg}/L}{4}$$

$$= \frac{0.41 \text{ mg}/L}{4}$$

$$= 0.103 \text{ mg}/L$$

YEAR 2, QUARTER 4

$$\text{Annual Running TTHM Average, mg}/L = \frac{0.07 \text{ mg}/L + 0.13 \text{ mg}/L + 0.11 \text{ mg}/L + 0.09 \text{ mg}/L}{4}$$

$$= \frac{0.40 \text{ mg}/L}{4}$$

$$= 0.10 \text{ mg}/L$$

EXAMPLE 14

An 18-inch diameter water main is to be flushed at a velocity of 4 ft/sec. What should be the reading on the flowmeter in gallons per minute?

Known	Unknown
Diameter, in = 18 in	Flow, GPM
Velocity, ft/sec = 4 ft/sec	

1. Calculate the cross-sectional area of the pipe in square feet.

$$\text{Area, sq ft} = \frac{(0.785)(\text{Diameter, in})^2}{144 \text{ sq in/sq ft}}$$

$$= \frac{(0.785)(18 \text{ in})^2}{144 \text{ sq in/sq ft}}$$

$$= 1.77 \text{ sq ft}$$

2. Determine the flow in the pipe in cubic feet per second (CFS). Use Q = AV.

$$\text{Flow, CFS} = (\text{Area, sq ft})(\text{Velocity, ft/sec})$$

$$= (1.77 \text{ sq ft})(4 \text{ ft/sec})$$

$$= 7.1 \text{ CFS}$$

3. Calculate the flowmeter reading in gallons per minute (GPM).

$$\text{Flow, GPM} = (\text{Flow, cu ft/sec})(7.48 \text{ gal/cu ft})(60 \text{ sec/min})$$

$$= (7.1 \text{ cu ft/sec})(7.48 \text{ gal/cu ft})(60 \text{ sec/min})$$

$$= 3,186 \text{ GPM}$$

EXAMPLE 15

During a 20-minute time span the water in a tank three feet in diameter increases by four feet. Calculate the total flow in gallons and the flow in gallons per minute.

Known	Unknown
Time, min = 20 min	1. Volume, gal
Diameter, ft = 3 ft	2. Flow, GPM
Depth, ft = 4 ft	

1. Calculate the total flow in gallons.

$$\text{Volume, gal} = (0.785)(\text{Diameter, ft})^2(\text{Depth, ft})(7.48 \text{ gal/cu ft})$$

$$= (0.785)(3 \text{ ft})^2(4 \text{ ft})(7.48 \text{ gal/cu ft})$$

$$= 211 \text{ gal}$$

2. Calculate the flow in gallons per minute.

$$\text{Flow, GPM} = \frac{\text{Volume, gal}}{\text{Time, min}}$$

$$= \frac{211 \text{ gal}}{20 \text{ min}}$$

$$= 10.6 \text{ GPM}$$

EXAMPLE 16

A water meter is tested in a laboratory. The meter reads 1.3 GPM and the actual flow was 1.33 GPM. Determine the meter accuracy as a percentage.

Known	Unknown
Meter Reading, GPM = 1.3 GPM	Meter Accuracy, %
Actual Flow, GPM = 1.33 GPM	

Determine the meter accuracy as a percentage.

$$\text{Meter Accuracy, \%} = \frac{(\text{Meter Reading, GPM})(100\%)}{\text{Actual Flow, GPM}}$$

$$= \frac{(1.3 \text{ GPM})(100\%)}{1.33 \text{ GPM}}$$

$$= 97.7\%$$

EXAMPLE 17

A water meter is tested in a laboratory. The meter reads 260 gallons. The actual flow was measured in a volumetric measuring tank three feet in diameter. During the meter test five feet of water were added to the tank. Determine the meter accuracy as a percentage.

Known	Unknown
Meter Reading, gal = 260 gal	Meter Accuracy, %
Diameter, ft = 3 ft	
Depth, ft = 5 ft	

1. Calculate the actual volume of water that flowed through the meter in gallons.

$$\text{Volume, gal} = (0.785)(\text{Diameter, ft})^2(\text{Depth, ft})(7.48 \text{ gal/cu ft})$$

$$= (0.785)(3 \text{ ft})^2(5 \text{ ft})(7.48 \text{ gal/cu ft})$$

$$= 264 \text{ gal}$$

2. Determine the meter accuracy as a percentage.

$$\text{Meter Accuracy, \%} = \frac{(\text{Meter Reading, gal})(100\%)}{\text{Actual Volume, gal}}$$

$$= \frac{(260 \text{ gal})(100\%)}{264 \text{ gal}}$$

$$= 98.5\%$$

EXAMPLE 18

A trench four feet wide, 1,100 feet long, and six feet deep is to be excavated for a water main. How many cubic feet and also cubic yards of soil have to be excavated for the water main?

Known	Unknown
Length, ft = 1,100 ft	1. Volume Excavated, cu ft
Width, ft = 4 ft	2. Volume Excavated, cu yd
Depth, ft = 6 ft	

1. Calculate the volume excavated in cubic feet.

$$\text{Volume, cu ft} = (\text{Length, ft})(\text{Width, ft})(\text{Depth, ft})$$

$$= (1,100 \text{ ft})(4 \text{ ft})(6 \text{ ft})$$

$$= 26,400 \text{ cu ft}$$

2. Calculate the volume excavated in cubic yards.

$$\text{Volume, cu yd} = \frac{\text{Volume, cu ft}}{27 \text{ cu ft/cu yd}}$$

$$= \frac{26,400 \text{ cu ft}}{27 \text{ cu ft/cu yd}}$$

$$= 978 \text{ cu yd}$$

EXAMPLE 19

A flowmeter indicates a flow rate of seven gallons per minute. How many gallons of water will flow through the meter in three hours and 30 minutes?

Known	Unknown
Flow Rate, GPM = 7 GPM	Flow Volume, gal
Time, hr and min = 3 hr and 30 min	

1. Convert three hours and 30 minutes to hours.

$$\text{Time, hr} = 3 \text{ hr} + \frac{30 \text{ min}}{60 \text{ min/hr}}$$

$$= 3 \text{ hr} + 0.5 \text{ hr}$$

$$= 3.5 \text{ hr}$$

2. Calculate the total flow volume of water in gallons.

$$\text{Flow Volume, gal} = (\text{Flow Rate, gal/min})(\text{Time, hr})(60 \text{ min/hr})$$

$$= (7 \text{ gal/min})(3.5 \text{ hr})(60 \text{ min/hr})$$

$$= 1,470 \text{ gal}$$

EXAMPLE 20

A chemical solution tank is 4.5 feet in diameter and 5 feet deep. How long will it take to fill this tank when water flows in at a rate of 3.6 gallons per minute?

Known	Unknown
Diameter, ft = 4.5 ft	Time to Fill, min
Depth, ft = 5 ft	
Flow Rate, GPM = 3.6 GPM	

1. Calculate the tank volume in gallons.

$$\text{Tank Volume, gal} = (0.785)(\text{Diameter, ft})^2(\text{Depth, ft})(7.48 \text{ gal/cu ft})$$

$$= (0.785)(4.5 \text{ ft})^2(5 \text{ ft})(7.48 \text{ gal/cu ft})$$

$$= 595 \text{ gal}$$

2. Calculate the time to fill the tank in minutes.

$$\text{Time to Fill, min} = \frac{\text{Tank Volume, gal}}{\text{Flow Rate, gal/min}}$$

$$= \frac{595 \text{ gal}}{3.6 \text{ gal/min}}$$

$$= 165 \text{ min}$$

3. Convert 165 minutes to hours and minutes.

$$\text{Time to Fill, hr} = \frac{165 \text{ min}}{60 \text{ min/hr}}$$

$$= 2.75 \text{ hr}$$

$$= 2 \text{ hr} + (0.75 \text{ hr})(60 \text{ min/hr})$$

$$= 2 \text{ hr } 45 \text{ min}$$

EXAMPLE 21

A cylindrical water storage tank is 30 feet in diameter and 15 feet high. Calculate the gallons of water per foot of depth and then the total gallons when the water is 11.6 feet deep.

Known	Unknown
Diameter, ft = 30 ft	1. Volume, gal/ft
Height, ft = 15 ft	2. Volume, gal
Depth, ft = 11.6 ft	(for Depth = 11.6 ft)

1. Calculate the volume of water in the tank per foot of depth.

$$\text{Volume, gal} = (\text{Area, sq ft})(\text{Depth, ft})(7.48 \text{ gal/cu ft})$$

$$\text{Volume, gal/ft} = (0.785)(\text{Diameter, ft})^2(7.48 \text{ gal/cu ft})$$

$$= (0.785)(30 \text{ ft})^2(7.48 \text{ gal/cu ft})$$

$$= 5,285 \text{ gal/ft}$$

2. Calculate the volume of water in the tank in gallons when the water is 11.6 feet deep in the tank.

$$\text{Volume, gal} = (\text{Volume, gal/ft})(\text{Depth, ft})$$

$$= (5,285 \text{ gal/ft})(11.6 \text{ ft})$$

$$= 61,306 \text{ gallons}$$

EXAMPLE 22

Estimate the flow from a pipe in gallons per minute. Water is flowing from a two-inch diameter pipe. The pipe is 36 inches above the point where the water hits the ground. The water hits the ground 60 inches (horizontal distance) from where the water leaves the pipe. See Figure A.3 for a sketch of this problem.

Known	Unknown
Diameter, in = 2 in	Flow, GPM
Height, in = 36 in	
Distance, in = 60 in	

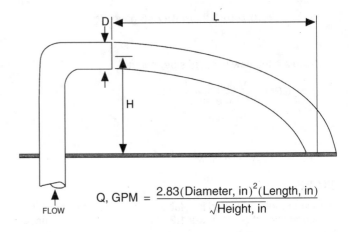

$$Q, \text{GPM} = \frac{2.83(\text{Diameter, in})^2(\text{Length, in})}{\sqrt{\text{Height, in}}}$$

Fig. A.3   Sketch of flow from a hydrant

Estimate the flow from the pipe in gallons per minute.

$$\text{Flow, GPM} = \frac{2.83(\text{Diameter, in})^2(\text{Distance, in})}{\sqrt{\text{Height, in}}}$$

$$= \frac{2.83(2 \text{ in})^2(60 \text{ in})}{\sqrt{36 \text{ in}}}$$

$$= 113 \text{ GPM}$$

A.134   Disinfection

EXAMPLE 23

How many gallons of five percent sodium hypochlorite will be needed to disinfect a well with a 15-inch diameter casing and well screen? The well is 180 feet deep and there are 75 feet of water in the well. Use an initial chlorine dose of 100 mg/L.

Known	Unknown
Hypochlorite, % = 5%	5% Hypochlorite, gal
Chlorine Dose, mg/L = 100 mg/L	
Diameter, in = 15 in	
Well Depth, ft = 180 ft	
Water Depth, ft = 75 ft	

1. Calculate the volume of water in the well in gallons.

$$\text{Water Vol,} \atop \text{gal} = \frac{(0.785)(\text{Diameter, in})^2(\text{Depth, ft})(7.48 \text{ gal/cu ft})}{144 \text{ sq in/sq ft}}$$

$$= \frac{(0.785)(15 \text{ in})^2(75 \text{ ft})(7.48 \text{ gal/cu ft})}{144 \text{ sq in/sq ft}}$$

$$= 688 \text{ gal}$$

2. Determine the pounds of chlorine needed.

$$\text{Chlorine, lbs} = (\text{Volume, M Gal})(\text{Dose, mg/}L)(8.34 \text{ lbs/gal})$$

$$= (0.000688 \text{ M Gal})(100 \text{ mg/}L)(8.34 \text{ lbs/gal})$$

$$= 0.57 \text{ lb Chlorine}$$

3. Calculate the gallons of five percent sodium hypochlorite solution needed.

$$\text{Sodium Hypochlorite} \atop \text{Solution, gal} = \frac{(\text{Chlorine, lbs})(100\%)}{(8.34 \text{ lbs/gal})(\text{Hypochlorite, \%})}$$

$$= \frac{(0.57 \text{ lb})(100\%)}{(8.34 \text{ lbs/gal})(5\%)}$$

$$= 1.37 \text{ gallons}$$

Use 1.4 gallons of five percent sodium hypochlorite to disinfect the well.

EXAMPLE 24

A new 10-inch diameter water main 650 feet long needs to be disinfected. An initial chlorine dose of 400 mg/L is expected to maintain a chlorine residual of over 300 mg/L during the three-hour disinfection period. How many gallons of 5.25 percent sodium hypochlorite solution will be needed?

Known		Unknown
Diameter of Pipe, in	= 10 in	5.25% Hypochlorite, gal
Length of Pipe, ft	= 650 ft	
Chlorine Dose, mg/L	= 400 mg/L	
Hypochlorite, %	= 5.25%	

1. Calculate the volume of water in the pipe in gallons.

$$\text{Pipe Volume,} \atop \text{gal} = \frac{(0.785)(\text{Diameter, in})^2(\text{Length, ft})(7.48 \text{ gal/cu ft})}{144 \text{ sq in/sq ft}}$$

$$= \frac{(0.785)(10 \text{ in})^2(650 \text{ ft})(7.48 \text{ gal/cu ft})}{144 \text{ sq in/sq ft}}$$

$$= 2,650 \text{ gal}$$

2. Determine the pounds of chlorine needed.

$$\text{Chlorine, lbs} = (\text{Volume, M Gal})(\text{Dose, mg/}L)(8.34 \text{ lbs/gal})$$

$$= (0.00265 \text{ M Gal})(400 \text{ mg/}L)(8.34 \text{ lbs/gal})$$

$$= 8.84 \text{ lbs Chlorine}$$

3. Calculate the gallons of 5.25 percent sodium hypochlorite solution needed.

$$\text{Sodium Hypochlorite} \atop \text{Solution, gal} = \frac{(\text{Chlorine, lbs})(100\%)}{(8.34 \text{ lbs/gal})(\text{Hypochlorite, \%})}$$

$$= \frac{(8.84 \text{ lbs})(100\%)}{(8.34 \text{ lbs/gal})(5.25\%)}$$

$$= 20.2 \text{ gallons}$$

Twenty gallons of 5.25 percent sodium hypochlorite solution should do the job.

EXAMPLE 25

A service storage reservoir has been taken out of service for inspection, maintenance, and repairs. The reservoir needs to be disinfected before being placed back on line. The reservoir is 30 feet in diameter and 8 feet deep. An initial chlorine dose of 100 mg/L is expected to maintain a chlorine residual of over 50 mg/L during the 24-hour disinfection period. How many gallons of 15 percent sodium hypochlorite solution will be needed?

Known		Unknown
Diameter, ft	= 30 ft	15% Hypochlorite, gal
Tank Depth, ft	= 8 ft	
Chlorine Dose, mg/L	= 100 mg/L	
Hypochlorite, %	= 15%	

1. Calculate the volume of water in the tank in gallons.

$$\text{Tank Volume,} \atop \text{gal} = (0.785)(\text{Diameter, ft})^2(\text{Depth, ft})(7.48 \text{ gal/cu ft})$$

$$= (0.785)(30 \text{ ft})^2(8 \text{ ft})(7.48 \text{ gal/cu ft})$$

$$= 42,277 \text{ gal}$$

2. Determine the pounds of chlorine needed.

$$\text{Chlorine, lbs} = (\text{Volume, M Gal})(\text{Dose, mg/}L)(8.34 \text{ lbs/gal})$$

$$= (0.042277 \text{ M Gal})(100 \text{ mg/}L)(8.34 \text{ lbs/gal})$$

$$= 35.3 \text{ lbs Chlorine}$$

3. Calculate the gallons of 15 percent sodium hypochlorite solution needed.

$$\text{Sodium Hypochlorite Solution, gal} = \frac{(\text{Chlorine, lbs})(100\%)}{(8.34 \text{ lbs/gal})(\text{Hypochlorite, \%})}$$

$$= \frac{(35.3 \text{ lbs})(100\%)}{(8.34 \text{ lbs/gal})(15\%)}$$

$$= 28.2 \text{ gallons}$$

Twenty-eight gallons of 15 percent sodium hypochlorite solution should do the job.

EXAMPLE 26

Calculate the actual chlorine dose in milligrams per liter if 200 gallons of a two percent sodium hypochlorite solution were used to treat two million gallons of water.

Known	Unknown
Volume Hypochl, gal = 200 gal	Chlorine Dose, mg/L
Volume Water, M Gal = 2 M Gal	
Hypochlorite, % = 2%	

1. Calculate the pounds of chlorine used.

$$\text{Chlorine, lbs} = \frac{(\text{Hypochlorite, gal})(8.34 \text{ lbs/gal})(\text{Hypochlorite, \%})}{100\%}$$

$$= \frac{(200 \text{ gal})(8.34 \text{ lbs/gal})(2\%)}{100\%}$$

$$= 33.36 \text{ lbs Chlorine}$$

2. Calculate the actual chlorine dose in milligrams per liter.

$$\text{Actual Dose, mg/L} = \frac{\text{Chlorine, lbs}}{(\text{Volume Water, M Gal})(8.34 \text{ lbs/gal})}$$

$$= \frac{33.36 \text{ lbs Chlorine}}{(2 \text{ M Gal})(8.34 \text{ lbs/gal})}$$

$$= 2.0 \text{ mg/L}$$

EXAMPLE 27

A deep-well turbine pump delivers 250 GPM against typical operating heads. If the desired chlorine dose is 2.5 mg/L, what should be the chlorine feed rate in pounds per day?

Known	Unknown
Flow, GPM = 250 GPM	Chlorine Feed, lbs/day
Dose, mg/L = 2.5 mg/L	

1. Convert flow from gallons per minute to million gallons per day.

$$\text{Flow, MGD} = \frac{(\text{Flow, GPM})(60 \text{ min/hr})(24 \text{ hr/day})}{1,000,000/\text{M}}$$

$$= \frac{(250 \text{ GPM})(60 \text{ min/hr})(24 \text{ hr/day})}{1,000,000/\text{M}}$$

$$= 0.36 \text{ MGD}$$

2. Calculate the chlorine feed rate in pounds of chlorine per day.

$$\text{Chlorine Feed, lbs/day} = (\text{Flow, MGD})(\text{Dose, mg/L})(8.34 \text{ lbs/gal})$$

$$= (0.36 \text{ MGD})(2.5 \text{ mg/L})(8.34 \text{ lbs/gal})$$

$$= 7.5 \text{ lbs/day}$$

EXAMPLE 28

Estimate the chlorine demand for a water in milligrams per liter if the chlorine dose is 2.6 mg/L and the chlorine residual is 0.4 mg/L.

Known	Unknown
Chlorine Dose, mg/L = 2.6 mg/L	Chlorine Demand, mg/L
Chlorine Residual, mg/L = 0.4 mg/L	

Estimate the chlorine demand in milligrams per liter.

$$\text{Chlorine Demand, mg/L} = \text{Chlorine Dose, mg/L} - \text{Chlorine Residual, mg/L}$$

$$= 2.6 \text{ mg/L} - 0.4 \text{ mg/L}$$

$$= 2.2 \text{ mg/L}$$

EXAMPLE 29

Estimate the gallons of hypochlorite pumped by a hypochlorinator if the hypochlorite solution is in a container with a diameter of three feet and the hypochlorite level drops 18 inches (1.5 feet) during a specific time period.

Known	Unknown
Diameter, ft = 3 ft	Hypochlorite Pumped, gal
Drop, ft = 1.5 ft	

Estimate the gallons of hypochlorite pumped.

$$\text{Hypochlorite, gal} = (\text{Container Area, sq ft})(\text{Drop, ft})(7.48 \text{ gal/cu ft})$$

$$= (0.785)(3 \text{ ft})^2(1.5 \text{ ft})(7.48 \text{ gal/cu ft})$$

$$= 79.3 \text{ gallons}$$

EXAMPLE 30

Estimate the desired strength (as a percent chlorine) of a hypochlorite solution being pumped by a hypochlorinator that delivers 80 gallons per day. The water being treated requires a chlorine feed rate of ten pounds of chlorine per day.

Known	Unknown
Hypochlorinator Flow, GPD = 80 GPD	Hypochlorite Strength, %
Chlorine Required, lbs/day = 10 lbs/day	

Estimate the desired hypochlorite strength as a percent chlorine.

$$\text{Hypochlorite Strength, \%} = \frac{(\text{Chlorine Required, lbs/day})(100\%)}{(\text{Hypochlorinator Flow, gal/day})(8.34 \text{ lbs/gal})}$$

$$= \frac{(10 \text{ lbs/day})(100\%)}{(80 \text{ GPD})(8.34 \text{ lbs/gal})}$$

$$= 1.5\%$$

EXAMPLE 31

How many gallons of water must be added to ten gallons of a five percent hypochlorite solution to produce a 1.5 percent hypochlorite solution?

Known		**Unknown**
Hypochlorite, gal	= 10 gal	Water Added, gal
Desired Hypochlorite, %	= 1.5%	
Actual Hypochlorite, %	= 5%	

Calculate the gallons of water that must be added to produce a 1.5 percent hypochlorite solution.

$$\text{Water Added, gal} \atop \text{(to hypochlorite solution)} = \frac{(\text{Hypo, gal})(\text{Hypo, \%}) - (\text{Hypo, gal})(\text{Des Hypo, \%})}{\text{Desired Hypo, \%}}$$

$$= \frac{(10 \text{ gal})(5\%) - (10 \text{ gal})(1.5\%)}{1.5\%}$$

$$= \frac{50 - 15}{1.5}$$

$$= 23.3 \text{ gallons}$$

A.14   BASIC CONVERSION FACTORS (METRIC SYSTEM)

LENGTH

100 cm	= 1 m	100 cm/m
3.281 ft	= 1 m	3.281 ft/m

AREA

2.4711 ac	= 1 ha*	2.4711 ac/ha
10,000 sq m	= 1 ha	10,000 sq m/ha

VOLUME

1,000 m*L*	= 1 liter	1,000 m*L*/*L*
1,000 *L*	= 1 cu m	1,000 *L*/cu m
3.785 *L*	= 1 gal	3.785 *L*/gal

WEIGHT

1,000 mg	= 1 gm	1,000 mg/gm
1,000 gm	= 1 kg	1,000 gm/kg

DENSITY

1 kg	= 1 liter	1 kg/*L*

PRESSURE

10.015 m	= 1 kg/sq cm	10.015 m/kg/sq cm
1 Pascal	= 1 N/sq m	1 Pa/N/sq m
1 psi	= 6,895 Pa	1 psi/6,895 Pa

FLOW

3,785 cu m/day	= 1 MGD	3,785 cu m/day/MGD
3.785 M*L*/day	= 1 MGD	3.785 M*L*/day/MGD

---

* hectare

A.15   TYPICAL WATER DISTRIBUTION SYSTEM PROBLEMS (METRIC SYSTEM)

A.150   Flows

EXAMPLE 1

Convert a flow of 500 gallons per minute to liters per second and cubic meters per day.

Known		**Unknown**
Flow, GPM = 500 GPM		1. Flow, liters/sec
		2. Flow, cu m/day

1. Convert the flow from 500 GPM to liters per second.

$$\text{Flow, liters/sec} = \frac{(\text{Flow, gal/min})(3.785 \text{ liters/gal})}{60 \text{ sec/min}}$$

$$= \frac{(500 \text{ gal/min})(3.785 \text{ liters/gal})}{60 \text{ sec/min}}$$

$$= 31.5 \text{ liters/sec}$$

2. Convert the flow from 500 GPM to cubic meters per day.

$$\text{Flow,} \atop \text{cu m/day} = \frac{(\text{Flow, GPM})(3.785 \text{ } L\text{/gal})(60 \text{ min/hr})(24 \text{ hr/day})}{1,000 \text{ } L\text{/cu m}}$$

$$= \frac{(500 \text{ GPM})(3.785 \text{ } L\text{/gal})(60 \text{ min/hr})(24 \text{ hr/day})}{1,000 \text{ } L\text{/cu m}}$$

$$= 2,725 \text{ cu m/day}$$

A.151   Chemical Doses

EXAMPLE 2

Determine the chlorinator setting in kilograms per 24 hours if 4,000 cubic meters of water per day are to be treated with a desired chlorine dose of 2.5 mg/*L*.

Known		**Unknown**
Flow, cu m/day	= 4,000 cu m/day	Chlorinator Setting, kg/24 hours
Chlorine Dose, mg/*L*	= 2.5 mg/*L*	

Determine the chlorinator setting in kilograms per 24 hours.

$$\text{Chlorinator Setting, kg/day} = \frac{(\text{Flow cu m/day})(\text{Dose, mg/}L)(1,000 \text{ } L\text{/cu m})}{(1,000 \text{ mg/gm})(1,000 \text{ gm/kg})}$$

$$= \frac{(4,000 \text{ cu m/day})(2.5 \text{ mg/}L)(1,000 \text{ } L\text{/cu m})}{(1,000 \text{ mg/gm})(1,000 \text{ gm/kg})}$$

$$= 10 \text{ kg/day}$$

EXAMPLE 3

Determine the actual dose or chemical feed in kilograms per day from a dry chemical feeder. A bucket placed under the chemical feeder weighed 100 grams empty and 1,400 grams after 30 minutes.

Known		**Unknown**
Empty Bucket, gm	= 100 gm	Chemical Feed, kg/day
Full Bucket, gm	= 1,400 gm	
Time to Fill, min	= 30 min	

Determine the chemical feed in kilograms of chemical per day.

$$\text{Chemical Feed, kg/day} = \frac{\text{Chemical Applied, kg}}{\text{Length of Application, days}}$$

$$= \frac{(1,400 \text{ gm} - 100 \text{ gm})(60 \text{ min/hr})(24 \text{ hr/day})}{(1,000 \text{ gm/kg})(30 \text{ min})}$$

$$= 62.4 \text{ kg/day}$$

EXAMPLE 4

Determine the chemical feed in kilograms of chlorine per day from a hypochlorinator. The hypochlorite solution is 1.4 percent or 14,000 mg chlorine per liter. During a test run the hypochlorinator delivered 400 mL during 12 minutes.

Known		Unknown
Cl Solution, %	= 1.4%	Chemical Feed, kg/day
Cl Conc, mg/L	= 14,000 mg/L	
Vol Pumped, mL	= 400 mL	
Time Pumped, min	= 12 min	

Calculate the chlorine fed by the hypochlorinator in kilograms of chlorine per day.

$$\text{Chlorine Feed, kg/day} = \frac{(\text{Cl Conc, mg/}L)(\text{Vol Pumped, m}L)(60 \text{ min/hr})(24 \text{ hr/day})}{(\text{Time Pumped, min})(1{,}000 \text{ m}L/L)(1{,}000 \text{ mg/gm})(1{,}000 \text{ gm/kg})}$$

$$= \frac{(14{,}000 \text{ mg/}L)(400 \text{ m}L)(60 \text{ min/hr})(24 \text{ hr/day})}{(12 \text{ min})(1{,}000 \text{ m}L/L)(1{,}000 \text{ mg/gm})(1{,}000 \text{ gm/kg})}$$

$$= 0.67 \text{ kg/day}$$

A.152 Distribution System Facilities

EXAMPLE 5

A pressure gage at a fire hydrant reads 350 kiloPascals (kPa). What is the pressure head on the gage at the fire hydrant in meters?

Known	Unknown
Pressure, kPa = 350 kPa	Pressure Head, m

Convert the pressure from kPa to meters of head.

$$\text{Pressure Head, m} = (\text{Pressure, kPa})(0.102 \text{ m/kPa})$$

$$= (350 \text{ kPa})(0.102 \text{ m/kPa})$$

$$= 35.7 \text{ m}$$

EXAMPLE 6

A pressure gage on a fire hydrant reads 50 meters of pressure head. What is the pressure in kiloPascals?

Known	Unknown
Pressure Head, m = 50 m	Pressure, kPa

Calculate the pressure head in kiloPascals.

$$\text{Pressure, kPa} = \frac{\text{Pressure Head, m}}{0.102 \text{ m/kPa}}$$

$$= \frac{50 \text{ m}}{0.102 \text{ m/kPa}}$$

$$= 490 \text{ kPa}$$

or

$$\text{Pressure, kPa} = (\text{Pressure Head, m})(9.79 \text{ kPa/m})$$

$$= (50 \text{ m})(9.79 \text{ kPa/m})$$

$$= 490 \text{ kPa}$$

EXAMPLE 7

A 100-mm diameter water main needs to be flushed with a water velocity of 1.5 meters per second. What should be the flow through the main in cubic meters per second and liters per second?

Known	Unknown
Pipe Diameter, mm = 100 mm	1. Flow, cu m/sec
Velocity, m/sec = 1.5 m/sec	2. Flow, L/sec

1. Calculate the pipe area in square meters.

$$\text{Area, sq m} = (0.785)(\text{Diameter, m})^2$$

$$= (0.785)(0.1 \text{ m})^2$$

$$= 0.00785 \text{ sq m}$$

2. Determine the flow in cubic meters per second.

$$\text{Flow, cu m/sec} = (\text{Area, sq m})(\text{Velocity, m/sec})$$

$$= (0.00785 \text{ sq m})(1.5 \text{ m/sec})$$

$$= 0.0118 \text{ cu m/sec}$$

3. Convert the flow from cubic meters per second to liters per second.

$$\text{Flow, liters/sec} = (\text{Flow, cu m/sec})(1{,}000 \text{ }L/\text{cu m})$$

$$= (0.0118 \text{ cu m/sec})(1{,}000 \text{ }L/\text{cu m})$$

$$= 11.8 \text{ }L/\text{sec}$$

EXAMPLE 8

Determine the actual leakage in liters per day per kilometer per millimeter of diameter from a new section of pipe if the leak rate during the test period was 200 liters per day. The section tested is 0.4 kilometer long and the pipe diameter is 300 millimeters.

Known		Unknown
Leak Rate, L/day	= 200 L/day	Actual Leakage, L/day/km-mm
Length, km	= 0.4 km	
Diameter, mm	= 300 mm	

Calculate the actual leakage from the test section in liters per day per kilometer per millimeter of pipe diameter.

$$\text{Actual Leakage, } L/\text{day/km-mm} = \frac{\text{Leak Rate, }L/\text{day}}{(\text{Length, km})(\text{Diameter, mm})}$$

$$= \frac{200 \text{ }L/\text{day}}{(0.4 \text{ km})(300 \text{ mm})}$$

$$= 1.67 \text{ liters per day/km-mm}$$

EXAMPLE 9

An underground rectangular water storage tank must be drained and cleaned. The tank is 10 meters long and 3 meters wide. The groundwater is 2 meters above the bottom of the tank. What is the force of the water pressure lifting upward on the bottom of the tank?

Known	Unknown
Length, m = 10 m	Uplift Force, kg
Width, m = 3 m	
Height, m = 2 m	

1. Calculate the area of the tank bottom.

   Area, sq m = (Length, m)(Width, m)

   = (10 m)(3 m)

   = 30 sq m

2. Calculate the pressure on the bottom of the tank in kilograms per square meter (kg/sq m).

   Pressure, kg/sq m = (Density of Water, kg/cu m)(Height, m)

   = (1,000 kg/cu m)(2 m)

   = 2,000 kg/sq m

3. Calculate the uplift force on the bottom of the tank.

   Uplift Force, kg = (Area, sq m)(Pressure, kg/sq m)

   = (30 sq m)(2,000 kg/sq m)

   = 60,000 kg

If the tank weighs less than 60,000 kg, the tank will float when emptied if the groundwater is two meters or higher above the bottom of the tank.

EXAMPLE 10 (Advanced Problem)

Estimate the C Factor for a 250-mm diameter water main when the flow is 5 MLD (mega or million liters per day) and the drop in pressure head elevation between two pressure gages 400 meters apart is two meters.

Known		Unknown
Diameter, mm	= 250 mm	C Factor
	= 0.25 m	
Flow, MLD	= 5 MLD	
Energy Loss, m	= 2 m	
Distance, m	= 400 m	

1. Calculate the slope.

   $$\text{Slope} = \frac{\text{Energy Loss, m}}{\text{Distance, m}}$$

   $$= \frac{2\text{ m}}{400\text{ m}}$$

   = 0.005 m/m

2. Calculate the C Factor.

   $$\text{C Factor} = \frac{\text{Flow, MLD}}{23.97(\text{Diameter, m})^{2.63}(\text{Slope})^{0.54}}$$

   $$= \frac{5\text{ MLD}}{23.97(0.25\text{ m})^{2.63}(0.005)^{0.54}}$$

   $$= \frac{5\text{ MLD}}{(23.97)(0.026)(0.057)}$$

   = 141

EXAMPLE 11 (Advanced Problem)

Estimate the C Factor for a 400-mm diameter pipe if a field test was conducted using a flow of 100 liters per second. The difference in elevation of the pressure heads between two fire hydrants 250 meters apart is 0.5 meter.

Known		Unknown
Diameter, mm	= 400 mm	C Factor
Flow, L/sec	= 100 L/sec	
Pres Head Dif, m	= 0.5 m	
Distance, m	= 250 m	

1. Determine the slope of the energy grade line.

   $$\text{Slope} = \frac{\text{Difference in Elev Pres Head, m}}{\text{Distance, m}}$$

   $$= \frac{0.5\text{ m}}{250\text{ m}}$$

   = 0.002 m/m

2. Calculate the C Factor.

   $$\text{C Factor} = \frac{\text{Flow, L/sec}}{277.8(\text{Diameter, m})^{2.63}(\text{Slope})^{0.54}}$$

   $$= \frac{100\text{ L/sec}}{277.8(0.4\text{ m})^{2.63}(0.002)^{0.54}}$$

   $$= \frac{100\text{ L/sec}}{(277.8)(0.0890)(0.03488)}$$

   = 115

A.153 Distribution System Operation and Maintenance

EXAMPLE 12

Calculate the quarterly average TTHM (Total Trihalomethane) for a water supply system in micrograms per liter based on the results of tests taken during a three-month period.

Test Number	1	2	3	4	5	6
TTHM, µg/L	140	100	80	160	150	90

Known	Unknown
Test Results Shown Above	Average TTHM, µg/L

$$\text{Avg TTHM, µg/L} = \frac{\text{Sum of Measurements, µg/L}}{\text{Number of Measurements}}$$

$$= \frac{140 + 100 + 80 + 160 + 150 + 90}{6}$$

$$= \frac{720\text{ µg/L}}{6}$$

= 120 µg/L

or

= 0.12 mg/L

EXAMPLE 13

Determine the annual running TTHM average using the quarterly results shown below.

Known	Unknown
Results of Quarterly TTHM Analyses	Annual Running TTHM Average, mg/L

YEAR 1

QUARTER	CONC, mg/L
1	0.07
2	0.09
3	0.12
4	0.08

YEAR 2

1	0.09
2	0.11
3	0.13
4	0.07

Calculate the annual running TTHM average for each quarter during Year 2. Use the quarterly TTHM average for the quarter being considered and the three quarters immediately before the one being considered.

YEAR 2, QUARTER 1

$$\text{Annual Running TTHM Average, mg/L} = \frac{\text{Sum of Measurements, mg/L}}{\text{Number of Measurements}}$$

$$= \frac{0.09\ mg/L + 0.08\ mg/L + 0.12\ mg/L + 0.09\ mg/L}{4}$$

$$= \frac{0.38\ mg/L}{4}$$

$$= 0.095\ mg/L$$

YEAR 2, QUARTER 2

$$\text{Annual Running TTHM Average, mg/L} = \frac{0.11\ mg/L + 0.09\ mg/L + 0.08\ mg/L + 0.12\ mg/L}{4}$$

$$= \frac{0.40\ mg/L}{4}$$

$$= 0.10\ mg/L$$

YEAR 2, QUARTER 3

$$\text{Annual Running TTHM Average, mg/L} = \frac{0.13\ mg/L + 0.11\ mg/L + 0.09\ mg/L + 0.08\ mg/L}{4}$$

$$= \frac{0.41\ mg/L}{4}$$

$$= 0.103\ mg/L$$

YEAR 2, QUARTER 4

$$\text{Annual Running TTHM Average, mg/L} = \frac{0.07\ mg/L + 0.13\ mg/L + 0.11\ mg/L + 0.09\ mg/L}{4}$$

$$= \frac{0.40\ mg/L}{4}$$

$$= 0.10\ mg/L$$

EXAMPLE 14

A 450-mm diameter water main is to be flushed at a velocity of 1.2 m/sec. What should be the reading on the flowmeter in liters per second?

Known	Unknown
Diameter, mm = 450 mm	Flow, L/sec
= 0.45 m	
Velocity, m/sec = 1.2 m/sec	

1. Calculate the cross-sectional area of the pipe in square meters.

$$\text{Area, sq m} = (0.785)(\text{Diameter, m})^2$$

$$= (0.785)(0.45\ m)^2$$

$$= 0.159\ sq\ m$$

2. Determine the flow in the pipe in liters per second.

$$\text{Flow, L/sec} = (\text{Area, sq m})(\text{Velocity, m/sec})(1,000\ L/cu\ m)$$

$$= (0.159\ sq\ m)(1.2\ m/sec)(1,000\ L/cu\ m)$$

$$= 191\ liters/sec$$

EXAMPLE 15

During a 20-minute time span the water in a tank one meter in diameter increases by 1.2 meters. Calculate the total flow in liters and the flow in liters per second.

Known	Unknown
Time, min = 20 min	1. Volume, liters
Diameter, m = 1 m	2. Flow, liters/sec
Depth, m = 1.2 m	

1. Calculate the total flow in liters.

$$\text{Volume, liters} = (0.785)(\text{Diameter, m})^2(\text{Depth, m})(1,000\ L/cu\ m)$$

$$= (0.785)(1\ m)^2(1.2\ m)(1,000\ L/cu\ m)$$

$$= 942\ liters$$

2. Calculate the flow in liters per second.

$$\text{Flow, L/sec} = \frac{\text{Volume, liters}}{(\text{Time, min})(60\ sec/min)}$$

$$= \frac{942\ liters}{(20\ min)(60\ sec/min)}$$

$$= 0.785\ liters/sec$$

EXAMPLE 16

A water meter is tested in a laboratory. The meter reads 0.08 liter per second and the actual flow was 0.083 liter per second. Determine the meter accuracy as a percentage.

Known	Unknown
Meter Reading, L/sec = 0.08 L/sec	Meter Accuracy, %
Actual Flow, L/sec = 0.083 L/sec	

Determine the meter accuracy as a percentage.

$$\text{Meter Accuracy, \%} = \frac{(\text{Meter Reading, } L/\text{sec})(100\%)}{\text{Actual Flow, } L/\text{sec}}$$

$$= \frac{(0.08\,L/\text{sec})(100\%)}{0.083\,L/\text{sec}}$$

$$= 96.4\%$$

EXAMPLE 17

A water meter is tested in a laboratory. The meter reads 1,550 liters. The actual flow was measured in a volumetric measuring tank one meter in diameter. During the meter test two meters of water were added to the tank. Determine the meter accuracy as a percentage.

Known	Unknown
Meter Reading, liters = 1,550 L	Meter Accuracy, %
Diameter, m = 1 m	
Depth, m = 2 m	

1. Calculate the actual volume of water that flowed through the meter in liters.

$$\text{Volume, liters} = (0.785)(\text{Diameter, m})^2(\text{Depth, m})(1{,}000\,L/\text{cu m})$$

$$= (0.785)(1\,\text{m})^2(2\,\text{m})(1{,}000\,L/\text{cu m})$$

$$= 1{,}570\ \text{liters}$$

2. Determine the meter accuracy as a percentage.

$$\text{Meter Accuracy, \%} = \frac{(\text{Meter Reading, } L)(100\%)}{\text{Actual Volume, } L}$$

$$= \frac{(1{,}550\,L)(100\%)}{1{,}570\,L}$$

$$= 98.7\%$$

EXAMPLE 18

A trench 1.5 meters wide, 320 meters long, and 2 meters deep is to be excavated for a water main. How many cubic meters of soil have to be excavated for the water main?

Known	Unknown
Length, m = 320 m	Volume Excavated, cu m
Width, m = 1.5 m	
Depth, m = 2 m	

Calculate the volume excavated in cubic meters.

$$\text{Volume, cu m} = (\text{Length, m})(\text{Width, m})(\text{Depth, m})$$

$$= (320\,\text{m})(1.5\,\text{m})(2\,\text{m})$$

$$= 960\ \text{cu m}$$

EXAMPLE 19

A flowmeter indicates a flow rate of two liters per second. How many liters of water will flow through the meter in 45 minutes?

Known	Unknown
Flow Rate, L/sec = 2 L/sec	Flow Volume, L
Time, min = 45 min	

Calculate the total flow of water in liters.

$$\text{Flow Volume, } L = (\text{Flow Rate, } L/\text{sec})(\text{Time, min})(60\ \text{sec/min})$$

$$= (2\ L/\text{sec})(45\ \text{min})(60\ \text{sec/min})$$

$$= 5{,}400\ \text{liters}$$

EXAMPLE 20

A chemical solution tank is 1.5 meters in diameter and two meters deep. How long will it take to fill this tank when water flows at a rate of two liters per second?

Known	Unknown
Diameter, m = 1.5 m	Time to Fill, min
Depth, m = 2 m	
Flow Rate, L/sec = 2 L/sec	

1. Calculate the volume in the tank in liters.

$$\text{Tank Volume, } L = (0.785)(\text{Diameter, m})^2(\text{Depth, m})(1{,}000\,L/\text{cu m})$$

$$= (0.785)(1.5\,\text{m})^2(2\,\text{m})(1{,}000\,L/\text{cu m})$$

$$= 3{,}533\ L$$

2. Calculate the time to fill the tank in minutes.

$$\text{Time to Fill, min} = \frac{\text{Tank Volume, liters}}{(\text{Flow Rate, } L/\text{sec})(60\ \text{sec/min})}$$

$$= \frac{3{,}533\,L}{(2\,L/\text{sec})(60\ \text{sec/min})}$$

$$= 29.4\ \text{min}$$

EXAMPLE 21

A cylindrical water storage tank is 10 meters in diameter and four meters high. Calculate the volume of water in cubic meters per meter of depth and then the total volume in cubic meters and megaliters (ML) when the water is 3.7 meters deep.

Known	Unknown
Diameter, m = 10 m	1. Volume, cu m/m for a depth of 3.7 m
Height, m = 4 m	2. Volume, cu m
Depth, m = 3.7 m	3. Volume, ML

1. Calculate the volume of water in the tank in cubic meters per meter of depth.

$$\text{Volume, cu m} = (\text{Area, sq m})(\text{Depth, m})$$

$$\text{Volume, cu m/m} = (0.785)(\text{Diameter, m})^2(1\ \text{m/1 m})$$

$$= (0.785)(10\,\text{m})^2(1\ \text{m/1 m})$$

$$= 78.5\ \text{cu m/m}$$

2. Calculate the volume of water in the tank in cubic meters when the water is 3.7 meters deep in the tank.

Volume, cu m = (Volume, cu m/m)(Depth, m)

= (78.5 cu m/m)(3.7 m)

= 290 cu m

3. Convert the volume of water in the tank from cubic meters to mega or million liters (ML).

$$\text{Volume, M}L = \frac{(\text{Volume, cu m})(1{,}000\ L/\text{cu m})}{1{,}000{,}000/\text{M}}$$

$$= \frac{(290\ \text{cu m})(1{,}000\ L/\text{cu m})}{1{,}000{,}000/\text{M}}$$

= 0.29 ML

EXAMPLE 22

Estimate the flow from a pipe in liters per second. Water is flowing from a 50-mm diameter pipe. The pipe is one meter above the point where the water hits the ground. The water hits the ground 2.3 meters (horizontal distance) from where the water leaves the pipe. See Figure A.3 on page 564 for a sketch of this problem.

Known	Unknown
Diameter, mm = 50 mm	Flow, L/sec
Height, m = 1 m	
Distance, m = 2.3 m	

Estimate the flow from the pipe in liters per second.

$$\text{Flow, }L/\text{sec} = \frac{1{,}739(\text{Diameter, m})^2(\text{Distance, m})}{\sqrt{\text{Height, m}}}$$

$$= \frac{1{,}739(0.050\ \text{m})^2(2.3\ \text{m})}{\sqrt{1\ \text{m}}}$$

= 10.0 L/sec

A.154 Disinfection

EXAMPLE 23

How many liters of five percent sodium hypochlorite will be needed to disinfect a well with a 0.5-meter diameter casing and well screen? The well is 60 meters deep and there are 25 meters of water in the well. Use an initial chlorine dose of 100 mg/L.

Known	Unknown
Hypochlorite, % = 5%	5% Hypochlorite, L
Chlorine Dose, mg/L = 100 mg/L	
Diameter, m = 0.5 m	
Well Depth, m = 60 m	
Water Depth, m = 25 m	

1. Calculate the volume of water in the well in liters.

Water Vol, liters = (0.785)(Diameter, m)^2(Depth, m)(1,000 L/cu m)

= (0.785)(0.5 m)^2(25 m)(1,000 L/cu m)

= 4,906 liters

2. Determine the grams of chlorine needed.

$$\text{Chlorine, gm} = \frac{(\text{Volume, }L)(\text{Dose, mg}/L)}{1{,}000\ \text{mg/gm}}$$

$$= \frac{(4{,}906\ L)(100\ \text{mg}/L)}{1{,}000\ \text{mg/gm}}$$

= 491 gm Chlorine

3. Calculate the liters of five percent sodium hypochlorite solution needed. One liter of water weighs 1,000 grams.

$$\begin{aligned}\text{Sodium Hypochlorite} \\ \text{Solution, liters}\end{aligned} = \frac{(\text{Chlorine, gm})(100\%)}{(1{,}000\ \text{gm}/L)(\text{Hypochlorite, \%})}$$

$$= \frac{(491\ \text{gm})(100\%)}{(1{,}000\ \text{gm}/L)(5\%)}$$

= 9.8 liters

Use ten liters of five percent sodium hypochlorite to disinfect the well.

4. ALTERNATE SOLUTION

Calculate the liters of five percent sodium hypochlorite solution needed. A five percent solution is the same as 50,000 mg per liter.

$$\begin{aligned}\text{Sodium Hypochlorite} \\ \text{Solution, liters}\end{aligned} = \frac{(\text{Volume, }L)(\text{Dose, mg}/L)}{\text{Hypochlorite Solution, mg}/L}$$

$$= \frac{(4{,}906\ L)(100\ \text{mg}/L)}{50{,}000\ \text{mg}/L}$$

= 9.8 liters

EXAMPLE 24

A new 250-mm diameter water main 200 meters long needs to be disinfected. An initial chlorine dose of 400 mg/L is expected to maintain a chlorine residual of over 300 mg/L during the three-hour disinfection period. How many liters of 5.25 percent (52,500 mg/L) sodium hypochlorite solution will be needed?

Known	Unknown
Diameter of Pipe, m = 0.25 m	5.25% Hypochlorite, L
Length of Pipe, m = 200 m	
Chlorine Dose, mg/L = 400 mg/L	
Hypochlorite, % = 5.25%	
Hypochlorite, mg/L = 52,500 mg/L	

1. Calculate the volume of water in the pipe in liters.

Pipe Vol, L = (0.785)(Diameter, m)^2(Length, m)(1,000 L/cu m)

= (0.785)(0.25 m)^2(200 m)(1,000 L/cu m)

= 9,813 L

2. Calculate the liters of 5.25 percent sodium hypochlorite solution needed.

$$\text{Sodium Hypochlorite Solution, } L = \frac{(\text{Volume, } L)(\text{Dose, mg/}L)}{\text{Hypochlorite Solution, mg/}L}$$

$$= \frac{(9,813\ L)(400\ \text{mg/}L)}{52,500\ \text{mg/}L}$$

$$= 75\ \text{liters}$$

Seventy-five liters of 5.25 percent sodium hypochlorite solution should do the job.

EXAMPLE 25

A service storage reservoir has been taken out of service for inspection, maintenance, and repairs. The reservoir needs to be disinfected before being placed back on line. The reservoir is 12 meters in diameter and 3 meters deep. An initial chlorine dose of 100 mg/L is expected to maintain a chlorine residual of over 50 mg/L during the 24-hour disinfection period. How many liters of 15 percent sodium hypochlorite solution will be needed?

Known		Unknown
Diameter, m	= 12 m	15% Hypochlorite, L
Tank Depth, m	= 3 m	
Chlorine Dose, mg/L	= 100 mg/L	
Hypochlorite, %	= 15%	
Hypochlorite, mg/L	= 150,000 mg/L	

1. Calculate the volume of water in the tank in liters.

$$\text{Tank Volume, } L = (0.785)(\text{Diameter, m})^2(\text{Depth, m})(1,000\ L/\text{cu m})$$

$$= (0.785)(12\ \text{m})^2(3\ \text{m})(1,000\ L/\text{cu m})$$

$$= 339,120\ L$$

2. Calculate the liters of 15 percent sodium hypochlorite solution needed.

$$\text{Sodium Hypochlorite Solution, } L = \frac{(\text{Volume, } L)(\text{Dose, mg/}L)}{\text{Hypochlorite Solution, mg/}L}$$

$$= \frac{(339,120\ L)(100\ \text{mg/}L)}{150,000\ \text{mg/}L}$$

$$= 226\ \text{liters}$$

Two hundred twenty-six liters of 15 percent sodium hypochlorite solution should do the job.

EXAMPLE 26

Calculate the actual chlorine dose in milligrams per liter if 800 liters of a two percent sodium hypochlorite solution were used to treat eight megaliters of water.

Known		Unknown
Volume Hypochl, L	= 800 L	Chlorine Dose, mg/L
Volume Water, ML	= 8 ML	
Hypochlorite, %	= 2%	
Hypochlorite, mg/L	= 20,000 mg/L	

Calculate the actual chlorine dose in milligrams per liter.

$$\text{Actual Dose, mg/}L = \frac{(\text{Hypochlorite, mg/}L)(\text{Volume Hypochl, } L)}{(\text{Volume Water, M}L)(1,000,000/\text{M})}$$

$$= \frac{(20,000\ \text{mg/}L)(800\ L)}{(8\ \text{M}L)(1,000,000/\text{M})}$$

$$= 2.0\ \text{mg/}L$$

EXAMPLE 27

A deep-well turbine pump delivers 15 liters per second against typical operating heads. If the desired chlorine dose is 2.5 mg/L, what should be the chlorine feed rate in kilograms per day and milligrams per second?

Known	Unknown
Flow, L/sec = 15 L/sec	1. Chlorine Feed, kg/day
Dose, mg/L = 2.5 mg/L	2. Chlorine Feed, mg/sec

1. Calculate the chlorine feed in kilograms per day.

$$\text{Chlorine Feed, kg/day} = \frac{(\text{Flow, }L/\text{sec})(\text{Dose, mg/}L)(60\ \text{sec/min})(60\ \text{min/hr})(24\ \text{hr/day})}{(1,000\ \text{mg/gm})(1,000\ \text{gm/kg})}$$

$$= \frac{(15\ L/\text{sec})(2.5\ \text{mg/}L)(60\ \text{sec/min})(60\ \text{min/hr})(24\ \text{hr/day})}{(1,000\ \text{mg/gm})(1,000\ \text{gm/kg})}$$

$$= 3.24\ \text{kg/day}$$

2. Calculate the chlorine feed in milligrams per second.

$$\text{Chlorine Feed, mg/sec} = (\text{Flow, }L/\text{sec})(\text{Dose, mg/}L)$$

$$= (15\ L/\text{sec})(2.5\ \text{mg/}L)$$

$$= 37.5\ \text{mg/sec}$$

EXAMPLE 28

Estimate the chlorine demand for a water in milligrams per liter if the chlorine dose is 2.6 mg/L and the chlorine residual is 0.4 mg/L.

Known	Unknown
Chlorine Dose, mg/L = 2.6 mg/L	Chlorine Demand, mg/L
Chlorine Residual, mg/L = 0.4 mg/L	

Estimate the chlorine demand in milligrams per liter.

$$\text{Chlorine Demand, mg/}L = \text{Chlorine Dose, mg/}L - \text{Chlorine Residual, mg/}L$$

$$= 2.6\ \text{mg/}L - 0.4\ \text{mg/}L$$

$$= 2.2\ \text{mg/}L$$

EXAMPLE 29

Estimate the liters of hypochlorite pumped by a hypochlorinator if the hypochlorite solution is in a container with a diameter of one meter and the hypochlorite level drops 45 centimeters during a specific time period.

Known	Unknown
Diameter, m = 1 m	Hypochlorite Pumped, liters
Drop, cm = 45 cm	

Estimate the liters of hypochlorite pumped.

$$\text{Hypochlorite, liters} = (\text{Container Area, sq m})(\text{Drop, m})(1{,}000\ L/\text{cu m})$$

$$= \frac{(0.785)(1\ m)^2(45\ cm)(1{,}000\ L/\text{cu m})}{100\ cm/m}$$

$$= 353 \text{ liters}$$

EXAMPLE 30

Estimate the desired strength (as a percent chlorine) of a hypochlorite solution being pumped by a hypochlorinator that delivers 300 liters per day. The water being treated requires a chlorine feed rate of five kilograms of chlorine per day.

Known	Unknown
Hypochlorinator Flow, L/day = 300 L/day	Hypochlorite Stength, %
Chlorine Required, kg/day = 5 kg/day	

Estimate the desired hypochlorite strength as a percent chlorine.

$$\text{Hypochlorite Strength, \%} = \frac{(\text{Chlorine Required, kg/day})(100\%)}{(\text{Hypochlorinator Flow, } L/\text{day})(1\ kg/L)}$$

$$= \frac{(5\ kg/\text{day})(100\%)}{(300\ L/\text{day})(1\ kg/L)}$$

$$= 1.67\%$$

EXAMPLE 31

How many liters of water must be added to 40 liters of a five percent hypochlorite solution to produce a 1.67 percent hypochlorite solution?

Known	Unknown
Hypochlorite, L = 40 L	Water Added, L
Desired Hypo, % = 1.67%	
Actual Hypo, % = 5%	

Calculate the liters of water that must be added to produce a 1.67 percent hypochlorite solution.

$$\text{Water Added, } L \text{ (to hypochlorite solution)} = \frac{(\text{Hypo, } L)(\text{Hypo, \%}) - (\text{Hypo, } L)(\text{Des Hypo, \%})}{\text{Des Hypo, \%}}$$

$$= \frac{(40\ L)(5\%) - (40\ L)(1.67\%)}{1.67\%}$$

$$= \frac{200 - 66.8}{1.67}$$

$$= 80 \text{ liters}$$

WATER ABBREVIATIONS

ac	acre		km	kilometer
ac-ft	acre-feet		kN	kilonewton
af	acre feet		kW	kilowatt
amp	ampere		kWh	kilowatt-hour
°C	degrees Celsius		L	liter
CFM	cubic feet per minute		lb	pound
CFS	cubic feet per second		lbs/sq in	pounds per square inch
Ci	Curie		m	meter
cm	centimeter		M	mega
cu ft	cubic feet		M	million
cu in	cubic inch		mg	milligram
cu m	cubic meter		MGD	million gallons per day
cu yd	cubic yard		mg/L	milligram per liter
°F	degrees Fahrenheit		min	minute
ft	feet or foot		mL	milliliter
ft-lb/min	foot-pounds per minute		mm	millimeter
g	gravity		N	Newton
gal	gallon		ohm	ohm
gal/day	gallons per day		Pa	Pascal
GFD	gallons of flux per square foot per day		pCi	picoCurie
gm	gram		ppb	parts per billion
GPD	gallons per day		ppm	parts per million
gpg	grains per gallon		psf	pounds per square foot
GPM	gallons per minute		psi	pounds per square inch
gr	grain		psig	pounds per square inch gage
ha	hectare		RPM	revolutions per minute
HP	horsepower		sec	second
hr	hour		sq ft	square feet
in	inch		sq in	square inches
k	kilo		W	watt
kg	kilogram			

WATER WORDS

A Summary of the Words Defined

in

WATER DISTRIBUTION SYSTEM
OPERATION AND MAINTENANCE,

WATER TREATMENT PLANT OPERATION,

and

SMALL WATER SYSTEM
OPERATION AND MAINTENANCE

PROJECT PRONUNCIATION KEY

by Warren L. Prentice

The Project Pronunciation Key is designed to aid you in the pronunciation of new words. While this key is based primarily on familiar sounds, it does not attempt to follow any particular pronunciation guide. This key is designed solely to aid operators in this program.

You may find it helpful to refer to other available sources for pronunciation help. Each current standard dictionary contains a guide to its own pronunciation key. Each key will be different from each other and from this key. Examples of the difference between the key used in this program and the *WEBSTER'S NEW WORLD COLLEGE DICTIONARY*[1] "Key" are shown below.

In using this key, you should accent (say louder) the syllable that appears in capital letters. The following chart is presented to give examples of how to pronounce words using the Project Key.

WORD	SYLLABLE				
	1st	2nd	3rd	4th	5th
acid	AS	id			
coliform	COAL	i	form		
biological	BUY	o	LODGE	ik	cull

The first word, *ACID*, has its first syllable accented. The second word, *COLIFORM*, has its first syllable accented. The third word, *BIOLOGICAL*, has its first and third syllables accented.

We hope you will find the key useful in unlocking the pronunciation of any new word.

Term	Project Key	Webster Key
acid	AS-id	aś id
coliform	COAL-i-form	kō′ lə fôrm
biological	BUY-o-LODGE-ik-cull	bī ə läj′ i kəl

---

[1] The WEBSTER'S NEW WORLD COLLEGE DICTIONARY, Fourth Edition, 1999, was chosen rather than an unabridged dictionary because of its availability to the operator. Other editions may be slightly different.

WATER WORDS

>GREATER THAN >GREATER THAN

DO >5 mg/*L* would be read as DO GREATER THAN 5 mg/*L*.

<LESS THAN <LESS THAN

DO <5 mg/*L* would be read as DO LESS THAN 5 mg/*L*.

A

ABC ABC

See **A**SSOCIATION OF **B**OARDS OF **C**ERTIFICATION.

ACEOPS ACEOPS

See **A**LLIANCE OF **CE**RTIFIED **OP**ERATOR**S**, LAB ANALYSTS, INSPECTORS, AND SPECIALISTS (ACEOPS).

atm atm

The abbreviation for atmosphere. One atmosphere is equal to 14.7 psi or 100 kPa.

AWWA AWWA

See **A**MERICAN **W**ATER **W**ORKS **A**SSOCIATION.

ABSORPTION (ab-SORP-shun) ABSORPTION

The taking in or soaking up of one substance into the body of another by molecular or chemical action (as tree roots absorb dissolved nutrients in the soil).

ACCOUNTABILITY ACCOUNTABILITY

When a manager gives power/responsibility to an employee, the employee ensures that the manager is informed of results or events.

ACCURACY ACCURACY

How closely an instrument measures the true or actual value of the process variable being measured or sensed.

ACID RAIN ACID RAIN

Precipitation which has been rendered (made) acidic by airborne pollutants.

ACIDIC (uh-SID-ick) ACIDIC

The condition of water or soil which contains a sufficient amount of acid substances to lower the pH below 7.0.

ACIDIFIED (uh-SID-uh-FIE-d) ACIDIFIED

The addition of an acid (usually nitric or sulfuric) to a sample to lower the pH below 2.0. The purpose of acidification is to "fix" a sample so it won't change until it is analyzed.

ACRE-FOOT ACRE-FOOT

A volume of water that covers one acre to a depth of one foot, or 43,560 cubic feet (1,233.5 cubic meters).

ACTIVATED CARBON ACTIVATED CARBON

Adsorptive particles or granules of carbon usually obtained by heating carbon (such as wood). These particles or granules have a high capacity to selectively remove certain trace and soluble materials from water.

ACUTE HEALTH EFFECT

ACUTE HEALTH EFFECT

An adverse effect on a human or animal body, with symptoms developing rapidly.

ADSORBATE (add-SORE-bait)

ADSORBATE

The material being removed by the adsorption process.

ADSORBENT (add-SORE-bent)

ADSORBENT

The material (activated carbon) that is responsible for removing the undesirable substance in the adsorption process.

ADSORPTION (add-SORP-shun)

ADSORPTION

The gathering of a gas, liquid, or dissolved substance on the surface or interface zone of another material.

AERATION (air-A-shun)

AERATION

The process of adding air to water. Air can be added to water by either passing air through water or passing water through air.

AEROBIC (AIR-O-bick)

AEROBIC

A condition in which atmospheric or dissolved molecular oxygen is present in the aquatic (water) environment.

AESTHETIC (es-THET-ick)

AESTHETIC

Attractive or appealing.

AGE TANK

AGE TANK

A tank used to store a known concentration of chemical solution for feed to a chemical feeder. Also called a DAY TANK.

AIR BINDING

AIR BINDING

The clogging of a filter, pipe or pump due to the presence of air released from water. Air entering the filter media is harmful to both the filtration and backwash processes. Air can prevent the passage of water during the filtration process and can cause the loss of filter media during the backwash process.

AIR GAP

AIR GAP

An open vertical drop, or vertical empty space, that separates a drinking (potable) water supply to be protected from another water system in a water treatment plant or other location. This open gap prevents the contamination of drinking water by backsiphonage or backflow because there is no way raw water or any other water can reach the drinking water supply.

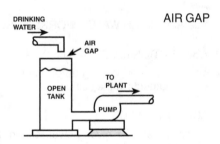

AIR PADDING

AIR PADDING

Pumping dry air (dew point −40°F) into a container to assist with the withdrawal of a liquid or to force a liquified gas such as chlorine out of a container.

AIR STRIPPING

AIR STRIPPING

A treatment process used to remove dissolved gases and volatile substances from water. Large volumes of air are bubbled through the water being treated to remove (strip out) the dissolved gases and volatile substances.

ALARM CONTACT

ALARM CONTACT

A switch that operates when some preset low, high or abnormal condition exists.

ALGAE (AL-gee)

ALGAE

Microscopic plants which contain chlorophyll and live floating or suspended in water. They also may be attached to structures, rocks or other submerged surfaces. Excess algal growths can impart tastes and odors to potable water. Algae produce oxygen during sunlight hours and use oxygen during the night hours. Their biological activities appreciably affect the pH, alkalinity, and dissolved oxygen of the water.

ALGAL (AL-gull) BLOOM

ALGAL BLOOM

Sudden, massive growths of microscopic and macroscopic plant life, such as green or blue-green algae, which can, under the proper conditions, develop in lakes and reservoirs.

ALGICIDE (AL-juh-SIDE)   ALGICIDE

Any substance or chemical specifically formulated to kill or control algae.

ALIPHATIC (AL-uh-FAT-ick) HYDROXY ACIDS   ALIPHATIC HYDROXY ACIDS

Organic acids with carbon atoms arranged in branched or unbranched open chains rather than in rings.

ALIQUOT (AL-li-kwot)   ALIQUOT

Representative portion of a sample. Often an equally divided portion of a sample.

ALKALI (AL-ka-lie)   ALKALI

Any of certain soluble salts, principally of sodium, potassium, magnesium, and calcium, that have the property of combining with acids to form neutral salts and may be used in chemical water treatment processes.

ALKALINE (AL-ka-LINE)   ALKALINE

The condition of water or soil which contains a sufficient amount of alkali substances to raise the pH above 7.0.

ALKALINITY (AL-ka-LIN-it-tee)   ALKALINITY

The capacity of water to neutralize acids. This capacity is caused by the water's content of carbonate, bicarbonate, hydroxide, and occasionally borate, silicate, and phosphate. Alkalinity is expressed in milligrams per liter of equivalent calcium carbonate. Alkalinity is not the same as pH because water does not have to be strongly basic (high pH) to have a high alkalinity. Alkalinity is a measure of how much acid must be added to a liquid to lower the pH to 4.5.

ALLIANCE OF CERTIFIED OPERATORS,   ALLIANCE OF CERTIFIED OPERATORS,
  LAB ANALYSTS, INSPECTORS,   LAB ANALYSTS, INSPECTORS,
  AND SPECIALISTS (ACEOPS)   AND SPECIALISTS (ACEOPS)

A professional organization for operators, lab analysts, inspectors, and specialists dedicated to improving professionalism; expanding training, certification, and job opportunities; increasing information exchange; and advocating the importance of certified operators, lab analysts, inspectors, and specialists. For information on membership, contact ACEOPS, 808 1st Avenue SE, Le Mars, IA 51031, phone (712) 548-4281 or e-mail: Info@aceops.org.

ALLUVIAL (uh-LOU-vee-ul)   ALLUVIAL

Relating to mud and/or sand deposited by flowing water. Alluvial deposits may occur after a heavy rainstorm.

ALTERNATING CURRENT (A.C.)   ALTERNATING CURRENT (A.C.)

An electric current that reverses its direction (positive/negative values) at regular intervals.

ALTITUDE VALVE   ALTITUDE VALVE

A valve that automatically shuts off the flow into an elevated tank when the water level in the tank reaches a predetermined level. The valve automatically opens when the pressure in the distribution system drops below the pressure in the tank.

AMBIENT (AM-bee-ent) TEMPERATURE   AMBIENT TEMPERATURE

Temperature of the surrounding air (or other medium). For example, temperature of the room where a gas chlorinator is installed.

AMERICAN WATER WORKS ASSOCIATION   AMERICAN WATER WORKS ASSOCIATION

A professional organization for all persons working in the water utility field. This organization develops and recommends goals, procedures and standards for water utility agencies to help them improve their performance and effectiveness. For information on AWWA membership and publications, contact AWWA, 6666 W. Quincy Avenue, Denver, CO 80235. Phone (303) 794-7711.

AMPERAGE (AM-purr-age)   AMPERAGE

The strength of an electric current measured in amperes. The amount of electric current flow, similar to the flow of water in gallons per minute.

AMPERE (AM-peer)   AMPERE

The unit used to measure current strength. The current produced by an electromotive force of one volt acting through a resistance of one ohm.

AMPEROMETRIC (am-PURR-o-MET-rick)   AMPEROMETRIC

A method of measurement that records electric current flowing or generated, rather than recording voltage. Amperometric titration is a means of measuring concentrations of certain substances in water.

AMPEROMETRIC (am-PURR-o-MET-rick) TITRATION

AMPEROMETRIC TITRATION

A means of measuring concentrations of certain substances in water (such as strong oxidizers) based on the electric current that flows during a chemical reaction. Also see TITRATE.

AMPLITUDE

AMPLITUDE

The maximum strength of an alternating current during its cycle, as distinguished from the mean or effective strength.

ANAEROBIC (AN-air-O-bick)

ANAEROBIC

A condition in which atmospheric or dissolved molecular oxygen is *NOT* present in the aquatic (water) environment.

ANALOG

ANALOG

The readout of an instrument by a pointer (or other indicating means) against a dial or scale.

ANALYZER

ANALYZER

A device which conducts periodic or continuous measurement of some factor such as chlorine, fluoride or turbidity. Analyzers operate by any of several methods including photocells, conductivity or complex instrumentation.

ANGSTROM (ANG-strem)

ANGSTROM

A unit of length equal to one-tenth of a nanometer or one-tenbillionth of a meter (1 Angstrom = 0.000 000 000 1 meter). One Angstrom is the approximate diameter of an atom.

ANION (AN-EYE-en)

ANION

A negatively charged ion in an electrolyte solution, attracted to the anode under the influence of a difference in electrical potential. Chloride ion (Cl^-) is an anion.

ANIONIC (AN-eye-ON-ick) POLYMER

ANIONIC POLYMER

A polymer having negatively charged groups of ions; often used as a filter aid and for dewatering sludges.

ANNULAR (AN-you-ler) SPACE

ANNULAR SPACE

A ring-shaped space located between two circular objects, such as two pipes.

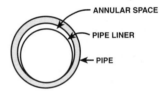

ANODE (an-O-d)

ANODE

The positive pole or electrode of an electrolytic system, such as a battery. The anode attracts negatively charged particles or ions (anions).

APPARENT COLOR

APPARENT COLOR

Color of the water that includes not only the color due to substances in the water but suspended matter as well.

APPROPRIATIVE

APPROPRIATIVE

Water rights to or ownership of a water supply which is acquired for the beneficial use of water by following a specific legal procedure.

APPURTENANCE (uh-PURR-ten-nans)

APPURTENANCE

Machinery, appliances, structures and other parts of the main structure necessary to allow it to operate as intended, but not considered part of the main structure.

AQUEOUS (A-kwee-us)

AQUEOUS

Something made up of, similar to, or containing water; watery.

AQUIFER (ACK-wi-fer)

AQUIFER

A natural underground layer of porous, water-bearing materials (sand, gravel) usually capable of yielding a large amount or supply of water.

ARCH   ARCH

(1) The curved top of a sewer pipe or conduit.

(2) A bridge or arch of hardened or caked chemical which will prevent the flow of the chemical.

ARTESIAN (are-TEE-zhun)   ARTESIAN

Pertaining to groundwater, a well, or underground basin where the water is under a pressure greater than atmospheric and will rise above the level of its upper confining surface if given an opportunity to do so.

ASEPTIC (a-SEP-tick)   ASEPTIC

Free from the living germs of disease, fermentation, or putrefaction. Sterile.

ASSOCIATION OF BOARDS OF CERTIFICATION   ASSOCIATION OF BOARDS OF CERTIFICATION
   (ABC)   (ABC)

An international organization representing over 150 boards which certify the operators of waterworks and wastewater facilities. For information on ABC publications regarding the preparation of and how to study for operator certification examinations, contact ABC, 208 Fifth Street, Ames, IA 50010-6259. Phone (515) 232-3623.

ASYMMETRIC (A-see-MET-rick)   ASYMMETRIC

Not similar in size, shape, form or arrangement of parts on opposite sides of a line, point or plane.

ATOM   ATOM

The smallest unit of a chemical element; composed of protons, neutrons and electrons.

AUDIT, WATER   AUDIT, WATER

A thorough examination of the accuracy of water agency records or accounts (volumes of water) and system control equipment. Water managers can use audits to determine their water distribution system efficiency. The overall goal is to identify and verify water and revenue losses in a water system.

AUTHORITY   AUTHORITY

The power and resources to do a specific job or to get that job done.

AVAILABLE CHLORINE   AVAILABLE CHLORINE

A measure of the amount of chlorine available in chlorinated lime, hypochlorite compounds, and other materials that are used as a source of chlorine when compared with that of elemental (liquid or gaseous) chlorine.

AVAILABLE EXPANSION   AVAILABLE EXPANSION

The vertical distance from the sand surface to the underside of a trough in a sand filter. This distance is also called FREEBOARD.

AVERAGE   AVERAGE

A number obtained by adding quantities or measurements and dividing the sum or total by the number of quantities or measurements. Also called the arithmetic mean.

$$\text{Average} = \frac{\text{Sum of Measurements}}{\text{Number of Measurements}}$$

AVERAGE DEMAND   AVERAGE DEMAND

The total demand for water during a period of time divided by the number of days in that time period. This is also called the average daily demand.

AXIAL TO IMPELLER   AXIAL TO IMPELLER

The direction in which material being pumped flows around the impeller or flows parallel to the impeller shaft.

AXIS OF IMPELLER   AXIS OF IMPELLER

An imaginary line running along the center of a shaft (such as an impeller shaft).

B

BOD (pronounce as separate letters) BOD

Biochemical **O**xygen **D**emand. The rate at which organisms use the oxygen in water while stabilizing decomposable organic matter under aerobic conditions. In decomposition, organic matter serves as food for the bacteria and energy results from its oxidation. BOD measurements are used as a measure of the organic strength of wastes in water.

BACK PRESSURE BACK PRESSURE

A pressure that can cause water to backflow into the water supply when a user's water system is at a higher pressure than the public water system.

BACKFLOW BACKFLOW

A reverse flow condition, created by a difference in water pressures, which causes water to flow back into the distribution pipes of a potable water supply from any source or sources other than an intended source. Also see BACKSIPHONAGE.

BACKSIPHONAGE BACKSIPHONAGE

A form of backflow caused by a negative or below atmospheric pressure within a water system. Also see BACKFLOW.

BACKWASHING BACKWASHING

The process of reversing the flow of water back through the filter media to remove the entrapped solids.

BACTERIA (back-TEAR-e-ah) BACTERIA

Bacteria are living organisms, microscopic in size, which usually consist of a single cell. Most bacteria use organic matter for their food and produce waste products as a result of their life processes.

BAFFLE BAFFLE

A flat board or plate, deflector, guide or similar device constructed or placed in flowing water or slurry systems to cause more uniform flow velocities, to absorb energy, and to divert, guide, or agitate liquids (water, chemical solutions, slurry).

BAILER (BAY-ler) BAILER

A 10- to 20-foot-long pipe equipped with a valve at the lower end. A bailer is used to remove slurry from the bottom or the side of a well as it is being drilled.

BASE-EXTRA CAPACITY METHOD BASE-EXTRA CAPACITY METHOD

A cost allocation method used by water utilities to determine water rates for various water user groups. This method considers base costs (O & M expenses and capital costs), extra capacity costs (additional costs for maximum day and maximum hour demands), customer costs (meter maintenance and reading, billing, collection, accounting) and fire protection costs.

BASE METAL BASE METAL

A metal (such as iron) which reacts with dilute hydrochloric acid to form hydrogen. Also see NOBLE METAL.

BATCH PROCESS BATCH PROCESS

A treatment process in which a tank or reactor is filled, the water is treated or a chemical solution is prepared, and the tank is emptied. The tank may then be filled and the process repeated.

BENCH SCALE TESTS BENCH SCALE TESTS

A method of studying different ways or chemical doses for treating water on a small scale in a laboratory.

BIOCHEMICAL OXYGEN DEMAND (BOD) BIOCHEMICAL OXYGEN DEMAND (BOD)

The rate at which organisms use the oxygen in water while stabilizing decomposable organic matter under aerobic conditions. In decomposition, organic matter serves as food for the bacteria and energy results from its oxidation. BOD measurements are used as a measure of the organic strength of wastes in water.

BIOLOGICAL GROWTH BIOLOGICAL GROWTH

The activity and growth of any and all living organisms.

BLANK BLANK

A bottle containing only dilution water or distilled water; the sample being tested is not added. Tests are frequently run on a *SAMPLE* and a *BLANK* and the differences are compared. The procedure helps to eliminate or reduce test result errors that could be caused when the dilution water or distilled water used is contaminated.

BOND

(1) A written promise to pay a specified sum of money (called the face value) at a fixed time in the future (called the date of maturity). A bond also carries interest at a fixed rate, payable periodically. The difference between a note and a bond is that a bond usually runs for a longer period of time and requires greater formality. Utility agencies use bonds as a means of obtaining large amounts of money for capital improvements.

(2) A warranty by an underwriting organization, such as an insurance company, guaranteeing honesty, performance, or payment by a contractor.

BONNET (BON-it)

The cover on a gate valve.

BOWL, PUMP

The submerged pumping unit in a well, including the shaft, impellers and housing.

BRAKE HORSEPOWER

(1) The horsepower required at the top or end of a pump shaft (input to a pump).

(2) The energy provided by a motor or other power source.

BREAKPOINT CHLORINATION

Addition of chlorine to water until the chlorine demand has been satisfied. At this point, further additions of chlorine will result in a free chlorine residual that is directly proportional to the amount of chlorine added beyond the breakpoint.

BREAKTHROUGH

A crack or break in a filter bed allowing the passage of floc or particulate matter through a filter. This will cause an increase in filter effluent turbidity. A breakthrough can occur (1) when a filter is first placed in service, (2) when the effluent valve suddenly opens or closes, and (3) during periods of excessive head loss through the filter (including when the filter is exposed to negative heads).

BRINELLING (bruh-NEL-ing)

Tiny indentations (dents) high on the shoulder of the bearing race or bearing. A type of bearing failure.

BUFFER

A solution or liquid whose chemical makeup neutralizes acids or bases without a great change in pH.

BUFFER CAPACITY

A measure of the capacity of a solution or liquid to neutralize acids or bases. This is a measure of the capacity of water for offering a resistance to changes in pH.

C

C FACTOR

A factor or value used to indicate the smoothness of the interior of a pipe. The higher the C Factor, the smoother the pipe, the greater the carrying capacity, and the smaller the friction or energy losses from water flowing in the pipe. To calculate the C Factor, measure the flow, pipe diameter, distance between two pressure gages, and the friction or energy loss of the water between the gages.

$$C \text{ Factor} = \frac{\text{Flow, GPM}}{193.75(\text{Diameter, ft})^{2.63}(\text{Slope})^{0.54}}$$

CT VALUE

Residual concentration of a given disinfectant in mg/L times the disinfectant's contact time in minutes.

CAISSON (KAY-sawn)

A structure or chamber which is usually sunk or lowered by digging from the inside. Used to gain access to the bottom of a stream or other body of water.

CALCIUM CARBONATE EQUILIBRIUM

A water is considered stable when it is just saturated with calcium carbonate. In this condition the water will neither dissolve nor deposit calcium carbonate. Thus, in this water the calcium carbonate is in equilibrium with the hydrogen ion concentration.

CALCIUM CARBONATE (CaCO₃) EQUIVALENT

CALCIUM CARBONATE (CaCO₃) EQUIVALENT

An expression of the concentration of specified constituents in water in terms of their equivalent value to calcium carbonate. For example, the hardness in water which is caused by calcium, magnesium and other ions is usually described as calcium carbonate equivalent. Alkalinity test results are usually reported as mg/L $CaCO_3$ equivalents. To convert chloride to $CaCO_3$ equivalents, multiply the concentration of chloride ions in mg/L by 1.41, and for sulfate, multiply by 1.04.

CALIBRATION

CALIBRATION

A procedure which checks or adjusts an instrument's accuracy by comparison with a standard or reference.

CALL DATE

CALL DATE

First date a bond can be paid off.

CAPILLARY ACTION

CAPILLARY ACTION

The movement of water through very small spaces due to molecular forces.

CAPILLARY FORCES

CAPILLARY FORCES

The molecular forces which cause the movement of water through very small spaces.

CAPILLARY FRINGE

CAPILLARY FRINGE

The porous material just above the water table which may hold water by capillarity (a property of surface tension that draws water upward) in the smaller void spaces.

CARCINOGEN (CAR-sin-o-JEN)

CARCINOGEN

Any substance which tends to produce cancer in an organism.

CATALYST (CAT-uh-LIST)

CATALYST

A substance that changes the speed or yield of a chemical reaction without being consumed or chemically changed by the chemical reaction.

CATALYZE (CAT-uh-LIZE)

CATALYZE

To act as a catalyst. Or, to speed up a chemical reaction.

CATALYZED (CAT-uh-LIZED)

CATALYZED

To be acted upon by a catalyst.

CATHODE (KA-thow-d)

CATHODE

The negative pole or electrode of an electrolytic cell or system. The cathode attracts positively charged particles or ions (cations).

CATHODIC (ca-THOD-ick) PROTECTION

CATHODIC PROTECTION

An electrical system for prevention of rust, corrosion, and pitting of metal surfaces which are in contact with water or soil. A low-voltage current is made to flow through a liquid (water) or a soil in contact with the metal in such a manner that the external electromotive force renders the metal structure cathodic. This concentrates corrosion on auxiliary anodic parts which are deliberately allowed to corrode instead of letting the structure corrode.

CATION (CAT-EYE-en)

CATION

A positively charged ion in an electrolyte solution, attracted to the cathode under the influence of a difference in electrical potential. Sodium ion (Na^+) is a cation.

CATIONIC POLYMER

CATIONIC POLYMER

A polymer having positively charged groups of ions; often used as a coagulant aid.

CAUTION

CAUTION

This word warns against potential hazards or cautions against unsafe practices. Also see DANGER, NOTICE, and WARNING.

CAVITATION (CAV-uh-TAY-shun)

CAVITATION

The formation and collapse of a gas pocket or bubble on the blade of an impeller or the gate of a valve. The collapse of this gas pocket or bubble drives water into the impeller or gate with a terrific force that can cause pitting on the impeller or gate surface. Cavitation is accompanied by loud noises that sound like someone is pounding on the impeller or gate with a hammer.

CENTRATE CENTRATE

The water leaving a centrifuge after most of the solids have been removed.

CENTRIFUGAL (sen-TRIF-uh-gull) PUMP CENTRIFUGAL PUMP

A pump consisting of an impeller fixed on a rotating shaft that is enclosed in a casing, and having an inlet and discharge connection. As the rotating impeller whirls the liquid around, centrifugal force builds up enough pressure to force the water through the discharge outlet.

CENTRIFUGE CENTRIFUGE

A mechanical device that uses centrifugal or rotational forces to separate solids from liquids.

CERTIFICATION EXAMINATION CERTIFICATION EXAMINATION

An examination administered by a state agency that water distribution system operators take to indicate a level of professional competence. In the United States, certification of water distribution system operators is mandatory.

CERTIFIED OPERATOR CERTIFIED OPERATOR

A person who has the education and experience required to operate a specific class of treatment facility as indicated by possessing a certificate of professional competence given by a state agency or professional association.

CHARGE CHEMISTRY CHARGE CHEMISTRY

A branch of chemistry in which the destabilization and neutralization reactions occur between stable negatively charged and stable positively charged particles.

CHECK SAMPLING CHECK SAMPLING

Whenever an initial or routine sample analysis indicates that a Maximum Contaminant Level (MCL) has been exceeded, *CHECK SAMPLING* is required to confirm the routine sampling results. Check sampling is in addition to the routine sampling program.

CHECK VALVE CHECK VALVE

A special valve with a hinged disc or flap that opens in the direction of normal flow and is forced shut when flows attempt to go in the reverse or opposite direction of normal flows.

CHELATING (key-LAY-ting) AGENT CHELATING AGENT

A chemical used to prevent the precipitation of metals (such as copper).

CHELATION (key-LAY-shun) CHELATION

A chemical complexing (forming or joining together) of metallic cations (such as copper) with certain organic compounds, such as EDTA (ethylene diamine tetracetic acid). Chelation is used to prevent the precipitation of metals (copper). Also see SEQUESTRATION.

CHLORAMINATION (KLOR-ah-min-NAY-shun) CHLORAMINATION

The application of chlorine and ammonia to water to form chloramines for the purpose of disinfection.

CHLORAMINES (KLOR-uh-means) CHLORAMINES

Compounds formed by the reaction of hypochlorous acid (or aqueous chlorine) with ammonia.

CHLORINATION (KLOR-uh-NAY-shun) CHLORINATION

The application of chlorine to water, generally for the purpose of disinfection, but frequently for accomplishing other biological or chemical results (aiding coagulation and controlling tastes and odors).

CHLORINATOR (KLOR-uh-NAY-ter) CHLORINATOR

A metering device which is used to add chlorine to water.

CHLORINE DEMAND CHLORINE DEMAND

Chlorine demand is the difference between the amount of chlorine added to water and the amount of residual chlorine remaining after a given contact time. Chlorine demand may change with dosage, time, temperature, pH, and nature and amount of the impurities in the water.

$$\text{Chlorine Demand, mg}/L = \text{Chlorine Applied, mg}/L - \text{Chlorine Residual, mg}/L$$

CHLORINE REQUIREMENT   CHLORINE REQUIREMENT

The amount of chlorine which is needed for a particular purpose. Some reasons for adding chlorine are reducing the number of coliform bacteria (Most Probable Number), obtaining a particular chlorine residual, or oxidizing some substance in the water. In each case a definite dosage of chlorine will be necessary. This dosage is the chlorine requirement.

CHLORINE RESIDUAL   CHLORINE RESIDUAL

The concentration of chlorine present in water after the chlorine demand has been satisfied. The concentration is expressed in terms of the total chlorine residual, which includes both the free and combined or chemically bound chlorine residuals.

CHLOROPHENOLIC (klor-o-FEE-NO-lick)   CHLOROPHENOLIC

Chlorophenolic compounds are phenolic compounds (carbolic acid) combined with chlorine.

CHLOROPHENOXY (KLOR-o-fuh-KNOX-ee)   CHLOROPHENOXY

A class of herbicides that may be found in domestic water supplies and cause adverse health effects. Two widely used chlorophenoxy herbicides are 2,4-D (2,4-Dichlorophenoxy acetic acid) and 2,4,5-TP (2,4,5-Trichlorophenoxy propionic acid (silvex)).

CHLORORGANIC (klor-or-GAN-ick)   CHLORORGANIC

Organic compounds combined with chlorine. These compounds generally originate from, or are associated with, life processes such as those of algae in water.

CHRONIC HEALTH EFFECT   CHRONIC HEALTH EFFECT

An adverse effect on a human or animal body with symptoms that develop slowly over a long period of time or that recur frequently.

CIRCLE OF INFLUENCE   CIRCLE OF INFLUENCE

The circular outer edge of a depression produced in the water table by the pumping of water from a well. Also see CONE OF INFLUENCE and CONE OF DEPRESSION.

[SEE DRAWING ON PAGE 589]

CIRCUIT   CIRCUIT

The complete path of an electric current, including the generating apparatus or other source; or, a specific segment or section of the complete path.

CIRCUIT BREAKER   CIRCUIT BREAKER

A safety device in an electric circuit that automatically shuts off the circuit when it becomes overloaded. The device can be manually reset.

CISTERN (SIS-turn)   CISTERN

A small tank (usually covered) or a storage facility used to store water for a home or farm. Often used to store rainwater.

CLARIFIER (KLAIR-uh-fire)   CLARIFIER

A large circular or rectangular tank or basin in which water is held for a period of time during which the heavier suspended solids settle to the bottom. Clarifiers are also called settling basins and sedimentation basins.

CLASS, PIPE AND FITTINGS   CLASS, PIPE AND FITTINGS

The working pressure rating, including allowances for surges, of a specific pipe for use in water distribution systems. The term is used for cast iron, ductile iron, asbestos cement and some plastic pipe.

CLEAR WELL   CLEAR WELL

A reservoir for the storage of filtered water of sufficient capacity to prevent the need to vary the filtration rate with variations in demand. Also used to provide chlorine contact time for disinfection.

COAGULANT AID   COAGULANT AID

Any chemical or substance used to assist or modify coagulation.

COAGULANTS (co-AGG-you-lents)   COAGULANTS

Chemicals that cause very fine particles to clump (floc) together into larger particles. This makes it easier to separate the solids from the water by settling, skimming, draining or filtering.

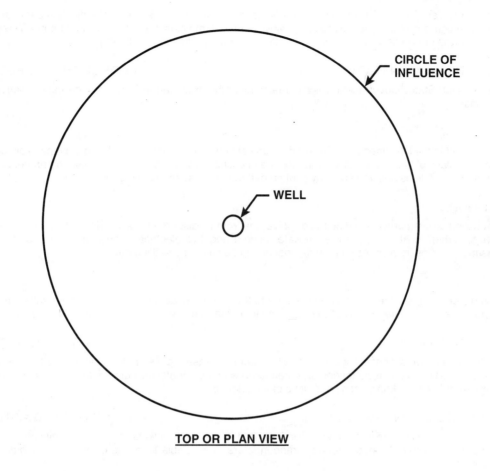

TOP OR PLAN VIEW

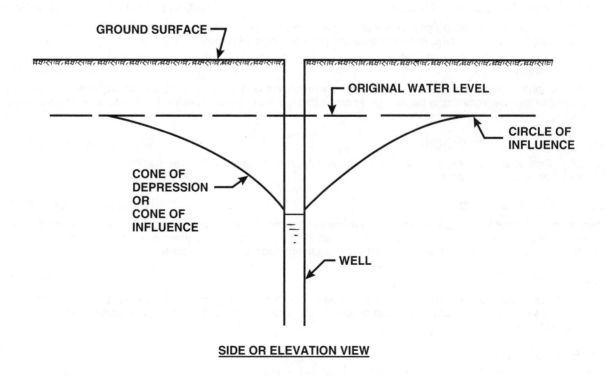

SIDE OR ELEVATION VIEW

CIRCLE OF INFLUENCE and CONE OF DEPRESSION/CONE OF INFLUENCE

COAGULATION (co-AGG-you-LAY-shun)   COAGULATION

The clumping together of very fine particles into larger particles (floc) caused by the use of chemicals (coagulants). The chemicals neutralize the electrical charges of the fine particles, allowing them to come closer and form larger clumps. This clumping together makes it easier to separate the solids from the water by settling, skimming, draining or filtering.

CODE OF FEDERAL REGULATIONS (CFR)   CODE OF FEDERAL REGULATIONS (CFR)

A publication of the United States Government which contains all of the proposed and finalized federal regulations, including environmental regulations.

COLIFORM (COAL-i-form)   COLIFORM

A group of bacteria found in the intestines of warm-blooded animals (including humans) and also in plants, soil, air and water. Fecal coliforms are a specific class of bacteria which only inhabit the intestines of warm-blooded animals. The presence of coliform bacteria is an indication that the water is polluted and may contain pathogenic (disease-causing) organisms.

COLLOIDS (CALL-loids)   COLLOIDS

Very small, finely divided solids (particles that do not dissolve) that remain dispersed in a liquid for a long time due to their small size and electrical charge. When most of the particles in water have a negative electrical charge, they tend to repel each other. This repulsion prevents the particles from clumping together, becoming heavier, and settling out.

COLOR   COLOR

The substances in water that impart a yellowish-brown color to the water. These substances are the result of iron and manganese ions, humus and peat materials, plankton, aquatic weeds, and industrial waste present in the water. Also see TRUE COLOR.

COLORIMETRIC MEASUREMENT   COLORIMETRIC MEASUREMENT

A means of measuring unknown chemical concentrations in water by measuring a sample's color intensity. The specific color of the sample, developed by addition of chemical reagents, is measured with a photoelectric colorimeter or is compared with "color standards" using, or corresponding with, known concentrations of the chemical.

COMBINED AVAILABLE CHLORINE   COMBINED AVAILABLE CHLORINE

The total chlorine, present as chloramine or other derivatives, that is present in a water and is still available for disinfection and for oxidation of organic matter. The combined chlorine compounds are more stable than free chlorine forms, but they are somewhat slower in disinfection action.

COMBINED AVAILABLE CHLORINE RESIDUAL   COMBINED AVAILABLE CHLORINE RESIDUAL

The concentration of residual chlorine that is combined with ammonia, organic nitrogen, or both in water as a chloramine (or other chloro derivative) and yet is still available to oxidize organic matter and help kill bacteria.

COMBINED CHLORINE   COMBINED CHLORINE

The sum of the chlorine species composed of free chlorine and ammonia, including monochloramine, dichloramine, and trichloramine (nitrogen trichloride). Dichloramine is the strongest disinfectant of these chlorine species, but it has less oxidative capacity than free chlorine.

COMBINED RESIDUAL CHLORINATION   COMBINED RESIDUAL CHLORINATION

The application of chlorine to water to produce combined available chlorine residual. This residual can be made up of monochloramines, dichloramines, and nitrogen trichloride.

COMMODITY-DEMAND METHOD   COMMODITY-DEMAND METHOD

A cost allocation method used by water utilities to determine water rates for the various water user groups. This method considers the commodity costs (water, chemicals, power, amount of water use), demand costs (treatment, storage, distribution), customer costs (meter maintenance and reading, billing, collection, accounting) and fire protection costs.

COMPETENT PERSON   COMPETENT PERSON

A competent person is defined by OSHA as a person capable of identifying existing and predictable hazards in the surroundings, or working conditions which are unsanitary, hazardous or dangerous to employees, and who has authorization to take prompt corrective measures to eliminate the hazards.

COMPLETE TREATMENT   COMPLETE TREATMENT

A method of treating water which consists of the addition of coagulant chemicals, flash mixing, coagulation-flocculation, sedimentation and filtration. Also called CONVENTIONAL FILTRATION.

COMPOSITE (come-PAH-zit) (PROPORTIONAL) SAMPLE

COMPOSITE (PROPORTIONAL) SAMPLE

A composite sample is a collection of individual samples obtained at regular intervals, usually every one or two hours during a 24-hour time span. Each individual sample is combined with the others in proportion to the rate of flow when the sample was collected. The resulting mixture (composite sample) forms a representative sample and is analyzed to determine the average conditions during the sampling period.

COMPOUND   COMPOUND

A pure substance composed of two or more elements whose composition is constant. For example, table salt (sodium chloride, NaCl) is a compound.

CONCENTRATION POLARIZATION   CONCENTRATION POLARIZATION

(1) A buildup of retained particles on the membrane surface due to dewatering of the feed closest to the membrane. The thickness of the concentration polarization layer is controlled by the flow velocity across the membrane.

(2) Used in corrosion studies to indicate a depletion of ions near an electrode.

(3) The basis for chemical analysis by a polarograph.

CONDITIONING   CONDITIONING

Pretreatment of sludge to facilitate removal of water in subsequent treatment processes.

CONDUCTANCE   CONDUCTANCE

A rapid method of estimating the dissolved solids content of a water supply. The measurement indicates the capacity of a sample of water to carry an electric current, which is related to the concentration of ionized substances in the water. Also called SPECIFIC CONDUCTANCE.

CONDUCTIVITY   CONDUCTIVITY

A measure of the ability of a solution (water) to carry an electric current.

CONDUCTOR   CONDUCTOR

A substance, body, device or wire that readily conducts or carries electric current.

CONDUCTOR CASING   CONDUCTOR CASING

The outer casing of a well. The purpose of this casing is to prevent contaminants from surface waters or shallow groundwaters from entering a well.

CONE OF DEPRESSION   CONE OF DEPRESSION

The depression, roughly conical in shape, produced in the water table by the pumping of water from a well. Also called the CONE OF INFLUENCE. Also see CIRCLE OF INFLUENCE.

[SEE DRAWING ON PAGE 589]

CONE OF INFLUENCE   CONE OF INFLUENCE

The depression, roughly conical in shape, produced in the water table by the pumping of water from a well. Also called the CONE OF DEPRESSION. Also see CIRCLE OF INFLUENCE.

[SEE DRAWING ON PAGE 589]

CONFINED SPACE   CONFINED SPACE

Confined space means a space that:

A. Is large enough and so configured that an employee can bodily enter and perform assigned work; and

B. Has limited or restricted means for entry or exit (for example, manholes, tanks, vessels, silos, storage bins, hoppers, vaults, and pits are spaces that may have limited means of entry); and

C. Is not designed for continuous employee occupancy.

(Definition from the Code of Federal Regulations (CFR) Title 29 Part 1910.146.)

CONFINED SPACE, CLASS "A"   CONFINED SPACE, CLASS "A"

A confined space that presents a situation that is immediately dangerous to life or health (IDLH). These include but are not limited to oxygen deficiency, explosive or flammable atmospheres, and/or concentrations of toxic substances.

(Definition from NIOSH, "Criteria for a Recommended Standard: Working in Confined Spaces.")

CONFINED SPACE, CLASS "B"   CONFINED SPACE, CLASS "B"

A confined space that has the potential for causing injury and illness, if preventive measures are not used, but not immediately dangerous to life and health.

(Definition from NIOSH, "Criteria for a Recommended Standard: Working in Confined Spaces.")

CONFINED SPACE, CLASS "C"   CONFINED SPACE, CLASS "C"

A confined space in which the potential hazard would not require any special modification of the work procedure.

(Definition from NIOSH, "Criteria for a Recommended Standard: Working in Confined Spaces.")

CONFINED SPACE, NON-PERMIT   CONFINED SPACE, NON-PERMIT

A non-permit confined space is a confined space that does not contain or, with respect to atmospheric hazards, have the potential to contain any hazard capable of causing death or serious physical harm.

CONFINED SPACE, PERMIT-REQUIRED
  (PERMIT SPACE)   CONFINED SPACE, PERMIT-REQUIRED
  (PERMIT SPACE)

A confined space that has one or more of the following characteristics:

- Contains or has a potential to contain a hazardous atmosphere,
- Contains a material that has the potential for engulfing an entrant,
- Has an internal configuration such that an entrant could be trapped or asphyxiated by inwardly converging walls or by a floor which slopes downward and tapers to a smaller cross section, or
- Contains any other recognized serious safety or health hazard.

(Definition from the Code of Federal Regulations (CFR) Title 29 Part 1910.146.)

CONFINING UNIT   CONFINING UNIT

A layer of rock or soil of very low hydraulic conductivity that hampers the movement of groundwater in and out of an aquifer.

CONSOLIDATED FORMATION   CONSOLIDATED FORMATION

A geologic material whose particles are stratified (layered), cemented or firmly packed together (hard rock); usually occurring at a depth below the ground surface. Also see UNCONSOLIDATED FORMATION.

CONSUMER CONFIDENCE REPORTS   CONSUMER CONFIDENCE REPORTS

An annual report prepared by a water utility to communicate with its consumers. The report provides consumers with information on the source and quality of their drinking water. The report is an opportunity for positive communication with consumers and to convey the importance of paying for good quality drinking water.

CONTACTOR   CONTACTOR

An electric switch, usually magnetically operated.

CONTAMINATION   CONTAMINATION

The introduction into water of microorganisms, chemicals, toxic substances, wastes, or wastewater in a concentration that makes the water unfit for its next intended use.

CONTINUOUS SAMPLE   CONTINUOUS SAMPLE

A flow of water from a particular place in a plant to the location where samples are collected for testing. This continuous stream may be used to obtain grab or composite samples. Frequently, several taps (faucets) will flow continuously in the laboratory to provide test samples from various places in a water treatment plant.

CONTROL LOOP   CONTROL LOOP

The path through the control system between the sensor, which measures a process variable, and the controller, which controls or adjusts the process variable.

CONTROL SYSTEM   CONTROL SYSTEM

An instrumentation system which senses and controls its own operation on a close, continuous basis in what is called proportional (or modulating) control.

CONTROLLER   CONTROLLER

A device which controls the starting, stopping, or operation of a device or piece of equipment.

CONVENTIONAL FILTRATION CONVENTIONAL FILTRATION

A method of treating water which consists of the addition of coagulant chemicals, flash mixing, coagulation-flocculation, sedimentation and filtration. Also called COMPLETE TREATMENT. Also see DIRECT FILTRATION and IN-LINE FILTRATION.

CONVENTIONAL TREATMENT CONVENTIONAL TREATMENT

See CONVENTIONAL FILTRATION. Also called COMPLETE TREATMENT.

CORPORATION STOP CORPORATION STOP

A water service shutoff valve located at a street water main. This valve cannot be operated from the ground surface because it is buried and there is no valve box. Also called a corporation cock.

CORROSION CORROSION

The gradual decomposition or destruction of a material by chemical action, often due to an electrochemical reaction. Corrosion may be caused by (1) stray current electrolysis, (2) galvanic corrosion caused by dissimilar metals, or (3) differential-concentration cells. Corrosion starts at the surface of a material and moves inward.

CORROSION INHIBITORS CORROSION INHIBITORS

Substances that slow the rate of corrosion.

CORROSIVE GASES CORROSIVE GASES

In water, dissolved oxygen reacts readily with metals at the anode of a corrosion cell, accelerating the rate of corrosion until a film of oxidation products such as rust forms. At the cathode where hydrogen gas may form a coating on the cathode and slow the corrosion rate, oxygen reacts rapidly with hydrogen gas forming water, and again increases the rate of corrosion.

CORROSIVITY CORROSIVITY

An indication of the corrosiveness of a water. The corrosiveness of a water is described by the water's pH, alkalinity, hardness, temperature, total dissolved solids, dissolved oxygen concentration, and the Langelier Index.

COULOMB (COO-lahm) COULOMB

A measurement of the amount of electrical charge carried by an electric current of one ampere in one second. One coulomb equals about 6.25×10^{18} electrons (6,250,000,000,000,000,000 electrons).

COUPON COUPON

A steel specimen inserted into water to measure the corrosiveness of water. The rate of corrosion is measured as the loss of weight of the coupon (in milligrams) per surface area (in square decimeters) exposed to the water per day. 10 decimeters = 1 meter = 100 centimeters.

COVERAGE RATIO COVERAGE RATIO

The coverage ratio is a measure of the ability of the utility to pay the principal and interest on loans and bonds (this is known as "debt service") in addition to any unexpected expenses.

CROSS CONNECTION CROSS CONNECTION

A connection between a drinking (potable) water system and an unapproved water supply. For example, if you have a pump moving nonpotable water and hook into the drinking water system to supply water for the pump seal, a cross connection or mixing between the two water systems can occur. This mixing may lead to contamination of the drinking water.

CRYPTOSPORIDIUM (CRIP-toe-spo-RID-ee-um) CRYPTOSPORIDIUM

A waterborne intestinal parasite that causes a disease called cryptosporidiosis (CRIP-toe-spo-rid-ee-O-sis) in infected humans. Symptoms of the disease include diarrhea, cramps, and weight loss. *Cryptosporidium* contamination is found in most surface waters and some groundwaters. Commonly referred to as "crypto."

CURB STOP CURB STOP

A water service shutoff valve located in a water service pipe near the curb and between the water main and the building. This valve is usually operated by a wrench or valve key and is used to start or stop flows in the water service line to a building. Also called a curb cock.

CURIE CURIE

A measure of radioactivity. One Curie of radioactivity is equivalent to 3.7×10^{10} or 37,000,000,000 nuclear disintegrations per second.

CURRENT   CURRENT

A movement or flow of electricity. Water flowing in a pipe is measured in gallons per second past a certain point, not by the number of water molecules going past a point. Electric current is measured by the number of coulombs per second flowing past a certain point in a conductor. A coulomb is equal to about 6.25×10^{18} electrons (6,250,000,000,000,000,000 electrons). A flow of one coulomb per second is called one ampere, the unit of the rate of flow of current.

CYCLE   CYCLE

A complete alternation of voltage and/or current in an alternating current (A.C.) circuit.

D

DBP   DBP

See **DISINFECTION BY-PRODUCT.**

DPD (pronounce as separate letters)   DPD

A method of measuring the chlorine residual in water. The residual may be determined by either titrating or comparing a developed color with color standards. DPD stands for N,N-diethyl-p-phenylene-diamine.

DANGER   DANGER

The word *DANGER* is used where an immediate hazard presents a threat of death or serious injury to employees. Also see CAUTION, NOTICE, and WARNING.

DANGEROUS AIR CONTAMINATION   DANGEROUS AIR CONTAMINATION

An atmosphere presenting a threat of causing death, injury, acute illness, or disablement due to the presence of flammable and/or explosive, toxic or otherwise injurious or incapacitating substances.

A. Dangerous air contamination due to the flammability of a gas or vapor is defined as an atmosphere containing the gas or vapor at a concentration greater than 10 percent of its lower explosive (lower flammable) limit.

B. Dangerous air contamination due to a combustible particulate is defined as a concentration greater than 10 percent of the minimum explosive concentration of the particulate.

C. Dangerous air contamination due to the toxicity of a substance is defined as the atmospheric concentration immediately hazardous to life or health.

DATEOMETER (day-TOM-uh-ter)   DATEOMETER

A small calendar disc attached to motors and equipment to indicate the year in which the last maintenance service was performed.

DATUM LINE   DATUM LINE

A line from which heights and depths are calculated or measured. Also called a datum plane or a datum level.

DAY TANK   DAY TANK

A tank used to store a chemical solution of known concentration for feed to a chemical feeder. A day tank usually stores sufficient chemical solution to properly treat the water being treated for at least one day. Also called an AGE TANK.

DEAD END   DEAD END

The end of a water main which is not connected to other parts of the distribution system by means of a connecting loop of pipe.

DEBT SERVICE   DEBT SERVICE

The amount of money required annually to pay the (1) interest on outstanding debts; or (2) funds due on a maturing bonded debt or the redemption of bonds.

DECANT   DECANT

To draw off the upper layer of liquid (water) after the heavier material (a solid or another liquid) has settled.

DECANT WATER   DECANT WATER

Water that has separated from sludge and is removed from the layer of water above the sludge.

DECHLORINATION (dee-KLOR-uh-NAY-shun)   DECHLORINATION

The deliberate removal of chlorine from water. The partial or complete reduction of residual chlorine by any chemical or physical process.

DECIBEL (DES-uh-bull) DECIBEL

A unit for expressing the relative intensity of sounds on a scale from zero for the average least perceptible sound to about 130 for the average level at which sound causes pain to humans. Abbreviated dB.

DECOMPOSITION, DECAY DECOMPOSITION, DECAY

The conversion of chemically unstable materials to more stable forms by chemical or biological action. If organic matter decays when there is no oxygen present (anaerobic conditions or putrefaction), undesirable tastes and odors are produced. Decay of organic matter when oxygen is present (aerobic conditions) tends to produce much less objectionable tastes and odors.

DEFLUORIDATION (de-FLOOR-uh-DAY-shun) DEFLUORIDATION

The removal of excess fluoride in drinking water to prevent the mottling (brown stains) of teeth.

DEGASIFICATION (DEE-GAS-if-uh-KAY-shun) DEGASIFICATION

A water treatment process which removes dissolved gases from the water. The gases may be removed by either mechanical or chemical treatment methods or a combination of both.

DELEGATION DELEGATION

The act in which power is given to another person in the organization to accomplish a specific job.

DEMINERALIZATION (DEE-MIN-er-al-uh-ZAY-shun) DEMINERALIZATION

A treatment process which removes dissolved minerals (salts) from water.

DENSITY (DEN-sit-tee) DENSITY

A measure of how heavy a substance (solid, liquid or gas) is for its size. Density is expressed in terms of weight per unit volume, that is, grams per cubic centimeter or pounds per cubic foot. The density of water (at 4°C or 39°F) is 1.0 gram per cubic centimeter or about 62.4 pounds per cubic foot.

DEPOLARIZATION DEPOLARIZATION

The removal or depletion of ions in the thin boundary layer adjacent to a membrane or pipe wall.

DEPRECIATION DEPRECIATION

The gradual loss in service value of a facility or piece of equipment due to all the factors causing the ultimate retirement of the facility or equipment. This loss can be caused by sudden physical damage, wearing out due to age, obsolescence, inadequacy or availability of a newer, more efficient facility or equipment. The value cannot be restored by maintenance.

DESALINIZATION (DEE-SAY-leen-uh-ZAY-shun) DESALINIZATION

The removal of dissolved salts (such as sodium chloride, NaCl) from water by natural means (leaching) or by specific water treatment processes.

DESICCANT (DESS-uh-kant) DESICCANT

A drying agent which is capable of removing or absorbing moisture from the atmosphere in a small enclosure.

DESICCATION (DESS-uh-KAY-shun) DESICCATION

A process used to thoroughly dry air; to remove virtually all moisture from air.

DESICCATOR (DESS-uh-KAY-tor) DESICCATOR

A closed container into which heated weighing or drying dishes are placed to cool in a dry environment in preparation for weighing. The dishes may be empty or they may contain a sample. Desiccators contain a substance, such as anhydrous calcium chloride, which absorbs moisture and keeps the relative humidity near zero so that the dish or sample will not gain weight from absorbed moisture.

DESTRATIFICATION (de-STRAT-uh-fuh-KAY-shun) DESTRATIFICATION

The development of vertical mixing within a lake or reservoir to eliminate (either totally or partially) separate layers of temperature, plant, or animal life. This vertical mixing can be caused by mechanical means (pumps) or through the use of forced air diffusers which release air into the lower layers of the reservoir.

DETECTION LAG DETECTION LAG

The time period between the moment a process change is made and the moment when such a change is finally sensed by the associated measuring instrument.

DETENTION TIME DETENTION TIME

(1) The theoretical (calculated) time required for a small amount of water to pass through a tank at a given rate of flow.

(2) The actual time in hours, minutes or seconds that a small amount of water is in a settling basin, flocculating basin or rapid-mix chamber. In storage reservoirs, detention time is the length of time entering water will be held before being drafted for use (several weeks to years, several months being typical).

$$\text{Detention Time, hr} = \frac{(\text{Basin Volume, gal})(24 \text{ hr/day})}{\text{Flow, gal/day}}$$

DEW POINT DEW POINT

The temperature to which air with a given quantity of water vapor must be cooled to cause condensation of the vapor in the air.

DEWATER DEWATER

(1) To remove or separate a portion of the water present in a sludge or slurry. To dry sludge so it can be handled and disposed of.

(2) To remove or drain the water from a tank or a trench.

DIATOMACEOUS (DYE-uh-toe-MAY-shus) EARTH DIATOMACEOUS EARTH

A fine, siliceous (made of silica) "earth" composed mainly of the skeletal remains of diatoms.

DIATOMS (DYE-uh-toms) DIATOMS

Unicellular (single cell), microscopic algae with a rigid (box-like) internal structure consisting mainly of silica.

DIELECTRIC (DIE-ee-LECK-trick) DIELECTRIC

Does not conduct an electric current. An insulator or nonconducting substance.

DIGITAL READOUT DIGITAL READOUT

Use of numbers to indicate the value or measurement of a variable. The readout of an instrument by a direct, numerical reading of the measured value. The signal sent to such readouts is usually an analog signal.

DILUTE SOLUTION DILUTE SOLUTION

A solution that has been made weaker, usually by the addition of water.

DIMICTIC (die-MICK-tick) DIMICTIC

Lakes and reservoirs which freeze over and normally go through two stratification and two mixing cycles within a year.

DIRECT CURRENT (D.C.) DIRECT CURRENT (D.C.)

Electric current flowing in one direction only and essentially free from pulsation.

DIRECT FILTRATION DIRECT FILTRATION

A method of treating water which consists of the addition of coagulant chemicals, flash mixing, coagulation, minimal flocculation, and filtration. The flocculation facilities may be omitted, but the physical-chemical reactions will occur to some extent. The sedimentation process is omitted. Also see CONVENTIONAL FILTRATION and IN-LINE FILTRATION.

DIRECT RUNOFF DIRECT RUNOFF

Water that flows over the ground surface or through the ground directly into streams, rivers, or lakes.

DISCHARGE HEAD DISCHARGE HEAD

The pressure (in pounds per square inch or psi) measured at the centerline of a pump discharge and very close to the discharge flange, converted into feet. The pressure is measured from the centerline of the pump to the hydraulic grade line of the water in the discharge pipe.

$$\text{Discharge Head, ft} = (\text{Discharge Pressure, psi})(2.31 \text{ ft/psi})$$

DISINFECTION (dis-in-FECT-shun) DISINFECTION

The process designed to kill or inactivate most microorganisms in water, including essentially all pathogenic (disease-causing) bacteria. There are several ways to disinfect, with chlorination being the most frequently used in water treatment. Compare with STERILIZATION.

DISINFECTION BY-PRODUCT (DBP)

DISINFECTION BY-PRODUCT (DBP)

A contaminant formed by the reaction of disinfection chemicals (such as chlorine) with other substances in the water being disinfected.

DISTILLATE (DIS-tuh-late)

DISTILLATE

In the distillation of a sample, a portion is collected by evaporation and recondensation; the part that is recondensed is the distillate.

DIVALENT (die-VAY-lent)

DIVALENT

Having a valence of two, such as the ferrous ion, Fe^{2+}. Also called bivalent.

DIVERSION

DIVERSION

Use of part of a stream flow as a water supply.

DRAFT

DRAFT

(1) The act of drawing or removing water from a tank or reservoir.

(2) The water which is drawn or removed from a tank or reservoir.

DRAWDOWN

DRAWDOWN

(1) The drop in the water table or level of water in the ground when water is being pumped from a well.

(2) The amount of water used from a tank or reservoir.

(3) The drop in the water level of a tank or reservoir.

DRIFT

DRIFT

The difference between the actual value and the desired value (or set point); characteristic of proportional controllers that do not incorporate reset action. Also called OFFSET.

DYNAMIC PRESSURE

DYNAMIC PRESSURE

When a pump is operating, the vertical distance (in feet) from a reference point (such as a pump centerline) to the hydraulic grade line is the dynamic head. Also see ENERGY GRADE LINE, STATIC HEAD, STATIC PRESSURE, and TOTAL DYNAMIC HEAD.

Dynamic Pressure, psi = (Dynamic Head, ft)(0.433 psi/ft)

E

EPA

EPA

United States **E**nvironmental **P**rotection **A**gency. A regulatory agency established by the U.S. Congress to administer the nation's environmental laws. Also called the U.S. EPA.

EDUCTOR (e-DUCK-ter)

EDUCTOR

A hydraulic device used to create a negative pressure (suction) by forcing a liquid through a restriction, such as a Venturi. An eductor or aspirator (the hydraulic device) may be used in the laboratory in place of a vacuum pump. As an injector, it is used to produce vacuum for chlorinators. Sometimes used instead of a suction pump.

EFFECTIVE RANGE

EFFECTIVE RANGE

That portion of the design range (usually from 10 to 90+ percent) in which an instrument has acceptable accuracy. Also see RANGE and SPAN.

EFFECTIVE SIZE (E.S.)

EFFECTIVE SIZE (E.S.)

The diameter of the particles in a granular sample (filter media) for which 10 percent of the total grains are smaller and 90 percent larger on a weight basis. Effective size is obtained by passing granular material through sieves with varying dimensions of mesh and weighing the material retained by each sieve. The effective size is also approximately the average size of the grains.

EFFLUENT

EFFLUENT

Water or other liquid—raw (untreated), partially or completely treated—flowing *FROM* a reservoir, basin, treatment process, or treatment plant.

EJECTOR

EJECTOR

A device used to disperse a chemical solution into water being treated.

ELECTROCHEMICAL REACTION

ELECTROCHEMICAL REACTION

Chemical changes produced by electricity (electrolysis) or the production of electricity by chemical changes (galvanic action). In corrosion, a chemical reaction is accompanied by the flow of electrons through a metallic path. The electron flow may come from an external source and cause the reaction, such as electrolysis caused by a D.C. (direct current) electric railway or the electron flow may be caused by a chemical reaction as in the galvanic action of a flashlight dry cell.

ELECTROCHEMICAL SERIES

ELECTROCHEMICAL SERIES

A list of metals with the standard electrode potentials given in volts. The size and sign of the electrode potential indicates how easily these elements will take on or give up electrons, or corrode. Hydrogen is conventionally assigned a value of zero.

ELECTROLYSIS (ee-leck-TRAWL-uh-sis)

ELECTROLYSIS

The decomposition of material by an outside electric current.

ELECTROLYTE (ee-LECK-tro-LITE)

ELECTROLYTE

A substance which dissociates (separates) into two or more ions when it is dissolved in water.

ELECTROLYTIC (ee-LECK-tro-LIT-ick) CELL

ELECTROLYTIC CELL

A device in which the chemical decomposition of material causes an electric current to flow. Also, a device in which a chemical reaction occurs as a result of the flow of electric current. Chlorine and caustic (NaOH) are made from salt (NaCl) in electrolytic cells.

ELECTROMOTIVE FORCE (E.M.F.)

ELECTROMOTIVE FORCE (E.M.F.)

The electrical pressure available to cause a flow of current (amperage) when an electric circuit is closed. Also called VOLTAGE.

ELECTROMOTIVE SERIES

ELECTROMOTIVE SERIES

A list of metals and alloys presented in the order of their tendency to corrode (or go into solution). Also called the GALVANIC SERIES. This is a practical application of the theoretical ELECTROCHEMICAL SERIES.

ELECTRON

ELECTRON

(1) A very small, negatively charged particle which is practically weightless. According to the electron theory, all electrical and electronic effects are caused either by the movement of electrons from place to place or because there is an excess or lack of electrons at a particular place.

(2) The part of an atom that determines its chemical properties.

ELEMENT

ELEMENT

A substance which cannot be separated into its constituent parts and still retain its chemical identity. For example, sodium (Na) is an element.

END BELLS

END BELLS

Devices used to hold the rotor and stator of a motor in position.

END POINT

END POINT

Samples of water or wastewater are titrated to the end point. This means that a chemical is added, drop by drop, to a sample until a certain color change (blue to clear, for example) occurs. This is called the *END POINT* of the titration. In addition to a color change, an end point may be reached by the formation of a precipitate or the reaching of a specified pH. An end point may be detected by the use of an electronic device such as a pH meter. The completion of a desired chemical reaction.

ENDEMIC (en-DEM-ick)

ENDEMIC

Something peculiar to a particular people or locality, such as a disease which is always present in the population.

ENDRIN (EN-drin)

ENDRIN

A pesticide toxic to freshwater and marine aquatic life that produces adverse health effects in domestic water supplies.

ENERGY GRADE LINE (EGL)

ENERGY GRADE LINE (EGL)

A line that represents the elevation of energy head (in feet) of water flowing in a pipe, conduit or channel. The line is drawn above the hydraulic grade line (gradient) a distance equal to the velocity head ($V^2/2g$) of the water flowing at each section or point along the pipe or channel. Also see HYDRAULIC GRADE LINE.

[SEE DRAWING ON PAGE 599]

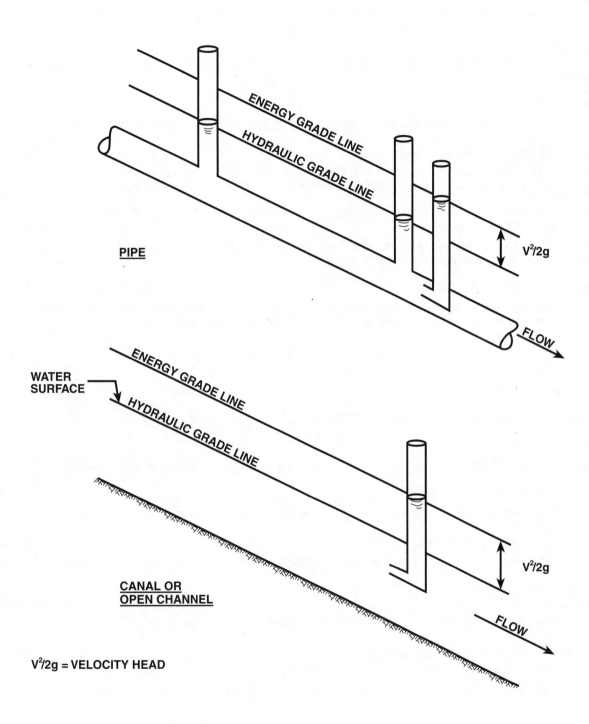

PIPE

WATER
SURFACE

ENERGY GRADE LINE

HYDRAULIC GRADE LINE

$V^2/2g$

CANAL OR
OPEN CHANNEL

FLOW

$V^2/2g$ = VELOCITY HEAD

ENERGY GRADE LINE and HYDRAULIC GRADE LINE

ENTERIC

Of intestinal origin, especially applied to wastes or bacteria.

ENTRAIN

To trap bubbles in water either mechanically through turbulence or chemically through a reaction.

ENZYMES (EN-zimes)

Organic substances (produced by living organisms) which cause or speed up chemical reactions. Organic catalysts and/or biochemical catalysts.

EPIDEMIC (EP-uh-DEM-ick)

A disease that occurs in a large number of people in a locality at the same time and spreads from person to person.

EPIDEMIOLOGY (EP-uh-DE-me-ALL-o-gee)

A branch of medicine which studies epidemics (diseases which affect significant numbers of people during the same time period in the same locality). The objective of epidemiology is to determine the factors that cause epidemic diseases and how to prevent them.

EPILIMNION (EP-uh-LIM-knee-on)

The upper layer of water in a thermally stratified lake or reservoir. This layer consists of the warmest water and has a fairly uniform (constant) temperature. The layer is readily mixed by wind action.

EQUILIBRIUM, CALCIUM CARBONATE

A water is considered stable when it is just saturated with calcium carbonate. In this condition the water will neither dissolve nor deposit calcium carbonate. Thus, in this water the calcium carbonate is in equilibrium with the hydrogen ion concentration.

EQUITY

The value of an investment in a facility.

EQUIVALENT WEIGHT

That weight which will react with, displace or is equivalent to one gram atom of hydrogen.

ESTER

A compound formed by the reaction between an acid and an alcohol with the elimination of a molecule of water.

EUTROPHIC (you-TRO-fick)

Reservoirs and lakes which are rich in nutrients and very productive in terms of aquatic animal and plant life.

EUTROPHICATION (you-TRO-fi-KAY-shun)

The increase in the nutrient levels of a lake or other body of water; this usually causes an increase in the growth of aquatic animal and plant life.

EVAPORATION

The process by which water or other liquid becomes a gas (water vapor or ammonia vapor).

EVAPOTRANSPIRATION (ee-VAP-o-TRANS-purr-A-shun)

(1) The process by which water vapor passes into the atmosphere from living plants. Also called TRANSPIRATION.

(2) The total water removed from an area by transpiration (plants) and by evaporation from soil, snow and water surfaces.

F

FACULTATIVE (FACK-ul-TAY-tive)

Facultative bacteria can use either dissolved molecular oxygen or oxygen obtained from food materials such as sulfate or nitrate ions. In other words, facultative bacteria can live under aerobic, anoxic, or anaerobic conditions.

FEEDBACK

The circulating action between a sensor measuring a process variable and the controller which controls or adjusts the process variable.

FEEDWATER

The water that is fed to a treatment process; the water that is going to be treated.

FINISHED WATER

Water that has passed through a water treatment plant; all the treatment processes are completed or "finished." This water is ready to be delivered to consumers. Also called PRODUCT WATER.

FIX, SAMPLE

A sample is "fixed" in the field by adding chemicals that prevent the water quality indicators of interest in the sample from changing before final measurements are performed later in the lab.

FIXED COSTS

Costs that a utility must cover or pay even if there is no demand for water or no water to sell to customers. Also see VARIABLE COSTS.

FLAGELLATES (FLAJ-el-LATES)

Microorganisms that move by the action of tail-like projections.

FLAME POLISHED

Melted by a flame to smooth out irregularities. Sharp or broken edges of glass (such as the end of a glass tube) are rotated in a flame until the edge melts slightly and becomes smooth.

FLOAT ON SYSTEM

A method of operating a water storage facility. Daily flow into the facility is approximately equal to the average daily demand for water. When consumer demands for water are low, the storage facility will be filling. During periods of high demand, the facility will be emptying.

FLOC

Clumps of bacteria and particulate impurities that have come together and formed a cluster. Found in flocculation tanks and settling or sedimentation basins.

FLOCCULATION (FLOCK-you-LAY-shun)

The gathering together of fine particles after coagulation to form larger particles by a process of gentle mixing.

FLUIDIZED (FLEW-id-I-zd)

A mass of solid particles that is made to flow like a liquid by injection of water or gas is said to have been fluidized. In water treatment, a bed of filter media is fluidized by backwashing water through the filter.

FLUORIDATION (FLOOR-uh-DAY-shun)

The addition of a chemical to increase the concentration of fluoride ions in drinking water to a predetermined optimum limit to reduce the incidence (number) of dental caries (tooth decay) in children. Defluoridation is the removal of excess fluoride in drinking water to prevent the mottling (brown stains) of teeth.

FLUSHING

A method used to clean water distribution lines. Hydrants are opened and water with a high velocity flows through the pipes, removes deposits from the pipes, and flows out the hydrants.

FLUX

A flowing or flow.

FOOT VALVE

A special type of check valve located at the bottom end of the suction pipe on a pump. This valve opens when the pump operates to allow water to enter the suction pipe but closes when the pump shuts off to prevent water from flowing out of the suction pipe.

FREE AVAILABLE RESIDUAL CHLORINE

That portion of the total available residual chlorine composed of dissolved chlorine gas (Cl_2), hypochlorous acid (HOCl), and/or hypochlorite ion (OCl^-) remaining in water after chlorination. This does not include chlorine that has combined with ammonia, nitrogen, or other compounds.

FREE RESIDUAL CHLORINATION   FREE RESIDUAL CHLORINATION

The application of chlorine to water to produce a free available chlorine residual equal to at least 80 percent of the total residual chlorine (sum of free and combined available chlorine residual).

FREEBOARD   FREEBOARD

(1) The vertical distance from the normal water surface to the top of the confining wall.

(2) The vertical distance from the sand surface to the underside of a trough in a sand filter. This distance is also called AVAILABLE EXPANSION.

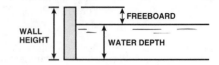

FRICTION LOSSES   FRICTION LOSSES

The head, pressure or energy (they are the same) lost by water flowing in a pipe or channel as a result of turbulence caused by the velocity of the flowing water and the roughness of the pipe, channel walls, or restrictions caused by fittings. Water flowing in a pipe loses head, pressure or energy as a result of friction losses. Also see HEAD LOSS.

FUNGI (FUN-ji)   FUNGI

Mushrooms, molds, mildews, rusts, and smuts that are small non-chlorophyll-bearing plants lacking roots, stems and leaves. They occur in natural waters and grow best in the absence of light. Their decomposition may cause objectionable tastes and odors in water.

FUSE   FUSE

A protective device having a strip or wire of fusible metal which, when placed in a circuit, will melt and break the electric circuit if heated too much. High temperatures will develop in the fuse when a current flows through the fuse in excess of that which the circuit will carry safely.

G

GIS   GIS

Geographic **I**nformation **S**ystem. A computer program that combines mapping with detailed information about the physical locations of structures such as pipes, valves, and manholes within geographic areas. The system is used to help operators and maintenance personnel locate utility system features or structures and to assist with the scheduling and performance of maintenance activities.

GAGE PRESSURE   GAGE PRESSURE

The pressure within a closed container or pipe as measured with a gage. In contrast, absolute pressure is the sum of atmospheric pressure (14.7 lbs/sq in) *PLUS* pressure within a vessel (as measured with a gage). Most pressure gages read in "gage pressure" or psig (**p**ounds per **s**quare **i**nch **g**age pressure).

GALVANIC CELL   GALVANIC CELL

An electrolytic cell capable of producing electric energy by electrochemical action. The decomposition of materials in the cell causes an electric (electron) current to flow from cathode to anode.

GALVANIC SERIES   GALVANIC SERIES

A list of metals and alloys presented in the order of their tendency to corrode (or go into solution). Also called the ELECTROMOTIVE SERIES. This is a practical application of the theoretical ELECTROCHEMICAL SERIES.

GALVANIZE   GALVANIZE

To coat a metal (especially iron or steel) with zinc. Galvanization is the process of coating a metal with zinc.

GARNET   GARNET

A group of hard, reddish, glassy, mineral sands made up of silicates of base metals (calcium, magnesium, iron and manganese). Garnet has a higher density than sand.

GAUGE, PIPE   GAUGE, PIPE

A number that defines the thickness of the sheet used to make steel pipe. The larger the number, the thinner the pipe wall.

GEOGRAPHIC INFORMATION SYSTEM (GIS)   GEOGRAPHIC INFORMATION SYSTEM (GIS)

A computer program that combines mapping with detailed information about the physical locations of structures such as pipes, valves, and manholes within geographic areas. The system is used to help operators and maintenance personnel locate utility system features or structures and to assist with the scheduling and performance of maintenance activities.

GEOLOGICAL LOG

A detailed description of all underground features discovered during the drilling of a well (depth, thickness and type of formations).

GEOPHYSICAL LOG

A record of the structure and composition of the earth encountered when drilling a well or similar type of test hole or boring.

GERMICIDE (GERM-uh-SIDE)

A substance formulated to kill germs or microorganisms. The germicidal properties of chlorine make it an effective disinfectant.

GIARDIA (gee-ARE-dee-ah)

A waterborne intestinal parasite that causes a disease called giardiasis (GEE-are-DIE-uh-sis) in infected humans. Symptoms of the disease include diarrhea, cramps, and weight loss. *Giardia* contamination is found in most surface waters and some groundwaters.

GIARDIASIS (GEE-are-DIE-uh-sis)

Intestinal disease caused by an infestation of *Giardia* flagellates.

GRAB SAMPLE

A single sample of water collected at a particular time and place which represents the composition of the water only at that time and place.

GRADE

(1) The elevation of the invert (or bottom) of a pipeline, canal, culvert, or similar conduit.

(2) The inclination or slope of a pipeline, conduit, stream channel, or natural ground surface; usually expressed in terms of the ratio or percentage of number of units of vertical rise or fall per unit of horizontal distance. A 0.5 percent grade would be a drop of one-half foot per hundred feet of pipe.

GRAVIMETRIC

A means of measuring unknown concentrations of water quality indicators in a sample by *WEIGHING* a precipitate or residue of the sample.

GRAVIMETRIC FEEDER

A dry chemical feeder which delivers a measured weight of chemical during a specific time period.

GREENSAND

A mineral (glauconite) material that looks like ordinary filter sand except that it is green in color. Greensand is a natural ion exchange material which is capable of softening water. Greensand which has been treated with potassium permanganate ($KMnO_4$) is called manganese greensand; this product is used to remove iron, manganese and hydrogen sulfide from groundwaters.

GROUND

An expression representing an electrical connection to earth or a large conductor which is at the earth's potential or neutral voltage.

H

HTH (pronounce as separate letters)

High **T**est **H**ypochlorite. Calcium hypochlorite or $Ca(OCl)_2$.

HARD WATER

Water having a high concentration of calcium and magnesium ions. A water may be considered hard if it has a hardness greater than the typical hardness of water from the region. Some textbooks define hard water as water with a hardness of more than 100 mg/L as calcium carbonate.

HARDNESS, WATER

A characteristic of water caused mainly by the salts of calcium and magnesium, such as bicarbonate, carbonate, sulfate, chloride and nitrate. Excessive hardness in water is undesirable because it causes the formation of soap curds, increased use of soap, deposition of scale in boilers, damage in some industrial processes, and sometimes causes objectionable tastes in drinking water.

HEAD
HEAD

The vertical distance (in feet) equal to the pressure (in psi) at a specific point. The pressure head is equal to the pressure in psi times 2.31 ft/psi.

HEAD LOSS
HEAD LOSS

The head, pressure or energy (they are the same) lost by water flowing in a pipe or channel as a result of turbulence caused by the velocity of the flowing water and the roughness of the pipe, channel walls, or restrictions caused by fittings. Water flowing in a pipe loses head, pressure or energy as a result of friction losses. The head loss through a filter is due to friction losses caused by material building up on the surface or in the top part of a filter. Also see FRICTION LOSSES.

HEADER
HEADER

A large pipe to which the ends of a series of smaller pipes are connected. Also called a MANIFOLD.

HEAT SENSOR
HEAT SENSOR

A device that opens and closes a switch in response to changes in the temperature. This device might be a metal contact, or a thermocouple which generates a minute electric current proportional to the difference in heat, or a variable resistor whose value changes in response to changes in temperature. Also called a TEMPERATURE SENSOR.

HECTARE (HECK-tar)
HECTARE

A measure of area in the metric system similar to an acre. One hectare is equal to 10,000 square meters and 2.4711 acres.

HEPATITIS (HEP-uh-TIE-tis)
HEPATITIS

Hepatitis is an inflammation of the liver caused by an acute viral infection. Yellow jaundice is one symptom of hepatitis.

HERBICIDE (HERB-uh-SIDE)
HERBICIDE

A compound, usually a manmade organic chemical, used to kill or control plant growth.

HERTZ
HERTZ

The number of complete electromagnetic cycles or waves in one second of an electric or electronic circuit. Also called the frequency of the current. Abbreviated Hz.

HETEROTROPHIC (HET-er-o-TROF-ick)
HETEROTROPHIC

Describes organisms that use organic matter for energy and growth. Animals, fungi and most bacteria are heterotrophs.

HIGH-LINE JUMPERS
HIGH-LINE JUMPERS

Pipes or hoses connected to fire hydrants and laid on top of the ground to provide emergency water service for an isolated portion of a distribution system.

HOSE BIB
HOSE BIB

Faucet. A location in a water line where a hose is connected.

HYDRATED LIME
HYDRATED LIME

Limestone that has been "burned" and treated with water under controlled conditions until the calcium oxide portion has been converted to calcium hydroxide ($Ca(OH)_2$). Hydrated lime is quicklime combined with water. $CaO + H_2O \rightarrow Ca(OH)_2$. Also called slaked lime. Also see QUICKLIME.

HYDRAULIC CONDUCTIVITY (K)
HYDRAULIC CONDUCTIVITY (K)

A coefficient describing the relative ease with which groundwater can move through a permeable layer of rock or soil. Typical units of hydraulic conductivity are feet per day, gallons per day per square foot, or meters per day (depending on the unit chosen for the total discharge and the cross-sectional area).

HYDRAULIC GRADE LINE (HGL)
HYDRAULIC GRADE LINE (HGL)

The surface or profile of water flowing in an open channel or a pipe flowing partially full. If a pipe is under pressure, the hydraulic grade line is at the level water would rise to in a small vertical tube connected to the pipe. Also see ENERGY GRADE LINE.

[SEE DRAWING ON PAGE 599]

HYDRAULIC GRADIENT
HYDRAULIC GRADIENT

The slope of the hydraulic grade line. This is the slope of the water surface in an open channel, the slope of the water surface of the groundwater table, or the slope of the water pressure for pipes under pressure.

HYDROGEOLOGIST (HI-dro-gee-ALL-uh-gist) HYDROGEOLOGIST

A person who studies and works with groundwater.

HYDROLOGIC (HI-dro-LOJ-ick) CYCLE HYDROLOGIC CYCLE

The process of evaporation of water into the air and its return to earth by precipitation (rain or snow). This process also includes transpiration from plants, groundwater movement, and runoff into rivers, streams and the ocean. Also called the WATER CYCLE.

HYDROLYSIS (hi-DROLL-uh-sis) HYDROLYSIS

(1) A chemical reaction in which a compound is converted into another compound by taking up water.

(2) Usually a chemical degradation of organic matter.

HYDROPHILIC (HI-dro-FILL-ick) HYDROPHILIC

Having a strong affinity (liking) for water. The opposite of HYDROPHOBIC.

HYDROPHOBIC (HI-dro-FOE-bick) HYDROPHOBIC

Having a strong aversion (dislike) for water. The opposite of HYDROPHILIC.

HYDROPNEUMATIC (HI-dro-new-MAT-ick) HYDROPNEUMATIC

A water system, usually small, in which a water pump is automatically controlled (started and stopped) by the air pressure in a compressed-air tank.

HYDROSTATIC (HI-dro-STAT-ick) PRESSURE HYDROSTATIC PRESSURE

(1) The pressure at a specific elevation exerted by a body of water at rest, or

(2) In the case of groundwater, the pressure at a specific elevation due to the weight of water at higher levels in the same zone of saturation.

HYGROSCOPIC (HI-grow-SKOP-ick) HYGROSCOPIC

Absorbing or attracting moisture from the air.

HYPOCHLORINATION (HI-poe-KLOR-uh-NAY-shun) HYPOCHLORINATION

The application of hypochlorite compounds to water for the purpose of disinfection.

HYPOCHLORINATORS (HI-poe-KLOR-uh-NAY-tors) HYPOCHLORINATORS

Chlorine pumps, chemical feed pumps or devices used to dispense chlorine solutions made from hypochlorites such as bleach (sodium hypochlorite) or calcium hypochlorite into the water being treated.

HYPOCHLORITE (HI-poe-KLOR-ite) HYPOCHLORITE

Chemical compounds containing available chlorine; used for disinfection. They are available as liquids (bleach) or solids (powder, granules, and pellets) in barrels, drums, and cans. Salts of hypochlorous acid.

HYPOLIMNION (HI-poe-LIM-knee-on) HYPOLIMNION

The lowest layer in a thermally stratified lake or reservoir. This layer consists of colder, more dense water, has a constant temperature and no mixing occurs.

I

ICR ICR

The Information Collection Rule (ICR) specifies the requirements for monitoring microbial contaminants and disinfection by-products (DBPs) by large public water systems (PWSs). It also requires large PWSs to conduct either bench- or pilot-scale testing of advanced treatment techniques.

IDLH IDLH

Immediately Dangerous to Life or Health. The atmospheric concentration of any toxic, corrosive, or asphyxiant substance that poses an immediate threat to life or would cause irreversible or delayed adverse health effects or would interfere with an individual's ability to escape from a dangerous atmosphere.

IMHOFF CONE

IMHOFF CONE

A clear, cone-shaped container marked with graduations. The cone is used to measure the volume of settleable solids in a specific volume (usually one liter) of water.

IMPELLER

IMPELLER

A rotating set of vanes in a pump or compressor designed to pump or move water or air.

IMPERMEABLE (im-PURR-me-uh-BULL)

IMPERMEABLE

Not easily penetrated. The property of a material or soil that does not allow, or allows only with great difficulty, the movement or passage of water.

INDICATOR (CHEMICAL)

INDICATOR (CHEMICAL)

A substance that gives a visible change, usually of color, at a desired point in a chemical reaction, generally at a specified end point.

INDICATOR (INSTRUMENT)

INDICATOR (INSTRUMENT)

A device which indicates the result of a measurement. Most indicators in the water utility field use either a fixed scale and movable indicator (pointer) such as a pressure gage or a movable scale and movable indicator like those used on a circular flow-recording chart. Also called a RECEIVER.

INFILTRATION (IN-fill-TRAY-shun)

INFILTRATION

The seepage of groundwater into a sewer system, including service connections. Seepage frequently occurs through defective or cracked pipes, pipe joints, connections or manhole walls.

INFLUENT

INFLUENT

Water or other liquid—raw (untreated) or partially treated—flowing *INTO* a reservoir, basin, treatment process, or treatment plant.

INFORMATION COLLECTION RULE (ICR)

INFORMATION COLLECTION RULE (ICR)

The Information Collection Rule (ICR) specifies the requirements for monitoring microbial contaminants and disinfection by-products (DBPs) by large public water systems (PWSs). It also requires large PWSs to conduct either bench- or pilot-scale testing of advanced treatment techniques.

INITIAL SAMPLING

INITIAL SAMPLING

The very first sampling conducted under the Safe Drinking Water Act (SDWA) for each of the applicable contaminant categories.

INJECTOR WATER

INJECTOR WATER

Service water in which chlorine is added (injected) to form a chlorine solution.

IN-LINE FILTRATION

IN-LINE FILTRATION

The addition of chemical coagulants directly to the filter inlet pipe. The chemicals are mixed by the flowing water. Flocculation and sedimentation facilities are eliminated. This pretreatment method is commonly used in pressure filter installations. Also see CONVENTIONAL FILTRATION and DIRECT FILTRATION.

INORGANIC

INORGANIC

Material such as sand, salt, iron, calcium salts and other mineral materials. Inorganic substances are of mineral origin, whereas organic substances are usually of animal or plant origin. Also see ORGANIC.

INORGANIC WASTE

INORGANIC WASTE

Waste material such as sand, salt, iron, calcium, and other mineral materials which are only slightly affected by the action of organisms. Inorganic wastes are chemical substances of mineral origin; whereas organic wastes are chemical substances of an animal or plant origin.

INPUT HORSEPOWER

INPUT HORSEPOWER

The total power used in operating a pump and motor.

$$\text{Input Horsepower, HP} = \frac{(\text{Brake Horsepower, HP})(100\%)}{\text{Motor Efficiency, \%}}$$

INSECTICIDE
INSECTICIDE

Any substance or chemical formulated to kill or control insects.

INSOLUBLE (in-SAWL-you-bull)
INSOLUBLE

Something that cannot be dissolved.

INTEGRATOR
INTEGRATOR

A device or meter that continuously measures and calculates (adds) a process rate variable in cumulative fashion; for example, total flows displayed in gallons, million gallons, cubic feet, or some other unit of volume measurement. Also called a TOTALIZER.

INTERFACE
INTERFACE

The common boundary layer between two substances such as water and a solid (metal); or between two fluids such as water and a gas (air); or between a liquid (water) and another liquid (oil).

INTERLOCK
INTERLOCK

An electric switch, usually magnetically operated. Used to interrupt all (local) power to a panel or device when the door is opened or the circuit is exposed to service.

INTERNAL FRICTION
INTERNAL FRICTION

Friction within a fluid (water) due to cohesive forces.

INTERSTICE (in-TUR-stuhz)
INTERSTICE

A very small open space in a rock or granular material. Also called a PORE, VOID, or void space. Also see VOID.

INVERT (IN-vert)
INVERT

The lowest point of the channel inside a pipe, conduit, or canal.

ION
ION

An electrically charged atom, radical (such as SO_4^{2-}), or molecule formed by the loss or gain of one or more electrons.

ION EXCHANGE
ION EXCHANGE

A water treatment process involving the reversible interchange (switching) of ions between the water being treated and the solid resin. Undesirable ions in the water are switched with acceptable ions on the resin.

ION EXCHANGE RESINS
ION EXCHANGE RESINS

Insoluble polymers, used in water treatment, that are capable of exchanging (switching or giving) acceptable cations or anions to the water being treated for less desirable ions.

IONIC CONCENTRATION
IONIC CONCENTRATION

The concentration of any ion in solution, usually expressed in moles per liter.

IONIZATION (EYE-on-uh-ZAY-shun)
IONIZATION

The splitting or dissociation (separation) of molecules into negatively and positively charged ions.

J

JAR TEST
JAR TEST

A laboratory procedure that simulates a water treatment plant's coagulation/flocculation units with differing chemical doses and also energy of rapid mix, energy of slow mix, and settling time. The purpose of this procedure is to *ESTIMATE* the minimum or ideal coagulant dose required to achieve certain water quality goals. Samples of water to be treated are commonly placed in six jars. Various amounts of chemicals are added to each jar, stirred and the settling of solids is observed. The dose of chemicals that provides satisfactory settling, removal of turbidity and/or color is the dose used to treat the water being taken into the plant at that time. When evaluating the results of a jar test, the operator should also consider the floc quality in the flocculation area and the floc loading on the filter.

JOGGING
JOGGING

The frequent starting and stopping of an electric motor.

JOULE (jewel)   JOULE

A measure of energy, work or quantity of heat. One joule is the work done when the point of application of a force of one newton is displaced a distance of one meter in the direction of the force. Approximately equal to 0.7375 ft-lbs (0.1022 m-kg).

K

KELLY   KELLY

The square section of a rod which causes the rotation of the drill bit. Torque from a drive table is applied to the square rod to cause the rotary motion. The drive table is chain or gear driven by an engine.

KILO   KILO

(1) Kilogram.

(2) Kilometer.

(3) A prefix meaning "thousand" used in the metric system and other scientific systems of measurement.

KINETIC ENERGY   KINETIC ENERGY

Energy possessed by a moving body of matter, such as water, as a result of its motion.

KJELDAHL (KELL-doll) NITROGEN   KJELDAHL NITROGEN

Nitrogen in the form of organic proteins or their decomposition product ammonia, as measured by the Kjeldahl Method.

L

LANGELIER INDEX (L.I.)   LANGELIER INDEX (L.I.)

An index reflecting the equilibrium pH of a water with respect to calcium and alkalinity. This index is used in stabilizing water to control both corrosion and the deposition of scale.

$$\text{Langelier Index} = pH - pH_s$$
where pH = actual pH of the water, and
pH_s = pH at which water having the same alkalinity and calcium content is just saturated with calcium carbonate.

LAUNDERING WEIR (LAWN-der-ing weer)   LAUNDERING WEIR

Sedimentation basin overflow weir. A plate with V-notches along the top to ensure a uniform flow rate and avoid short-circuiting.

LAUNDERS (LAWN-ders)   LAUNDERS

Sedimentation basin and filter discharge channels consisting of overflow weir plates (in sedimentation basins) and conveying troughs.

LEAD (LEE-d)   LEAD

A wire or conductor that can carry electric current.

LEATHERS   LEATHERS

O-rings or gaskets used with piston pumps to provide a seal between the piston and the side wall.

LEVEL CONTROL   LEVEL CONTROL

A float device (or pressure switch) which senses changes in a measured variable and opens or closes a switch in response to that change. In its simplest form, this control might be a floating ball connected mechanically to a switch or valve such as is used to stop water flow into a toilet when the tank is full.

LINDANE (LYNN-dane)   LINDANE

A pesticide that causes adverse health effects in domestic water supplies and also is toxic to freshwater and marine aquatic life.

LINEARITY (LYNN-ee-AIR-it-ee)   LINEARITY

How closely an instrument measures actual values of a variable through its effective range; a measure used to determine the accuracy of an instrument.

LITTORAL (LIT-or-al) ZONE   LITTORAL ZONE

(1) That portion of a body of fresh water extending from the shoreline lakeward to the limit of occupancy of rooted plants.

(2) The strip of land along the shoreline between the high and low water levels.

LOGARITHM (LOG-a-rith-m)   LOGARITHM

The exponent that indicates the power to which a number must be raised to produce a given number. For example: if $B^2 = N$, the 2 is the logarithm of N (to the base B), or $10^2 = 100$ and $\log_{10} 100 = 2$. Also abbreviated to "log."

LOGGING, ELECTRICAL   LOGGING, ELECTRICAL

A procedure used to determine the porosity (spaces or voids) of formations in search of water-bearing formations (aquifers). Electrical probes are lowered into wells, an electric current is induced at various depths, and the resistance measured of various formations indicates the porosity of the material.

LOWER EXPLOSIVE LIMIT (LEL)   LOWER EXPLOSIVE LIMIT (LEL)

The lowest concentration of gas or vapor (percent by volume in air) that explodes if an ignition source is present at ambient temperature. At temperatures above 250°F the LEL decreases because explosibility increases with higher temperature.

M

M or MOLAR   *M* or MOLAR

A molar solution consists of one gram molecular weight of a compound dissolved in enough water to make one liter of solution. A gram molecular weight is the molecular weight of a compound in grams. For example, the molecular weight of sulfuric acid (H_2SO_4) is 98. A one *M* solution of sulfuric acid would consist of 98 grams of H_2SO_4 dissolved in enough distilled water to make one liter of solution.

MBAS   MBAS

Methylene - **B**lue - **A**ctive **S**ubstances. These substances are used in surfactants or detergents.

MCL   MCL

Maximum **C**ontaminant **L**evel. The largest allowable amount. MCLs for various water quality indicators are specified in the National Primary Drinking Water Regulations (NPDWR).

MCLG   MCLG

Maximum **C**ontaminant **L**evel **G**oal. MCLGs are health goals based entirely on health effects. They are a preliminary standard set but not enforced by EPA. MCLs consider health effects, but also take into consideration the feasibility and cost of analysis and treatment of the regulated MCL. Although often less stringent than the corresponding MCLG, the MCL is set to protect health.

mg/*L*   mg/*L*

See MILLIGRAMS PER LITER, mg/*L*.

MPN (pronounce as separate letters)   MPN

MPN is the **M**ost **P**robable **N**umber of coliform-group organisms per unit volume of sample water. Expressed as a density or population of organisms per 100 m*L* of sample water.

MSDS   MSDS

See **MATERIAL SAFETY DATA SHEET**.

MACROSCOPIC (MACK-row-SKAWP-ick) ORGANISMS   MACROSCOPIC ORGANISMS

Organisms big enough to be seen by the eye without the aid of a microscope.

MANDREL (MAN-drill)   MANDREL

A special tool used to push bearings in or to pull sleeves out.

MANIFOLD   MANIFOLD

A large pipe to which the ends of a series of smaller pipes are connected. Also called a HEADER.

MANOMETER (man-NAH-mut-ter)   MANOMETER

An instrument for measuring pressure. Usually, a manometer is a glass tube filled with a liquid that is used to measure the difference in pressure across a flow measuring device such as an orifice or a Venturi meter. The instrument used to measure blood pressure is a type of manometer.

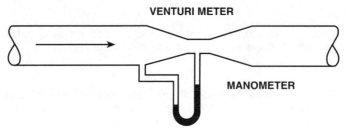

MATERIAL SAFETY DATA SHEET (MSDS)   MATERIAL SAFETY DATA SHEET (MSDS)

A document which provides pertinent information and a profile of a particular hazardous substance or mixture. An MSDS is normally developed by the manufacturer or formulator of the hazardous substance or mixture. The MSDS is required to be made available to employees and operators whenever there is the likelihood of the hazardous substance or mixture being introduced into the workplace. Some manufacturers are preparing MSDSs for products that are not considered to be hazardous to show that the product or substance is *NOT* hazardous.

MAXIMUM CONTAMINANT LEVEL (MCL)   MAXIMUM CONTAMINANT LEVEL (MCL)
See MCL.

MEASURED VARIABLE   MEASURED VARIABLE

A characteristic or component part that is sensed and quantified (reduced to a reading of some kind) by a primary element or sensor.

MECHANICAL JOINT   MECHANICAL JOINT

A flexible device that joins pipes or fittings together by the use of lugs and bolts.

MEG   MEG

A procedure used for checking the insulation resistance on motors, feeders, bus bar systems, grounds, and branch circuit wiring. Also see MEGGER.

MEGGER (from megohm)   MEGGER

An instrument used for checking the insulation resistance on motors, feeders, bus bar systems, grounds, and branch circuit wiring. Also see MEG.

MEGOHM   MEGOHM

Meg means one million, so 5 megohms means 5 million ohms. A megger reads in millions of ohms.

MENISCUS (meh-NIS-cuss)   MENISCUS

The curved surface of a column of liquid (water, oil, mercury) in a small tube. When the liquid wets the sides of the container (as with water), the curve forms a valley. When the confining sides are not wetted (as with mercury), the curve forms a hill or upward bulge. When a meniscus forms in a measuring device, the top of the liquid level of the sample is determined by the bottom of the meniscus.

MESH   MESH

One of the openings or spaces in a screen or woven fabric. The value of the mesh is usually given as the number of openings per inch. This value does not consider the diameter of the wire or fabric; therefore, the mesh number does not always have a definite relationship to the size of the hole.

MESOTROPHIC (MESS-o-TRO-fick)   MESOTROPHIC

Reservoirs and lakes which contain moderate quantities of nutrients and are moderately productive in terms of aquatic animal and plant life.

METABOLISM (meh-TAB-uh-LIZ-um)   METABOLISM

(1)   The biochemical processes in which food is used and wastes are formed by living organisms.

(2)   All biochemical reactions involved in cell formation and growth.

METALIMNION (MET-uh-LIM-knee-on)   METALIMNION

The middle layer in a thermally stratified lake or reservoir. In this layer there is a rapid decrease in temperature with depth. Also called the THERMOCLINE.

METHOXYCHLOR (meth-OXY-klor)   METHOXYCHLOR

A pesticide which causes adverse health effects in domestic water supplies and is also toxic to freshwater and marine aquatic life. The chemical name for methoxychlor is 2,2-bis(p-methoxyphenol)-1,1,1-trichloroethane.

METHYL ORANGE ALKALINITY   METHYL ORANGE ALKALINITY

A measure of the total alkalinity in a water sample. The alkalinity is measured by the amount of standard sulfuric acid required to lower the pH of the water to a pH level of 4.5, as indicated by the change in color of methyl orange from orange to pink. Methyl orange alkalinity is expressed as milligrams per liter equivalent calcium carbonate.

MICROBIAL (my-KROW-bee-ul) GROWTH   MICROBIAL GROWTH

The activity and growth of microorganisms such as bacteria, algae, diatoms, plankton and fungi.

MICRON (MY-kron)   MICRON

μm, Micrometer or Micron. A unit of length. One millionth of a meter or one thousandth of a millimeter. One micron equals 0.00004 of an inch.

MICROORGANISMS (MY-crow-OR-gan-IS-zums)   MICROORGANISMS

Living organisms that can be seen individually only with the aid of a microscope.

MIL   MIL

A unit of length equal to 0.001 of an inch. The diameter of wires and tubing is measured in mils, as is the thickness of plastic sheeting.

MILLIGRAMS PER LITER, mg/L MILLIGRAMS PER LITER, mg/L

A measure of the concentration by weight of a substance per unit volume. For practical purposes, one mg/L of a substance in fresh water is equal to one part per million parts (ppm). Thus a liter of water with a specific gravity of 1.0 weighs one million milligrams. If water contains 10 milligrams of calcium, the concentration is 10 milligrams per million milligrams, or 10 milligrams per liter (10 mg/L), or 10 parts of calcium per million parts of water, or 10 parts per million (10 ppm).

MILLIMICRON (MILL-uh-MY-kron)   MILLIMICRON

A unit of length equal to $10^{-3}\mu$ (one thousandth of a micron), 10^{-6} millimeters, or 10^{-9} meters; correctly called a nanometer, nm.

MOLAR   MOLAR

See **M** for MOLAR.

MOLE   MOLE

The molecular weight of a substance, usually expressed in grams.

MOLECULAR WEIGHT   MOLECULAR WEIGHT

The molecular weight of a compound in grams is the sum of the atomic weights of the elements in the compound. The molecular weight of sulfuric acid (H_2SO_4) in grams is 98.

Element	Atomic Weight	Number of Atoms	Molecular Weight
H	1	2	2
S	32	1	32
O	16	4	64
			98

MOLECULE (MOLL-uh-KULE)   MOLECULE

The smallest division of a compound that still retains or exhibits all the properties of the substance.

MONOMER (MON-o-MER)   MONOMER

A molecule of low molecular weight capable of reacting with identical or different monomers to form polymers.

MONOMICTIC (mo-no-MICK-tick)   MONOMICTIC

Lakes and reservoirs which are relatively deep, do not freeze over during the winter months, and undergo a single stratification and mixing cycle during the year. These lakes and reservoirs usually become destratified during the mixing cycle, usually in the fall of the year.

MONOVALENT   MONOVALENT

Having a valence of one, such as the cuprous (copper) ion, Cu^+.

MOST PROBABLE NUMBER (MPN)   MOST PROBABLE NUMBER (MPN)

See MPN.

MOTILE (MO-till)   MOTILE

Capable of self-propelled movement. A term that is sometimes used to distinguish between certain types of organisms found in water.

MOTOR EFFICIENCY   MOTOR EFFICIENCY

The ratio of energy delivered by a motor to the energy supplied to it during a fixed period or cycle. Motor efficiency ratings will vary depending upon motor manufacturer and usually will be near 90.0 percent.

MUDBALLS   MUDBALLS

Material that is approximately round in shape and varies from pea-sized up to two or more inches in diameter. This material forms in filters and gradually increases in size when not removed by the backwashing process.

MULTI-STAGE PUMP   MULTI-STAGE PUMP

A pump that has more than one impeller. A single-stage pump has one impeller.

N

N or NORMAL   *N* or NORMAL

A normal solution contains one gram equivalent weight of reactant (compound) per liter of solution. The equivalent weight of an acid is that weight which contains one gram atom of ionizable hydrogen or its chemical equivalent. For example, the equivalent weight of sulfuric acid (H_2SO_4) is 49 (98 divided by 2 because there are two replaceable hydrogen ions). A one *N* solution of sulfuric acid would consist of 49 grams of H_2SO_4 dissolved in enough water to make one liter of solution.

NESHTA (formerly NETA)   NESHTA

See **N**ATIONAL **E**NVIRONMENTAL, **S**AFETY & **H**EALTH **T**RAINING **A**SSOCIATION (NESHTA).

NETA   NETA

See **N**ATIONAL **E**NVIRONMENTAL, **S**AFETY & **H**EALTH **T**RAINING **A**SSOCIATION (NESHTA).

NIOSH (NYE-osh)   NIOSH

The **N**ational **I**nstitute of **O**ccupational **S**afety and **H**ealth is an organization that tests and approves safety equipment for particular applications. NIOSH is the primary federal agency engaged in research in the national effort to eliminate on-the-job hazards to the health and safety of working people. The NIOSH Publications Catalog, Seventh Edition, NIOSH Pub. No. 87-115, lists the NIOSH publications concerning industrial hygiene and occupational health. To obtain a copy of the catalog, write to National Technical Information Service (NTIS), 5285 Port Royal Road, Springfield, VA 22161. NTIS Stock No. PB88-175013, price, $141.00, plus $5.00 shipping and handling per order.

NOM (NATURAL ORGANIC MATTER)   NOM (NATURAL ORGANIC MATTER)

Humic substances composed of humic and fulvic acids that come from decayed vegetation.

NPDES PERMIT

National Pollutant Discharge Elimination System permit is the regulatory agency document issued by either a federal or state agency which is designed to control all discharges of pollutants from point sources and storm water runoff into U.S. waterways. NPDES permits regulate discharges into navigable waters from all point sources of pollution, including industries, municipal waste-water treatment plants, sanitary landfills, large agricultural feedlots and return irrigation flows.

NPDWR

National Primary Drinking Water Regulations.

NSDWR

National Secondary Drinking Water Regulations.

NTU

Nephelometric Turbidity Units. See TURBIDITY UNITS (TU).

NAMEPLATE

A durable metal plate found on equipment which lists critical operating conditions for the equipment.

NATIONAL ENVIRONMENTAL, SAFETY & HEALTH TRAINING ASSOCIATION (NESHTA)
(formerly NATIONAL ENVIRONMENTAL TRAINING ASSOCIATION (NETA))

A professional organization devoted to serving the environmental trainer and promoting better operation of waterworks and pollution control facilities. For information on NESHTA membership and publications, contact NESHTA, 5320 North 16th Street, Suite 114, Phoenix, AZ 85016-3241. Phone (602) 956-6099.

NATIONAL ENVIRONMENTAL TRAINING ASSOCIATION (NETA)

See NATIONAL ENVIRONMENTAL, SAFETY & HEALTH TRAINING ASSOCIATION (NESHTA).

NATIONAL INSTITUTE OF OCCUPATIONAL SAFETY AND HEALTH

See NIOSH.

NATIONAL PRIMARY DRINKING WATER REGULATIONS

Commonly referred to as NPDWR.

NATIONAL SECONDARY DRINKING WATER REGULATIONS

Commonly referred to as NSDWR.

NEPHELOMETRIC (NEFF-el-o-MET-rick)

A means of measuring turbidity in a sample by using an instrument called a nephelometer. A nephelometer passes light through a sample and the amount of light deflected (usually at a 90-degree angle) is then measured.

NEWTON

A force which, when applied to a body having a mass of one kilogram, gives it an acceleration of one meter per second per second.

NITRIFICATION (NYE-truh-fuh-KAY-shun)

An aerobic process in which bacteria oxidize the ammonia and organic nitrogen in water into nitrite and then nitrate.

NITROGENOUS (nye-TRAH-jen-us)

A term used to describe chemical compounds (usually organic) containing nitrogen in combined forms. Proteins and nitrate are nitrogenous compounds.

NOBLE METAL

A chemically inactive metal (such as gold). A metal that does not corrode easily and is much scarcer (and more valuable) than the so-called useful or base metals. Also see BASE METAL.

NOMINAL DIAMETER

NOMINAL DIAMETER

An approximate measurement of the diameter of a pipe. Although the nominal diameter is used to describe the size or diameter of a pipe, it is usually not the exact inside diameter of the pipe.

NONIONIC (NON-eye-ON-ick) POLYMER

NONIONIC POLYMER

A polymer that has no net electrical charge.

NON-PERMIT CONFINED SPACE

NON-PERMIT CONFINED SPACE

See CONFINED SPACE, NON-PERMIT.

NONPOINT SOURCE

NONPOINT SOURCE

A runoff or discharge from a field or similar source. A point source refers to a discharge that comes out the end of a pipe.

NONPOTABLE (non-POE-tuh-bull)

NONPOTABLE

Water that may contain objectionable pollution, contamination, minerals, or infective agents and is considered unsafe and/or unpalatable for drinking.

NONVOLATILE MATTER

NONVOLATILE MATTER

Material such as sand, salt, iron, calcium, and other mineral materials which are only slightly affected by the actions of organisms and are not lost on ignition of the dry solids at 550°C. Volatile materials are chemical substances usually of animal or plant origin. Also see INORGANIC WASTE and VOLATILE MATTER or VOLATILE SOLIDS.

NORMAL

NORMAL

See *N* or NORMAL.

NOTICE

NOTICE

This word calls attention to information that is especially significant in understanding and operating equipment or processes safely. Also see CAUTION, DANGER, and WARNING.

NUTRIENT

NUTRIENT

Any substance that is assimilated (taken in) by organisms and promotes growth. Nitrogen and phosphorus are nutrients which promote the growth of algae. There are other essential and trace elements which are also considered nutrients.

O

ORP (pronounce as separate letters)

ORP

Oxidation-**R**eduction **P**otential. The electrical potential required to transfer electrons from one compound or element (the oxidant) to another compound or element (the reductant); used as a qualitative measure of the state of oxidation in water treatment systems. ORP is measured in millivolts, with negative values indicating a tendency to reduce compounds or elements and positive values indicating a tendency to oxidize compounds or elements.

OSHA (O-shuh)

OSHA

The Williams-Steiger **O**ccupational **S**afety and **H**ealth **A**ct of 1970 (OSHA) is a federal law designed to protect the health and safety of industrial workers and also the operators of water supply systems and treatment plants. The Act regulates the design, construction, operation and maintenance of water supply systems and water treatment plants. OSHA also refers to the federal and state agencies which administer the OSHA regulations.

OCCUPATIONAL SAFETY AND
  HEALTH ACT OF 1970

OCCUPATIONAL SAFETY AND
HEALTH ACT OF 1970

See OSHA.

ODOR THRESHOLD

ODOR THRESHOLD

The minimum odor of a water sample that can just be detected after successive dilutions with odorless water. Also called THRESHOLD ODOR.

OFFSET

OFFSET

The difference between the actual value and the desired value (or set point); characteristic of proportional controllers that do not incorporate reset action. Also called DRIFT.

OHM   OHM

The unit of electrical resistance. The resistance of a conductor in which one volt produces a current of one ampere.

OLFACTORY (ol-FAK-tore-ee) FATIGUE   OLFACTORY FATIGUE

A condition in which a person's nose, after exposure to certain odors, is no longer able to detect the odor.

OLIGOTROPHIC (AH-lig-o-TRO-fick)   OLIGOTROPHIC

Reservoirs and lakes which are nutrient poor and contain little aquatic plant or animal life.

OPERATING PRESSURE DIFFERENTIAL   OPERATING PRESSURE DIFFERENTIAL

The operating pressure range for a hydropneumatic system. For example, when the pressure drops below 40 psi the pump will come on and stay on until the pressure builds up to 60 psi. When the pressure reaches 60 psi the pump will shut off.

OPERATING RATIO   OPERATING RATIO

The operating ratio is a measure of the total revenues divided by the total operating expenses.

ORGANIC   ORGANIC

Substances that come from animal or plant sources. Organic substances always contain carbon. (Inorganic materials are chemical substances of mineral origin.) Also see INORGANIC.

ORGANICS   ORGANICS

(1) A term used to refer to chemical compounds made from carbon molecules. These compounds may be natural materials (such as animal or plant sources) or manmade materials (such as synthetic organics). Also see ORGANIC.

(2) Any form of animal or plant life. Also see BACTERIA.

ORGANISM   ORGANISM

Any form of animal or plant life. Also see BACTERIA.

ORGANIZING   ORGANIZING

Deciding who does what work and delegating authority to the appropriate persons.

ORIFICE (OR-uh-fiss)   ORIFICE

An opening (hole) in a plate, wall, or partition. An orifice flange or plate placed in a pipe consists of a slot or a calibrated circular hole smaller than the pipe diameter. The difference in pressure in the pipe above and at the orifice may be used to determine the flow in the pipe.

ORTHOTOLIDINE (or-tho-TOL-uh-dine)   ORTHOTOLIDINE

Orthotolidine is a colorimetric indicator of chlorine residual. If chlorine is present, a yellow-colored compound is produced. This reagent is no longer approved for chemical analysis to determine chlorine residual.

OSMOSIS (oz-MOE-sis)   OSMOSIS

The passage of a liquid from a weak solution to a more concentrated solution across a semipermeable membrane. The membrane allows the passage of the water (solvent) but not the dissolved solids (solutes). This process tends to equalize the conditions on either side of the membrane.

OUCH PRINCIPLE   OUCH PRINCIPLE

This principle says that as a manager when you delegate job tasks you must be **O**bjective, **U**niform in your treatment of employees, **C**onsistent with utility policies, and **H**ave job relatedness.

OVERALL EFFICIENCY, PUMP   OVERALL EFFICIENCY, PUMP

The combined efficiency of a pump and motor together. Also called the WIRE-TO-WATER EFFICIENCY.

OVERDRAFT   OVERDRAFT

The pumping of water from a groundwater basin or aquifer in excess of the supply flowing into the basin. This pumping results in a depletion or "mining" of the groundwater in the basin.

OVERFLOW RATE   OVERFLOW RATE

One of the guidelines for the design of settling tanks and clarifiers in treatment plants. Used by operators to determine if tanks and clarifiers are hydraulically (flow) over- or underloaded. Also called SURFACE LOADING.

$$\text{Overflow Rate, GPD/sq ft} = \frac{\text{Flow, gallons/day}}{\text{Surface Area, sq ft}}$$

OVERHEAD   OVERHEAD

Indirect costs necessary for a water utility to function properly. These costs are not related to the actual production and delivery of water to consumers, but include the costs of rent, lights, office supplies, management and administration.

OVERTURN   OVERTURN

The almost spontaneous mixing of all layers of water in a reservoir or lake when the water temperature becomes similar from top to bottom. This may occur in the fall/winter when the surface waters cool to the same temperature as the bottom waters and also in the spring when the surface waters warm after the ice melts. This is also called "turnover."

OXIDATION (ox-uh-DAY-shun)   OXIDATION

Oxidation is the addition of oxygen, removal of hydrogen, or the removal of electrons from an element or compound. In the environment, organic matter is oxidized to more stable substances. The opposite of REDUCTION.

OXIDATION-REDUCTION POTENTIAL (ORP)   OXIDATION-REDUCTION POTENTIAL (ORP)

The electrical potential required to transfer electrons from one compound or element (the oxidant) to another compound or element (the reductant); used as a qualitative measure of the state of oxidation in water treatment systems. ORP is measured in millivolts, with negative values indicating a tendency to reduce compounds or elements and positive values indicating a tendency to oxidize compounds or elements.

OXIDIZING AGENT   OXIDIZING AGENT

Any substance, such as oxygen (O_2) or chlorine (Cl_2), that will readily add (take on) electrons. The opposite is a REDUCING AGENT.

OXYGEN DEFICIENCY   OXYGEN DEFICIENCY

An atmosphere containing oxygen at a concentration of less than 19.5 percent by volume.

OXYGEN ENRICHMENT   OXYGEN ENRICHMENT

An atmosphere containing oxygen at a concentration of more than 23.5 percent by volume.

OZONATION (O-zoe-NAY-shun)   OZONATION

The application of ozone to water for disinfection or for taste and odor control.

P

PCBs   PCBs
See **POLYCHLORINATED BIPHENYLS**.

pCi/*L*   pCi/*L*

picoCurie per liter. A picoCurie is a measure of radioactivity. One picoCurie of radioactivity is equivalent to 0.037 nuclear disintegrations per second.

pcu (PLATINUM COBALT UNITS)   pcu (PLATINUM COBALT UNITS)

Platinum cobalt units are a measure of color using platinum cobalt standards by visual comparison.

PMCL   PMCL

Primary **M**aximum **C**ontaminant **L**evel. Primary MCLs for various water quality indicators are established to protect public health.

PPM   PPM
See **PARTS PER MILLION**.

PSIG   PSIG

Pounds per **S**quare **I**nch **G**age pressure. The pressure within a closed container or pipe measured with a gage in pounds per square inch. See GAGE PRESSURE.

PACKER ASSEMBLY   PACKER ASSEMBLY

An inflatable device used to seal the tremie pipe inside the well casing to prevent the grout from entering the inside of the conductor casing.

PALATABLE (PAL-uh-tuh-bull)   PALATABLE

Water at a desirable temperature that is free from objectionable tastes, odors, colors, and turbidity. Pleasing to the senses.

PARSHALL FLUME   PARSHALL FLUME

A device used to measure the flow in an open channel. The flume narrows to a throat of fixed dimensions and then expands again. The rate of flow can be calculated by measuring the difference in head (pressure) before and at the throat of the flume.

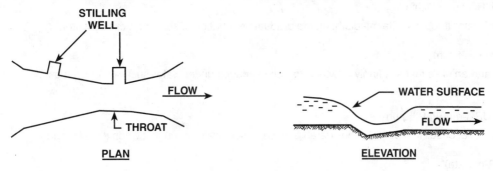

PARTICLE COUNT   PARTICLE COUNT

The results of a microscopic examination of treated water with a special particle counter which classifies suspended particles by number and size.

PARTICLE COUNTER   PARTICLE COUNTER

A device which counts and measures the size of individual particles in water. Particles are divided into size ranges and the number of particles is counted in each of these ranges. The results are reported in terms of the number of particles in different particle diameter size ranges per milliliter of water sampled.

PARTICLE COUNTING   PARTICLE COUNTING

A procedure for counting and measuring the size of individual particles in water. Particles are divided into size ranges and the number of particles is counted in each of these ranges. The results are reported in terms of the number of particles in different particle diameter size ranges per milliliter of water sampled.

PARTICULATE (par-TICK-you-let)   PARTICULATE

A very small solid suspended in water which can vary widely in size, shape, density, and electrical charge. Colloidal and dispersed particulates are artificially gathered together by the processes of coagulation and flocculation.

PARTS PER MILLION (PPM)   PARTS PER MILLION (PPM)

Parts per million parts, a measurement of concentration on a weight or volume basis. This term is equivalent to milligrams per liter (mg/L) which is the preferred term.

PASCAL   PASCAL

The pressure or stress of one newton per square meter. Abbreviated Pa.

     1 psi = 6,895 Pa = 6.895 kN/sq m = 0.0703 kg/sq cm

PATHOGENIC (PATH-o-JEN-ick) ORGANISMS   PATHOGENIC ORGANISMS

Organisms, including bacteria, viruses or cysts, capable of causing diseases (giardiasis, cryptosporidiosis, typhoid, cholera, dysentery) in a host (such as a person). There are many types of organisms which do *NOT* cause disease. These organisms are called non-pathogenic.

PATHOGENS (PATH-o-jens)   PATHOGENS

Pathogenic or disease-causing organisms.

PEAK DEMAND

The maximum momentary load placed on a water treatment plant, pumping station or distribution system. This demand is usually the maximum average load in one hour or less, but may be specified as the instantaneous load or the load during some other short time period.

PERCENT SATURATION

The amount of a substance that is dissolved in a solution compared with the amount dissolved in the solution at saturation, expressed as a percent.

$$\text{Percent Saturation, \%} = \frac{\text{Amount of Substance That Is Dissolved} \times 100\%}{\text{Amount Dissolved in Solution at Saturation}}$$

PERCOLATING (PURR-co-LAY-ting) WATER

Water that passes through soil or rocks under the force of gravity.

PERCOLATION (PURR-co-LAY-shun)

The slow passage of water through a filter medium; or, the gradual penetration of soil and rocks by water.

PERIPHYTON (pair-e-FI-tawn)

Microscopic plants and animals that are firmly attached to solid surfaces under water such as rocks, logs, pilings and other structures.

PERMEABILITY (PURR-me-uh-BILL-uh-tee)

The property of a material or soil that permits considerable movement of water through it when it is saturated.

PERMEATE (PURR-me-ate)

(1)   To penetrate and pass through, as water penetrates and passes through soil and other porous materials.

(2)   The liquid (demineralized water) produced from the reverse osmosis process that contains a *LOW* concentration of dissolved solids.

PERMIT-REQUIRED CONFINED SPACE
   (PERMIT SPACE)

See CONFINED SPACE, PERMIT-REQUIRED (PERMIT SPACE).

PESTICIDE

Any substance or chemical designed or formulated to kill or control animal pests. Also see INSECTICIDE and RODENTICIDE.

PET COCK

A small valve or faucet used to drain a cylinder or fitting.

pH (pronounce as separate letters)

pH is an expression of the intensity of the basic or acidic condition of a liquid. Mathematically, pH is the logarithm (base 10) of the reciprocal of the hydrogen ion activity.

$$\text{pH} = \text{Log}\frac{1}{\left[H^+\right]}$$

The pH may range from 0 to 14, where 0 is most acidic, 14 most basic, and 7 neutral. Natural waters usually have a pH between 6.5 and 8.5.

PHENOLIC (fee-NO-lick) COMPOUNDS

Organic compounds that are derivatives of benzene.

PHENOLPHTHALEIN (FEE-nol-THAY-leen) ALKALINITY

The alkalinity in a water sample measured by the amount of standard acid required to lower the pH to a level of 8.3, as indicated by the change in color of phenolphthalein from pink to clear. Phenolphthalein alkalinity is expressed as milligrams per liter of equivalent calcium carbonate.

PHOTOSYNTHESIS (foe-toe-SIN-thuh-sis)   PHOTOSYNTHESIS

A process in which organisms, with the aid of chlorophyll (green plant enzyme), convert carbon dioxide and inorganic substances into oxygen and additional plant material, using sunlight for energy. All green plants grow by this process.

PHYTOPLANKTON (FI-tow-PLANK-ton)   PHYTOPLANKTON

Small, usually microscopic plants (such as algae), found in lakes, reservoirs, and other bodies of water.

PICO   PICO

A prefix used in the metric system and other scientific systems of measurement which means 10^{-12} or 0.000 000 000 001.

PICOCURIE   PICOCURIE

A measure of radioactivity. One picoCurie of radioactivity is equivalent to 0.037 nuclear disintegrations per second.

PITLESS ADAPTER   PITLESS ADAPTER

A fitting which allows the well casing to be extended above ground while having a discharge connection located below the frost line. Advantages of using a pitless adapter include the elimination of the need for a pit or pump house and it is a watertight design, which helps maintain a sanitary water supply.

PLAN VIEW   PLAN VIEW

A diagram or photo showing a facility as it would appear when looking down on top of it.

PLANKTON   PLANKTON

(1) Small, usually microscopic, plants (phytoplankton) and animals (zooplankton) in aquatic systems.

(2) All of the smaller floating, suspended or self-propelled organisms in a body of water.

PLANNING   PLANNING

Management of utilities to build the resources and financial capability to provide for future needs.

PLUG FLOW   PLUG FLOW

A type of flow that occurs in tanks, basins or reactors when a slug of water moves through a tank without ever dispersing or mixing with the rest of the water flowing through the tank.

POINT SOURCE   POINT SOURCE

A discharge that comes out the end of a pipe. A nonpoint source refers to runoff or a discharge from a field or similar source.

POLARIZATION   POLARIZATION

The concentration of ions in the thin boundary layer adjacent to a membrane or pipe wall.

POLE SHADER   POLE SHADER

A copper bar circling the laminated iron core inside the coil of a magnetic starter.

POLLUTION   POLLUTION

The impairment (reduction) of water quality by agricultural, domestic, or industrial wastes (including thermal and radioactive wastes) to a degree that has an adverse effect on any beneficial use of water.

POLYANIONIC (poly-AN-eye-ON-ick)   POLYANIONIC

Characterized by many active negative charges especially active on the surface of particles.

POLYCHLORINATED BIPHENYLS   POLYCHLORINATED BIPHENYLS

A class of organic compounds that cause adverse health effects in domestic water supplies.

POLYELECTROLYTE (POLY-ee-LECK-tro-lite)   POLYELECTROLYTE

A high-molecular-weight (relatively heavy) substance having points of positive or negative electrical charges that is formed by either natural or manmade processes. Natural polyelectrolytes may be of biological origin or derived from starch products and cellulose derivatives. Manmade polyelectrolytes consist of simple substances that have been made into complex, high-molecular-weight substances. Used with other chemical coagulants to aid in binding small suspended particles to larger chemical flocs for their removal from water. Often called a POLYMER.

POLYMER (POLY-mer)
POLYMER

A long chain molecule formed by the union of many monomers (molecules of lower molecular weight). Polymers are used with other chemical coagulants to aid in binding small suspended particles to larger chemical flocs for their removal from water.

PORE
PORE

A very small open space in a rock or granular material. Also called an INTERSTICE, VOID, or void space. Also see VOID.

POROSITY
POROSITY

(1) A measure of the spaces or voids in a material or aquifer.

(2) The ratio of the volume of spaces in a rock or soil to the total volume. This ratio is usually expressed as a percentage.

$$\text{Porosity, \%} = \frac{(\text{Volume of Spaces})(100\%)}{\text{Total Volume}}$$

POSITIVE BACTERIOLOGICAL SAMPLE
POSITIVE BACTERIOLOGICAL SAMPLE

A water sample in which gas is produced by coliform organisms during incubation in the multiple tube fermentation test. See Chapter 11, Laboratory Procedures, "Coliform Bacteria," in *WATER TREATMENT PLANT OPERATION*, Volume I, for details.

POSITIVE DISPLACEMENT PUMP
POSITIVE DISPLACEMENT PUMP

A type of piston, diaphragm, gear or screw pump that delivers a constant volume with each stroke. Positive displacement pumps are used as chemical solution feeders.

POSTCHLORINATION
POSTCHLORINATION

The addition of chlorine to the plant effluent, *FOLLOWING* plant treatment, for disinfection purposes.

POTABLE (POE-tuh-bull) WATER
POTABLE WATER

Water that does not contain objectionable pollution, contamination, minerals, or infective agents and is considered satisfactory for drinking.

POWER FACTOR
POWER FACTOR

The ratio of the true power passing through an electric circuit to the product of the voltage and amperage in the circuit. This is a measure of the lag or lead of the current with respect to the voltage. In alternating current the voltage and amperes are not always in phase; therefore, the true power may be slightly less than that determined by the direct product.

PRECHLORINATION
PRECHLORINATION

The addition of chlorine at the headworks of the plant *PRIOR TO* other treatment processes mainly for disinfection and control of tastes, odors, and aquatic growths. Also applied to aid in coagulation and settling.

PRECIPITATE (pre-SIP-uh-TATE)
PRECIPITATE

(1) An insoluble, finely divided substance which is a product of a chemical reaction within a liquid.

(2) The separation from solution of an insoluble substance.

PRECIPITATION (pre-SIP-uh-TAY-shun)
PRECIPITATION

(1) The process by which atmospheric moisture falls onto a land or water surface as rain, snow, hail, or other forms of moisture.

(2) The chemical transformation of a substance in solution into an insoluble form (precipitate).

PRECISION
PRECISION

The ability of an instrument to measure a process variable and repeatedly obtain the same result. The ability of an instrument to reproduce the same results.

PRECURSOR, THM (pre-CURSE-or)
PRECURSOR, THM

Natural organic compounds found in all surface and groundwaters. These compounds *MAY* react with halogens (such as chlorine) to form trihalomethanes (tri-HAL-o-METH-hanes) (THMs); they *MUST* be present in order for THMs to form.

PRESCRIPTIVE (pre-SKRIP-tive)
PRESCRIPTIVE

Water rights which are acquired by diverting water and putting it to use in accordance with specified procedures. These procedures include filing a request (with a state agency) to use unused water in a stream, river or lake.

PRESENT WORTH PRESENT WORTH

The value of a long-term project expressed in today's dollars. Present worth is calculated by converting (discounting) all future benefits and costs over the life of the project to a single economic value at the start of the project. Calculating the present worth of alternative projects makes it possible to compare them and select the one with the largest positive (beneficial) present worth or minimum present cost.

PRESSURE CONTROL PRESSURE CONTROL

A switch which operates on changes in pressure. Usually this is a diaphragm pressing against a spring. When the force on the diaphragm overcomes the spring pressure, the switch is actuated (activated).

PRESSURE HEAD PRESSURE HEAD

The vertical distance (in feet) equal to the pressure (in psi) at a specific point. The pressure head is equal to the pressure in psi times 2.31 ft/psi.

PRESTRESSED PRESTRESSED

A prestressed pipe has been reinforced with wire strands (which are under tension) to give the pipe an active resistance to loads or pressures on it.

PREVENTIVE MAINTENANCE UNITS PREVENTIVE MAINTENANCE UNITS

Crews assigned the task of cleaning sewers (for example, balling or high-velocity cleaning crews) to prevent stoppages and odor complaints. Preventive maintenance is performing the most effective cleaning procedure, in the area where it is most needed, at the proper time in order to prevent failures and emergency situations.

PRIMARY ELEMENT PRIMARY ELEMENT

(1) A device that measures (senses) a physical condition or variable of interest. Floats and thermocouples are examples of primary elements. Also called a SENSOR.

(2) The hydraulic structure used to measure flows. In open channels, weirs and flumes are primary elements or devices. Venturi meters and orifice plates are the primary elements in pipes or pressure conduits.

PRIME PRIME

The action of filling a pump casing with water to remove the air. Most pumps must be primed before start-up or they will not pump any water.

PROCESS VARIABLE PROCESS VARIABLE

A physical or chemical quantity which is usually measured and controlled in the operation of a water treatment plant or an industrial plant.

PRODUCT WATER PRODUCT WATER

Water that has passed through a water treatment plant. All the treatment processes are completed or finished. This water is the product from the water treatment plant and is ready to be delivered to the consumers. Also called FINISHED WATER.

PROFILE PROFILE

A drawing showing elevation plotted against distance, such as the vertical section or *SIDE* view of a pipeline.

PRUSSIAN BLUE PRUSSIAN BLUE

A blue paste or liquid (often on a paper like carbon paper) used to show a contact area. Used to determine if gate valve seats fit properly.

PUMP BOWL PUMP BOWL

The submerged pumping unit in a well, including the shaft, impellers and housing.

PUMPING WATER LEVEL PUMPING WATER LEVEL

The vertical distance in feet from the centerline of the pump discharge to the level of the free pool while water is being drawn from the pool.

PURVEYOR (purr-VAY-or), WATER PURVEYOR, WATER

An agency or person that supplies water (usually potable water).

PUTREFACTION (PEW-truh-FACK-shun) PUTREFACTION

Biological decomposition of organic matter, with the production of foul-smelling and -tasting products, associated with anaerobic (no oxygen present) conditions.

Q

QUICKLIME QUICKLIME

A material that is mostly calcium oxide (CaO) or calcium oxide in natural association with a lesser amount of magnesium oxide. Quicklime is capable of combining with water, that is, becoming slaked. Also see HYDRATED LIME.

R

RADIAL TO IMPELLER RADIAL TO IMPELLER

Perpendicular to the impeller shaft. Material being pumped flows at a right angle to the impeller.

RADICAL RADICAL

A group of atoms that is capable of remaining unchanged during a series of chemical reactions. Such combinations (radicals) exist in the molecules of many organic compounds; sulfate (SO_4^{2-}) is an inorganic radical.

RANGE RANGE

The spread from minimum to maximum values that an instrument is designed to measure. Also see EFFECTIVE RANGE and SPAN.

RANNEY COLLECTOR RANNEY COLLECTOR

This water collector is constructed as a dug well from 12 to 16 feet (3.5 to 5 m) in diameter that has been sunk as a caisson near the bank of a river or lake. Screens are driven radially and approximately horizontally from this well into the sand and the gravel deposits underlying the river.

[SEE DRAWING ON PAGE 623]

RATE OF RETURN RATE OF RETURN

A value which indicates the return of funds received on the basis of the total equity capital used to finance physical facilities. Similar to the interest rate on savings accounts or loans.

RAW WATER RAW WATER

(1) Water in its natural state, prior to any treatment.

(2) Usually the water entering the first treatment process of a water treatment plant.

REAERATION (RE-air-A-shun) REAERATION

The introduction of air through forced air diffusers into the lower layers of the reservoir. As the air bubbles form and rise through the water, oxygen from the air dissolves into the water and replenishes the dissolved oxygen. The rising bubbles also cause the lower waters to rise to the surface where oxygen from the atmosphere is transferred to the water. This is sometimes called surface reaeration.

REAGENT (re-A-gent) REAGENT

A pure chemical substance that is used to make new products or is used in chemical tests to measure, detect, or examine other substances.

RECARBONATION (re-CAR-bun-NAY-shun) RECARBONATION

A process in which carbon dioxide is bubbled into the water being treated to lower the pH. The pH may also be lowered by the addition of acid. Recarbonation is the final stage in the lime-soda ash softening process. This process converts carbonate ions to bicarbonate ions and stabilizes the solution against the precipitation of carbonate compounds.

RECEIVER RECEIVER

A device which indicates the result of a measurement. Most receivers in the water utility field use either a fixed scale and movable indicator (pointer) such as a pressure gage or a movable scale and movable indicator like those used on a circular flow-recording chart. Also called an INDICATOR.

RECORDER RECORDER

A device that creates a permanent record, on a paper chart or magnetic tape, of the changes in a measured variable.

REDUCING AGENT REDUCING AGENT

Any substance, such as base metal (iron) or the sulfide ion (S^{2-}), that will readily donate (give up) electrons. The opposite is an OXIDIZING AGENT.

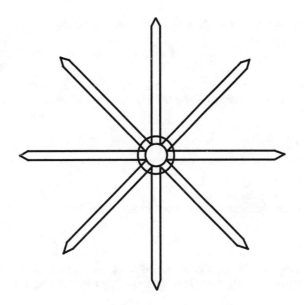

PLAN VIEW OF COLLECTOR PIPES

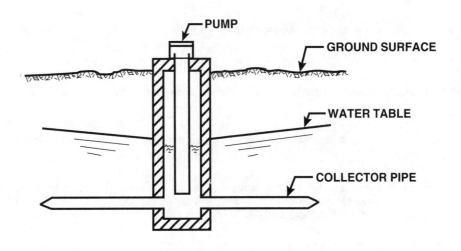

ELEVATION VIEW

RANNEY COLLECTOR

REDUCTION (re-DUCK-shun)

REDUCTION

Reduction is the addition of hydrogen, removal of oxygen, or the addition of electrons to an element or compound. Under anaerobic conditions (no dissolved oxygen present), sulfur compounds are reduced to odor-producing hydrogen sulfide (H_2S) and other compounds. The opposite of OXIDATION.

REFERENCE

REFERENCE

A physical or chemical quantity whose value is known exactly, and thus is used to calibrate instruments or standardize measurements. Also called a STANDARD.

REGULATORY NEGOTIATION

REGULATORY NEGOTIATION

A process whereby the U.S. Environmental Protection Agency acts on an equal basis with outside parties to reach consensus on the content of a proposed rule. If the group reaches consensus, the US EPA commits to propose the rule with the agreed upon content.

RELIQUEFACTION (re-LICK-we-FACK-shun)

RELIQUEFACTION

The return of a gas to the liquid state; for example, a condensation of chlorine gas to return it to its liquid form by cooling.

REPRESENTATIVE SAMPLE

REPRESENTATIVE SAMPLE

A sample portion of material or water that is as nearly identical in content and consistency as possible to that in the larger body of material or water being sampled.

RESIDUAL CHLORINE

RESIDUAL CHLORINE

The concentration of chlorine present in water after the chlorine demand has been satisfied. The concentration is expressed in terms of the total chlorine residual, which includes both the free and combined or chemically bound chlorine residuals.

RESIDUE

RESIDUE

The dry solids remaining after the evaporation of a sample of water or sludge. Also see TOTAL DISSOLVED SOLIDS.

RESINS

RESINS

See ION EXCHANGE RESINS.

RESISTANCE

RESISTANCE

That property of a conductor or wire that opposes the passage of a current, thus causing electric energy to be transformed into heat.

RESPIRATION

RESPIRATION

The process in which an organism uses oxygen for its life processes and gives off carbon dioxide.

RESPONSIBILITY

RESPONSIBILITY

Answering to those above in the chain of command to explain how and why you have used your authority.

REVERSE OSMOSIS (oz-MOE-sis)

REVERSE OSMOSIS

The application of pressure to a concentrated solution which causes the passage of a liquid from the concentrated solution to a weaker solution across a semipermeable membrane. The membrane allows the passage of the water (solvent) but not the dissolved solids (solutes). The liquid produced is a demineralized water. Also see OSMOSIS.

RIPARIAN (ri-PAIR-ee-an)

RIPARIAN

Water rights which are acquired together with title to the land bordering a source of surface water. The right to put to beneficial use surface water adjacent to your land.

RODENTICIDE (row-DENT-uh-SIDE)

RODENTICIDE

Any substance or chemical used to kill or control rodents.

ROTAMETER (RODE-uh-ME-ter)

ROTAMETER

A device used to measure the flow rate of gases and liquids. The gas or liquid being measured flows vertically up a tapered, calibrated tube. Inside the tube is a small ball or bullet-shaped float (it may rotate) that rises or falls depending on the flow rate. The flow rate may be read on a scale behind or on the tube by looking at the middle of the ball or at the widest part or top of the float.

ROTOR   ROTOR

The rotating part of a machine. The rotor is surrounded by the stationary (non-moving) parts (stator) of the machine.

ROUTINE SAMPLING   ROUTINE SAMPLING

Sampling repeated on a regular basis.

S

SCADA (ss-KAY-dah) SYSTEM   SCADA SYSTEM

Supervisory **C**ontrol **A**nd **D**ata **A**cquisition system. A computer-monitored alarm, response, control and data acquisition system used by drinking water facilities to monitor their operations.

SCFM   SCFM

Cubic **F**eet of air per **M**inute at **S**tandard conditions of temperature, pressure, and humidity (0°C, 14.7 psia, and 50% relative humidity).

SDWA   SDWA

See **S**AFE **D**RINKING **W**ATER **A**CT.

SMCL   SMCL

Secondary **M**aximum **C**ontaminant **L**evel. Secondary MCLs for various water quality indicators are established to protect public welfare.

SNARL   SNARL

Suggested **N**o **A**dverse **R**esponse **L**evel. The concentration of a chemical in water that is expected not to cause an adverse health effect.

SACRIFICIAL ANODE   SACRIFICIAL ANODE

An easily corroded material deliberately installed in a pipe or tank. The intent of such an installation is to give up (sacrifice) this anode to corrosion while the water supply facilities remain relatively corrosion free.

SAFE DRINKING WATER ACT   SAFE DRINKING WATER ACT

Commonly referred to as SDWA. An Act passed by the U.S. Congress in 1974. The Act establishes a cooperative program among local, state and federal agencies to ensure safe drinking water for consumers. The Act has been amended several times, including the 1980, 1986, and 1996 Amendments.

SAFE WATER   SAFE WATER

Water that does not contain harmful bacteria, or toxic materials or chemicals. Water may have taste and odor problems, color and certain mineral problems and still be considered safe for drinking.

SAFE YIELD   SAFE YIELD

The annual quantity of water that can be taken from a source of supply over a period of years without depleting the source permanently (beyond its ability to be replenished naturally in "wet years").

SALINITY   SALINITY

(1) The relative concentration of dissolved salts, usually sodium chloride, in a given water.

(2) A measure of the concentration of dissolved mineral substances in water.

SANITARY SURVEY   SANITARY SURVEY

A detailed evaluation and/or inspection of a source of water supply and all conveyances, storage, treatment and distribution facilities to ensure protection of the water supply from all pollution sources.

SAPROPHYTES (SAP-row-FIGHTS)   SAPROPHYTES

Organisms living on dead or decaying organic matter. They help natural decomposition of organic matter in water.

SATURATION   SATURATION

The condition of a liquid (water) when it has taken into solution the maximum possible quantity of a given substance at a given temperature and pressure.

SATURATOR (SAT-you-RAY-tore)   SATURATOR

A device which produces a fluoride solution for the fluoridation process. The device is usually a cylindrical container with granular sodium fluoride on the bottom. Water flows either upward or downward through the sodium fluoride to produce the fluoride solution.

SCHEDULE, PIPE   SCHEDULE, PIPE

A sizing system of numbers that specifies the I.D. (inside diameter) and O.D. (outside diameter) for each diameter pipe. The schedule number is the ratio of internal pressure in psi divided by the allowable fiber stress multiplied by 1,000. Typical schedules of iron and steel pipe are schedules 40, 80, and 160. Other forms of piping are divided into various classes with their own schedule schemes.

SCHMUTZDECKE (sh-moots-DECK-ee)   SCHMUTZDECKE

A layer of trapped matter at the surface of a slow sand filter in which a dense population of microorganisms develops. These microorganisms within the film or mat feed on and break down incoming organic material trapped in the mat. In doing so the microorganisms both remove organic matter and add mass to the mat, further developing the mat and increasing the physical straining action of the mat.

SECCHI (SECK-key) DISC   SECCHI DISC

A flat, white disc lowered into the water by a rope until it is just barely visible. At this point, the depth of the disc from the water surface is the recorded Secchi disc transparency.

SEDIMENTATION (SED-uh-men-TAY-shun)   SEDIMENTATION

A water treatment process in which solid particles settle out of the water being treated in a large clarifier or sedimentation basin.

SEIZE UP   SEIZE UP

Seize up occurs when an engine overheats and a part expands to the point where the engine will not run. Also called "freezing."

SENSITIVITY (PARTICLE COUNTERS)   SENSITIVITY (PARTICLE COUNTERS)

The smallest particle a particle counter will measure and count.

SENSOR   SENSOR

A device that measures (senses) a physical condition or variable of interest. Floats and thermocouples are examples of sensors. Also called a PRIMARY ELEMENT.

SEPTIC (SEP-tick)   SEPTIC

A condition produced by bacteria when all oxygen supplies are depleted. If severe, the bottom deposits produce hydrogen sulfide, the deposits and water turn black, give off foul odors, and the water has a greatly increased chlorine demand.

SEQUESTRATION (SEE-kwes-TRAY-shun)   SEQUESTRATION

A chemical complexing (forming or joining together) of metallic cations (such as iron) with certain inorganic compounds, such as phosphate. Sequestration prevents the precipitation of the metals (iron). Also see CHELATION.

SERVICE PIPE   SERVICE PIPE

The pipeline extending from the water main to the building served or to the consumer's system.

SET POINT   SET POINT

The position at which the control or controller is set. This is the same as the desired value of the process variable. For example, a thermostat is set to maintain a desired temperature.

SEWAGE   SEWAGE

The used household water and water-carried solids that flow in sewers to a wastewater treatment plant. The preferred term is WASTEWATER.

SHEAVE (SHE-v)   SHEAVE

V-belt drive pulley which is commonly made of cast iron or steel.

SHIM   SHIM

Thin metal sheets which are inserted between two surfaces to align or space the surfaces correctly. Shims can be used anywhere a spacer is needed. Usually shims are 0.001 to 0.020 inch thick.

SHOCK LOAD   SHOCK LOAD

The arrival at a water treatment plant of raw water containing unusual amounts of algae, colloidal matter, color, suspended solids, turbidity, or other pollutants.

SHORT-CIRCUITING   SHORT-CIRCUITING

A condition that occurs in tanks or basins when some of the flowing water entering a tank or basin flows along a nearly direct pathway from the inlet to the outlet. This is usually undesirable since it may result in shorter contact, reaction, or settling times in comparison with the theoretical (calculated) or presumed detention times.

SIMULATE   SIMULATE

To reproduce the action of some process, usually on a smaller scale.

SINGLE-STAGE PUMP   SINGLE-STAGE PUMP

A pump that has only one impeller. A multi-stage pump has more than one impeller.

SLAKE   SLAKE

To mix with water so that a true chemical combination (hydration) takes place, such as in the slaking of lime.

SLAKED LIME   SLAKED LIME

See HYDRATED LIME.

SLOPE   SLOPE

The slope or inclination of a trench bottom or a trench side wall is the ratio of the vertical distance to the horizontal distance or "rise over run." Also see GRADE (2).

2 VERTICAL

1 HORIZONTAL

2:1 SLOPE

SLUDGE (sluj)   SLUDGE

The settleable solids separated from water during processing.

SLURRY (SLUR-e)   SLURRY

A watery mixture or suspension of insoluble (not dissolved) matter; a thin, watery mud or any substance resembling it (such as a grit slurry or a lime slurry).

SOFT WATER   SOFT WATER

Water having a low concentration of calcium and magnesium ions. According to U.S. Geological Survey guidelines, soft water is water having a hardness of 60 milligrams per liter or less.

SOFTWARE PROGRAMS   SOFTWARE PROGRAMS

Computer programs; the list of instructions that tell a computer how to perform a given task or tasks. Some software programs are designed and written to monitor and control water distribution systems and treatment processes.

SOLENOID (SO-luh-noid)   SOLENOID

A magnetically (electric coil) operated mechanical device. Solenoids can operate small valves or electric switches.

SOLUTION   SOLUTION

A liquid mixture of dissolved substances. In a solution it is impossible to see all the separate parts.

SOUNDING TUBE   SOUNDING TUBE

A pipe or tube used for measuring the depths of water.

SPAN   SPAN

The scale or range of values an instrument is designed to measure. Also see RANGE.

SPECIFIC CAPACITY

A measurement of well yield per unit depth of drawdown after a specific time has passed, usually 24 hours. Typically expressed as gallons per minute per foot (GPM/ft or cu m/day/m).

SPECIFIC CAPACITY TEST

A testing method used to determine the adequacy of an aquifer or well by measuring the specific capacity.

SPECIFIC CONDUCTANCE

A rapid method of estimating the dissolved solids content of a water supply. The measurement indicates the capacity of a sample of water to carry an electric current, which is related to the concentration of ionized substances in the water. Also called CONDUCTANCE.

SPECIFIC GRAVITY

(1) Weight of a particle, substance, or chemical solution in relation to the weight of an equal volume of water. Water has a specific gravity of 1.000 at 4°C (39°F). Particulates in raw water may have a specific gravity of 1.005 to 2.5.

(2) Weight of a particular gas in relation to the weight of an equal volume of air at the same temperature and pressure (air has a specific gravity of 1.0). Chlorine has a specific gravity of 2.5 as a gas.

SPECIFIC YIELD

The quantity of water that a unit volume of saturated permeable rock or soil will yield when drained by gravity. Specific yield may be expressed as a ratio or as a percentage by volume.

SPOIL

Excavated material such as soil from the trench of a water main.

SPORE

The reproductive body of an organism which is capable of giving rise to a new organism either directly or indirectly. A viable (able to live and grow) body regarded as the resting stage of an organism. A spore is usually more resistant to disinfectants and heat than most organisms. Gangrene and tetanus bacteria are common spore-forming organisms.

SPRING LINE

Theoretical center of a pipeline. Also, the guideline for laying a course of bricks.

STALE WATER

Water which has not flowed recently and may have picked up tastes and odors from distribution lines or storage facilities.

STANDARD

A physical or chemical quantity whose value is known exactly, and thus is used to calibrate instruments or standardize measurements. Also called a REFERENCE.

STANDARD DEVIATION

A measure of the spread or dispersion of data.

STANDARD METHODS

STANDARD METHODS FOR THE EXAMINATION OF WATER AND WASTEWATER, 20th Edition. A joint publication of the American Public Health Association (APHA), American Water Works Association (AWWA), and the Water Environment Federation (WEF) which outlines the accepted laboratory procedures used to analyze the impurities in water and wastewater. Available from American Water Works Association, Bookstore, 6666 West Quincy Avenue, Denver, CO 80235. Order No. 10079. Price to members, $168.00; nonmembers, $213.00; price includes cost of shipping and handling.

STANDARD SOLUTION

A solution in which the exact concentration of a chemical or compound is known.

STANDARDIZE

To compare with a standard.

(1) In wet chemistry, to find out the exact strength of a solution by comparing it with a standard of known strength. This information is used to adjust the strength by adding more water or more of the substance dissolved.

(2) To set up an instrument or device to read a standard. This allows you to adjust the instrument so that it reads accurately, or enables you to apply a correction factor to the readings.

STARTERS (MOTOR) STARTERS (MOTOR)

Devices used to start up large motors gradually to avoid severe mechanical shock to a driven machine and to prevent disturbance to the electrical lines (causing dimming and flickering of lights).

STATIC HEAD STATIC HEAD

When water is not moving, the vertical distance (in feet) from a specific point to the water surface is the static head. (The static pressure in psi is the static head in feet times 0.433 psi/ft.) Also see DYNAMIC PRESSURE and STATIC PRESSURE.

STATIC PRESSURE STATIC PRESSURE

When water is not moving, the vertical distance (in feet) from a specific point to the water surface is the static head. The static pressure in psi is the static head in feet times 0.433 psi/ft. Also see DYNAMIC PRESSURE and STATIC HEAD.

STATIC WATER DEPTH STATIC WATER DEPTH

The vertical distance in feet from the centerline of the pump discharge down to the surface level of the free pool while no water is being drawn from the pool or water table.

STATIC WATER LEVEL STATIC WATER LEVEL

(1) The elevation or level of the water table in a well when the pump is not operating.

(2) The level or elevation to which water would rise in a tube connected to an artesian aquifer, basin, or conduit under pressure.

STATOR STATOR

That portion of a machine which contains the stationary (non-moving) parts that surround the moving parts (rotor).

STERILIZATION (STARE-uh-luh-ZAY-shun) STERILIZATION

The removal or destruction of all microorganisms, including pathogenic and other bacteria, vegetative forms and spores. Compare with DISINFECTION.

STETHOSCOPE STETHOSCOPE

An instrument used to magnify sounds and convey them to the ear.

STORATIVITY (S) STORATIVITY (S)

The volume of groundwater an aquifer releases from or takes into storage per unit surface area of the aquifer per unit change in head. Also called the storage coefficient.

STRATIFICATION (STRAT-uh-fuh-KAY-shun) STRATIFICATION

The formation of separate layers (of temperature, plant, or animal life) in a lake or reservoir. Each layer has similar characteristics such as all water in the layer has the same temperature. Also see THERMAL STRATIFICATION.

STRAY CURRENT CORROSION STRAY CURRENT CORROSION

A corrosion activity resulting from stray electric current originating from some source outside the plumbing system such as D.C. grounding on phone systems.

SUBMERGENCE SUBMERGENCE

The distance between the water surface and the media surface in a filter.

SUBSIDENCE (sub-SIDE-ence) SUBSIDENCE

The dropping or lowering of the ground surface as a result of removing excess water (overdraft or overpumping) from an aquifer. After excess water has been removed, the soil will settle, become compacted and the ground surface will drop and can cause the settling of underground utilities.

SUCTION LIFT SUCTION LIFT

The *NEGATIVE* pressure [in feet (meters) of water or inches (centimeters) of mercury vacuum] on the suction side of the pump. The pressure can be measured from the centerline of the pump *DOWN TO* (lift) the elevation of the hydraulic grade line on the suction side of the pump.

SUPERCHLORINATION (SUE-per-KLOR-uh-NAY-shun) SUPERCHLORINATION

Chlorination with doses that are deliberately selected to produce free or combined residuals so large as to require dechlorination.

SUPERNATANT (sue-per-NAY-tent)   SUPERNATANT

Liquid removed from settled sludge. Supernatant commonly refers to the liquid between the sludge on the bottom and the scum on the water surface of a basin or container.

SUPERSATURATED   SUPERSATURATED

An unstable condition of a solution (water) in which the solution contains a substance at a concentration greater than the saturation concentration for the substance.

SURFACE LOADING   SURFACE LOADING

One of the guidelines for the design of settling tanks and clarifiers in treatment plants. Used by operators to determine if tanks and clarifiers are hydraulically (flow) over- or underloaded. Also called OVERFLOW RATE.

$$\text{Surface Loading, GPD/sq ft} = \frac{\text{Flow, gallons/day}}{\text{Surface Area, sq ft}}$$

SURFACTANT (sir-FAC-tent)   SURFACTANT

Abbreviation for surface-active agent. The active agent in detergents that possesses a high cleaning ability.

SURGE CHAMBER   SURGE CHAMBER

A chamber or tank connected to a pipe and located at or near a valve that may quickly open or close or a pump that may suddenly start or stop. When the flow of water in a pipe starts or stops quickly, the surge chamber allows water to flow into or out of the pipe and minimize any sudden positive or negative pressure waves or surges in the pipe.

[SEE DRAWING ON PAGE 631]

SUSPENDED SOLIDS   SUSPENDED SOLIDS

(1) Solids that either float on the surface or are suspended in water, wastewater, or other liquids, and which are largely removable by laboratory filtering.

(2) The quantity of material removed from water in a laboratory test, as prescribed in *STANDARD METHODS FOR THE EXAMINATION OF WATER AND WASTEWATER*, and referred to as Total Suspended Solids Dried at 103–105°C.

T

TCE   TCE

See **TR**I**CH**LORO**ET**HANE.

TDS   TDS

See **TO**TAL **D**ISSOLVED **S**OLIDS.

THM   THM

See **TR**I**H**ALO**M**ETHANES.

THM PRECURSOR   THM PRECURSOR

See PRECURSOR, THM.

TAILGATE SAFETY MEETING   TAILGATE SAFETY MEETING

Brief (10 to 20 minutes) safety meetings held every 7 to 10 working days. The term *TAILGATE* comes from the safety meetings regularly held by the construction industry around the tailgate of a truck.

TELEMETRY (tel-LEM-uh-tree)   TELEMETRY

The electrical link between the transmitter and the receiver. Telephone lines are commonly used to serve as the electrical line.

TEMPERATURE SENSOR   TEMPERATURE SENSOR

A device that opens and closes a switch in response to changes in the temperature. This device might be a metal contact, or a thermocouple that generates minute electric current proportional to the difference in heat, or a variable resistor whose value changes in response to changes in temperature. Also called a HEAT SENSOR.

THERMAL STRATIFICATION (STRAT-uh-fuh-KAY-shun)   THERMAL STRATIFICATION

The formation of layers of different temperatures in a lake or reservoir. Also see STRATIFICATION.

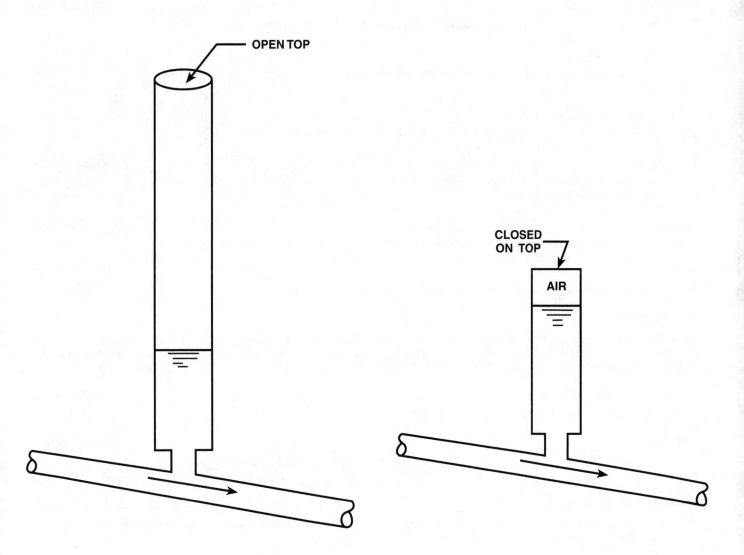

OPEN TOP

CLOSED ON TOP

AIR

TYPES OF SURGE CHAMBERS

THERMOCLINE (THUR-moe-KLINE)   THERMOCLINE

The middle layer in a thermally stratified lake or reservoir. In this layer there is a rapid decrease in temperature with depth. Also called the METALIMNION.

THERMOCOUPLE   THERMOCOUPLE

A heat-sensing device made of two conductors of different metals joined at their ends. An electric current is produced when there is a difference in temperature between the ends.

THICKENING   THICKENING

Treatment to remove water from the sludge mass to reduce the volume that must be handled.

THRESHOLD ODOR   THRESHOLD ODOR

The minimum odor of a water sample that can just be detected after successive dilutions with odorless water. Also called ODOR THRESHOLD.

THRESHOLD ODOR NUMBER (TON)   THRESHOLD ODOR NUMBER (TON)

The greatest dilution of a sample with odor-free water that still yields a just-detectable odor.

THRUST BLOCK   THRUST BLOCK

A mass of concrete or similar material appropriately placed around a pipe to prevent movement when the pipe is carrying water. Usually placed at bends and valve structures.

TIME LAG   TIME LAG

The time required for processes and control systems to respond to a signal or to reach a desired level.

TIMER   TIMER

A device for automatically starting or stopping a machine or other device at a given time.

TITRATE (TIE-trate)   TITRATE

To *TITRATE* a sample, a chemical solution of known strength is added drop by drop until a certain color change, precipitate, or pH change in the sample is observed (end point). Titration is the process of adding the chemical reagent in small increments (0.1 − 1.0 milliliter) until completion of the reaction, as signaled by the end point.

TOPOGRAPHY (toe-PAH-gruh-fee)   TOPOGRAPHY

The arrangement of hills and valleys in a geographic area.

TOTAL CHLORINE   TOTAL CHLORINE

The total concentration of chlorine in water, including the combined chlorine (such as inorganic and organic chloramines) and the free available chlorine.

TOTAL CHLORINE RESIDUAL   TOTAL CHLORINE RESIDUAL

The total amount of chlorine residual (value for residual chlorine, including both free chlorine and chemically bound chlorine) present in a water sample after a given contact time.

TOTAL DISSOLVED SOLIDS (TDS)   TOTAL DISSOLVED SOLIDS (TDS)

All of the dissolved solids in a water. TDS is measured on a sample of water that has passed through a very fine mesh filter to remove suspended solids. The water passing through the filter is evaporated and the residue represents the dissolved solids. Also see SPECIFIC CONDUCTANCE.

TOTAL DYNAMIC HEAD (TDH)   TOTAL DYNAMIC HEAD (TDH)

When a pump is lifting or pumping water, the vertical distance (in feet) from the elevation of the energy grade line on the suction side of the pump to the elevation of the energy grade line on the discharge side of the pump.

TOTAL ORGANIC CARBON (TOC)   TOTAL ORGANIC CARBON (TOC)

TOC measures the amount of organic carbon in water.

TOTALIZER   TOTALIZER

A device or meter that continuously measures and calculates (adds) a process rate variable in cumulative fashion; for example, total flows displayed in gallons, million gallons, cubic feet, or some other unit of volume measurement. Also called an INTEGRATOR.

TOXAPHENE (TOX-uh-FEEN)   TOXAPHENE

A chemical that causes adverse health effects in domestic water supplies and also is toxic to freshwater and marine aquatic life.

TOXIC (TOX-ick)   TOXIC

A substance which is poisonous to a living organism.

TRANSDUCER (trans-DUE-sir)   TRANSDUCER

A device that senses some varying condition measured by a primary sensor and converts it to an electrical or other signal for transmission to some other device (a receiver) for processing or decision making.

TRANSMISSION LINES   TRANSMISSION LINES

Pipelines that transport raw water from its source to a water treatment plant. After treatment, water is usually pumped into pipelines (transmission lines) that are connected to a distribution grid system.

TRANSMISSIVITY (TRANS-miss-SIV-it-tee)   TRANSMISSIVITY

A measure of the ability to transmit (as in the ability of an aquifer to transmit water).

TRANSPIRATION (TRAN-spur-RAY-shun)   TRANSPIRATION

The process by which water vapor is released to the atmosphere by living plants. This process is similar to people sweating. Also see EVAPOTRANSPIRATION.

TREMIE (TREH-me)   TREMIE

A device used to place concrete or grout under water.

TRICHLOROETHANE (TCE) (try-KLOR-o-ETH-hane)   TRICHLOROETHANE (TCE)

An organic chemical used as a cleaning solvent that causes adverse health effects in domestic water supplies.

TRIHALOMETHANES (THMs) (tri-HAL-o-METH-hanes)   TRIHALOMETHANES (THMs)

Derivatives of methane, CH_4, in which three halogen atoms (chlorine or bromine) are substituted for three of the hydrogen atoms. Often formed during chlorination by reactions with natural organic materials in the water. The resulting compounds (THMs) are suspected of causing cancer.

TRUE COLOR   TRUE COLOR

Color of the water from which turbidity has been removed. The turbidity may be removed by double filtering the sample through a Whatman No. 40 filter when using the visual comparison method.

TUBE SETTLER   TUBE SETTLER

A device that uses bundles of small-bore (2 to 3 inches or 50 to 75 mm) tubes installed on an incline as an aid to sedimentation. The tubes may come in a variety of shapes including circular and rectangular. As water rises within the tubes, settling solids fall to the tube surface. As the sludge (from the settled solids) in the tube gains weight, it moves down the tubes and settles to the bottom of the basin for removal by conventional sludge collection means. Tube settlers are sometimes installed in sedimentation basins and clarifiers to improve particle removal.

TUBERCLE (TOO-burr-cull)   TUBERCLE

A protective crust of corrosion products (rust) which builds up over a pit caused by the loss of metal due to corrosion.

TUBERCULATION (too-BURR-cue-LAY-shun)   TUBERCULATION

The development or formation of small mounds of corrosion products (rust) on the inside of iron pipe. These mounds (tubercles) increase the roughness of the inside of the pipe thus increasing resistance to water flow (decreases the C Factor).

TURBID   TURBID

Having a cloudy or muddy appearance.

TURBIDIMETER   TURBIDIMETER

See TURBIDITY METER.

TURBIDITY (ter-BID-it-tee)

The cloudy appearance of water caused by the presence of suspended and colloidal matter. In the waterworks field, a turbidity measurement is used to indicate the clarity of water. Technically, turbidity is an optical property of the water based on the amount of light reflected by suspended particles. Turbidity cannot be directly equated to suspended solids because white particles reflect more light than dark-colored particles and many small particles will reflect more light than an equivalent large particle.

TURBIDITY METER

An instrument for measuring and comparing the turbidity of liquids by passing light through them and determining how much light is reflected by the particles in the liquid. The normal measuring range is 0 to 100 and is expressed as Nephelometric Turbidity Units (NTUs).

TURBIDITY UNITS (TU)

Turbidity units are a measure of the cloudiness of water. If measured by a nephelometric (deflected light) instrumental procedure, turbidity units are expressed in nephelometric turbidity units (NTU) or simply TU. Those turbidity units obtained by visual methods are expressed in Jackson Turbidity Units (JTU) which are a measure of the cloudiness of water; they are used to indicate the clarity of water. There is no real connection between NTUs and JTUs. The Jackson turbidimeter is a visual method and the nephelometer is an instrumental method based on deflected light.

TURN-DOWN RATIO

The ratio of the design range to the range of acceptable accuracy or precision of an instrument. Also see EFFECTIVE RANGE.

U

UNCONSOLIDATED FORMATION

A sediment that is loosely arranged or unstratified (not in layers) or whose particles are not cemented together (soft rock); occurring either at the ground surface or at a depth below the surface. Also see CONSOLIDATED FORMATION.

UNIFORMITY COEFFICIENT (U.C.)

The ratio of (1) the diameter of a grain (particle) of a size that is barely too large to pass through a sieve that allows 60 percent of the material (by weight) to pass through, to (2) the diameter of a grain (particle) of a size that is barely too large to pass through a sieve that allows 10 percent of the material (by weight) to pass through. The resulting ratio is a measure of the degree of uniformity in a granular material such as filter media.

$$\text{Uniformity Coefficient} = \frac{\text{Particle Diameter}_{60\%}}{\text{Particle Diameter}_{10\%}}$$

UPPER EXPLOSIVE LIMIT (UEL)

The point at which the concentration of a gas in air becomes too great to allow an explosion upon ignition due to insufficient oxygen present.

V

VARIABLE COSTS

Costs that a utility must cover or pay that are associated with the production and delivery of water. The costs vary or fluctuate on the basis of the volume of water treated and delivered to customers (water production). Also see FIXED COSTS.

VARIABLE FREQUENCY DRIVE

A control system that allows the frequency of the current applied to a motor to be varied. The motor is connected to a low-frequency source while standing still; the frequency is then increased gradually until the motor and pump (or other driven machine) are operating at the desired speed.

VARIABLE, MEASURED

A factor (flow, temperature) that is sensed and quantified (reduced to a reading of some kind) by a primary element or sensor.

VARIABLE, PROCESS

A physical or chemical quantity which is usually measured and controlled in the operation of a water treatment plant or an industrial plant.

VELOCITY HEAD

The energy in flowing water as determined by a vertical height (in feet or meters) equal to the square of the velocity of flowing water divided by twice the acceleration due to gravity ($V^2/2g$).

VENTURI METER

A flow measuring device placed in a pipe. The device consists of a tube whose diameter gradually decreases to a throat and then gradually expands to the diameter of the pipe. The flow is determined on the basis of the difference in pressure (caused by different velocity heads) between the entrance and throat of the Venturi meter.

NOTE: Most Venturi meters have pressure sensing taps rather than a manometer to measure the pressure difference. The upstream tap is the high pressure tap or side of the manometer.

VENTURI METER

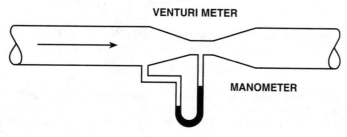

MANOMETER

VISCOSITY (vis-KOSS-uh-tee)

A property of water, or any other fluid, which resists efforts to change its shape or flow. Syrup is more viscous (has a higher viscosity) than water. The viscosity of water increases significantly as temperatures decrease. Motor oil is rated by how thick (viscous) it is; 20 weight oil is considered relatively thin while 50 weight oil is relatively thick or viscous.

VOID

A pore or open space in rock, soil or other granular material, not occupied by solid matter. The pore or open space may be occupied by air, water, or other gaseous or liquid material. Also called an INTERSTICE, PORE, or void space.

VOLATILE (VOL-uh-tull)

(1)   A volatile substance is one that is capable of being evaporated or changed to a vapor at relatively low temperatures. Volatile substances also can be partially removed by air stripping.

(2)   In terms of solids analysis, volatile refers to materials lost (including most organic matter) upon ignition in a muffle furnace for 60 minutes at 550°C. Natural volatile materials are chemical substances usually of animal or plant origin. Manufactured or synthetic volatile materials such as ether, acetone, and carbon tetrachloride are highly volatile and not of plant or animal origin. Also see NONVOLATILE MATTER.

VOLATILE ACIDS

Fatty acids produced during digestion which are soluble in water and can be steam-distilled at atmospheric pressure. Also called organic acids. Volatile acids are commonly reported as equivalent to acetic acid.

VOLATILE LIQUIDS

Liquids which easily vaporize or evaporate at room temperature.

VOLATILE MATTER

Matter in water, wastewater, or other liquids that is lost on ignition of the dry solids at 550°C.

VOLATILE SOLIDS

Those solids in water or other liquids that are lost on ignition of the dry solids at 550°C.

VOLTAGE

The electrical pressure available to cause a flow of current (amperage) when an electric circuit is closed. Also called ELECTROMOTIVE FORCE (E.M.F.).

VOLUMETRIC

A measurement based on the volume of some factor. Volumetric titration is a means of measuring unknown concentrations of water quality indicators in a sample *BY DETERMINING THE VOLUME* of titrant or liquid reagent needed to complete particular reactions.

VOLUMETRIC FEEDER   VOLUMETRIC FEEDER

A dry chemical feeder which delivers a measured volume of chemical during a specific time period.

VORTEX   VORTEX

A revolving mass of water which forms a whirlpool. This whirlpool is caused by water flowing out of a small opening in the bottom of a basin or reservoir. A funnel-shaped opening is created downward from the water surface.

W

WARNING   WARNING

The word *WARNING* is used to indicate a hazard level between *CAUTION* and *DANGER*. Also see CAUTION, DANGER, and NOTICE.

WASTEWATER   WASTEWATER

A community's used water and water-carried solids (including used water from industrial processes) that flow to a treatment plant. Storm water, surface water, and groundwater infiltration also may be included in the wastewater that enters a wastewater treatment plant. The term "sewage" usually refers to household wastes, but this word is being replaced by the term "wastewater."

WATER AUDIT   WATER AUDIT

A thorough examination of the accuracy of water agency records or accounts (volumes of water) and system control equipment. Water managers can use audits to determine their water distribution system efficiency. The overall goal is to identify and verify water and revenue losses in a water system.

WATER CYCLE   WATER CYCLE

The process of evaporation of water into the air and its return to earth by precipitation (rain or snow). This process also includes transpiration from plants, groundwater movement, and runoff into rivers, streams and the ocean. Also called the HYDROLOGIC CYCLE.

WATER HAMMER   WATER HAMMER

The sound like someone hammering on a pipe that occurs when a valve is opened or closed very rapidly. When a valve position is changed quickly, the water pressure in a pipe will increase and decrease back and forth very quickly. This rise and fall in pressures can cause serious damage to the system.

WATER PURVEYOR (purr-VAY-or)   WATER PURVEYOR

An agency or person that supplies water (usually potable water).

WATER TABLE   WATER TABLE

The upper surface of the zone of saturation of groundwater in an unconfined aquifer.

WATERSHED   WATERSHED

The region or land area that contributes to the drainage or catchment area above a specific point on a stream or river.

WATT   WATT

A unit of power equal to one joule per second. The power of a current of one ampere flowing across a potential difference of one volt.

WEIR (weer)   WEIR

(1)   A wall or plate placed in an open channel and used to measure the flow of water. The depth of the flow over the weir can be used to calculate the flow rate, or a chart or conversion table may be used to convert depth to flow.

(2)   A wall or obstruction used to control flow (from settling tanks and clarifiers) to ensure a uniform flow rate and avoid short-circuiting.

WEIR (weer) DIAMETER   WEIR DIAMETER

Many circular clarifiers have a circular weir within the outside edge of the clarifier. All the water leaving the clarifier flows over this weir. The diameter of the weir is the length of a line from one edge of a weir to the opposite edge and passing through the center of the circle formed by the weir.

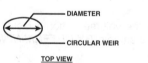

TOP VIEW

CROSS SECTION

WEIR LOADING

A guideline used to determine the length of weir needed on settling tanks and clarifiers in treatment plants. Used by operators to determine if weirs are hydraulically (flow) overloaded.

$$\text{Weir Loading, GPM/ft} = \frac{\text{Flow, GPM}}{\text{Length of Weir, ft}}$$

WELL ISOLATION ZONE

The surface or zone surrounding a water well or well field, supplying a public water system, with restricted land uses to prevent contaminants from a not permitted land use to move toward and reach such water well or well field. Also see WELLHEAD PROTECTION AREA (WHPA).

WELL LOG

A record of the thickness and characteristics of the soil, rock and water-bearing formations encountered during the drilling (sinking) of a well.

WELLHEAD PROTECTION AREA (WHPA)

The surface and subsurface area surrounding a water well or well field, supplying a public water system, through which contaminants are reasonably likely to move toward and reach such water well or well field. Also see WELL ISOLATION ZONE.

WET CHEMISTRY

Laboratory procedures used to analyze a sample of water using liquid chemical solutions (wet) instead of, or in addition to, laboratory instruments.

WHOLESOME WATER

A water that is safe and palatable for human consumption.

WIRE-TO-WATER EFFICIENCY

The combined efficiency of a pump and motor together. Also called the OVERALL EFFICIENCY.

WYE STRAINER

A screen shaped like the letter Y. The water flows in at the top of the Y and the debris in the water is removed in the top part of the Y.

X

(NO LISTINGS)

Y

YIELD

The quantity of water (expressed as a rate of flow—GPM, GPH, GPD, or total quantity per year) that can be collected for a given use from surface or groundwater sources. The yield may vary with the use proposed, with the plan of development, and also with economic considerations. Also see SAFE YIELD.

Z

ZEOLITE

A type of ion exchange material used to soften water. Natural zeolites are siliceous compounds (made of silica) which remove calcium and magnesium from hard water and replace them with sodium. Synthetic or organic zeolites are ion exchange materials which remove calcium or magnesium and replace them with either sodium or hydrogen. Manganese zeolites are used to remove iron and manganese from water.

ZETA POTENTIAL

In coagulation and flocculation procedures, the difference in the electrical charge between the dense layer of ions surrounding the particle and the charge of the bulk of the suspended fluid surrounding this particle. The zeta potential is usually measured in millivolts.

ZONE OF AERATION ZONE OF AERATION

The comparatively dry soil or rock located between the ground surface and the top of the water table.

ZONE OF SATURATION ZONE OF SATURATION

The soil or rock located below the top of the groundwater table. By definition, the zone of saturation is saturated with water. Also see WATER TABLE.

ZOOPLANKTON (ZOE-PLANK-ton) ZOOPLANKTON

Small, usually microscopic animals (such as protozoans), found in lakes and reservoirs.

SUBJECT INDEX

NOTES

NOTES

NOTES

NOTES

NOTES

NOTES